Selected Titles in This Series

Volume

3 Xia-Xi Ding and Tai-Ping Liu, Editors
Nonlinear Evolutionary Partial Differential Equations
1997

2.2 William H. Kazez, Editor
Geometric Topology
1997

2.1 William H. Kazez, Editor
Geometric Topology
1997

1 B. Greene and S.-T. Yau, Editors
Mirror Symmetry II
1997

Nonlinear Evolutionary Partial Differential Equations

AMS/IP

Studies in Advanced Mathematics

Volume 3

Nonlinear Evolutionary Partial Differential Equations

International Conference
on Nonlinear Evolutionary Partial
Differential Equations
June 21–25, 1993
Beijing, People's Republic of China

Xia-Xi Ding and Tai-Ping Liu, Editors

American Mathematical Society • International Press

Shing-Tung Yau, Managing Editor

1991 *Mathematics Subject Classification.* Primary 35L65, 35K57, 58F07, 35Qxx;
Secondary 76-xx.

Library of Congress Cataloging-in-Publication Data

International Conference on Nonlinear Evolutionary Partial Differential Equations (1993 : Peking, China)

Nonlinear evolutionary partial differential equations : International Conference on Nonlinear Evolutionary Partial Differential Equaitons, June 21–25, 1993, Beijing, People's Republic of China / Xia-Xi Ding, Tai-Ping Liu, editors.

p. cm. — (AMS/IP studies in advanced mathematics, ISSN 1089-3288 ; v. 3)

Includes bibliographical references.

ISBN 0-8218-0661-0 (acid-free paper)

1. Differential equations, Nonlinear—Congresses. 2. Evolution equations—Congresses. I. Ting, Hsia-hsi, 1928– . II. Liu, Tai-Ping, 1945– . III. Title. IV. Series: AMS/IP studies in advanced mathematics : no. 3.

QA377.I564 1993

515′.353—dc20

96-44773

CIP

∞ The paper used in this book is acid-free and falls within the guidelines
established to ensure permanence and durability.

10 9 8 7 6 5 4 3 2 1   02 01 00 99 98 97

Contents

Parallel Invited Talks

Contributions

Preface

An International Conference on Nonlinear Evolutionary Partial Differential Equations(ICEPDE) was held in Beijing, June 21-25, 1993. It was attended by over 180 scientists, 140 from China and over 40 from abroad. There were 102 research reports presented at the conference of which 9 were planary invited lectures by senior scientists, 36 were half-hour parallel invited talks and 57 were short presentations. This Proceeding contains a selection of these papers.

The topics for the conference was selected, after long deliberation, because of its importance in the natural sciences as well as its mathematical interest. The topics discussed included conservation laws, dispersion waves, Einstein's theory of gravitation, reaction-diffusion equations, solution of the Navier-Stokes equations, etc. New results on these subjects were presented, and they are described in this volume.

The participants engaged enthusiastically in discussion and other informal interactions. It's especially valuable that many young scientists were able to participate to meet foreign scientists and to report on their own work. For many of the foreign scientists this was their first opportunity to learn of the developments by Chinese scientists. The conference was successful from both the scientific and cultural viewpoints.

Academia Sinica sponsors the Conference. Two of its Institutes, The Institute of Applied Mathematics and the Wuhan Institute of Mathematical Physics, supplied the local organization works. The State Science and Technology Commission of China and the National Science Foundation of China provided financial support. It is a pleasure to thank the members of these organizations, the local organizing committee and the others who helped make the conference such a success.

April, 1994

Xia-Xi Ding Tai-Ping Liu

AMS/IP Studies in Advanced Mathematics
Volume **3**, 1997

MULTISCALE STRUCTURE FOR HYPERBOLIC WAVES

BRIAN BOSTON, JAMES GLIMM[2-6], JOHN W. GROVE[1-4],
RICHARD HOLMES, AND QIANG ZHANG[5-6]

ABSTRACT. The hyperbolic, or conservative, description of nonlinear waves is recognized to be incomplete, and appeal to a physically more complete set of equations may be needed to resolve ambiguities, such as nonuniqueness of solutions, at the hyperbolic level. Here we consider a very simple example of a hyperbolic wave, namely a fluid interface, and show that it has potentially a very complicated structure. We survey recent progress of the authors and co-workers in understanding the structure of a mixing zone, as a diffused boundary or interface between two fluids, in several cases for which standard linear diffusion, with a growth rate $O(t^{1/2})$, is incorrect. The methods use a mixture of theoretical analysis, numerical simulation by the front tracking method, and comparison to experimental and field data.

Keywords: Front tracking, Mach reflection, Richtmyer-Meshkov

1. INTRODUCTION

According to classical theories of linear diffusion, the mixing zone associated with an interface between two fluids will grow at a rate of $O(t^{1/2})$. The derivation of this property is based on solutions to the heat equation, and ultimately upon an infinite scale separation between the microscopic disturbance causing the diffision and the macroscopic scale of the observation. Not all problems allow such a scale separation. We consider three examples with no such length scale separation, and often with multiple scale disturbances, for which the mixing rates differ from the $O(t^{1/2})$ law. In the cases examined, significant progress has been recently obtained concerning such anomalous mixing rates [8, 5, 3, 3, 9, 10, 11, 4].

The three problems we consider are:

[1]Suppoited in part by the National Science Foundation Grant no. DMS-9057429.
[2]Supported in part by the U.S. Army Research Office through the Mathematical Sciences Institute of Cornell University under subcontract to SUNY Stony Brook, ARO contract number DAAL03-91-C-0027.
[3]Supported in part by the National Science Foundation Grant no. DMS-9201581.
[4]Supported in part by the U. S. Army Research Office, grant no. DAAL03-92-G-0185.
[5]Supported in part by the Applied Mathematics Subprogram of the U.S. Department of Energy DE-FG02-90ER25084.
[6]Supported in part by the Oak Ridge National Laboratory subcontract 19X-SJ067V.

1. The Richtmyer-Meshkov problem of an interface between fluids of distinct density, which is accelerated impulsively (by a shock wave).
2. The Rayleigh-Taylor problem of a continuously accelerated interface.
3. Dispersive mixing of passive contaminants for flow in porous media (*i.e.*, groundwater).

There are a number of other examples to which our methods should apply, such as the Taylor-Saffman and Kelvin-Helmholtz interfacial instabilities.

Our methods use a combination of theory, simulation, and comparison to experimental and field data. The theory is based upon Riemann solutions for nonlinear conservation laws, linearized instability analysis, fractals and renormalization group methods. The simulations are based on front tracking. For the problems studied, the computations by this method appear to be superior to those obtained by other methods. The main results of the authors and co-workers include a consistent prediction of mixing rates, by these various methods, for the problems considered.

2. Front Tracking

For the nonlinear conservation law

$$u_t + \nabla \cdot F(u) = 0 \ , \tag{2.1}$$

we suppose that the Jacobian matrix

$$A = \frac{\partial F}{\partial u} \tag{2.2}$$

has real eigenvalues

$$\lambda_1 \leq \cdots \leq \lambda_n \ . \tag{2.3}$$

In other words 2.1 is hyperbolic. The basic building block for wave interactions and for front tracking (along with most modern finite difference methods for 2.1), is the Riemann problem. This is the Cauchy problem with scale invariant initial data.

$$u(t=0,x) = \begin{cases} u_L, & \text{if } x < 0 \\ u_R, & \text{if } x > 0 \end{cases} \tag{2.4}$$

The solution of 2.1-2.4, i.e., the Riemann solution, is typically composed of a certain number of elementary waves, separated by constant states. These waves are either shock waves, satisfying a jump relation $s[u] = [F]$ or rarefaction waves solving the ODE $du/d\xi = r_i(u)$, where $\xi = x/t$ and r_i is the ith right eigenvector of A. These same waves appear and interact as part of a general solution of 2.1.

The idea of front tracking is to give the solution discontinuities, such as those associated with shock waves, an explicit computational degree of freedom. The front tracking discretization contains the explicit location of the discontinuity, as well as a pair of state values $u_\pm$ at the discontinuity, in addition to regular grid values for every mesh block. This results in a more accurate as well as a more stable solution. (See Figure 1).

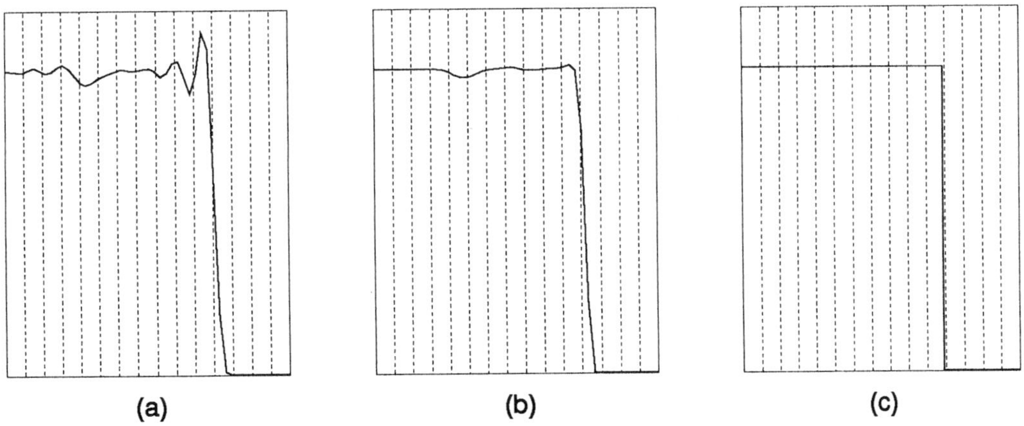

FIGURE 1. Pressure profiles of a shock computed by (a) the Lax-Wendroff scheme, (b) a second order Godunov method and (c) front tracking. Note the discontiuity in the front tracking profile. The vertical lines are three computational cells apart. Both Lax-Wendroff and the second order Godunov scheme have an entropy error from the initialization of the shock. This is most clearly seen behind the shock in (b).

The Front Tracking algorithm is a method of constructing a discrete solution $u(t+\Delta t, x)$, given $u(t,x)$. To construct $u(t+\Delta t, x)$, we advance the regular grid states and front states separately. The regular grid states are constructed by finite differences, while the front states are constructed by the solution of Riemann problems. Coupling between these two systems of states is enforced using the method of characteristics. In Figure 2 we show a propagated front, with information from the interior reaching the new front location along characteristic rays.

We illustrate the power of front tracking by applying it to the problem of shock wave reflection by a ramp, shown schematically in Figure 3. This problem allows a number of different solution patterns as the ramp angle, θ_w, and the incident shock strength are varied. Of considerable interest is a region of apparent nonuniqueness, where simple (shock polar) theories allow both a regular reflection and a Mach reflection configuration, in an interval $\theta \in [\theta_e, \theta_N]$. The von Neumann or mechanical equilibrium criterion states that Mach reflection will occur as long as it is an allowed solution, i.e. for $\theta \leq \theta_N$. Recent computations of B. Boston (presented here) come very close to realizing this conjecture numerically.

The example presented here (Figure 4) consists of a shock wave in air (modeled as a perfect gas with $\gamma = 1.4$) colliding with an oblique ramp. The gas ahead of the shock wave was stagnant at an ambient pressure of 0.3 bars and a temperature of 300° K, with an incident shock Mach number of 2.0. The interaction was modeled with a ramp angle θ_w of 49°. A computational grid of 256×256

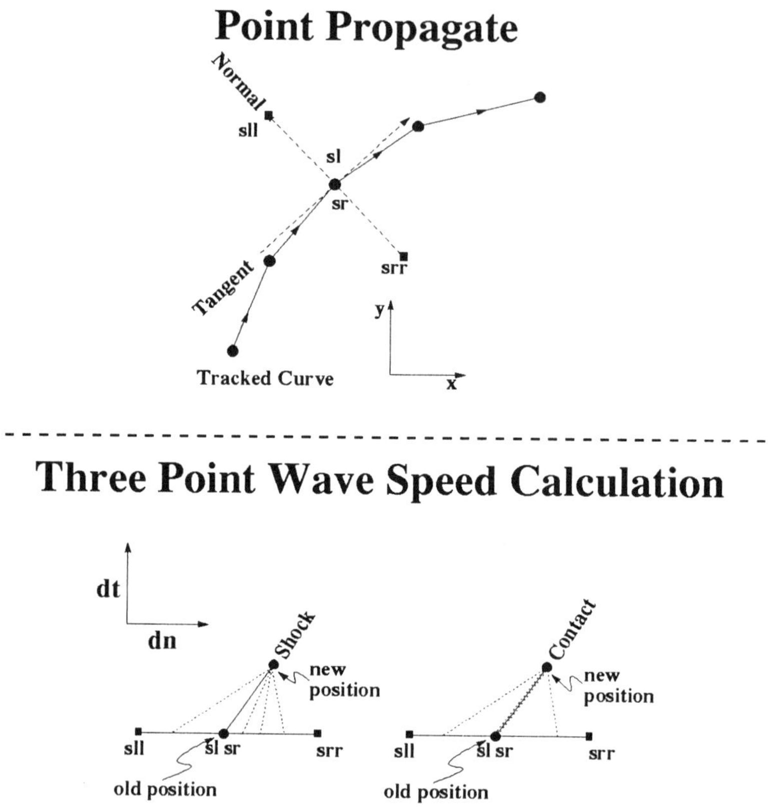

FIGURE 2. A schematic depiction of the propagation of a tracked front and the coupling of the front states to the interior states via the method of characteristics.

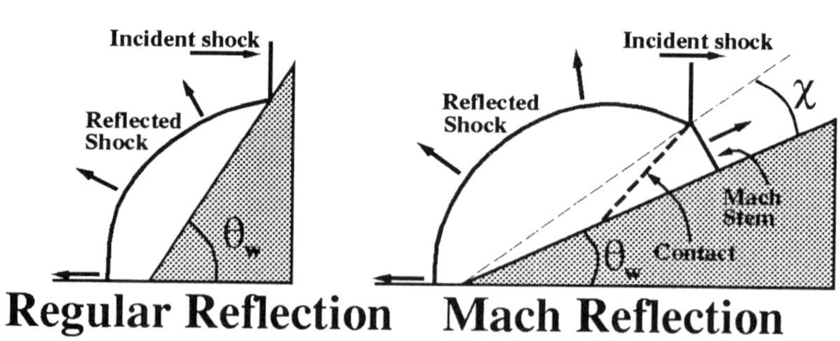

FIGURE 3. A schematic of regular and Mach reflection. Here, χ is the Mach node trajectory angle, and θ_w is the ramp angle. The incident shock is moving from the left to the right.

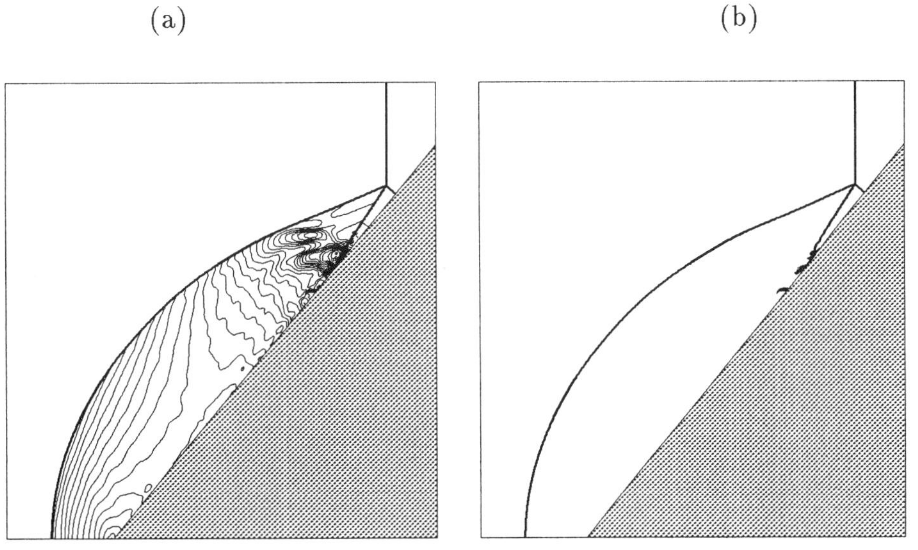

FIGURE 4. (a) Density contours for the simulation of Mach reflection. Here the Mach triple point trajectory angle $\chi = 1.6^\circ$. (b) The tracked wave fronts from the same computation.

zones was used for this simulation, which was conducted on a Sun Microsystems Sparcstation2 with 64 megabytes of RAM. The run required about 20 CPU hours. Our investigations have shown that grids as coarse as 100×100 would resolve the basic structure (front locations), and much of the interior structure of the reflected shock bubble. This represents a tremendous saving in both time and computing resources when compared with other simulations of this problem achieving comparable resolution.

The approach to the transition between these two configurations is best understood by plotting $\chi = \chi(\theta)$ *vs.* θ. In Figure 5 , we show these plots. Note that laboratory experiments, Godunov calculations on a Cray and front tracking calculations on a Sun workstation all agree, within their domain of definition. The experiments resolve a triple point angle down to $\chi \approx 2.0^\circ$. The best available Cray computation using the Godunov method, automatic mesh refinement, and a limited version of front tracking achieves $\chi \approx 0.9^\circ$. The (full) front tracking computations of B. Boston reach $\chi \approx 0.1^\circ$. Extrapolating to $\chi = 0$ gives a transition point θ_T very near, but slightly above θ_N, thus lending support to the conjecture that θ_N is the transition point for the inviscid limit.

3. Shock Accelerated Fluid Mixing

In this problem, a shock wave is incident on an interface separating gases of different densities, and possibly different equations of state. If the gas interface is not perfectly planar, a mixing zone is created, which grows nearly linearly in time, due to differential vorticity deposited by the shock wave as it crosses the interface. Consider the single mode case in which the initial interface position

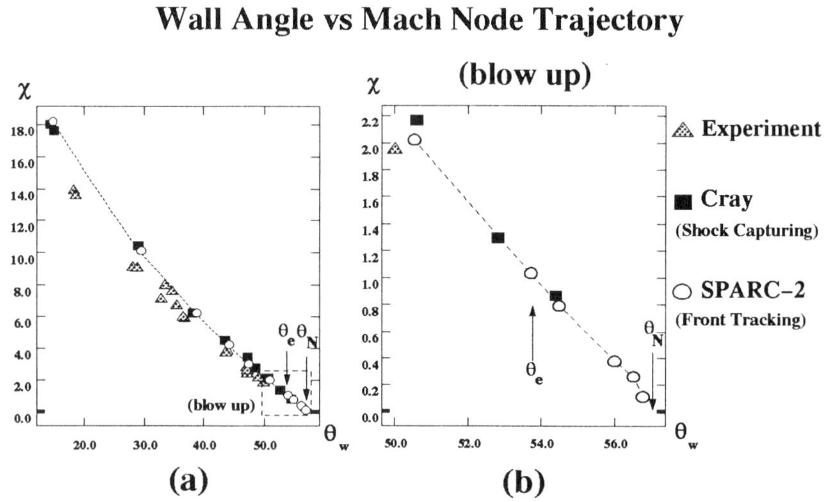

FIGURE 5. A comparison of the Mach triple point trajectory for experiments, fine grid shock capturing computations, and front tracking. We note that both numerical methods are in substantial agreement with each other, and are close to the experimental measurements. θ_e and θ_N are conjectured values for transition, according to the mechanical equilibrium and the detachment transition criteria.

has the shape of a single sine wave of amplitude $A_0 = A(0)$. The result will be to produce a nonlinear single mode perturbation of amplitude $A(t)$. The object of the study is to predict the behavior of $A(t)$.

If A_0 is small, the problem can be solved by expansion in powers of A_0. When only linear terms are retained, a PDE for $A(t)$ results, which can be integrated numerically, as was first done by Richtmyer [7].

The result of the shock-interface interaction is a transmitted shock wave and a reflected wave, which can either be a shock or a rarefaction wave. Richtmyer's linear solution, valid only for the case of a reflected shock wave, was recently extended to the reflected rarefaction wave case [8]. The linear solution assumes that the growing disturbance has the shape of a pure sine wave. This assumption will only be valid for small times. Although the linear model, and especially an approximation to it known as the impulsive model, is commonly used for late times, there is no scientific basis to ensure validity of this procedure. We find, in contrast, that there is excellent agreement between front tracking simulations and the linear model, but only for times much smaller than experimental times.

The major result of this section is the approximate agreement between numerical simulation and experiment. This is the first such agreement known to the authors, since simulations for this problem were first attempted. The calculations were based on the front tracking method, and are described in [5]. The main results are summarized in Figure 6.

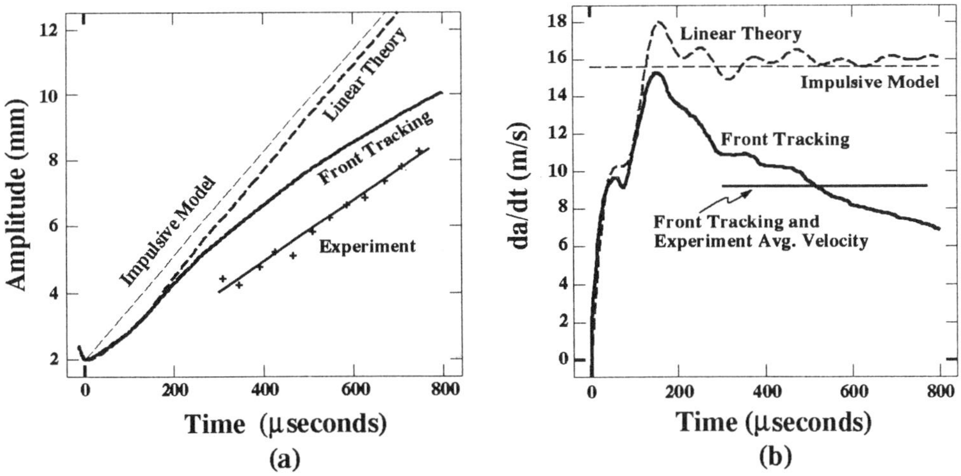

FIGURE 6. Comparison of Front Tracking simulation with experiment [2], linear theory and the impulsive model. Front Tracking agrees with the linear model for early times and agrees approximately with experiment over the times of experimental measurement.

4. Fluid Mixing by Continuous Acceleration: The Rayleigh-Taylor Problem

Here we consider an interface between fluids of different densities, $\rho_1 \neq \rho_2$, and an accelerating force, such as gravity in the unstable direction, i.e. pointing from the heavy fluid to the light. The unstable interface gives rise to a mixing zone, and we consider in particular the height $h(t)$ measured from the undisturbed heavy fluid to the initial interface position. The interface is given an initial small random disturbance, using randomly chosen fourier coefficents [3]. There are two main results of our program involving simulation based on the Front Tracking Method and a simple theoretical analysis based on the renormalization group. The results concern

$$h(t) \cong \alpha A g t^2 \tag{4.1}$$

where

$$A = \frac{\rho_1 - \rho_2}{\rho_1 + \rho_2} \tag{4.2}$$

is the dimensionless density ratio and α is a dimensionless constant to be determined.

The first main result is for the incompressible case. Consistent experimental data [6], results from numerical simulations and theoretical results give for incompressible or nearly incompressible flow,

$$\alpha_{incomp} \simeq 0.06 \tag{4.3}$$

We note in particular that the theoretical, renormalization group, approach has no adjustable parameters, in contrast to most macroscopic multiphase or turbulence computations.

The second main result displays the dependence of α on compressibility. There are no experiments in this case. Both Front Tracking simulation and the renormalization group based theory show a significant dependence of α on compressiblity, with values of α_{comp} more than double the .06 value of α_{incomp} [3].

5. Dispersion of Passive Tracers in Porous Media Flow

We find three main regimes for dispersivity. The origin of dispersion for flow in porous media is geological heterogeneity, especially of the permeability. The heterogeneity typically occurs on all length scales, including those of the observation, and so the hypothesis, necessary for classical theories, of a large scale separation between disturbance and observation, is commonly violated.

The first dispersion flow-regime is classical (Fickian) diffusion, with a mixing width proportional to $t^{1/2}$. In general, we consider a power law mixing length $\ell(t) \simeq t^\gamma$. In the second, anomalous mixing regime, $1/2 < \gamma < 1$, and in a sense the dispersion is dominated by a transient response to large length scale. In this case, the transient and long time asymptotic mixing rate has been determined by several consistent approaches: simple and renormalized perturbation theory and the ensemble average of direct simulation of multiple realizations for a stochastic heterogeneous geology.

A final mixing regime is identified which is essentially divergent at large distance (non-renormalizable) in the sense that contributions from higher order perturbation theory are more divergent than lower order ones. Finite results are obtained by use of an infrared cutoff, but those results are nonunique, and characterize the cutoff as much as the stochastic partial differential whose asymptotics is "described" by mixing exponents. See [10, 11] for further details.

Time dependent stochastic partial differential equations with time as well as spatially correlated random velocity can be studied in this framework, and the same three regions emerge [11]. The Kolmogorov point, identified in [1], on the basis of its association with the 5/3 exponent, lies in the non-uniqueness (non-renormalizable) region. The properties of this point, including the value of the exponent and a phase transition or discontinuity in the functional form of the exponent, result from specific choices of the cutoff, including a phase transiton which occurs in the functional form of the cutoff in [1].

References

1. M. Avellaneda and A. Majda. Mathematical models with exact renormalization for turbulent transport. *Comm. Math. Phys.*, 131:381–429, 1990.
2. R. Benjamin, D. Besnard, and J. Haas. Shock and reshock of an unstable interface. LANL report LA-UR 92-1185, Los Alamos National Laboratory, 1993.
3. Y. Chen, Y. Deng, J. Glimm, G. Li, D. H. Sharp, and Q. Zhang. A renormalization group scaling analysis for compressible two-phase flow. *Phys. Fluids A*, 5(11):2929–2937, 1993.

4. J. Glimm, W. B. Lindquist, F. Pereira, and Q. Zhang. A theory of macrodispersion for the scaleup problem. In *Flow in Porous Media*, 1993. to appear.
5. J. Grove, R. Holmes, D. H. Sharp, Y. Yang, and Q. Zhang. Quantitative theory of Richtmyer-Meshkov instability. In *Phys. Rev. Lett.*, pages 3473–3476.
6. K. I. Read. Experimental investigation of turbulent mixing by Rayleigh-Taylor instability. *Physica D*, 12:45, 1984.
7. R. D. Richtmyer. Taylor instability in shock acceleration of compressible fluids. *Comm. Pure Appl. Math*, 13:297–319, 1960.
8. Y. Yang, Q. Zhang, and D. H. Sharp. Small amplitude theory of Richtmyer-Meshkov instability. *Physics of Fluids A*, Submitted, 1993.
9. Q. Zhang. A multi-length scale theory of the anomalous mixing length growth for tracer flow in heterogeneous porous media. *Journal of Statistical Physics*, 66:485–501, 1991.
10. Q. Zhang. The asymptotic scaling behavior of mixing induced by a random velocity field. In *Adv. in Appl. Math.*, 1994. to appear.
11. Q. Zhang and J. Glimm. Inertial range scaling of laminar shear flow as a model of turbulent transport. *Comm. Math. Phys.*, 146:217–229, 1992.

DEPARTMENT OF APPLIED MATHEMATICS AND STATISTICS, STATE UNIVERSITY OF NEW YORK AT STONY BROOK, STONY BROOK, NY 11794-3600

E-mail address: boston@ams.sunysb.edu, glimm@ams.sunysb.edu, grove@ams.sunysb.edu, holmes@ams.sunysb.edu zhang@ams.sunysb.edu,

AMS/IP Studies in Advanced Mathematics
Volume 3, 1997

Euler-Poisson-Darboux Equations and Hyperbolic Conservation Laws

Gui-Qiang Chen

1. Euler-Poisson-Darboux Equations

The classical Euler-Poisson-Darboux equation, proposed by Euler [20] in 1790, is of the form

$$\eta_{w_1 w_2} - \frac{\alpha}{w_1 - w_2}\eta_{w_1} + \frac{\beta}{w_1 - w_2}\eta_{w_2} = 0, \tag{1.1}$$

where α and β are constants. The main feature of this equation is the singularity of the coefficients of the first-order terms along the line $w_1 = w_2$. This is the Fuchsian type singularity. For the case $\alpha = \beta$, the coordinate transformation:

$$(w_1, w_2) \longrightarrow (s, \tau) \equiv (\frac{w_1 + w_2}{2}, \frac{w_1 - w_2}{2}) \tag{1.2}$$

transforms (1.1) into

$$\eta_{ss} - \eta_{\tau\tau} - \frac{2\beta}{\tau}\eta_\tau = 0. \tag{1.3}$$

There is another similar type equation with the form

$$\eta_{ss} + \eta_{\tau\tau} + \frac{2\beta}{\tau}\eta_\tau = 0, \tag{1.4}$$

sometimes so called the Beltrami equation in some literature.

Such an equation has a close connection with many important linear partial differential equations arising from wave theory, fluid dynamics as well as geometry. The spherically symmetric solution

$$\eta(r,t) = \eta(x,t), \quad r = |x|, \; x \in \mathbf{R}^n,$$

This work was supported in part by ONR grant N00014-91-J-1384, NSF grant DMS 9207080, and Alfred P. Sloan Foundation Fellowship, USA

of the wave equation

$$\eta_{tt} - \Delta\eta = 0, \quad x \in \mathbf{R}^n, \tag{1.5}$$

satisfies

$$\eta_{tt} - \eta_{rr} - \frac{n-1}{r}\eta_r = 0. \tag{1.6}$$

The Tricomi equation

$$y\eta_{xx} + \eta_{yy} = 0 \tag{1.7}$$

is a mixed hyperbolic-elliptic equation. This equation is hyperbolic when $y < 0$ and elliptic when $y > 0$. Under the coordinate transformations

$$(x, y) \longrightarrow (s, \tau) \equiv (x, \frac{2}{3}(-y)^{\frac{3}{2}}), \quad y < 0,$$

and

$$(x, y) \longrightarrow (s, \tau) \equiv (x, \frac{2}{3}y^{\frac{3}{2}}), \quad y > 0,$$

this equation is transformed into

$$\eta_{ss} - \eta_{\tau\tau} - \frac{1}{3\tau}\eta_\tau = 0, \quad y < 0, \tag{1.8}$$

and

$$\eta_{ss} + \eta_{\tau\tau} + \frac{1}{3\tau}\eta_\tau = 0, \quad y > 0, \tag{1.9}$$

respectively. We refer the reader to [2,65,66,11] for detailed discussions for such a connection.

Such a connection has attracted great attentions from mathematicians for two centuries. Poisson [53] first solved (1.1) for a special case, and then Darboux [13] completely solved (1.1) to study the corresponding surface problems. Riemann, Volterra, Beltrami, and many others also made important contributions in studying this equation and its connection with many other important partial differential equations. The general solution of the equation (1.1) for $w_1 \leq w_2$ has the forms

$$\begin{aligned} \eta(w_1, w_2) = &\int_{w_1}^{w_2} (w_2 - s)^{-\alpha}(s - w_1)^{-\beta}\phi(s)ds \\ &+ (w_2 - w_1)^{1-\alpha-\beta}\int_{w_1}^{w_2} (w_2 - s)^{\beta-1}(s - w_1)^{\alpha-1}\psi(s)ds, \end{aligned} \tag{1.10}$$

for the case $0 < \alpha + \beta < 1$, and

$$\eta(w_1, w_2) = \int_{w_1}^{w_2} (w_2 - s)^{\beta-1}(s - w_1)^{-\beta}\phi(s)ds + \int_{w_1}^{w_2} (w_2 - s)^{\beta-1}(s - w_1)^{-\beta} ln\frac{(w_2 - s)(s - w_1)}{w_1 - w_2}\psi(s)ds, \tag{1.11}$$

for the case $\alpha + \beta = 1$, respectively. Along the line $w_1 = w_2$, the solution $\eta(w_1, w_2)$ has the singularity $(w_2 - w_1)^{1-\alpha-\beta}$ for the case $0 < \alpha + \beta < 1$, and $ln(w_2 - w_1)$ for the case $\alpha + \beta = 1$, which are Fuchsian type singularities.

Early in the present century, many problems arising from mechanics and other areas involved the Euler-Poisson-Darboux type equation with nonconstant function $\alpha = \alpha(w_1, w_2)$ and $\beta = \beta(w_1, w_2)$ that are regular along the line $w_1 = w_2$. Such a study was made by many authors including A. Weinstein, J. Lions, A. Erdélyi, and many Chinese mathematicians (see [2,65,66,11] for the references). On the other hand, the theory of linear equations with multiple characteristics is also well developed. An important feature of such equations is loss of differentiability [14], which leads to ill-posedness in Sobolev spaces and to the use of Gevrey classes [27]. Another feature is that sign conditions on the subprincipal symbol play an important role (cf. [24,52,31]).

The corresponding singular equations arising from the connection with hyperbolic systems of conservation laws, in general, are much more singular. The typical form of such equations is

$$\eta_{w_1 w_2} - \frac{\alpha(\frac{w_1}{w_2})}{w_1 - w_2}\eta_{w_1} + \frac{\beta(\frac{w_1}{w_2})}{w_1 - w_2}\eta_{w_2} = 0. \tag{1.12}$$

This equation has an extra singularity, high singularity of α and β at the origin, besides the singularity on the line $w_1 = w_2$.

2. Hyperbolic Conservation Laws

Quasilinear systems of conservation laws in one dimension take the following form

$$\partial_t U + \partial_x F(U) = 0, \quad U \in \mathbf{R}^n. \tag{2.1}$$

Such a system is hyperbolic if the $n \times n$ matrix $\nabla F(U)$ has n real eigenvalues $\lambda_j(U)$ and linearly independent eigenvectors $r_j(U), 1 \le j \le n$. Denote

$$\mathcal{D} \equiv \{U : \lambda_i(U) = \lambda_j(U), \quad i \ne j, 1 \le i, j \le n\}, \tag{2.2}$$

as the degenerate set. If the set $\mathcal{D}$ is empty, then this system is strictly hyperbolic and, otherwise, is nonstrictly hyperbolic. Such a set allows a degree of interaction, or nonlinear resonance, between different characteristic modes, which is missing in the strictly hyperbolic case, and causes an analytic difficulty due to the degeneracy. A point $U_* \in \mathcal{D}$ is hyperbolic degenerate if $\nabla F(U_*)$ is diagonalizable and, otherwise, is parabolic degenerate.

Since F is a nonlinear function, solutions of the Cauchy problem for (2.1) (even starting from smooth initial data) generally develop singularities in a finite time, and then the solutions become discontinuous functions. This situation reflects the physical phenomenon of breaking of waves and development of shock waves. For this reason, attention focuses on solutions in the space of discontinuous functions, where one can not directly use the classical analytic techniques that predominate in the theory of partial differential equations of other types.

To overcome this difficulty, a natural idea is to construct approximate solutions $U^\epsilon(x,t)$ to (2.1) using shock capturing methods and then to study the convergence and consistency of the approximate solutions to (2.1). There are two typical shock capturing methods to construct approximate solutions $U^\epsilon(x,t)$:

(1) Viscosity method: $U^\epsilon(x,t)$ are governed by the following Cauchy problem

$$\begin{cases} \partial_t U + \partial_x F(U) = \epsilon \partial_x (D(U)\partial_x U), \\ U|_{t=0} = U_0^\epsilon(x) \end{cases} \tag{2.3}$$

with a properly selected and nonnegative matrix;

(2) Finite difference method: $U^\epsilon(x,t)$ are governed by the following difference equations

$$\begin{cases} D_t U + D_x F(U) = 0, \\ U(x,0) = U_0^\epsilon(x), \end{cases} \tag{2.4}$$

with the space mesh length ϵ and the time mesh length Δt determined by the Courant-Friedrichs-Levy condition [36,55].

The viscosity method is motivated from fluid dynamics: The compressible gas flow determined by the system of compressible Euler equations, which is a typical example of (2.1), can be in general regarded as the limit of the gas flows determined by the system of the Navier-Stokes equations when the effect of the viscosity and heat conductivity vanishes in an appropriate sense of topology. The difference methods come from the fact that numerical computations for systems of conservation laws are limited to calculations of discrete difference equations and Cauchy mesh data. In this context, a good numerical regularization will produce a good numerical approximation to hyperbolic conservation laws. Examples of this approach include the Lax-Friedrichs scheme [36], the Glimm scheme [25], and the Godunov scheme [29]. The key issue is whether the approximate solutions converge in an appropriate topology and the limit function is consistent with the system (2.1). Solving this issue involves two aspects: one is to construct good approximate solutions and another one needs the compactness analysis in an appropriate topology.

For strictly hyperbolic and genuinely nonlinear systems of conservation laws with C^2 flux functions: $F \in C^2$ and $\nabla\lambda_j(U) \cdot r_j(U) \neq 0, U \in \mathbf{R}^n, 1 \leq j \leq n$. DiPerna [18] proved that any uniformly bounded viscosity solution sequence must be compact in L^∞ no matter whether corresponding approximate initial data sequence is compact or not and the limit function generated from the vanishing viscosity method is consistent with (2.1). An extension of this theorem to the linear degenerate case can be found in Serre [61]. An alternative proof of the DiPerna's theorem was given by Morawetz [47]. The $L^p, 1 < p \leq \infty$, uniformly estimates for viscosity approximate solutions were discussed by Dafermos [12]. Also see [39] for the convergence of uniformly L^p bounded viscosity approximate solutions for the system of elasticity.

If the flux function F is not C^2 or the genuine nonlinearity fails, the situation is different. Greenberg and Rascle [28] showed by an example that the loss of C^2 smoothness for the flux function can produce oscillation. For the linearly degenerate case, the initial oscillation will propagate along the linearly degenerate characteristic fields in the entropy weak solution (cf. [6,61,54]). In these cases, one cannot expect the convergence of viscous approximate solutions. The next issue is whether the degenerate set $\mathcal{D}$ affects the limiting behavior of approximate solutions $U^\epsilon(x,t)$ to conservation laws when ϵ tends to zero.

Recent decades, nonstrictly hyperbolic systems of conservation laws have arisen from many important fields such as continuous mechanics including the vacuum problem, multiphase flows in porous media, MHD, and elasticity. On the other hand, nonstrict hyperbolicity of systems is generic in some sense. For example, for three space dimensional systems of conservation laws, Lax [38] indicated that systems with $2(mod\,4)$ equations must be nonstrictly hyperbolic. The result can be extended to systems with $\pm 2, \pm 3, \pm 4 (mod\,8)$ equations [22]. Then the plane wave solutions of such systems are governed by corresponding one-dimensional hyperbolic systems with $\mathcal{D} \neq \emptyset$.

3. Connection: Entropy and Young Measure

The connection between the Euler-Poisson-Darboux equations and nonstrictly hyperbolic systems of conservation laws is the entropy (see [37]).

A pair of functions $(\eta, q) : \mathbf{R}^n \to \mathbf{R}^2$ is called an entropy-entropy flux pair if they satisfy the following linear hyperbolic system

$$\nabla q(U) = \nabla \eta(U) \nabla F(U). \tag{3.1}$$

The function $\eta(U)$ is called an entropy function. Clearly, any C^1 solution satisfies

$$\partial_t \eta(U) + \nabla \eta(U) \nabla F(U) \partial_x U = 0, \tag{3.2}$$

or

$$\partial_t \eta(U) + \partial_x q(U) = 0. \tag{3.3}$$

For the BV solution, the second term in (3.2) has no meaning in the classical sense because of the multiplication of the Radon measure with the discontinuous function. If $\eta(U)$ is an entropy, then the left side of (3.2) becomes the left side of (3.3) that makes sense even for L^∞ solutions in the sense of distribution. An L^∞ function $U(x,t)$ is called an entropy weak solution [37] if

$$\partial_t \eta(U) + \partial_x q(U) \leq 0, \tag{3.4}$$

in the sense of distribution for any convex entropy-entropy flux pair.

Assume that the system is endowed with globally defined Riemann invariants $w_i(U), 1 \leq i \leq n$, satisfying

$$\nabla w_i(U) \cdot \nabla F(U) = \lambda_i(U) \nabla w_i(U). \tag{3.5}$$

The necessary and sufficient condition for the existence of the Riemann invariants w_i for strictly hyperbolic systems is the well-known Frobenius condition:

$$l_i\{r_j, r_k\} = 0, \quad \text{for any } j, k \neq i,$$

where l_i denotes the left eigenvector corresponding to λ_i and $\{\cdot,\cdot\}$ is the Poisson bracket of vector fields in U-space. For $n = 2$, the Frobenius condition always holds. For smooth solution $U(x,t)$, the corresponding Riemann invariants $w_i, 1 \leq i \leq n$, satisfy the transport equations

$$\partial_t w_i + \lambda_i(w) \partial_x w_i = 0,$$

which indicate that w_i is invariant along the i-th characteristic fields.

Taking the inner product of (3.5) with the right eigenvectors r_i of ∇F produces the characteristic form

$$(\lambda_i \nabla \eta - \nabla q) \cdot r_i = 0,$$

that is,

$$\lambda_i \eta_{w_i} = q_{w_i}. \tag{3.6}$$

For $n = 2$, the linear hyperbolic system (3.5) is equivalent to (3.6) and the system (3.6) can be reduced to the following linear second-order hyperbolic equation:

$$\eta_{w_1 w_2} + \frac{\lambda_{2w_1}}{\lambda_2 - \lambda_1}\eta_{w_2} - \frac{\lambda_{1w_2}}{\lambda_2 - \lambda_1}\eta_{w_1} = 0. \tag{3.7}$$

For the case $\mathcal{D} = \emptyset$, the equation (3.7) is, in general, regular and hyperbolic for which the Goursat problem is well posed in the coordinates of Riemann invariants. The entropy space is infinite-dimensional and is represented by two families $\alpha(w_1)$ and $\beta(w_2)$ of functions of one variable. However, for the case $\mathcal{D} \neq \emptyset$, the situation is much more complicated because of the singularity of the functions $\frac{\lambda_{2w_1}}{\lambda_2 - \lambda_1}$ and $\frac{\lambda_{1w_2}}{\lambda_2 - \lambda_1}$ on the set $(w_1(\mathcal{D}), w_2(\mathcal{D})) \in \mathbf{R}^2$. The typical form of such equations is

$$\eta_{w_1 w_2} - \frac{\alpha(\frac{w_1}{w_2})}{w_1 - w_2}\eta_{w_1} + \frac{\beta(\frac{w_1}{w_2})}{w_1 - w_2}\eta_{w_2} = 0. \tag{3.8}$$

This equation has an extra singularity of α and β at the origin, besides the singularity on the line $w_1 = w_2$. The issue is if there exist nontrivial regular solutions of the singular equation such as (3.8).

The connection between the compactness problem of the approximate solutions and the entropy is the Young measure via the compensated compactness ideas first developed by Tartar and Murat [64,49] and a related observation by Ball [1].

The Young measure is a useful tool for studying the limiting behavior of measurable function sequences. For an arbitrary sequence of measurable maps

$$U^\epsilon : \mathbf{R}^m \to \mathbf{R}^n, \quad \|U^\epsilon\|_{L^\infty} \leq M < \infty,$$

that converges in the weak-star topology of L^∞ to a function U,

$$w^* - \lim_{\epsilon \to 0} U^\epsilon = U,$$

there exists a subsequence (still labeled) U^ϵ and a family of Young measures

$$\nu_y \in \mathrm{Prob}(\mathbf{R}^n), \quad \mathrm{supp}\nu_y \subset \{\lambda : |\lambda| \leq M\}, \quad y \in \mathbf{R}^m,$$

such that for all continuous functions g,

$$w^* - \lim_{\epsilon \to 0} g(U^\epsilon(y)) = \int_{\mathbf{R}^n} g(\lambda) d\nu_y(\lambda) \equiv < \nu_y, g(\lambda) >,$$

for almost all points $y \in \mathbf{R}^m$. In particular, U^ϵ converges strongly to U if and only if the Young measure ν_y at almost all points y is equal to a Dirac mass concentrated at $U(y)$.

In the most cases, one can estimate (cf. [18,5]) that the approximate solution $U^\epsilon(x,t)$ generated by the shock capturing methods for (3.1) satisfying

(1) $\|U^\epsilon\|_{L^\infty} \le M < \infty$.

(2) For C^2 entropy-entropy flux pair $(\eta_i, q_i), i = 1, 2$, determined by (3.6) and (3.7),

$$\partial_t \eta_i(U^\epsilon) + \partial_x q_i(U^\epsilon) \quad \text{compact in } H^{-1}_{\text{loc}}. \tag{3.9}$$

Then there exist a subsequence (still labeled) U^ϵ and a family of Young measures $\{\nu_{x,t} : \mathbf{R}^2_+ \to \text{Prob}(\mathbf{R}^2)\}, \text{supp}\nu_{x,t} \subset \{\lambda : |\lambda| \le M\}$, such that, for any continuous function g,

$$w^* - \lim_{\epsilon \to 0} g(U^\epsilon(x,t)) = \int_{\mathbf{R}^2_+} g(\lambda) d\nu_{x,t}(\lambda) \equiv < \nu_{x,t}, g(\lambda) >, \tag{3.10}$$

and that, for any C^2 entropy-entropy flux pair determined by (3.6) and (3.7),

$$\left\langle \nu_{x,t}, \begin{vmatrix} \eta_1(\lambda) & q_1(\lambda) \\ \eta_2(\lambda) & q_2(\lambda) \end{vmatrix} \right\rangle = \begin{vmatrix} < \nu_{x,t}, \eta_1(\lambda) > & < \nu_{x,t}, q_1(\lambda) > \\ < \nu_{x,t}, \eta_2(\lambda) > & < \nu_{x,t}, q_2(\lambda) > \end{vmatrix} \quad a.e. \ (x,t) \in \mathbf{R}^2_+. \tag{3.11}$$

This commutativity relation is derived from (3.9), the Young measure representation theorem for the measurable function sequence, and a basic continuity theorem for the 2×2 determinant in the weak topology (cf. [63,5]). This indicates that proving the compactness of the approximate solutions $U^\epsilon(x,t)$ is equivalent to solving the functional equation (3.11) for the Young measure $\nu_{x,t}$ for all possible C^2 entropy-entropy flux pair determined by (3.6) and (3.7). If one clarifies the structure of the Young measure $\nu_{x,t}(\lambda)$, one can understand the limiting behavior of the approximate solutions. For example, if one can prove that almost all Young measures $\nu_{x,t}, (x,t) \in \mathbf{R}^2_+$, are Dirac masses, then one can conclude the strong convergence of the approximate solutions $U^\epsilon(x,t)$ almost everywhere. Therefore, the structure of the Young measures is determined by the C^2 solutions of the Euler-Poisson-Darboux type equation (3.7). The larger the set of C^2 solutions to the Euler-Poisson-Darboux type entropy equation (3.7) is, the easier solving the functional equation (3.11) for the Young measures is.

4. Applications

Now we apply this connection to address whether the degenerate set $\mathcal{D}$ affects the limiting behavior of approximate solutions $U^\epsilon(x,t)$ to conservation laws when ϵ tends to zero. To make my points more precisely, I will focus on two of the most important systems in this area. The first one is the system of isentropic gas dynamics, which is of parabolic degeneracy.

4.1. Parabolic Degenerate Systems

The system of isentropic Euler equations are of the form

$$\left\{\begin{array}{l}\partial_t\rho+\partial_x m=0,\\ \partial_t m+\partial_x(\frac{m^2}{\rho}+p(\rho))=0,\end{array}\right. \tag{4.1}$$

where ρ, m, and p are the density, the mass, and the pressure, respectively. For the polytropic gas, $p(\rho)=\rho^\gamma/\gamma$, where γ is the adiabatic exponent and $1<\gamma\le 5/3$ for usual gases. In this case, the flux function is not C^2.

One can check that $\mathcal{D}=\{\rho=0\}$, which is the vacuum state, and the degeneracy is parabolic. An interesting mathematical issue is whether the dissipation limit is still stable near the vacuum.

The eigenvalues of (4.1) are

$$\lambda_1=\frac{m}{\rho}-\rho^\theta,\quad \lambda_2=\frac{m}{\rho}+\rho^\theta,\quad \theta=\frac{\gamma-1}{2},$$

and corresponding Riemann invariants of (4.1) are

$$w_1=\frac{m}{\rho}-\frac{\rho^\theta}{\theta},\quad w_2=\frac{m}{\rho}+\frac{\rho^\theta}{\theta}.$$

Then the entropy equation (3.7) in the Riemann coordinates is the following Euler-Poisson-Darboux equation

$$\eta_{w_1w_2}+\frac{\alpha}{w_2-w_1}(\eta_{w_2}-\eta_{w_1})=0,\quad \alpha=\frac{3-\gamma}{2(\gamma-1)}. \tag{4.2}$$

Since this system is the most typical one in fluid dynamics, the study of this system has an extensive history dating back to the work of Riemann [56], where a special Cauchy problem, so-called Riemann problem, was solved. Zhang and Guo [68] established an existence theorem of the global solutions to this system with a class of initial data by using the characteristic method. Nishida [50] obtained the first large data existence theorem with locally finite total variation for the case $\gamma=1$ using Glimm's scheme [25]. Large-data theorems have also been obtained for general $\gamma > 1$ in the case where the initial data with locally finite total variation are restricted to prevent the development of cavities (e.g. [51,17,41]). The difficult point in bounding total variation norm at low densities is that the coupling between characteristic fields increases as the density decreases. This difficulty is a reflection of the fact that the strict hyperbolicity fails at the vacuum: $\mathcal{D} \neq \emptyset$. Using the compensated compactness ideas developed by Tartar and Murat [64,49], DiPerna [19] established a global existence theorem with L^∞ bounded data for $\gamma = 1 + \frac{2}{2m+1}, m \geq 2$ integers, with the aid of the viscosity method.

X. Ding, P. Luo, and I [15] first succeeded in proving the convergence of the Lax-Friedrichs scheme for this system in the case $\gamma = 3/2$ with the aid of compensated compactness and an analysis of the Euler-Poisson-Darboux equation (4.2) with $\alpha = 3/2$. Then we established a compensated compactness framework for approximate solutions to the Euler equations for the general case $1 < \gamma \leq 5/3$ (cf. [4,5]). Uniform boundedness of the approximate solutions coupling with H^{-1} compactness of corresponding entropy dissipation measures implies the compactness of the approximate solutions in L^∞. This means that the degenerate set $\mathcal{D}$ and the loss of C^2 differentiability of the flux function do not affect the compactness of approximate solutions for this case.

The proof is based on a careful analysis of weak entropy determined by (4.2) and a detailed study of regularities of the family of the Young measures $\nu_{x,t}$ determined by the approximate solutions and restricted by a commutativity relation (3.11). The basic motivation is that the commutativity relation (3.11) represents an imbalance of regularity and that the operator on the left is more regular than one on the right as a result of cancellation, which forces the measure $\nu_{x,t}$ to reduce to a point mass. The essential difficulty is that only a subspace of entropy pairs, weak entropy pairs, can be used in the relation (3.11). By a careful analysis and complicated calculation, we constructed several classes of useful weak entropy pairs containing the Hilbert transformation functions by using fractional derivative techniques and analyzed their behavior near the line $w_1 = w_2$. The key strategy is to choose the weak entropy pairs whose leading term is coercive with respect to the 2×2 determinant and to show that the coercive behavior guarantees that the derivative of $\nu_{x,t}$ vanishes except at one point.

Then this compactness framework is applied to prove the convergence of the Lax-Friedrichs scheme, the Godunov scheme, the viscosity approximate solutions, the MUSCL type schemes, the entropy flux-splitting schemes to this system (cf. [5, 16, 10, 9]). Recently, this framework has been applied to the initial-boundary problem for this system [63], the Euler equations coupled with the Poisson equation that model the hydrodynamic behavior of semiconductors [69,46], and the global solutions of the compressible Euler equations with geometrical structure including transonic nozzle flow, cylindrically symmetric flow, spherically symmetric flow, and symmetric rotating flow [7].

4.2. Hyperbolic Degenerate Systems

The typical example with hyperbolic degeneracy is the gradient quadratic flux system, which is umbilic degenerate, the most singular case:

$$\partial_t U + \partial_x (dC(U)) = 0, \quad U = (u, v)^{\top} \in \mathbf{R}^2, \tag{4.3}$$

where

$$C(U) = \frac{1}{3} a u^3 + b u^2 v + u v^2, \tag{4.4}$$

and a and b are two real parameters.

Such systems are generic in the following sense. For any smooth nonlinear flux function, take its Taylor expansion about the isolated umibilic point. The first three terms including the quadratic terms determines the local behavior of the hyperbolic singularity near the umbilic point. The hyperbolic degeneracy enables us to eliminate the linear term by a coordinate transformation to obtain the system with a homogeneous quadratic polynomial flux. Such a polynomial flux contains some inessential scaling parameters. There is a nonsingular linear coordinate transformation to transform the above system into (4.3)-(4.4), first studied by Marchesin, Isaacson, Plohr, and Temple and in a more satisfactory form by Schaeffer and Shearer [58]. From the viewpoint of group theory, such a reduction from six to two parameters is natural: for the six dimensional space of quadratic mappings acted by the four dimensional group $GL(2, \mathbf{R})$, one expects the generic orbit to have codimension two.

The Riemann solutions for such systems were discussed by Isaacson, Marchesin, Paes-Leme, Plohr, Schaeffer, Shearer, Temple, and others (cf. [32,33,59,60]). Two new shock waves, the overcompressive shock wave and the undercompressive shock wave, were discovered that are quite different from gas dynamical shock. The overcompressive shock can be easily understood using the Lax entropy condition [37]. It is known that there is a traveling wave solution connecting the left state and the right state of the undercompressive shock for the artificial viscosity system. Stability of such traveling waves for the overcompressive shock and the undercompressive shock has been studied (cf. [42,43]). The next issues are whether the compactness of the corresponding approximate solutions is affected by the viscosity matrix as the viscosity parameter goes to zero, to understand the sensitivity of the undercompressive shock wave with respect to the viscosity matrix, and whether corresponding global existence of entropy weak solutions can be obtained as a corollary from this effort.

The global existence of weak solutions to the Cauchy problem for a special case of such quadratic flux systems was solved by Kan [34] using the viscosity method. A different proof was given independently to the same problem by Lu [44].

In the papers [8], P.-T. Kan and I established an L^∞ compactness framework for sequences of approximate solutions to general hyperbolic systems with umbilic degeneracy specially including (2.2)-(2.3). Under this framework the approximate solution sequences, which are apriori bounded in L^∞ and which produce correct entropy dissipations, lead to the compactness of the corresponding Riemann invariant sequences. This means that the viscosity matrix does not affect the compactness of the corresponding uniformly bounded Riemann invariant sequences. One of the principal difficulties associated with such systems is the general lack of enough classes of entropy functions that can be verified to satisfy certain weak compactness conditions. This is due to possible singularities of entropy functions near the regions of nonstrict hyperbolicity. The analysis leading to the compactness involves two steps:

In the first step, we constructed regular entropy functions governed by a highly singular entropy equation, the Euler-Poisson-Darboux type equation,

$$\eta_{w_1 w_2} - \frac{\alpha(\frac{w_1}{w_2})}{w_2 - w_1}\eta_{w_1} + \frac{\beta(\frac{w_1}{w_2})}{w_2 - w_1}\eta_{w_2} = 0. \tag{4.5}$$

There are two main difficulties. The first is that the coefficients of the entropy equation are multiple-valued functions near the umbilic points in the Riemann invariant coordinates. This difficulty is overcome by a detailed analysis of the singularities of the Riemann function of the entropy equation. This analysis involves a study of a corresponding Euler-Poisson-Darboux equation and requires very complicated estimates and calculations. An appropriate choice of Goursat data leads to a cancellation of singularities and regular entropies in the Riemann invariant coordinates are obtained. The second difficulty is that the nonlinear corresponding between the physical state coordinates and the Riemann invariant coordinates is, in general, irregular. A regular entropy function in the Riemann invariant coordinates is usually no longer regular in the physical coordinates. We overcame this by a detailed analysis of the correspondence between these two coordinates.

In the second step, we studied the structure of the Young measures associated with the approximate sequences, and proved that the support of the Young measures lies in finite isolated points or separate lines in the Riemann invariant coordinates. This is achieved by a delicate use of Serre's technique [61] and regular entropy functions, constructed in the first step, in the Tartar-Murat commutation equation [64] for Young measures associated with the approximate solution sequences.

This compactness framework is successfully applied to prove the convergence of the Lax-Friedrichs scheme, the Godunov scheme, and the viscosity method for the quadratic flux systems. Some corresponding existence theorems of global entropy solutions for such systems are established. The compactness is achieved by reducing the support of the corresponding Young measures to a Dirac mass in the physical space.

References

1. J.-M. Ball, *Convexity conditions and existence theorems in nonlinear elasticity*, Arch. Rat. Mech. Anal. **63** (1977), 337-403.
2. A. V. Betsadze, *Equations of the Mixed Type*, Macmillan Company, New York, 1964.
3. T. Chang and L. Hsiao, *The Riemann Problem and Interaction of Waves in Gas Dynamics*, Longman, Avon, 1989.
4. G.-Q. Chen, *Convergence of the Lax-Friedrichs scheme for isentropic gas dynamics*, Acta Math. Sci. **6** (1986), 75-120; **8** (1988), 243-276 (in Chinese).
5. G.-Q. Chen, *The compensated compactness method and the system of isentropic gas dynamics*, MSRI Preprint 00527-91 Mathematical Science Research Institute Berkeley (1990).
6. G.-Q. Chen, *The method of quasidecoupling to conservation laws*, Arch. Rat. Mech. Anal. **121** (1992), 131-185.
7. G.-Q. Chen and J. Glimm, In preparation (1993).
8. G.-Q. Chen and P.-T. Kan, *Hyperbolic conservation laws with umbilic degeneracy I, II*, Preprint (1993).
9. G.-Q. Chen and Ph. Le Floch, *Flux splittings for hyperbolic conservation laws*, C. R. Acad. Sci. Paris, Série I **317** (1993).
10. G.-Q. Chen and J.-G. Liu, *Convergence of second-order schemes for isentropic gas dynamics*, Math. Comp. **6** (1993), 607-627.
11. R. Courant and K.O. Friedrichs, *Supersonic flow and shock waves*, Springer-Verlag, New York, 1948.
12. C.M. Dafermos, *Estimates for conservation laws with little viscosity*, SIAM J. Math. Anal. **18** (1987), 409-421.
13. G. Darboux, *Lecons Sur La Theorie Generalle des Surfaces*, T. II,, Paris, 1914.
14. E. De Giorgi, *Un esempio di non-unicità della soluzione del problema di Cauchy*, Università di Roma, Rendiconti di Mathematica **14** (1955), 382-387.
15. X. Ding, G.-Q. Chen, and P. Luo, *Convergence of the Lax-Friedrichs scheme for the isentropic gas dynamics (I)*, Acta Mathematica Scientia **5**, 415-432; **7** (1987), 467-480.
16. X. Ding, G.-Q. Chen, and P. Luo, *Convergence of the fractional step Lax-Friedrichs scheme and Godunov scheme for the isentropic system of gas dynamics*, Commun. Math. Phys. **121** (1989), 63-84.
17. Ding Shia-Shi, Zhang-Tung, Wang Ching-Hua, Hsiao-Ling, and Li Tsai-Chung, *A study of the global solutions for quasilinear hyperbolic systems of conservation laws*, Scientica Sinica **16** (1973), 317-335.
18. R. DiPerna, *Convergence of approximate solutions to conservation laws*, Arch. Rat. Mech. Anal. **82** (1983), 27–70.

19. R. DiPerna, *Convergence of the viscosity method for isentropic gas dynamics*, Commun. Math. Phys. **91** (1983), 1-30.
20. L. Euler, *Institutions Calculi Integralis, III*, Petropolis, 1790.
21. L. C. Evans, *Weak Convergence Methods for Nonlinear Partial Differential Equations*, CBMS Lecture Notes of AMS, 1990.
22. S. Friedlands, J. W. Robbin, and J. Sylvester, *On the crossing rule*, Comm. Pure Appl. Math. **37** (1984), 19-64.
23. H. Frid and M. M. Santos, *Nonstrictly hyperbolic systems of conservation laws of the conjugate type*, Preprint (1993).
24. M. Froissart, *Hyperbolic Equations and Waves*, Battelle Seattle 1968 Recontres, Springer-Verlag, Berlin Heidelberg, 1970.
25. J. Glimm, *Solutions in the large for nonlinear hyperbolic systems of equations*, Comm. Pure Appl. Math. **18** (1965), 697-715.
26. J. Glimm and A. Majda, *Multidimensional Hyperbolic Problems and Computations*, IMA Vol. 29, Springer-Verlag, New York, 1991.
27. M. Gevrey, *Sur la nature analytique des solutions des équations aux dérivavées partielles*, Annales École Norm. Sup. **35** (1917), 129-189.
28. J. Greenberg and M. Rascle, *Time periodic solutions to systems of conservation laws*, Preprint 1992.
29. S. Godunov, *A difference method for numerical calculation of discontinuous solutions of the equations of hydrodynamics*, Mat. Sb. **47(89)** (1959), 271-360.
30. C.-H. Gu and others, *Discontinuous initial value problems for systems of quasilinear hyperbolic equations I, II, III,*, Acta Math. Sinica **11** (1961), 314-323, 324-327; **12** (1962), 132-143.
31. L. Hörmander, *The Analysis of Linear Partial Differential Operators II*, Springer-Verlag, Berlin Heidelberg, 1983.
32. E. Isaacson, D. Marchesin, B. Plohr, and B. Temple, *The Riemann problem near a hyperbolic singularity: the classification of solutions of quadratic Riemann problem (I)*, SIAM J. Appl. Math. **48** (1988), 1-24.
33. E. Isaacson and B. Temple, *The classification of solutions of quadratic Riemann problem (II)-(III)*, SIAM J. Appl. Math. **48** (1988), 1287-13011, 1302-1318.
34. P.-T Kan, *Convergence of the viscosity method for a system of nonstrictly hyperbolic conservation laws*, Ph.D. thesis, New York University (1989).
35. B.L. Keyfitz, *Some elementary connections among nonstrictly hyperbolic conservation laws*, Contemporary Mathematics **60** (1987), 67-77.
36. P.D. Lax, *Weak solutions of nonlinear hyperbolic equations and their numerical computation*, Comm. Pure Appl. Math **7** (1954), 159-193.
37. P.D. Lax, *Shock waves and entropy*, In: Contributions to Nonlinear Functional Analysis, ed. E. A. Zarantonello, Academic Press, New York, 1971, pp. 603-634.
38. P.D. Lax, *The multiplicity of eigenvalues*, Bull. AMS **6** (1982), 213-314.
39. L.W. Lin, *Global solutions to a first-order quasilinear equation without the convex condition*, Acta Sci. Nat. Jilin Univ. **2** (1979), 17-26 (in Chinese).
40. P.-X. Lin, *Young measures and application of compensated compactness to system of nonlinear elasticity*, Preprint, Heriot-Watt University, 1989.
41. T.-P. Liu, *Quasilinear hyperbolic system*, Comm. Math. Phys. **68** (1979), 141–172.
42. T.-P. Liu and K. Zumbrun, *Stability of an undercompressive shock*, Preprint (1993).
43. T.-P. Liu and Z. Xin, *Stability of viscous shock waves associated with a system of nonstrictly hyperbolic conservation laws*, Comm. Pure Appl. Math. **45** (1992), 361-388.
44. Y.-G. Lu, *Convergence of the viscosity method for a nonstrictly hyperbolic system*, Acta Mathematica Sinica **12** (1992).
45. T. Makino, K. Mizohata, and S. Ukai, *Global weak solutions of the compressible Euler equations with spherical symmetry I, II*, Japan J. Industrial Appl. Math. **9** (1992), 431-449.
46. P. Marcati and R. Natalini, *Weak solutions to a hydrodynamic model for semiconductors: the Cauchy problem*, Preprint Quaderno No. 2, IAC, Italy (1993).
47. C.S. Morawetz, *An alternative proof of DiPerna's theorem*, Comm. Pure Appl. Math. **44** (1991), 1081–1090.

48. C.S. Morawetz, *On a weak solution for a transonic flow problem*, Comm. Pure Appl. Math. **38** (1985), 797-818.
49. F. Murat, *L'injection du cone positif de H^{-1} dans $W^{-1,q}$ est compacte pour tout $q < 2$*, J. Math. Pures Appl. **60** (1981), 309-322.
50. T. Nishida, *Global solution for an initial-boundary-value problem of a quasilinear hyperbolic system*, Proc. Jap. Acad. **44** (1968), 642-646.
51. T. Nishida and J. Smoller, *Solutions in the large for some nonlinear hyperbolic conservation laws*, Comm. Pure Appl. Math. **26** (1973), 183-200.
52. Y. Ohya, *On E. E. Levi's functions for hyperbolic equations with triple characteristics*, Comm. Pure Appl. Math. **25** (1972), 257-263.
53. S. D. Poisson, *Memoire sur lintegration des equations lineaires aux derivees partielles*, Journal de L'école Polytechnique **12** (1823), 19.
54. M. Rascle, *On the static and dynamic study of oscillations for some nonlinear hyperbolic systems of conservation laws*, Anal. Non Linéaire **8** (1991), 333-350.
55. R.D. Richtmyer and K. W. Morton, *Difference Methods for Initial Value Problems*, 2nd ed., Wiley-Interscience, New York, 1967.
56. B. Riemann, *Uber die Fortpflanzung ebener Luftwellen von endlicher Schwingungsweite*, Koenig. Gesell. Wiss., Goettingen **8**, (1860), 43.
57. B. Rubino, *On the vanishing viscosity approximation to the Cauchy problem for a 2×2 system of conservation laws*, Ann. Inst. H. Poincare Anal. Non Linéaire (to appear).
58. D. Schaeffer and M. Shearer, *The classification of 2×2 systems of nonstrictly hyperbolic conservation laws, with application to oil recovery*, Comm. Pure Appl. Math. **40** (1987), 141-178.
59. D. Schaeffer and M. Shearer, *Riemann problems for nonstrictly hyperbolic 2×2 systems of conservation laws*, Trans. AMS. **304** (1987), 267-306.
60. M. Shearer, D. Schaeffer, D. Marchesin, and P.J. Paes-Leme, *Solution of the Riemann problem for a prototype 2×2 system of nonstrictly hyperbolic conservation laws*, Arch. Rat. Mech. Anal. **97** (1987), 299-320.
61. D. Serre, *La compacité par compensation pour les systémes hyperboliques non linéaires de deux équations à une dimension d'espace*, J. Math. Pures Appl. **65** (1986), 423-468.
62. J. Smoller, *Shock Waves and Reaction-Diffusion Equations*, Springer-Verlag, New York, 1982.
63. S. Takeno, *Initial boundary problems for isentropic gas dynamics*, Proc. Royal Soc. Edinburgh **120A** (1992), 1-23.
64. L. Tartar, *Compensated compactness and applications to partial differential equations*, Research Notes in Mathematics, Nonlinear Analysis and Mechanics, ed. R. J. Knops, Vol. 4, Pitman Press, New York, 1979.
65. Wu Xin-Mou, *Equations of Mathematical Physics*, Chinese Advanced Education Press, Beijing, 1956 (in Chinese).
66. Yang Guang-Jin, *The Euler-Poisson-Darboux Equations*, Yuannan Education Press, Yuannan, 1989 (in Chinese).
67. Ying Lung-an and Teng Zhen-huan, *Hyperbolic Systems of Conservation Laws and Numerical Methods*, Academic Press, Beijing, 1991 (in Chinese).
68. Zhang Tong and Guo Yu-Fa, *A class of initial-value problem for systems of aerodynamic equations*, Acta Math. Sinica **15** (1965), 386-396.
69. B. Zhang, *Convergence of the Godunov scheme for a simplified one-dimensional hydrodynamic model for semiconductor devices*, Commun. Math. Phys. (1993).

Department of Mathematics, The University of Chicago
5734 S. University Avenue
Chicago, IL 60637
USA

AMS/IP Studies in Advanced Mathematics
Volume **3**, 1997

ENTROPY FOR HYPERBOLIC SYSTEMS OF CONSERVATION LAWS IN SEVERAL SPACE DIMENSIONS

by

C.M. DAFERMOS[1]

Abstract

The paper discusses how involutions may compensate for the absence of convexity in entropies associated with hyperbolic systems of conservation laws arising in classical physics.

1 Introduction

Nonlinear hyperbolic systems of conservation laws

$$\partial_t U + \sum_{\alpha=1}^{m} \partial_\alpha G^\alpha(U) = 0 \tag{1.1}$$

in several space dimensions will be a source of challenging mathematical problems for a long time to come. The theory will have to cope with the fundamental difficulties already encountered in the study of the Cauchy problem in one space dimension, namely the breakdown of smooth solutions and nonuniqueness within the class of solutions with shocks. In several dimensions, however, these difficulties are exacerbated by a number of complications [Ma].

Only L^2 norms are scale invariant under wave focussing or radiation from a point source. Consequently, the function class BV which is serving so effectively in one space dimension, seems to be inadequate for systems in several dimensions, unless the Jacobian matrices DG^α of the fluxes commute [Df2, Ra]. Until the issue of existence is finally settled, we should keep our

[1]Partially supported by the National Science Foundation under grant # DMS-9208284, by the Army Research Office under contract # DAAH04-93-G-0125, and by the Office of Naval Research under contract # N00014-92-J-1481.

options open by considering solutions in a very broad function class, like L^∞.

Any effort to construct solutions, as in one space dimension, by tracking waves would face serious obstacles: Solutions of multidimensional Riemann problems [CH] indicate that exceedingly intricate wave patterns emerge from even simple wave interactions.

Still another, somewhat unexpected, difficulty with systems of conservation laws in several space dimensions is induced by symmetry. Virtually all such systems of physical interest, in particular those arising in nonlinear elastodynamics, nonlinear electrodynamics, magnetohydrodynamics, general relativity, relativistic hydrodynamics, etc., exhibit symmetries partly manifested in the presence of constraints

$$\sum_{\alpha=1}^{m} \partial_\alpha P^\alpha(U) = 0 \tag{1.2}$$

that are automatically satisfied by solutions, so long as they are compatible with the initial data, and thus dubbed *involutions* [Bo1, Bo2, Df1]. Those of us with faith in the unity of Physics and Mathematics may trust that, ultimately, this special structure will prove helpful in the analysis of the systems; at the same time, however, we do recognize it as a source of complication. For example, symmetry often induces massive failure of strict hyperbolicity, with dire consequences already familiar in one space dimension. Another difficulty arises from the interaction between invariance and entropy principles.

Entropy inequalities

$$\partial_t \eta(U) + \sum_{\alpha=1}^{m} \partial_\alpha q^\alpha(U) \leq 0 \tag{1.3}$$

for the systems of conservation laws of Continuum Physics derive from the Clausius-Duhem inequality [TN] and express, explicitly or implicitly, the Second Law of Thermodynamics. Such inequalities are regarded by physicists as statements of stability. Ample supportive (mathematical) evidence for that view is available, provided that the entropy function $\eta(U)$ is convex. As a result, the requirement of convexity is often embodied in the definition itself of entropy [La]. The trouble is that in several space dimensions convexity of the entropy function is often incompatible with physical invariance (see Section 2).

Here we will experiment with the idea that the entropy function need not be convex. For a proper entropy it would suffice that, along trajectories generated by (1.1), the map $t \mapsto \eta(U(\cdot,t))$ be lower semicontinuous in L^∞ weak*. We shall give evidence that, for stability considerations, lack of convexity in the entropy function may be compensated by the presence of involutions; thus physical symmetry provides the cure for the ill it has caused.

The results of this paper may be extended in a routine manner to the case where the conservation laws and/or the system of involutions and/or the entropy inequality contain lower order "source" terms.

2 L^∞ Solutions and Proper Entropies

We consider a system of conservation laws (1.1). The state vector U takes values in a ball $\mathcal{O}$ centered at the origin in $I\!R^n$ and for $\alpha = 1,\cdots,m$, G^α are given, smooth functions from $\mathcal{O}$ to $I\!R^n$. The space variable x takes values in $I\!R^m$ and ∂_α denotes $\partial/\partial x^\alpha$. The time variable t takes values in some interval $[0,\omega)$ and ∂_t stands for $\partial/\partial t$.

The system is hyperbolic, i.e., for each unit vector ν in $I\!R^m$ and every fixed U in $\mathcal{O}$, the $n \times n$ matrix

$$\Lambda(U;\nu) = \sum_{\alpha=1}^{m} \nu_\alpha DG^\alpha(U) \tag{2.1}$$

has real eigenvalues $\lambda_1(U;\nu) \leq \cdots \leq \lambda_n(U;\nu)$ and linearly independent eigenvectors $R_1(U;\nu),\cdots,R_n(U;\nu)$.

As a model for the ensuing discussion we consider the system of (nonlinear) elastodynamics in Lagrangian coordinates:

$$\begin{cases} \partial_t F^i_\alpha = \partial_\alpha v^i & , \quad i = 1,\cdots,m; \quad \alpha = 1,\cdots,m \\ \partial_t v^i = \sum_{\alpha=1}^{m} \partial_\alpha T^\alpha_i(F) & , \quad i = 1,\cdots,m, \end{cases} \tag{2.2}$$

where v denotes velocity, F is the deformation gradient, and T stands for the Piola-Kirchhoff stress, which is the gradient of the strain energy function σ, i.e.,

$$T^\alpha_i(F) = \frac{\partial \sigma(F)}{\partial F^i_\alpha} \ , \quad i = 1,\cdots,m; \quad \alpha = 1,\cdots,m. \tag{2.3}$$

The principle of *material frame indifference* [TN] ordains that σ be invariant under rigid rotations:

$$\sigma(OF) = \sigma(F) \text{ for all orthogonal matrices } O. \tag{2.4}$$

The system (2.2) is hyperbolic if for every fixed F and any unit vector ν in $I\!R^m$ the *acoustic tensor* [TN]

$$Q_{ij}(F;\nu) = \sum_{\alpha=1}^{m}\sum_{\beta=1}^{m} \frac{\partial^2\sigma(F)}{\partial F^i_\alpha \partial F^j_\beta}\nu_\alpha\nu_\beta \tag{2.5}$$

is positive definite. The $m^2 + m$ characteristic speeds are the $2m$ square roots of the m eigenvalues of Q, and zero, with multiplicity $m^2 - m$.

A second example is provided by Maxwell's equations:

$$\begin{cases} \partial_t B = -\operatorname{curl} E \\ \partial_t D = \operatorname{curl} H \end{cases} \tag{2.6}$$

where B is the magnetic induction, D is the electric displacement, E is the electric field and H is the magnetic field, interrelated by constitutive equations

$$E = E(B,D) \ , \ H = H(B,D) \tag{2.7}$$

satisfying the *lossless condition*:

$$\frac{\partial H}{\partial D} = \frac{\partial E}{\partial B}. \tag{2.8}$$

We prescribe initial data

$$U(x,0) = U_0(x) \ , \ x \in I\!R^m, \tag{2.9}$$

where U_0 is in $L^\infty(I\!R^m;\mathcal{O})$, and consider (weak) solutions U of (1.1), (2.9) in $L^\infty(I\!R^m \times [0,\omega);\mathcal{O})$, that is

$$\int_0^\omega \int_{I\!R^m} [\partial_t\phi U + \sum_{\alpha=1}^{m} \partial_\alpha\phi G^\alpha(U)]dxdt + \int_{I\!R^m} \phi(x,0)U_0(x)dx = 0 \tag{2.10}$$

holds for all Lipschitz continuous test functions ϕ with compact support in $I\!R^m \times [0,\omega)$. For convenience, we will be assuming throughout that $U_0(\cdot)$ and $U(\cdot,t)$ have compact support in $I\!R^m$, for all t in $[0,\omega)$.

Proposition 2.1. *After redefining, if necessary, U on a set of measure zero, the map $t \mapsto U(\cdot, t)$ is continuous in L^∞ weak*.*

Proof. Fix $0 < s < \tau < \omega$. Let $\chi \in C_0^\infty(\mathbb{R}^m)$. For positive small δ, ε, we apply (2.10) with $\phi(x,t) = \chi(x)\theta(t)$, where $\theta(t) = 0$ for $0 \leq t < s - \delta, \theta(t) = \delta^{-1}(t-s) + 1$ for $s - \delta \leq t < s, \theta(t) = 1$ for $s \leq t < \tau, \theta(t) = \varepsilon^{-1}(\tau - t) + 1$ for $\tau \leq t < \tau + \varepsilon$, and $\theta(t) = 0$ for $\tau + \varepsilon \leq t < \omega$. This yields

$$\begin{aligned}(2.11)\qquad & \frac{1}{\delta}\int_{s-\delta}^{s}\int_{\mathbb{R}^m} \chi(x)U(x,t)dxdt - \frac{1}{\varepsilon}\int_{\tau}^{\tau+\varepsilon}\int_{\mathbb{R}^m} \chi(x)U(x,t)dxdt \\ & + \int_s^\tau \int_{\mathbb{R}^m} \sum_{\alpha=1}^m \partial_\alpha \chi(x) G^\alpha(U(x,t))dxdt = O(\delta) + O(\varepsilon).\end{aligned}$$

From (2.11) it follows that we may modify U on a set of measure zero so that

$$(2.12)\qquad \int_{\mathbb{R}^m} \chi(x)U(x,\tau)dx = \lim_{\varepsilon \to 0} \frac{1}{2\varepsilon}\int_{\tau-\varepsilon}^{\tau+\varepsilon}\int_{\mathbb{R}^m} \chi(x)U(x,t)dxdt$$

holds for any τ in $(0,\omega)$ and all χ in $L^1(\mathbb{R}^m)$ and this normalization renders the map $t \mapsto U(\cdot,t)$ continuous in L^∞ weak*. The same argument shows that as $t \downarrow 0$ $U(\cdot,t)$ converges to $U_0(\cdot)$ in L^∞ weak*. The proof is complete.

Entropy inequalities (1.3) are employed to weed out physically inadmissible solutions of (1.1). We want all Lipschitz solutions to be admissible and since these are time-reversible they must satisfy (1.3) as an equality. This leads to the standard conditions

$$(2.13)\qquad Dq^\alpha = D\eta DG^\alpha \ , \ \alpha = 1, \cdots, m,$$

relating the entropy η with the entropy flux q. The compatibility conditions on η resulting from (2.13) are

$$(2.14)\qquad H\eta DG^\alpha = (DG^\alpha)^T H\eta \ , \ \alpha = 1, \cdots, m,$$

where $H\eta$ denotes the Jacobian matrix of η.

In the context of L^∞ solutions, (1.3) should be interpreted as

$$(2.15)\qquad \int_0^\omega \int_{\mathbb{R}^m} [\partial_t \psi \eta(U) + \sum_{\alpha=1}^m \partial_\alpha \psi q^\alpha(U)]dxdt + \int_{\mathbb{R}^m} \psi(x,0)\eta(U_0(x))dx \geq 0$$

for all nonnegative, Lipschitz continuous test functions ψ with compact support in $\mathbb{R}^m \times [0,\omega)$.

For the system (2.2) of elastodynamics, the natural choice is

$$\eta(F,v) = \frac{1}{2}|v|^2 + \sigma(F) \ , \quad q^\alpha(F,v) = -\sum_{i=1}^{m} v^i T_i^\alpha(F), \tag{2.16}$$

i.e., "entropy" is mechanical energy and "entropy flux" is the rate of work.

For the system (2.6) of electromagnetism

$$\eta(B,D) = \frac{1}{4\pi}\int\{H\cdot dB + E\cdot dD\} \ , \quad q(B,D) = \frac{1}{4\pi}E\times H, \tag{2.17}$$

that is "entropy" is electromagnetic field energy and "entropy flux" is the Poynting vector.

For inequalities (1.3) to induce stability, conditions are needed on the entropy. The thesis of this paper is that the natural assumption on η is provided by

Definition 2.1. The entropy η for the system (1.1) is *proper* if along the trajectories of solutions $U(\cdot,t)$ of (1.1) the map $t\mapsto\eta(U(\cdot,t))$ is lower semicontinuous in L^∞ weak*.

In view of Proposition 2.1, every convex entropy is proper. The assumption that entropy is convex is natural in many physical models (for instance elastic fluids) and has become standard in the theory [La]. In many situations, however, especially in several space dimensions, convexity of the entropy is contrary to experience or even in direct conflict with physical laws. For example, in the case of elastodynamics it is well known that convexity of the entropy η in (2.16), i.e., convexity of the strain energy function σ, is generally incompatible with the principle of material frame indifference (2.4) and also contrary to physical experience (it would rule out backling, for instance).

The aim of this paper is to discuss nonconvex proper entropies and their implications to stability.

3 Involutions

We consider a system (1.1) of hyperbolic conservation laws endowed with a system (1.2) of involutions. For $\alpha = 1,\cdots,m, P^\alpha$ is defined on $\mathcal{O}$ and takes values in $\mathbb{R}^k$. Every (weak) solution of (1.1) satisfies automatically (1.2), for all t, so long as it does so at $t = 0$. Here we shall restrict attention to the special but very common (in the applications) case of linear involutions [Df1]:

$$\sum_{\alpha=1}^{m} A^{\alpha}\partial_{\alpha}U = 0, \tag{3.1}$$

where $A^{\alpha}, \alpha = 1, \cdots, m$, are constant $k \times n$ matrices.

Both our model systems (2.2) and (2.6) have linear involutions, namely,

$$\partial_{\beta}F^{i}_{\alpha} - \partial_{\alpha}F^{i}_{\beta} = 0 \ , \ i = 1, \cdots, m \ ; \ \alpha, \beta = 1, \cdots, m \tag{3.2}$$

and

$$divB = 0 \ , \ divD = 0, \tag{3.3}$$

respectively. Nevertheless, there are examples of systems, arising in Physics, with nonlinear involutions. For example, if one formulates elastodynamics in Eulerian, rather than Lagrangian, coordinates, the involutions (3.2) become

$$\sum_{j=1}^{m} \partial_j[(\det F)^{-1}(F^{j}_{\beta}F^{i}_{\alpha} - F^{j}_{\alpha}F^{i}_{\beta})] = 0 \ , \ i = 1, \cdots, m \, ; \, \alpha, \beta = 1, \cdots, m. \tag{3.4}$$

A collection $A^{\alpha}, \alpha = 1, \cdots, m$, of $k \times n$ matrices generates a system of involutions (3.1) for (1.1) if

$$A^{\alpha}G^{\beta}(U) + A^{\beta}G^{\alpha}(U) = S^{\alpha}A^{\beta}U + S^{\beta}A^{\alpha}U \ , \ \ \alpha, \beta = 1, \cdots, m, \tag{3.5}$$

where $S^{\alpha}, \alpha = 1, \cdots, m$, are constant $k \times k$ matrices (cf. [Bo2, Df1]). This follows immediately from the observation that when (3.5) holds the k-vector field $V(x,t)$ on the left-hand side of (3.1), with U any (weak) solution of (1.1), satisfies the linear transport equation

$$\partial_t V + \sum_{\alpha=1}^{m} S^{\alpha}\partial_{\alpha}V = 0. \tag{3.6}$$

Condition (3.5) is satisfied for our models of elastodynamics (2.2), (3.2) and electromagnetism (2.6), (3.3), in both cases with $S^{\alpha} = 0, \alpha = 1, \cdots, m$. The equations of general relativity are cited [Bo2] as an example of a system with linear involutions in which (3.5) hold for $S^{\alpha} \neq 0$. Also note that the vector field $V(x,t)$ on the left-hand side of the nonlinear involutions (3.4) of elastodynamics in Eulerian coordinates satisfies the transport equation

$$\partial_t V + \sum_{j=1}^{m} \partial_j(v^j V) = 0. \tag{3.7}$$

Assuming (3.5) holds, we set, for each unit vector ν in $\mathbb{R}^m$,

$$N(\nu) = \sum_{\alpha=1}^{m} \nu_\alpha A^\alpha, \tag{3.8}$$

$$M(\nu) = \sum_{\alpha=1}^{m} \nu_\alpha S^\alpha. \tag{3.9}$$

Recalling the notation (2.1), it follows easily from (3.5) that

$$N(\nu)\Lambda(U;\nu) = M(\nu)N(\nu). \tag{3.10}$$

In particular, any eigenvector $R(U;\nu)$ of $\Lambda(U;\nu)$ must lie in the kernel of $N(\nu)$, unless $N(\nu)R(U;\nu)$ is an eigenvector of $M(\nu)$, with eigenvalue $\lambda(U;\nu)$. Since $M(\nu)$ is a constant matrix, the latter may occur only when $\lambda(U;\nu)$ and $R(U;\nu)$ are independent of U (linearly degenerate characteristic field). Note that solutions $U = u(\nu \cdot x - \lambda t)R(\nu)$ of (1.1) generated by any linearly degenerate characteristic field with $N(\nu)R(\nu) \neq 0$ are suppressed by the involutions (3.1).

The situation with shocks is quite analogous. Shock waves propagating in the direction of the unit vector ν in $\mathbb{R}^m$ with speed s satisfy the Rankine-Hugoniot jump condition

$$\sum_{\alpha=1}^{m} \nu_\alpha [G^\alpha(U)] = s[U]. \tag{3.11}$$

Combining (3.11), (3.5), (3.8) and (3.9) we obtain

$$M(\nu)N(\nu)[U] = sN(\nu)[U]. \tag{3.12}$$

One possibility is that the amplitude $[U]$ lies in the kernel of $N(\nu)$. This will include, in particular, all shocks associated with solutions of (1.1) satisfying the involutions. Indeed, the Rankine-Hugoniot jump conditions of (3.1) are

$$N(\nu)[U] = \sum_{\alpha=1}^{m} \nu_\alpha A^\alpha [U] = 0. \tag{3.13}$$

The other possibility for (3.12) is that $N(\nu)[U]$ is a (nonzero) eigenvector of $M(\nu)$ with eigenvalue s. Again this may occur only when s is constant along its shock curve, i.e., the shock is associated with a linearly degenerate characteristic field.

The above discussion motivates the following

Definition 3.1. The *wave cone* $\mathcal{K}$ in $\mathbb{R}^n$ of the system of involutions (3.1) is the union of the kernels of $N(\nu)$ over all unit vectors ν in $\mathbb{R}^m$.

In particular, as shown above, the wave cone contains every nondegenerate characteristic direction as well as the amplitude of all nondegenerate shocks.

For our model system (2.2) a simple calculation shows that the wave cone $\mathcal{K}$ consists of all vectors $(\xi \otimes \nu, w)$ in $\mathbb{R}^{m^2+m}$ with arbitrary ξ, ν, w in $\mathbb{R}^m$.

4 Proper Entropies in the Presence of Linear Involutions

Here we assume that our system (1.1) is endowed with k linear involutions (3.1) and seek to characterize proper entropies, in the sense of Definition 2.1. The same question arises under a different guise in the Calculus of Variations and is being addressed there in the framework of the theory of Compensated Compactness [Da, Ta].

Definition 4.1. An entropy η for the system (1.1) endowed with linear involutions (3.1) is *A-quasiconvex* if

$$\int_{\mathcal{C}} \eta(\overline{U})dx \leq \int_{\mathcal{C}} \eta(U)dx \tag{4.1}$$

holds for every hypercube $\mathcal{C}$ in $\mathbb{R}^m$, each fixed $\overline{U}$ in $\mathcal{O}$ and any $U \in L^\infty(\mathbb{R}^m; \mathcal{O})$ such that

$$\sum_{\alpha=1}^{m} A^\alpha \partial_\alpha U = 0, \tag{4.2}$$

$$\int_{\mathcal{C}} [U(x) - \overline{U}]dx = 0. \tag{4.3}$$

The above concept of A-quasiconvexity was introduced by Dacorogna [Da] as an extension of the notion of quasiconvexity in the sense of Morrey [Mo]. It admits a very natural interpretation in the present context as a statement that, among all fields with the same "mass", entropy is extremized at the equilibrium state, in the spirit of classical Thermodynamics. Every convex η is A-quasiconvex (Proposition 2.2 in [Da]). An immediate corollary of Definition 2.1, Proposition 2.1, and Theorem 2.3 in [Da] is

Proposition 4.1. *Any A-quasiconvex entropy η for the system* (1.1) *with linear involutions* (3.1) *is proper.*
In fact, A-quasiconvexity would also be a necessary condition for a proper entropy if the only constraints on U were those induced by the involutions (3.1) (Theorem 2.1 in [Da]).

A very important consequence of A-quasiconvexity is (cf. [Ta] or Theorem 3.1 in [Da]):

Proposition 4.2. *If η is A-quasiconvex, then it is convex in the direction of the wave cone $\mathcal{K}$ (Definition 3.1), i.e., for each fixed $\overline{U} \in \mathcal{O}$ and $\xi \in \mathcal{K}$ the function $\eta(\overline{U} + s\xi)$ of the real variable s is convex.*

Note that, in general, it is not easy to determine whether a given entropy is A-quasiconvex while convexity in the direction of $\mathcal{K}$ may be easily tested. It turns out that convexity in the direction of the wave cone is equivalent to A-quasiconvexity when η is quadratic (Theorem 4.1 in [Da]) but not in general.

To illustrate the above, let us consider our model system (2.2) of elastodynamics. Recalling the discussion at the end of Section 3, we conclude that the entropy in (2.16) will be convex in the direction of the wave cone if and only if the strain energy function $\sigma(F)$ is rank-one convex (Legendre-Hadamard condition), i.e., the acoustic tensor (2.5) is nonnegative definite. Rank-one convexity is necessary but not generally sufficient for quasiconvexity [Sv]. A sufficient condition for quasiconvexity is polyconvexity in the sense of Ball [Ba]: σ is *polyconvex* (in three dimensions, $m = 3$) if

$$\sigma(F) = W(F, F^*, \det F) \tag{4.4}$$

where F^* denotes the adjucate of F, $F^* = (\det F)F^{-1}$, and W is a convex function. Thus, if σ is polyconvex, the entropy η in (2.16) is proper.

5 Proper Entropies, Involutions and Stability

We consider the system (1.1) endowed with linear involutions (3.1) and an A-quasiconvex (and thereby proper) entropy η which is uniformly convex in the direction of the wave cone $\mathcal{K}$ (recall Proposition 4.2), i.e.,

$$\xi^T H\eta(U)\xi \geq \mu|\xi|^2 \ , \ \xi \in \mathcal{K} \ , \ U \in \mathcal{O}, \tag{5.1}$$

where $H\eta$ denotes the Hessian matrix of η and $\mu > 0$. For simplicity, we assume (3.5) holds with $S^\alpha = 0, \alpha = 1, \cdots, m$, and that the dimension of the null space of $\Lambda(U;\nu)$ equals the rank of $N(\nu)$, for any unit vector ν is $I\!R^m$ (recall (3.10)).

It was shown in [Df1] that, when they exist, Lipschitz solutions of (1.1) are stable within the class of BV solutions which satisfy inequality (1.3) with η a proper entropy. Here we will establish stability of Lipschitz solutions within the even broader class of L^∞ solutions.

Proposition 5.1. *Let $\overline{U}$ and U be solutions of* (1.1) *on $I\!R^m \times [0,\omega)$ taking values in a compact subset of $\mathcal{O}$ and vanishing outside a bounded cylinder $\mathcal{B} \times [0,\omega)$. Assume $\overline{U}$ is (uniformly) Lipschitz while U is in L^∞ and satisfies the entropy inequality* (1.3). *Furthermore,*

$$\sum_{\alpha=1}^{m} A^\alpha \partial_\alpha [U(x,0) - \overline{U}(x,0)] = 0, \tag{5.2}$$

$$\int_{I\!R^m} [U(x,0) - \overline{U}(x,0)] dx = 0. \tag{5.3}$$

Then there are positive constants κ, C, c, γ such that if the oscillation of U over any ball of arbitrarily small radius does not exceed κ we have, for $\tau \in [0,\omega)$,

(5.4)
$$\int_{I\!R^m} |U(x,\tau) - \overline{U}(x,\tau)|^2 dx \leq C(1+\tau)\exp(\gamma\tau + c\tau^2) \int_{I\!R^m} |U(x,0) - \overline{U}(x,0)|^2 dx.$$

Proof. It is very similar with the proof of Theorem 4.1 in [Df1] (see also [Di]) and so here we shall only provide a sketch, emphasizing the points where switching from BV to L^∞ may make a difference.

For any vectors $\overline{U}$ and U in $\mathcal{O}$ we set

$$h(U,\overline{U}) = \eta(U) - \eta(\overline{U}) - D\eta(\overline{U})[U - \overline{U}], \tag{5.5}$$

$$f^\alpha(U,\overline{U}) = q^\alpha(U) - q^\alpha(\overline{U}) - D\eta(\overline{U})[G^\alpha(U) - G^\alpha(\overline{U})], \tag{5.6}$$

$$Z^\alpha(U,\overline{U}) = G^\alpha(U) - G^\alpha(\overline{U}) - DG^\alpha(\overline{U})[U-\overline{U}]. \tag{5.7}$$

We fix any nonnegative, Lipschitz continuous test function ψ with compact support in $I\!R^m \times [0,\omega)$, evaluate h and f^α along the two solutions $U(x,t), \overline{U}(x,t)$ and use (2.15) (note that (2.15) holds as an equality for $\overline{U}$, because $\overline{U}$ is Lipschitz continuous) to infer

(5.8)

$$\int_0^\omega \int_{I\!R^m} [\partial_t\psi h(U,\overline{U}) + \sum_{\alpha=1}^m \partial_\alpha\psi f^\alpha(U,\overline{U})]dxdt + \int_{I\!R^m} \psi(x,0)h(U(x,0),\overline{U}(x,0))dx$$

$$\geq -\int_0^\omega \int_{I\!R^m} [\partial_t\psi D\eta(\overline{U})[U-\overline{U}] + \sum_{\alpha=1}^m \partial_\alpha\psi D\eta(\overline{U})[G^\alpha(U) - G^\alpha(\overline{U})]]dxdt$$

$$-\int_{I\!R^m} \psi(x,0)D\eta(\overline{U}(x,0))[U(x,0)-\overline{U}(x,0)]dx.$$

We apply (2.10), using the appropriate components of the (Lipschitz continuous) vector field $\psi D\eta(\overline{U})$ as test function ϕ, which yields

$$\int_0^\omega \int_{I\!R^m} [\partial_t[\psi D\eta(\overline{U})][U-\overline{U}] + \sum_{\alpha=1}^m \partial_\alpha[\psi D\eta(\overline{U})][G^\alpha(U) - G^\alpha(\overline{U})]]dxdt \tag{5.9}$$

$$+\int_{I\!R^m} \psi(x,0)D\eta(\overline{U}(x,0))[U(x,0)-\overline{U}(x,0)]dx = 0.$$

We recall that $\overline{U}$ is a classical solution of (1.1) and use (2.14) to get

(5.10)

$$\partial_t D\eta(\overline{U}) = \partial_t\overline{U}^T H\eta(\overline{U}) = -\sum_{\alpha=1}^m \partial_\alpha\overline{U}^T G^\alpha(\overline{U})H\eta(\overline{U}) = -\sum_{\alpha=1}^m \partial_\alpha\overline{U}^T H\eta(\overline{U})G^\alpha(\overline{U})$$

so that, by virtue of (5.7),

$$\partial_t D\eta(\overline{U})[U-\overline{U}] + \sum_{\alpha=1}^m \partial_\alpha D\eta(\overline{U})[G^\alpha(U) - G^\alpha(\overline{U})] \tag{5.11}$$

$$= \sum_{\alpha=1}^m \partial_\alpha\overline{U}^T H\eta(\overline{U})Z^\alpha(U,\overline{U}).$$

In (5.8) we choose a test function ψ such that $\psi = 1$ on $\mathcal{B} \times [0,\tau], \psi(x,t) = 1 - \varepsilon^{-1}(t-\tau)$ for $x \in \mathcal{B}, \tau < t \leq \tau + \varepsilon$, and $\psi(x,t) = 0$ for $t > \tau + \varepsilon$. Combining (5.8), (5.9), and (5.11) we obtain

(5.12)
$$\frac{1}{\varepsilon}\int_{\tau}^{\tau+\varepsilon}\int_{\mathbb{R}^m} h(U,\overline{U})dxdt - \int_{\mathbb{R}^m} h(U(x,0),\overline{U}(x,0))dx$$
$$\leq -\int_0^{\tau+\varepsilon}\int_{\mathbb{R}^m}\psi\sum_{\alpha=1}^m \partial_\alpha \overline{U}^T H\eta(\overline{U})Z^\alpha(U,\overline{U})dxdt.$$

We let $\varepsilon \downarrow 0$. Recalling that $\overline{U}$ is Lipschitz continuous, that η is proper (Definition 2.1) and Proposition 2.1, we conclude

(5.13)
$$\int_{\mathbb{R}^m} h(U(x,\tau),\overline{U}(x,\tau))dx \leq \int_{\mathbb{R}^m} h(U(x,0),\overline{U}(x,0))dx$$
$$-\int_0^\tau\int_{\mathbb{R}^m}\sum_{\alpha=1}^m \partial_\alpha \overline{U}^T H\eta(\overline{U})Z^\alpha(U,\overline{U})dxdt.$$

Once (5.13) has been established, we may demonstrate (5.4) by retracing, step by step, the proof of Theorem 4.1 in [Df1]. The key point is to show that when the local oscillation of $\overline{U}$ is sufficiently small,

(5.14)
$$\int_{\mathbb{R}^m} h(U(x,\tau),\overline{U}(x,\tau))dx \geq \delta\|U(\cdot,\tau)-\overline{U}(\cdot,\tau)\|_{L^2}^2 - c_1\|U(\cdot,\tau)-\overline{U}(\cdot,\tau)\|_{H^{-1}}^2.$$

The proof of (5.14) hinges on that $U, \overline{U}$ satisfy the involutions (3.1) and η is convex in the direction of the wave cone. The last term on the right-hand side of (5.14) may be estimated by

(5.15)
$$\|U(\cdot,\tau)-\overline{U}(\cdot,\tau)\|_{H^{-1}}$$
$$\leq \|U(\cdot,0)-\overline{U}(\cdot,0)\|_{H^{-1}} + \int_0^\tau \|\partial_t[U(\cdot,t)-\overline{U}(\cdot,t)]\|_{H^{-1}}dt$$
$$= \|U(\cdot,0)-\overline{U}(\cdot,0)\|_{H^{-1}} + \int_0^\tau \|\sum_{\alpha=1}^m \partial_\alpha[G^\alpha(U(\cdot,t)) - G^\alpha(\overline{U}(\cdot,t))]\|_{H^{-1}}dt$$
$$\leq \|U(\cdot,0)-\overline{U}(\cdot,0)\|_{L^2} + \sum_{\alpha=1}^m\int_0^\tau \|G^\alpha(U(\cdot,t)) - G^\alpha(\overline{U}(\cdot,t))\|_{L^2}dt.$$

Finally, (5.4) follows from (5.13), (5.14), (5.15) and the generalized Gronwall inequality. This completes the sketch of the proof of Proposition 5.1.

References

[Ba] J.M. Ball, Convexity conditions and existence theorems in nonlinear elasticity. Arch. Rat. Mech. Anal. **63** (1977), 337-403.

[Bo1] G. Boillat, Symétrization des systèmes d' équations aux dérivées partielles avec densité d' énergie convexe et contraintes. C.R. Acad. Sc. Paris **295** (1982), 551-554.

[Bo2] G. Boillat, Involutions des systèmes conservatifs. C.R. Acad. Sc. Paris **307** (1988), 891-894.

[CH] T. Chang, and L. Hsiao, *The Riemann Problem and Interaction of Waves in Gas Dynamics.* Essex : Longman 1989.

[Da] B. Dacorogna, *Weak Continuity and Weak Lower Semicontinuity of Non-Linear Functionals.* Lecture Notes in Math. No. 922. Berlin : Springer-Verlag 1982.

[Df1] C.M. Dafermos, Quasilinear hyperbolic systems with involutions. Arch. Rat. Mech. Anal. **94** (1986), 373-389.

[Df2] C.M. Dafermos, L^p stability for systems of conservation laws in several space dimensions (to appear).

[Di] R.J. DiPerna, Uniqueness of solutions to hyperbolic conservation laws. Indiana U. Math. J. **28** (1979), 137-188.

[La] P.D. Lax, Shock waves and entropy. *Contributions to Functional Analysis*, ed. E.A. Zarantonello. New York : Academic Press 1971, pp. 603-634.

[Ma] A. Majda, *Compressible Fluid Flow and Systems of Conservation Laws in Several Space Variables.* New York: Springer-Verlag 1984.

[Mo] C.B. Morrey, Quasiconvexity and the lower semicontinuity of multiple integrals, Pacific J. Math. **2** (1952), 25-53.

[Ra] J. Rauch, *BV* estimates fail for most quasilinear hyperbolic systems in dimensions greater than one. Commun. Math. Phys. **106** (1986), 481-484.

[Sv] V. Šverák, Rank-one convexity does not imply quasiconvexity. *Proc. Roy. Soc. Edinburgh* **120** (1992), 185-189.

[Ta] L. Tartar, Compensated compactness and applications to partial differential equations. *Nonlinear Analysis and Mechanics* Vol. IV, ed. R. J. Knops. London : Pitman 1979, pp. 136-212.

[TN] C. Truesdell, and W. Noll, *The Nonlinear Field Theories of Mechanics. Handbuch der Physik III/3*, ed. S. Flügge. Berlin : Springer-Verlag 1965.

Brown University
Providence, RI 02912 USA

AMS/IP Studies in Advanced Mathematics
Volume **3**, 1997

Some Results on Nonlinear Evolution Equations[1]

XIA-XI DING

§1. INTRODUCTIONS

Our research work on nonlinear evolution equations began from 1958. In order to solve some practical problems in water wave and gas dynamics, we need to study the theoretical basis of shock waves. Motivated by the famous paper of E.Hopf on Burgers equation published in 1950, i.e. the paper "The partial differential equation $u_t + uu_x = \mu u_{xx}$, CPAM, 3(1950), our work has been developed along two lines on these years. Firstly, corresponding to the case $\mu = 0$, we have been devoted to hyperbolic equations and systems. We have worked on this direction for several decades. Secondly, corresponding to the case $\mu > 0$, we investigate parabolic equations and systems. The work in this direction includes parabolic equation with strong nonlinearity, reaction and diffusion equations, Navier-Stokes equations, and so on.

In the following *I* would like to introduce some work of mine and my students.

§2. NONLINEAR HYPERBOLIC EQUATIONS.

The research work on nonlinear hyperbolic equations has been developed on two lines. One is the study of Riemann problem and the interactions of elementary waves, either in one dimensional case or in two dimensional case. Professor Zhang Tung and Xiao Ling and their students have done a lot in this direction [1]. They will give lecture on this topic in this conference. Another one is the study of the general initial value problem with different approaches, such as difference methods, viscosity method, etc. For difference methods we have ever used Lax-Friedrichs scheme, Godunov scheme, Glimm scheme. Using Glimm scheme we proved some global existence theorem for certain special kind of initial value problems; including the interaction of shock wave and rarefaction wave for the

[1]Partially supported by grants from National science Foundation

isentropic gas dynamical systems. In 1985-1986, using compensated compactness principle we (Ding, Chen, Luo) proved the convergence of the Lax-Friedrichs scheme for isentropic gas dynamics equatuions with general initial values including vacuums [3][4][5] . Then we got the global existence theorems for isentropic gas case, i.e. the adiabatic exponent $1 < \gamma \leq \frac{5}{3}$. Afterwards we generalized the above result to the case of Godunov scheme and non-homogeneous equation [6]. After this work, we still have some people in China working continuously in the viscosity method and compensated compactness theory. For example, we consider how to avoid the use of measure theory in the compensated compactness throry. We (Chen, Lu) have got some results in this respect. In the case of single conservation law we succeeded in get the positive result at the same time as that was done by Tartar. Also we get [11] the global existence theorem for a combustion model

$$\begin{cases} u_t + qz + f(u)_x = 0, \\ z_t + kg(u)z = 0, \end{cases}$$

when the initial value $(u_0, z_0) \in L_p$.

In the applications of compensated compactness principle we get other results as well. For examples we proved some global existence theorems for the non-strictly hyperbolic system [9]

$$\begin{cases} v_t + (uv)_x = 0, \\ u_t + (\frac{1}{2}u^2 + p(v))_x = 0. \quad P(v) = \int_0^v s(s+\delta)^{\gamma-3} ds, \gamma > 3, \end{cases}$$

and the quadratic conservation laws

$$\begin{cases} v_t + \frac{1}{2}(3u^2 + v^2)_x = 0 \\ v_t + uv_x = 0. \end{cases}$$

We also get results on the interactions of elementary waves of these systems.

I would like to say it is professor Lu Yunguang mainly working in these topics.

Besides these *I* should still talk about some work by Wu Yonghui on the global solutions of the two-dimentional imcompressible Euler equations.

Many people worked on the initial value problem of the system

$$\begin{cases} \frac{\partial \vec{u}}{\partial t} - \varepsilon \triangle \vec{u} + (\vec{u} \cdot \nabla)\vec{u} = -\nabla p, \\ div\ \vec{u} = 0, \\ \vec{u}(x, 0) = \vec{u}_0(x), div\ \vec{u}_0 = 0. \end{cases}$$

For examples:

(1) When the initial curl $\omega_0 = \frac{\partial u_{02}}{\partial x_1} - \frac{\partial u_{01}}{\partial x_2} \in L^p(R^2) \cap L^1(R^2), (p > 1)$, Majda and Diperna proved the global existence of the solutions of the above system;

(2) When $\omega_0 \in L^1(R^2) \cap H^{-1}_{loc}(R^2), \omega_0(x) \geq 0, u_0(x) \in L_2(R^2)$,, J. M. Delort proved the analogous global existence theorem;

(3) When $\omega_0 \in L(Log^+.L)$,, Supp ω_0 is bounded, $u_0 \in L^2(R^2)$, Chae Dongho proved the analogous global existence theorem.

Using Orlicz space theory, Wu Yonghui proved the analogous global existence theorem for $u_0(x) \in L^2(R^2), \omega_0 \in L(log^+L)^\alpha, (\alpha > \frac{1}{2})$ in his thesis, which is an improvement to all of the above results.

Recently we [23] studied the Riemann problem and general initial value problem of the system

$$\begin{cases} u_t + uu_x = 0, \\ \omega_t + u\omega_x = 0. \end{cases}$$

where we adopted Lebesgue-Stieltjes integration in the definition of generalized solutions in stead of using Lebesgue integral.

§3. Nonlinear parabolic equations

It was pointed in Hopf's paper that Burgers equation is only a model of Navier-Stokes equations. The true subject which need to be investigated thoroughly should be the 3-dimensional or its generalized n-dimensional Navier-Stokes equations:

$$\begin{cases} \vec{u}_t + (\vec{u}.\nabla)\vec{u} = -\nabla p + \gamma\triangle\vec{u}, \\ div\ \vec{u} = 0, \quad \vec{u} = (u_1, ..., u_n). \end{cases} \tag{N}$$

When $n = 2$, they take the form

$$\begin{cases} \frac{\partial u_i}{\partial t} + u_j \frac{\partial u_i}{\partial x_i} = -\frac{\partial p}{\partial x_i} + \gamma\triangle u_i, \\ \sum \frac{\partial u_i}{\partial x_i} = 0, \quad i = 1, 2, \end{cases} \tag{3.1}$$

or

$$\begin{cases} \frac{\partial u_i}{\partial t} + \frac{\partial(u_i u_j)}{\partial x_j} = -\frac{\partial p}{\partial x_i} + \gamma\triangle u_i, \\ \sum \frac{\partial u_i}{\partial x_i} = 0, \quad i = 1, 2. \end{cases} \tag{3.2}$$

In order to simplify the study, we omit the presure p and the equation of continuity, i.e,

$$\frac{\partial u_i}{\partial t} + u_j \frac{\partial u_i}{\partial x_j} = \gamma\triangle u_i, \quad i = 1, 2, \tag{3.3}$$

or

$$\frac{\partial u_i}{\partial t} + \frac{\partial(u_i u_j)}{\partial x_j} = \gamma \triangle u_i, \quad i = 1, 2. \tag{3.4}$$

For (3.3), Ding and Wu [12] proved the global existence theorem for its initial value problem. For the similar form (3.4), it is very difficult to prove the global existence theorem. Luo proved a global existence theorem for (3.4) in [13] when the initial value is small in some sense.

In [14] Teman discussed the large time behavior of solutions of the initial boundany value problem for system (3.1) in a bounded domain, i.e. for initial values in L_2 with zero divergence there always exists a global attractor and one can get the estimates of its Hausdauff and Fractal dimensions. Ding and Wu [15] have generalized this result to some unbounded domains such as strips. But for general unbounded domains, this is somehow very difficult, since one cannot get global estimates independent of initial values and also need to treat the problem that the Sobolev imbedding is not compact. So we consider a weaker problem, i.e. we consider Navier-Stokes equations with linear damp

$$\begin{cases} \frac{\partial u_i}{\partial t} + \sum \frac{\partial u_i u_j}{\partial x_j} + \lambda u_i = -\frac{\partial p}{\partial x_i} + \gamma \triangle u_i + f_i(x), \\ \sum \frac{\partial u_i}{\partial x_i} = 0. \end{cases} \tag{3.5}$$

In [16] we study the pure initial value problem:$u_i(x,0) = \phi_i(x)$.

In order to deal with the lack of compactness in Sobolev's imbedding we shall use the weighted Lebesgue L_p class $L_{p.l}$, i.e.

$$L_{p \cdot l} \equiv \{\varphi | (1 + x^2)^{\frac{l}{2}} \varphi \in L_p\}.$$

Firstly we prove the global existence throrem of the initial value problem of (3.5). Here we take the original Leray's iteration with some technical change. We hope it may be a goal to attack the higher-dimensional problem of Navier-Stokes equations.

Put

$$\begin{cases} \frac{\partial u_i^{(n)}}{\partial t} + \sum \frac{\partial (u_i^{(n-1)} u_j^{(N-1)}}{\partial x_j} + \lambda u_i^{(n-1)} = -\frac{\partial p^{(n-1)}}{\partial x_i} + \gamma \triangle u_i^{(n)} + f_i(x), \\ u_i(x,0) = \Phi_i(x), \end{cases} \tag{3.6}$$

or its equivalent system of integral equations

$$u_i^{(n)} = T(t)\Phi_i(x) + \int_0^t T(t-s)[-\frac{\partial p^{(n-1)}}{\partial x_i} + \sum (\frac{\partial}{\partial x_i} u_i^{(n-1)} u_j^{(n-1)} - \lambda u_i^{(n-1)} + f_i] ds, \tag{3.7}$$

where

$$T(t)\Phi = \frac{1}{4\pi t}\iint_{-\infty}^{\infty} \Phi(\varsigma)e^{-\frac{(x-\varsigma)^2}{4t}}d\varsigma.$$

Without loss of generality we may suppose div $f = 0$. Then we have

$$-\triangle p^{(n-1)} = \sum \frac{\partial^2}{\delta x_i \partial x_j}(u_i{}^{(n-1)} u_j^{(n-1)}),$$

Hence

$$p^{(n-1)} = C_i R_i R_j u_i^{(n-1)} u_j^{(n-1)},$$

where R_i is the Risez transform

$$R_i f(x) = C_0 \int_{R^2} \frac{y_i}{|y|^3} f(x-y)dy, \quad i = 1, 2,$$

$c_0 = \frac{2}{\pi}$.

Using the iteration (3.6) or (3.7), we can get solution in $C(0, T; H^2)$ for any $T > 0$.

If the initial value $\Phi(x) = (\Phi_1(x), \Phi_2(x)) \in H, i.e. \Phi_i(x) \in L_2$, div $\Phi_i(x) = 0$, we can get the existence of solutions in $L^\infty([0,\infty), H) \cap ([0,\infty), V)$ where

$$V \equiv \{\Phi \in H^1, div\Phi = 0\}.$$

If we suppose $\Phi_i(x) \in L_{2,1}$ again, then we can get $u(x,t) \in L^\infty([0,\infty), L_{2,1}$. From these results we can prove the existence of global attractor and get the dimension estimates.

Theorem. If $\Phi(x) \cap L_{2,1}(R^2) \in H$, then there exists a global attractor $A \equiv \omega(B)$ of the initial value problem of (3.5), and we have the estimates of Hausdauff dimension $d_H(A)$ and Fractal dimension $d_F(A)$ of A:

$$d_H(A) \le \frac{c_L}{4\lambda^2}|f|^2 + 1,$$

$$d_F(A) \le (\frac{c_L}{2\lambda^2}|f|^2 + 1),$$

where the constant c_L is the constant in Lieb-Thiring ineguality, $B \in V$ is the absorbing set of (3.5), $\omega(B)$ is the so-called ω set of B.

From the above we need consider the initial value problem with initial data in the weighted Lebesgue class $L_{r,\gamma}$ instead of L_p sometimes. We have developed a theory on this kind of problem. For example, we consider the Cauchy problem of the single superlinear conservation law with viscosity

$$\begin{cases} \frac{\partial u}{\partial t} + \sum \frac{\partial f_i(u)}{\partial x_i} + f_0(u) = \Delta u, \\ u(x,0) = \varphi(x) \in L_{p,\gamma}. \end{cases}$$

and get the solvability of the problem, where the solution $u \in L^\infty([0,\infty), L_{p,r})$[18]. Now if $f_i(u) \equiv 0, \gamma = 0, f_0 = |u|^{r-1}u$, the problem reduces to the case which Fujita and Weissler considered.

For n-dimensional Navier-Stokes equations [N] we have the following result[17].

Theorem. If $\Phi_i(x) \in L_{q,k\alpha}$, $||\Phi||_{q,k\alpha}$ is sufficiently small, then there exists a unique global solution of the Cauchy problem of [N] with $u_i(x,t) \in C([0,\infty), L_{q,k\alpha})$, moreover

$$u_i(x,t) \in C([0,\infty), L^\alpha_{2q}),$$

$$||u||_{q,k\alpha} + t^{\frac{n}{4q}}||u||_{2q,\alpha} < c,$$

where c is independent of $t, q = \frac{n}{1-\mu\alpha} \geq n$, k,and α, satisfy:

(i) If $0 < k < 1$, then $1 - \frac{n}{q} \leq \alpha < n-1$,,

(ii) If $k > 1$, then $\frac{1-\frac{n}{q}}{2-\lambda} \leq \alpha < \frac{n-1}{k}$. The last inequality implies

$$k < \frac{2(n-1)}{n-\frac{n}{q}} \leq 2.$$

Similarily to the reaction-diffusion equation, Ding and Yan [19] studied the Cauchy problem of the semilinear Stokes equation

$$\begin{cases} \frac{\partial u_i}{\partial t} - \Delta u_i = -\nabla p + |u_i|^{r-1}u_i, \quad i = 1,2, \\ \sum \frac{\partial u_i}{\partial x_i} = 0, \\ u_i(x,0) = \Phi_i(x). \end{cases}$$

By taking $u_1 = \psi_{x_2}$ and $u_2 = -\psi_{x_1}$, the above system reduces to

$$\begin{cases} \frac{\partial \Delta\psi}{\partial t} = \Delta^2\psi + \frac{\partial}{\partial x_2}(|\psi_{x_2}|^{r-1}\psi_{x_2}) - \frac{\partial}{\partial x_1}(|\psi_{x_1}|^{r-1}\psi_{x_1}), \\ \Delta\psi|_{t=0} = \Delta\psi_0. \end{cases}$$

for which we have

Theorem: If $r > 2$, for sufficiently small $\lambda > 0$ and $||\Delta\psi_0||_{q,k\lambda}$ there exists a global solution ψ of the above problem, and $||\Delta\psi||_{p,k\lambda} + t^b||\Delta\psi||_{p,\lambda} < C$, where C is independent of t,

$$q = \frac{2r}{r+2}, \quad b = \frac{1}{r} - \frac{1}{p} + \frac{1}{2},$$

$$\frac{2(r+1)}{3+r-k\lambda} < p < \frac{2r(r+1)}{2+r+r(r+2)}$$

$$1 \le k \le r+1, \quad 0 \le \lambda < \frac{r-2}{rk},$$

In the proof of all of the above results we need two lemmas in weighted Lebesque class $L_{p,l}$.

Lemma 1. (Generalized Young inequality) If $l < n(1-\frac{1}{p}), h < n(1-\frac{1}{q}), h+l \ge 0, \frac{1}{r} < \frac{1}{p'} + \frac{1}{q'}$, then we have

$$\||x|^{-k} f\|_{q'} \le \||x|^l q\|_p \||x|^{-h-l} G\|_r, (p, r \ge 1)$$

where f is the convolution of g and G, i.e.

$$f(x) = \int G(x-y) g(y) dy,$$

$$G = G(|x|) = |x|^\mu k(|x|) \quad k(|x|), \quad is \quad nondecreasing.$$

Lemma 2. (Weighted Calderon and Zygmund theorem)
If the Fourier transform of $k(x)$ is $\tilde{k}(x)$

$$\|\tilde{k}(x)\|_\infty \le c, \quad |k(x)| \le \frac{c}{|x|^n},$$

$$|k(x) - k(x-y)| \le \frac{c(y)}{|x|^{n+1}}, \quad when \quad |y| < \frac{|x|}{2},$$

$$T_f = \int k(x-y) f(y) dy,$$

then

$$\||x|^l T f\|_p \le c_p \||x|^l f\|_p.$$

where $-\frac{n}{p} < l < n(1-\frac{1}{p}), p > 1$..

Corresponding to the semilinear heat equation considered by Fujita and Weissler, Ding and *Zhao*[20] considered the Cauchy problem

$$\begin{cases} u_t = \Delta u + |u|^{r-1} u - u, \\ u(x,0) = \Phi(x), \end{cases}$$

in $L^\infty([0,\infty), L_{p,l})$.

Ding and Wang[21] studied the L^∞ solvability and blow-up property of this equation. Here we need a well-known result of elliptic equation.

For $1 < r < \frac{n+2}{n-2}$, the problem

$$\begin{cases} -\triangle u = u^r - u, & x \in R^n, \\ u(x) > 0, u(x) \to 0, |x| \to \infty, \end{cases}$$

has a unique positive solution, denoted by $\bar{u}(x)$..

We have

Theorem. 1. If $\Phi(x) \leq \bar{u}(x)$, then there exists a unique $L_p(1 \leq p \leq +\infty)$ global solution $u(x,t)$,, and $\lim_{t\to\infty} u(x,t) = 0, \forall x$..

2. If $\Phi(x) \geq \bar{u}(x), \Phi(x) \not\equiv \bar{u}(x)$ then there does not exist any L_p global solution $(2 < p \leq \infty)$.

References

[1]. Chang Tung and Hsiao Ling; Longman Scieutific and Technical 1989.

[2]. Ding Xiaxi and others ; Scientia sinica V.16. 1973.

[3]. Ding Xiaxi, Chen Guiqiang, and Luo peizhu, Acta Math. Scientia 7 : 4, 1987.

[4]. Ding Xiaxi, Chen Guiqiang, and Luo Peizhu, Acta Math. Scientia, 8 : 1, 1988.

[5]. Chen Guiqiang, Acta Math. Scientia 8 : 3, 1988.

[6]. Ding Xiaxi, Chen Guiqiang and Luo Peizhu, Comm, Math, phy. (121)1989.

[7]. Chen Guiqiang and Lu Yunguang, Chinese Science Bulletin 34(1989).

[8]. Chen Guiqiang and Lu Yunguang, Acta Math.Sci. 8 : 1(1988).

[9]. Lu Yunguang; Comm. Math. phy. (1992).

[10]. Lu Yunguang and Zhao Huijiang, Nonlinear Analyais 17 : 2(1990).

[11]. Lu Yungnang, Proceeding Royal society Edinbury 120 (1992).

[12]. Ding Xiaxi and Wu Yonghui, Acta Math. Sci., 13 : 2(1993).

[13]. Luo Peizhu and Ding Zhong, Acta Math. Sci., 13 : 1(1993).

[14]. Temam R. Applied Math. Sciences, Springor-Verlag, New York, Vol.68.

[15]. Ding Xiaxi and Wu Yonghui, to appear in Acta Math. Scientia.

[16]. Ding Xiaxi (to appear)

[17]. Ding Xiaxi (to appear)

[18]. Ding Xiaxi, IMA preprint series $ 558 August (1989), University of Minnesota.

[19]. Ding Xiaxi and Yan Guozhang, Acta Math, Sci., 13 : 2(1993).

[20]. DIng Xiaxi and Zhao Huijiang, Acta Math,Sci., 11 : 3(1991).

[21]. Wang Ming-Xin and Ding Xiaxi, Scientia Sinica 36 : 4(1993)

[22]. Wu Yong-Hui, Thesis (1993).

[23]. Xiaxi Ding, Preprint, No. 167, (1993), Math. Dept. University of Jyvaskyla, Finland.

Institute of Applied Mathematics
Academia Sinica
100080 Beijing, P.R.China

Wuhan Institute of Mathematical Sciences
Academia Sinica
430071, Wuhan, P.R. China

AMS/IP Studies in Advanced Mathematics
Volume **3**, 1997

THE ZERO DISPERSION LIMIT, A DETERMINISTIC ANALOGUE OF TURBULENCE

PETER D. LAX

1. Introduction

The usual theories of turbulence are *stochastic*, i.e., they regard turbulent flows as ensembles of flows generated by some stochastic mechanism. Many such mechanisms have been envisaged: a random force field impinging on a fluid, or a random distribution of initial data. Perhaps the most intriguing randomization is suggested by the following speculation: there is a critical time at which the flow under study breaks down, in the sense of losing most of its smoothness. The flow can be continued beyond breakdown time as a *generalized solution* of the governing equations. Unlike smooth solutions, generalized solutions are not uniquely determined by their initial data; so the continuations as generalized solutions beyond breakdown time could constitute an ensemble of flows. To be sure, this ensemble would have to be provided with a physically plausible measure.

Whatever the source of the ensemble, the quantities of interest from the mathematical, physical, or engineering point of view – velocity, pressure, temperature, etc. – are obtained as ensemble averages. The existence of these averages has to be established, and laws have to be derived governing their evolution in time. It is here that the nonlinearity of the equations complicates matters: products of quantities of interest, say velocity time pressure, appear in the equations, and their ensemble average is *not* equal to the product of the ensemble average of the factors. Thus the equations governing the time evolution of ensemble averages deal with the ensemble averages of all functions of the physically interesting quantities, a very large set of functions; in fact we surmise that an even larger set of ensemble averages would have to be included.

Even if we could find such an autonomous system of equations, it would be so huge that it would be useless for numerical purposes. This leaves one with the task of finding approximations to the exact equations, involving only a finite number of averages, whose solutions are close enough for engineering purposes to the solution of the exact equations.

Deterministic theories are, on other hand, based on the dependence of solutions on physically relevant parameters in the equations, such as viscosity, heat conduction coefficient, etc. Suppose that as one – or several – of these parameters tend to zero, or some other critical value, a solution of the flow equation with prescribed initial data becomes, in a certain portion of space-time, more and more oscillatory, at wave lengths that tend to zero, and amplitudes that do not tend to zero. Such an oscillatory family of functions cannot converge strongly, but could converge in a weak sense. If it does, the weak limits can be regarded as analogues of ensemble averages. This paper is devoted to exploring this analogy.

The tasks for such a deterministic theory are strikingly similar to the stochastic case. The existence of weak limits has to be established, and laws have to be derived that govern their time evolution. Again, nonlinearity rears up to complicate this task: the weak limit of products differs from the product of weak limits of the factors; thus we are forced to deal with the weak limits of all functions of the flow variables, and in fact, as we shall show, with an even larger class of weak limits. The resulting system of equations would be so huge as to be useless for the extracting of numerical information, and would have to be approximated by a finite system.

The main example we shall deal with is a dispersive approach to the equations of compressible flow. But first we discuss two examples of dispersive approximations to a simplified model of compressible flow, given by

$$u_t + uu_x = 0. \tag{1.1}$$

This can be written as a conservation law

$$u_t + \frac{1}{2}(u^2)_x = 0. \tag{1.2}$$

2. The Zero Dispersion Limit of the KdV Equation as an Approximation of (1.1)

The equation in question is

$$u_t + uu_x + \varepsilon u_{xxx} = 0. \tag{2.1}$$

Denote by $u(x,t;\varepsilon)$ the solution of this equation with prescribed initial values

$$u(x,0;\varepsilon) = u_0(x). \tag{2.2}$$

For each smooth initial function u_0 there is a critical value t_{crit} such that for $t < t_{\text{crit}}$, equation (1.1) has a smooth solution with initial value u_0, but that as t tends to t_{crit}, the solution $u(x,t)$ loses its smoothness at least at one point x. The author and David Levermore have shown in [11] that for $t < t_{\text{crit}}$, $u(x,t;\varepsilon)$ tends uniformly to the smooth solution $u(x,t)$ of (1.1). For $t > t_{\text{crit}}$, $u(x,t;\varepsilon)$ becomes oscillatory in the neighborhood of those points x where $u(x,t_{\text{crit}})$ is singular; the wavelength of these oscillations is of order $\sqrt{\varepsilon}$. It was shown in [11] that as ε tends to zero, $u(x,t;\varepsilon)$ converges weakly in these oscillatory regions to a limit $\overline{u}$. This weak limit $\overline{u}$ is *not* a solution of the conservation law (1.2); this can be seen by writing (2.1) in conservation form:

$$u_t + \left(\frac{1}{2}u^2\right)_x + \varepsilon u_{xxx} = 0, \tag{2.3}$$

and taking the limit of each term in the sense of the theory of distributions: the first term tends to $\overline{u}_t$, and the third term tends to zero. It was shown in [11] that as ε tends to zero, u^2 converges weakly to a limit, which we denote as $\overline{u^2}$. Then the distribution limit of the second term in (2.3) is $\left(\frac{1}{2}\overline{u}^2\right)_x$, so that we obtain the limit equation

$$\overline{u}_t + \left(\frac{1}{2}\overline{u^2}\right)_x = 0. \tag{2.4}$$

As is well known, when u and u^2 both converge weakly but not strongly, the weak limit of the latter is *not* the square of the former, but is larger:

$$\overline{u^2} > \overline{u}^2. \tag{2.5}$$

Therefore (2.4) is not the conservation law (1.2) but an equation containing two functions $\overline{u}$ and $\overline{u^2}$.

The t derivative of $\overline{u^2}$ can be expressed by multiplying (2.1) by $2u$ and writing it in conservation form:

$$2uu_t + 2u^2u_x + 2\varepsilon uu_{xxx} = (u^2)_t + \frac{2}{3}(u^3)_x + \varepsilon(u^2)_{xxx} - 3\left(\varepsilon(u_x)^2\right)_x. \tag{2.6}$$

Then we take the limit of this equation in the sense of the theory of distributions as ε tends to zero. The first term tends to $\left(\overline{u^2}\right)_t$, and the second term to $\frac{2}{3}\left(\overline{u^3}\right)_x$, where $\overline{u^3}$ is the weak limit of u^3. The next term tends to zero. To determine the limit of the last term we note that since $u(x,t;\varepsilon)$ is oscillatory with wave length $\sqrt{\varepsilon}$, u_x is expected to behave like $1/\sqrt{\varepsilon}$, and therefore $\varepsilon(u_x)^2$ could have a weak limit; we denote this weak limit by $\overline{w}$. So we get from (2.6) that

$$\left(\overline{u^2}\right)_t + \frac{2}{3}\left(\overline{u^3}\right)_x - 3\overline{w}_x = 0. \tag{2.7}$$

Note that $\overline{w}$ is *not* attainable as weak limit of some function of u.

The KdV equation (2.1) has an infinite number of conservation forms, all obtainable by multiplying (2.1) by a suitable polynomial in u and its partial derivatives. They are of the form

$$F_t + W_x = 0,$$

where F and W are polynomials as above. One can take the weak limits of these and obtain partial differential equations analogous to (2.7), but these do not constitute an autonomous infinite system of partial differential equations. To determine the time evolution of $\overline{u}$ we proceeded differently in [11]; because of the complete integrability of the KdV equation, the solution $u(x,t;\varepsilon)$ of (2.1) can be expressed explicitly in terms of the initial data

(2.2) by the scattering transform. This explicit expression contains ε as a parameter; we take its weak limit as ε tends to zero, obtaining $\overline{u}(x,t)$ in terms of the minimum of a quadratic functional, subject to inequality side conditions. Further analysis of this minimum problem leads to an expression for $\overline{u}$ in terms of a solution of a hyperbolic system for a finite number of functions. The simplest of these equations have been first derived by Whitham; see [15]. The number of functions needed to describe $\overline{u}$ increases with t; see also [3].

The more difficult periodic case has been studied by Stephanos Venakides in [14].

3. A Dispersive Discrete Approximation of Equation (1.1)

In this section we describe a discrete approximation to (1.1), studied in [4], where the continuum is coarse-grained in cells of length Δ, u is replaced by its averages over the cells, and its derivative u_x is replaced by a centered difference quotient. Denoting the average of u over the k-th cell by u_k we obtain the following system of ordinary differential equations:

$$\frac{d}{dt}u_k + u_k\frac{u_{k+1}-u_{k-1}}{2\Delta} = 0. \tag{3.1}$$

The centered difference quotient is an approximation of the following kind:

$$\frac{u_{k+1}-u_{k-1}}{2\Delta} \simeq u_x + \frac{\Delta^2}{6}u_{xx};$$

setting this into (3.1) gives

$$u_t + uu_x + \frac{\Delta^2}{6}uu_{xxx} \simeq 0,$$

which bears a resemblance to the KdV equation (2.1), at least for u positive, with $\varepsilon \sim (\Delta^2/6)u$.

Denote by $u(x,t;\Delta)$ the solution of (3.1), defined to be piecewise constant over intervals of length Δ :

$$u(x,t;\Delta) = u_k(t), \qquad \left(k-\frac{1}{2}\right)\Delta < x < \left(\frac{k+1}{2}\right)\Delta, \tag{3.2}$$

with prescribed initial values (2.2). In [4], Jonathan Goodman and the author have investigated the behavior of $u(x,t;\Delta)$ as Δ tends to zero. We have proved that for $t < t_{\text{crit}}$, $u(x,t;\Delta)$ tends uniformly to the smooth solution of (1.1), (2.2). For $t > t_{\text{crit}}$, numerical calculations indicate that $u(x,t;\Delta)$ is oscillatory in a portion of x,t space, with wave length prportional to Δ. Numerical calculations indicate that as Δ tends to zero, $u(x,t;\Delta)$ converges weakly in these oscillatory regions to a limit $\overline{u}$, and $\overline{u}$ does not satisfy the conservation law (1.2).

The conservation law satisfied by $\overline{u}$ can be derived by rewriting (3.1) as follows:

$$\frac{d}{dt}u + \frac{u(x+\Delta)u(x) - u(x)u(x-\Delta)}{2\Delta} = 0.$$

The distribution limit of the first term is $\overline{u}_t$, and of the second term $\left(\frac{1}{2}\overline{u}_2\right)_x$, where $\overline{u}_2$ is weak limit

$$\overline{u}_2 = w - \lim u(x)u(x+\Delta). \tag{3.3}$$

So we obtain

$$\overline{u}_t + \left(\frac{1}{2}\overline{u}_2\right)_x = 0, \tag{3.4}$$

an analogue of equation (2.4). There is, however, a difference between the two; we have seen earlier that $\overline{u^2} > \overline{u}^2$, whereas no such relation holds between $\overline{u^2}$ and $\overline{u}^2$. In fact, our numerical calculations suggest, at least in one case, $\overline{u}_2 < \overline{u}^2$.

Since (3.4) is a single equation for two unknown functions, it is unsuitable for the determination of $\overline{u}$. To do that, we show in [4] that if $u_k(t)$ are positive at $t = 0$, they remain positive for all t. Then we introduce as new variable

$$a_k = \sqrt{u_k};$$

the equations satisfied by the a_k, can be obtained by setting $u_k = a_k^2$ in (3.1):

$$\frac{d}{dt}a_k + a_k\frac{a_{k+1}^2 - a_{k-1}^2}{4\Delta} = 0.$$

This equation has been shown by Kac and van Moerbeke to be completely integrable (see [8]), and therefore can be solved. This makes it possible to study the limit $\Delta \to 0$.

4. The Equations of Compressible Flow and von Neumann's Difference Scheme

The equation of compressible flow in one space dimension can be written in the following Lagrangian form:

$$u_t + p_x = 0, \tag{4.1}$$

$$V_t - u_x = 0, \tag{4.2}$$

$$e_t + pu_x = 0. \tag{4.3}$$

Here u is velocity, p is pressure, V specific volume, and e internal energy, connected with p and V via an equation of state:

$$e = e(p, V). \tag{4.4}$$

The first equation is conservation of momentum, the second conservation of mass, the third the work equation. Multiplying the first equation by u and adding it to the third yields conservation of energy:

$$\left(e + \frac{1}{2}u^2\right)_t + (up)_z = 0. \tag{4.5}$$

Multiplying the second equation by p and adding it to the third yields the conservation of entropy:

$$S_t = 0, \tag{4.6}$$

where $S = S(p, V)$ is entropy.

In 1944 von Neumann proposed solving these equations by a centered difference scheme (see [12]), where the thermodynamic quantities p, V, e were carried on a rectangular mesh, and u on another mesh displaced from the first one by $\Delta x/2$ in the x direction and $\Delta t/2$ in the t direction; see (4.8)-(4.10). As initial data he took

$$u(x, 0) = \begin{cases} -1 & x > 0 \\ 1 & x < 0 \end{cases}, \qquad V(x, 0) = 1,\ p(x, 0) = 1. \tag{4.7}$$

The exact solution consists of two shocks emerging from the origin, propagating to the right and to the left, containing between them an expanding region where $u = 0$. Numerical calculations carried out at the Ballistic Research Laboratory at Aberden, Maryland, showed oscillations on mesh scale. Von Neumann interpreted the oscillations in velocity as heat energy converted from kinetic energy by the irreversible action of the shock; he conjuctured that as $\Delta x, \Delta t$ tend to zero, the approximate solutions converge weakly but not strongly, and that the weak limit is the exact solution of the equations of compressible flow.

The difference scheme of von Neumann, because of the centering of the difference quotients, is dispersive; it is this quality that is responsible for the oscillatory nature of the solutions. As we have seen in the case of the KdV equation, weak limits of oscillatory solutions of nonlinear conservation equations do not satisfy the original conservation equation; therefore I have strogly doubted the second part of von Neumann's conjecture, and said so in [9]. Thereafter, in [7], Tom Hou and the author have made a systematic investigation of von Neumann's scheme, mainly by careful numerical studies but also by some analysis, and came to the following conclusions:

Extend, as in (3.2), the solutions of the difference equation to all x and t as picewise constant.

(a) For those values of t for which the equations of compressible flow (4.1)-(4.3) have a smooth solution for all x, u, V, p converge uniformly to these smooth solutions as Δx tends to zero and $\Delta t/\Delta x = \lambda$ remains a constant, less than the maximum value permitted by the CFL condition.

(b) For t greater than the value described in (a) the sequences u, V, p converge weakly, but not strongly, to limits $\overline{u}, \overline{V}, \overline{p}$. These weak limits $\overline{u}, \overline{V}, \overline{p}$ appear to be independent of $\lambda = \Delta t/\Delta x$.

(c) $\overline{u}, \overline{V}$ and $\overline{p}$ satisfy the conservation law of momentum and mass, but fail to satisfy the law of conservation of energy.

It is instructive to elaborate (c). We write out von Neumann's difference equations:

$$\frac{u_k^{n+1/2} - u_k^{n-1/2}}{\Delta t} + \frac{p_{k+1/2}^n - p_{k-1/2}^n}{\Delta x} = 0, \tag{4.8}$$

$$\frac{V_{k+1/2}^{n+1} - V_{k+1/2}^n}{\Delta t} + \frac{u_{k+1}^{n+1/2} - u_k^{n+1/2}}{\Delta x} = 0, \tag{4.9}$$

$$\frac{e_{k+1/2}^{n+1} - e_{k+1/2}^n}{\Delta t} + p_{k+1/2}^{n+1/2} \frac{\left(u_{k+1}^{n+1/2} - u_k^{n+1/2}\right)}{\Delta x} = 0, \tag{4.10}$$

where

$$p_{k+1/2}^{n+1/2} = \frac{p_{k+1/2}^n + p_{k+1/2}^{n+1}}{2}. \tag{4.11}$$

We can take the distribution limits of equations (4.8) and (4.9); we obtain

$$\overline{u}_t + \overline{p}_x = 0, \qquad \overline{V}_t - \overline{u}_x = 0. \tag{4.12}$$

Before we can take limit of (4.10) we have to turn it into a conservation law. This can be accomplished by an ingenuous trick of Trulio and Trigger: add equations $(4.8)_n$ and $(4.8)_{n+1}$ and divide by two; we get

$$\frac{u_k^{n+3/2} - u_k^{n-1/2}}{2\Delta t} + \frac{p_{k+1/2}^{n+1/2} - p_{k-1/2}^{n+1/2}}{\Delta x} = 0, \tag{4.13}$$

where $p^{n+1/2}$ is defined by (4.11). We multiply (4.13) by $u_k^{n+1/2}$; using the abbreviation

$$K_k^n = \frac{1}{2} u_k^{n+1/2} u_k^{n-1/2} \tag{4.14}$$

we can write the resulting relation as

$$\frac{K_k^{n+1} - K_k^n}{\Delta t} + u_k^{n+1/2} \frac{(p_{k+1/2}^{n+1/2} - p_{k-1/2}^{n+1/2})}{\Delta x} = 0.$$

We add this to (4.10) and get

$$\frac{E_k^{n+1} - E_k^n}{\Delta t} + \frac{(u_{k+1} p_{k+1/2} - u_k p_{k-1/2})^{n+1/2}}{\Delta x} = 0, \tag{4.15}$$

where

$$E_k^n = e_{k+1/2}^n + K_k^n. \tag{4.16}$$

We now take the distribution limit of equation (4.15); we get, using (4.16), that

$$\left(\overline{e} + \overline{K}\right)_t + (\overline{up})_x = 0. \tag{4.17}$$

This looks like the conservation law of energy, but it is not really. For the energy equation ought to state that

$$\left(e\left(\overline{p}, \overline{V}\right) + \frac{1}{2}\overline{u}^2\right)_t + (\overline{u}\,\overline{p})_x = 0,$$

which is different from (4.17), just as (1.2) is different from (2.4) and (3.4).

More analytical work can be done in the extreme case $\lambda = 0$, where von Neumann's scheme reduces to the system of ordinary differential equations

$$\frac{d}{dt}u_k + \frac{p_{k+1/2} - p_{k-1/2}}{\Delta} = 0, \tag{4.18}$$

$$\frac{d}{dt}V_{k+1/2} - \frac{u_{k+1} - u_k}{\Delta} = 0, \tag{4.19}$$

and similarly for the work equation. When the initial values of V, p are isentropic, they stay isentropic for all t. In this case p is a function of V in the equation of state:

$$p = f(V), \tag{4.20}$$

and the work equation is not needed.

5.

In this section we show how to interpret the system of equations (4.18), (4.19) describing the motion of a chain of identical particles linked by identical springs, with a spring law (4.20); in fact von Neumann was aware of this interpretation.

We introduce of Eulerian coordinates $X_k(t)$ on the semidiscretised level well. To simplify notation we denote differentiation with respect to t by a dot

$$\dot{} = \frac{d}{dt}. \tag{5.1}$$

Then (4.18) and (4.19) can be rerwitten as

$$\dot{u}_k = \frac{p_{k-1/2} - p_{k+1/2}}{\Delta}, \tag{5.2}$$

$$\dot{V}_{k+1/2} = \frac{u_{k+1} - u_k}{\Delta}. \tag{5.3}$$

We define $X_k(t)$ by

$$\dot{X}_k = u_k. \tag{5.4}$$

Set (4.16) into (4.15), obtaining,

$$\dot{V}_{k+1/2} = \frac{\dot{X}_{k+1} - \dot{X}_k}{\Delta}; \tag{5.5}$$

if the initial values of X_k are so chosen that

$$V_{k+1/2} = \frac{X_{k+1} - X_k}{\Delta} \tag{5.6}$$

at $t = 0$, then it follows from (5.5) that (5.6) holds for all t as well. In the recursion relation (5.6), $X_0(0)$ may be chosen arbitrarily, say zero.

Set (5.4) into equation (5.2); the result is

$$\ddot{X}_k = \frac{p_{k-1/2} - p_{k+1/2}}{\Delta}.$$

Setting (5.6) into the equation of state (4.20), we can rewrite the above as

$$\ddot{X}_k + \left[f\left(\frac{X_{k+1} - X_k}{\Delta}\right) - f\left(\frac{X_k - X_{k-1}}{\Delta}\right)\right] \Big/ \Delta = 0. \tag{5.7}$$

This can be looked upon as a semidiscrete approximation to the nonlinear wave equation. It can be given a different interpretation after an appropriate change of scale:

Set

$$\frac{X_k}{\Delta} = q_k, \tag{5.8}$$

and define a rescaled time s that is related to t by

$$s = \frac{t}{\Delta}; \tag{5.9}$$

then

$$\frac{d}{dt} = \frac{1}{\Delta}\frac{d}{ds}\,. \tag{5.10}$$

Setting this into (4.19) and denoting $\frac{d}{ds}$ by $'$, we get

$$q_k'' + f(q_{k+1} - q_k) - f(q_k - q_{k-1}) = 0. \tag{5.11}$$

We recognize (5.11) as the equation governing the vibrations of an anharmonic chain of particles with unit mass, linked by identical springs to their two nearest neighbors. Here $q_k(t)$ is the distance at time t of the k-th mass from the origin, and $f(d)$ the negative of the force exerted on each by two particles separated by distance d and linked by a spring. The force $-f$ is a monotonically increasing function of d, assumed to be zero when the two masses are a unit distance apart.

Equation (5.11) is the semidiscrete version of von Neumann's original formulation of his method [12].

There is an extensive literature in physics about lattice vibrations, see e.g. those quoted in [5]. In [5] Holian and Straub investigated the so-called shock problem for lattices: particles are set in motion by giving them constant velocity directed toward origin:

$$q_k(0) = 0, \qquad q_k'(0) = \begin{cases} a & k < 0 \\ 0 & k = 0\,, \\ -a & k > 0 \end{cases} \tag{5.12}$$

here a is some positive quantity.

Using the relations (5.4)-(5.9) we can transform this initial value problem for lattice vibration governed by equation (4.23) into an initial value problem for the semidiscrete approximations (4.14), (4.15) of the isentropic flow equations in Lagrangian coordinates. The initial values corresponding to (4.24) are

$$V_{k+1/2}(0) = 1, \qquad u_k = \begin{cases} a & k < 0 \\ 0 & k = 0\,; \\ -a & k > 0 \end{cases} \tag{5.13}$$

this is the discretized version of the continuum Riemann initial value problem

$$V(\xi,0) = 1, \qquad u(\xi,0) = \begin{cases} a & \xi < 0 \\ -a & \xi > 0 \end{cases}. \tag{5.13)$'$}$$

The rescaling (5.8) and (5.9) show that the large time limit $s \to \infty$ for the lattice shock problem (5.11), (5.12) is equivalent with the $\Delta \to 0$ limit for the semidiscrete approximation (5.2), (5.3) for the Riemann shock problem (5.13)′, provided that the spring force is identified as pressure.

In their numerical experiments, Holian and Straub solved equation (5.11) by a centered difference scheme in time; so effectively they carried out von Neumann's scheme for isentropic gas. They included in their studies a variety of physically interesting force laws associated with the names of Lennard-Jones, Morse and Toda. Their main observation is that the motion produced by the initial data (5.12) consists of two shock waves, travelling in opposite direction, emerging from the origin. These shocks are dispersive in the sense that behind the shock front the q_k perform large amplitude oscillations; many details of these observations are described.

In a subsequent paper [6], Holian, Flaschka and McLaughlin took advantage of the complete integrability of the Toda lattice, characterized by the exponential force law

$$f(V) = e^{-V}. \tag{5.14}$$

By analyzing the motion of the particles as a slowly varying wave train of solitons they reached a number of significant conclusions:

(i) They derived a formula for the shock-speed as function of a.

(ii) They showed the existence of a critical value of a, which is $a = 1$ for the force law (5.14), such that for weak shocks, i.e. $a < 1$, there is a region, emanating from the origin with speed less than shock speed, in which u tends uniformly to zero. As a approaches 1, this region disappears. For $a > 1$, u is oscillatory everywhere behind the shock; there is a region in which the oscillation is in the binary mode, i.e., the displacement of adjacent particles is equal and opposite.

(iii) They observed that qualitative features of the behavior described in (ii) hold for order anharmonic lattices.

In their numerical experiments described in [5] Holian and Straub have found a critical speed for the von Neumann shock problem in all chains, no matter what the force law. A mathematical proof for this is utterly lacking, except for the completely integrable Toda chain; we don't even understand intuitively why chains behave differently under subcritical and supercritical impact. I would like to close this talk by drawing an analogy with other systems that show critical behavior.

Consider dissapative dynamical systems driven by external forces. When the force is small or moderate, such a dynamical system settles down to a steady state. If however the force exceeds a critical value, the system abandons the steady state and settles into periodic motion. The intuitive explanation of this kind of behavior is that excessive rate of energy input strains the ability of the system in its steady state to dissipate that energy.

Any finite portion of a vibrating lattice can be regarded as a dissipative system; the dissipative mechanism is radiation of energy from the endpoints of the lattice. The rate of energy input is proportional to the square of the velocity of particles impacting on the lattice. Whatever a mathematical argument demonstrating criticality can be based on this picture remains to be seen.

6. Summary

There are points of similarity between the examples described in Section 2 through 4 and turbulent flows:

(a) The coexistence of smooth and oscillatory regions in space-time of solutions of (2.1), (3.1), and (4.8)-(4.11) bear a resemlance to intermittent turbulence.

(b) The Whitham equations and their generalizations can serve as a model for equations governing ensemble averages of turbulent flows.

There are major differences between dispersive and turbulent flows. The dispersive flows studied here have oscillations on a single scale, whereas turbulent flows contain a hierarchy of scales; the interaction of these scales is at the center of the problem of describing turbulence.

This talk is based on the articles [7] and [10] in the Bibliography. For further reading on the zero dispersion limit see [9].

Acknowledgement. This research was supported partly by the Applied Mathematical Sciences Program of the Department of Energy under Grant DE-FG0288ER25053.

References

[1] Venakides, S., Deift, P., and Oba, R., *The Toda Shock Problem*, Comm. Pure Appl. Math., **44**(1991), 1171-1242.

[2] Flaschka, H., *The Toda lattice I*, Phys. Rev., **B9** (1974), 1924-1925; *II*, Prog. Theoret. Phys., **51**(1974), 703-716.

[3] Flaschka, H., Forest, G., and McLaughlin, D.W., *Multiphase averaging and the inverse spectral solution of the Korteweg-de Vries equation*, Comm. Pure Appl. Math., **33**(1980), 739-784.

[4] Goodman, J., and Lax, P.D., *Dispersive difference schemes I*, Comm. Pure Appl. Math., **41**(1988), 591-613.

[5] Holian, B.L., and Straub, G.K., *Molecular dynamics of shock waves in one-dimensional chains*, Phys. Rev., **B18**(1978), 1593-1608.

[6] Holian, B.L., Flaschka, H., and McLaughlin, D.W., *Shock waves in the Toda lattice: Analysis*, Phys. Rev., **A24**(1981), 2595-2623.

[7] Hou, T., and Lax, P.D., *Dispersive approximations in fluid dynamics*, Comm. Pure Appl. Math., **44**(1991), 1-40.

[8] Kac, M., and van Moerbeke, P., *On an explicitly solvable system of nonlinear differential equations related to the Toda lattice*, Adv. in Math., **16**(1975), 160-169.

[9] Lax, P.D., Levermore, C.D., and Venakides, S., *The generation and propagation of oscillations in dispersive IVP's and their limiting behavior*, Important Developments in Soliton Theory 1980-1990, T. Fokas and V.E. Zakharov, eds. Springer-Verlag, Berlin, 1992.

[10] Lax, P.D., *The zero dispersion limit, a deterministic analogue of turbulence*, Comm. Pure Apl. Math., Vol.XLIV, 1047-1056.

[11] Lax, P.D., and Levermore, C.D., *The small dispersion limit of the Korteweg-de Vries equation*, Comm. Pure Appl. Math., **36**(1983), I, 253-290, II, 571-593, III, 809-829.

[12] von Neumann, J., *Proposal and analysis of a new numerical method in the treatment of hydrodynamical shock problem, VI*, collected works, Pergamon, London, 1963.

[13] Venakides, S., *The Zero Dispersion limit of the periodic KdV equation*, AMS Transactions 301 (1987), 189-226.

[14] Whitham, G.B., *Linear and Nonlinear Waves*, Wiley, New York, 1974.

COURANT INSTITUTE

AMS/IP Studies in Advanced Mathematics
Volume 3, 1997

Nonlinear Waves for Viscous Conservation Laws

TAI-PING LIU
(STANFORD UNIVERSITY)

Abstract

Recent progresses on the qualitative understanding of viscous conservation laws are described. Several important phsycal models are of the form of viscous conservation laws, such as the compressible Navier-Stokes equations, MHD and visco-elasticity equations. There are diffusion waves, shock waves, viscous contact discontinuities, expansion waves and others. Thus far, the study of shock waves and diffusion waves is most substantial. We will concentrate on these two types of waves. There are three types of shock waves: classical (Lax), undercompressive and overcompressive shock waves. Two types of diffusion waves: linear and nonlinear (Burgers or generalized Burgers), exist. To treat wave patterns consisting of distinct wave types, an approach of pointwise estimates has been developed in recent years. We will describe this new approach when applying to physical models.

1. Introduction

Consider viscous conservation laws

$$u_t + f(u)_x = (B(u)_x)_x, \tag{1.1}$$

where $u(x,t)$ is an n-vector, $-\infty < x < \infty, t \geq 0$. We assume that the associated inviscid conservation laws are hyperbolic:

$$u_t + f(u)_x = 0, \tag{1.2}$$

$$f'(u)r_i(u) = \lambda_i(u)r_i(u), i = 1, 2, \cdots, n,$$
$$\lambda_1(u) \leq \lambda_2(u) \leq \cdots \leq \lambda_n(u). \tag{1.3}$$

Much of the progress is made for the simplified situation of artificial viscosity:

$$u_t + f(u)_x = u_{xx}. \tag{1.1'}$$

In the case the system (1.2) is strictly hyperbolic, that is, the characteristic speeds $\lambda_i(u), i = 1, \cdots, n$, are distinct, we have classical (Lax) shocks. For the simplified situation that the i-characteristic field is genuinely nonlinear, [Lax],

$$\nabla\lambda_i(u).r_i(u) \neq 0, \tag{g.nl}$$

there exist i-shocks satisfying the admissibility condition

$$\lambda_i(u_-) > s > \lambda_i(u_+). \tag{E}$$

where u_- and u_+ are the values of the shock at $x = -\infty$ and $x = \infty$, respectively. See [Chang-Hsiao] and references therein for shocks for non-genuinely nonlinear characteristic fields. These shocks occur in the important physical situations such as gas flows. For such a shock, the two sets of n characteristics $\lambda_1(u_\pm), \cdots, \lambda_n(u_\pm)$, have the property that n+1 of the corresponding characteristic curves impinge on the shock, while the rest $n-1$ leave the shock. When (1.2) is not strictly hyperbolic, there are shocks which are less (or more) compressive than the classical ones. That is, less (or more) than n+1 characteristics impinge on the shock. We call these shocks undercompressive (or overcompressive) shocks. In Section3, 4 and 5 we describe the distinct behaviours of these three types of shocks for the viscous conservation laws.

To understand the behaviour of shocks we need to study the diffusion waves which dissipate around a constant state. For genuinely nonlinear fields, Burgers or generalized Burgers equations govern the nonlinear diffusion waves. Otherwise, we have linear diffusion waves, governed by heat equation. In Section 2 we describe a recent result on pointwise estimates for diffusion waves for the general viscous conservation laws (1.1).

Finally, in Section5 we make some concluding remarks on general nonlinear waves for the viscous conservation laws. We also raise open questions.

2. Diffusion waves

For physical systems the viscosity matrix B(u) in (1.1) is usually not uniformly positive when the flux is diagonalized. This makes the system not uniformly parabolic. In order to make the system dissipative, certain coupling of different characteristic fields needs to be assumed. This issue has been studied systematically by [Kawashima] for perturbation of constant states. Linearize (1.1) around the zero state and diagonalize the flux matrix to get

$$u_t + \Lambda u_x = D u_{xx}. \tag{2.1}$$

Here Λ is the diagonal matrix formed by the characteristic speed $\lambda_i(0)$ and D is viscosity matrix B(0) in the coordinate of characteristic directions $r_i(0)$. For (2.1) to be dissipative, the matrix need to be nonnegative definite and the diagonal elements positive. Other important assumption on the viscosity matrix is that there exists a dissipative entropy for the nonlinear system. These properties are met for the compressible Navier-Stokes, viscous MHD and elasticity equations. In the case the inviscid system is not strictly hyperbolic, the block matrix in D corresponding to the same eigen value of Λ must be positive definite. Thus if

$$\lambda_k = \cdots = \lambda_l, \text{ for some } 1 \leq k \leq l \leq n, \tag{2.2}$$

then it is required that the submatrix of D

$$D' = [D_{ij}], \quad k \leq i, j \leq l,$$

is positive definite. Diffusion waves for (1.1) are time-asymptotic solutions. To construct them we first consider the quadratic approximation of the flux function $f(u)$

$$f(u) = f(0) + f'(0)u + A(u, u) + O(1)|u|^3,$$

for an $n \times n$ matrix A. For simplicity, we assume that $f'(0)$ has already been diagonalized and equals Λ. The diffusion waves are constructed for each distinct eigenvalue. In (2.2) we let the diffusion waves to be the solutions of

$$w_t + \lambda_k(0) w_x + A'(w, w)_x = D' w_{xx}. \tag{2.3}$$

Here w is a $l - k$ vector and A' is a $(l - k) \times (l - k)$ matrix

$$A' = [A_{ij}], \quad k \leq i, j \leq l.$$

In the simplified situation when the inviscid system is strictly hyperbolic, $k = l$, (2.3) is the Burgers equation or heat equation, depending on whether λ_k is genuinely nonlinear or not. When $k < l$, (2.3) may be called the generalized Burgers equations. It has self-similar solutions just as the heat and Burgers equations.

The study of diffusion waves as the time-asymptotic states for perturbation of constant states has been studied using energy method, [Liu2],[Kawashima]. Recently Zeng and the author [Liu-Zeng, to appear] have succeeded in obtaining the pointwise estimates for the convergence of general solutions to the diffusion waves. Our approach consists of pointwise estimates of the Green function of the linearized system (2.1) and the coupling of nonlinear waves pertaining to different characteristic directions. The Green function in general contain δ function because the system is not uniformly parabolic, but hyperbolic-parabolic.

3. Classical (Lax) shocks

The study of the behavior of the perturbation of shock waves involves also the diffusion waves because characteristics leaving the shocks carry diffusion waves. Consider, for instance, the perturbation of a p-shock, $1 \leq p \leq n$. Since we assume the shock to be classical, there are $p-1$ characteristic leaving the shock from the left and $n-p-1$ leaving from the right. Thus a perturbation produces a family of $n-1$ diffusion waves. The ones leaving from the left diffuse about the constant state u_- and take values in the characteristic directions $r_i(u_-), i = 1, \cdots, p-1$. Similar happening occurs for those leaving from the right. The perturbation also produce a translation, say of the amount x_0, of the shock creating a flux proportional to $x_0(u_+ - u_-)$. These happening contributes to the conservation laws, [Liu2],

$$\int_{-\infty}^{\infty} v(x,0)dx = \sum_{i=1}^{p-1} c_i r_i(u_-) + \sum_{i=p+1}^{n} c_i r_i(u_+) + x_0(u_+ - u_+) \tag{3.1}$$

This yields the strength c_i of diffusion waves and the amount of translation x_0 caused by the perturbation $v(x,0)$. The ability to determine the shock location through the conservation laws is crucial for the energy method to work for the stability analysis,[Goodman], [Matsumura-Nishihara], [Liu2], [Szepessy-Xin].

For nonclassical shocks, one cannot determine the shock location through (3.1) because the number of diffusion waves leaving the shock is not $n-1$ and so the vectors on the right-hand side of (3.1) do not form a basis of the state space R^n. It turns out that the sense of stability for the nonclassical shocks differs from that for the classical ones. Also, the energy method needs to be sharpened. A more powerful method is being developed by the author to obtain the pointwise estimates. The first step has been taken in [Liu4] for diffusion waves. It is being completed for the classical shocks,[Liu, to appear], and is being applied to other types of waves.

4. Overcompressive Shocks

For an overcompressive shock (u_-, u_+) there are more than $n+1$ characteristics impinging on the shock and less than $n-1$ characteristics leaving it. The system of ordinary differential equations

$$-s(u-u_\pm)+f(u)-f(u_\pm)=B(u)u' \tag{4.1}$$

satisfied by the travelling waves for (1.1) have more than one connecting orbit. In fact, if there are $n+1+k$ characteristics impinging on the shock then the number of unstable nodes at $x=-\infty$ plus the number of stable nodes at $x=\infty$ equal $n+k+1$ and there are in general k-parameters family of solutions to (4.1). This is in contrast to the classical case, where there is an unique connecting orbit. s is coupled with the fact that a perturbation of a shock profile would produce $n-1-k$ diffusion waves, since there are $n-1-k$ characteristics leaving the shock.

The above considerations yield the following scenario: When a shock profile is perturbed, the time-asymptotic state consists of $n-1-k$ diffusion waves propagating away from the shock plus one of the shock profile, properly translated, among the k-parameters of connecting orbits. This scenario is plausible because it is supported by the following consideration of conservation laws: Consider the perturbation of shock profile $\phi(x-st)$

$$u(x,0)=\phi(x)+v(x,0). \tag{4.2}$$

Then it follows from (1.1) that

$$\int_{-\infty}^{\infty} v(x,t)dx=\int_{-\infty}^{\infty} v(x,0)dx, \tag{4.3}$$

that is, we have n conservation laws. We argue that it is possible to find another shock profile $\psi(x+x_0-st)$, properly translated by x_0 and $n-1-k$ diffusion waves θ_i with proper strengh $c_1,\cdots,c_{n-1-k}$, cf (3.1), so that the integral of solution minus these waves is zero:

$$\int_{-\infty}^{\infty}[u(x,t)-\psi(x+x_0-st)-\sum_{i=1}^{n-1-k}\theta_i]dx=0. \tag{4.4}$$

This is so because the integral is an n-vector and there are also n-parameters to choose, namely, the k-parameter of shock profiles, the translation x_0 of the profile, and the strenghs $c_1,\cdots,c_{n-1-k}$, of the $n-1-k$ diffusion waves $\theta_1,\cdots,\theta_{n-1-k}$.

The above argument indicates that a time-asymptotic state can be chosen so as not to contradict the conservation laws (4.3). For this we may require the perturbation to be small, since the k-parameter families of shock profiles may stay bounded in some sense and so the profile ψ exists only under this smallness assumption. It is therefore conjectured that the above time-asymptotic state indeed represents the large-time state of the solution when the perturbation is sufficiently small. This conjecture has been proved for simple model [Freistuhler-Liu].

So far we have argued that an overcompressive viscous profile is stable upon small perturbation in the sense that the perturbation yields another nearby profile with the same given end states at $x = \pm\infty$. This differs from the classical case, where the perturbation yields the same and unique profile. There is another important difference from the classical shocks. When the perturbation exceeds certain amount another profile ψ may not exist to satisfy the conservation laws (4.4). By straightforward dimensional analysis, this is equivalent to saying that for a given perturbation, no matter how small, ψ may not exist to satisfy (4.4) if the viscosity matrix $B(u)$ is small. When this happens, an overcompressive shock is stable in the aformentioned sense, but not uniformly with respect to the strengh of the viscosities. In other words, it should not be regarded as an hyperbolic, inviscid, admissible shock. This criterion is a right one. It is stronger than the usually called "viscosity criterion" of calling the inviscid shock admissible when there exist shock profiles. It is also more stringent than the requirement that the corresponding viscous profile is nonlinearly stable.

In [Liu3] a rotationally invariant model for MHD and nonlinear elasticity is shown to have the property of stability not uniformly with respect to viscosity. This is expected for the full MHD and elasticity models.

5. Undercompressive Shocks

Because the number of characteristic impinging on an undercompressive shock is less than $n+1$, the connecting orbit for the system of ordinary differential equations (4.1) is saddle-saddle connection. This is non-generic in the sense that if the end states $u_\pm$ are perturbed, still maitaining the jump (Rankine-Hugoniot) condition, the connection disappears in general. Nevertheless, such waves exist in general flow and is generic in the flowing sense: To solve the Riemann problem for the hyperbolic system, undercompressive shock is often needed. It is possible to include the wave from the viscous point of view because if one of the end state, say u_-, is fixed, then the right state always exists for varying the shock speed so as to have connecting orbit. Unlike the classical case, the right state is unique and not the whole Hugoniot curve.

In spite of the precarious sense of the genericity of undercompressive shocks, its viscous profile is nonlinearly stable. A perturbation now produces more than

$n-1$ diffusion waves, because there are more than $n-1$ characteristics leaving the shock. There is also the time-asymptotic location of the shock to determine. This implies that the n conservation laws (4.3) are not sufficient to determine the more than n parameters, namely, the strenghs of the diffusion waves and the location of the shock. In other words, it appears that the problem is under-determined. This is deceptive. It simply means that the shock location cannot be determined by the conservation laws, but depends on the initial data in a more nonlinear way.

The inability to determine the shock location through the conservation laws (4.3) creats technical difficulties for the stability analysis. The author's recent introduction of pointwise estimates for perturbation of general nonlinear waves becomes necessary for the stability analysis of undercompressive shocks. The pointwise estimates yield the degree of nonlinear coupling of nonlinear waves pertaining to different characteristic fields. In the case of the perturbation of an undercompressive shock it allows us to estimate the coupling of the diffusion waves and the shock. This makes it possible trace the location of the shock. A prerequisite for this approach to work is the adequate knowledge of the Green functions for the linearized system. For the classical shock for (1.1)' this is being done by the author. For particular undercompressive shock, it is being done by Zumbrun and the author [Liu-Zumbrun, to appear].

An important difference between the classical and undercompressive shocks is that the undercompressive shocks' existence, though not their stability, depends sensitively on the form of the viscosity matrix $B(u)$, since they are saddle-saddle connections. Because they are stable, each one should be regarded as admissible hyperbolic shock. This has the important consequence that the hyperbolic theory for (1.2) depends on the form of the viscosity, and there are more than one hyperbolic theory. This is very different from the other types of shock waves.

The above nonuniqueness of hyperbolic theory also occurs in several other important physical situations. For instance, in combustion theory, the weak detonations and deflagrations are undercompressive. As a consequence, the only unique inviscid theory, that of Chapman-Juoget [Courant-Friedrichs], precludes these waves. For a simple combustion model one can show that if dissipation is included then an one-parameter of hyperbolic theory emerges which includes the weak combustion waves,[Liu-Tan-Ying]. Another example concerns the addition of dispersion for the MHD model, [Wu]. Again, the hyperbolic theory seems to depend on the form of dissipation and dispersion present.

Concluding Remarks

One of the outstanding problem is to understand the behavior of solutions to the viscous conservation laws (1.1) for varying strengh of the viscosities. A particular problem in this regard is the zero dissipation problem, the study of the convergence of the solutions as the viscosity matrix $B(u)$ tends to zero. This problem has been studied sucessfully using the theory of compensated compactness for a number of 2x2 systems, [Ding-Chen-Luo] (see also the article of G.C.Chen in this volume). The present geometric approach works for the zero dissipation problem, but so far only for particular initial data, though the ideas are applicable to general systems, [Hoff-Liu]. Besides the advantage of applicable to general systems, the present approach has an important advantage of having understanding of the behavior, besides the convergence, of the solutions in the process of zero dissipation.

As of now the understanding of viscous shocks and diffusion waves is much more than for other types of waves, principally the viscous expansion waves and contact discontinuities. There is some works on the later using the energy method. There is need to study the pointwise estimates for their perturbation. This is lacking right now. The contact discontinuities are important because of their instability. This is well-studied for the incompressible flow, which possesses only the contact discontinuities, in fact. It is interesting to consider wave patterns containing all types of waves. Since different type of waves behaves very differently, it remains a challenging problem to deal with all the waves present in a general solution.

Another outstanding problem is the well-posedness of the initial value problem for the hyperbolic conservation laws (1.2). The main problem is to find the appropriate topology for the stability. So far the attempts to prove the uniqueness use the $L_2(x) - norm$. This is a convenient choice because the $L_2 - norm$ is natural for functional-analytic approach of making uses of adjoint operator and entropy functions. On the other hand, it can be seen that the solution operator is not stable with respect to the L_2. In fact, for dissipative system, say the heat equation, the solution is dissipative in L_2, but for solutions which are compressive, it is unstable in the $L_2 - norm$. For the hyperbolic conservation laws, a general solution contains both compressive and expansive waves. The well-known N-waves, for example, is dissipative,[Liu1]. When compressive waves form shocks, the solution operator is unstable in L_2. It is an outstanding problem to find a suitable norm and a new approach for the stability analysis. For the scalar conservation law, the $L_1 - norm$ is the right norm.

The above two problems are closely related in the sense that a well-posed theory for the hyperbolic system would provide a clear goal for a parabolic theory which depends not so strongly on the strength of the viscosities as the present one does. At present one often uses the characteristic method to study the behavior of the solutions of the hyperbolic systems. [Glimm], [Glimm-Lax], [Liu1]. For such an approach the natural function space is the functions of bounded variation. For the parabolic system, energy method is natural and one obtains a well-posed theory in the Sobolev spaces. Our pointwise approach for the parabolic systems may be viewed as a small step in the effort of bridging the gap, still quite big, of the persent two distinct thinkings for respective systems.

References

[1] Chang, T. and Hsiao, L., The Riemann Problem and Interaction of Waves in Gas Dynamics, Longman Scientific and Technical, 1989.

[2] Courant, R., and Friedrichs, K.,Supersonic Flows and Shock Waves, Springer, 1948.

[3] Ding, X.,Chen, G. and Luo, P., Convergence of the fractional step Lax-Friedrichs scheme for insentropic system of gas dynamics, Comm. Math. Phys. 121(1989)63-81.

[4] Freistuhler, H. and Liu, T.-P., Nonlinear stability of overcompressive shock waves in a rotationally invariant system of viscous conservation laws, Comm. Math. Phys., 153(1993)147-158.

[5] Hoff, D. and Liu, T.-P., The inviscid limit for the Navier-Stokes equations of compressible isentropic flow with shock data, Indiana Univ. J.,38(1989)861-915.

[6] Glimm, J., Solution in the large for nonlinear hyperbolic systems of equations, Comm. Pure Appl. Math. 18(1965)697-715.

[7] Glimm, J. and Lax, P.D., Decay of Solutions to Hyperbolic Conservation Laws, Memior, Amer. Math. Soc.101(1970).

[8] Goodman, J., Nonlinear stability of viscous profiles for conservation laws, Arch. Rational Mech Anal., 95(1986)325-344.

[9] Kawashima, Large-time behaviors of solutions to hyperbolic-parabolic systems of conservation laws and applications, Proc. Roy. Soc. Edinburgh 106A(1987)169-194.

10] Lax, P.D., Hyperbolic conservation laws, II, Comm. Pure Appl. Math., 10(1957)537-566.

[11] Liu, T.-P.,1, Adimissible Solutions for Hyperbolic Conservation Laws, Memior, Amer. Math. Soc. 240(1981).

[12] Liu, T.-P.,2, Nonlinear Stability of Shock Waves for Viscous Conservation Laws, Memior, Amer. Math. Soc., 328 (1985).

[13] Liu, T.-P.,3, on viscousity criterion for hyperbolic conservation laws, Viscous Profiles and Numerical Methods for shock Waves, ed. Shearer, M., Philadelphia SIAM (1991)105-114.

[14] Liu, T.-P.4, Interaction of nonlinear hyperbolic waves, Nonlinear Analysis, Ed. Liu, F.-C. and Liu, T.-P., Academia Sinica, Taipei, R.O.C., World Scientific (1989)171-183.

[15] Liu, T.-P., Tan, D., and Ying, L., Time-asymptotic states for a combustion model. (to appear.)

[16] Matsumura, A. and Nishihara, K., On stability of travelling wave solutions of a one-dimensional model system for compressible viscous gas.

[17] Szepessy, A. and Xin, Z., Nonlinear stability of viscous shock waves, Arch. Rational Mech. Anal. (to appear.)

[18] Wu, C.C., New theory of MHD shock waves, Viscous Profiles and Numerical Methods for Shock Waves, ed. Shearer, M., Philadelphia SIAM (1991)209-236.

AMS/IP Studies in Advanced Mathematics
Volume 3, 1997

Singular perturbation problems to a combustion equation in very long cylindrical domains

M. Mimura, K. Sakamoto, S.-I. Ei

Abstract

By combining two timing method with singular perturbation techniques, it is shown that a combution model with a small parameter ϵ in very long cylindrical domains is formally reduced to a one dimensional problem as ϵ tends to zero. This enables us to study more qualitatively spatio-temporal pattern dynamics in combustion processes from application view points.

1 Introduction

Variety of spatio and/or temporal patterns are observed in combustion processes with or without supply of fuel. In order to theoretically understand such pattern formation, several mathematical models have been proposed so far. Among them, we focus our attention to a thermal and diffusive equation which describes combustion processes in a single step exothermic reaction. The equation is given by the following two component system for u and v :

$$\left\{\begin{array}{lcl} u_t &=& \Delta_y u + u_{xx} + vf(u) \\ \\ v_t &=& d(\Delta_y v + v_{xx}) - \epsilon v f(u) \end{array}\right. , \quad t > 0, \quad (x,y) \in \Omega_L = (0,L) \times \Omega, \tag{1}$$

where u and v denote respectively the (nondimensionalized) absolute temperature and concentration of a fuel. Δ_y is the Laplacian operator in y, and Ω_L is a cylindrical domain with the section Ω in $\mathbf{R}^n$ and the length L. The nonlinearity $f(u)$ takes the form

$$f(u) = \exp[u/(1+u/\alpha)], \quad \alpha > 0,$$

which is called the Arrhenius rate in irreversible chemical kinetics (Frank-Kamenetzky [5], for instance). The paremeters d and ϵ are positive constants and especially ϵ is assumed to be sufficiently small but non zero (Sattinger [8]).

The initial and boundary conditions are

$$(u,v)(0,x,y) = (u_i,v_i)(x,y) \geq 0, \quad (x,y) \in \Omega_L, \tag{2}$$

$$\left\{\begin{array}{l} (u,v)(t,x,y) = (u_0,v_0)(y) \geq 0, \quad t > 0, \quad (x,y) \in \Gamma_0 = \{x = 0, \ y \in \Omega\}, \\ \\ (u,v)(t,x,y) = (u_1,v_1)(y) \geq 0, \quad t > 0, \quad (x,y) \in \Gamma_1 = \{x = L, \ y \in \Omega\}, \end{array}\right. \tag{3}$$

$$u = 0, \;\; d\partial v/\partial \nu = \epsilon h(\beta - v), \;\; t > 0, \;\; (x, y) \in \Gamma = \{0 < x < L, \; y \in \partial\Omega\}, \tag{4}$$

where ν is the outward normal unit vector on the boundary $\partial\Omega$. The condition on v in (4) indicates that the in- or out-flux of the fuel, with the flux rate ϵh through the boundary Γ, depends on the difference between β and the value of v on the boundary. h and β are both positive but may depend on x and y. In particular, when $h = 0$, there is no fuel supply through Γ. The compatibility conditions on $(u_0, v_0), (u_1, v_1)$ and h are required on the boundaries $\Gamma_0, \; \Gamma_1$. Instead of (3), we may impose the following boundary conditions :

$$\left\{ \begin{array}{rcl} u(t,x,y) & = & 0 \\ \partial v(t,x,y)/\partial x & = & 0 \end{array} \right. , \;\; t > 0, \;\; (x,y) \in \Gamma_0 \cup \Gamma_1.$$

In the previous paper [4], as a special case of (1), the authors have considered

$$\left\{ \begin{array}{rcl} u_t & = & \Delta_y u + v f(u) \\ v_t & = & d\Delta_y v - \epsilon v f(u) \end{array} \right. , \;\; t > 0, \;\; y \in \Omega \tag{5}$$

with the boundary conditions

$$u = 0, \;\; d\partial v/\partial \nu = \epsilon h(\beta - v), \;\; t > 0, \; y \in \partial\Omega \tag{6}$$

and have investigated the effect of β on dynamics of solutions to (5),(6) by using fast-slow dynamics methods and have shown that there exists a spatio-temoral periodic solution for suitalbe values of β. It should be emphasized that this solution is regarded as an infinite dimensional relaxation oscillation.

The purpose of this paper is to understand the effect of the shape of Ω, which is the cross-section of the cylindrical domain Ω_L, as well as the length L, on the dynamics of solutions to the problem (1)-(4).

In order to do this, we consider the situation in which the cylindrical domain Ω_L is very long in the $x-$direction in the sense that

(A1) $\qquad L^2 = 1/\epsilon\delta,$

where δ is of order $O(1)$ compared with ϵ. Thus, by transforming the x variable by x/L, (1) can be written as

$$\left\{ \begin{array}{rcl} u_t & = & \epsilon\delta u_{xx} + (\Delta_y u + v f(u)) \\ v_t & = & \epsilon\mu v_{xx} + (d\Delta_y v - \epsilon v f(u)) \end{array} \right. , \;\; t > 0, \;\; (x,y) \in \Omega_1 = (0,1) \times \Omega, \tag{7}$$

where $\mu = \delta d$. Correspondingly, the initial and boundary conditions (2) - (4) become

$$(u, v)(0, x, y) = (u_i, v_i)(x, y) \geq 0, \;\; (x, y) \in \Omega_1, \tag{8}$$

$$\begin{cases} (u,v)(t,x,y) = (u_0,v_0)(y) \geq 0, \ \ t>0, \ \ (x,y) \in \Gamma_0 = \{x=0, \ y \in \Omega\}, \\ (u,v)(t,x,y) = (u_1,v_1)(y) \geq 0, \ \ t>0, \ \ (x,y) \in \Gamma_1 = \{x=1, \ y \in \Omega\}, \end{cases} \tag{9}$$

$$u=0, \ \ d\partial v/\partial \nu = \epsilon h(\beta - v), \ \ t>0, \ \ (x,y) \in \Gamma = \{0<x<1, \ y \in \partial\Omega\}. \tag{10}$$

Thus, the problem with which we are concerned in this paper is (7)-(10) with $\epsilon > 0$ sufficiently small.

2 The perturbed Gelfand problem

In this section, we consider the following boundary value problem with parameter $\lambda > 0$:

$$\Delta_y \phi + \lambda f(\phi) = 0, \ \ y \in \Omega, \qquad \phi = 0, \ \ y \in \partial\Omega. \tag{11}$$

This problem, called *the perturbed Gelfand problem*, has been extensively investigated by many authors. One fundamental result is stated in the following :

Theorem1 (Dancer [3]). *Let $\Omega = \{y \in \mathbf{R}^n | \ |y| < R\}$. For any $\lambda > 0$, (11) with $\alpha > 0$ has at least one and at most finitely many positive solutions $\phi(y;\lambda)$.*

From the view point of qualitative property of solutions, it is important to know the effect of λ on the number of positive solutions $\phi(y;\lambda)$. For a finite interval ($n = 1$) or a disk ($n = 2$), if $\alpha > 0$ is small, the solution branch is monotone increasing with $\lambda > 0$, that is, there is a unique solution $\phi(y;\lambda)$ for any fixed $\lambda > 0$, while if $\alpha > 0$ is rather large, the branch takes S shape structure, that is, there are three subbranches, a small (S-), a middle (M-), and a large (L-) ones with two turning points at $\lambda = \lambda_*, \ \lambda^*$. The corresponding solutions ϕ_S, ϕ_M, and ϕ_L have the following properties :

(i) $\phi_S(y;\lambda) < \phi_M(y;\lambda) < \phi_L(y;\lambda)$, *for* $\lambda \in (\lambda_*, \lambda^*)$ *and* $y \in \Omega$.

(ii) For $0 < \lambda < \lambda' < \lambda^*$, $\phi_S(y;\lambda) < \phi_S(y;\lambda')$, $y \in \Omega$.

(iii) For $\lambda_* < \lambda < \lambda'$, $\phi_L(y;\lambda) < \phi_L(y;\lambda')$, $y \in \Omega$.

(iv) The $S-$branch and $L-$branch are stable, while the $M-$branch is unstable with respect to the time dependent problem of (11).

On the other hand, in higher dimensions ($n \geq 3$), the situation is drastically different, depending on the value of α. In fact, when $n = 3$, the global branch of solutions of (11) does not necessarily takes S shape structure but exhibits double S-shape or more complex ones for suitable values of α. The reader should be refered to [1].

Hereafter, we assume

(A2) *The solution branch of* (11) *takes the S-shape structure satisfying* (i)-(iv) *in the above.*

3 Fast-Slow dynamics problem.

Using the fast-slow dynamics method (see [4], for instance), we study the time evolution of the solutions to the problem (7)-(10). To do so, we first consider the case when $\epsilon = 0$. The resulting problem is

$$\begin{cases} u_t = \Delta_y u + v f(u) \\ v_t = d\Delta_y v \end{cases}, \quad t > 0, \quad (x, y) \in \Omega_1 \tag{12}$$

with the initial and boundary conditions

$$(u, v)(0, x, y) = (u_i, v_i)(x, y) \geq 0, \quad (x, y) \in \Omega_1, \tag{13}$$

$$\begin{cases} (u, v)(t, x, y) = (u_0, v_0)(y) \geq 0, \quad t > 0, \quad (x, y) \in \Gamma_0 = \{x = 0, \ y \in \Omega\}, \\ (u, v)(t, x, y) = (u_1, v_1)(y) \geq 0, \quad t > 0, \quad (x, y) \in \Gamma_1 = \{x = 1, \ y \in \Omega\}, \end{cases} \tag{14}$$

$$u = 0, \quad d\partial v/\partial\nu = \epsilon(\beta - v), \quad t > 0, \quad (x, y) \in \Gamma = \{0 < x < L, \ y \in \partial\Omega\}. \tag{15}$$

By the second equation in (12), one can easily find that $v(t, x, y)$ converges to the spatial average of the initial data $v_i(x, y)$, that is,

$$\lim_{t\to\infty} v(t, x, y) = |\Omega|^{-1} \int_\Omega v_i(x, y) dy = \lambda_i(x).$$

Therefore, the first equation in (12) becomes asymptotically autonomous and the $\omega-$limit set of the solution is contained in the $\omega-$limit set of the following problem :

$$\begin{cases} u_t = \Delta_y u + \lambda_i(x) f(u), \quad t > 0, \quad y \in \Omega, \\ u = 0, \quad t > 0, \quad y \in \partial\Omega. \end{cases} \tag{16}$$

Since $u_i(x, y) \geq 0$, we immediately know that for arbitrarily fixed $x \in (0, 1)$, $u(t, x, y)$ converges to one of the stable positive solutiions of (11) with $\lambda = \lambda_i(x)$, unless the initial data $u_i(y)$ is on the stable manifold of $\phi_M(\,\cdot\,; \lambda_i(x))$. We thus expect the following asymptotic behavior, depending upon the position of $x \in (0, 1)$,

$$\lim_{t\to\infty} (u, v)(t, x, y) = \begin{cases} (\phi_S(y; \lambda_i(x)), \lambda_i(x)) \\ \qquad or \\ (\phi_L(y; \lambda_i(x)), \lambda_i(x)) \end{cases}, \quad y \in \Omega, \tag{17}$$

which is called the *fast dynamics* for $(u, v)(t, x, y)$. At this stage, we should remark two points : (i) there possibly occurs *internal transition layers* in the $x-$direction on $(0, 1)$, because of the bistability property of ϕ_S and ϕ_L, and (ii) $v(t, x, y)$ becomes spatially constant in y. This implies that the fast dynamics for v is not valid for large time t.

In order to obtain the slow dynamics of v for large time t, we integrate over Ω the second equation in (7) and using the boundary conditions in (10), we obtain

$$\epsilon^{-1}\partial/\partial t \int_\Omega v dy = \int_{\partial\Omega} h(\beta - v)dy + \int_\Omega (\mu v_{xx} - v f(u))dy. \tag{18}$$

Since $v(t, x, y)$ is spatially constant in $y-$direction for large time t, we may replace it by, say, $\lambda(t, x)$. Moreover, (17) indicates $u(t, x, y) = \phi(y; \lambda(t, x))$, where ϕ stands for either ϕ_S or ϕ_L, depending on the position x. To make the discussion simple, we consider the situation in which there is $x = s(t)$, which is called an *interfacial point* in $(0, 1)$, such that :

$$u(t, x, y) = \phi(y; \lambda(t, x)) = \begin{cases} \phi_L(y; \lambda(t, x)), & 0 < x < s(t), \\ \phi_S(y; \lambda(t, x)), & s(t) < x < 1. \end{cases} \tag{19}$$

To assure the above relation, we assume that $u_0(y) = \phi_L(y; \lambda_i(0))$ and $u_1(y) = \phi_S(y; \lambda_i(1))$. Consequently, substituting $(u, v) = (\phi(y; \lambda(t, x)), \lambda(t, x))$ into (18), we obtain the one dimensional scalar reaction-diffusion equation for $\lambda(t, x)$.

$$\epsilon^{-1}\partial\lambda/\partial t = \mu\lambda_{xx} + h^*(\beta - \lambda) - \lambda \int_\Omega f(\phi(y; \lambda))dy,$$

where $h^* = h|\partial\Omega||\Omega|^{-1}$. Therefore, the equation for λ has the form :

$$\epsilon^{-1}\partial\lambda/\partial t = \mu\lambda_{xx} + \begin{cases} G_L(\lambda), & t > 0, \ 0 < x < s(t), \\ G_S(\lambda), & t > 0, \ s(t) < x < 1, \end{cases} \tag{20}$$

where G_L, G_S are defined by

$$G_L(\lambda) = h^*(\beta - \lambda) - \lambda \int_\Omega f(\phi_L(y; \lambda))dy,$$

$$G_S(\lambda) = h^*(\beta - \lambda) - \lambda \int_\Omega f(\phi_S(y; \lambda))dy.$$

The initial and boundary conditions are plausible to take

$$\lambda(0, x) = \lambda_i(x), \ 0 < x < 1, \tag{21}$$

$$\lambda(t, 0) = |\Omega|^{-1} \int_\Omega v_0(y)dy, \ \lambda(t, 1) = |\Omega|^{-1} \int_\Omega v_1(y)dy. \tag{22}$$

The dynamics of solutions $(u, v) = (\phi(y; \lambda(t, x)), \lambda(t, x))$ to (20)-(22) is called the *slow dynamics* of $(u(t, x, y), v(t, x, y))$. However, the initial and boundary value problem (20)-(22) has not yet been well formulated, because the internal layer position $x = s(t)$ is not known *a priori*. We therefore have to derive an evolution equation which describes the dynamics of $s(t)$. This will be done in the next section.

4 Singular perturbation problems.

In order to derive an evolutional equation for $s(t)$, we use the well known singular perturbation argument. Becuause of the existence of the interfacial point $x = s(t)$ in the fast dynamics, we expect that $u(t,x,y)$ exhibits an internal layer in the vicinity of $x = s(t)$ when $\epsilon > 0$ is suffciently small. To study the dynamics of such a layer, we introduce the stretching coordinate ξ via $\xi = (x - s(t))/\sqrt{\epsilon}$. The resulting equation is

$$\begin{cases} \tilde{u}_t - (1/\sqrt{\epsilon})s_t\tilde{u}_\xi &= \delta\tilde{u}_{\xi\xi} + \Delta_y\tilde{u} + \tilde{v}f(\tilde{u}) \\ \tilde{v}_t - (1/\sqrt{\epsilon})s_t\tilde{v}_\xi &= \mu\tilde{v}_{\xi\xi} + d\Delta_y\tilde{v} - \epsilon\tilde{v}f(\tilde{u}) \\ -s(t)/\sqrt{\epsilon} < &\xi < (1 - s(t))/\sqrt{\epsilon},\ t > 0,\ y \in \Omega, \end{cases} \tag{23}$$

where $\tilde{u}(\xi) = u(s(t) + \sqrt{\epsilon}\xi)$ and $\tilde{v}(\xi) = v(s(t) + \sqrt{\epsilon}\xi)$. The equations in (23) suggest us to also introduce the stretching coordinate in time by $\tau = \sqrt{\epsilon}(t - t_0)$. Then, (23) becomes as follows :

$$\begin{cases} \sqrt{\epsilon}U_\tau - s_\tau U_\xi &= \delta U_{\xi\xi} + \Delta_y U + Vf(U) \\ \sqrt{\epsilon}V_\tau - s_\tau V_\xi &= \mu V_{\xi\xi} + d\Delta_y V - \epsilon V f(U) \\ -s(t_0 + \sqrt{\epsilon}\tau)/\sqrt{\epsilon} < &\xi < (1 - s(t_0 + \sqrt{\epsilon}\tau))/\sqrt{\epsilon},\ t > 0,\ y \in \Omega, \end{cases} \tag{24}$$

where $U(\tau) = \tilde{u}(t_0 + \sqrt{\epsilon}\tau)$ and $V(\tau) = \tilde{v}(t_0 + \sqrt{\epsilon}\tau)$. If we assume :

$$\lim_{\epsilon\to 0} v(t,x,y) = \lim_{\epsilon\to 0} v(t_0 + \sqrt{\epsilon}\tau, s(t_0 + \sqrt{\epsilon}\tau) + \sqrt{\epsilon}\xi, y) = \lambda(t_0, s(t_0)),$$

then the first equation in (24) reduces to a scalar equation for u only :

$$-s_\tau(t_0)U_\xi(0,\xi,y) = \delta U_{\xi\xi}(0,\xi,y) + \Delta_y U(0,\xi,y) + \lambda(t_0, s(t_0))f(U(0,\xi,y), \tag{25}$$

for $\xi \in \mathbf{R},\ y \in \Omega$ with the boundary conditions

$$\begin{cases} U(0,\xi,y) &= 0,\ \xi \in \mathbf{R},\ y \in \partial\Omega, \\ U(0,-\infty,y) &= \phi_L(y;\lambda(t_0,s(t_0))),\ y \in \Omega, \\ U(0,\infty,y) &= \phi_S(y;\lambda(t_0,s(t_0))),\ y \in \Omega. \end{cases} \tag{26}$$

The following lemma plays a key role to the solution of the problem (25),(26).

Lemma 1 (Vega [10]). *Consider the following nonlinear eigenvalue problem for a fixed $\lambda \in (\lambda_*, \lambda^*)$ in the cylindrical domain $\Omega_\infty = \{(z,y) \in \mathbf{R} \times \Omega\}$:*

$$-cu_z = \delta u_{zz} + \Delta_y u + \lambda f(u),\ (z,y) \in \Omega_\infty \tag{27}$$

with the boundary conditions

$$\begin{cases} u(-\infty, y) = \phi_L(y;\lambda), \quad u(\infty, y) = \phi_S(y;\lambda), \quad y \in \Omega, \\ u = 0, z \in \mathbf{R}, \quad y \in \partial\Omega. \end{cases} \tag{28}$$

Then there is a unique $c(\lambda)$ *such that* (27),(28) *has a solution* $U^*(z, y; \lambda)$ *which is monotone in* z.

Remark. *The sign of* $c(\lambda)$ *is determined by that of* $H(\phi_L(\cdot;\lambda)) - H(\phi_S(\cdot;\lambda))$, *where*

$$H(u) = \int_\Omega [|\nabla u|^2 - 2\lambda F(u)]dy, \quad F(u) = \int_0^u f(s)ds.$$

We thus find that $c(\lambda) > 0$ *for* $\lambda > \lambda_*$ *but close to* λ_* *and that* $c(\lambda) < 0$ *for* $\lambda < \lambda^*$ *but close to* λ^*. *It is numerically confirmed that* $c(\lambda)$ *is a monotone decreasing function of* λ.

Now comparing (25) with (27), we arrive at

$$ds(t_0)/d\tau = c(\lambda(t_0, s(t_0)).$$

Consequently, as the limiting system of (7) as $\epsilon \to 0$, we arrive the following free boundary problem for $(\lambda(t,x), s(t))$:

$$\epsilon^{-1}\partial\lambda/\partial t = \mu\lambda_{xx} + \begin{cases} G_L(\lambda), \quad t > 0, \quad 0 < x < s(t), \\ G_S(\lambda), \quad t > 0, \quad s(t) < x < 1, \end{cases} \tag{29}$$

$$\epsilon^{-1/2} ds/dt = c(\lambda(t, s(t))), \quad t > 0, \tag{30}$$

with the initial and boundary conditions

$$\lambda(0, x) = \lambda_i(x), \quad 0 < x < 1, \quad s(0) = s_i, \quad 0 < s_i < 1, \tag{31}$$

$$\lambda(t, 0) = |\Omega|^{-1} \int_\Omega v_0(y)dy, \quad \lambda(t, 1) = |\Omega|^{-1} \int_\Omega v_1(y)dy, \quad t > 0. \tag{32}$$

As for the continuity condition on λ, it is required to have $\lambda \in C^1([0,1])$ (see [2]). Therefore, the problem for $(\lambda(t,x), s(t))$ can be well formulated as (29)-(32).

If we could obtain a solution $(\lambda(t,x), s(t))$ of (29)-(32), we find that the solution (u, v) of (7) with (8)-(10) is approximated by $(u, v) = (\phi(y; \lambda(t,x)), \lambda(t,x))$ for sufficiently small $\epsilon > 0$. This representation indicates that the time profile of the solution (u, v) is given by $\lambda(t,x)$, while the spatial profile is determined by $\phi(y; \lambda(t,x))$. The effects of the shape of the domain Ω and the length in the $x-$direction of the cylindrical domain are, respectively, encoded in $\phi(y; \lambda(t,x))$ and $\lambda(t,x)$. However, the rigorous justification of this reduction in the limit $\epsilon \to 0$ remains unsolved.

5 Reduction to a one dimensional problem.

In the previous sections, we find that the original problem in an $(n+1)-$dimensional cylindrical domain is formally reduced to the one dimensional free boundary problem when $\epsilon > 0$ is sufficiently close to zero. From theoretical (as well as numerical) view points, this reduction enables us to treat the problem more qualitatively. However, even so, the situation is not so simple, because of the difficulty of solving the free boundary problem. In order to avoid this difficulty, we propose a usual one dimensional reaction-diffusion system for two components (u, v). The system is given by

$$\begin{cases} u_t - \epsilon\delta u_{xx} &= -au + vf(u) &= F(u,v) \\ v_t - \epsilon\mu v_{xx} &= \epsilon[h^*(\beta - v) - vf(u)] &= \epsilon G(u,v), \end{cases} \tag{33}$$

for $t > 0,\ \ 0 < x < 1$, where $h^* = h|\partial\Omega|/|\Omega|$, and u and v denote the temperature and the concentration of fuel.

The system (33) is intuitively derived from the original systems (7) with (8)-(10) as follows : The term $-au$ in the first equation implies the effect of heat radiation related to the Dirichlet boundary condition in (10) and the term $h^*(\beta - v)$ is the fuel supply related to the Robin conditions in (10). As in Fig.1, if α is small, the nullcline of F is monotone increasing with u, while if α is rather large, it takes S-shape with one local minimum and one local maximum points $u = u_*,\ u^*\ (u_* < u^*)$, while that of G is monotone decreasing. We note that there is only one intersecting point of $G = 0 = F$ in the $(u, v)-$plane. Of course the location of the intersecting point changes depending on the value of β. When α is large, we write three branches of $\{F(u, v) = 0\}$ as $u = \phi_S(v),\ \phi_M(v)$ and $\phi_L(v)$.

The initial and boundary conditions are given by

$$(u, v)(0, x) = (u_i, v_i)(x) \geq 0, 0 < x < 1, \tag{34}$$

$$\begin{cases} (u,v)(t,0) &= (u_0, v_0) \geq 0,\ \ t > 0, \\ (u,v)(t,1) &= (u_1, v_1) \geq 0,\ \ t > 0, \end{cases} \tag{35}$$

where (u_0, v_0) and (u_1, v_1) are assumed to satisfy $u_0 = \phi_L(v_0)$ and $u_1 = \phi_S(v_1)$. For sufficiently small $\epsilon > 0$, the singular perturbation argument similar to the one in the previous section shows that, in a short time, u tends to $\phi_S(v_i(x))$ and $\phi_L(v_i(x)$ so that there appear internal transition layers at some points in $0 < x < 1$. For simplicity, we consider the situation where there is one layer at $x = s(t)$ in the limit $\epsilon \to 0$ such that $u = \phi_L(v), 0 < x < s(t),\ u = \phi_S(v), s(t) < x < 1$. Substituting this relation into the second equation in (33), we obtain

$$\epsilon^{-1}\partial v/\partial t = \begin{cases} \mu v_{xx} + G_L(v), & t > 0,\ \ 0 < x < s(t), \\ \mu v_{xx} + G_S(v), & t > 0,\ \ s(t) < x < 1, \end{cases} \tag{36}$$

where $G_L(v) = g(\phi_L(v), v)$ and $G_L(v) = g(\phi_L(v), v)$, which are qualitatively the same as the ones in (30). The equation describing the time evolution of $s(t)$ is derived in a way similar to the one in (29),

$$\epsilon^{-1/2} ds(t)/dt = c(v(t, s(t))), \quad t > 0, \tag{37}$$

where $c(v)$ is the velocity of the travelling front solution $u(z)$ $(z = x - c\sigma)$ of the following reaction-diffusion equation for v fixed :

$$u_\sigma = \delta u_{xx} + F(u, v), \quad \sigma > 0, \quad x \in \mathbf{R} \tag{38}$$

with the boundary conditions

$$u(\sigma, -\infty) = \phi_L(v), \qquad u(\sigma, \infty) = \phi_S(v). \tag{39}$$

Here we note that $u = \phi_L(v)$ and $u = \phi_S(v)$ satisfy $F(u, v) = 0$. It can be shown that the velocity $c(v)$ is a monotone decreasing function of v, which is the similar property to that of $c(\lambda)$, as was stated in Remark 1 in the previous section. Thus we obtain the limiting system of the one dimensional problem as $\epsilon \to 0$ as (36),(37). The justification of this reduction was discussed in [2].

Consequently, we can say that the original problem in very long cylindrical domain is qualitatively equal to the one dimensional problem (33)-(35) in the singular limit $\epsilon \downarrow 0$. This allows us to consider the one dimensional problem (33)-(35) as an approximation to the original one for sufficiently small $\epsilon > 0$. The justification of this statement has been numerically confirmed, although it has not yet been proved rigorously. Theoretical analyses of the one-dimensinal problem, with complementary numerical simulations, have revealed that a variety of spatio and/or temporal patterns are exhibited in cylindrical domains, depending on the shape of the section and the length of the cylinder. It will be reported in a forthcoming paper [7].

References

[1] Bebernes, J. and Eberly, D. : *Mathematical Problems from Combustion Theory,* Springer-Verlag, New York (1989)

[2] Chen, X.-Y. : Dynamics of interfaces in reaction-diffusion systems, Hiroshima Math. J. 21(1991) 47 -83

[3] Dancer, E.N. : On the structure of solutions of an equation in catalysis theory when a parameter is large, J. Diff. Equations, 37(1980) 404-437

[4] Ei, S.-I. and Mimura, M. : Relaxation oscillations in combustion models of thermal self-ignition, J. Dynamics and Diff. Equations, 4(1992) 191-229

[5] Frank-Kamenetzky, D.A. : *Diffusion and Heat Exchange in Chemical Kinetics* (transtated by N. Thon), Princeton Univ. Press, Princeton, N.J. (1955)

[6] Henry, D. :*Geometric Theory of Semilinear Parabolic Equations.* Lecture Notes in Mathematics, 840, Springer-Verlag, New York (1881)

[8] Mimura, M., Nagayama, M. and Sakamoto,K. : Numerical treatment on spatio-temporal patterns arising in a combustion equation in cylindrical domains, in preparation

[9] Sattinger, D.F. : A nonlinear parabolic system in the theory of combustion. Q. Appl. Math. 33 (1975) 47-61

[10] Vega, J.M. : Traveling wavefronts of reaction-diffusion equations in a cylindrical domains. Comm. P.D.Es. 18(1993) 505-531

(M.Mimura) Department of Mathematical Sciences,
University of Tokyo, Tokyo, Japan

(K.Sakamoto) Department of Mathematics,
Faculty of Science, Hiroshima University, Higashi-Hiroshima, Japan

(S.-I. Ei) Graduate School of Integrated Science,
Yokohama City University, Yokohama, Japan

AMS/IP Studies in Advanced Mathematics
Volume 3, 1997

PARTICLE- LIKE SOLUTIONS OF THE EINSTEIN- YANG-MILLS EQUATIONS AND THE EXISTENCE OF AN EXOTIC BLACK-HOLE SOLUTION

by

J.A. SMOLLER AND A.G. WASSERMAN[1]

The coupled Einstein/Yang-Mills (EYM) equations with SU(2) gauge group are, (c.f. [1-4]),

$$R_{ij} - \frac{1}{2}Rg_{ij} = \sigma T_{ij}, and \quad d^* F_{ij} = 0.$$

Here T_{ij} is the stress-energy tensor associated to the su(2)-valued Yang-Mills curvature 2-form F_{ij}, and $R_{ij} - \frac{1}{2}Rg_{ij}$ is the Einstein tensor computed with respect to the sought-for metric g_{ij}. For static, stationary, spherically symmetric solutions of the above equations, the line element for the metric can be written as

$$ds^2 = -T(r)^{-2}dt^2 + A(r)^{-1}dr^2 + r^2(d\theta^2 + sin^2\theta d\phi^2),$$

and the curvature 2-form as

$$F = w'\tau_1 dr d\theta + w'\tau_2 dr(sin\theta d\phi) + (1 - w^2)\tau_3 d\theta(sin\theta d\phi).$$

Here $(T(r), A(r))$ and $w(r)$ are the unknown metric and connection coefficients, respectively, and τ_1, τ_2, τ_3, the Pauli matrices, form a basis for the Lie algebra su(2). In this set-up, the EYM equations become the following system of ordinary differential equations, [1-4]:

[1]Research of both authors supported by the NSF, Contract # DMS-89-05205

$$r^2Aw'' + r\left[(1-A) - \frac{(1-w^2)^2}{r}\right]w' + w(1-w^2) = 0$$

$$rA' + (1+2w'^2)A = 1 - \frac{(1-w^2)^2}{r^2}$$

$$2rA\frac{T'}{T} = \frac{(1-w^2)^2}{r^2} + (1-2w'^2)A - 1.$$

Notice that the T-equation decouples from the first two equations. The problem is to find solutions that have the following asymptotic behavior at infinity:

$$A(r) \simeq 1 - \frac{\mu}{r}, w^2(r) \simeq 1, and \quad T(r)^{-2} \simeq 1 - \frac{\mu}{r}.$$

where μ is a positive constant, the "total mass".

There are two main difficulties with these equations: First, the equations are singular if $r = 0$ or if $A = 0$; the $r = 0$ singularity is remedied by a careful choice of initial conditions; namely, $w(0) = 1$, $w'(0) = 0, A(0) = 1$. The existence theorem proved in [2] says that solutions are parametrized by $w''(0) = -\lambda$. Second, the equations are fiercly non- linear. Defining

$$\Phi(r) = r(1-A) - \frac{(1-w^2)^2}{r}, and \quad u(r) = (1-w^2),$$

equation (1) becomes

$$r^2Aw'' + \Phi(r)w' + u(r)w = 0,$$

just a tame second order linear homogeneous ode. To really make use of this point of view , we must make sure that $A(r) > 0$ and $|w'(r)| < \infty$; i. e. , we must prevent "crash".

Set

$$\Gamma = \{(w, w', A, r) \in \Re^4 : w^2 \leq 1, r > 0, 0 < A \leq 1\};$$

then then equations define a "flow" on Γ. Solutions we seek must lie in Γ.

Now crash is a real danger; in fact, we have the following result.

Proposition 1 ([2]): If $\lambda \geq 2$, the solution crashes with in the region $w > 0$. If $\lambda \leq 1$, then no crash occurs provided $w^2 \leq 1$.

On the other hand, it is proved in [2] that if $\Phi(r) > 0$, then no crash can occur. (From (1), we see that $\Phi(r) > 0$ represents a "friction" term). Now even though $\Phi(r) > 0$ is really $\Phi(r, w, A)$, we can still prove general things about $\Phi(r)$; namely, we have

Proposition 2 ([3,4]): If $r \geq 1+\epsilon$, or $w^2(r)$ is sufficiently close to 1, then $\Phi(r) > 0$. In this case, there are constants $\eta > 0, m > 0$ such that

$$A(r) \geq \eta \quad and \quad |w'(r)| \leq m.$$

Thus $\Phi > 0$ positive prevents crash.

The fundamental result which we need is the following "no crash" result.

Porposition 3 ([3]): Let

$$\Lambda_n(r) = (w_n(r), w_n'(r), A_n(r), r), a \leq r \leq b_n$$

be a sequence of orbit segments in Γ, with uniformly bounded rotation; i.e., there is an $M > 0$ such that

$$\theta_n(r) \equiv \frac{w_n'(r)}{w_n(r)} \geq -M, \theta_n(a) \geq -\frac{\pi}{2}.$$

Suppose too that $\Lambda_n(b_n) \to P = (\bar{w}, \bar{w}', \bar{A}, b) \in \Gamma$. Then the (backwards) orbit through P lies in Γ for $a \leq r \leq b$, and has rotation $\geq -M$.

One application of the no-crash result is

Theorem 4 ([3]) : There are an infinite number of particle- like solutions of the EYM equations. The nth solution has rotation number $-n\pi$, and each solution has finite total mass; i.e.,

$$\lim_{r\to\infty} \mu_n(r) \equiv lim_{n\to\infty} r(1 - A_n(r)) = \mu_n < \infty.$$

Sketch of proof. Let $\bar{\lambda}$ be the infinum of the crashing λ's, and let $\lambda_n \to \bar{\lambda}, \lambda_n < \bar{\lambda}$. Consider the orbit segments $\Lambda_n(r), 0 \leq r \leq 2$ of the $\lambda_n - orbits$. By passing to a subsequence , we may assume that $\Lambda_n(2)$ converges to a point $p \in \Gamma$, because $1 \geq A_n(2) \geq \eta > 0, w_n^2(2) \leq 1$, and $|w_n'(w)| \leq m$. The (backwards) orbit through P must crash because nearby orbits crash; therefore by the no-crash proposition, the $\lambda_n - orbits$ have unbounded rotation.

The significance of these solutions is that the Yang- Mills repulsive force balances the attractive gravitational force, and prevents the formation of singularities in 4- dimensional spacetime. Furthermore, covariant quantities , which represent masses, come directly out of our solutions; they are not arbitrary constants.

Another application of these ideas is to the existence of black -hole solutions of the EYM equations. These are solutions defined for all $r \geq r_h$, where $A(r_h) = 0$, which have the asymptotic behavior (1). The singularity in the w-equation at

$r = r_h$ is again remedied by a careful choice of initial conditions; in this case we require $\Phi(r_h)w' + u(r_h)w = 0$. This condition describes different types of curves in the $w - w'$ plane for $r_h > 1, r_h = 1$ and $r_h < 1$. In all cases the theorem is the same:

Theorem 5 ([4]): For any even horizon $r_h > 0$, there exists an infinite number of black-hole solutions the the EYM equations. Moreover the masser external to the black holes are uniformly bounded.

These solution provide counterexamples to to the Wheeler "n0- hair" conjecture.

We now describe a new type of black hole solution to the EYM soluations; a so-called "exotic" black hole. This solution cannot be investigated numerically.

Let λ_n be the parameter of a particle-like solution, (as described above) , with rotation number $-n\pi$, and let $\Lambda_n(r)$ be the associated orbit. Passing to a subsequence if necessary , $\lim_{n\to\infty} \Lambda_n(2) = P \in \Gamma$, because $0 < \eta \leq A_n(2) \leq 1, w_n^2(2) \leq 1, |w_n'(2)| \leq m$. The orbit through $P, \Lambda(r) = (w(r), w'(r), A(r), r)$, is the solution that we wish to investigate.

First, as $r \to \infty, \Lambda(r)$ connects to a "rest" point, $(w, w') = (\pm 1, 0)$. Namely, it cannot crash because $r > 0$, and it cannot exit Γ via $w^2 = 1$, because then the λ_n orbits, for large n, would have to exit Γ, by continuous dependence . Then any orbit staying in Γ for all r must connect to a rest point $(\pm 1, 0)$, see [2,3].

Next, going backwards in r; i.e. , for $r < 2$, the orbit $\Lambda(r)$ must crash! Otherwise, for large n , the λ_n − *orbits* would have bounded rotation. In fact ,$A(1) = 0$; i.e. , the crash occurs at $r = 1$. This vanishing of the metric coeficient A defines a black hole.

Now more can be said about the crash of the exotic orbit (c.f.[5]) ; namely the following properties hold:

(i) $\lim_{r\to 1_+} w(r) = 0$

(ii) $\overline{\lim}_{r\to 1_+} w'(r) = +\infty, \quad \underline{\lim}_{r\to 1_+} w'(r) = -\infty$

(iii) For any $\epsilon > 0$, the exotic orbit has infinite rotation in the $1 < r < 1 + \epsilon$ Thus near $r = 1, w(r)$ behaves qualitatively like $(r-1)sin\frac{1}{r-1}$.

Now there is nothing special about $r = 2$; for all $r > 1$, $\lim_{n\to\infty} \Lambda_n(r) = \Lambda(r)$. What about $r < 1$? Since $\lim_{n\to\infty} \lambda_n = \bar{\lambda} \leq 2$, it follows that $\Lambda_n(r) \to \bar{\Lambda}(r)$, where $\bar{\Lambda}(r)$ denotes the orbit starting at $r = 0$, corresponding to the parameter value $\bar{\lambda}$. In other words, the exotic black hole solution is a (pseudo) continuation of the limit of particle -like solutions. (At $r = 1, A$ and T are C' functions, but w is only continuous.) More details concerning this exotic black hold solution can be found in [5].

References

1. Bartnik, R.,and McKinnon, J. , Particle- like solutions of the Einstein/Yang-Mills equations, Phys. Rev. Lett. 61, 141-144, (1988).
2. Smoller, J., Wasserman, A.,Yan, S.-T. , and McLeod, J. , Smooth static solutions of the Einstein/Yang-Mills equations, Comm. Math. Phys . 143,115-147, (1991).
3. Smoller, J., and Wasserman, A., Existence of infinitely- many smooth, static, global solutions of the Einstein/Yang-Mills equations, Comm. Math. Phys. 151, 303-325, (1993).
4. Smoller, J., and Wasserman, A.,Yan, S.-T., Existence of infinitely- many black -hole solutions of the Einstein/Yang-Mills equations, Comm. Math. Phys. 154, 377-401, (1993).
5. Smoller, J., and Wasserman, A., An investigation of the limiting behavior of particle- like solutions to the Einstein/Yang- Mills equations, and a new black hole solution, Comm. Math. Phys, (to appear)

The University of Michigan
Ann Arbor, MI 48109-1003
USA

AMS/IP Studies in Advanced Mathematics
Volume 3, 1997

The Large Time Behavior of Cauchy Problem for A Dissipative Benjamin-Ono Type Equation

Zhou Yulin and Guo Boling

1 Introduction

The study of turbulent behavior is one of the most important problems in plasma physics. [1] present a model for studying plasma and fluid turbulence that is amenable to an exact soliton like nonlinear mode decompsition. The basic dynamical model used is the dissipative Benjamin-Ono (DBO) equation, which is a nonlinear wave equation of internal wave in a two-fluid system:

$$u_t + \mu H u_x + \beta H u_{xx} - \nu u_{xx} + 2uu_x = 0 \tag{1.1}$$

where the Hilbert transform H is defined as

$$Hf(x) = \frac{1}{\pi} P \int_{-\infty}^{\infty} \frac{f(x')}{x' - x} dx' \tag{1.2}$$

and P denotes the principle value of the integral. The growth $\mu H(u_x)$ is the same as tha growth term in a nonlinear ion acoustic wave which is caused by energy transfer from electrons to the ion wave. The dissipation νu_{xx} is from kinematic viscosity.

The purpose of this work is to estiblish the existence of the global attractor and the estimates of its Hausdorff and fractal dimensions for the Cauchy problem of the general equation

$$u_t - \alpha u_{xx} + 2uu_x + \beta H u_{xx} + \delta H u_x = g(u) + h(x) \tag{1.3}$$

$$u|_{t=0} = u_0(x), \quad x \in R \tag{1.4}$$

of the dissipative Benjamin-Ono type equation, where $\alpha > 0$, β δ are real constants.

It is well known that the existence of the global attractor and the estimate of its dimension hinge crucially upon the compactness of ω-limit sets of trajectories. In the previous study of nonlinear parabolic dynamics, this compactness requirement is trivially satisfied and the trajectories themselves are precompac. For an evolution equation of functions defined on an unbounded Euclidean domain, this precompactness property dose not hold true. It is our aim in this note to present a method of the weighted function space to overcome this difficulty. It needs to make a priori estimates uniformly for $t \in R^+$ in weighted space to construct the semigroup and its attractor.

2 Existence of global smooth solution and a priori estimate uniformly for t 2 R^+

Let $H_{0,\gamma}$ be a Hilbert space with the norm

$$\|u\|_{0,\gamma}^2 = \int |u(x)|^2(1+|x|^2)^\gamma dx \tag{2.1}$$

and let $H_{l,\gamma}$ $(l=1,2,\cdots)$ be weighted Sobolev spaces with the norms

$$\|u\|_{l,\gamma}^2 = \sum_{|\alpha|\le l} \|\partial^\alpha u\|_{0,\gamma}^2 \tag{2.2}$$

where $\partial^\alpha = \partial_1^{\alpha_1}\cdots\partial_n^{\alpha_n}$, $\partial_i = \frac{\partial}{\partial x_i}$, $|\alpha| = \alpha_1+\cdots+\alpha_n$.

It is evident that, as $\gamma>0$, function $u(x)$ which belong to $H_{0,\gamma}$ are decreasing at infinity faster than the functions from $L_2(R^n)$.

We use in the estimations for the solution of (1.3) (1.4) the function $\phi(x)$,

$$\phi(x) = \phi_\epsilon(x) = (1+|\epsilon x|^2)^\gamma \tag{2.3}$$

where $0<\epsilon\le 1$ and ϵ is sufficiently small. The function $\phi(x)$ is in the definition of norm in $H_{0,\gamma}$ with $\epsilon=1$. For $\epsilon\ne 0$, the following norms are equivalent:

$$C_\epsilon^{-1}\|\phi_\epsilon^{\frac12}u\| \le \|u\|_{0,\gamma} \le C_\epsilon\|\phi_\epsilon^{\frac12}u\| \tag{2.4}$$

(Here and bolow $\|\cdot\|$ denotes $\|\cdot\|_{0,0}$). The estimate

$$|\nabla\phi| \le C\epsilon\phi \tag{2.5}$$

holds. It is easy to verify the inequalities

$$C^{-1}\|u\|_{l,\gamma} \le \|\phi_\epsilon^{\frac12}\|_l \le \|u\|_{l,\gamma}, \quad l=1,2. \tag{2.6}$$

Here and blow $\|\cdot\|_l$ denotes $\|\cdot\|_{l,0}$.

It is similar to the proof of [2], we have

Theorem 2.1 Suppose that the following conditions are satisfied:

(1) $\alpha>0$, $\beta>0$, $\delta\le 0$

(2) $g(u)\in C^m$, $g'(u)\le b=$ constant, $g(0)=0$

(3) $u_0(x)\in H_{m,\gamma}$, $h(x)\in H_{m,\gamma}$.

Then there exists a unique global solution $u(x,t)$ of the Cauchy problem (1.3) (1.4):

$$u(x,t)\in L^\infty(0,T;\ H_{m,\gamma}(R)).$$

Theorem 2.2 Suppose that the following conditions are satisfied:

(1) $\alpha>0$, $\beta>0$, $\delta\le 0$

(2) $g(u)\in H^M$, $g'(u)\le b=$ constant <0, $g(0)=0$

(3) $u_0(x)\in H_{M,\gamma}$, $h(x)\in H_{M,\gamma}$.

Then for the smooth solution of Cauchy problem (1.3)(1.4) there are estimates

$$\|u_{x^M}(\cdot,t)\|_{L_2(R)} \le K\left\{e^{\frac{3}{4}bt}\|u_0\|_{H^M(R)} + \|h\|_{H^M(R)}\right\} \tag{2.7}$$

for any $t \in R^+$, $M = 0, 1, 2, \cdots$ where K is a constant dependent on $\alpha > 0$, $\beta \ge 0$ and as $M \ge 1$ also dependent on the norm $\sup_{t\in R^+}\|u(\cdot,t)\|_{L_2(R)}$.

Now we give a priori estimates uniformly for $t \in R^+$ for the solution of the Cauchy problem (1.3)(1.4) in weighted Sobolev space.

Lemma 2.1 (A. P. Calderon [3]). Let $A: R \to R$ be a C^∞ function. Then the operator $[H;A]\partial_x$ maps $L_2(R)$ into $L_2(R)$ with

$$\|[H,A]\partial_x f\| \le C\|A'f\| \tag{2.8}$$

where H is the Hilbert transform, i.e.

$$Hf(x) = \frac{1}{\pi}P\int_{-\infty}^{\infty}\frac{f(y)}{x-y}dy = F^{-1}(\mathrm{i}\ \mathrm{sgn}\xi\,\hat{f}(\xi))$$

with "$\wedge$" and F^{-1} denoting the Fourier transform and its invers respectively. $[H\partial_x, A]$ denotes

$$[H\partial_x, A]f = [H,A]f_x + H(A'f),$$

and

$$[A,B]f = ABf - BAf$$

is the commutator of singular integral operator A and B.

Lemma 2.2 Suppose that the following conditions are satisfied:

(1) $g(u) \in C^1$, $g(0) = 0$, $g'(u) \le -b$, $b > 0$.

(2) $\alpha > |\beta| + \frac{|\delta|}{2}$, $b > \frac{3}{2}|\delta|$.

(3) $u_0(x) \in H_{0,\gamma}$, $h(x) \in H_{0,\gamma}$.

Then for the smooth solution of Cauchy problem (1.3)(1.4) we have

$$\|u(\cdot,t)\|_{0,\gamma} \le C_\epsilon(e^{-\frac{|\delta|}{2}t}\|u_0\|_{0,\gamma} + \frac{2}{b|\delta|}(1 - e^{-\frac{1}{2}|\delta|t})\|h\|_{0,\gamma}),$$

$$\overline{\lim_{t\to\infty}}\|u(\cdot,t)\|_{0,\gamma} \le C_\epsilon\frac{2}{b|\delta|}\|h\|_{0,\gamma}^2 = E_0, \tag{2.9}$$

$$\overline{\lim_{t\to\infty}}\frac{1}{t}\int_0^t\|u(\cdot,\tau)\|_{1,\gamma}^2 d\tau \le E_0$$

Proof. Multiplying (1.3) by $\phi_\epsilon(x)u$ and integrating with respect to x, we obtain

$$(\phi_\epsilon u, u_t - \alpha H u_{xx} + 2uu_x + \beta H u_{xx} + \delta H u_x - g(u) - h(x)) = 0. \tag{2.10}$$

Here we have

$$(\phi_\epsilon u, u_t) = \frac{1}{2}\frac{d}{dt}\|\phi_\epsilon^{\frac{1}{2}}u(\cdot,t)\|^2,$$

$$(\phi_\epsilon u, -\alpha u_{xx}) = \alpha(\phi'_\epsilon u, u_x) + \alpha(\phi_\epsilon u_x, u_x),$$

$$|(\phi_\epsilon' u, u_x)| \le \gamma\epsilon(||\phi_\epsilon^{\frac{1}{2}} u||^2 + ||\phi_\epsilon^{\frac{1}{2}} u_x||^2),$$

$$|(\phi_\epsilon u, 2uu_x)| = \frac{2}{3}|(\phi_\epsilon, (u^3)_x)| = \frac{2}{3}|(\phi_\epsilon', u^3)| \le \frac{4}{3}\gamma\epsilon||u||_\infty||\phi_\epsilon^{\frac{1}{2}} u||^2,$$

$$(\phi_\epsilon u, \beta H u_{xx}) = -\beta(\phi_\epsilon' u, H u_x) - \beta(\phi_\epsilon u_x, H u_x),$$

$$|\beta(\phi_\epsilon' u, H u_x)| \le |\beta|\gamma\epsilon(||\phi_\epsilon^{\frac{1}{2}} u||^2 + ||\phi_\epsilon^{\frac{1}{2}} H u_x||^2),$$

$$|\beta(\phi_\epsilon u_x, H u_x)| \le \frac{|\beta|}{2}(||\phi_\epsilon^{\frac{1}{2}} u_x||^2 + ||\phi_\epsilon^{\frac{1}{2}} H u_x||^2).$$

In view of Lemma 2.1, it follows

$$||\phi_\epsilon^{\frac{1}{2}} H u_x||^2 \le ||\phi_\epsilon^{\frac{1}{2}} u_x||^2 + c\epsilon||\phi_\epsilon^{\frac{1}{2}} u||^2. \tag{2.11}$$

Hence

$$|(\phi_\epsilon u, \beta H u_{xx})| \le |\beta|(1+\gamma\epsilon)||\phi_\epsilon^{\frac{1}{2}} u_x||^2 + \frac{|\beta|}{2}\epsilon[c + 2\gamma(1+c\epsilon)]||\phi_\epsilon^{\frac{1}{2}} u||^2,$$

$$|(\phi_\epsilon u, \delta H u_x)| \le |\delta|||\phi_\epsilon^{\frac{1}{2}} u||||\phi_\epsilon^{\frac{1}{2}} H u_x|| \le \frac{|\delta|}{2}[(1+c\epsilon)||\phi_\epsilon^{\frac{1}{2}} u||^2 + ||\phi_\epsilon^{\frac{1}{2}} u_x||^2],$$

$$(g(u), \phi_\epsilon u) \le -b||\phi_\epsilon^{\frac{1}{2}} u||^2,$$

$$(\phi_\epsilon u, h) \le ||\phi_\epsilon^{\frac{1}{2}} u||^2||\phi_\epsilon^{\frac{1}{2}} h||^2 \le \frac{b}{2}||\phi_\epsilon^{\frac{1}{2}} u||^2 + \frac{1}{2b}||\phi_\epsilon^{\frac{1}{2}} h||^2.$$

Thus from (2.10) we have

$$\frac{1}{2}\frac{d}{dt}||\phi_\epsilon^{\frac{1}{2}} u||^2 + [\alpha - \alpha\gamma\epsilon - |\beta|(1+\gamma\epsilon) - \frac{|\delta|}{2}]||\phi_\epsilon^{\frac{1}{2}} u_x||^2$$

$$+[b - \alpha\gamma\epsilon - \frac{4}{3}\gamma\epsilon||u||_\infty - \frac{|\beta|}{2}\epsilon(c + 2\gamma(1+c\epsilon)) - \frac{|\delta|}{2}(c\epsilon+1) - \frac{b}{2}]||\phi_\epsilon^{\frac{1}{2}} u||^2$$

$$\le \frac{1}{2b}||\phi_\epsilon^{\frac{1}{2}} h||^2.$$

For small ϵ and from Theorem 2.2, $\sup_{t\in R_+} ||u(\cdot,t)||_\infty \le \infty$ and the conditions of the lemma, we get

$$\frac{d}{dt}||\phi_\epsilon^{\frac{1}{2}} u||^2 + \frac{|\delta|}{2}||\phi_\epsilon^{\frac{1}{2}} u||^2 \le \frac{1}{2b}||\phi_\epsilon^{\frac{1}{2}} h||^2. \tag{2.12}$$

From the inequalities (2.4) and (2.12), it follows (2.9).

Lemma 2.3 If $||\phi_\epsilon^{\frac{1}{2}} u_{x^{j-1}}|| \le M$ $(j = 1, 2, \cdots)$, then we have

$$||\phi_\epsilon^{\frac{1}{2}} u_{x^j}|| \leq \epsilon_1 ||\phi_\epsilon^{\frac{1}{2}} u_{x^{j+1}}|| + C(M, \epsilon_1), \tag{2.13}$$

where $\epsilon_1 > 0$ is sufficiently small and C is a constant depending on M and $\epsilon_1 > 0$.

Proof. By using Sobolev's inequality,

$$\begin{aligned}
||\phi_\epsilon^{\frac{1}{2}} u_{x^j}|| &= ||(\phi_\epsilon^{\frac{1}{2}} u_{x^{j-1}})_x - (\phi_\epsilon^{\frac{1}{2}})' u_{x^{j-1}}|| \\
&\leq ||(\phi_\epsilon^{\frac{1}{2}} u_{x^{j-1}})_x|| + \gamma\epsilon ||\phi_\epsilon^{\frac{1}{2}} u_{x^{j-1}}|| \leq ||\phi_\epsilon^{\frac{1}{2}} u_{x^{j-1}}||^{\frac{1}{2}} ||(\phi_\epsilon^{\frac{1}{2}} u_{x^{j-1}})_{xx}||^{\frac{1}{2}} + \gamma\epsilon ||\phi_\epsilon^{\frac{1}{2}} u_{x^{j-1}}|| \\
&\leq \frac{\epsilon_1}{2} ||(\phi_\epsilon^{\frac{1}{2}} u_{x^{j-1}})_{xx}|| + (\frac{1}{2\epsilon_1} + \gamma\epsilon) M \\
&\leq \frac{\epsilon_1}{2} \{ ||\phi_\epsilon^{\frac{1}{2}} u_{x^{j+1}}|| + 2||(\phi_\epsilon^{\frac{1}{2}})' u_{x^j}|| + ||(\phi_\epsilon^{\frac{1}{2}})'' u_{x^{j-1}}|| \} + (\frac{1}{2\epsilon_1} + \gamma\epsilon) M \\
&\leq \frac{\epsilon_1}{2} ||\phi_\epsilon^{\frac{1}{2}} u_{x^{j+1}}|| + \epsilon_1 \gamma\epsilon ||\phi_\epsilon^{\frac{1}{2}} u_{x^j}|| + [\frac{1}{8}\gamma\epsilon^2 \epsilon_1 (2\gamma + 1) + \frac{1}{2\epsilon_1} + \gamma\epsilon] M. \qquad (2.14)
\end{aligned}$$

Taking $\epsilon_1 \gamma \epsilon = \frac{1}{2}$, we get the inequality (2.13) immediately.

Lemma 2.4 Suppose that the conditions of Lemma 2.2 are satisfied and assume that $u_0(x) \in H_{1,\gamma}$, $h(x) \in H_{1,\gamma}$. Then for the smooth solution $u(x,t)$ of the Cauchy problem (1.3)(1.4) we have

$$||u(\cdot, t)||_{0,\gamma} \leq C_\epsilon [e^{-\frac{|\delta|}{2}t} ||u_0||_{0,\gamma} + \frac{2}{b|\delta|}(1 - e^{-\frac{1}{2}|\delta| t})(||h||_{1,\gamma} + ||u||_{0,\gamma}). \tag{2.15}$$

$$\overline{\lim_{t\to\infty}} ||u(\cdot, t)||_{1,\gamma} \leq C_\epsilon \frac{2}{b|\delta|} (||h||_{1,\gamma} + ||u||_{0,\gamma}) = E_1 \tag{2.16}$$

$$\overline{\lim_{t\to\infty}} \frac{1}{t} \int_0^t ||u(\cdot, t)||_{2,\gamma}^2 dt \leq E_1'. \tag{2.17}$$

Proof. Differentiating (1.3) with respect to x, then taking the inner product with $\phi_\epsilon u_x$, we get

$$\begin{aligned}
(\phi_\epsilon u_x, u_{tx} &- \alpha u_{xxx} + 2(u u_x)_x + \beta H u_{xxx} \\
&+ \delta H u_{xx} - g'(u) u_x - h'(x)) = 0. \qquad (2.18)
\end{aligned}$$

Here

$$(\phi_\epsilon u_x, u_{xt}) = \frac{1}{2}\frac{d}{dt} ||\phi_\epsilon^{\frac{1}{2}} u_x||^2,$$

$$(\phi_\epsilon u_x, -\alpha u_{xxx}) = \alpha ||\phi_\epsilon^{\frac{1}{2}} u_{xx}||^2 + \alpha(\phi_\epsilon' u_x, u_{xx}),$$

$$\begin{aligned}
|(\phi_\epsilon' u_x, u_{xx})| &\leq ||(\phi_\epsilon')^{\frac{1}{2}} u_x|| ||(\phi_\epsilon')^{\frac{1}{2}} u_{xx}|| \\
&\leq C\epsilon^2 ||\phi_\epsilon^{\frac{1}{2}} u_x|| ||\phi_\epsilon^{\frac{1}{2}} u_{xx}||
\end{aligned}$$

$$\leq \frac{C}{2}\epsilon^2(\|\phi_\epsilon^{\frac{1}{2}}u_x\|^2+\|\phi_\epsilon^{\frac{1}{2}}u_{xx}\|^2);$$

$$|(\phi_\epsilon u_x,(2uu_x)_x)|\leq 2C\epsilon\|u\|_\infty\|\phi_\epsilon^{\frac{1}{2}}u_x\|^2+\eta_1\|\phi_\epsilon^{\frac{1}{2}}u_{xx}\|^2+\frac{1}{2\eta_1}\|\phi_\epsilon^{\frac{1}{2}}uu_x\|^2$$

$$\leq \eta_1\|\phi_\epsilon^{\frac{1}{2}}u_{xx}\|^2+2C\epsilon\|u\|_\infty\|\phi_\epsilon^{\frac{1}{2}}u_x\|^2+\frac{\|u\|_\infty^2}{\eta_1}[\epsilon_1^2\|\phi_\epsilon^{\frac{1}{2}}u_{xx}\|^2+C_1^2(\|\phi_\epsilon^{\frac{1}{2}}u\|,\epsilon_1)],$$

$$|(\phi_\epsilon u_x,\beta Hu_{xxx})|\leq\beta|(\phi_\epsilon' u_x,Hu_{xx})|+\beta|(\phi_\epsilon u_{xx},Hu_{xx})|$$

$$\leq C\beta\epsilon(\|\phi_\epsilon^{\frac{1}{2}}u_x\|^2+\|\phi_\epsilon^{\frac{1}{2}}Hu_{xx}\|^2)+\frac{\beta}{2}(\|\phi_\epsilon^{\frac{1}{2}}u_{xx}\|^2+\|\phi_\epsilon^{\frac{1}{2}}Hu_{xx}\|^2)$$

$$\leq C\beta\epsilon\|\phi_\epsilon^{\frac{1}{2}}u_x\|^2+\frac{\beta}{2}\|\phi_\epsilon^{\frac{1}{2}}u_{xx}\|^2+(\frac{\beta}{2}+C\epsilon\beta)(\|\phi_\epsilon^{\frac{1}{2}}u_{xx}\|^2+C\epsilon\|\phi_\epsilon^{\frac{1}{2}}u_x\|^2)$$

$$=(\beta+C\epsilon\beta)\|\phi_\epsilon^{\frac{1}{2}}u_{xx}\|^2+C\epsilon(\frac{3\beta}{2}+C\epsilon\beta)\|\phi_\epsilon^{\frac{1}{2}}u_x\|^2,$$

$$|(\phi_\epsilon u_x,\delta Hu_{xx})|\leq\delta\|\phi_\epsilon^{\frac{1}{2}}u_x\|\|\phi_\epsilon^{\frac{1}{2}}Hu_{xx}\|$$

$$\leq\eta_2\|\phi_\epsilon^{\frac{1}{2}}Hu_{xx}\|^2+\frac{\delta^2}{4\eta_2}\|\phi_\epsilon^{\frac{1}{2}}u_x\|^2$$

$$\leq\eta_2\|\phi_\epsilon^{\frac{1}{2}}u_{xx}\|^2+(\eta_2C\epsilon+\frac{\delta^2}{4\eta_2})\|\phi_\epsilon^{\frac{1}{2}}u_x\|^2$$

$$\leq\eta_2\|\phi_\epsilon^{\frac{1}{2}}u_{xx}\|^2+(\eta_2C\epsilon+\frac{\delta^2}{4\eta_2})(\epsilon_1^2\|\phi_\epsilon^{\frac{1}{2}}u_{xx}\|^2+C_1^2)$$

$$\leq[\eta_2+\epsilon_1^2(\eta_2C\epsilon+\frac{\delta^2}{4\eta_2})]\|\phi_\epsilon^{\frac{1}{2}}u_{xx}\|^2+(\eta_2C\epsilon+\frac{\delta^2}{4\eta_2})C_1^2\leq\eta_3\|\phi_\epsilon^{\frac{1}{2}}u_{xx}\|^2+C_2,$$

$$(\phi_\epsilon u_x,g'(u)u_x)\leq -b\|\phi_\epsilon^{\frac{1}{2}}u_x\|^2,$$

$$(\phi_\epsilon u_x,h')\leq\frac{b}{2}\|\phi_\epsilon^{\frac{1}{2}}u_x\|^2+\frac{1}{2b}\|\phi_\epsilon^{\frac{1}{2}}h'\|^2.$$

Hence from (2.18) we have

$$\frac{1}{2}\frac{d}{dt}\|\phi_\epsilon^{\frac{1}{2}}u_x\|^2+(\alpha-\frac{C}{2}\epsilon^2-\eta_1-\frac{2\epsilon_1^2\|u\|_\infty^2}{\eta_1}-|\beta|-C|\beta|\epsilon-\eta_3)\|\phi_\epsilon^{\frac{1}{2}}u_{xx}\|^2$$

$$+(b-\frac{b}{2}-\frac{C}{2}\epsilon^2-2C\epsilon\|u\|_\infty-C\epsilon|\beta|(\frac{3}{2}+C\epsilon))\|\phi_\epsilon^{\frac{1}{2}}u_x\|^2$$

$$\leq \frac{1}{2b}||\phi_\epsilon^{\frac{1}{2}} h'||^2 + C_3(||\phi_\epsilon^{\frac{1}{2}} u||^2). \tag{2.19}$$

For given small η_1 and η_2, then choosing sufficiently small ϵ and ϵ_1, from (2.19) it follows (2.15)-(2.17).

Similarly we have

Lemma 2.5 Suppose that the conditions of Lemma 2.4 are satisfied and assume that $u_0(x) \in H^1(R) \cap H_{1,\gamma+\delta_1}$, $h(x) \in H_{1,\gamma+\delta_1}$. Then for the smooth solution $u(x,t)$ of the Cauchy problem (1.3)(1.4) we have

$$\sup_{t\in R_+} ||u(\cdot,t)||_{1,\gamma+\delta_1} \leq E_{1,\delta_1}, \tag{2.20}$$

where the constant $\delta_1 > 0$.

Lemma 2.6 Suppose that the conditions of Lemma 2.4 are satisfied and assume that $u_0(x) \in H_{2,\gamma}$, $h(x) \in H_{2,\gamma}$. Then for the smooth solution $u(x,t)$ of the Cauchy problem (1.3)(1.4), we have

$$||u(\cdot,t)||_{2,\gamma} \leq C_\epsilon(e^{-\frac{|\delta|}{2}t}||u_0||_{2,\gamma} + \frac{2}{b|\delta|}(1 - e^{-\frac{1}{2}|\delta|t})(||h||_{2,\gamma} + ||u||_{1,\gamma}). \tag{2.21}$$

$$\overline{\lim_{t\to\infty}}||u(\cdot,t)||_{2,\gamma} \leq C_\epsilon \frac{2}{b|\delta|}(||h||_{2,\gamma} + ||u||_{1,\gamma}) = E_2 \tag{2.22}$$

$$\overline{\lim_{t\to\infty}} \frac{1}{t}\int_0^t ||u(\cdot,t)||_{3,\gamma}^2 dt \leq E_2'. \tag{2.23}$$

Proof. Differentiating (1.3) with respect to x two times, then taking the inner product with $\phi_\epsilon u_{xx}$, we get

$$(\phi_\epsilon u_{xx}, u_{xxt} - \alpha u_{xxxx} + 2(uu_x)_{xx} + \beta H u_{xxxx}$$
$$+\delta H u_{xxx} - (g'(u)u_x)_x - h'') = 0. \tag{2.24}$$

Here we have

$$(\phi_\epsilon u_{xx}, u_{xxt}) = \frac{1}{2}\frac{d}{dt}||\phi_\epsilon^{\frac{1}{2}} u_{xx}||^2,$$

$$(\phi_\epsilon u_{xx}, -\alpha u_{xxxx}) = \alpha((\phi_\epsilon u_{xx})_x, u_{xxx}) = \alpha(\phi_\epsilon' u_{xx}, u_{xxx}) + \alpha||\phi_\epsilon^{\frac{1}{2}} u_{xxx}||^2,$$

$$|(\phi_\epsilon' u_{xx}, u_{xxx})| \leq C\epsilon(||\phi_\epsilon^{\frac{1}{2}} u_{xx}||^2 + ||\phi_\epsilon^{\frac{1}{2}} u_{xxx}||^2),$$

$$(\phi_\epsilon u_{xx}, (2uu_x)_{xx}) = -(\phi_\epsilon' u_{xx} + \phi_\epsilon u_{xxx}, 2u_x^2 + 2uu_{xx}),$$

$$|(\phi_\epsilon' u_{xx}, 2u_x^2 + 2uu_{xx})| \leq C\epsilon[||\phi_\epsilon^{\frac{1}{2}} u_{xx}||^2 + ||\phi_\epsilon^{\frac{1}{2}} u_x^2||^2 + ||\phi_\epsilon^{\frac{1}{2}} uu_{xx}||^2]$$

$$\leq C\epsilon[||\phi_\epsilon^{\frac{1}{2}} u_{xx}||^2 + ||u_x||_\infty^2||\phi_\epsilon^{\frac{1}{2}} u_x||^2 + ||u||_\infty||\phi_\epsilon^{\frac{1}{2}} u_{xx}||^2] \leq C\epsilon[||\phi_\epsilon^{\frac{1}{2}} u_{xx}||^2 + 1],$$

$$|(\phi_\epsilon u_{xxx}, 2u_x^2 + 2uu_{xx})| \leq |(\phi_\epsilon u_{xxx}, 2u_x^2)| + |(\phi_\epsilon u_{xxx}, 2uu_{xx})|$$

$$\leq 2\|u_x\|_\infty \|\phi_\epsilon^{\frac{1}{2}} u_{xxx}\| \|\phi_\epsilon^{\frac{1}{2}} u_x\| + 2\|u\|_\infty \|\phi_\epsilon^{\frac{1}{2}} u_{xx}\| \|\phi_\epsilon^{\frac{1}{2}} u_{xxx}\| \leq \epsilon_1 \|\phi_\epsilon^{\frac{1}{2}} u_{xxx}\|^2 + C_1.$$

Here the boundness of $\sup_{t\in R_+} \|u_x(\cdot,t)\|_{L_\infty(R)}$, $\sup_{t\in R_+} \|u(\cdot,t)\|_{L_\infty(R)}$ and $\sup_{t\in R_+} \|\phi_\epsilon^{\frac{1}{2}} u_x(\cdot,t)\|_{L_2(R)}$ have been used by means of Theorem 2.2 and Lemma 2.5 respectively. Furthermore we have

$$|(\phi_\epsilon u_{xx}, \beta H u_{xxxx})| \leq |\beta|\, |(\phi_\epsilon' u_{xx}, H u_{xxx})| + |\beta|\, |(\phi_\epsilon u_{xxx}, H u_{xxx})|$$

$$\leq C\epsilon(\|\phi_\epsilon^{\frac{1}{2}} u_{xx}\|^2 + \|\phi_\epsilon^{\frac{1}{2}} H u_{xxx}\|^2) + \frac{|\beta|}{2}(\|\phi_\epsilon^{\frac{1}{2}} u_{xxx}\|^2 + \|\phi_\epsilon^{\frac{1}{2}} H u_{xxx}\|^2)$$

$$\leq (|\beta| + C\epsilon)\|\phi_\epsilon^{\frac{1}{2}} u_{xxx}\|^2 + C_1\epsilon \|\phi_\epsilon^{\frac{1}{2}} u_{xx}\|^2.$$

It is similar to the proof of Lemma 2.5,

$$|(\phi_\epsilon u_{xx}, \delta H u_{xxx})| \leq \eta_1 \|\phi_\epsilon^{\frac{1}{2}} u_{xxx}\|^2 + C_2.$$

And also

$$(\phi_\epsilon u_{xx}, (g'(u)u_x)_x) = (\phi_\epsilon u_{xx}, g''(u)u_x^2 + g'(u)u_{xx})$$

$$\leq -b\|\phi_\epsilon^{\frac{1}{2}} u_{xx}\|^2 + \|g''(u)\|_\infty \|u_x\|_\infty \|\phi_\epsilon^{\frac{1}{2}} u_{xx}\| \|\phi_\epsilon^{\frac{1}{2}} u_x\|$$

$$\leq (-b + C\epsilon)\|\phi_\epsilon^{\frac{1}{2}} u_{xx}\|^2 + C_3,$$

$$|(\phi_\epsilon u_{xx}, h'')| \leq \frac{b}{2}\|\phi_\epsilon^{\frac{1}{2}} u_{xx}\|^2 + \frac{1}{2b}\|\phi_\epsilon^{\frac{1}{2}} h''\|^2.$$

Hence from (2.24) we obtain

$$\frac{1}{2}\frac{d}{dt}\|\phi_\epsilon^{\frac{1}{2}} u_{xx}\|^2 + (\alpha - C\epsilon - \epsilon_1 - \beta - \epsilon_2 - \eta_1)\|\phi_\epsilon^{\frac{1}{2}} u_{xxx}\|^2$$

$$+(\frac{b}{2} - C_5\epsilon)\|\phi_\epsilon^{\frac{1}{2}} u_{xx}\|^2 \leq \frac{1}{2b}\|\phi_\epsilon^{\frac{1}{2}} h''\|^2 + C_4. \tag{2.25}$$

As ϵ, ϵ_1, ϵ_2 and η_1 are suitably small, from inequality (2.25) we can get (2.21) (2.22) (2.23).

Lemma 2.7 Suppose that $u_1(x,0)$, $u_2(x,0) \in H_{1,\gamma}$, $\alpha > |\beta|$, $g'(u) \leq b$, $b =$const. and $u_1(x,t)$, $u_2(x,t)$ be any two smooth solutions of the Cauchy problem (1.3)(1.4). Then the inequality

$$\|u_1(\cdot,t) - u_2(\cdot,t)\|_{1,\gamma}^2 \leq C_0\|u_1(\cdot,0) - u_2(\cdot,0)\|_{1,\gamma}^2, \quad t \in [0,T] \tag{2.26}$$

holds, where the constant C_0 depends only on T.

Proof. First, we estimate $\|u_1(\cdot,t) - u_2(\cdot,t)\|_{0,\gamma}$. Let $w(x,t) = u_1(x,t) - u_2(x,t)$, subtracting the equation (1.3) for u_1 and u_2, we get

$$(u_1 - u_2)_t - \alpha(u_1 - u_2)_{xx} + 2u_1u_{1x} - 2u_2u_{2x} + \beta H(u_1 - u_2)_{xx}$$

$$+\delta H(u_1 - u_2)_x = g(u_1) - g(u_2). \tag{2.27}$$

Multiplying this equation by $\phi_\epsilon(u_1 - u_2)$, and integrating with respect to t and x, we deduce that

$$\frac{1}{2}\frac{d}{dt}\|\phi_\epsilon^{\frac{1}{2}} w(\cdot,t)\|^2\Big|_0^t + \alpha((\phi_\epsilon w)_x, w_x) + (\phi_\epsilon w, (2u_1u_{1x} - 2u_2u_{2x}))$$

$$+(\beta H w_{xx}, \phi_\epsilon w) + (\delta H w_x, \phi_\epsilon w) = (g(u_1) - g(u_2), \phi_\epsilon w)$$

$$\leq b\|\phi_\epsilon^{\frac{1}{2}} w(\cdot,t)\|^2. \tag{2.28}$$

Here we have

$$\alpha((\phi_\epsilon w)_x, w_x) = \alpha(\phi_\epsilon' w, w_x) + \alpha\|\phi_\epsilon^{\frac{1}{2}} w_x\|^2,$$

$$|\alpha(\phi_\epsilon' w, w_x)| \leq C\epsilon\|\phi_\epsilon^{\frac{1}{2}} w\|\,\|\phi_\epsilon^{\frac{1}{2}} w_x\| \leq \epsilon\|\phi_\epsilon^{\frac{1}{2}} w_x\|^2 + \frac{C^2\epsilon}{4}\|\phi_\epsilon^{\frac{1}{2}} w\|^2,$$

$$2u_1u_{1x} - 2u_2u_{2x} = 2u_1w_x + 2u_{2x}w,$$

$$|(\phi_\epsilon w, 2u_1w_x + 2u_{2x}w)| \leq |((\phi_\epsilon u_1)_x, w^2)| + ((2\phi_\epsilon u_{2x}, w^2)|$$

$$\leq C\epsilon\|u_1\|_\infty\|\phi_\epsilon^{\frac{1}{2}} w\|^2 + \|u_{1x}\|_\infty\|\phi_\epsilon^{\frac{1}{2}} w\|^2 + 2\|u_{2x}\|\,\|\phi_\epsilon^{\frac{1}{2}} w\|^2 \leq C_1\|\phi_\epsilon^{\frac{1}{2}} w\|^2,$$

$$|(\beta H w_{xx}, \phi_\epsilon w)| = |\beta(H w_x, (\phi_\epsilon w)_x)|$$

$$\leq |\beta|C\epsilon\|\phi_\epsilon^{\frac{1}{2}} H w_x\|\,\|\phi_\epsilon^{\frac{1}{2}} w\| + |\beta|\|\phi_\epsilon^{\frac{1}{2}} w_x\|\,\|\phi_\epsilon^{\frac{1}{2}} H w_x\|$$

$$\leq (|\beta| + \epsilon)\|\phi_\epsilon^{\frac{1}{2}} w_x\|^2 + C_2\|\phi_\epsilon^{\frac{1}{2}} w\|^2,$$

$$|(\delta H w_x, \phi_\epsilon w)| \leq |\delta|\|\phi_\epsilon^{\frac{1}{2}} w\|\,\|\phi_\epsilon^{\frac{1}{2}} H w_x\| \leq \epsilon_1\|\phi_\epsilon^{\frac{1}{2}} w_x\|^2 + C_3\|\phi_\epsilon^{\frac{1}{2}} w\|^2.$$

Thus from (2.28) we have

$$\frac{1}{2}\frac{d}{dt}\|\phi_\epsilon^{\frac{1}{2}} w(\cdot,t)\|^2 + (\alpha - 2\epsilon - |\beta| - \epsilon_1)\|\phi_\epsilon^{\frac{1}{2}} w_x\|^2$$

$$\leq (b + \frac{C^2\epsilon}{4} + C_1 + C_2 + C_3)\|\phi_\epsilon^{\frac{1}{2}} w\|^2.$$

For small ϵ and ϵ_1, $\alpha - 2\epsilon - |\beta| - \epsilon_1 > 0$. By use of Gronwall's inequality, we get

$$\|\phi_\epsilon^{\frac{1}{2}} w(t)\|^2 \leq e^{C_0't}\|\phi_\epsilon^{\frac{1}{2}} w(0)\|^2,$$

where $C_0' = 2(b + \frac{C^2\epsilon}{4} + C_1 + C_2 + C_3)$. Let $C_0 = Ce^{C_0'T}$, we get

$$\|u_1(t) - u_2(t)\|_{0,\gamma}^2 \leq C_0\|u_1(0) - u_2(0)\|_{0,\gamma}^2, \quad t \in [0,T].$$

Similarly, we can get (2.26).

Lemma 2.8 (A. V. Babin, M. I. Vishik)[4] Let a set B be bounded in $H_{1,\gamma+\delta}$, $\delta > 0$ and in $H_{2,\gamma}$. Then B is compact in $H_{1,\gamma}$.

3 Existence of global attractor and its dimensions

Theorem 3.1 (R. Temam [5]) We assume that H is a metric space and that the nonlinear operators $S(t)$ of H into itself for $t \geq 0$ satisfy

$$S(t+s) = S(t) \cdot S(s) \ , \quad s,t \geq 0, \quad S(0) = I \text{ (Identity in } H). \tag{3.1}$$

And also $S(t)$ are continuous and uniformly compact for t large, that means for every bounded set B there exists t_0, which may depends on B, such that $\bigcup\limits_{t \geq t_0} S(t)B$ is relatively compact in H. We also assume that there exist an open set $U \subset H$ and a bounded set B of U such that B is absorbing in U. Then the $\omega-$limit set of B: $A = \omega(B) = \cup_{t \geq t_0} S(t)B$ is a compact attractor, which attracts the bounded sets of U. It is the maximal bounded attractor in U. Furthermore, if H is a Banach space and U is convex and connected, then A is connected too.

By use of a priori integral estimates and fixed point argument, we have

Theorem 3.2 Suppose that the following conditions are satisfied:

$$(1) \quad \alpha > 0, \quad \beta > 0, \quad \delta \leq 0, \quad \alpha > \beta - \frac{\delta}{2}.$$

$$(2) \quad g(u) \in C^2, \quad g'(u) \leq -b, \quad b > 0 \text{ is a constant}, \quad g(0) = 0,$$

and

$$b + \frac{3}{2}\delta > 0.$$

$$(3) \quad u_0(x) \in H_{2,\gamma} \ , \quad h(x) \in H_{2,\gamma}.$$

Then there esxists a global attractor A of Cauchy problem (1.3)(1.4), i.e., there is a set A, such that

(1) A is compact in $H_{1,\gamma}$;

(2) $S_t A = A, \quad \forall t \geq 0$;

(3) $\lim\limits_{t \to \infty} \text{dist } (S_t B, A) = 0$, for any boundedset $B \in H_{1,\gamma}$,

where

$$\text{dist } (x,y) = \sup_{x \in X} \inf_{y \in Y} \|x - y\|_{1,\gamma}.$$

and $S_t u_0$ is a semi-group operator generated by the Cauchy problem (1.3)(1.4).

Proof. On account of the result of Theorem 3.1, we shall prove this theorem by checking the conditions in Theorem 3.1. Under the assumptions of the theorem, from Theorem 2.1 we know that exists a semi-group operator generated by the Cauchy problem (1.3)(1.4). Thus we set the Banach space $H = H_{1,\gamma}$ and $S_t : H_{1,\gamma} \to H_{1,\gamma}$. From Lemma 2.7, it is a continuous (nonlinear) operator from $H_{1,\gamma}$ into itself. Let

$$B = \{u(\cdot,t) \in H_{1,\gamma}(R) \ , \ \|u\|_{1,\gamma} \leq 2E_1\} \tag{3.2}$$

is the bounded absorbing set of the semi-group operator S_t from Lemma 2.4. $B_1 = S_tB$ is bounded in $H_{2,\gamma}$ by Lemma 2.6. From Lemma 2.5 and Lemma 2.8, the absorbing set B_1 is compact in $H_{1,\gamma}$. Hence the ω-limit set of B, $A = \omega(B)$ is a compact attractor for the Cauchy problem (1.3), (1.4).

In order to establish the upper bounds of Hausdorff and fractal dimensions for the global attractor of the Cauchy problem (1.3)(1.4), we need the following linear variation corresponding to the problem (1.3)(1.4).

$$v_t + L(u(t))v = 0, \tag{3.3}$$

$$v(0) = v_0, \tag{3.4}$$

where

$$L(u(t))v = -\alpha v_{xx} + 2uv_x + 2u_xv + \beta Hv_{xx} + \delta Hv_x - g'(u)v. \tag{3.5}$$

Since the solution of problem (1.3)(1.4) is sufficiently smooth, we can easily prove that the linear problem (3.3) and (3.4) has a global smooth solution as long as the initial data mildly smooth, i.e., there is a solution operator G_t such that $v(t) = G_tv_0$.

It can be verified that the semi-group operator S_t for any $t \geq 0$ is uniformly differentiable on A in the metric of $H_{0,0} = L_2$. Namely, the Frechet derivative $S_t'(u_0)$ exists, and $G_tv_0 = S_t'(u_0)v_0$. In fact, we set

$$w(t) = S_t(u_0 + v_0) - S_t(u_0) - G_t(u_0)v_0 = u_1(t) - u(t) - v(t).$$

Thus we have

$$\begin{aligned}\partial_t w(t) &= L_1(u_1(t)) - L_1(u(t)) + L(u(t))v(t)\\ &= L_1(u(t) + v(t) + w(t)) - L_1(u(t)) + L(u(t))v(t),\end{aligned} \tag{3.6}$$

$$w(0) = 0, \tag{3.7}$$

where $u_t = L_1(u)$ is the operator form of equation (1.3).

Theorefore, (3.6) can be rewritten in the form

$$\partial_t w + L(u(t))w = \Lambda_0(u, v, w), \tag{3.8}$$

where

$$\Lambda_0(u, v, w) = L_1(u(t) + v(t) + w(t)) - L_1(u(t)) + L(u(t))(v(t) + w(t)). \tag{3.9}$$

By applying the theory of linear differential equations, we have the following L_2 estimate

$$\|w(t)\|^2 \leq C\|v_0\|^2. \tag{3.10}$$

This implies the semi-group operator S_t can be differentiated in $L_2(R)$.

Denote by $v_1(t)$, $v_2(t)$, $\cdots, v_J(t)$ the solution of linear equation (3.3) corresponding respectively to the initial data $v_1(0) = \xi_1, \cdots, v_J(0) = \xi_J$, here $\xi = (\xi_1, \xi_2, \cdots, \xi_J) \in L_2$, and by the simple computation [6], we can deduce that

$$\frac{d}{dt}\|v_1(t) \wedge \cdots \wedge v_J(t)\|^2 + 2T_r(L(u(t))Q_J)\|v_1(t) \wedge \cdots \wedge v_J(t)\|^2 = 0. \tag{3.11}$$

where $L(u(t)) = L(S_t u_0)$ is a linear map: $v \to L(u(t))v$, "$\wedge$" denotes the exterior product, Tr the trace of operator, and $Q_J(t)$ the orthographic projection of space $L_2(R)$ to the spaning subspace generated by $v_1(t), \cdots, v_J(t)$. Therefore from (3.11) we can obtain the change of the volume $\wedge_{j=0}^J \xi$ of the J dimensional cube by

$$\omega_J(t) = \sup_{u_0 \in A} \sup_{\xi_j \in L_2, \xi_j \leq 1} \|v_1(t) \wedge \cdots \wedge v_J(t)\|^2_{\wedge^j L_2}$$

$$\leq \sup_{u_o \in A} \exp\left(-\int_0^t \inf(\operatorname{Tr} L(S_\tau u_0), Q_j(\tau))d\tau.\right. \tag{3.12}$$

Noticing the result in [6], we know that ω_j is sub-exponented with respect to t, i.e.

$$\omega_j(t+t') \leq \omega_j(t)\omega_j(t') \ , \quad t, t' \geq 0. \tag{3.13}$$

Hence we have

$$\lim_{t\to\infty} \omega_j(t)^{\frac{1}{t}} = \pi_j \ , \quad (j = 1, 2, \cdots, J) \tag{3.14}$$

and

$$\pi_j \leq \exp(-q_J) \tag{3.15}$$

where

$$q_J = \overline{\lim_{t\to\infty}} \left(\inf \frac{1}{t} \int_0^t \inf(\operatorname{Tr} L(S_\tau u_0) \cdot Q_J(\tau))d\tau.\right. \tag{3.16}$$

Definition 3.1 The Hausdorff measure of a set X is defined by

$$n_H(X, d) = \lim_{\epsilon \to 0} n_H(X, d, \epsilon) = \sup_{\epsilon > 0} n_H(X, d, \epsilon),$$

where

$$n_H(X, d, \epsilon) = \inf \sum_i r_i^d$$

the infimum is takes over the balls that cover the set X with the radius $r_i \leq \epsilon$. The Hausdorff dimension of the set X is defined by a number $d_H(X) \in [0, \infty)$, such that

$$n_H(X, d) = 0, \quad \text{for } d > d_H(X)$$

and

$$n_H(X, d) = \infty, \quad \text{for } d < d_H(X).$$

Definition 3.2 The fractal dimension is defined by the number

$$d_F(X) = \lim_{\epsilon \to 0} \sup \frac{\log n_X(\epsilon)}{\log(1/\epsilon)},$$

where $n_X(\epsilon)$ denotes the smallest number of the balls that cover the set X with their radiuses less than or equal to ϵ.

From the results of [5], we see that

$$d_F(X) = \inf\{d > 0, \quad n_F(X, d) = 0\}$$

where

$$n_F(X, d) = \lim_{\epsilon \to 0} \sup(\epsilon^d n_X(\epsilon)).$$

Since $n_F(X, d) \geq n_H(X, d)$, thus we have

$$d_H(X) \leq d_F(X).$$

Theorem 3.3 [**6**] Let A be the attractor of nonlinear evolution equation (such as Navier Stokes equation, equation (1.3) etc.) that is bounded in $H^1(R)$. Then if $q_J > 0$ for some J, the Hausdorff dimension of X is no more than J, and its fractal dimension is less than or equal to

$$J \left(H \max_{1 \leq l \leq J} \frac{-q_l}{q_J} \right). \tag{3.17}$$

Lemma 3.1 (Sobolev-Lieb-Thiring Inequality, [5]) Let ϕ_j, $1 \leq j \leq N$ be a finite family of $H^m(R^n)$ which is orthonornal in $L_2(R^n)$ and set, for almost every $x \in R^n$

$$\rho(x) = \sum_{j=1}^{N} |\phi_j(x)|^2.$$

Then there exists a constant k_0 such that

$$\left(\int_{R^n} \rho(x)^{p/(p-1)} dx \right)^{2m(p-1)/n} \leq k_0 \sum_{j=1}^{N} \int_{R^n} |D^m \phi_j|^2 dx, \tag{3.18}$$

where the constant k_0 depends on m, n, p and it is independent of the ϕ_j and of N.

Theorem 3.4 Under the conditions of Theorem 3.3, then the Hausdorff and fractal dimensions of the global attrator A of the Cauchy problem (1.3) (1.4) are finite, and

$$d_H(A) \leq J_0, \quad d_F(A) \leq 2J_0, \tag{3.19}$$

where J_0 is the smallest integer, such that

$$J_0 \geq \left(b - \frac{\delta^2}{2\alpha} \right)^{-1} \alpha^{-1/2} \left(\frac{2^{3/2}}{3^{3/2}} k_0^{2/3} \|u_x\|_{L_{3/2}}^{3/2} \right). \tag{3.20}$$

Proof. Suppose that $\{\phi_1, \cdots, \phi_J\}$ is an orthogonal basis of the subspace $Q_J L_2(R)$, we have

$$\mathrm{Tr}(L(u(t))Q_J)$$

$$= \sum_{j=1}^{J} \{(-\alpha \phi_{jxx} + 2u\phi_{jx} + 2u_x \phi_j + \beta H \phi_{jxx} + \delta H \phi_{jx} - g'(u)\phi_j, \, \phi_j)\}$$

$$= \sum_{j=1}^{J}\{\alpha\|\phi_{jx}\|^2 - (u_x, \phi_j^2) + 2(u_x, \phi_j^2) + \delta(H\phi_{jx}, \phi_j) - (g'(u)\phi_j, \phi_j)\}.$$

By Sobolev-Lieb-Thiring inequality (3.18) it follows

$$\int_R \rho^3(x)dx \leq k_0 \sum_{j=1}^{J} \int_R |\phi_{jx}|^2 dx.$$

Hence we have

$$\sum_{j=1}^{J}(u_x, \phi_j^2) = (u_x, \rho(x)) \leq \|u_x\|_{L_{3/2}}\|\rho\|_{L_3}$$

$$\leq \|u_x\|_{L_{3/2}}(k_3 \sum_{j=1}^{J} \int |\phi_{jx}|^2 dx)^{1/3} \leq k_0^{1/3}\|u_x\|_{L_{3/2}}(\sum_{j=1}^{J} \|\phi_{jx}\|^2)^{1/3}.$$

And also

$$\left|\sum_{j=1}^{J}(\delta H\phi_{jx}, \phi_j)\right| = \left|\sum_{j=1}^{J}(\delta H\phi_j, \phi_{jx})\right| \leq \delta \sum_{j=1}^{J} \|\phi_{jx}\| \, \|H\phi_j\|$$

$$= \delta \sum_{j=1}^{J} \|\phi_{jx}\| \, \|\phi_j\|, \leq \frac{\alpha}{2} \sum_{j=1}^{J} \|\phi_{jx}\|^2 + \frac{\delta^2}{2\alpha} J,$$

where $\|\phi_j\|^2 = 1$ for $j = 1, \cdots, J$. Hence we get

$$\operatorname{Tr}(L(u(t))Q_J) \geq \frac{\alpha}{2} \sum_{j=1}^{J} \|\phi_{jx}\|^2 - k_0^{1/3}\|u_x\|_{L_{3/2}} \sum_{j=1}^{J} \|\phi_{jx}\|^{1/3} + \left(b - \frac{\delta^2}{2\alpha}\right) J.$$

Let $s = (\sum_{j=1}^{J} \|\phi_{jx}\|^2)^{1/3}$, $\quad f(s) = \frac{\alpha}{2}s^2 - k_0^{1/3}\|u_x\|_{L_{3/2}} s + \left(b - \frac{\delta^2}{2\alpha}\right) J.$

For $f'(s) = \frac{3\alpha}{2}s^2 - k_0^{1/3}\|u_x\|_{L_{3/2}} = 0,$

there is $s_m = \frac{2^{1/2}k_0^{1/6}}{(3\alpha)^{1/2}}\|u_x\|_{L_{3/2}}^{1/2}.$

Let us choose J, such that

$$f(s_m) = \frac{\alpha}{2}s_m^3 - k_0^{1/3}\|u_x\|_{L_{3/2}} s_m + \left(b - \frac{\delta^2}{2\alpha}\right) J > 0.$$

This gives

$$\left(b - \frac{\delta^2}{2\alpha}\right) J > \frac{2^{3/2}k_0^{2/3}}{3^{3/2}\alpha^{1/2}}\|u_x\|_{L_{3/2}}^{3/2}.$$

References

[1] S. Qian, Y. C. Lee and H. H. Chen, A Study of Nonlinear Dynamical Models of Plasma Turbluence, Phys. Fluids B 1 (1), 1989, 87-100.

[2] Zhou Yulin, Guo Boling, Global Solutions and Their Large-time Behavior of Cauchy Problem for Equations of Deep Water Type, Preprnt 1993.

[3] A. P. Calderon, Commutators of Singular Integral Operators, Proc. Nat. Acad. Sci., USA 53 (1965), 1092-1099.

[4] A. V. Bain and M. I. Vishik, Attrators of Partial Differential Evolution Equations in a Unbounded Domain, Proc. Roy. Soc. Edinburgh, 116A (1990), 221-243.

[5] R. Temam, Infinite Dimensional Systems in Mechanics and Physics, Springer-Verlag, Berlin New York, 1988.

[6] P. Constantin, C. Fioas and R. Temam, Attractor representing tubulent flows, Mem. Amer. Math. Sci., (3), 53 (1985), 314.

[7] Zhou Yulin and Guo Boling, Initial Value Problems for a Nonlinear Singular Integral-Differential Equation of Deep Water, Proceedings of a Symposium held in Tianjin, June 23-July 5, 1986, Lecture Notes in Mathematics, 1306, S. S. Chen (ed.), Partial Differential Equations.

[8] Rafael Jose Iorio, Jr., On the Cauchy Problem for the Benjamin- Ono Equation, Comm. in Partial Differential Eqs., 11 (1986), 1031-1084.

[9] L. Abdelouhab, J. L. Bona, M. Felland, J. C. Saut, Nonlocal Models for Nonlinear Dispersive Waves, Physics D 40 (1989), 360-392.

Institute of Applied Physics and
Computational Mathematics
Bejing P. O. Box 8009
100088, Chain

AMS/IP Studies in Advanced Mathematics
Volume 3, 1997

Stability of Boundary Integral Methods for Water Waves

J. Thomas Beale, Thomas Y. Hou and John Lowengrub

Abstract. We consider stability of several boundary integral methods for water waves. The numerical study of fluid interfaces is a difficult task because the underlying problem is very singular and sensitive to small perturbations. This makes it difficult to distinguish the numerical instability from a physical one. Here we summarize our work in studying stability of boundary integral methods for water waves[6]. Our study shows that the numerical method must satisfy a strong compatibility constraint between the quadrature rule for the velocity integral and the derivative rule in order to be stable. Violation of this compatibility condition will lead to numerical instabilities.

1. Introduction

Many physically interesting problems involve propagation of free surfaces in fluids. Water waves, boundaries between immiscible fluids, vortex sheets, and Hele-Shaw flows are some of the better known examples. Accurate simulation of these free surfaces in fluids presents a problem of considerable difficulty because the underlying physical problem is very singular and is sensitive to small perturbations. Straightforward numerical discretization of these interface problems may lead to numerical instabilities.

In this paper, we are mainly concerned with stability of boundary integral methods for water waves, although most of our analytical techniques apply to other fluid interface problems, see[7]. The boundary integral method has been one of the most common approaches in solving these interfacial problems, see, e.g. [2]. Recent applications using boundary integral formulation include [1, 2, 4, 10, 11, 13, 14, 16], The advantage of using boundary integral methods is to reduce the two-dimensional problem into an one-dimensional problem involving the interface position only, consequently avoiding the difficulty of differentiating discontinuous fluid quantities across the fluid interface. However, numerical simulations using boundary integral methods are susceptible to numerical instabilities. As shown by our stability analysis, boundary integral methods are generically unstable unless they satisfy a certain compatibility condition between the choice of quadrature rule for the velocity integral and the choice of spatial derivative.

The key in analyzing stability of these methods is to identify the most singular(or the leading order) contributions of the methods, and to see how various terms balance one another. To this end, we need to study the symbols of discrete singular operators, such as the discrete Hilbert transform and its variants. In order to satisfy the compatibility condition between the quadrature rule for the integral and the derivative rule, we need to apply certain Fourier filtering to the interface variables. The amount of filtering is determined by the quadrature in approximating the velocity integral and the discrete derivative operators being used. With this modification, we can prove stability of the methods. This modification is especially crucial in removing numerical instabilities for finite order derivative operators. We show in Section 4 that without this modification, the schemes using finite order derivative operators are numerically unstable.

We remark that Baker and Nachbin have also independently analyzed stability of several boundary integral method for vortex sheets with surface tension [3]. Their approach is to linearize the solution around equilibrium. The resulting linearized equations have constant coefficients which can be analyzed by Fourier analysis. However, as our purpose is to prove convergence, our analysis requires much more subtle energy estimates and has to take into account balances of singular terms that are normally dropped out in a near equilibrium linearization.

The organization of the paper is as follows. In Section 2, we present the stability analysis for the spatial discretization for water waves without surface tension. The case with surface tension is given in Section 3. In Section 4, we show that without the appropriate modifications in the boundary integral methods, the methods are unstable. A linear stability analysis around equilibrium is given. The result is confirmed by our numerical experiments.

2. Water waves without Surface Tension.

We consider the classical 2-D water wave problem. That is, we consider the evolution of an incompressible, inviscid and irrotational fluid with a free surface under influence of gravity. We intend this section to be illustrative and the reader is referred to [6] for additional details. The boundary integral formulation of the water wave problem is well known; we refer to [2] for a complete derivation. In 2-D, the free surface is just a curve,$\Gamma = (x(\alpha,t), y(\alpha,t))$ where α is a parametrization of the curve. Using complex variables, Γ can be written as $z(\alpha,t) = x(\alpha,t) + iy(\alpha,t)$. Moreover, we assume that the interface is periodic in the horizontal direction, i.e.$z(\alpha + 2\pi, t) = 2\pi + z(\alpha,t)$. Thus, it is enough to consider the interface over one period. The interface can be interpreted as a vortex sheet. Assuming the velocity goes to a constant far from the interface, the complex velocity off Γ can be written as

$$u(z) - iv(z) = \frac{1}{4\pi i}\int_{-\pi}^{\pi} \gamma(\alpha')cot(\frac{z - z(\alpha')}{2})d\alpha', \tag{I}$$

where γ is the unnormalized vortex sheet strength, $\gamma(\alpha+2\pi) = \gamma(\alpha)$ and $z \neq z(\alpha)$. The representation of the velocity given by (I) has a tangential discontinuity across Γ and γ is a measure of its jump. The normal component of the velocity is continuous across Γ, however. We take the normal velocity of the interface to be that of the fluid to satisfy the kinematic condition. Consquently, the normal velocity of Γ is given *uniquely* by the normal component of the limiting value of (I) at Γ. Here, we use the usual convention of choosing the tangential velocity to be that of the fluid. We move the interface position z by the velocity of the interface :

$$z_t{}^* = \frac{1}{4\pi i}\int_{-\pi}^{\pi} \gamma(\alpha')cot(\frac{z(\alpha) - z(\alpha')}{2})d\alpha' + \frac{\gamma(\alpha)}{2z_\alpha(\alpha)} \equiv u(\alpha) - iv(\alpha), \tag{1}$$

where z^* is the complex conjugate of z. It remains to consider the evolution of γ. This is can be done in terms of a velocity potential ϕ : $(u, v) = \nabla\phi.\gamma$ and ϕ are related through the kinetic condition

$$\phi_\alpha = \frac{\gamma}{2} + Re\left[\frac{z_\alpha(\alpha)}{4\pi i}\int_{-\pi}^{\pi} \gamma(\alpha')cot(\frac{z(\alpha) - z(\alpha')}{2})d\alpha'\right] \tag{2}$$

This is a Fredholm integral equation of the second kind for γ and is invertible with a bounded inverse, see[2]. In practice, an iterative method will be used to invert the integral equation for γ once the discretization is specified. The evolution of the velocity potential is given by Bernoulli's equation

$$\phi_t - \frac{1}{2}(u^2 + v^2) + gy = 0, \tag{3}$$

in the frame moving with the fluid. Equations (1)-(3) completely determine the evolution of the fluid interface.

We now consider the discretization of Equations (1)-(3). It turns out that straightfoward numerical discretizations of (1)-(3) are generically unstable! The reason for this is two-fold and will become clear in the analysis. First, a certain compatibility is required between the choice of quadrature rule for the velocity integrals and the choice of spatial derivative for the z_α and ϕ_α terms. This compatibility ensures that a delicate balance of terms on the continuous level is preserved on the discrete level and is crucial for maintaining numerical stability. Second, the periodicity of the numerical solution introduces aliasing errors which affect the balance of terms on the discrete level.

We now show how standard schemes can be modified so that they become numerically stable. We begin by considering discrete derivative operators. Define a derivative operator as

$$(\hat{D_h f})_k = ik\rho(kh)\hat{f}_k, \quad k = -N/2+1, ..., N/2,$$

where $\hat{f}_k$ is the discrete Fourier transform of f. The choice of $\rho(x)$ varies depending on what kind of derivative operators is used. For example, $\rho_2(kh) = sin(kh)/kh$ for the second order centered differencing; $\rho_c(kh) = 3sin(kh)/(kh(2+cos(kh)))$ for the cubic spline approximation. Note that $\rho(kh) \geq 0$. It is clear that for the $r-th$ order derivative operator, $\rho(kh)$ satisfies

$$|1-\rho(kh)| \leq C(kh)^r$$

If D_h corresponds to the spectral derivative, $\rho(kh)$ plays the role of Fourier smoothing. In that case, ρ should satisfy (i) $\rho \geq 0, \rho(\pm\pi) = 0$; (ii) $\rho(x) = 1$ for $|x| \leq \lambda\pi$ for some $0 < \lambda < 1$; (iii) ρ is smooth. Condition (ii) is to ensure spectral accuracy.

We denote by S_h as the pseudo-spectral derivative operator without smoothing. This corresponds to the case of $\rho(kh) \equiv 1$ in the definition of D_h. We also introduce a smoothing operator for the interface variable:

$$\hat{z_k^p} = \rho(kh)\hat{z}_k$$

It is cleal that z^p is an $r-th$ order approximation to z if ρ corresponds to the $r-th$ order derivative operator.

To approximate the velocity integral, we use the alternating trapezoidal rule:

$$\int_{-\pi}^{\pi} \gamma(\alpha')cot(\frac{z(\alpha_i - z(\alpha')}{2})d\alpha' \simeq \sum_{\substack{j=-N/2+1 \\ (j-i) odd}}^{N/2} \gamma_j \cdot cot(\frac{z_i - z_j}{2})2h,$$

where we have used the notations that $z_j \simeq z(\alpha_j)$ and $\gamma_j \simeq \gamma(\alpha_j)$. The advantage of using the alternating trapezoidal quadrature is that the approximation is spectrally accurate, see e.g. [14,8].

Now we can present our numerical algorithm for the water wave equations(1)-(3) as follows:

$$\frac{dz_i^*}{dt} = \frac{1}{4\pi i} \sum_{\substack{j=-N/2+1 \\ (j-i)odd}}^{N/2} \gamma_j cot(\frac{z_i^p - z_j^p}{2})(2h) + \frac{\gamma_i}{2(D_n z)_i} \equiv u_i - iv_i, \tag{4}$$

$$\frac{d\phi_i}{dt} = \frac{1}{2}(u_i^2 + v_i^2) - gy_i, \tag{5}$$

$$D_h\phi_i = \frac{\gamma_i}{2} + Re\left[\frac{D_h z_i}{4\pi i} \sum_{\substack{j=-N/2+1 \\ (j-i)odd}}^{N/2} \gamma_j cot(\frac{z_i^p - z_j^p}{2})2h\right], \tag{6}$$

where $\hat{z_k}^p = \rho(kh)\hat{z}_k$. In practice, the solution to (6), γ, is obtained iteratively.

Remark. The Fourier filtering z^p in (4) and (6) is to balance the high frequency errors introduced by D_h. This will become apparent in the analysis. The choice of such filtering is sharp. If we use finite order derivative operators or Fourier smoothing for the spectral derivative, the use of z^p is necessary for stability(i.e. for avoiding numerical ill-posedness). We do not need to apply filtering on γ because γ defined by (6) has been filtered implicitly through D_h and z^p.

One of the main results of this paper is the following convergence theorem:

Theorem 1: Convergence of Boundary Integral Methods

Assume that $z(\cdot,t), \phi(\cdot,t) \in C^{m+2}[0,2\pi]$ and $\gamma(\cdot,t) \in C^{m+1}[0,2\pi]$ for $m \geq 3$, and $|z(\alpha,t) - z(\beta,t)| \geq c|\alpha - \beta|$ for $0 \leq t \leq T$ and $c > 0$. Furthermore, assume that

$$(u_t, v_t) \cdot \vec{N} - (0,-g) \cdot \vec{N} \geq c_0 > 0. \tag{7}$$

Here (u,v) is the Lagrangian velocity, $\vec{N}$ is the normal vector to the interface, pointing out of the fluid region; and c_0 is some constant. Then if D_h corresponds to a $r-th$ order derivative approximation, we have for $h \leq h_0(T)$

$$||z(t) - z(\cdot,t)||_{l^2} \leq C(T)h^r, \quad ||\phi(t) - \phi(\cdot,t)||_{l^2} \leq C(T)h^r,$$

$$||\gamma(t) - \gamma(\cdot,t)||_{l^2} \leq C(T)h^{r-1},$$

where $||\cdot||_{l^2}$ is the discrete L^2 norm defined by $||z||_{l^2}^2 = \sum_{j=1}^N |z_j|^2 h$. If D_h corresponds to the spectral approximation, we have the same convergence result as above except for replacing h^r by h^m.

We remark that the condition (7) simply means that the interface is not accelerating downward, normal to itself, as rapidly as the normal acceleration of gravity. If (7) is violated, it would generate the Rayleigh-Taylor instability as if water were above the interface. It can be viewed as a natural generalization of the criterion of Taylor[15]. In the following, we will outline the key ingredients of the proof.

Consistency

The spectral accuracy of the alternating quadrature rule follows from the fact that the integrand has a simple pole which can be desingularized due to the oddness of the principal singular term. First we rewrite the velocity integral as

$$\int_{-\pi}^{\pi} \gamma(\alpha')cot(\frac{z(\alpha)-z(\alpha')}{2})d\alpha' = \int_{-\pi}^{\pi} [\gamma(\alpha')cot(\frac{z(\alpha)-z(\alpha')}{2}) + \frac{\gamma(\alpha)}{z_\alpha(\alpha)}cot(\frac{\alpha'}{2})]d\alpha'$$
$$\equiv \int_{-\pi}^{\pi} F(\alpha,\alpha')d\alpha'.$$

Note that the desingularized kernel F has a removable singularity, i.e. F is smooth in α'. It is well known that the trapezoidal rule gives spectrally accurate approximation to integrals with smooth periodic integrands. We have

$$\int_{-\pi}^{\pi} \gamma(\alpha')cot(\frac{z(\alpha_i)-z(\alpha')}{2})d\alpha'$$
$$= \sum_{j-1}^{N} F(\alpha_i,\alpha_j)h + O(h^m)$$
$$= \sum_{j\neq i} \frac{\gamma(\alpha_j)}{z(\alpha_i)-z(\alpha_j)}h + hF(\alpha_i,\alpha_i) + O(h^m).$$

Now use the Richardson extrapolation to eliminate the first order term. This will give rise the alternating point quadrature rule, and its spectral accuracy follows.

Stability

In this section, we concentrate on proving the linear stability of the scheme given in (4)-(6). Given the linear stability, nonlinear stability can be obtained relatively easily by using the smallness of the error and an induction argument. The reader is referred to [6] for detail.

Analytically, it is more convenient to deal with the $1/z$ kernel than the $cot(z/2)$ kernel. This is done as follows:

$$\frac{1}{2}cot(\frac{z}{2}) = \frac{1}{z} + \sum_{k=1}^{\infty} \frac{2z}{z^2-(2k\pi)^2},$$

where the sum is interpreted symmetrically. This converges absolutely away from $z = 2k\pi$, for any integer k. Using this and the discrete periodicity $z_{j+N} = 2\pi + z_j$, and $\gamma_{j+N} = \gamma_j$, we get

$$\frac{1}{4\pi i}\sum_{\substack{j=-N/2+1\\(j-i)odd}}^{N/2}\gamma_j cot(\frac{z_i{}^p - z_j{}^p}{2})2h = \frac{1}{4\pi i}\sum_{(j-i)odd}\frac{\gamma_j}{z_i{}^p - z_j{}^p}2h.$$

To obtain stability of the method, we need to estimate

$$\left[\frac{1}{4\pi i}\sum_{(j-i)odd}\frac{\gamma_j}{z_i{}^p - z_j{}^p}2h + \frac{\gamma_i}{2D_h z}\right]$$
$$-\left[\frac{1}{4\pi i}\sum_{(j-i)odd}\frac{\gamma(\alpha_j)}{z^p(\alpha_i) - z^p(\alpha_j)}2h + \frac{\gamma(\alpha_i)}{2D_h z(\alpha_i)}\right]$$

in terms of $z_j - z(\alpha_j)$ and $\gamma_j - \gamma(\alpha_j)$. Define the variations for z, γ and ϕ as follows:

$$\dot{z}_j = z_j - z(\alpha_j), \dot{\gamma}_j = \gamma_j - \gamma(\alpha_j), \dot{\phi}_j = \phi_j - \phi(\alpha_j).$$

Introduce a $m-th$ order smoothing operator A_{-m} which satisfies the following property:

$$||D_h{}^k(A_{-m}(\dot{z}_j))||_{l^2} \leq C||\dot{z}||_{l^2}, ||A_{-m}(S_h^k\dot{z}_j))||_{l^2} \leq C|\dot{z}||_{l^2}, for \quad 0 \leq k \leq m.$$

In particular, when $m = 0, A_0(\dot{z})$ denotes a bounded operator in l^2

$$||A_0(\dot{z}_j))||_{l^2} \leq C||\dot{z}||_{l^2}.$$

With these notations, we now outline the stability analysis.

The key idea in our analysis is to identify the leading order terms that are essential for stability of the method. To illustrate the idea, we choose to study the variation in the velocity integral

$$\dot{E}_i \equiv \frac{1}{4\pi i}\sum_{(j-i)odd}\frac{\gamma_j}{z_i{}^p - z_j{}^p}2h - \frac{1}{4\pi i}\sum_{(j-i)odd}\frac{\gamma(\alpha_j)}{z^p(\alpha_i) - z^p(\alpha_j)}2h.$$

The linearized variations are given by

$$\begin{aligned}\dot{E}_i^L &\equiv \frac{1}{2\pi i}\sum_{(j-i)odd}\frac{\dot{\gamma}_j}{z(\alpha_i)^p - z(\alpha_j)^p}2h - \frac{1}{2\pi i}\sum_{(j-i)odd}\frac{\gamma(\alpha_j)(\dot{z}_i^p - \dot{z}_j^p)}{(z(\alpha_i)^p - z(\alpha_j)^p)^2}2h\\ &= \frac{1}{2\pi i}\sum_{(j-i)odd}\frac{\dot{\gamma}_j}{z(\alpha_i) - z(\alpha_j)}2h - \frac{1}{2\pi i}\sum_{(j-i)odd}\frac{\gamma(\alpha_j)(\dot{z}_i^p - \dot{z}_j^p)}{(z(\alpha_i) - z(\alpha_j))^2}2h + A_{-1}(\dot{z})\\ &\quad + A_{-1}(\dot{\gamma}),\end{aligned}$$

since $z^p = z + O(h^r)$. To identify the most singular terms, we use the Taylor expansion to obtain the most singular symbols

$$\frac{1}{z(\alpha_i) - z(\alpha_j)} = \frac{1}{z_\alpha(\alpha_i)(\alpha_i - \alpha_j)} + f(\alpha_i, \alpha_j),$$
$$\frac{\gamma(\alpha_j)}{(z(\alpha_i) - z(\alpha_j))^2} = \frac{\gamma(\alpha_i)}{z_\alpha{}^2(\alpha_i)(\alpha_i - \alpha_j)^2} + \frac{D(\alpha_i)}{(\alpha_i - \alpha_j)} + g(\alpha_i, \alpha_j),$$

where $f(\alpha_i, \alpha_j)$ and $g(\alpha_i, \alpha_j)$ are smooth function of α_i and α_j, and

$$D(\alpha_i) = \frac{(-\gamma_\alpha(\alpha_i) z_\alpha(\alpha_i) + \gamma(\alpha_i) z_{\alpha\alpha}(\alpha_i))}{z_\alpha{}^3(\alpha_i)}$$

This nuturally leads us to studying properties of the discrete Hilbert transform

$$H_h(\dot{\gamma}_i) \equiv \frac{1}{\pi} \sum_{(j-i)odd} \frac{\dot{\gamma}_j}{\alpha_i - \alpha_j} 2h.$$

Note that there is no unique way to define the discrete Hilbert transform. This particular version of the discrete Hilbert transform is determined by our quadrature rule in approximating the velocity integral. Define another discrete operator Λ_h as follows:

$$\Lambda_h(\dot{z}_i) \equiv \frac{1}{\pi} \sum_{(j-i)odd} \frac{\dot{z}_i - \dot{z}_j}{(\alpha_i - \alpha_j)^2} 2h.$$

We define the corresponding continuous singular operators as follows:

$$H(\dot{\gamma}) \equiv \frac{1}{\pi} \int \frac{\dot{\gamma}(\alpha')}{\alpha - \alpha'} d\alpha', \Lambda(\dot{z}) \equiv \frac{1}{\pi} \int \frac{\dot{z}(\alpha) - \dot{z}(\alpha')}{(\alpha - \alpha')^2} d\alpha'.$$

It is easy to show that

$$\Lambda(\dot{z}) = H(D_\alpha \dot{z}), \tag{8}$$

where D_α is the continuous derivative operator.

It turns out that in order to maintain numerical stabilities of the boundary integral method, the quadrature rule for the velocity integral and the derivative rule for the spatial derivative D_α must satisfy a compatibility condition similar to (8). That is, given a quadrature rule, which defines a corresponding discrete operators H_h and Λ_h, and a derivative rule D_h, they must satisfy a compatibility condition similar to (8):

$$\Lambda_h(\dot{z}_i) = H_h(D_h)(\dot{z}_i), \tag{9}$$

for $\dot{z}$ satisfying $\hat{\dot{z}}_0 = \hat{\dot{z}}_{N\backslash 2} = 0$. If (9) is violated, it would generate a singular operator of the form $(\Lambda_h - H_h(D_h))(\dot{z})$ in the error equations. This will generate numerical instability, see Section 4.

In the case of the alternating trapezoidal quadrature rule,the discrete Hilbert transform defined above has properties that are surprisingly similar to those of the continuum counterpart. We summarize its properties below:

$$\hat{H_h f_k} = -i \quad sgn(k)\hat{f}_k, \tag{10}$$

$$\Lambda_h f_i = H_h(S_h f_i), \tag{11}$$

for f satisfying $\hat{f}_0 = \hat{f}_{N\backslash 2=0}$, where S_h is the pseudo-spectral derivative operator without smoothing. Thus the compatibility condition (9) would imply that a spectral derivative operator, S_h, should be used to obtain numerical stabilities.

Actually, we can relax the compatibility condition (9) to allow non-spectral derivative operators if we use numerical filtering in the approximations of the velocity integral. It is sufficient to impose the compatibility condition

$$\Lambda_h(\dot{z}_i^p) = H_h(D_h)(\dot{z}_i). \tag{12}$$

Thus, if D_h has the Fourier symbol, $(\hat{D}_h)_k = ik\rho(kh)$, then the filtering $\dot{z}^p$ has the form $\hat{\dot{z}}_k = \rho(kh)\hat{z}_k$. This explains why we need to filter z in (4) and (6) when we approximate the velocity integral.

Using (10)-(11), we obtain the leading order contributions to $\dot{E}^L$ as

$$\begin{aligned} \dot{E}_i^L &= \frac{1}{2iz_\alpha(\alpha_i)} H_h(\dot{\gamma}_i) - \frac{\gamma(\alpha_i)}{2iz_\alpha(\alpha_i)^2} H_h S_h(\dot{z}_i^p) \\ &= \frac{1}{2iz_\alpha(\alpha_i)} H_h(\dot{\gamma}_i) - \frac{\gamma(\alpha_i)}{2iz_\alpha(\alpha_i)^2} H_h D_h(\dot{z}_i), \end{aligned}$$

where we have used the identity $H_h S_h(\dot{z}^p) = H_h D_h(\dot{z})$..

The Fourier filtering is also needed to eliminate aliasing errors in the lower order terms. Typically, the lower order terms are of the form $\sum_{(j-i)odd} f(\alpha_i, \alpha_j)\dot{z}_j 2h$ for a smooth function f. In the continuous case, we have for any $\dot{z} \in L^2$

$$\int f(\alpha, \alpha')\dot{z}(\alpha')d\alpha' = A_{-m}(\dot{z}),$$

where m is the degree of regularities of f. This is no longer true at the discrete level due to aliasing errors associated with the alternating point quadrature rule.

For example, if we let $\dot{z}_j = e^{i(N/2-1)\alpha'}$ and $f(\alpha,\alpha') = \frac{g(\alpha)-g(\alpha_j)}{\alpha-\alpha'}$ with $g(\alpha) = e^{i2\alpha}$. Then we have from (10) that

$$\sum_{(j-i)odd} f(\alpha_i,\alpha_j)\dot{z}_j 2h = g(\alpha_i)H_h(\dot{z}) - H_h(g\dot{z})$$

$$= (e^{i2\alpha_i})\cdot(-ie^{i(N/2-1)\alpha_i}) - ie^{i(-N/2+1)\alpha_i} = -2ie^{i(-N/2+1)\alpha_i},$$

which is of course no smoother than $\dot{z}$. This is a result of aliasing errors at the highest frequencies and is why we must use the Fourier smoothing to eliminate the aliasing errors in the high modes. With the p smoothing, we can show that [6]

$$\sum_{(j-i)odd} f(\alpha_i,\alpha_j)\dot{z}_j^p 2h = A_{-1}(\dot{z}). \tag{13}$$

If $\rho(x)$ satisfies in addition that $\rho'(\pm\pi) = 0$, then we have

$$\sum_{(j-i)odd} f(\alpha_i,\alpha_j)\dot{z}_j^p 2h = A_{-2}(\dot{z}). \tag{14}$$

Now it follows from the above estimates that

$$\dot{E}_i^L = \frac{1}{2iz_\alpha(\alpha_i)}H_h(\dot{\gamma}_i) - \frac{\gamma(\alpha_i)}{2iz_\alpha(\alpha_i)^2}H_hD_h(\dot{z}_i) + A_0(\dot{z}_i) + A_{-1}(\dot{\gamma}_i) + R_h(\dot{\gamma}_i), \tag{15}$$

where $R_h(\dot{\gamma}) = \sum_{(j-i)odd} f(\alpha_i,\alpha_j)\dot{\gamma}_j$ for some smooth function f. Combining (15) with error estimates on $\gamma_i/(2D_hz_i)$, we obtain our stability estimates for the complex velocity $\frac{d\dot{z}^*}{dt}$, i.e.

$$\frac{d\dot{z}_i^*}{dt} = \frac{1}{2z_\alpha(\alpha_i)}(I - iH_h)[\dot{\gamma}_i - \frac{\gamma(\alpha_i)}{z_\alpha(\alpha_i)}D_h\dot{z}_i] + A_0(\dot{z}) + A_{-1}(\dot{\gamma}_i) + R_h(\dot{\gamma}_i) + O(h^r). \tag{16}$$

Since $\dot{\gamma}$ is related to $D_h\dot{\phi}$, $D_h\dot{z}$ and $H_hD_h\dot{z}$ through variation of (6),we can show that $R_h(\dot{\gamma}) = A_0(\dot{z},\dot{\phi})$ using (13) and $A_{-1}(\dot{\gamma}) = A_0(\dot{z},\dot{\phi})$.

To complete our estimates, we need to estimate the commutator $[H_h, f]$ and discrete produce rule:

$$[H_h, f]\dot{z}^p = fH_h(\dot{z}^p) - H_h(f\dot{z}^p) = A_{-1}(\dot{z}), \tag{17}$$

$$D_h(f(\alpha_i)\dot{z}_i) = f(\alpha_i)D_h(\dot{z}_i) + \dot{z}_i^q f_\alpha(\alpha_i) + hA_0(\dot{z}_i) \tag{18}$$

where f is a smooth function, $\hat{\dot{z}}_k^q = \hat{\dot{z}}_k q(kh)$ and $q(x) = \frac{d}{dx}(\rho(x)x)$. The precise form of the lower order term $\dot{z}_i^q f_\alpha(\alpha_i)$ is crucial in designing a stable discretization for curvature when surface tension is included.

We remark that (17)-(18) would not be valid due to aliasing errors if we replace $\dot{z}^p$ by $\dot{z}$ in (17) or use S_h in (18) without smoothing. In practical computations, aliasing errors only come into effect when the physical solution is under-resolved, i.e. there are significant contributions from high frequency components. Therefore, the numerical instability due to aliasing errors is not as severe as the one caused by violating the compatibility condition. Violation of the compatibility condition can cause severe numerical instabilities even when a physical solution is very smooth and well-resolved,see Section 4.

Another key idea is to project the error equations into the local normal and tangent coordinates. This plays the role of performing a linearization around a "local equilibrium solution". Notice that the motion of the interface is governed only by its normal velocity component. The tangential component of the interface velocity only changes the frame on which we parametrize the interface.Therefore, we expect that the most important contribution in the stability analysis comes from terms that are associated with the normal component of the interface variables. Denote by $\vec{N}$ and $\vec{T}$ the local unit normal and tangential vectors respectively. With some manipulations, we can show that the error equations are reduced to

$$(\dot{x}_i^N)_t = \sigma(\alpha_i) H_h D_h \dot{F}_i + A_0(\dot{x}_i^N, \dot{\delta}_i) + A_0(\dot{F}_i) + O(h^r) \tag{19}$$

$$(\dot{\delta}_i)_t = A_0(\dot{x}_i^N, \dot{\delta}_i) + A_0(\dot{F}_i) + O(h^r) \tag{20}$$

$$(\dot{F}_i)_t = -c(\alpha_i)\dot{x}_i^N + A_{-1}(\dot{x}_i^N, \dot{\delta}_i, \dot{F}_i) + O(h^r), \tag{21}$$

where $\dot{x}^N$ and $\dot{x}^T$ are the normal and tangential components of $(\dot{x}, \dot{y})$ respectively, $\dot{\delta} = \dot{x}^T + H_h \dot{x}^N$, and $\dot{F} = \dot{\phi} - u\dot{x} - v\dot{y}$, and $c(\alpha) = (u_t, v_t + g) \cdot \vec{N}$. Thus we have reduced the error estimates for the full nonlinear, nonlocal water wave equations into a simple linear and almost local system for the variation quantities. The lower order terms are nonlocal, but they are smoother than the principal linearized terms. This simplification helps us identify which terms are most important and what kind of balances we should look for among different singular terms in our energy estimates.

Recall that $\Lambda_h = H_h D_h$. It is important to note that Λ_h is a positive and symmetric operator. In fact Λ_h has the discrete Fourier symbol $|k|\rho(kh)$. Using the sign condition, $c(\alpha) > 0$, and energy estimates for equations (19)-(21), we can prove convergence of the methods.

3. The Case with Surface Tension

The effect of surface tension is to introduce a discontinuity in pressure across the interface proportional to the (mean) curvature. The pressure is larger below when the interface is curved toward the lower region. Surface tension plays an important role in understanding many physically interesting phenomena such as pattern formation in Hele-Shaw cells, the formation of fluid droplets, the motion of capillary waves on free surfaces, and the noise generation at the ocean surface [12].

As before, we assume that the fluid occupies the lower region. As it was shown in [15], in the case with surface tension, the direction of gravity plays no role in the stability estimates. Consequently the analysis also applies to the case with fluid above the interface. The earlier form of Bernoulli's equation (3) is now replaced by

$$\phi_t = \frac{1}{2}(u^2 + v^2) - gy + \tau\kappa. \tag{22}$$

Here κ is the curvature and the constant τ is the coefficient of surface tension.

In the case with surface tension, we need to use the sharper estimate (14) to suppress the aliasing errors. This is because ϕ now contains second order derivatives of z. In order to close the energy estimates, we require a $A_{-1}(\dot{F})$ in (19) and (20). This amounts to having a better control of the aliasing errors arising from the alternating quadrature rule. This requirment will be satisfied if we impose in addition that $\rho'(\pm\pi) = 0$. For the spectral derivative approximation, we can always require that the Fourier smoothing multiplier $\rho(x)$ satisfies this property. The numerical scheme is now given by

$$\frac{dz_i^*}{dt} = \frac{1}{4\pi i} \sum_{(j-i)odd} \gamma_j cot(\frac{z_i^p - z_j^p}{2})2h + \frac{\gamma_i}{2(D_h z_i)} \equiv \omega_i = u_i - iv_i \tag{23}$$

$$\frac{d\phi_i}{dt} = \frac{1}{2}(u_i^2 + v_i^2) - gy_i + \tau\kappa_i \tag{24}$$

$$D_h\phi_i = \frac{\gamma_i}{2} + Re[\frac{D_h z_i}{4\pi i} \sum_{(j-i)odd} \gamma_j cot(\frac{z_i^p - z_j^p}{2})2h] \tag{25}$$

$$\kappa_i = \frac{D_h x_i^q D_h^2 y_i - D_h y_i^q D_h^2 x_i}{((D_h x_l^q)^2 + (D_h y_i^q)^2)^3/2}, \tag{26}$$

where $\hat{x}_i^p = \rho(kh)\hat{x}_k$, and $\hat{x}_i^q = q(kh)\hat{x}_k, q(x) = \frac{d}{dx}(x\rho(x))$. Note that if $\rho(kh)$ corresponds to a $r-th$ order derivative, then

$$|\rho(kh) - 1| \leq C(kh)^r, \quad |q(kh) - 1| \leq C(kh)^r.$$

Hence z^q and z^p are $r-th$ order approximations to z.

For finite order approximations, the natural filtering associated with D_h is not strong enough since $\rho'(\pi) \neq 0$ for all finite order derivative operators. Therefore we need to apply an additional filtering before we take the finite order derivative in equations (23) and (25). That is, we should replace D_h by $\tilde{D}_h$ in (23) and (25), where $\tilde{D}_h z_i = D_h z_i^s$ and $\hat{z}_k^s = \hat{z}_k s(kh)$. The Fourier multiplier s satisfies

$$|1 - s(kh)| \leq C(kh)^r, s \geq 0, and \quad s(\pm\pi) = 0.$$

The derivative operators in (26) remain unchanged. To maintain the balances of high frequency errors, we need to modify the filtering operator accordingly. That is $\hat{z}_k^p$ now becomes $\rho(kh)s(kh)\hat{z}_k$. With these modifications, the algorithm is exactly the same as (23)-(26). We can prove the following result:

Theorem 2. *Assume that $z(\cdot,t), \phi(\cdot,t) \in C^{m+2}[0, 2\pi]$, and $\gamma(\cdot,t) \in C^{m+1}[0,2\pi]$ for $t \leq T$ and $m \geq 4$. If D_h corresponds to a r-th order approximation, then we have*

$$||z(\cdot,t) - z(t)||_{H_h^1} \leq C(T)h^{r-1},$$

$$||\phi(\cdot,t) - \phi(t)||_{H_h^{1/2}} \leq C(T)h^{r-1},$$

$$||\gamma(\cdot,t) - \gamma(t)||_{H_h^{1/2}} \leq C(T)h^{r-2},$$

where $||\phi||_{H_h^1}^2 = \sum_{|k|\leq N/2} (1 + |k|\rho(kh))^2 |\hat{\phi}_k|^2$. *If D_h corresponds to the spectral derivative, the result is the same by replacing r by m.*

Remark The use of z^q in (26) is to balance the high mode error due to aliasing in the spectral approximation or in finite order accurate approximation. It is determined by the precise form of the discrete product rule:

$$D_h(fz) = fD_h z + (D_h f)z^q + hA_0(z),$$

as given by (18).

To illustrate how the algorithm looks like in a concrete example, we consider the case of second order finite difference approximation D_h:

$$D_h f_i = \frac{f_{i+1} - f_{i-1}}{2h}.$$

The corresponding $\rho(kh)$ is given by $\rho(x) = (sinx)/x$, so that $q(x) = cos(x)$. Clearly, we have

$$\hat{z_k^q} = cos(kh)\hat{z_k} = [\frac{e^{-ikh} + e^{ikh}}{2}]\hat{z}_k,$$

which implies that

$$z_l^q = \frac{z_{i+1} + z_{i-1}}{2}$$

As a result, we obtain

$$D_h z_i^q = \frac{z_{i+2} - z_{i-2}}{4h} = D_{2h} z_i, D_h^2 z_i = \frac{z_{i+2} - 2z_i + z_{i-2}}{h^2}$$

This simply says that we should use every other grid points in discretizing the curvature κ.

The proof of Theorem 2 relies on a key estimate on the variation of the curvature. We can show that the linear part of $\dot{k}$ has the simple expression in terms of the normal component of $\dot{z}$ almost exactly as that in the continuum case [5]

$$\dot{\kappa} = \sigma(\alpha_i) D_h[\sigma(\alpha_i) D_h \dot{x}_i^N] + A_0(\dot{z}) \tag{27}$$

In proving (27), we have repeatedly used the discrete product rule

$$D_h(fz) = f D_h z + (D_h f) z^q + h A_0(z).$$

Combining the estimate for $\dot{\kappa}$ with the previous estimates obtained in the case without surface tension, we arrive at

$$(\dot{x}_i^N)_t = \sigma(\alpha_i) H_h D_h \dot{F}_i + A_0(\dot{x}_i^N, \dot{\delta}_i) + A_{-1}(\dot{F}_i) + O(h^r) \tag{28}$$

$$(\dot{\delta}_i)_t = A_0(\dot{x}_i^N, \dot{\delta}_i) + A_{-1}(\dot{F}_i) + O(h^r) \tag{29}$$

$$(\dot{F}_i)_i = \tau\sigma(\alpha_i) D_h[\sigma(\alpha_i) D_h \dot{x}_i^N] + A_0(\dot{z}) - c(\alpha_i)\dot{x}_i^N + A_{-1}(\dot{F}) + O(h^r) > \tag{30}$$

Energy estimates on (28)-(30) will imply convergence of the method.

One should note that the leading order contribution in (30) comes from the surface tension. The term with the sign condition $c(\alpha_i)$ becomes a lower order term in this case. Therefore it plays no role in the leading order stability analysis. This is in contrast with the case without surface tension where the sign condition for $c(\alpha)$ is crucial in determining the linear stability of the interface perturbation.

Remark. The presence of surface tension introduces a stiff ODE system through the curvature term. As a consequence, the time step stability constraint is very severe if an explicit time discretization is used. For example, the time step stability constraint for Hele-Shaw flows is given by $\Delta t \leq Ch^3$. To maintain stability, we are forced to use very small time step $\triangle t$. This has been one of the major difficulties in numerical simulations of interfacial flows with surface tension. Recently, Hou, Lowengrub and Shelley [9] have been able to remove this severe time step stability constraint by reformulating the problem in terms of the arclength metric and the tangent angle variables, and using an implicit time discretization. This reformulation eliminates the high order time step stability constraint leaving only a first order constraint to be satisfied for stability. Many interfacial problems which are difficult to solve previously are now solvable using our method.

4.Examples of Unstable Schemes

In this section, we present a class of unstable schemes based on an equivalent boundary integral formulation. We will demonstrate the numerical instability by performing a von Neumann stability analysis around equilibrium (also see [3]). Consider a boundary integral formulation that uses the dipole strength μ. The vortex sheet strength is defined in terms of μ by the relationship $\gamma = \mu_\alpha$. The evolution equations for z and ϕ are the same as before. The only difference is in the kinetic condition between ϕ and μ.

$$\phi = \frac{\mu}{2} + Re[\frac{1}{4\pi i}\int_{-\pi}^{\pi} \mu(\alpha')z_\alpha(\alpha')cot[\frac{z(\alpha)-z(\alpha')}{2}]d\alpha']. \tag{31}$$

It is clear that equation (2) is obtained by differentiating the relation (31) with respect to α and integrating by parts. A natural numerical approximation to the above equations is given by

$$\frac{dz_i^*}{dt} = \frac{1}{4\pi i}\sum_{(j-i)odd} \gamma_j cot(\frac{z_i - z_j}{2})2h + \frac{\gamma_i}{2(D_h z_i)} \equiv u_i - iv_i, \tag{32}$$

$$\frac{d\phi_i}{dt} = \frac{1}{2}(u_i^2 + v_i^2) - gy_i + \tau\kappa_i, \tag{33}$$

$$\phi_i = \frac{\mu_i}{2} + Re[\frac{1}{4\pi i}\sum_{(j-i)odd} \mu_j D_h z_j cot(\frac{z_i - z_j}{2})2h],$$

$$\gamma_i = D_h\mu_i, \kappa_i = \frac{D_h x_i D_h^2 y_i - D_h y_i D_h^2 x_i}{((D_h x_i)^2 + (D_h y_i)^2)^{3/2}}. \tag{34}$$

Here D_h can be any finite order derivative approximation. As before, we denote its Fourier symbol as $\hat{D_h f_k} = ik\rho(kh)\hat{f}_k$.

Now we perform a linear stability analysis around equilibrium. Let

$$z = \alpha + \dot{z}; \phi = \dot{\phi}, \mu = 1 + \dot{\mu}.$$

Substituting the above expressions into the discrete equations (32)-(34), and using (10)-(11), we obtain to the leading order

$$\frac{d}{dt}\dot{z}^* = \frac{1}{2i}H_h(D_h\dot{\mu}) + D_h\frac{\dot{\mu}}{2}, \tag{35}$$

$$\frac{d}{dt}\dot{\phi} = -g\dot{y} + \tau D_h^2\dot{y}, \tag{36}$$

$$\dot{\phi} = \frac{\dot{\mu}}{2} + \frac{1}{2}(H_h(D_h\dot{y}) - H_h(S_h\dot{y})), \tag{37}$$

where H_h is the discrete Hilbert transform and S_h is the pseudo-spectral derivative operator without smoothing. This system of equations has constant coefficients in the Fourier space. The eigenvalues of the resulting system in the Fourier space give the growth rates of $\dot{x}, \dot{y}, \dot{\phi}$. They are explicitly given by

$$\lambda_1 = 0; \lambda_2, \lambda_3 = (k^2\rho(1-\rho) \pm \sqrt{k^4\rho^2(1-\rho)^2 - 16[g|k|\rho + \tau(|k|\rho)^3]})/2,$$

where τ is the surface tension coefficient. This is the standard nomal mode analysis. Baker and Nachbin [3] independently used this approach to study a number of discretizations for vortex sheets with surface tension. Notice that if $\rho \neq 0$, 1, then

$$\lambda_2, \lambda_3 \simeq O(k^2)$$

for large k. This indicates the numerical high mode instability is even stronger than that of Kelvin-Helmholtz! This indicates that except for the spectral derivative operator, all other discretizations are linearly unstable.It is also clear that the instability is caused by violating the compatibility condition in (37), i.e. $H_hD_h - \Lambda_h \neq 0$ (see(11)). If the Fourier filtering is used as in (4) and (6), then this term vanishes and one can easily see that the modified method is stable. On the other hand, aliasing instabilities cannot be seen from this linear stability analysis around the equilibrium solution because there is no mode mixing for constant coefficient problems.

The above analysis is also confirmed by our numerical experiment. We tested the above algorithms for the second order finite difference, fourth order finite difference, and cubic spline derivative operators [6]. In our calculations, we choose a sinusoidal perturbation to the equilibrium solution. Surface tension is set to be zero. If a spectral derivative is used,the numerical solution is smooth and exists for long times. In Figure 1, we plot the numerical interface position obtained using the spectral derivative operator at different times. Clearly the numerical solution is stable and smooth. In Figure 2a, we plot the numerical vortex sheet strength obtained using the second order derivative operator. We can see that there is a critical time beyond which the numerical solution starts to produce numerical oscillations. The oscillations grow rapidly with respect to time. If we further reduce the mesh size, the numerical oscillations will appear earlier. In Figure 2b, we examine the growth of the Fourier coefficients for the vortex sheet strength. We show the log plot of spectrum at different times ranging from $t = 0.008$ to $t = 0.22$. For t small, we see that the spectrum decays exponentially. But due to numerical roundoff errors, the computer can not represent accurately the Fourier coefficients smaller than 10^{-16}. These roundoff errors are amplified by the numerical instability in time. As we can see, the roundoff errors at the high modes are amplified the fastest. The ordering of the curves is in the increasing order of time. Only after the time $t = 0.2$, we begin to see the effect of this numerical instability in the physical space. Numerical oscillations start to grow rapidly in time. Very similar behaviors are observed for the fourth order finite difference and the cubic spline derivative approximations. However, with the modifications given by (23)-(26), the scheme using the second order derivative operator is stable. The numerical calculations presented in Figure 3 confirm this ($\tau = 0.01$).

Acknowledgements. JTB acknowledges the partial support from NSF grant DMS-9102782. TYH acknowledges partial support from a NSF grant DMS-9003202, a Sloan Foundation Research Fellowship, and an Air Force of Scientific Research grant AFOSR-90-0090. JSL acknowledges the support of the National Science Foundation through a Mathematical Sciences Postdoctoral Fellowship, the University of Minnesota Army High Performance Computing Research Center.

Bibliography

[1] C. Anderson, "*A Vortex Method for Flows with Slight Density Variations,*" J. Comp. Phys., v. 61, 1985, p. 417.

[2] G. Baker, D. Meiron and S. Orszag, "*Generalized Methods for Free Surface Flow Problems,* " J. Fluid Mech., v. 123, 1982, p. 477-501.

[3] G. Baker and A. Nachbin, "*Stable Methods for Vortex Sheet Motion with Surface Tension,* " preprint (1993), submitted to J. Comp. Phys.

[4] G. Baker and M. Shelley, "*On the Connection between Thin Vortex Layers and Vortex Sheets,*" J. Fluid Mech., v. 215, 1990, p. 161.

[5] J. T. Beale, T. Y.Hou and J. Lowengrub, "*Growth Rates for the Linear Motion of Fluid Interfaces Away from Equilibrium,*" Comm. Pure Appl.

Math., v. 46, 1993, pp. 1269-1301.

[6] J. T. Beale, T. Y. Hou, and J. Lowengrub, *"Convergence of Boundary Integral Methods for Water Waves,"* preprint (1993), submitted to SIAMJ. Numer. Anal.

[7] J. T. Beale, T. Y. Hou, and J. Lowengrub, *"On the Well-Posedness of Two Fluid Interfacial Flows with Surface Tension,"* in **Singularities in Fluid, Plasmas and Optics,**. NATO ASI Series C: Mathematical and Physical Sciences, Vol. 404, 1993, pp. 11-38, Ed. Russel Caflisch and George Papanicolaou.

[8] T. Y. Hou, J. Lowengrub, and R. Krasny, *"Convergence of a Point Vortex Method for Vortex Sheets,"* SIAMJ. Numer. Anal., v. 28, 1991, p. 308-320.

[9] T. Y. Hou, J. Lowengrub, and M. Shelley, *"Removing the Stiffness from Interfacial Flows with Surface Tension"* perprint (1993), submitted to J. Comp. Phys.

[10] R. M. Kerr,*"Simulation of Rayleigh-Taylor Flows Using Vortex Blobs,"* J. Comp. Phys., v. 76, 1988, p. 48-84.

[11] R. Krasny,*"Desingularization of Periodic Vortex Sheet Rollup,"* J. Comp. Phys., v. 65, 1986, p. 65.

[12] A. Prospeeritti, L. A. Crum and H. C. Pumphrey,*"The Underwater Noise of Rain,"* J. Geophys. Res., v. 94, 1989, p. 323.

[13] D. Pullin,*"Numerical Studies of Surface Tension effects in Nonlinear Kelvin-Helmholtz and Rayleigh-Taylor Instabilities,"* J. Fluid Mech., v. 119, 1982, p. 507.

[14] M. Shelley,*" A Study of Singularity Formation in Vortex Sheet Motion by a Spectrally Accurate Method,"* J. Fluid Mech.m., v. 244, 1992, p. 493.

[15] G. I. Taylor, *"The Instability of Liquid Surfaces when Accelerated in a Directional Perpendicular to their Plances,"* Proc. Roy. Soc. A, v. 201, 1950, p. 192-196.

[16] G. Tryggvason,*" Numerical Simulations of the Rayleigh Taylor Instabilities,"* J. Comp. Phys., v. 75, 1988, p. 253-282.

Dept. of Mathematics, Duke University, Durham, NC 27708.
Applied Mathematics,California Institute of Technology,Pasadena,CA91125.
School of Mathematics, University of Minnesota, Minneapolis, MN 55455.

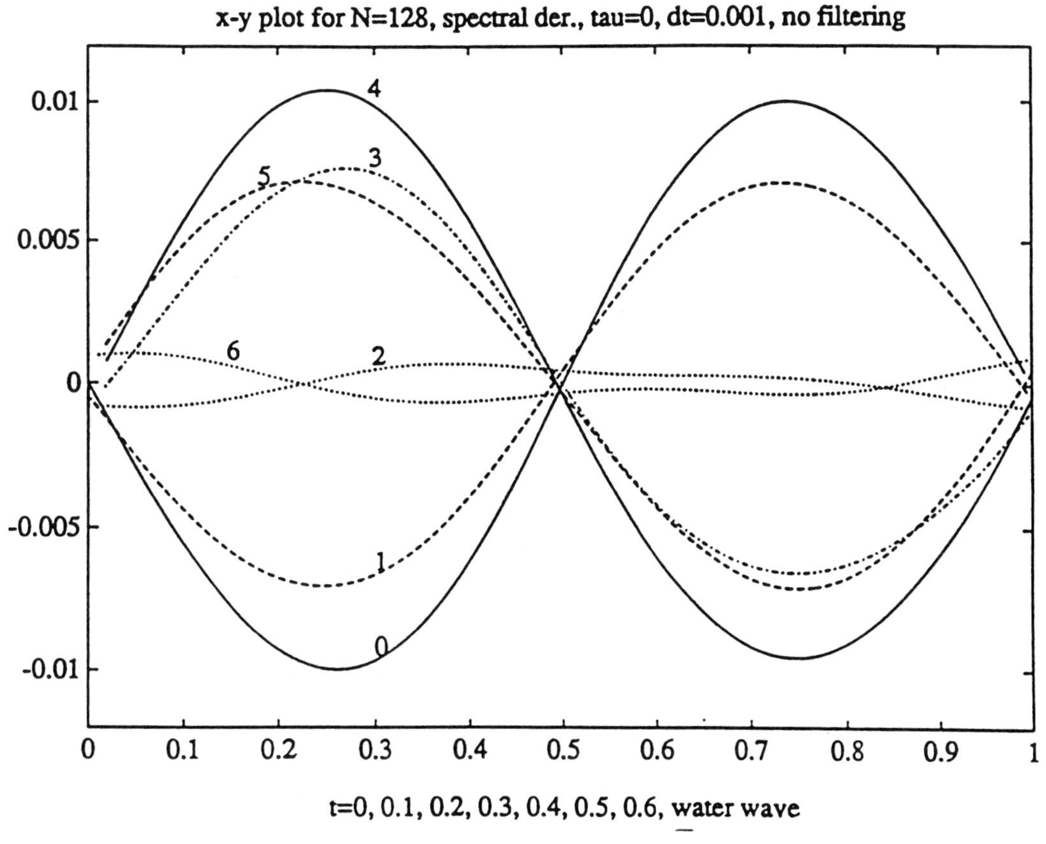
x-y plot for N=128, spectral der., tau=0, dt=0.001, no filtering
0.01
0.005
0
-0.005
-0.01
0
0.1
0.2
0.3
0.4
0.5
0.6
0.7
0.8
0.9
1
4
3
5
6
2
1
0
t=0, 0.1, 0.2, 0.3, 0.4, 0.5, 0.6, water wave

Figure 1

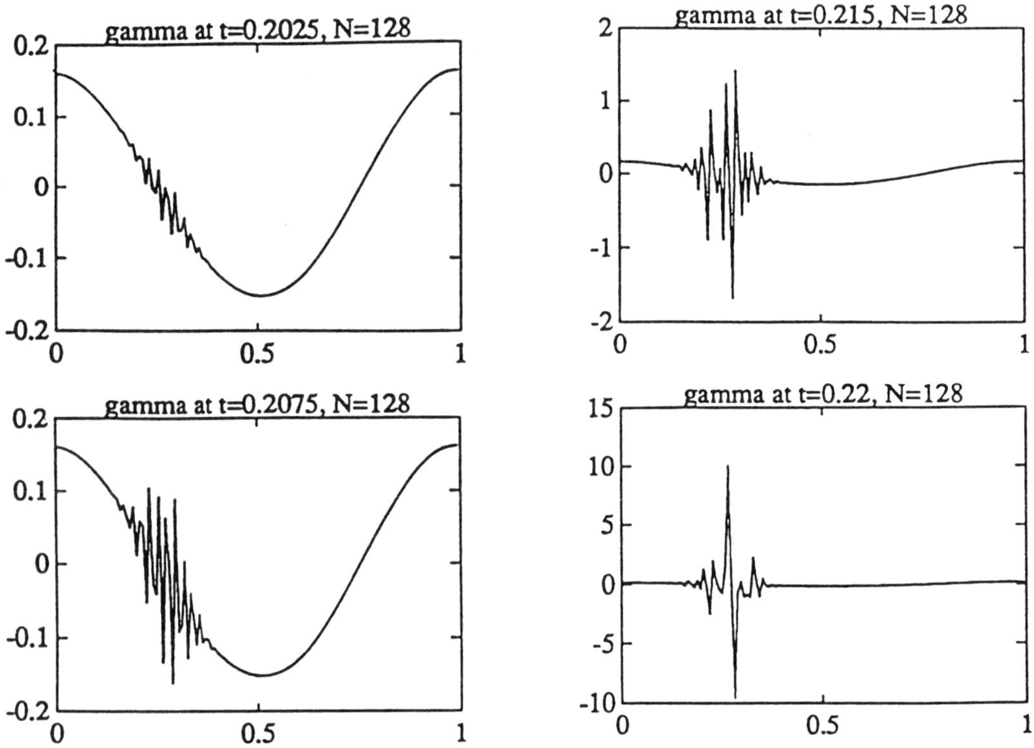

Figure 2a

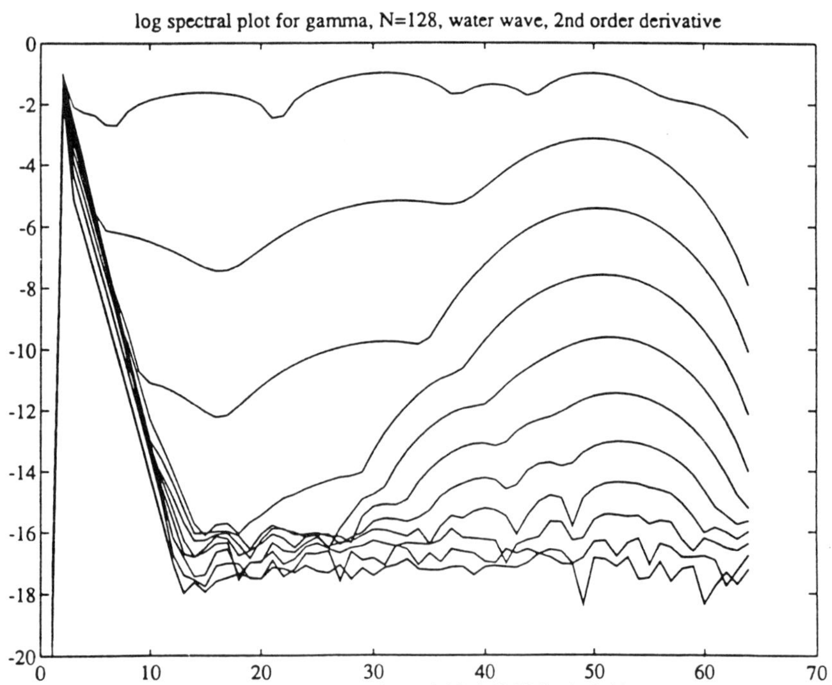

Figure 2b

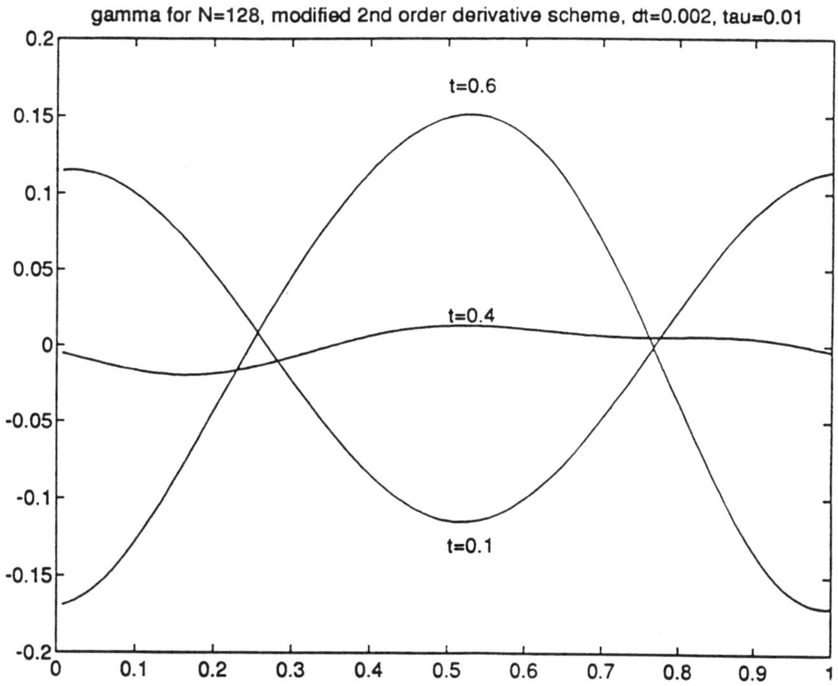

Figure 3a

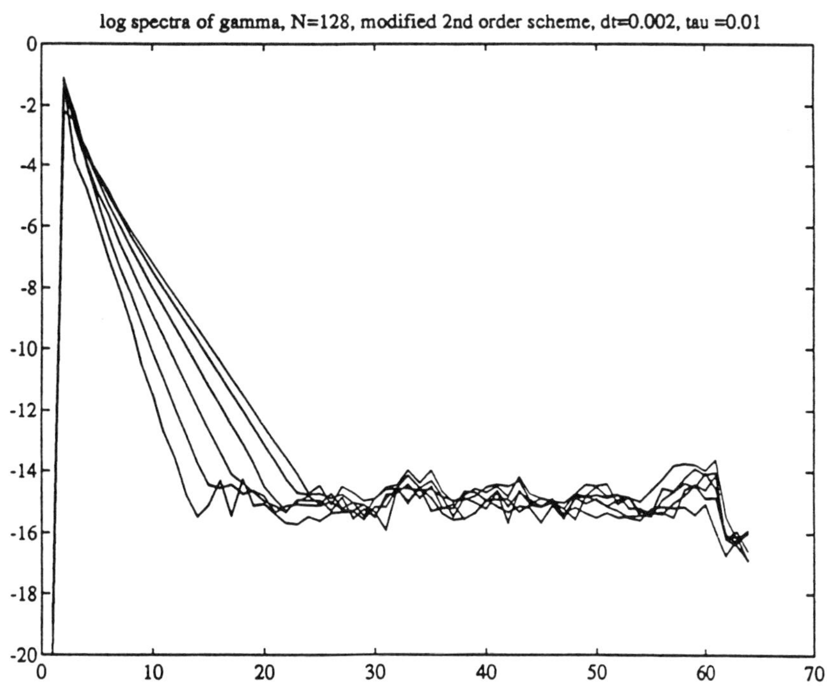

Figure 3b

AMS/IP Studies in Advanced Mathematics
Volume 3, 1997

ON THE COMPACTLY SUPPORTED SOLUTIONS OF KPP OR FISCHER TYPE EQUATIONS

Z.BIRO, R.KERSNER[1]

Computer and Automation Institute,
Hungarian Academy of Sciences
H-1518, Budapest, POB 63, Hungary

ABSTRACT. We investigate the supprot properties of sulution to a KPP type equation with compactly supported data

Keywords: degenerate diffusion, KPP equation, positivity, localization

1. INTRODUCTION

By several reasons (see e.g. [M], in particular Chapter 11, [C] and the references there) a natural generalization of Kolmogorov-Petrovsky-Piscunov [KPP] or Fischer (1937) equation

$$u_t = u_{xx} + u(1-u) \tag{KPP}$$

is the equation

$$u_t = (u^m)_{xx} + s_0 u^p - a_0 u^q \tag{1}$$

where m, p, q, s_0 and a_0 are positive constants, $x \in \Re$ –the space–, $t > 0$ the time variables. If $u(x,t)$ represents the density at x and in t, mu^{m-1} is the (density-dependent) diffusion coefficient, $s_0 u^{p-1}$ and $a_0 u^{q-1}$ are the (density-dependent) source and absorption intensities. The initial data

[1]Supported in part by Research Fund OTKA T-4374

$$u(x,0) = u_0(x) \tag{2}$$

is positive on $(-l,l), l > 0$, equal to zero for $|x| > l$ and is continuous.

The study of travelling-wave solution, i.e. the solutions of the form $f(x - \lambda t)$, of [KPP] and the related equation have become an essential part of the analysis of processes described by these models. Since the paper of [KPP] several fundamental results have been obtained about the existence, stability etc. of such waves (Aronson, Weinberger, Fife, McLeod, Hadeler , Rothe, Uchiyama, Kanel, Sattinger, Bramson etc.; a rather complete list of references can be found in [GK]. In these investigations the phase-plane analysis of the corresponding ODE, the classical Gevrey's results and the maximum principle have played a vital role.

A complete theory of travelling waves for more general then (1) equations was constructed in [GK] by using the so-called integral- equation method. Applying the results of [GK] to equation (1), this method gives the following necessary and sufficient condition (we are interested in travelling-waves $f(\xi) = f(x - \lambda t)$ such that $f(-\infty) = (\frac{s_0}{a_0})^{\frac{1}{p-q}} \equiv M_0, f(+\infty) = 0)$:

The equation (1) has a monotonic global travelling-wave solution with a speed λ if and only if the integral equation

$$\Theta(s) = \lambda s - \int_0^s \frac{m\tau^{m-1}(s_0\tau^p - a_0\tau^q)}{\Theta(\tau)} d\tau \tag{3}$$

has a nonnegative, continuous solution $\Theta(s), s.t.\Theta(M_0) = 0$. If such a $\Theta(a)$ exists, then $f(\xi)$ is defined by

$$\int_0^{f(\xi)} \frac{ms^{m-1}}{\Theta(s)} ds = -\xi \tag{4}$$

We remark that in the case of $m > 1(e.g.)0 < f(\xi) < M_0$ does not necessarily hold unlike in (KPP) and the condition $f(+\infty) = 0(or f(-\infty) = M_0$ does not exclude the possibility of $f \equiv 0$ for $\xi \geq \xi_0(or \quad f \equiv M_0 \quad for \quad \xi \leq \xi_1)$. In fact, such a type of the behaviour of $f(\xi)$ depends on whether the integral (4) converges at 0 and M_0 or not. The following example ($[A], [N_2]$) is an illustration for that: the function

$$u_1(x,t) = \left[1 - exp\left(\frac{x - \frac{t}{\sqrt{2}}}{\sqrt{2}}\right)\right]_+$$

solves the equation $u_t = (uu_x)_x + u(1-u)$. One can see that $u_1 \equiv 0$ for $x \geq \frac{t}{\sqrt{2}}$.

In this note we begin to study solutions other than travelling-waves: the solutions with compact (in x) support (the waves like $u_1(x,t)$ could be compactly supported from the right or left only). By analogy with the theory of porous medium equation ($m > 1, a_0 = s_0 = 0 \quad in (1)$) we expect that there exists two continuous curves ς^+ and ς^- on the (x,t) plane such that $\varsigma^+(0) = l, \varsigma^-(0) = -l, u(x,t) > 0$ for $\varsigma^+(t) < x < \varsigma^-(t)$ and $u = 0$ otherwise.

Definition. We say that problem (1), (2) has the positivity property if $\forall x_0 \in \Re \exists t_0(x_0) \geq 0$ s.t. $\forall t > t_0, u(x_0,t) > 0$("$|\varsigma^\pm(t)| \to \infty$").. The problem (1),(2) has the localization property if $\exists L > 0 : u(x,t) \equiv 0$ for $|x| \geq L, t > 0$("$|\varsigma^\pm(t) \leq L$")

Remark. We have implicitly supposed that $\varsigma^\pm(t)$ exists for all $t > 0$. Of course, this is not the case for all combinations of (m,p,q). Sometimes $u(x,t)$ can also blow up in finite time, etc.

2. POSITIVITY, EXAMPLES

In the sequel we shall always assume that

$$(5) \qquad m > 1, \quad p \geq 1 \quad and \quad q \geq 1.$$

Theorem 1. Let $q > m$. Then, the problems (1) , (2) has the positivity property, independently on p.

A sketch of the proof.

We are going to show that $u(x,t)$ lies above a function $w(x,t)$ which has two interfaces ς_1^+ and ς_1^- such that $|\varsigma_1^\pm(t)| \to \infty$ when $t \to \infty$. From this it follows that $|\varsigma^\pm(t)) \to \infty$ has to hold as well. A natural candidate ("Barenblatt-Pattle") for such a $w(x,t)$ might be the function

$$w(x,t) = (kt + k_0)^{-\alpha} \left[a - \frac{x^2}{(kt+k_0)^\beta} \right]_+^\omega$$

with positive constants α, β, w, a, k and k_0, to be chosen later. In order to have a "subsolution", we must have

$$(6) \quad \mathcal{L}w := -w_t + (w^m)_{xx} + s_0 w^p - a_0 w^q > 0 \quad for \quad w > 0, u_0(x) \geq w(x,0).$$

After introducing the notation

$$\Theta = kt + k_0, A = a - \frac{x^2}{(kt+k_0)^\beta}$$

we calculate $\mathcal{L}w$:

$$\mathcal{L}w = k\alpha\Theta^{-a-1}A^{\omega} - k\beta\omega x^2\Theta^{-\alpha-\beta-1}A^{\omega-1}$$
$$+4\omega m(\omega m-1)x^2\Theta^{-\alpha m-2\beta}A^{\omega m-2} - 2\omega m\Theta^{-\alpha m-\beta}A^{\omega m-1}$$
$$+s_0\Theta^{-\alpha p}A^{\omega p} - a_0\Theta^{-\alpha q}A^{\omega q} := I_1 + I_2 + I_3 + I_4 + I_5 + I_6.$$

If we write

$$I_1 + I_4 + I_6$$
$$= \Theta^{-\alpha-1}A^{\omega}[k\alpha - 2\omega m\Theta^{-\alpha(m-1)-\beta+1}A^{\omega(m-1)-1} - a_0\Theta^{\alpha(q-1)+1}A^{\omega(q-1)}]$$

and

$$I_3 + I_2 = \omega x^2\Theta^{-\alpha m-2\beta}A^{\omega m-2}[4m(\omega m-1) - k\beta\Theta^{\alpha(m-1)+\beta-1}A^{1-\omega(m-1)}]$$

we can see–because Θ goes to infinity when t does and A can go to zero when x^2 close to $a(kt+k_0)^\beta$ – – that the following has to hold

$$\alpha(m-1) = 1-\beta \quad and \quad \omega = \frac{1}{m-1}. \tag{8}$$

The condition $-\alpha(q-1)+1 \leq 0$ is also necessary; with (8) it gives

$$-\frac{1-\beta}{m-1}(q-1)+1 \leq 0 \quad i.e. \quad \beta \leq \frac{q-m}{q-1}.$$

which is possible only if $m < q$ (in order to have $\alpha > 0$ one must have $\beta < 1$). In order to have $\mathcal{L}w > 0$, one also must have

$$k\alpha > 2\omega m \quad and \quad 4m(\omega m-1) \geq k\beta \quad or$$

$$\frac{2m}{(m-1)^2}(1-\beta) < k \leq \frac{4m}{\beta(m-1)}.$$

Such a k exists for small $\beta < \beta_0$. Taking $\beta = min(\beta_0, \frac{q-m}{q-1})$ we can see that $I_1 + I_4 + I_6 > 0$ for large k_0 and also $u_0(x) \geq w(x,0)$ for large k_0 and small $a > 0$. This yields the required properties.

Using the same ideas, the following assertion can be proven.

Theorem 2. Let $q > p$. Then the problem (1),(2) has the positivity property for any $m > 1$. If $q = p$, the same can be said, provided $s_0 \geq a_0$

Example. Consider the equation(1) with $m = q, p = 1, s_0 = 1, a_0 = \lambda$ i.e.

$$u_t = (u^m)_{xx} + u(1 - \lambda u^{m-1}) \tag{9}$$

with $u_0(x)$ having compact support:

$$u(x,0) = u_0(x) = f_0^{-\frac{1}{m-1}} \left[\rho_0 - \frac{1}{g_0} \left(ch \left(\frac{m-1}{m} \sqrt{\lambda} x \right) - 1 \right) \right]_+^{\frac{1}{m-1}}, \tag{10}$$

where f_0, g_0 and ρ_0 are arbitrary positive numbers. The solution of problem (9),(10) is

$$u(x,t) = \left(\frac{e^{(m-1)t}}{f(t)} \right)^{\frac{1}{m-1}} \left[\rho_0 - \frac{1}{g(t)} \left(ch \left(\frac{m-1}{m} \sqrt{\lambda} x \right) - 1 \right) \right]_+^{\frac{1}{m-1}}, \tag{11}$$

where the function $g(t)$ and $f(t)$ are defined by

$$\int_{g_0}^{g(t)} \left[s \left(s + \frac{2}{\rho_0} \right) \right]^{\frac{m-1}{2}} ds = \frac{c_0}{m-1} \left(e^{(m-1)t} - 1 \right)$$

and

$$f(t) = \lambda \rho_0 (e^{(m-1)t} - 1) + \frac{\lambda (m-1)^2}{m} \int_0^{\frac{1}{m-1}(e^{(m-1)t}-1)} \frac{ds}{g(s)} + f_0$$

It is easy to see that

$$g(0) = g_0 > 0, g'(t) > 0, g(t) \sim C_1 e^{\frac{m-1}{m} t} \quad for \quad t \gg 1 and$$

$$f(0) = f_0 > 0, f'(t) > 0, f(t) \sim \lambda \rho_0 e^{(m-1)t} \quad for \quad t \gg 1.$$

The function (11) has compact support for any $t > 0$. The equation for $\varsigma^{\pm}(t)$ can be written as

$$\rho_0 g(t) = ch \left(\frac{m-1}{m} \sqrt{\lambda} x \right) - 1.$$

For large t (large x) one has

$$e^{\frac{m-1}{m} t} \sim C_2 e^{\frac{m-1}{m} \sqrt{\lambda} x}, \quad so \quad \varsigma^+ \sim x \sim \frac{1}{\sqrt{\lambda}} t + C_3$$

and we can see that the free boundaries are asymptotically straight lines.

If x is from any bounded domain, we have the uniform convergence for $t \to \infty$:

$$u(x,t) \sim \left(\frac{e^{(m-1)t}}{\lambda \rho_0 e^{(m-1)t}}\right)^{\frac{1}{m-1}} \rho_0^{\frac{1}{m-1}} = \left(\frac{1}{\lambda}\right)^{\frac{1}{m-1}}$$

as it could be expected ($\lambda^{-\frac{1}{m-1}}$ is a solution to (9)).
Remark. If $\lambda < 0$ i.e. $\lambda = -|\lambda|$ in (9), (10) and (11) then, using the formula

$$ch\left(\frac{m-1}{m}\sqrt{-|\lambda|}x\right) = cos\left(\frac{m-1}{m}\sqrt{|\lambda|}x\right)$$

we shall have an interesting blow-up solution with a bounded and periodic initial condition.
Remark. We thind (but it is not so easy to show) that the solution(11) has been found earlier in $[N1]$ for $m = 2$.

3. LOCALIZATION, EXAMPLES

It is known ([K], [BKP], [CMM]) that the problem (1),(2) with the source term absent ($s_0 = 0$) has the localization property iff $q < m$. This fact is independent of the "size" of $u_0(x)$–max$u_0(x)$, measure of supp$u_0(x)$, etc. In the presence of source term ($s_0 > 0$) the situation alters. Some signs of that can be seen from [ACP], where the Cauchy-Dirichlet problem was treated.

We recall that the constant $M_0 = (\frac{a_0}{s_0})^{\frac{1}{p-q}}$ is a solution to (1) . In the first proposition the measure of supp$u_0(x)$ does not play any role but max$u_0(x)$ does: it has to strictly be less than M_0. In the second one max$u_0(x)$ can be greater than M_0 but less than some $M_1 > M_0$; at the same time suppu_0 has to remain in a fixed domain.

Theorem 3. Let $m > q, p \geq q$ and max$u_0 < M_0$. Then, the problem (1),(2) has the localization property.

Proof. It follows that $u(x,t) \leq M_0$ by the comparison principle. In the domain $\Re_+^2 = (0,\infty) \times (0,\infty)$ consider the auxiliary function

$$v(x,t) = v(x) = b\left(1 - \frac{x}{L}\right)_+^{\omega}$$

where the positive numbers ω, b and L have to be chosen. If $b >$max$u_0(x)$, then it is possible to choose $L > 0$ so large that the inequality $u_0(x) \leq v(x,0) = v(x)$ should be fulfilled . For $v(x) > 0$ we have

(12)
$$\mathcal{L}v \equiv (v^m)_{xx} + s_0 v^p - a_0 v^q$$
$$= v^q \left[-a_0 + s_0 b^{p-q} \left(1 - \frac{x}{L}\right)^{\omega(p-q)} + \omega m(\omega m - 1) b^{m-1} L^{-2} \left(1 - \frac{x}{L}\right)^{\omega(m-q)2-2} \right]$$

Let $\omega = \frac{2}{m-q}$. This is possible because $m > q$. The quantity in [ ] can be made negative (by choosing L even larger, if needed) iff $-a_0 + s_0 b^{p-q} < 0$ i.e. $b < (\frac{a_0}{s_0})^{\frac{1}{p-q}}$. Such a $b > max u_0(x)$ exists by the condition of the theorem. Now we can see that $(v^m)_x$ is a continuous function at $1 = \frac{x}{L}$ (because $\omega m = \frac{2m}{m-q} > 1$) so $v(x)$ is a generalized supersolution to (1). By the comparison argument, $u(x,t) \leq v(x)$ in $\Re^2_+$ in particular $u(x,t) \equiv 0$ for $x \geq L$. Using the function $v(x) = b(1+\frac{x}{L})^\omega$ in $\Re^2_- = (-\infty, 0) \times (0, \infty)$ one can show that $u(x,t) \equiv 0$ also for $x \leq -L$.

In order to be able to formulate our next result, we introduce some notation:

$$M_0 = \left(\frac{a_0}{s_0}\right)^{\frac{1}{p-q}}, M_1 = M_0 \left(\frac{m+p}{m+q}\right)^{\frac{1}{p-q}},$$

$$l_0 = M_1^{\frac{m-q}{2}} \left(\frac{m(m+q)}{2a_0}\right)^{\frac{1}{2}} \int_0^1 \frac{t^{\frac{m-q}{2}-1}}{\sqrt{1-t^{p-q}}} dt$$

$$= \frac{\sqrt{\pi}}{p-q} \left(\frac{m(m+q)}{2a_0}\right)^{\frac{1}{2}} \left(\frac{m+p}{m+q}\frac{a_0}{s_0}\right)^{\frac{m-q}{2(p-q)}} \frac{\Gamma(\frac{m-q}{2(p-q)})}{\Gamma(\frac{m+p-q}{2(p-q)})}.$$

The functions f and F are given by

$$f(s) = s_0 s^p - a_0 s^q \quad and \quad F(s) = \int_0^s \tau^{m-1} f(\tau) d\tau.$$

Obviously, $F(0) = F(M_1) = 0$ and $F(s) < 0$ for $s \in (0, M_1)$. Now we can define a function $v(x)$ by

$$\sqrt{\frac{m}{2}} \int_{v(x)}^{M_1} \frac{s^{m-1}}{\sqrt{-F(s)}} ds = |x| \quad for \quad x \in (-l_0, l_0) \quad and$$

$$v(x) \equiv 0 \quad for \quad |x| \geq l_0$$

Theorem 4. If $q < min(m,p)$ and $u_0(x) \leq v(x)$, then the problem (1),(2) has the *localization property*.

Proof. It can be checked that $v(x)$ is a (classical) solution to (1) when $x \in (-l_0, l_0), v(\pm l_0) = 0$. It follows from $(v^m)_x = \pm\sqrt{-2mF(v)}$ that at the points $x = \pm l_0$ the "flux" $(v^m)_x$ is continuous, so $v(x)$ is also a weak solution to (1) in $\Re$, which gives $u(x,t) \le v(x)$ in $\Re^2_+$. But $v(x) \equiv 0$ for $|x| \ge l_0$.

Example. If $m = p > q$ then $l_0 = \frac{\pi}{2}\frac{m}{\sqrt{s_0}}$ and

$$v(x) = \left(\frac{2m}{m+q}\frac{a_0}{s_0}\right)^{\frac{1}{m-q}} cos^{\frac{2}{m-q}}\left(\frac{\sqrt{s_0}}{m}x\right) \quad for \quad x \in (-l_0, l_0)$$

and $v(x) \equiv 0$ for $|x| \ge l_0$. It is interesting to note that l_0 does not depend on a_0 and q i.e. on the absorption term.

Example. As before, let $m = p$ but $q = 1$. Separating the variables and using the previous example one can find a non stationary solution to equation(1):

$$w(x,t) = \frac{M_1}{M_0}\left[\frac{s_0}{a_0} + \frac{1 - \omega_0^{m-1}\frac{s_0}{a_0}e^{a_0(m-1)t}}{\omega_0^{m-1}}\right]^{-\frac{1}{m-1}} cos^{\frac{2}{m-1}}\left(\frac{\sqrt{s_0}}{m}x\right)$$

for $x \in (-l_0, l_0), w \equiv 0$ for $|x| \ge l_0$. One can see from this formula that

If $w_0 > M_0$ then the solution blows up in finite time (cf.[GDE], [SGKM])

If $w_0 = M_0$ we have a steady state and

If $w_0 < M_0$ then the solution goes to zero exponentially as $t \to \infty$.

REFERENCES

[A] Aronson, D.G., Density-dependent Interaction-Diffusion Systems, In: Dynamics and Modelling of Reactive Systems, Academic Press (1980), 161 -176.

[ACP] Aronson, D.G.,Crandall, M.G., Peletier, L. A. , Stabilization of solution of a degenerated nonlinear diffusion problem, Nonlinear Analysis, Theory, Meth. Appl. 6 10 (1982), 1001-1002.

[BKP] Bertsch, M., Kersner, R. , Peletier, L. M. , Positivity versus localization in degenerate diffusion equation, Nonlinear Analysis, Theory, Meth. Appl **9** 9 (1985), 987-1008.

[CMM] Chen, X.Y., Matano, H., Mimura, M. , Finite- point extinction and continuity of Interfaces in a nonlinear diffusion equation with strong absorption, preprint 1990, pp. 1-38, to appear in J. Rein. Angew. Math..

[C] Choudhury, S.R., Painleve analysis and partial integrability of a calss of reaction-diffusion equations, Nonlinear Analysis, Theory, Meth. Appl. **18** 5 (1992), 445-459.

[GDE] Galaktionov, V.A., Dorodnitsyn, V.A., Elenin, G.G., Kurdyumov, S.P., Samarskii, A.A., Quasilinear heat conduction equation with source: blow-up, localization, symmetry, exact solution, asymptotics (in Russian), Modern Mathematical Problems, New Achievements, vol. 28, VINITI AN SSSR, Moscow, 1986, pp. 95-206; English transl. in Journal of Soviet mathematics 41 5 (1988), 1222-1292, Plenum Corp..

[GK] Gilding, B.H., Kersner, R., Travelling waves in nonlinear diffusion-advection-reaction, preprint 1993 Twente University, pp. 1-105.

[K] Kalashnikov, A.S., The nature of propogation of perturbation in processes that can be described by quasilinear degenerate parabolic equation, Trudy Semin, Petrovsky 1 (1975), 135-144.

[KPP] Kolmogorov, A., Petrovsky, I., Piscunov, N., Etude de l'equation de la diffusion avec croissance de la quantite de matiere et son application a un probleme biologique, Bulletin Univ. Moscou, Ser. Internationale, Math., Mec. 1 (1937), 1-25.

[M] Murray, J.D., Mathematical Biology, Springer Verlag, Berlin, Heidelberg, N.Y., 1989.

[N1] Newman, W.I., Some exact solution to a non-linear diffusion problem in population genetics and combustion, J. Theoret. Biol. 85 (1980) , 325-334.

[N2] Newman, W.I., The long time behavior of the solution to a nonlinear diffusion problem in population genetics and combustion, J. Theoret. Biol. 104 (1983), 473-484.

[SGKM] Samarskii, A.A., Galaktionov, V.A., Kurdyumov, S.P., Mikhailov, A.P. , Blow up in problems for quasilinear parabolic equations (in Russian), Nauka, Moscow (1987) ; English Translation: Walter de Gruyter, Berlin (1994), to appear.

AMS/IP Studies in Advanced Mathematics
Volume **3**, 1997

Delta Waves in Solutions of Conservation Laws

Tung Chang (Tong Zhang)[†,*] and Shuli Yang[*]

Abstract

It is well known that the elementary waves of nonlinear hyperbolic conservation laws involve shock waves (abbr. S), rarefaction waves (abbr. R) and contact discontinuities (abbr. J). A new kind of elementary waves, delta waves, will be discussed in the present paper.

Keywords: delta wave, delta-shock, conservation law, 2-D Burger's equation Euler equation.

1. Let us consider the Riemann problem of following system

$$\begin{cases} u_t + (u^2)_x = 0 & (1.1) \\ v_t + (uv)_x = 0 & (1.2) \end{cases}$$

$$(u,v)(x,0) = \begin{cases} (u^+, v^+) & x > 0 \\ (u^-, v^-) & x < 0 \end{cases} \tag{1.3}$$

It is easy to know that the two characteristics of (1.1-2) are

$$\Gamma_1 : \quad \frac{dx}{dt} = u, \quad \frac{v}{u} = const. \tag{1.4}$$

and

$$\Gamma_2 : \quad \frac{dx}{dt} = 2u, \quad u = const. \tag{1.5}$$

† Partly supported by National Fundamental Research Program of State Commission of Science and Technology and Academia Sinica

* Partly supported by NSFC

Since the system (1.1-2) is decoupled, we can obtain the entropy solution u(x,t) from conservation law (1.1), and there are two configurations of u(x,t) as shown in figure 1.1.

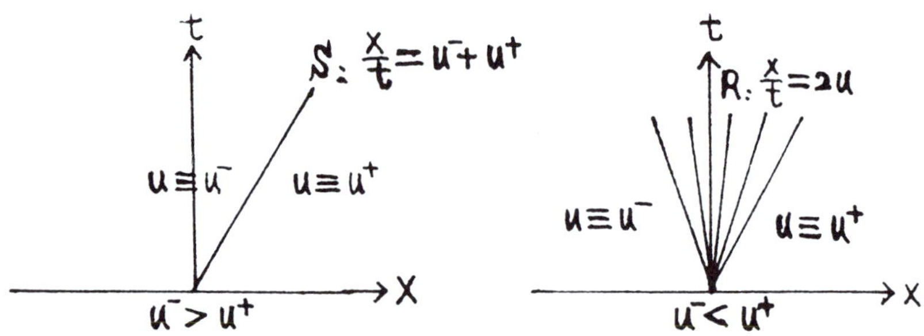

Fig 1.1

Substituting u(x,t) into (1.2), we obtain a linear equation of v(x,t). The discontinuity of v and/or u at origin (x,t)=(0,0) cause a J of v. Corresponding each configuration of u(x,t), there are three subconfigurations of v(x,t). Let us go into the details

1.1 Case $0 < u^- < u^+$ (figure 1.2.1)

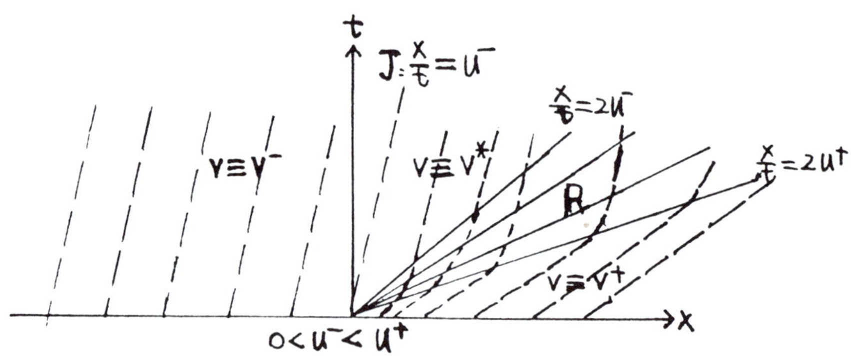

Fig. 1.2.1

According to the given u(x,t), we may obtain the Γ_1 (broken lines) field of v(x,t), the Γ_1 from $(x,t) = (x \leq 0)$ is $dx/dt = u^- < u^- + u^+$, and the Γ_1 from $(x,t)(x > 0)$ is $dx/dt = u^+ < u^- + u^+$, which will penetrates the area of R and is bent by R, after Γ_1 penetrate through out R, we have $v = v^* = v^+u^-/u^+$. So $x/t = u^-$ is a J of v.

Finally the solution is J+R.

1.2 Case $u^- < u^+ < 0$ (figure 1.2.2)

Analogously, the solution is R+J.

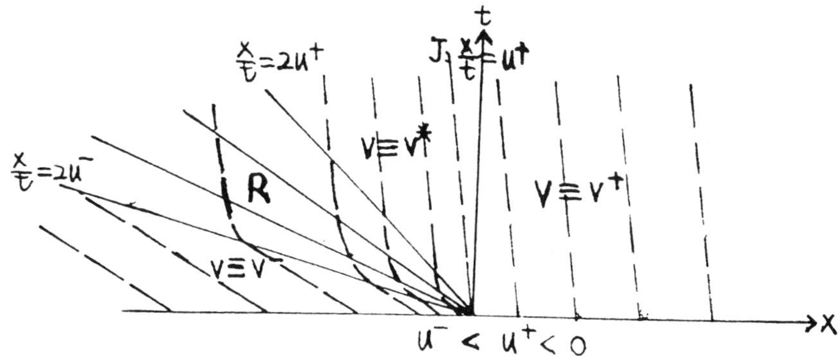

Fig. 1.2.2

1.3 Case $u^- < 0 < u^+$ (figure 1.2.3)

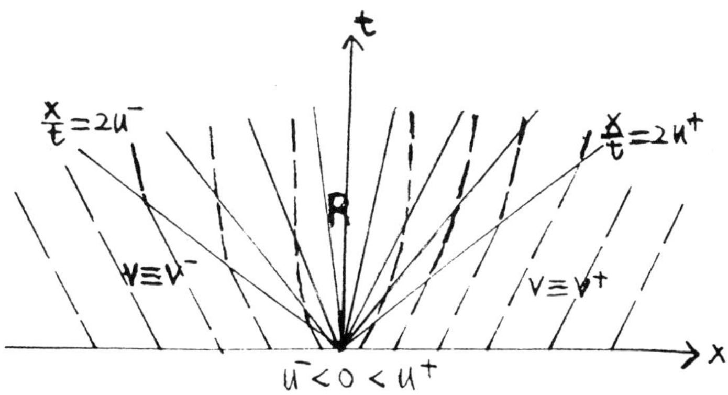

Fig. 1.2.3

Since all Γ_1 can not penetrate through out R, and they will parallel to t axis as $t \to \infty$, the solution is R.

1.4, Case $u^- > u^+ > 0$ (figure 1.2.4)

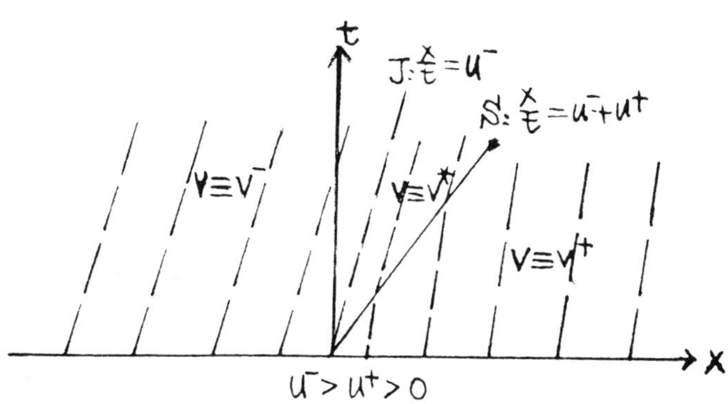

Fig. 1.2.4

The solution is J+S, it is easy to check the fact that S satisfies the Renkine Hugoniot condition of (1.1-2) iff $v_l/u_l = v_r/u_r$, where (u_l, v_l), (u_r, v_r) are the values of (u,v) on left and right banks of S respectively.

1.5 Case $0 > u^- > u^+$ (figure 1.2.5)
Analogously, the solution is S+J.

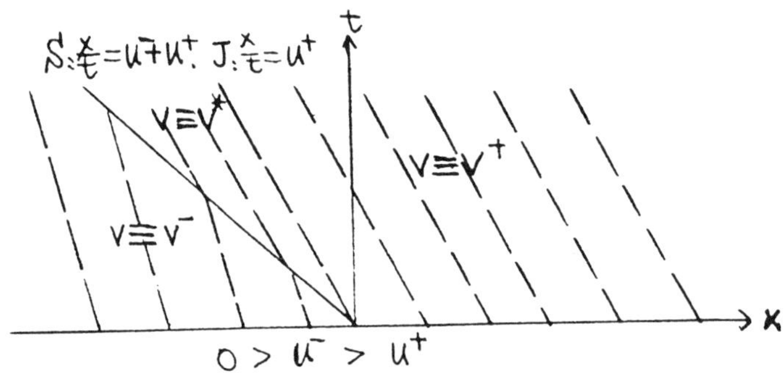

Fig. 1.2.5

1.6 Case $u^- > 0 > u^+$ (figure 1.2.6)

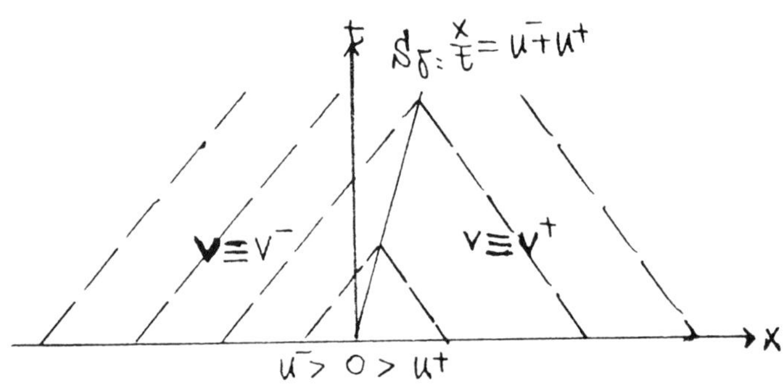

Fig. 1.2.6

Generally speaking, $v^+/u^+ \neq v^-/u^-$ which means that the discontinuity at $x/t = u^- + u^+$ does not satisfy the Rankine Hugoniot condition, i.e. the solution does not satisfy (1.1) in distribution sense unless some delta function in solution v is supported at the discontinuity line $x/t = u^- + u^+$.

2. Three papers have discussed delta waves in one dimensional case.
Korchinski studied the Riemann problem for system

$$\begin{cases} u_t + (\frac{1}{2}u^2)_x = 0 \\ v_t + (\frac{1}{2}uv)_x = 0 \end{cases} \tag{2.1}$$

in his Ph.D thesis in 1977. Basing on the numerical result and characteristic analysis, he constructed his unique solution of the Riemann problem, and called the overcompressive shock singular shock in the case $u^- \geq 0 \geq u^+$. Generalized delta functions (which are slightly more complicated than Dirac delta function) were defined and were used to prove his singular shocks satisfying (2.1) in the sense of distributions. The need for generalized delta functions is to overcome the problem of multiplying a delta function in v with a bounded discontinuous function u. It is a pity that his result has not received its due recognition, his paper has not been published so far and he even disappears in the mathematical community.

In 1991, Joseph solved the Riemann problem of system

$$\begin{cases} u_t + (u^2/2)_x = 0 \\ v_t + (uv)_x = 0 \end{cases} \tag{2.2}$$

By using the explicit formula of Riemann solutions (u^ϵ, v^ϵ) of system

$$\begin{cases} u_t + (u^2/2)_x = \frac{1}{2}\epsilon u_{xx} \\ v_t + (uv)_x = \frac{1}{2}\epsilon v_{xx} \end{cases} \tag{2.3}$$

he proved the existence of limit

$$\lim_{\epsilon \to 0}(u^\epsilon, v^\epsilon) = (u^0, v^0)$$

and defined (u^0, v^0) being Riemann solutions of (2.2). He clarified that a usual delta measure concentrates along line $x/t = (u^- + u^+)/2$ in the case of $u^- > u^+$.

3. In 1990, Tan, Y. and Z. both analytically and numerically constructed solutions to the Riemann problem of a two dimensional Burger's equations.

$$\begin{cases} u_t + (u^2)_x + (uv)_y = 0 \\ v_t + (uv)_x + (v^2)_y = 0 \end{cases} \tag{3.1}$$

$$(u, v)(x, y, 0) = (u_i, v_i) \quad \textit{in ith quadrant } (i = 1, 2, 3, 4) \tag{3.2}$$

We found that no classical weak solutions exist for certain values of initial data (u_i, v_i), and distributions of the form of Dirac delta-shock waves were found necessary and were used as parts in our solutions. Then, to investigate the validity of delta-shock waves, Tan, Zheng and Z. considered the Riemann problem of the reduced one dimensional model

$$\begin{cases} u_t + (u^2)_x = 0 \\ v_t + (uv)_x = 0 \end{cases} \tag{3.2}$$

independently from Korchinski.

By using of characteristic analysis in phase space (u,v), we obtained the unique solutions with six configurations (figure 1.2.1-6). In the case of $0 \in [u^+, u^-]$, the solution is just a delta-shock wave. A delta-shock wave (abbr. S_δ) is a generalization of an ordering shock wave. Speaking informally, it consists of a discontinuity line x=x(t) of (u,v), a Dirac delta function of v concentrates along the discontinuity line x=x(t). Delta-shock waves are more compressive than ordinary shock waves in the sense that more characteristics enter delta-shock waves.

Precisely, the S_δ of (3.2) is defined as follows:

$$(u,v) = \begin{cases} (u_l, v_l) & x < \sigma t \\ (\sigma, \frac{\rho(t)}{\sqrt{1+\sigma^2}}\delta) & x = \sigma t \\ (u_r, v_r) & x > \sigma t \end{cases} \tag{3.3}$$

where $\sigma = u_l + u_r$, $\rho(t) = (u_l v_r - u_r v_l)t/\sqrt{1+\sigma^2}$ and δ is a Dirac delta function supported at $x = \sigma t (a < t < b)$, δ is defined by:

$$< \delta, \psi(x,t) > = \int_a^b \psi(\sigma t, t)\sqrt{1+\sigma^2}dt \tag{3.4}$$

for all $\psi \in C_c(I\!R^2_+)$.

With this understanding, [TZZ] proved that delta-shock waves satisfy (3.2) in the sense of distributions.

Furthermore, [TZZ] proved that S_δ is stable in the w^* topology of $C_c^\infty(I\!R)$ for viscosity perturbation as follows

$$\begin{cases} u_t + (u^2)_x = \epsilon u_{xx} \\ v_t + (uv)_x = 0 \end{cases} \tag{3.5}$$

4. The system (1.1-2) is nonstrictly hyperbolic. For strictly hyperbolic system

$$\begin{cases} u_t + (u^2)_x = 0 \\ v_t + [(2u+1)u]_x = 0 \end{cases} \tag{3.6}$$

It is easy to know that delta-shock solutions appear in the case of

$$u^+ + 1 < u^- \tag{3.7}$$

More generally, the results in [TZZ] can be extended to

$$\begin{cases} u_t + F(u)_x = 0 \\ v_t + (g(u)v)_x = 0 \end{cases} \tag{3.8}$$

where F(u) is assumed to be convex and g(u) is smooth and increasing. The quantities σ and $\rho(t)$ in (3.3) should be changed to be

$$\sigma = [F]/[u]$$
$$\rho(t) = (\sigma[v] - [g(u)v])t$$

We refer the interested reader to Ouyang's paper for more details.

5. We will be back to discuss delta waves for two-dimensional system of conservation laws.

Zheng and Z. considered 2-D Riemann problems in gas dynamics as follows

$$\begin{cases} \rho_t + (\rho u)_x + (\rho v)_y = 0 \\ (\rho u)_t + (\rho u^2 + p)_x + (\rho uv)_y = 0 \\ (\rho v)_t + (\rho uv)_x + (\rho v^2 + p)_y = 0 \\ (\rho(e + \frac{u^2+v^2}{2})_t + (\rho u(h + \frac{u^2+v^2}{2}))_x + (\rho v(h + \frac{u^2+v^2}{2}))_y = 0 \end{cases} \tag{5.1}$$

$$(u, v, p, \rho)(x, y, 0) = (u_i, v_i, p_i, \rho_i) \ in \ ith \ quadrant \ (i = 1, 2, 3, 4) \tag{5.2}$$

where $e = p/(\gamma - 1)\rho$, $h = e + p/\rho$ and (u_i, v_i, p_i, ρ_i) are constant states, which are assumed to be that each pair of neighbouring ones can be connected by exactly one of planar waves of $\overset{\rightleftarrows}{S}$, $\overset{\rightleftarrows}{R}$ and $J^{\pm}$, and $J^{\pm}$ is defined by

$$J^{\pm} : \ curl(u, v)|_J = \pm\infty$$

The substantial difficulty of the problem is that how these four planar waves of $\overset{\rightleftarrows}{S}$, $\overset{\rightleftarrows}{R}$ and $J^{\pm}$ projecting from initial data interact and match together in a cone with vertex at origin in (x,y,t) space.

The classifications according to the different combinations of these four planar waves have been clarified (fundamentally by [ZZ] and eventually by Chen, C. and Y.). The Characteristic analysis in both phase space (u,v) and physical space $(\xi, \eta)(\xi = x/t, \eta = y/t)$ have been done case by case for selfsimilar solution $(u, v, p, \rho)(\xi, \eta)$. The conjecture for global structure of the solution for each case have been formed by [ZZ].

To approach the proof of the conjectures, Tan, Y. and Z. set p and ρ being constants in conservation laws of momentum and obtained (3.1). With the same framework used in gas dynamics, we solved the Riemann problem both analytically and numerically. The most interesting thing is that delta-shock waves appear in some solutions.

Let us have a look at the process of appearence of delta-shock in the case of four Js with different signs (figure 5.1).

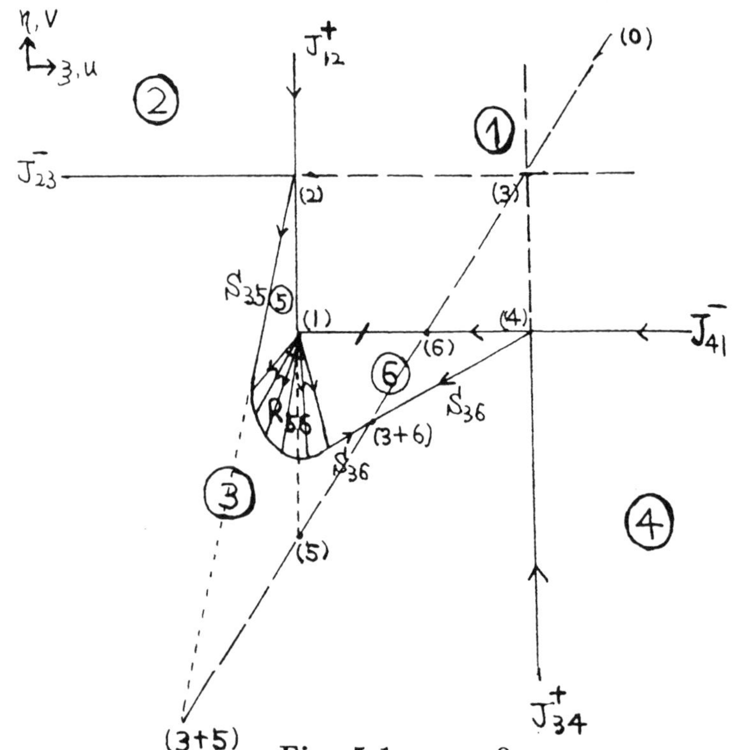

Fig. 5.1 $u_3 < 0$

In figure 5.1, the plane (u,v) coincides with plane (ξ, η), the broken lines are located in plane (u,v) and solid lines, in plane (ξ, η). (i) denotes (u_i, v_i) i=1,2,...,6. (0) denotes origin of both planes. $u_3 > 0$, $v_3 < 0$. ⓘ denotes constant state $(u, v) \equiv (u_i, v_i)$. All waves and the characteristic lines are oriented to point their singular points respectively. Four Js come from infinity ($|\xi|^2 + |\eta|^2 = \infty$, i.e. t=0), J_{12} and J_{23} meet at (2) and interact to form J_{15} and S_{35}, where (5) is the inersection point of line (1)(2) and line (0)(3). Analogously, J_{34} and J_{41} meet at (4), and interact to form J_{16} and S_{36}, where (6) is the intersection point of line (1)(4) and (0)(3).

J_{15} and J_{16} interact at (1) to form R_{56}. Then S_{35} penetrates R_{56} and changes to be S_{36}. At last two S_{36} match at their singular point (3+6).

Setting $u_3 \to 0+0$ in figure 5.1, we have $(5) \to (u_1, -\infty)$, and S_{35} coincides with J_{15}. It is easy to prove that

$$\int_{S_{35}}^{J_{15}} v_5 d\xi = v_2 - \eta = \lim_{u_3 \to 0-0} \int_{S_{35}}^{J_{15}} v_5 d\xi$$

So, a Dirac delta wave with density $v_2 - \eta$ is supported at the discontinuity line (2)(1) which we call delta-shock (abbr. S_δ denoted by the line – . – in plane (ξ, η)).

The solution for $u_3 = 0$, $v_3 < 0$ is shown in figure 5.2.

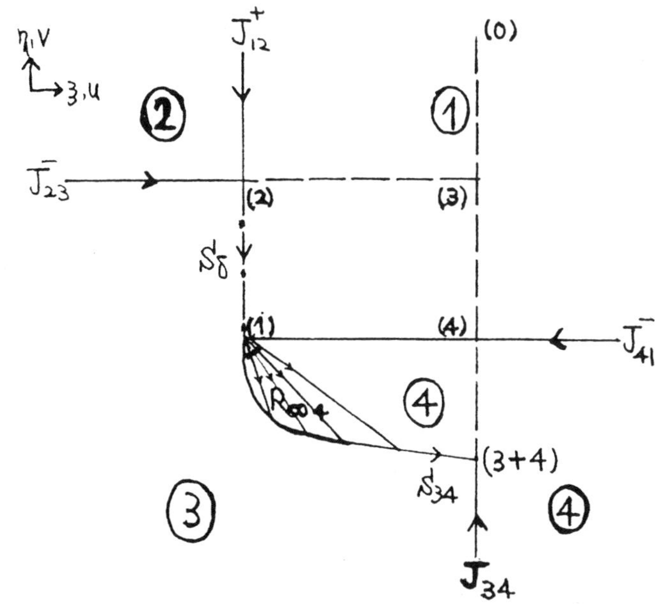

Fig. 5.2

The solution in case of $u_2 < 0 < u_3$ and $v_3 < 0$ is shown in figure 5.3.

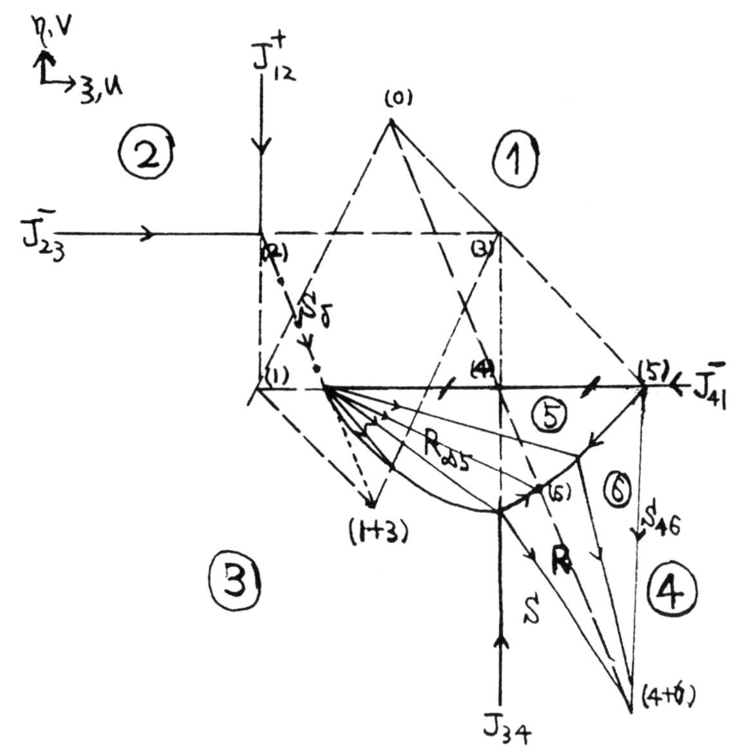

Fig. 5.3

The solution in case of $u_2 < 0 < u_3$, $v_3 > 0$ is shown in figure 5.4.

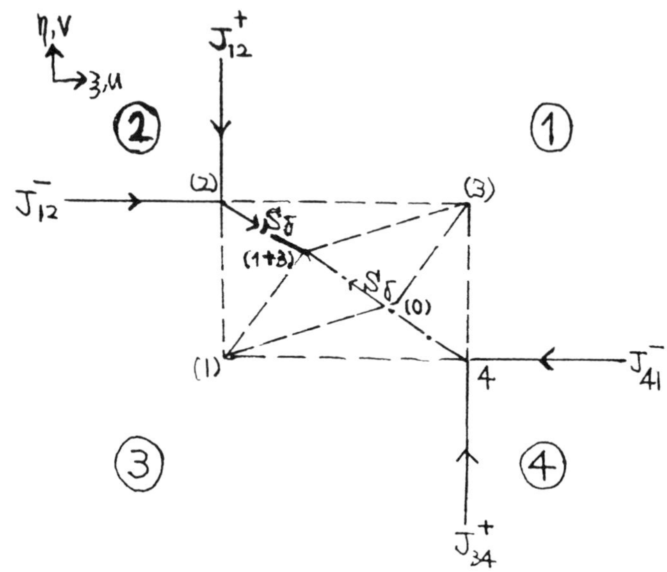

Fig. 5.4

Definition: if a discontinuity line $\eta = \eta(\xi)$ of solution $(u, v)(\xi, \eta)$ of (3.1-2) satisfying

(i)

$$\frac{d\xi}{d\eta} = \frac{\eta - (v_l + v_r)}{\xi - (u_l + u_r)} = \sigma \tag{5.3}$$

(ii) all of four characteristic lines at both banks do not go out of $\eta = \eta(\xi)$, we call it a delta-shock wave of (3.1-2).

In figure 5.4, it is easy to prove that when J_{12} and J_{23} interact at (2), there is no classical solution there, and a delta-shock wave S_δ is formed to connected ① and ③. The S_δ will end at its singular point (1+3) and match another S_δ starting from (4) there.

6. Reminding all fact described above, we are going to observe what may happen in gas dynamics.

We consider the case of four J_s with different signs with the same framework used in the last paragraph (figure 6.1).

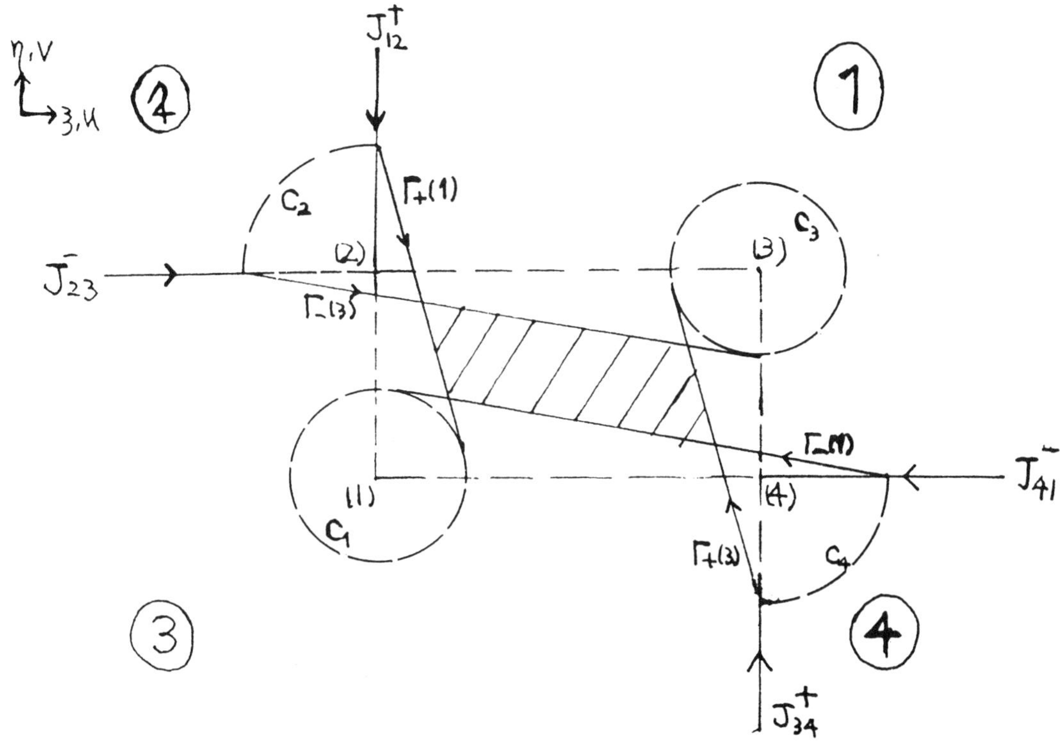

Fig. 6.1

The selfsimilar solution $(u, v, p, \rho)(\xi, \eta)$ is a pseudo steady flow, which is a transonic flow. Let us do some characteristic analysis. It is proved in [ZZ] that for constant state ①, the sonic curve is a circle C_1 with center (1), the flow is subsonic inside the circle and supersonic, outside the circle, two wave characteristics $\Gamma_\pm$ are the two half rays of the tangent lines of the sonic circle and the flow characteristic Γ_0 are the straight lines through (i). $\Gamma_\pm$ and Γ_0 are oriented to point to C_1. At infinity of plane (ξ, η), we have boundary condition ① + J^+_{12} + ② + J^-_{23} + J^-_{23} + ③ + J^+_{34} + ④ + J^-_{41}.

Extending them from infinity, J^+_{12} + ② + J^-_{23} will reach C_2, J^+_{34}+④+J^-_{41} will reach $C_4^\cdot$. ① will be bounded by J^+_{12} $+\Gamma_+(1)$ + $\Gamma_-(1)$+J^-_{41}, and ③ will be bounded by J^-_{23} $+\Gamma_-(3)$+$\Gamma_+(3)$+J^+_{34}. The determinacy domains of ① and ③ will overlap in the shaded domain in figure 6.1. [CCY] conjectured that the overlapping induces two shocks (figure 6.2).

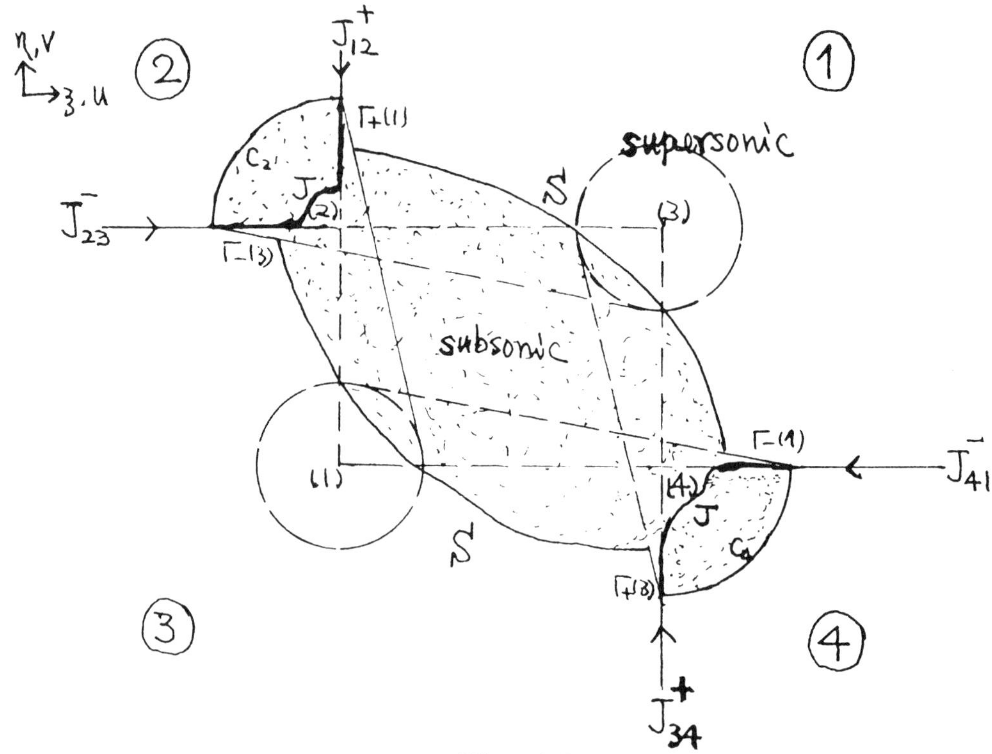

Fig. 6.2

The most interesting is that wether there exist delta waves in the solution. Figure 6.3-10 are the numerical experiment with MmB scheme, which was designed by Wu and Y. It is a nonsplitting second order accurate scheme and has high resolution nonoscillatory properties.

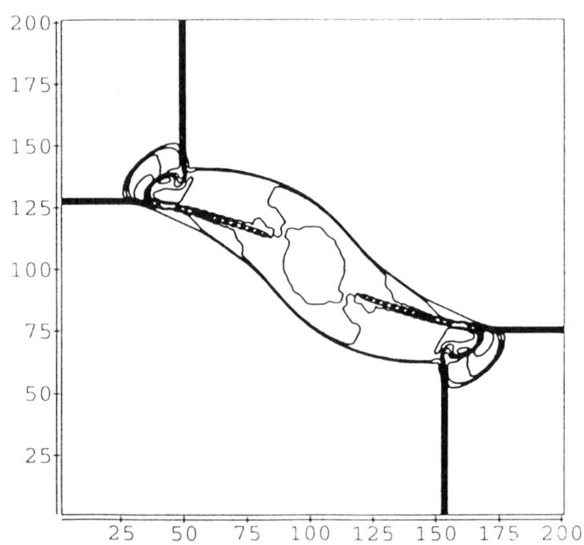

Fig. 6.3 Contour Lines for Density

The table of initial data (see figure 6.3-6):

$u_1 = -2.$	$u_2 = -2.$	$u_3 = 2.0$	$u_4 = 2.0$
$v_1 = -1.$	$v_2 = 1.0$	$v_3 = 1.0$	$v_4 = -1.$
$\rho_1 = 0.5$	$\rho_2 = 1.5$	$\rho_3 = 0.5$	$\rho_4 = 1.5$
$p_1 = 1.0$	$p_2 = 1.0$	$p_3 = 1.0$	$p_4 = 1.0$

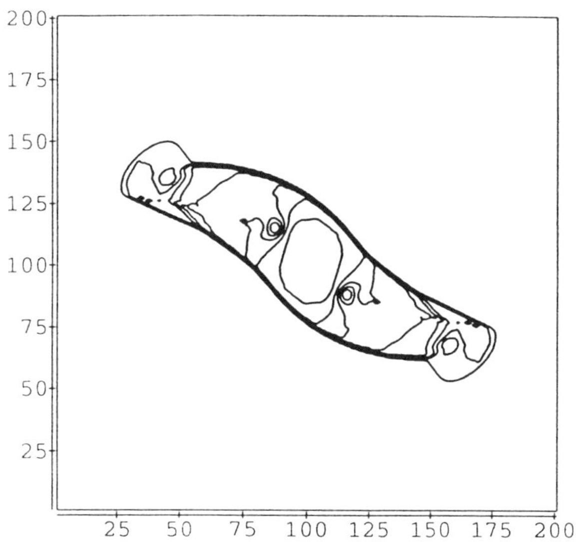

Fig. 6.4 Contour Lines for Pressure

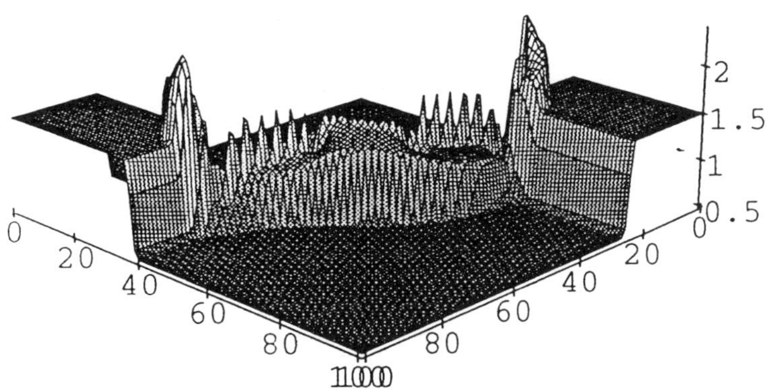

Fig. 6.5 Three Dimensional Figure for Density

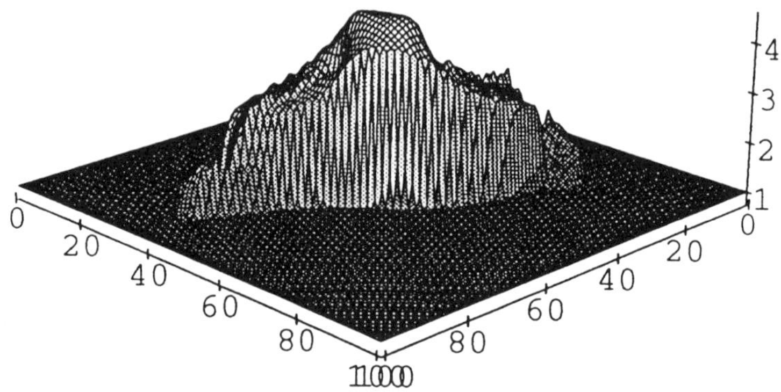

Fig. 6.6 Three Dimensional Figure for Pressure

The table of initial data (see figure 6.7-10):

$u_1 = -3.$	$u_2 = -3.$	$u_3=3.0$	$u_4=3.0$
$v_1 = -1.$	$v_2=1.0$	$v_3=1.0$	$v_4 = -1.$
$\rho_1=0.5$	$\rho_2=1.5$	$\rho_3=0.5$	$\rho_4=1.5$
$p_1=1.0$	$p_2=1.0$	$p_3=1.0$	$p_4=1.0$

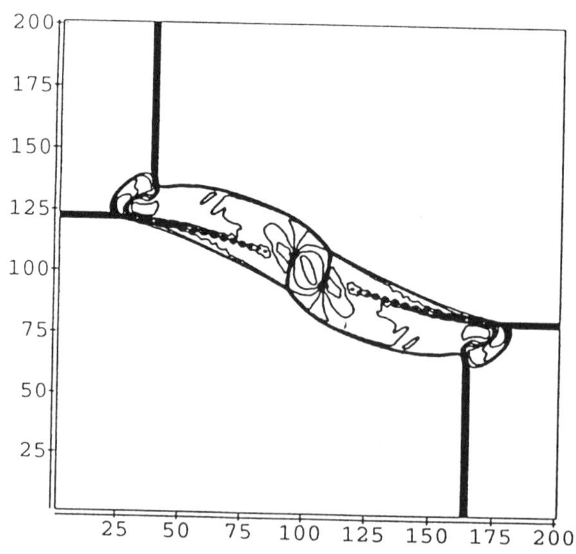

Fig. 6.7 Contour Lines for Density

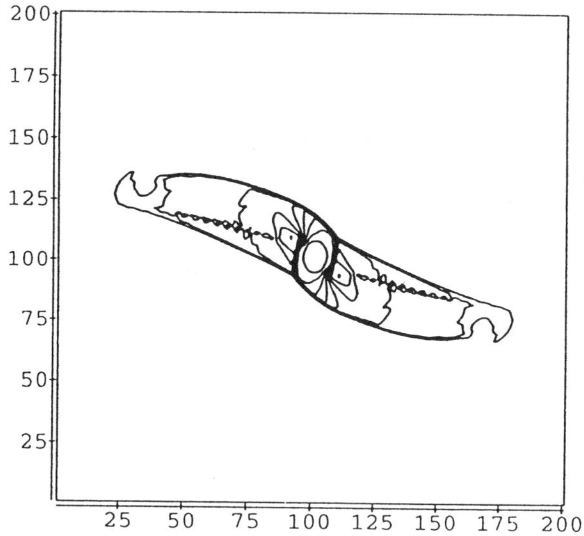

Fig. 6.8 Contour Lines for Pressure

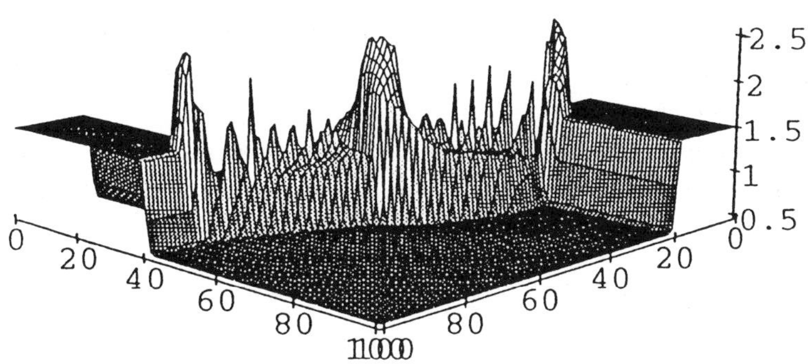

Fig. 6.9 Three Dimensional Figure for Density

In the supersonic domain and at the boundary of supersonic domain, everything is just the same as [CCY] analyzed and conjectured. In the subsonic domain between the two shocks, it is very clear that there exist two pressure and density waves which are quite different from S and J. The shapes of of these two waves in three dimensional figures like mountain chain standing on a plateau.

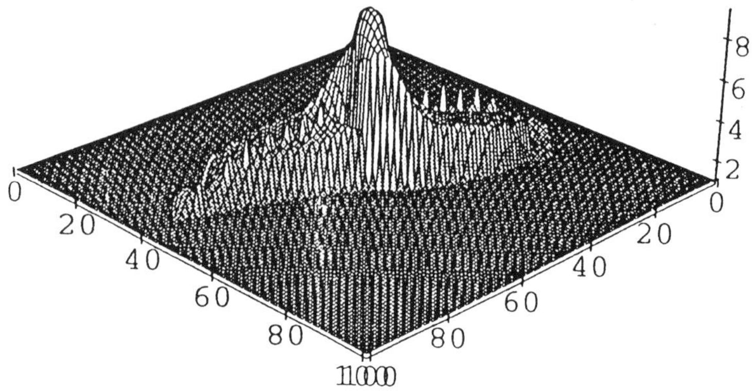

Fig. 6.10 Three Dimensional Figure for Pressure

There may be two explainations to the mountain chains in the numerical solutions. They are either the phenomena of the solutions themselves or errors of computations. According to our experiments we judge that they may be delta waves of the solutions.

There exist some natural phenomena which can not be explained by known theory so far. For examples, the air and sea disasters in Bermuda Triangle. If there exist delta waves in fluid mechanics, the disasters in Bermuda Triangle will be explained very well.

It is necessary to do more numerical experiments with our initial data by using various schemes to check if the same phenomena will appear.

References

[CCY] Tung Chang, Guiqiang Chen and Shuli Yang, On the Riemann problem for two-dimensional Euler equations, I. Interaction of shocks and rarefaction waves, II. Interaction of contact discontinuities, to appear.

[CT] Tung Chang and Dechun Tan, Two-dimensional Riemann problem for a hyperbolic system of conservation laws, Mathematica Acta Scientia, 11(1991), 4, 369-392.

[Jo] K. T. Joseph, A Riemann problem whose viscosity solutions contain $\delta-$ measures, Asymptotic Analysis, 7(1993), 105-120.

[Ko] D. J. Korchinski, Solutions of a Riemann problem for a 2×2 system of conservation laws with no classical weak solution, Thesis, Adelphi University, 1977.

[Ou] Yong Ouyang, Nonclassical solutions of Riemann problems for a hyperbolic system of conservation laws, Master thesis. Institute of Mathematics, Academia Sinica, 1992.

[TZ] Dechun Tan and Tong Zhang, Two dimensional Riemann problem for a hyperbolic system of nonlinear conservation laws, (I) Interaction of contact discontinuities, (II) Interaction of rarefaction waves and contact discontinuities, accepted for publication in J. D. E.

[TZZ] Dechun Tan, Tong Zhang and Yuxi Zheng, Delta-shock waves as limits of vanishing viscosity for hyperbolic systems of conservation laws, accepted for publication in J. D. E.

[WY] Huamo Wu and Shuli Yang, MmB – A new class of accurate high resolution schemes for conservation laws in two dimensions, IMPACT of Computing in Science and Engineering, 1(1989), 217-259.

[YZ] Shuli Yang and Tong Zhang, MmB difference solutions of 2-D Riemann problems for a 2×2 hyperbolic system of conservation laws, IMPACT of Computing in Science and Engineering, 3, 146-180(1991).

[ZZ] Tong Zhang and Yuxi Zheng, Conjecture on the structure of solutions of the Riemann problem for two-dimensional gas dynamics systems, SIAM J. Math. Anal., 21(1990), 593-630.

Tung Chang (Tong Zhang)
Institute of Mathematics
Academia Sinica
Beijing, 100080, China

Shuli Yang
Institute of Applied Mathematics
Academia Sinica
Beijing 100080, China

AMS/IP Studies in Advanced Mathematics
Volume **3**, 1997

SOLUTION TO M-D RIEMANN PROBLEMS FOR QUASILINEAR HYPERBOLIC SYSTEM OF PROPORTIONAL CONSERVATION LAWS

CHEN SHUXING

§ 1. INTRODUCTION.

In this paper we are going to discuss the general multidimensional Riemann problem for the system

$$\begin{pmatrix} u \\ v \end{pmatrix}_t + \begin{pmatrix} ug_1(u,v) \\ vg_1(u,v) \end{pmatrix}_x + \begin{pmatrix} ug_2(u,v) \\ vg_2(u,v) \end{pmatrix}_y = 0 \tag{1.1}$$

with the initial data

$$(u,v)|_{t=0} = (u_0(x,y), v_0(x,y)) \tag{1.2}$$

where $(u_0(x,y), v_0(x,y))$ has discontinuity on three curves issuing from the origin. That means, starting from the origin there are three smooth curves Γ_i, which devides the plane $\{t = 0\}$ to three angular shape domains Ω_i. And $(u_0(x,y), v_0(x,y))$ generally has discontinuity on Γ_i. The problem is essentially multidimensional in the sense that such a problem can not be regarded as a pertubation of an one-dimensional problem. Generally the solution of this problem may have a complicated flowery singularity structure issuing from the origin. The purpose of this paper is to construct the solution to such M-D Riemann problems with giving structure of singularities of the solution.

M-D Riemann problem for hyperbolic system was first put forward by Courant and Friedrichs in [1]. This problem is closely related to various problems in gas dynamics such as the interaction of shock waves, Mach reflection and refraction of shocks. However, for the Euler system the multidimensional Riemann problem is too difficult, so that there is not any essential progress in this subject (see [2]). Correspondingly, in the scalar

---

supported by NEC and NSF of China

equation case the Riemann problem has been disccussed by D.Wagner [3], W.B.Lindquist [4], T.Zhang & L.Xiao [5]. Besides, in [6,7] the self-similar solution of M-D Riemann problems for a 2X2 system are also studied when the initial data are piecewise constants.

The system (1.1) is a nonstrictly hyperbolic system, which arises in oil recovery problems, elastic theory and magnetohydrodynamics. The character of (1.1) is that as a description of conservation laws the components of flux are proportional, hence it is called system of proportional conservation laws. In one space-dimensional case Kerfitz & Kranzer [8], G-Q Chen [9] and T-P Liu [10] discussed Riemann problem and give the entropy condition for its weak solutions. We notice that the system (1.1) is symmetric hyperbolic if and only if $g_1(u,v), g_2(u,v)$ only depend on $r = (u^2+v^2)^{\frac{1}{2}}$. In this case the local existence of solution to Cauchy problem with smooth initial data is well known. The fact is a basis to treat the Cauchy problems with discontinuous initial data.

In §2 we shall give some basic facts for the system (1.1) as the preparation of our further discussions. Then we are going to solve our problem in three steps. In §3 we solve the Cauchy problem, which is a pertubation of 1-D problem. In §4 we consider the self-similar solutions of frozen problem of (1.1),(1.2), and give a detailed description of their singularities. Then in §5 we treat the generalized M-D Riemann problems (1.1),(1.2) as a pertubation of the case we consider in the last section. The conclusion is that the local singularity structure near the origin is the same as that for the frozen problem, and the existence of the solution to (1.1),(1.2) is also related to the existence for the corresponding frozen problem.

§ 2. Preparations

Let us first indicate some basic property of the system (1.1). Denote $r = (u^2+v^2)^{\frac{1}{2}}$ and $\theta = \arctan(v/u)$. It is easy to see that if g_1 and g_2 are only the functions of r, then the system is symmetric one. Moreover, we also have the following conclusion:

Lemma 1 If the system (1.1) is symmetric, then g_1 and g_2 must only depend on r.

Proof The system (1.1) can be written as

$$\frac{\partial}{\partial t}\begin{pmatrix} u \\ v \end{pmatrix} + A(u,v)\frac{\partial}{\partial x}\begin{pmatrix} u \\ v \end{pmatrix} + B(u,v)\frac{\partial}{\partial y}\begin{pmatrix} u \\ v \end{pmatrix} = 0 \tag{2.1}$$

where

$$A(u,v) = \begin{pmatrix} g_1 + u(g_1)_u & u(g_1)_v \\ v(g_1)_u & g_1 + v(g_1)_v \end{pmatrix}, B(u,v) = \begin{pmatrix} g_2 + u(g_2)_u & u(g_2)_v \\ v(g_2)_u & g_2 + v(g_2)_v \end{pmatrix}$$

Therefore, if the system is symmetric, we must have

$$u(g_i)_v = v(g_i)_u \qquad i = 1, 2 \tag{2.2}$$

Instead u and v we take r and θ as the arguments of g_i, then (2.2) implies $(g_i)_\theta = 0$. This means g_i only depends on r. q.e.d.

For the system (1.1) two eigenvalues of $\xi A + \eta B - \lambda I = 0$ are $\lambda_1 = g_1\xi + g_2\eta$, $\lambda_2 = g_1\xi + g_2\eta + r(g_1\xi + g_2\eta)'$. These two eigenvalues coinside if $r(g_1\xi + g_2\eta)' = 0$. Therefore, the system is nonstrictly hyperbolic. In the sequel we always assume $r \neq 0$. However, it is impossible to avoid the case $(g_1\xi + g_2\eta)' = 0$ completely.

When g_1 and g_2 are functions only depending on r, the system (1.1) can be reduced to a diagonal form:

$$(\frac{\partial}{\partial t} + g_1\frac{\partial}{\partial x} + g_2\frac{\partial}{\partial y})\theta = 0 \tag{2.3}$$

$$\frac{\partial r}{\partial t} + \frac{\partial(rg_1)}{\partial x} + \frac{\partial(rg_2)}{\partial y} = 0 \tag{2.4}$$

Hence the smooth solution of Cauchy problem for (1.1) can be obtained by integration along characteristics and iteration.

Let us turn to weak solutions of (1.1). Notice that under the rotation transformation of coordinate system

$$x_1 = x\cos\alpha + y\sin\alpha, y_1 = -x\sin\alpha + y\cos\alpha \tag{2.5}$$

(1.1) takes the form

$$\frac{\partial}{\partial t}\begin{pmatrix} u \\ v \end{pmatrix} + \frac{\partial}{\partial x_1}\begin{pmatrix} uG_1 \\ vG_1 \end{pmatrix} + \frac{\partial}{\partial y_1}\begin{pmatrix} uG_2 \\ vG_2 \end{pmatrix} = 0 \tag{2.6}$$

where $G_1 = g_1\cos\alpha + g_2\sin\alpha, G_2 = -g_1\sin\alpha + g_2\cos\alpha$. This means the form of system (1.1) keeps invariant under rotation.

Consider the solution having the form $u = u(h(t, x, y)), v = v(h(t, x, y))$, which is called simple wave. Substitute the form into (1.3), we have

$$(h_t + Ah_x + Bh_y)U' = 0 \tag{2.7}$$

where $U = (u, v)$, and U' represents the derivative of U as a function $U(h)$. Therefore, the direction (h_t, h_x, h_y) is characteristic, and $h = const.$ are characteristic surfaces. Notice that the simple wave should be formed by second characteristic surfaces, then $h(t,x,y)$ satisfies

$$h_t + (rg_1)_r h_x + (rg_2)_r h_y = 0 \tag{2.8}$$

Meanwhile, from (2.3) we know that the equality $v/u = const.$ always holds in the domain occupied by simple wave.

Consider the solution of (1.1) having discontinuity on the surface π with the normal $\vec{n} = (\sigma, \xi, \eta)$ at a point P. By a suitable rotation we may derive the Rankine-Hugoniot condition from one space-dimensional case. Assume that $\sigma > 0$ and the normal $\vec{n}$ points from " + " side of π to its " − " side, then

$$\sigma = \frac{u_+G_+ - u_-G_-}{u_+ - u_-} = \frac{v_+G_+ - v_-G_-}{v_+ - v_-} \tag{2.9}$$

where $G = g_1\xi + g_2\eta$, $(*)_\pm$ means the values of $(*)$ on two sides of the surface π. From (2.9) we obtain

$$(G_+ - G_-)(v_+u_- - u_+v_-) = 0.$$

So there are two possibilities: $G_+ = G_-$ corresponding to contact discontinuity, and $v_- : u_- = v_+ : u_+$ corresponding to shock. If the surface π bearing discontinuity is a shock, then under the assumption of monotonicity of G and $(rG)_r$ the entropy condition is

$$\theta_+ = \theta_-, G(r_-) > G(r_+) \tag{2.10}$$

(see [8-10]).

Lemma 2 If (u,v) is an admissible entropy solution of (1.1), G and $(rG)_r$ are monotonic increasing, where $G = g_1 \cos\alpha + g_2 \sin\alpha$, then the solution is also an admissible entropy solution of (2.3),(2.4).

Proof Obviously, in the domain where the solution is smooth, (1.1) is equivalent to (2.3),(2.4). Near any surface carring contact discontinuity r is continuous, so (1.1) and (2.3),(2.4) are also equivalent. Furthermore, near shock waves θ is continuous, and (2.6) is equivalent to $\sigma = \dfrac{r_+G_+ - r_-G_-}{r_+ - r_-}$, hence the only fact we have to verify is the equivalence of entropy condition for two systems. Indeed, by a coordinate rotation (2.5) the system (2.4) is transformed into the form

$$\frac{\partial r}{\partial t} + \frac{\partial(rG_1)}{\partial x_1} + \frac{\partial(rG_2)}{\partial y_1} = 0 \tag{2.11}$$

If the surface π is a shock, whose normal at P is $(n_t, n_x, n_y) = (\sigma, \cos\alpha, \sin\alpha)$, orientating from " + " side to " − " side, then the entropy condition for (2.11) is $(rG_1)_r(r_-) > (rG_1)_r(r_+)$. This is just the second relation of (2.10) in view of the monotonicity of functions G_1 and $(rG_1)_r$. q.e.d.

§ 3. PURTUBATION OF 1-D SOLUTION.

Next we construct the solution of Cauchy problem of (1.1) with initial data having discontinuity on a smooth curve. Assume that the initial data are

$$(u(0,x,y),v(0,x,y)) = \begin{cases} (u_1(x,y),v_1(x,y)) & \text{on the left of} \quad \Gamma \\ (u_2(x,y),v_2(x,y)) & \text{on the right of} \quad \Gamma \end{cases} \tag{3.1}$$

The data are discontinuous on a curve $\Gamma : x = g(y)$ satisfying $h(0) = 0$. If the discontinuous curve Γ of initial data on the plane (x, y) is not straight line, then the solution will not be a self-similar one. At a point P on Γ, the tangential line of Γ is denoted by $\ell : x = h'(0)y$. The Cauchy problem of (1.1) with the initial data

$$(u,v) = \begin{cases} (u_1(0,0),v_1(0,0)) & \text{on the left of } \ell \\ (u_2(0,0),v_2(0,0)) & \text{on the right of } \ell \end{cases} \tag{3.2}$$

is called frozen problem of (1.1),(3.1).

The frozen problem (1.1),(3.2) is 1-D problem indeed. By using a coordinate transformation (2.5) with $\alpha = \tan^{-1} h'(0)$, the problem is transformed into (2.6),(3.2) with ℓ being $x_1 = 0$. Hence we may seek the solution independent of y_1, and (2.6),(3.2) is reduced to

$$\frac{\partial}{\partial t}\begin{pmatrix} u \\ v \end{pmatrix} + \frac{\partial}{\partial x_1}\begin{pmatrix} uG_1 \\ vG_1 \end{pmatrix} = 0 \tag{3.3}$$

$$(u,v) = \begin{cases} (u_1(0,0),v_1(0,0)) & x_1 > 0 \\ (u_2(0,0),v_2(0,0)) & x_1 < 0 \end{cases} \tag{3.4}$$

The problem (3.3),(3.4) has been discussed in [8-10]. If G_1 satisfies $G_{1r} > 0, (rG_1)_{rr} > 0$, then the solution satisfying strong entropy criterion exists uniquely. However, due to the arbitrariness of α G_{1r} and $(rG_1)_{rr}$ may change their sign. Hence the case of nonexistence may often occurs. For instance, when $(rg)_{rr}$ changes its sign, the non-existence may be derived from the non-convex case for scalar nonlinear equation. Such non-existence has been also noticed by D.Wagner in [3].

As shown in [8-10], the solution of (3.3),(3.4) is composed of two waves issuing from $x_1 = 0$: J and R, or J and S. Therefore the solution of the above frozen problem (1.1),(3.2) is also composed of two waves issuing from ℓ.

Lemma 3 Assume that the solution of the frozen problem (1.1),(3.2) has two elementary waves issuing from ℓ, then near the point P the solution of (1.1),(3.1) has the same singularity structure.

Proof Let us consider the case "$J-S$" and "$J-R$" respectively. The system (1.1) is symmetric hyperbolic, then we can solve the Cauchy problems by using the data on two sides of Γ. These solutions are uniquely determined up to the first characteristics by (u_1, v_1) or to the second characteristics by (u_2, v_2). The solutions in these domain are denoted by (u_a, v_a) and (u_b, v_b) respectively. The contact discontinuity surface issuing from Γ is a characteristic surface of first class, it can also be determined by the initial data (u_1, v_1), and then denoted by $J : x = j(t, y)$. Moreover, J is composed of first characteristic curve. On each curve $\theta = \arctan(v_b/u_b)$ is invariant, hence the value (u_b, v_b) on the right side of J is known. By integration along characteristics for (2.3),(2.4) we may determine the solution (u_b, v_b) in between first and second characteristic surface issuing from Γ

By the initial data (u_2, v_2) the solution (u_c, v_c) can be determined up to the second characteristic surfaces issuing from Γ. Since the solution of the frozen problem has $J-S$ structure then by the entropy condition we have $r_1(0) > r_2(0)$, therefore the domain of (u_b, v_b) and (u_c, v_c) overlaps. In the overlapped wedged shape domain we determine the function $\phi(t, x, y)$ by

$$\phi_t(u_c - u_b) + \phi_x((ug_1)_c - (ug_1)_b) + \phi_y((ug_2)_c - (ug_2)_b) = 0 \tag{3.7}$$

with initial data

$$\phi(0, x, y) = x - h(y)$$

Solving $\phi(t, x, y) = 0$ we obtain $x = s(t, y)$, then global solution with $J-S$ structure issuing from Γ is

$$(u(t,x,y), v(t,x,y)) = \begin{cases} (u_a(t,x,y), v_a(t,x,y)) & \text{if } x < j(t,y) \\ (u_b(t,x,y), v_b(t,x,y)) & \text{if } j(t,y) < x < s(t,y) \\ (u_c(t,x,y), v_c(t,x,y)) & \text{if } s(t,y) < x \end{cases}$$

Next we study the $J-R$ structure issuing from a curve Γ on the plane (x, y). We still assume that the equation $x = h(y)$ of Γ satisfies $h(0) = 0$, and the initial data on the both sides of Γ has the same form as (3.2), but at any point near the origin the frozen initial data determines a $J-R$ structure. In this case we first determine the contact J and the solution $(u_b(t, x, y), v_b(t, x, y))$ in the domain right to J by using data (u_1, v_1), and determine the solution $(u_c(t, x, y), v_c(t, x, y))$ up to the second characteristics by using data (u_2, v_2) as we just did for $J-S$ structure, then we determine the centre wave issuing from Γ.

Right to the contact J we have $v_b/u_b = v_2/u_2$. Define functions

$$\begin{aligned}\tilde{u}(\theta, y) &= \theta u_b(0, g(y), y) + (1-\theta)u_2(g(y), y), \tilde{v}(\theta, y)\\ &= \theta v_b(0, g(y), y) + (1-\theta)v_2(g(y), y)\end{aligned}$$

For any $\theta \in [0,1]$ and any y near to 0, the second characteristic curve

$$\frac{dt}{ds} = 1, \frac{dx}{ds} = (rg_1)_r(\tilde{u}(\theta, y), \tilde{v}(\theta, y)), \frac{dy}{ds} = (rg_2)_r(\tilde{u}(\theta, y), \tilde{v}(\theta, y))$$

starting from $(0, g(y), y)$ can be determined. They form a wedged shape domain Ω_R bounded by surfaces $x = \pi_1(t, y)$ and $x = \pi_2(t, y)$, and the solution (u_r, v_r) can also be determined according to the form (2.3),(2.4). Therefore, the solution with $J - R$ structure issuing from Γ is

$$(u(t,x,y), v(t,x,y)) = \begin{cases} (u_a(t,x,y), v_a(t,x,y)) & \text{if } x < j(t,y) \\ (u_b(t,x,y), v_b(t,x,y)) & \text{if } j(t,y) < x < \pi_1(t,y) \\ (u_r(t,x,y), v_r(t,x,y)) & \text{if } \pi_1(t,y) < x < \pi_0(t,y) \\ (u_c(t,x,y), v_c(t,x,y)) & \text{if } \pi_0(t,y) < x \end{cases}$$

Remark We emphasize here that all above calculation is continuous with respect to initial data, because the solution of an ordinary equation depends on its initial data continuously, and each step of algebraic calculation is also continuous.

§ 4. Self-similar solution.

In order to treat the essentially M-D problem we also need to study the self-similar solution of the corresponding frozen problem first. The initial data for the frozen problem is

$$(u(0,x,y), v(0,x,y)) = (u_i, v_i), \quad \text{if} \quad \theta_{i-1} < \theta < \theta_i, \quad i = 1,2,3 \tag{4.1}$$

where θ_3 equals θ_0. To study self-similar solutions we introduce variables $\xi = x/t, \eta = y/t$. Correspondingly, the system (1.1) becomes

$$\begin{aligned} -\xi\frac{\partial u}{\partial \xi} - \eta\frac{\partial u}{\partial \eta} + \frac{\partial}{\partial \xi}(ug_1) + \frac{\partial}{\partial \eta}(ug_2) = 0 \\ -\xi\frac{\partial v}{\partial \xi} - \eta\frac{\partial v}{\partial \eta} + \frac{\partial}{\partial \xi}(vg_1) + \frac{\partial}{\partial \eta}(vg_2) = 0 \end{aligned} \tag{4.2}$$

The system has two eigenvalues:

$$\lambda_1 = \frac{\eta - g_2}{\xi - g_1}, \quad \lambda_2 = \frac{\eta - (rg_2)_r}{\xi - (rg_1)_r}$$

Let us denote the curve $\xi = g_1(r), \eta = g_2(r)$ by B_1, and the curve $\xi = (rg_1)_r, \eta = (rg_2)_r$ by B_2. At the point (ξ, η) the two characteristic directions orientate to $(g_1(r_0), g_2(r_0))$ or $((rg_1)_r(r_0), (rg_2)_r(r_0))$, where $r_0 = (u(\xi,\eta)^2 + v(\xi,\eta)^2))^{\frac{1}{2}}$. Besides, the slope of shock connecting the states "-" and "+" satisfies $\sigma = \dfrac{\eta - (r_+ g_{2+} - r_- g_{2-})/(r_+ - r_-)}{\xi - (r_+ g_{1+} - r_- g_{1-})/(r_+ - r_-)}$. Assume that the tangent of the shock intersect B_2 transversally only once, then by a coordinate transformation of (ξ, η), we may let η-axis coincides the direction of the tangent of the shock at (ξ, η). Then $g_1(r_-) > g_1(r_+)$ means $(rg_1)_r(r_-) > (rg_1)_r(r_+)$, which implies two second characteristics on both sides point to the shock. Therefore, at this point on the shock the entropy condition is equivalent to "three characteristic are incoming and one is outgoing".

To consider the wave graph of the self-similar solution we notice that the solution near infinity is easily determined in view of the property of finite propagation speed. On the other hand, near the origin many iteractions arise, and locally in (ξ, η) plane there are five kinds of interaction of waves, i.e. $R\otimes S$,$R\otimes J$,$S\otimes S$,$S\otimes J$ and $J\otimes J$, while the case $R\otimes R$ is trivial, because in this case two simple waves will be combined as one wave. Next let us treat the five interaction cases respectively. Without loss of generality we may assume all coming waves are straight ones.

(i) $R \otimes S$. Suppose that on (ξ, η) plane the shock S meets a simple wave R at the point Q. The solution in between R and S is denoted by (u_+, v_+), it is (u_-, v_-) on the outside of S, while it is (u_R, v_R) in the domain of simple wave. Then after interaction S will cut the simple wave R and change direction, while R stops propagating. The equation of the shock after the interaction is

$$\frac{d\eta}{d\xi} = \frac{\eta - (r_+ g_{2+} - r_- g_{2-})/(r_+ - r_-)}{\xi - (r_+ g_{1+} - r_- g_{1-})/(r_+ - r_-)}$$

with the initial condition $\xi|_{\eta=\eta_Q} = \xi_Q$.

(ii) $R \otimes J$. After interaction the contact discontinuous line becomes a curved contact penetrating into the simple wave, while the simple wave runs across the contact and continue propagating. In the process of interaction the all second characteristics forming the simple wave keep their direction, and the equation of the curved contact is

$$\frac{d\eta}{d\xi} = \frac{\eta - g_2(r_R)}{\xi - g_1(r_R)}$$

(iii) $S \otimes S$. If shock S_1 intersects shock S_2 at Q, the solution between S_1 and S_2 is (u_2, v_2), the solution on the outside of S_1 is (u_1, v_1), and it is (u_3, v_3) on the outside of S_2. Since S_1 is the line connecting (ξ, η) and $(((rg_1)(r_2) - (rg_1)(r_1))/(r_2 - r_1), ((rg_2)(r_2) - (rg_2)(r_1))/(r_2 - r_1))$, S_2 is the line connecting (ξ, η) and $(((rg_1)(r_3) - (rg_1)(r_2))/(r_3 - r_2), ((rg_2)(r_3) - (rg_2)(r_2))/(r_3 - r_2))$.

The produced shock S_3 is the line connecting (ξ, η) and $(((rg_1)(r_3) - (rg_1)(r_1))/(r_3 - r_1), ((rg_2)(r_3) - (rg_2)(r_1))/(r_3 - r_1))$, which is in between S_1 and S_2.

(iv) $S \otimes J$. In this case S and J run across each other. If we use notations as above, then S keeps its direction and J change its direction from "pointing to $(g_1(r_2), g_2(r_2))$" to "pointing to $(g_1(r_3), g_2(r_3))$".

(v) $J \otimes J$. Two contacts are denoted by J_1 and J_2, and other notations are as above Because $r = (u^2 + v^2)^{\frac{1}{2}}$ is invariant and both J_1 and J_2 orientate to $(g_1(r_2), g_2(r_2))$, hence J_1, J_2 must meet at $(g_1(r_2), g_2(r_2))$.

Let us turn to the global wave graph – the singularity structure of the solution in the whole plane (ξ, η). Generally, there are two waves coming from infinity in each direction θ_i, they are "J" and "S" or "J" and "R". If in each direction $\theta = \theta_i$ there are two waves coming from infinity, then we call the problem has a complete singularity structure. Therefore, from global viewpoint we essentially have four kinds of complete construction that is $JS - JR - JR, JS - JS - JR, JR - JR - JR, JS - JS - JS$.

Let us take the case $JS - JR - JR$ to show the method to construct the wave graph under the assumption

$$g_i(r) \text{ and } (rg_i)_r(r) \text{ are positive and monotone for } i = 1, 2 \qquad (4.3)$$

The Cartesian plane is regarded both (ξ, η) plane and (u, v) plane. Draw the curve $B_1 : \xi = g_1(r), \eta = g_2(r)$ and the curve $B_2 : \xi = (rg_1)_r(r), \eta = (rg_2)_r(r)$ for $r > 0$. Denote $(u_i^2 + v_i^2)^{\frac{1}{2}}$ by r_i, (u_i, v_i) by P_i, $(g_1(r), g_2(r))$ by $T(r)$, $((rg_1)_r(r), (rg_2)_r(r))$ by $Q(r)$. Due to the assumption (4.3) we know $T(r)$ and $Q(r)$ are one-to-one map from $(0, \infty)$ to B_1 and B_2, and B_1, B_2 are monotone curve in the first quadrant. Besides, we also assume :

When r is in $(\min r_i, \max r_i)$, the rays issuing from $Q(r)$ with direction θ_i intersect B_2 transversally only once. (4.4)

Therefore, without lose of generality we may assume B_1, B_2 are in between $\theta = \theta_1$ and $\theta = \theta_2$ when $r \to \infty$. Now if there are $J_{12'}$ and $S_{2'2}$ coming in θ_1 direction, $J_{2"3}$ and $R_{2"2}$ coming in θ_2 direction, $J_{1'3}$ and $R_{1'1}$ coming in θ_3 direction from infinity. According to the hypothesis $R_{2"2}$ is above B_2, while $S_{2'2}, R_{1'1}$ are under B_2. The shock $S_{2'2}$ orientating to the point $Q = (\frac{(rg_1)(r_2) - (rg_1)(r_1)}{r_2 - r_1}, \frac{(rg_2)(r_2) - (rg_2)(r_1)}{r_2 - r_1},)$ meets $R_{1'1}$ at some point, then as the discussion in the case $S \otimes R$, $S_{2'2}$ cuts $R_{1'1}$ and becomes a curved shock $\tilde{S}$ because of interaction. Since the state U in the centre wave $R_{1'1}$ is in between $P_{1'}$ and P_1, then the tangent of $\tilde{S}$, orienating the point on B_2 in between Q_1 and Q_2. The shock $\tilde{S}$ finally stops at Q_2, while $R_{1'1}$ runs up to B_2 except the part cut by $\tilde{S}$. On the other hand, $R_{2'2}$ stops at the interval Q_2Q_3 on B_1.

Afterwards, we complete the graph of all contact discontinuity. $J_{12'}$ may change its direction in the interaction $J \otimes R$ and becomes $J_{1'2"}$ after interaction , then all contact discontinuities $J_{1'2"}, J_{2"3}, J_{1'3}$ propagate up to B_1 and reach the point T_3. The whole process of construction is shown in Fig. 1. Furthermore, in figures 2–4 we have also given the global wave graphs in other three cases.

The self-similar solution of (4.2),(4.1) is stable in measure under pertubation of constant initial data. That means, for any fixed number ϵ and R, if we slightly change the values of u_i, v_i and θ_i (it is called as constant pertubation in the sequel), then the measure of the set $(t, x, y) \in D_R : |U - \tilde{U}| > \epsilon$ is small, where D_R is the set $(t^2 + x^2 + y^2)^{\frac{1}{2}}$, U and $\tilde{U}$ are the solutions of (4.2) satisfying the initial data (4.1) and purtubated one.

The proposition can be easily proved if there are two elementary waves coming from infinity in every direction $\theta = \theta_i$ for $i = 1, 2, 3$. In fact, since the steps to construct wave graphs are composed of algebraic calculation and solving initial value problems of ordinary equations, it is obvious that the place of all waves continuously depends on initial data u_i, v_i and θ_i. For any fixed number R and ϵ, we may find a number δ, so that under the pertubation with $|u_i - \tilde{u}_i| + |v_i - \tilde{v}_i| + |\theta_i - \tilde{\theta}_i| < \delta$ all possible shocks and contact discontinuities move a distance less than a given number μ, which approaches zero as $\delta \to 0$. That means there is a set E with measure less than $C(R)\mu$, such that the solution U outside E varies continuously. Therefore, if δ is sufficietly small, we have

$$\underset{(t,x,y) \in D_R}{mes} \{|U - \tilde{U}| > \epsilon\} < C(R)\mu.$$

This is the required stability.

If there is only one elementary wave coming from infinity in some direction $\theta = \theta_i$, new waves may increase under pertubation. However, the strength of these new waves are weak, and the solution is only slightly pertubated in measure. The details are left to readers.

§ 5. Pertubation of self-similar solution.

Now let us consider the solution of (1.1),(1.2). On the initial plane the discontiuous curves of the initial data form a distorted letter "Y".

We are going to seek the solution, local with respect to time t. By the result in the above section we only need to look for the solution near the origin, say in the domain $\omega_\delta : t + |x| + |y| < \delta$ with small δ. By scaling we may assume that three legs Γ_1, Γ_2 and Γ_3 of the letter Y are sufficiently near to straight lines.

In fact, if $(u(t,x,y), v(t,x,y))$ is a solution of (2.1), then for any α, the pair of functions $(u(\alpha t, \alpha x, \alpha y), v(\alpha t, \alpha x, \alpha y))$ also satisfies (2.1). But the transformation $T : t = \delta t_1, x = \delta x_1, y = \delta y_1$ transforms the neighbourhood in space (t,x,y) ω_δ into ω_1 in space (t_1, x_1, y_1). Since in the domain $\{|x| + |y| < \delta\}$ the distance between Γ_i and ℓ_i is $O(\delta^2)$, then in the domain $\{|x_1| + |y_1| < 1\}$ the distance between $T\Gamma_i$ and $T\ell_i$ is $O(\delta)$. Therefore, the three legs of "Y" in the bounded domain will be almost straight, if we take δ be sufficiently small.

In the sequel we denote t_1, x_1, y_1 by t, x, y again, and assume Γ_i is sufficiently near to its tangential line ℓ_i. Besides, without loss of generality we assume the equation of ℓ_i is $\theta (= \theta_i) = \frac{2}{3} i\pi - \frac{\pi}{2}$. The original problem (1.1),(1.2) is a pertubation of the correponding frozen problem, whose solution is self-similar one. For definiteness we still take the case $JS - JR - JR$ as an example to show how to construct non-self-similar solutions. The singularity structure of self-similar solution in $JS - JR - JR$ case has been shown in the figure 1 . In the space (t, x, y) the singularity structure of the solution is the projection of this structure with respect to the origin. For our convenience we use the same letter to denote the curves bearing singularities in the space (ξ, η) and the surfaces as the projection of these curves in the space (t, x, y), while we add tilde to every letter to denote corresponding pertubated wave.

Near the origin $\Gamma_i : x = h_i(y)$ is a pertubation of straight line, and $u_i(x,y) \sim u_i(0,0)$, $v_i(x,y) \sim v_i(0,0)$. Denote the initial data by

$$(u, v)|_{t=0} = (\phi_i(x,y), \psi_i(x,y)) \quad in \quad \Omega_i \qquad \text{for } i = 1,2,3$$

On Γ_3 : we define

$$\phi_b(x,y) = r_3(x,y) u_1(x,y) / r_1(x,y), \psi_b(x,y) = r_3(x,y) v_1(x,y) / r_1(x,y),$$

$\Phi(\alpha,x,y)=\alpha\phi_b(x,y)+(1-\alpha)\phi_1(x,y),\ \Psi(\alpha,x,y)=\alpha\psi_b(x,y)+(1-\alpha)\psi_1(x,y)$

where $r_i(x,y)=(\phi_i(x,y)^2+\psi_i(x,y)^2)^{\frac{1}{2}}$. The second characteristics issuing from Γ_3 is

$$x=(rg_1)_r(r(\alpha,h_3(\beta),\beta))t+h_3(\beta), y=(rg_2)_r(r(\alpha,h_3(\beta),\beta))t+\beta,$$

where $r(\alpha,h_3(\beta),\beta)=(\phi(\alpha,h_3(\beta),\beta)^2+\psi(\alpha,h_3(\beta),\beta)^2)^{\frac{1}{2}}$. When α runs from 0 to 1, β runs from -1 to 0, we obtain a family of second characteristics with two parameters. These characteristics form a center wave bounded by π_1 and π_2 issuing from Γ_3. On each characteristics $(u,v)=(\Phi(\alpha,h_3(\beta),\beta),\Psi(\alpha,h_3(\beta),\beta))$ (it is denoted by (u_R,v_R)). The centre wave extends up to the surface B_2.

By the same way we construct a centre wave bounded by π_3 and π_4 issuing from Γ_2 extending up to B_2.

In the direction θ_1 we solve (2.4) by using the initial data (ϕ_1,ψ_1) and (ϕ_2,ψ_2). The solution determined by initial data on the two sides of Γ_1 is defined in two domains respectively, but these two domains overlap each other. The overlapped domain is bounded by the second characteristic surfaces π_5,π_6 issuing from Γ_1, and both π_5 and π_6 pass the ray $Q_2: x=(rg_1)_r(r_2(0,0))t, y=(rg_2)_r(r_2(0,0))t$. Solve an initial problem of ordinary differential equation

$$\phi_t(u_2-u_b)+\phi_x((ug_1)_2-(ug_1)_b)+\phi_y((ug_2)_2-(ug_2)_b)=0 \qquad (5.1)$$

$$\phi(0,x,y)=y-x\tan\theta_1$$

where (u_b,v_b) equals (u_1,v_1) if the point (t,x,y) is on the right side of π_2, and it equals (u_R,v_R) on the left of π_2, we obtain a shock $\tilde{S}_{2'2}$, which cut a part of the centre wave $\tilde{R}_{1'1}$, and stops at the ray Q_2. Correspondingly, the solution on the both sides of $\tilde{S}_{2'2}$ has been determined.

Next we draw the contact discontinuity surfaces $\tilde{J}_{12'},\tilde{J}_{1'3},\tilde{J}_{2"3}$ as the pertubation of $J_{12'},J_{1'3},J_{2"3}$. These three contact discontinuities are formed by the first characteristics $(t(s),x(s),y(s))$ satisfying

$$\frac{dt}{ds}=1, \frac{dx}{ds}=g_1(r), \frac{dy}{ds}=g_2(r) \qquad (5.2)$$

which has a unique solution with given data. The surface $\tilde{J}_{12'}$ runs across $\tilde{R}_{1'1}$ and $\tilde{R}_{2"2}$, becomes $\tilde{J}_{1'2"}$, and then meets $\tilde{J}_{1'3},\tilde{J}_{2"3}$ at a curve satisfying (5.2) with the initial data

$$(t,x,y)|_{s=0}=(1,g_1(r_3),g_2(r_3))$$

where $r_3 = r_3(0, 0)$.

The solution of (1.1),(1.2) in all domains bounded by above-constructed waves are also determined in the mean time. In such a way we establish the existence of the solution. Similarly, the above method also works, if the waves coming from infinity determined by the initial data are $JS - JS - JR, JR - JR - JR, JS - JS - JS$. Summing up, we have

Theorem 1 Assume that (4.3),(4.4) are satisfied, and for the frozen problem there are two waves bearing singularity coming from infinity in each direction $\theta = \theta_i$, then there exists a solution of the frozen problem (1.1),(4.1). Moreover, there is a solution of the problem (1.1),(1.2) for small t, and the solution has similar local flowery singularity structure near the origin like its frozen problem.

Notice that all calculation in this section is continuous, then combining the discussion in the end of §3 and §4 we may also confirm

Theorem 2 Under the assumptions of Theorem 1, the local solution of (1.1),(1.2) is stable under any small pertubation of initial data (1.2).

References

[1] R.Courant and K.O.Friedrichs, *Supersonic Flow and Shock Waves*, Interscience Publishers Inc., New York, 1948.

[2] J.Glimm, *The Interaction of Nonlinear Hyperbolic Waves*, Comm. Pure Appl. Math. **41** (1988), 569-590.

[3] D.Wagner, *The Riemann problem in two space dimensions for a single conservation laws*, Math. Ann. **14** (1983), 534-559.

[4] B.Lindquist, *The scalar Riemann problem in two spatial dimensions: Piecewise smoothness of solutions and its breakdawn*, SIAM J. Anal. **14** (1986), 1178-1197.

[5] T.Zhang and L.Xiao, *The Riemann problem and interaction of waves in gas dynamics*, Pitman Monograph and Survey in Pure and Applied Math. 41, 1989.

[6] D.C.Tan and T.Zhang, *Two dimensional Riemann problem for a hyperbolic system of nonlinear conservation laws (I): Four J cases (preprint).*

[7] Chen Shuxing, *Structure and wave graph of solutions to M-D Riemann problems for a 2X2 quasilinear hyperbolic system*, ICTP preprint 92/347.

[8] B.L.Kerfitz and H.C.Kranzer, *A system of non-strictly hyperbolic conservarion laws arising in elasticity theory*, arch. Rat. Mech. Anal. **72** (1980), 219-241.

[9] G.Q.Chen, *Hyperbolic systems of conservation laws with a symmetry*, Comm. in PDEs **16** (1991), 1461-1487.

[10] T-P Liu, *On the viscosity criterion for hyperbolic conservation laws*, priprint.

Institute of Mathematics
Fudan University
Shanghai 200433, China

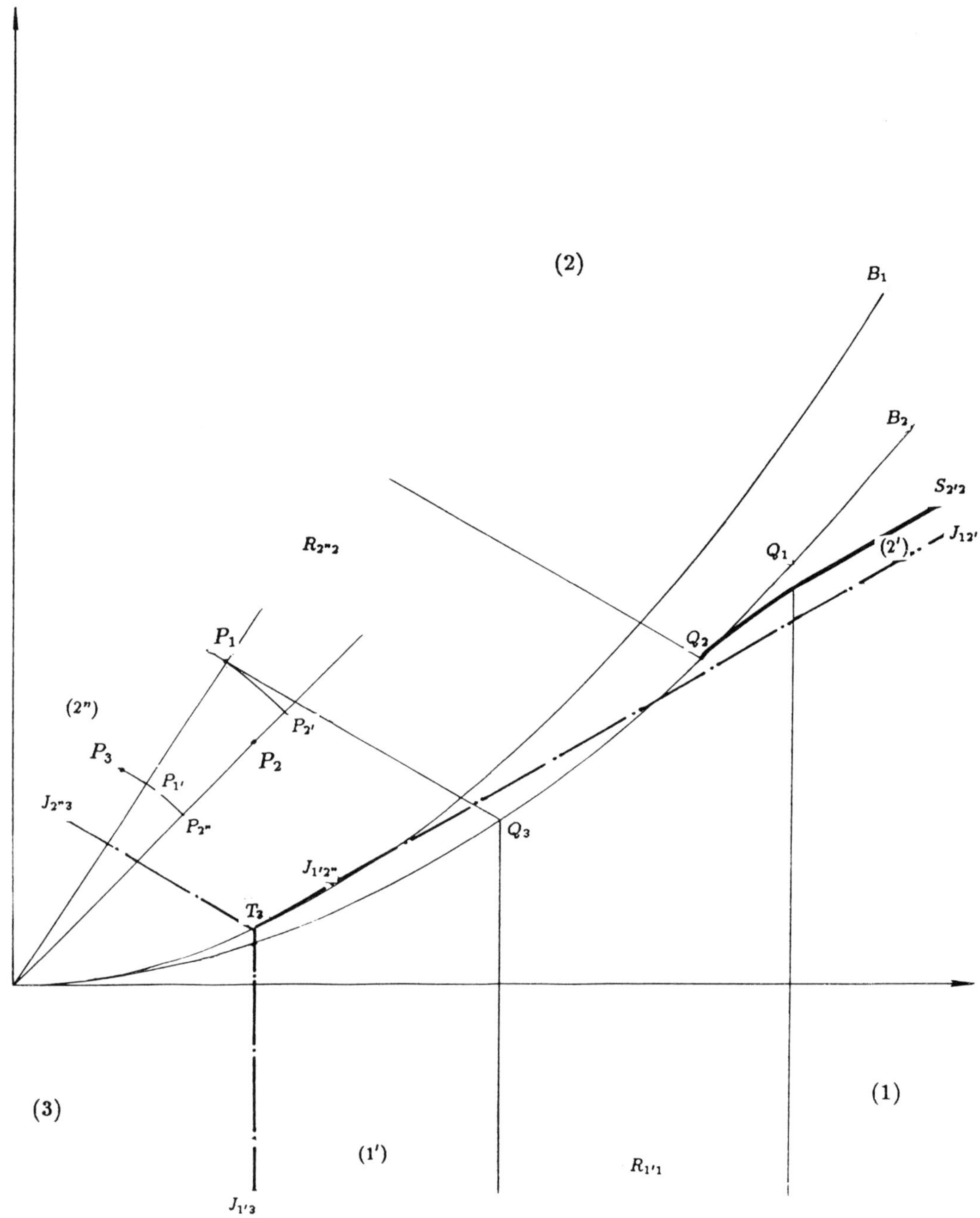
(2)
B_1
B_2
$S_{2'2}$
$J_{12'}$
$R_{2''2}$
Q_1
(2')
Q_2
P_1
(2")
$P_{2'}$
P_3
P_2
$P_{1'}$
$J_{2''3}$
$P_{2''}$
Q_3
$J_{1'2''}$
T_3
(3)
(1)
(1')
$R_{1'1}$
$J_{1'3}$

Fig. 1 $JS - JR - JR$

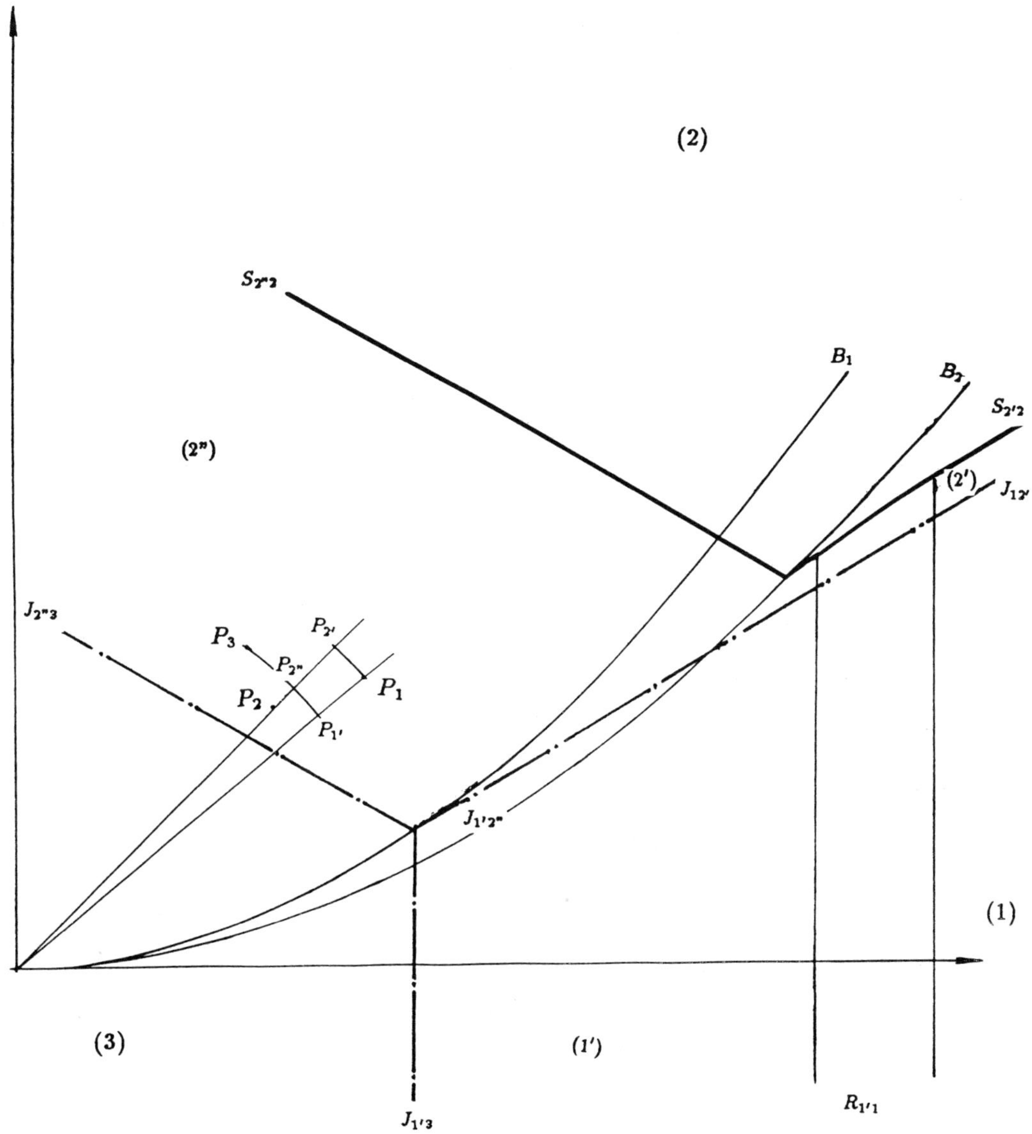

Fig. 2 $JS - JS - JR$

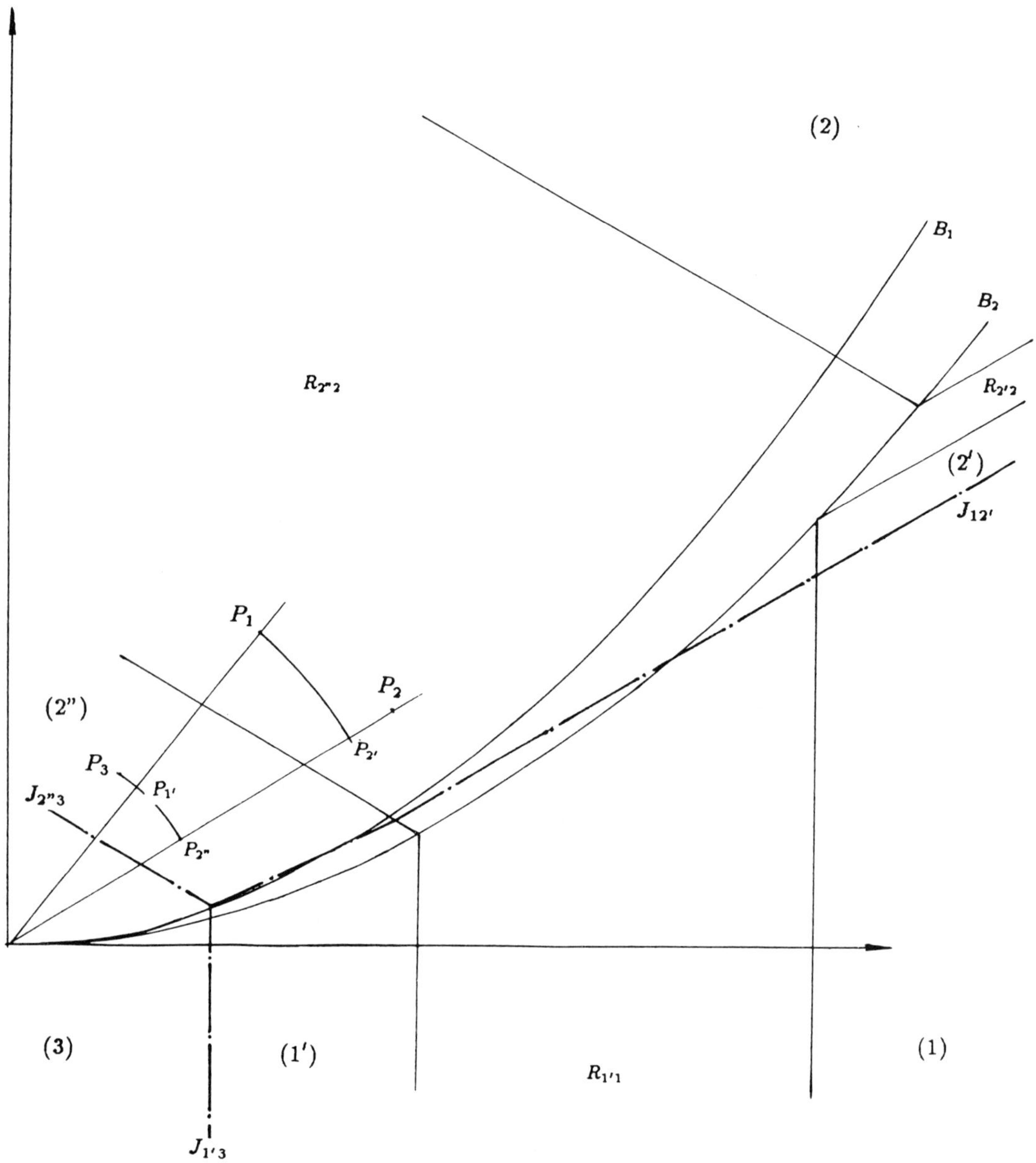

Fig. 3 $JR - JR - JR$

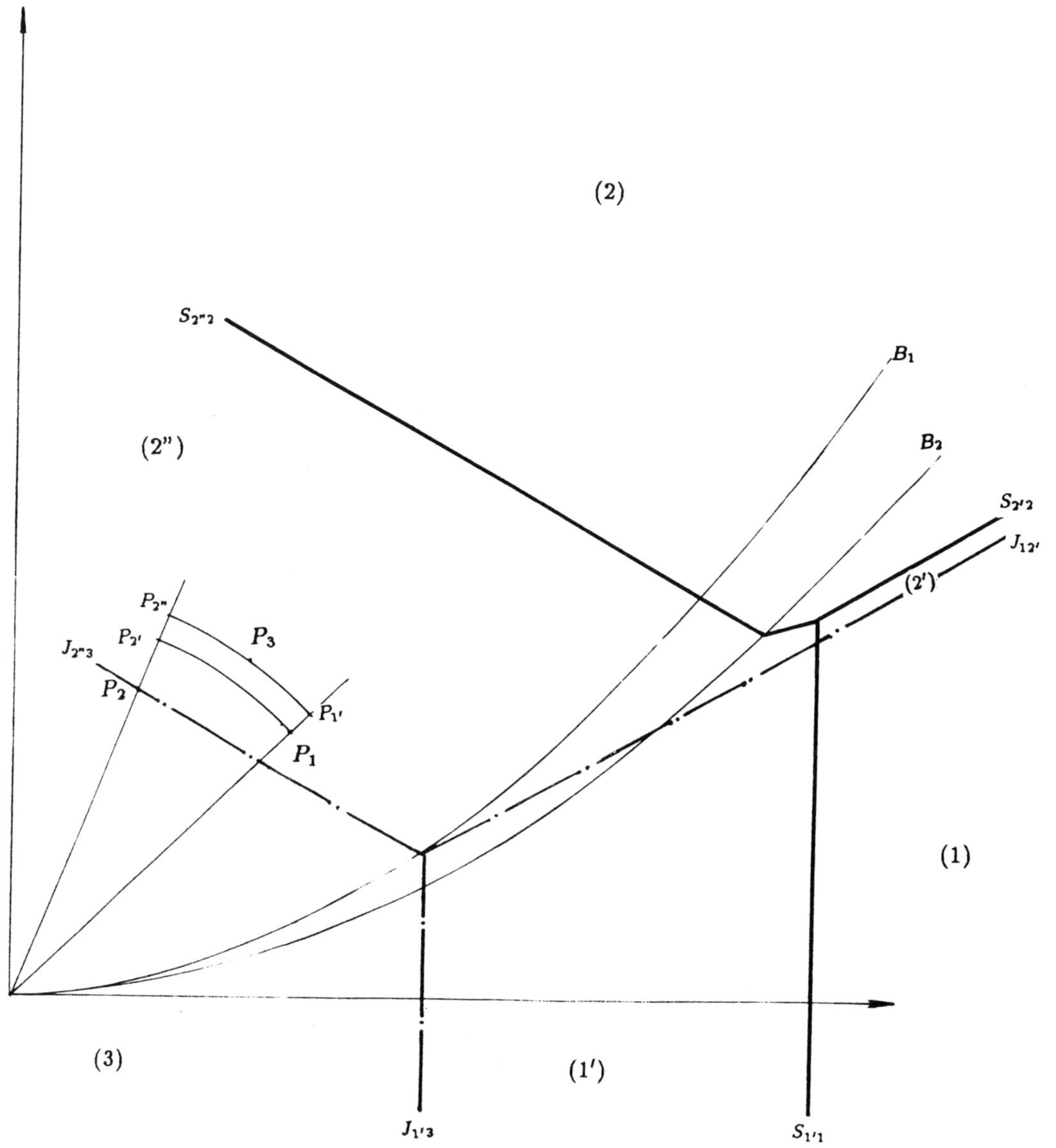

Fig. 4 $JS - JS - JS$

AMS/IP Studies in Advanced Mathematics
Volume **3**, 1997

CONTRIBUTIONS TO THE MATHEMATICAL THEORY OF MAGNETOHYDRODYNAMIC SHOCK WAVES

HEINRICH FREISTÜHLER

0. Abstract. Intermediate magnetohydrodynamic shock waves can exist and be conditionally stable in the presence of appropriate dissipative mechanisms. The zero dissipation limit is however "non-uniform." The dissipation-free case has a self-consistent theory in which intermediate shocks do not occur. Analogous properties hold for a wide class of systems of conservation laws, including those of nonlinear elasticity.

1. The equations and some basic properties

1.1. The full equations. Classical MHD is governed by the partial differential equations

$$
\begin{aligned}
\rho_t + \operatorname{div}(\rho \mathbf{V}^\top) &= 0\\
(\rho\mathbf{V})_t + \operatorname{div}(\rho\mathbf{V}\mathbf{V}^\top + \Pi) &= \mu\Delta\mathbf{V} + (\zeta-\mu)\operatorname{grad}(\operatorname{div}\mathbf{V}^\top)\\
\mathbf{B}_t + \operatorname{div}(\mathbf{B}\mathbf{V}^\top - \mathbf{V}\mathbf{B}^\top) &= \nu\Delta\mathbf{B}\\
\mathcal{E} + \operatorname{div}(\mathbf{V}^\top(\mathcal{E}I + \Pi)) &= \mu\operatorname{div}(\mathbf{V}^\top(\operatorname{grad}\mathbf{V} + (\operatorname{grad}\mathbf{V})^\top))\\
&\quad + (\zeta - 2\mu)\operatorname{div}(\mathbf{V}^\top\operatorname{div}(\mathbf{V}^\top))\\
&\quad + \nu\operatorname{div}(\mathbf{B}^\top(\operatorname{grad}\mathbf{B} - (\operatorname{grad}\mathbf{B})^\top))\\
&\quad + \kappa\Delta\vartheta\\
\operatorname{div}(\mathbf{B}^\top) &= 0.
\end{aligned}
\tag{1.1}
$$

These equations describe an ionized quasineutral fluid of density $\rho > 0$ and temperature $\vartheta > 0$ which moves under the influence of a magnetic field $\mathbf{B} \in \mathbb{R}^3$ at a velocity $\mathbf{V} \in \mathbb{R}^3$ through physical state space $\mathbb{R}^3$, in the course of time $t \in \mathbb{R}$. In (1.1),

$$\Pi = (p + \frac{1}{2}|\mathbf{B}|^2)I - \mathbf{B}\mathbf{B}^\top \in \mathbb{R}^{3\times 3}$$

denotes the total pressure tensor, and

$$\mathcal{E} = \rho(\frac{1}{2}|\mathbf{V}|^2 + e) + \frac{1}{2}|\mathbf{B}|^2$$

the total energy, with $p > 0$ the scalar fluiddynamic pressure, $e > 0$ the fluiddynamic internal energy. The two dissipation coefficients $\mu \geq 0$ and $\zeta \geq 0$ correspond to the intrinsic viscosity of the fluid; more precisely, $\mu = \zeta_1, \zeta = \zeta_2 + \frac{4}{3}\zeta_1$

with $\zeta_1, \zeta_2 \geq 0$ the so-called first and second viscosity coefficients of the fluid. The two remaining coefficients $\nu \geq 0$ and $\kappa \geq 0$ denote the electrical resistivity and the thermal conductivity of the fluid.

1.2. Equation of state. The variables ρ, p, ϑ, e which describe the internal state of the fluid, are intrinsically related with each other through the equation of state

$$e = \epsilon(\tau, \eta)$$

and the identities

$$\rho = \tau^{-1}, \quad p = -\epsilon_\tau(\tau, \eta), \quad \vartheta = \epsilon_\eta(\tau, \eta),$$

where $\tau > 0$ denotes the specific volume and $\eta \in \mathbb{R}$ the entropy of the fluid; the internal energy ϵ is required to satisfy the conditions

$$-\epsilon_\tau > 0, \quad \epsilon_\eta > 0, \quad D^2\epsilon > 0, \quad -\epsilon_{\tau\eta} > 0, \quad -\epsilon_{\tau\tau\tau} > 0. \tag{1.2}$$

The first two of these requirements amount to the positivity of pressure and density, the third to the concavity of physical entropy η (= convexity of the mathematical entropy $-\eta$), and the fourth and fifth are "Weyl's conditions." An important special case is the ideal gas

$$\epsilon(\tau, \eta) = c_v \exp(\frac{\eta}{c_v})\tau^{1-\gamma} \tag{1.3}$$

with constants $c_v > 0$ and $\gamma > 1$ denoting the specific heat at constant volume and the adiabatic exponent of the gas, respectively. For the ideal gas,

$$p = (\gamma - 1)c_v \rho \vartheta \quad \text{and} \quad e = c_v \vartheta. \tag{1.4}$$

1.3. Plane Waves. For solutions spatially depending only on one variable $x \in \mathbb{R}$ (an affine coordinate on $\mathbb{R}^3$), Equations (1.1) read

$$\begin{aligned}
\rho_t + (\rho v)_x &= 0 \\
(\rho v)_t + (\rho v^2 + p + \frac{1}{2}|b|^2)_x &= \zeta v_{xx} \\
(\rho \mathbf{w})_t + (\rho v \mathbf{w} - a\mathbf{b})_x &= \mu \mathbf{w}_{xx} \\
\mathbf{b}_t + (v\mathbf{b} - a\mathbf{w})_x &= \nu \mathbf{b}_{xx} \\
\mathcal{E} + ((\mathcal{E} + p + \frac{1}{2}(|\mathbf{b}|^2 - a^2)) - a\mathbf{b}\cdot\mathbf{w})_x &= \\
&\zeta(vv_x)_x + \mu(\mathbf{w}\cdot\mathbf{w}_x)_x + \nu(\mathbf{b}\cdot\mathbf{b}_x)_x + \kappa\vartheta_{xx}
\end{aligned} \tag{1.5}$$

and

$$a \equiv \text{const}, \tag{1.6}$$

where $v \in \mathbb{R}, \mathbf{w} \in \mathbb{R}^2$ and $a \in \mathbb{R}, \mathbf{b} \in \mathbb{R}^2$ are the longitudinal and transverse components of $\mathbf{V}$ and $\mathbf{B}$, respectively.

1.4. Wave speeds and non-strict hyperbolicity. For $\zeta = \mu = \nu = \kappa = 0$, Equations (1.5) constitute a 7×7 hyperbolic system of conservation laws. Its seven characteristic speeds $\lambda_{-3} \le \lambda_{-2} \le \lambda_{-1} \le \lambda_0 \le \lambda_1 \le \lambda_2 \le \lambda_3$ are of the form

$$\lambda_0 = v, \quad \lambda_{\pm 1} = v \pm c_-, \quad \lambda_{\pm 2} = v \pm c_a, \quad \lambda_{\pm 3} = v \pm c_+$$

with the fast and slow magnetoacoustic speeds $c_+ \ge c_- \ge 0$ and the Alfvén speed $c_a \ge 0$ given by

$$\begin{aligned} c_\pm^2 &= \frac{1}{2}[(c_s^2 + \rho^{-1}(a^2+b^2)) \pm \sqrt{(c_s^2+\rho^{-1}(a^2+b^2))^2 - 4c_s^2\rho^{-1}a^2}] \\ c_a^2 &= \rho^{-1}a^2, \end{aligned}$$

where c_s^2 is the squared sound speed,

$$c_s^2 = \pi_\rho(\rho, \eta)$$

with $\pi(\rho,\eta) = p = -\epsilon_\tau(\tau,\eta)$. From now on, we assume that

$$a \neq 0.$$

Obviously,

$$0 < c_- < c_a < c_+ \quad \text{if } \mathbf{b} \neq 0 \tag{1.7}$$

and

$$\begin{aligned} &0 < c_- = c_a < c_+ \quad \text{if } \mathbf{b} = 0 \text{ and } \Delta_a > 0 \\ &0 < c_- = c_a = c_+ \quad \text{if } \mathbf{b} = 0 \text{ and } \Delta_a = 0 \\ &0 < c_- < c_a = c_+ \quad \text{if } \mathbf{b} = 0 \text{ and } \Delta_a < 0 \end{aligned} \tag{1.8}$$

with

$$\Delta_a = \rho c_s^2 - a^2.$$

System (1.3) is thus strictly hyperbolic at every point with $\mathbf{b} \neq 0$, and non-strictly hyperbolic at every point with $\mathbf{b} = 0$. Typically, all three cases in (1.8) occur, with Δ_a vanishing along a smooth manifold which separates its complement in state space into two open sets where $\Delta_a > 0$ and $\Delta_a < 0$, respectively. E. g.,

$$\Delta_a = \gamma p - a^2$$

for an ideal gas.

1.5. Remark. For the classical knowlegde presented in this section, cf. the textbooks [31,2,28,7].

2. Ideal MHD shock waves and their viscous profiles

2.1 Jump condition and possible shocks [31,25]. Abbreviating (1.3) as

$$u_t + (f(u))_x = (D(u,\delta)u_x)_x \tag{2.1}$$

with $\delta = (\zeta, \mu, \nu, \kappa)$, the Rankine-Hugoniot conditions read

$$-su + f(u) = q \tag{2.2}$$

with parameters s (the shock speed) and q (the "relative flux"). For an open and dense set Q of values $(q,s) \in \mathbb{R}^7 \times \mathbb{R}$, Eqs. (2.2) have up to four solutions u_0, u_1, u_2, u_3 satisfying

$$\begin{aligned} s &< \lambda_{-3}(u_0) \\ \lambda_{-3}(u_1) < s &< \lambda_{-2}(u_1) \\ \lambda_{-2}(u_2) < s &< \lambda_{-1}(u_2) \\ \lambda_{-1}(u_3) < s &< \lambda_0(u_3) \end{aligned} \tag{2.3}$$

or

$$\begin{aligned} s &> \lambda_3(u_0) \\ \lambda_3(u_1) > s &> \lambda_2(u_1) \\ \lambda_2(u_2) > s &> \lambda_1(u_2) \\ \lambda_1(u_3) > s &> \lambda_0(u_3). \end{aligned} \tag{2.3$'$}$$

As cases (2.3), (2.3)$'$ differ only by a direction reversal $x \mapsto -x$, we restrict attention to (2.3) without loss of generality. The four states u_0, u_1, u_2, u_3 allow for various ideal shock waves

$$u(x,t) = \begin{cases} u^-, & x < st \\ u^+, & x > st, \end{cases} \tag{2.4}$$

namely the twelve species

$$u^- = u_i, \quad u^+ = u_j, \quad i,j \in \{0,1,2,3\}, i \neq j$$

which we briefly refer to as "$i \to j$." To each of these shocks, we assign its

$$\text{"type" } k = j - i = \#\{\text{incoming characteristics}\} - 7.$$

2.2. Physical entropy condition [31,25]. Entropy increases with the index:

$$\eta(u_0) < \eta(u_1) < \eta(u_2) < \eta(u_3);$$

in other words

$$\text{2nd law of thermodynamics} \quad \Leftrightarrow \quad i < j.$$

This restricts our further attention to species of positive type, namely shocks $0 \to 1$, $2 \to 3$ of type 1, which are classical Laxian shocks associated with the genuinely nonlinear fast and slow modes c_+, c_-, respectively, and the non-classical or "intermediate" shocks $0 \to 2$, $1 \to 3$ of type 2, $0 \to 3$ of type 3, and $1 \to 2$ of type 1.

2.3. Classical linear stability analysis [1] finds the classical shocks $0 \to 1$, $2 \to 3$ to be stable, the intermediate shocks $0 \to 2$, $1 \to 3$, $0 \to 3$, $1 \to 2$ to be unstable.

2.4. Small amplitude ideal intermediate shock waves

2.4.1. Theorem. *Consider a state u_* where $c_- < c_a = c_+$ ($c_- = c_a < c_+$) $\{c_- = c_a = c_+\}$. Then there exist shocks (2.4) of species $0 \to 1$, $0 \to 2$, $1 \to 2$ ($2 \to 3$, $1 \to 3$, $1 \to 2$) $\{0 \to 1$, $2 \to 3$, $0 \to 2$, $1 \to 3$, $0 \to 3$, $1 \to 2\}$ which assume values in an arbitrarily small neighborhood of u_*.*

Proof: The Hugoniot map

$$(u, s) \mapsto (f(u) - su, s)$$

has a singularity of cusp type (cusp type) {folded handkerchief type} [4,5] at the point (u_*, s_*) with $s_* = v_* - c_a(u_*)$. For details, see [20].

2.5. Existence of classical shock waves, and non-existence of intermediate shock waves, in the presence of dissipation

A solution of (2.1) of the form

$$u(x,t) = \phi(x - st)$$

with

$$\lim_{\xi \to \pm\infty} \phi(\xi) = u^{\pm}$$

is called a traveling wave shock with boundary states u^-, u^+, or a viscous profile for the ideal shock (2.4). The existence and shape of such solutions depends on $u^{\pm}$ (or (q, s) in (2.2)) as well as on $\delta = (\zeta, \mu, \nu, \kappa)$.

2.5.1. For all slow and fast shocks, viscous profiles exist for all values of $\delta = (\zeta, \mu, \nu, \kappa) \in (0, \infty)^4$. Cf. [25,9,10].

2.5.2. For all intermediate shocks, viscous profiles do not exist for a certain range (depending on the shock) of δ. Cf. [25,9,10].

2.6. Existence of intermediate shock waves in the presence of dissipation

2.6.1. Every intermediate shock of species $0 \to 3$ which is purely gasdynamic has a viscous profile for arbitrary $\delta \in (0, \infty)^4$. This follows directly from [26]; an early reference for this immediate observation is [2]. Note that every gasdynamic shock can be made into an $0 \to 3$ intermediate shock by superimposing a magnetic field $B = (a, b)$ with $b = 0$ and a in an appropriate range.

Let us call a shock "nondegenerate" if it is not purely gasdynamic and non-characteristic (i. e., $s \neq \lambda_i(u^\pm), i = -3, \dots, 3$).

2.6.2. For certain $\delta = (\zeta, \mu, \nu, \kappa)$ with $\zeta, \nu > 0, \mu, \kappa = 0$, there exist non-degenerate intermediate traveling wave shocks of species $0 \to 2$, $1 \to 3$, $0 \to 3$; for certain other such δ, such shocks of type $1 \to 2$ exist. Cf. [31,30].

2.6.3. There exist $\delta \in (0, \infty)^4$ for which intermediate traveling wave shocks of species $1 \to 2$ exist. Cf. [33].

2.6.4. Theorem. *For arbitrary nondegenerate intermediate ideal shocks of species* $0 \to 2$, $1 \to 3$, $0 \to 3$ $\{1 \to 2\}$ *in an ideal gas, viscous profiles exist for an open range* $\{$*a hypersurface*$\}$ *of values* $\delta \in (0, \infty)^4$. *More precisely, each such shock of type* k *has a* $(k-1)$*-parameter family of profiles associated with it, for each such* δ.

Proof: A careful analysis of the $\mu = \kappa = 0$ case is combined with a method of Szmolyan [34] from geometric singular perturbation theory. See [24].

2.6.5. Theorem. *For arbitrary* $\delta \in (0, \infty)^4$, *there exist non-degenerate intermediate traveling wave shocks of species* $0 \to 3$, $0 \to 2$, $1 \to 2$ *of large amplitude. Cf. [14,15].*

Proof: For the $0 \to 3$ case and the $0 \to 2$ case, this follows immediately by perturbing purely gasdynamic shocks or "switch-on" shocks, respectively. For the $1 \to 2$ case, it follows by perturbing from 180^0 contact discontinuities. See [14,15].

2.6.6. Theorem. *Consider a state* u_* *where* $c_- < c_a = c_+$ $(c_- = c_a < c_+)$ $\{c_- = c_a = c_+\}$. *Then, for arbitrarily chosen dissipation* $\delta \in (0, \infty)^4$, *there exist non-degenerate traveling wave shocks of species* $0 \to 1$, $0 \to 2$, $1 \to 2$ $(2 \to 3$, $1 \to 3$, $1 \to 2)$ $\{0 \to 1$, $2 \to 3$, $0 \to 2$, $1 \to 3$, $0 \to 3$, $1 \to 2\}$ *which assume values in an arbitrarily small neighborhood of* u_*.

Proof: The existence of the corresponding ideal shock waves of these respective types is due to 2.5. The existence of the profile orbits follows by considering the restriction of the flow of

$$D(u, \delta)u' = f(u) - su - q$$
$$q' = 0$$

to a center manifold at the point (u_*, q_*) with $q_* = f(u_*) - s_* u_*$ and using the well-known [25] Lyapunov function. For details, see [20].

3. Non-uniformity of vanishing viscosity approximation

3.1. Stability notions. Consider *some* hyperbolic system of conservation laws

$$u_t + (f(u))_x = 0, \tag{3.1}$$

together with a corresponding family of extensions thereof:

$$u_t + (f(u))_x = D[u]. \tag{3.2}$$

Here, f is an appropriate smooth mapping from a fixed and, for concreteness, compact subset U of $\mathbb{R}^n$ to $\mathbb{R}^n$. D in (3.2) ranges in some class $\mathcal{D}$, a collection of appropriately defined and interpreted *"dissipative"* operators. $\overline{\mathcal{D}} := \mathcal{D} \cup \{0\}$ is supposed to be a metric space, with metric d. Speaking of $\mathcal{D}$ reflects the fact that realistic continuum mechanical models are typically endowed with whole families of different conceivable or physically realizable dissipation mechanisms. A known or sought function u solving (3.1) or (3.2) depends on $x \in \mathbb{R}$ and $t \in [0,T]$, $T > 0$, and has values in U. For (3.1), consider solutions u, which are (at least) measurable and assume measurable initial data u_0, i. e.,

$$\int_0^T \int_{-\infty}^{\infty} \psi_t u + \psi_x f(u) dx dt \; + \int_{-\infty}^{\infty} u_0 \psi dx = 0, \psi \in C_0^\infty(\mathbb{R} \times [0,T));$$

for (3.2), do analogously, with u such that also $D[u]$ makes appropriate sense. Let $\mathcal{L}_0, \mathcal{L}$ be the sets of (Lebesgue-) measurable U-valued functions defined on $\mathbb{R}, \mathbb{R} \times [0,T]$, respectively. Consider $\mathcal{L}_0, \mathcal{L}$ as metric spaces with metrics l_0, l defined through

$$l_0(w_0, \tilde{w}_0) = \sup_{y \in \mathbb{R}} \int_{|x-y| \leq 1} |w_0(x) - \tilde{w}_0(x)| dx,$$

$$l(w, \tilde{w}) = \sup_{y \in \mathbb{R}} \int_0^T \int_{|x-y| \leq 1} |w(x,t) - \tilde{w}(x,t)| dx dt.$$

For a solution $u \in \mathcal{L}$, of (3.1) or (3.2), let $I(u) \in \mathcal{L}_0$ denote its initial values.

3.1.1. Definition. *Consider a solution $u^* \in \mathcal{L}$ of (1), with initial data $I(u^*) = u_0^* \in \mathcal{L}_0$.*

(i) *u^* is dynamically stable if a continuous map $S_0 : u_0 \mapsto u \in \mathcal{L}$ exists at least on a neighborhood of u_0^* in $\mathcal{L}_0$ such that each $u = S_0(u_0)$ solves (1) with data u_0 (i.e., $I(S_0(u_0)) = u_0$) and, especially, $S_0(u_0^*) = u^*$.*

(ii) *u^* is a viscosity limit (with respect to $\mathcal{D}$) if there exists a sequence $\{u^k, D_k\}_{k \in \mathbb{N}} \subset \mathcal{L} \times \mathcal{D}$ such that u_k solves (3.2) with $D = D_k$, and $l(u_k, u^*) + l_0(I(u_k), u_0^*) + d(D_k, 0) \to 0$ when $k \to \infty$.*

(iii) *u^* is a uniformly stable viscosity limit (with respect to $\mathcal{D}$) if a continuous map $S : (u_0, D) \mapsto u \in \mathcal{L}$ exists at least on a neighborhood $\{(u_0, D) \in \mathcal{L} \times \overline{\mathcal{D}}; l_0(u_0, u_0^*) < \varepsilon, d(D,0) < \varepsilon\}, \varepsilon > 0$, of $(u_0^*, 0)$ such that each $u = S(u_0, D)$ solves (3.2) with the respective D and data u_0 and, especially, $S(u_0^*, 0) = u^*$.*

3.2. Non-uniform viscosity limits in MHD

3.2.1. Theorem. *Fix any quadruple* $\delta_1 = (\zeta_1, \mu_1, \nu_1, \kappa_1) \in (0, \infty)^4$ *of positive dissipation coefficients and consider the vanishing viscosity approximation of (1.5) with* $\delta = (\zeta, \mu, \nu, \kappa) = 0$ *("ideal equations") by (1.5) with* $\delta = \alpha\delta_1$, $\alpha > 0$. *Assume the general property that for each fixed Riemann data and each* $\alpha > 0$, *there is a unique solution* u^α *which converges, as* $\alpha \searrow 0$, *to a piecewise continuous centered entropy solution of the ideal equations. Then the ideal intermediate shock waves observed in 2.4—as far as not being purely gasdynamic—cannot be uniformly stable viscosity limits.*

Proof: Analogous to a proof in [16]. See [20].
Remark: Gasdynamic (i. e., $\mathbf{b} = 0$) $0 \to 3$ intermediate shock waves seem to be uniformly stable viscosity limits.

3.2.2. Remark. An analogous result holds in elasticity. Cf. [20]. This finding complements considerations, e. g., in [3] on the mere existence of traveling waves in that context.

3.2.3. Remark. Note that 3.2.1, 3.2.2 document a violent local-in-time instability of vanishing viscosity approximation. The perception of this phenomenon, including 3.2.5 below, underlay my writing [12,16].

We state some analogous results—sharper and at the same time easier to understand intuitively—for isotropic Keyfitz-Kranzer models, i. e., systems of the form

$$u_t + (\varphi(|u|)u)_x = 0, \tag{3.3}$$

with parabolic counterparts

$$u_t + (\varphi(|u|)u)_x = \mu u_{xx}, \quad \mu > 0, \tag{3.4}$$

where $\varphi \in C^2(\mathbb{R}, \mathbb{R})$ is even and u takes values in $U = \{w \in \mathbb{R}^n; |w| \le \overline{r}\}$ with constants $\overline{r} > 0$ and $n \ge 2$. Systems (3.3),(3.4) are directly related to continuum mechanical systems. In particular, the case $\varphi(r) = r^2$ describes almost field-parallel propagation of small nonlinear MHD waves [8] (as well as small nonlinear transverse waves in an isotropic elastic medium). Note first that (3.4) has a reasonable solution theory.

3.2.4. Theorem. *For each* $\mu > 0$ *and any bounded and measurable initial data, (3.4) possesses a unique bounded and (for* $t > 0$*) smooth solution on* $\mathbb{R} \times [0, \infty)$.

Proof: Invariant regions. See [18].

The follwoing two theorems hold at least for φ with $(\partial^2(\varphi(r)r)/\partial r^2)|_{r=0} \ne 0$. Regarding vanishing viscosity approximation, we point out

3.2.5. Theorem. *There are data for which (i) there exists a vanishing viscosity limit solution of (3.3) (on any strip $\mathbb{R} \times [0,T], T > 0$), while (ii) no uniformly stable vanishing viscosity limit solution of (3.3) exists (on any such strip). (iii) In at least some cases of such data, there are at least two different vanishing viscosity limit solutions.—If, e. g., $\varphi(r) = r^2$, then for any step data*

$$u_0^*(x) = \begin{cases} u_l = \alpha_l e, & x < 0 \\ u_r = \alpha_r e = \alpha u_l, & x > 0 \end{cases} \qquad |e| = 1, \alpha_l \neq 0, \tag{3.5}$$

with $\alpha \leq 0$, properties (i) and (ii) hold. The property mentioned under (iii) holds at least if $-1 < \alpha < -\frac{1}{2}$.

Proof: See [12,16,18].

3.2.6. Theorem. *There exist solutions with non-Riemann data which are viscosity limits, but are not stable as such.*

Proof: For, e. g., $\varphi(r) = r^2$, such a solution is given by

$$u(x,t) = \begin{cases} (1,0), & |x-t| > 1 \\ (-1,0), & |x-t| < 1. \end{cases}$$

For (the simple) details, see [20].

3.2.7. Remark. Our varying the viscosity shows that the instability of intermediate shock waves observed in the ideal framework (2.3) is well compatible with their persistence (2.6.1-2.6.6) in the dissipative context.

3.2.8. Remark. For a numerical study of non-uniform vanishing viscosity approximation, see [22,23].

3.2.9. Remark. Wu, partly with co-authors, [35,36,37,6,29], has studied type 2 intermediate shock waves numerically. In these studies, the dissipation coefficients were fixed.

4. Conditional infinite time stability in the presence of dissipation

For traveling shock wave solutions

$$u^*(x,t) = \phi^*(x - st), \quad \phi^*(\pm\infty) = u^\pm \tag{4.1}$$

of a "viscous system" of n conservation laws

$$u_t + (f(u))_x = (B(u)u_x)_x \tag{4.2}$$

with hyperbolic inviscid part ($B = 0$), large time stability can be technically defined (for the present purpose) as follows.

4.1. Definition. *A traveling viscous shock wave solution (4.1) of (4.2) is called stable for infinite time if and only if with some appropriate norm $\|\cdot\|$ and some $\varepsilon > 0$, the following holds for any given sufficiently smooth perturbation $\overline{u}_0 : \mathbb{R} \longrightarrow \mathbb{R}^n$. If $\|\overline{u}_0\| < \varepsilon$, then the solution u of (4.2) with data*

$$u(x,0) = \phi^*(x) + \overline{u}_0(x), \quad x \in \mathbb{R}, \tag{4.3}$$

exists for all times $t > 0$ and converges in the sense

$$\lim_{t\to\infty} \sup_{x\in\mathbb{R}} |u(x,t) - \phi(x - st)| = 0 \tag{4.4}$$

to another viscous shock wave of profile ϕ with the same end states $\phi(\pm\infty) = u^\pm$.

The following two theorems establish infinite time stability of intermediate MHD shock waves in the framework of the two model systems

$$u_t + (|u|^2 u)_x = \mu u_{xx} \tag{4.5}$$

where $\mu > 0$ and u ranges in $\mathbb{R}^2$, and

$$\begin{aligned} y_t + (zy)_x &= \zeta y_{xx} \\ z_t + \frac{1}{2}(|y|^2 + z^2)_x &= \mu z_{xx}, \end{aligned} \tag{4.6}$$

where $\zeta, \mu > 0$ and (y, z) ranges in $\mathbb{R}^2 \times \mathbb{R}$. These two systems were proposed in [11] (see also [17]); they model the MHD equations (1.5) near points with double ($c- = c_a < c_+$ or $c_- < c_a = c_+$) and triple ($c- = c_a = c_+$) eigenvalues, respectively. (It turned out later that the cubic model (4.5) had been found and derived by Cohen and Kulsrud [8] as early as in 1974.)

Theorem 4.2. *Certain non-classical shock waves in model (4.5), which correspond to type 2 intermediate MHD shocks, are stable for infinite time.*

Proof: See [21].

Theorem 4.3. *Certain non-classical shock waves in model (4.6), which correspond to type 3 intermediate MHD shocks, are stable for infinite time.*

Proof: See [20].

4.4. Remark. Infinite time stability may be conditional in the sense that the relation

$$\int_{-\infty}^{\infty} \phi(x) - \phi^*(x)dx = \int_{-\infty}^{\infty} \overline{u}_0(x)dx$$

yields an a priori bound M on the $\mathcal{L}^1$-norm of perturbations $\overline{u}_0$ which permit (4.3), (4.4). The infinite-time stability of all shocks considered in 4.2, 4.3—except for the "purely Burgers", i. e., $y^{\pm} = 0$, ones for (4.6)—is conditional in this sense. The fact that the bound M scales with the viscosity, i. e., with the factor α in

$$u_t + (f(u))_x = \alpha(B(u)u_x)_x,$$

reflects the local-in-time non-uniformity observed earlier, i. e., violation of 3.1.1 (iii).

5. Hyperbolic theory

As in a number of previous publications, I point out again that despite the non-uniformity of vanishing viscosity approximation, the inviscid case seems to have its independent and self-consistent solution theory. In particular, one has the following two theorems.

5.1. Theorem. *The inviscid (i. e., $\mu = 0$) version of system 4.5 possesses an $\mathcal{L}^1_{loc}$-continuous solution operator. The solutions obtained via this operator do not contain non-classical shock waves.*

Proof: See [19]. The same can be proved analogously for the inviscid (i. e., $\zeta = \mu = 0$) version of system 4.6.

5.2. Theorem. *Consider a point u_* with double eigenvalues (i. e., $c- = c_a < c_+$ or $c_- < c_a = c_+$). If u_l, u_r are in a sufficiently small neigborhood of u_*, then there exists a centered, piecewise continuous solution u of the ideal MHD equations assuming Riemann data*

$$u(x,0) = \begin{cases} u_l, & x < 0 \\ u_r, & x > 0. \end{cases}$$

Under the two conditions (i) u has values in an appropriate neigborhood of u_*, and (ii) all discontinuities in u are "stable", this solution is unique.

Here, "stable" means either stable in the sense of linear stability analysis (condition "(LS)") or $\mathcal{L}^1_{loc}$-limit of such. For the proof—and this notion of stability—, cf. [13]. It is an open—and apparently difficult—problem to decide whether this solution operator extends to non-Riemann data.

REFERENCES

1. A. Akhiezer, G. Liubarskii, and R. Polovin, *The stability of shock waves in magnetohydrodynamics*, Sov. Phys. JETP 35 (1959), 507-511.
2. J. E. Anderson, "Magnetohydrodynamic shock waves," MIT Press: Cambridge MA 1963.
3. S. S. Antman and R. Malek-Madani, *Traveling waves in nonlinearly viscoelastic media and shock structure in elastic media*, Q. Appl. Math. 46 (1988), 77-93.
4. V. I. Arnol'd, "Singularity theory," Cambridge 1981.
5. T. Bröcker, "Differentiable germs and catastrophes," Cambridge 1975.
6. M. Brio and C. C. Wu, *An upwind differencing scheme for the equations of magnetohydrodynamics*, J. Comput. Phys. 75 (1988), 400-428.
7. H. Cabannes, "Theoretical magnetofluiddynamics," New York 1970.
8. R. H. Cohen and R. M. Kulsrud, *Nonlinear evolution of parallel-propagating hydromagnetic waves*, Phys. Fluids 17 (1974), 2215-2225.
9. C. C. Conley and J. Smoller, *On the structure of magnetohydrodynamic shock waves*, Commun. Pure Appl. Math. 28 (1974), 367-375.
10. C. C. Conley and J. Smoller, *On the structure of magnetohydrodynamic shock waves, II*, J. Math. Pures Appl. 54 (1975), 429-444.
11. H. Freistühler, "Anomale Schocks, strukturell labile Lösungen und die Geometrie der Rankine-Hugoniot-Bedingungen", Doctoral Thesis, Ruhr-Universität Bochum 1987.
12. H. Freistühler, *Instability of vanishing viscosity approximation for hyperbolic systems of conservation laws with rotational invariance*, J. Differ. Equations 87 (1990), 205-226.
13. H. Freistühler, *Rotational degeneracy of hyperbolic systems of conservation laws*, Arch. Rational Mech. Anal. 113 (1991), 39-64.
14. H. Freistühler, *Linear degeneracy and shock waves*, Math. Z. 207 (1991), 583-596.
15. H. Freistühler, *Some remarks on the structure of intermediate magnetohydrodynamic shocks*, J. Geophys. Res. 95 (1991), 3825–3827.
16. H. Freistühler, *Dynamical stability and vanishing viscosity: a case study of a nonstrictly hyperbolic system*, Commun. Pure Appl. Math. 45 (1992), 561-582.
17. H. Freistühler, *Hyperbolic systems of conservation laws with rotationally equivariant flux function*, Mat. Aplic. Comp. 11 (1992), 167-188.
18. H. Freistühler, *Non-uniformity of vanishing viscosity approximation*, Appl. Math. Lett. 6 (1993), 35-41.
19. H. Freistühler, *On the Cauchy problem for a class of hyperbolic systems of conservation laws*, J. Differ. Equations, to appear in 1994.
20. H. Freistühler, *Small amplitude intermediate MHD shock waves*, preprint Aachen 1993.
21. H. Freistühler and T.-P. Liu, *Nonlinear stability of overcompressive shock waves in a rotationally invariant system of viscous conservation laws*, Commun. Math. Phys. 153 (1993), 147-158.
22. H. Freistühler and E. B. Pitman, *A numerical study of a rotationally degenerate hyperbolic system. Part I: The Riemann problem*, J. Comput. Phys. 100 (1992), 306-321.
23. H. Freistühler and E. B. Pitman, *A numerical study of a rotationally degenerate hyperbolic system. Part II: The Cauchy problem*, SIAM J. Num. Anal., to appear.
24. H. Freistühler and P. Szmolyan, *Existence and bifurcation of viscous profiles for all intermediate magnetohydrodynamic shock waves*, SIAM J. Math. Anal., to appear.
25. P. Germain, *Contribution à la théorie des ondes de choc en magnétodynamique des fluides*, Off. Nat. Etud. Aéronaut., Publ. 97 (1959).
26. D. Gilbarg, *The existence and limit behavior of the one-dimensional shock layer*, Amer. J. Math. 73 (1951), 256-274.
27. M. Hesaraaki, *The structure of shock waves in magnetohydrodynamics*, Amer. Math. Soc. Mem. 302, 1984.
28. A. Jeffrey and T. Taniuti, "Nonlinear wave propagation," Academic Press: New York-London 1964.

29. C. F. Kennel, R. D. Blandford, and C. C. Wu, *Structure and evolution of small amplitude intermediate shock waves*, Phys. Fluids B 2 (1990), 253-276.
30. A. Kulikovskii and G. Liubimov, *On the structure of an inclined magnetohydrodynamic shock wave*, Appl. Math. Mech. (= engl. transl. of Prikladnaya Matematika i Mekhanika) 25 (1961), 171-179.
31. A. Kulikovskii and G. Liubimov, "Magnetohydrodynamics," Addison-Wesley: Reading 1965.
32. T.-P. Liu and C.-H. Wang, *On a non-strictly hyperbolic system of conservation laws*, J. Differ. Equations 57 (1985), 1-14.
33. K. Mischaikow and H. Hattori, *On the existence of intermediate magnetohydrodynamic shock waves*, J. Dynamics Diff. Eqs. 2 (1990), 252-281.
34. P. Szmolyan, *Transversal heteroclinic and homoclinic orbits in singular perturbation problems*, J. Differential Equations 92 (1992), 252-281..
35. C. C. Wu, *On MHD intermediate shocks*, Geophys. Res. Lett. 14 (1987), 668-671.
36. C. C. Wu, *The MHD intermediate shock interaction with an intermediate wave: Are intermediate shocks physical?*, J. Geophys. Res. 93 (1988), 987-993.
37. C. C. Wu, *Formation, structure, and stability of MHD intermediate shocks*, J. Geophys. Res. 95 (1990), 8149-8175.

Institut für Mathematik
Rheinisch-Westfälische Technische Hochschule Aachen
D-52056 Aachen, Germany

AMS/IP Studies in Advanced Mathematics
Volume **3**, 1997

NECESSARY AND SUFFICIENT CONDITIONS OF EXTINCTION AND BLOW-UP ON SOME PARABOLIC EQUATIONS

Yonggeng Gu

ABSTRACT

In this paper we are intresting the necessary and sufficient conditions of extinction and blow-up on the nonlinear growth power to the semilinear parabolic equations with initial-boundary value. The other initial-boundary value problems, as semilinear parabolic equations of high order, quasilinear parabolic equations ect.$\cdots$ will be studied in this paper also.

§ 1. INTRODUCTION

Suppose Ω be a bounded domain in the Euclid space $\mathbb{R}^n$ with the smooth boundary $\partial\Omega, n \geq 2$. We are intresting the initial-boundary value problem on $Q_T = \Omega \times (0,T), T > 0$ as following

$$\begin{cases} \dfrac{\partial u_t}{\partial t} - \Delta u + \lambda |u|^{\gamma-1}u = 0, & \text{in } Q_T, \\ u(x,t)|_{\partial\Omega\times(0,T)} = 0, & \\ u(x,0) = u_0(x), & \text{in } \Omega. \end{cases} \tag{1.1}$$

Hence γ and λ are given constants will be determined later, $u_0(x) \not\equiv 0$ is the initial datum and the operater $\Delta = \sum_{i=1}^{n} \frac{\partial^2}{\partial x_i^2}$ denotes the Laplacian.

It is well know there are lot of extensive work on the extinction and blow-up of positive solutions to the problem (1.1) (see [5, 6, 10]). The purpose of this paper is to establish the necessary and sufficient conditions of extinction and blow-up on the nonlinear growth power γ to the problem (1.1), where the sign of solutions may change in the domain Q_T. The method used in this paper is the estimate of integral norm of solutions. Nodoutedly, this method is more general and can be use to another problems, as equations of high order, qusilenear equations and system of equations (such as Navier-Stokes equation), in which the maximum principle may be not valid.

Now let us introduce some definitions and lemmas.

Definition 1.1. Suppose $u(x,t)$ is the solution of problem (1.1) , if there is a positive number $T_0 \in (0,\infty)$, such that

$$\begin{cases} u(x,t) \not\equiv 0. & (x,t) \in Q_{T_0}, \\ u(x,t) \equiv 0. & (x,t) \in Q_\infty \setminus Q_{T_0}. \end{cases}$$

Then we call the solution $u(x,t)$ is extinction in the finite time (or in short extinction).

Definition 1.2. Suppose $u(x,t)$ is the solution of problem (1.1), if there is a positive number $T_1 \in (0,\infty)$, such that

$$\lim_{t\to T_1^-} \|u(\cdot,t)\| = \infty.$$

Then we call the solution $u(x,t)$ is blow-up in the finite time under the norm $\|\cdot\|$ (or in short blow-up).

Lemma 1.3[6]. Suppose $\tau \in (0,1)$ is a constant, the function $y(t) \in C^1(0,\infty)$ satisfies

$$\begin{cases} \dfrac{dy}{dt} \le c_0 y^\tau, & y(t) \ge 0, \quad t \in (0,\infty) \\ y(0) = y_0 > 0. \end{cases}$$

Then the following estimate is valid

$$\begin{cases} y(t) \le (y_0^{1-\tau} - c_0 t)^{\frac{1}{1-\tau}}, & t \in [0,T_0), \\ y(t) = 0, & t \in [T_0,\infty), \end{cases}$$

if $c_0 > 0$, and

$$y(t) \le (y^{1-\tau} - c_0 t)^{\frac{1}{(1-\tau)}}, \quad t \ge 0,$$

if $c_0 \le 0$, where $T_0 > 0$ is a constant independent of $y(t)$.

Lemma 1.4[6]. Suppose the function $y(t) \in C^1(0,\infty)$, satisfies

$$\frac{dy}{dt} + c_1 y \ge c_2, \quad y(t) \ge 0, \quad t \in (0,\infty).$$

Then the following inequality is valid

$$y(t) \geq c, t \in (T_1, \infty),$$

where c_1, c_2, c and T_1 are positive constants independent of $y(t)$.

Lemma 1.5[10]. Suppose $k > 1$ is a constant, the function $y(t) \in C^1(0,\infty)$ satisfies

$$\begin{cases} \dfrac{dy}{dt} \geq c_3 y^k, \quad y(t) \geq 0, \quad t \in (0,\infty), \\ y(0) = y_0 > 0, \end{cases}$$

then the inequality

$$y(t) \geq (y_0^{1-k} - c_4 t)^{\frac{1}{1-k}}, \quad t \in (0, T_2)$$

is valid, where c_3, c_4 and T_2 are positive constants independent of $y(t)$.

Theorem A. Suppose $\lambda > 0$ is given, and $u_0(x) \in W_0^{1,2}(\Omega) \bigcap L^{\gamma+1}(\Omega), u_0(x) \not\equiv 0$, then the necessary and sufficient condition of extinction of solution $u(x,t)$ to the problem (1.1) is $\gamma \in (0,1)$.

Theorem B. Suppose $\lambda < 0$ is given, and $u_0(x) \in W_0^{1,2}(\Omega) \bigcap L^{\gamma+1}(\Omega)$ such that

$$\frac{1}{2}\| \nabla u_0\|_2^2 + \frac{\lambda}{\gamma+1}\|u_0\|_{\gamma+1}^{\gamma+1} \leq 0, \quad u_0(x) \not\equiv 0.$$

then the necessary and sufficient condition of blow-up of solution $u(x,t)$ to the problem (1.1) is $\gamma \in (1,\infty)$. Here $W_0^{l,p}(\Omega)$ (or $(W^{l,p}(\Omega))$ denote the Sobolev space, $l \geq 1, p > 1$, and $W^{0,p}(\Omega) = L^p(\Omega), \|\cdot\|_p$ is the norm in $L^p(\Omega)$, i.e.

$$\|u\|_p = (\int_\Omega |u(x)|^p dx)^{1/p}.$$

§ 2. THE SUFFICIENT CONDITIONS OF EXTINCTION

In this section we will discuss the sufficient conditions of extinction on some parabolic equations. At first we have

Theorem 2.1. Suppose $\lambda > 0, \gamma \in (0,1), u_0 \in L^2(\Omega)$, and $u_0 \not\equiv 0$, then the solution $u(x,t)$ of problem (1.1) will be extinction, and $\exists T_0, c > 0$, independent of the solution $u(x,t)$, the estimate

$$\begin{cases} \|u(\cdot,t)\|_2 \leq (\|u_0\|_2^{2-k} - ct), & t \in [0,T_0), \\ \|u(\cdot,t)_2 \equiv 0, & t \in [T_0,\infty), \end{cases} \tag{2.1}$$

holds, where

$$k=\frac{[2N(1-\gamma)+4(1+\gamma)]}{[N(1-\gamma)+4]}.$$

Proof. Multiplying two sides of equation of problem (1.1) by the solution $u(x,t)$, we have

$$\frac{1}{2}\frac{d}{dt}\|u\|_2^2+\|\nabla u\|_2^2+\lambda\|u\|_{1+\gamma}^{1+\gamma}=0. \tag{2.2}$$

By the Gagliardo-Nirenberg inequality[8], for some $\theta\in(0,1)$ we see

$$\|u\|\leq c\|u\|_{1+\gamma}^{1-\theta}\|\nabla u\|_2^{\theta}, \tag{2.3}$$

where the constant $c>0$, independent of $u(x,t)$, (Hereafter we denote by c generic positive constants independent of $u(x,t)$). Let

$$\theta=\frac{N(1-\gamma)}{N(1-\gamma)+2(1+\gamma)},$$

we see $\theta\in(0,1)$, because $\gamma\in(0,1)$. For $\forall\varepsilon>0,\exists C_\varepsilon>0$, using the Hölder inequality in (2.3) it is valid

$$\|u\|_2^k\leq C_\varepsilon\|u\|_{1+\gamma}^{1+\gamma}+\varepsilon\|\nabla u\|_2^2, \tag{2.4}$$

where

$$k=\frac{2N(1-\gamma)+4(1+\gamma)}{N(1-\gamma)+4}.$$

We see $k\in(1,2)$, because $\gamma\in(0,1)$. The formula (2.4), and (2.2) imply

$$\frac{d}{dt}\|u\|_2^2\leq -c\|u\|_2^k. \tag{2.5}$$

Let $y(t)=\|u(\cdot,t)\|_2^2$, by Lemma 1.3. we can obtain the estimate (2.1) immediately, and the proof of Theorem 2.1 is completed.

Next, we will discuss the extinction conditions for another initial boundary value problems.

Example 1. Let us consider the initial-boundary value problem of high order equations

$$\begin{cases}\dfrac{\partial u}{\partial t}+(-\Delta)^m u+\lambda|u|^{\gamma-1}u=0, & (x,t)\in Q_T,\\ \dfrac{\partial^i u}{\partial\nu}|_{\partial\Omega\times(0,T)}=0, & i=0,1,\cdots,m-1,\\ u(x,0)=u_0(x), & x\in\Omega,\end{cases} \tag{2.6}$$

where λ, γ are constants and $m \geq 1$ is positive integer, , ν is the unit outward normal on the boundary $\partial\Omega$ of domain $\Omega \subset \mathbb{R}^n$.

From the problem (2.6), we have the so-called energy equality

$$\frac{1}{2}\|u\|_2^2 + \| \nabla^m u\|_2^2 + \lambda\|u\|_{\gamma+1}^{\gamma+1} = 0. \tag{2.7}$$

For simply, we take $m = 2$ only, in this case, we see $u \in W_0^{2,2}(\Omega)$. Applying the Gagliardo-Nirenberg inequality and interpolation inequality, one derives

$$\|u\|_2^{1-\frac{\theta}{2}} \leq c\|u\|_{1+\gamma}^{1-\theta}\| \Delta u\|_2^{\frac{\theta}{2}}, \tag{2.8}$$

where the constant θ is given in (2.3). Let

$$k_1 = \frac{4N(1-\gamma)+8(1+\gamma)}{N(1-\gamma)+8},$$

we obtain

$$\|u\|_2^{k_1(1-\frac{\theta}{2})} \leq c\|u\|_{1+\gamma}^{1+\gamma} + \delta\| \Delta u\|_2^2, \tag{2.9}$$

where $\delta > 0$ is arbitrarily. From formula (2.9), together with (2.7), it follows

$$\frac{d}{dt}\|u\|_2^2 \leq -c\|u\|_2^h, \tag{2.10}$$

where $h = k_1(1-\theta) = [2N(1-\gamma)^2 + 8(1+\gamma)]/(N(1-\gamma)+8)$. If $\gamma \in (0,1)$, we have $h \in (1,2)$. Applying Lemma 1.3, $\exists T_0 > 0, c > 0$ such that

$$\begin{cases} \|u(\cdot,t)\|_2 \leq (\|u_0\|_2^{2-h} - ct)^{\frac{1}{2-h}}, & t \in [0,T_0), \\ \|u(\cdot,t)\|_2 = 0, & t \in [T_0,\infty). \end{cases} \tag{2.11}$$

For reason given, we have

Theorem 2.2. Suppose $\lambda > 0, \gamma \in (0,1), u_0 \in L^2(\Omega)$, and $u_0(x) \not\equiv 0$, then the solution $u(x,t)$ of problem (2.6) will be extinction and the estimate (2. 11) holds.

Example 2. Let us consider the intial-boundary value problem of qusilinear equations

$$\begin{cases} \dfrac{\partial u}{\partial t} - \Delta_p u + \lambda|u|^{\gamma-1}u = 0, & (x,t) \in Q_T, \\ u|_{\partial\Omega\times(0,T)} = 0, & \\ u(x,0) = u_0(x), & x \in \Omega, \end{cases} \tag{2.12}$$

where $p > 1, \lambda, \gamma$ are constants and

$$\Delta_p u = \sum_{i=1}^{N} \frac{\partial}{\partial x_i}\left(\left|\frac{\partial u}{\partial x_i}\right|^{p-2}\frac{\partial u}{\partial x_i}\right)$$

is the p-Laplacian. For $\forall q \geq 1 + \frac{N(p-2)}{p}$, we have the energy equality

$$\frac{1}{q+1}\frac{d}{dt}\|u\|_{q+1}^{q+1} + \frac{qp^p}{(q+p-1)^p}\| \nabla u^{\frac{q+pp-1}{p}} \|_p^p + \lambda\|u\|_{\gamma+q}^{\gamma+q} = 0. \tag{2.13}$$

On the analogy of Theorem 2.2. we have

Theorem 2.3. Suppose i) $\lambda > 0, p \geq 2, \gamma \in (0,1)$, or ii) $\lambda \geq 0, p \in (1,2), \gamma \geq 0$, then for $\forall u_0(x) \in L^{q+1}(\Omega), q \geq 1 + \frac{N(p-2)}{p}, u_0(x) \not\equiv 0, x \in \Omega$; the solution $u(x,t)$ of problem (2.12) will be extinction and the estimate like (2.1) is valid.

Example 3. Let us consider the initial boundary value problem of Navier-Stokes equation

$$\begin{cases} \dfrac{\partial \vec{u}}{\partial t} - \mu \Delta \vec{u} + (\vec{u}\cdot\Delta)\vec{u} + \lambda|\vec{u}|^{\gamma-1}\vec{u} = -\nabla P, \quad \nabla\vec{u} = 0, \quad (x,t) \in Q_T, \\ \vec{u}|_{\partial\Omega\times(0,T)} = 0, \\ \vec{u}(x,0) = \vec{u}_0(x), \qquad\qquad x \in \Omega, \end{cases} \tag{2.14}$$

where $\mu > 0, \lambda, \gamma$ are constants, $\vec{u}(x,t) = (u^1(x,t), u^2(x,t), \cdots, u^n(x,t))^\tau$ is the vector and P is the pressure. For problem (2.14), we have

Theorem 2.4. Suppose, $\lambda \geq 0, \gamma \in (0,1), u_0^i(x) \in L^2(\Omega), u_0^i(x) \not\equiv 0, i \in \{1,2,\cdots,N\}$, then the solution of problem (2.14) will be extinction and the estimate like (2.1) is valid.

§ 3. THE NECESSSARY CONDITIONS OF EXTINCTION

In this section, we will establish the neccssary conditions of extinction of solutions for problems (1.1), (2.1) and (2.13). First we have

Theorem 3.1. Suppose $\lambda > 0, \gamma \geq 1, u_0(x) \in W_0^{1,2}(\Omega)\bigcap L^{\gamma+1}(\Omega), u_0(x) \not\equiv 0$, then the solution $u(x,t)$ of problem (1.1) can not be extinction in the finite time, and then $\exists\widetilde{T}_0 > 0, \rho_0 > 0$, and $c > 0$, the estimate

$$\|u(\cdot,t)\|_2 \geq c\exp(-\rho_0^{1-\gamma}t), \quad t > \widetilde{T}_0, \tag{3.1}$$

holds

Proof. Before all, it is clear that the solution of problem (1.1) is not extinction when $\gamma = 1$. So we need only consider the case $\gamma > 1$. Let

$$\begin{cases} v(x,t) = \rho e^s u(x, \rho^{\gamma-1} s), \\ s = \rho^{-(\gamma-1)} t, \end{cases} \tag{3.2}$$

where $\rho > 0$ will be determinate later. By the transformation (3.2), the problem (1.1) can be writed as follows

$$\begin{cases} \frac{\partial v}{\partial s} - \rho^{\gamma-1} \triangle v + \lambda e^{-(\rho-1)s} |v|^{\gamma-1} v - v = 0, & (x,s) \in Q_s, \\ v|_{\partial\Omega \times (0,s)} = 0, & \\ v(x,0) = \rho u_0(x), & x \in \Omega. \end{cases} \tag{3.3}$$

Multiplying the equation of (3.3) by v and $\frac{\partial v}{\partial s}$ respectively, and integrating on Ω, we see

$$\frac{1}{2}\frac{d}{ds}\|v\|_2^2 + \rho^{\gamma-1}\| \nabla v\|_2^2 - \lambda e^{-(\gamma-1)s}\|v\|_{\gamma+1}^{\gamma+1} + \|v\|_2^2 = 0 \tag{3.4}$$

and

$$\begin{aligned} &\|\frac{\partial v}{\partial s}\|_2^2 + \frac{d}{ds}[\frac{1}{2}\rho^{\gamma-1}\| \nabla v\|_2^2 + \frac{\lambda}{\lambda+1} e^{-(\gamma-1)s}\|v\|_{\gamma+1}^{\gamma+1} + \|v\|_2^2] \\ &= -\frac{\lambda}{\lambda+1}(\gamma-1)e^{-(\gamma-1)s}\|v\|_{\gamma+1}^{\gamma+1} \le 0. \end{aligned} \tag{3.5}$$

So

$$\begin{aligned} &\frac{1}{2}\rho^{\gamma-1}\| \nabla v\|_2^2 + \frac{\lambda}{\lambda+1} e^{-(\gamma-1)s}\|v\|_{\gamma+1}^{\gamma+1} - \frac{1}{2}\|v\|_2^2 \\ &\le \rho^{\gamma+1}(\frac{1}{2}\| \nabla u_0\|_2^2 + \frac{\lambda}{\lambda+1}\|u_0\|_{\gamma+1}^{\gamma+1}) - \frac{1}{2}\|u_0\|_2^2. \end{aligned}$$

Due to $\gamma + 1 > 2$ and $u_0 \in W_0^{1,2}(\Omega) \bigcap L^{\gamma+1}(\Omega)$, we may choose $\rho = \rho_0$ small enough such that

$$E_0 \equiv \rho_0^{\gamma+1}(\frac{1}{2}\| \nabla u_0\|_2^2 + \frac{\lambda}{\lambda+1}\|u_0\|_{\gamma+1}^{\gamma+1}) - \frac{1}{2}\|u_0\|_2^2 < 0.$$

So we obtain

$$-\lambda e^{-(\gamma+1)s}\|v\|_{\gamma+1}^{\gamma+1} \ge -\frac{\gamma+1}{2}\rho_0^{\gamma-1}\| \nabla v\|_2^2 - \frac{\gamma+1}{2}\|v\|_2^2 - (\lambda+1)E_0. \tag{3.6}$$

From (3.1) and (3.6), it is clear that

$$\frac{d}{ds}\|v\|_2^2 + (\gamma-1)\|v\|_2^2 \ge -2(\gamma+1)E_0 > 0. \tag{3.7}$$

By Lemma 1.4. we have the estimate

$$\|v\|_2 \geq c, \;\; s > s_0,$$

where the constats c, s_0 are positive number independent of $v(x,s)$. Take account of the transformation (3.2), the estimate (3.1) follows.

In view of that the solution of problem (1.1) is not extinction when $\gamma \leq 0$, Theorem A in §1 was proved immediately by Theorem 2.1 and Theorem 3.1.

In the same way as Theorem 3.1. for the problem (2.6), we have

Theorem 3.2. Suppose $\lambda > 0, \gamma \geq 1, u_0(x) \in W_0^{m,2}(\Omega)\bigcap L^{\gamma+1}(\Omega)$, and $u_0(x) \not\equiv 0$, then the solution $u(x,t)$ of problem (2.6) is not extinction in the finite time, and the estimate like (3.1) holds

For the problem (2.13), we have

Theorem 3.3. Suppose $\lambda > 0, \rho \geq 2, \gamma \geq 1, u_0(x) \in W_0^1, p(\Omega)\bigcap L^{\gamma+1}(\Omega)$, $u_0(x) \not\equiv 0$, then the solution $u(x,t)$ of problem (2.13) is not extinction in the finite time, and then the estimate like (3.1) holds.

Proof. At first, under the transformation (3.2), we can write the problem (3.13) as

$$\begin{cases} \dfrac{\partial v}{\partial s} - \rho^{\gamma-p+1}e^{-(p-2)s}\,\triangle_p\, v + \lambda e^{(1-\gamma)s}|v|^{\gamma-1}v - v = 0, \quad (x,s) \in Q_s, \\ v|_{\partial\Omega\times(0,s)} = 0, \\ v(x,0) = \rho u_0(x), \end{cases} \tag{3.8}$$

Form (3.8) we obtain

$$\frac{1}{2}\frac{d}{ds}\|v\|_2^2 + \rho^{\gamma-p+1}e^{-(p-1)s}\|\bigtriangledown v\|_p^p + \lambda e^{-(\gamma-1)s}\|v\|_{\gamma+1}^{\gamma+1} - \|v\|_2^2 = 0,$$

and

$$\|\frac{\partial v}{\partial s}\|_2^2 + \frac{d}{ds}[\frac{1}{p}\rho^{\gamma-p+1}e^{-(p-2)s}\|\bigtriangledown v\|_p^p + \frac{\lambda}{\lambda+1}(\gamma-1)e^{-(\gamma-1)s}\|v\|_{\gamma+1}^{\gamma+1} - \frac{1}{2}\|v\|_2^2]$$
$$= -[\frac{p-2}{p}\rho^{\gamma-p+1}e^{-(p-2)s}\|\bigtriangledown v\|_p^p + \frac{\lambda}{\lambda+1}e^{-(\gamma-1)s}\|v\|_{\gamma+1}^{\gamma+1} \leq 0,$$

because $p \geq 2$, and $\gamma > 1$. So we arrive

$$\frac{1}{p}\rho^{\gamma-p+1}e^{-(p-2)s}\|\bigtriangledown v\|_p^p + \frac{\lambda}{\lambda+1}e^{-(\gamma-1)s}\|v\|_{\gamma+1}^{\gamma+1} - \frac{1}{2}\|v\|_2^2$$
$$\leq \frac{1}{p}\rho^{\gamma+1}\|\bigtriangledown u_0\|_p^p + \frac{\lambda}{\lambda+1}\rho^{\gamma+1}e^{-(\gamma-1)s}\|u_0\|_{\gamma+1}^{\gamma+1} - \frac{1}{2}\rho_0^2\|u_0\|_2^2.$$

Let

$$E_{0,p} = \frac{1}{p}\rho_0^{\gamma+1}\| \nabla u_0\|_p^p + \frac{p}{p+1}\rho_0^{\gamma+1}\|u_0\|_{\gamma+1}^{\gamma+1} - \frac{1}{2}\rho_0^2\|u_0\|_2^2.$$

Now let us choose ρ_0 small enough such that $E_{0,p} < 0$ (note that $\gamma + 1 > 2$) and then this theorem was proved to follow suit the proof of Theorem 3.1.

§ 4. BLOW-UP

In this section we will give the prove of Theorem B in §1. Before all, we have

Theorem 4.1. Suppose $\lambda < 0, \gamma > 1, u_0(x) \in W_0^{1,2}(\Omega) \bigcap L^{\gamma+1}(\Omega), u_0(x) \not\equiv 0$, and

$$\frac{1}{2}\| \nabla u_0\|_2^2 + \frac{\lambda}{\lambda+1}\|u_0\|_{\gamma+1}^{\gamma+1} \le 0,$$

then the solution of problem (1.1) will be blow-up, i.e. $\exists T_1 > 0, c > 0$ and $c_1 > 0$, the estimate

$$\|u(\cdot,t)\|_2 \ge c(c_1 - t)^{\frac{1}{1-\gamma}}, \ t < T_1 \tag{4.1}$$

is valid.

Proof. From problem (1.1), it is easy to see that

$$\frac{1}{2}\frac{d}{dt}\|u\|_2^2 + \| \nabla u\|_2^2 + \lambda\|u\|_{\gamma+1}^{\gamma+1} = 0. \tag{4.2}$$

and

$$\|u_t\|_2^2 + \frac{d}{dt}[\frac{1}{2}\| \nabla u\|_2^2 + \frac{\lambda}{\gamma+1}\|u\|_\lambda^{\gamma+1} = 0. \tag{4.3}$$

From (4.3) and the assumption, we see

$$\frac{1}{2}\| \nabla u\|_2^2 + \frac{\lambda}{\gamma+1}\|u\|_{\gamma+1}^{\gamma+1} \le \frac{1}{2}\| \nabla u_0\|_2^2 + \frac{\lambda}{\gamma+1}\|u_0\|_{\gamma+1}^{\gamma+1} \le 0. \tag{4.4}$$

Take account of (4.2) and (4.4), we obtain the estimate

$$\frac{d}{dt}\|u\|_2^2 \ge c\|u\|_{\gamma+1}^{\gamma+1} \ge c\|u\|_2^{\gamma+1}, \ c > 0. \tag{4.5}$$

Using Lemma 1.5., the estimate (4.1) is valid.

Theorem 4.2. Suppose $\lambda < 0, \gamma \in (0,1], u_0 \in L^2(\Omega)$, then the solution $u(x,t)$ of problem (1.1) is not blow-up, and $\exists c_0, c, c_1 > 0$ the estimate

$$\|u(\cdot,t)\|_2 \le \begin{cases} c(c_1+t)^{\frac{1}{1-\gamma}}, & \gamma \in (0,1), \\ ce^{c_0}t, & \gamma = 1, \end{cases} \tag{4.6}$$

is valid for all $t \geq 0$.

Proof. From problem (1.1) we see

$$\frac{1}{2}\frac{d}{dt}\|u\|_2^2 + \| \nabla u\|_2^2 + \lambda\|u\|_{\gamma+1}^{\gamma+1} = 0. \tag{4.7}$$

Due to $\lambda < 0, \gamma \in (0,1]$, it follows

$$\frac{d}{dt}\|u\|_2^2 \leq -\lambda\|u\|_{\gamma+1}^{\gamma+1} \leq c\|u\|_1^{\gamma+1}. \tag{4.8}$$

Take account of (4.8) and Lemma 1.3., we get the estimate (4.6) immediately.

Sum up Theorem 4.1. and Theorem 4.2., Theorem B holds.

Remark. A result to be analogous to Theorem A and Theorem B is valid for problem (2.6) and (2.12).

REFERENCES

[1] R.A.Adams, Sobolev Spaces, New York, Academic Press 1975.

[2] Q.X.Ye and Z.Y.Li, An Introduction to Reaction Diffusions, Science Press, Beijing, 1990 (in Chinese).

[3] Y.G.Gu, P.Z.Luo and X.X.Ding, generalized solutions of strongly nonlinear parabolic equations, *Scientist Sinica*, **26**:11(1983), 1129–1143.

[4] A.Friedman & M.A.Herrero, Extinction properties of semlinear heat equations with strong absorption, *J.of Math. & Appl.* **124** (1987), 530–546.

[5] M.Tsutsumi, On solutions of some dully nonlinear degenerate parabolic equations with absorption, *J.of Math.Anal.& Appl.***132** (1983), 187–212.

[6] H.Brezis and A.Friedman, Estimates on the support of solutions of parabolic variational inequathtes, *Illinois J.Math.* **20** (1976), 82–98.

[7] A.Friedman, Partial Differential Equations of Parabolic Type, Krigger, Malabar, Fla, 1983.

[8] L.L.Martinson, The finite velocity of propagation of thermal perturbation in media with constant thermal conductivity, *Zh.Vychist. Mat. Fi.*, **16**(1976), 1233–1241.

Institute of System Science, Academis simica, Beijing 100080 China

AMS/IP Studies in Advanced Mathematics
Volume 3, 1997

ANALYTIC SOLUTIONS OF BELOUSOV-ZHABOTINSKII REACTION

GUO BEN-YU CHEN ZHI-XIONG

I. Introduction

The Belousov-Zhabotinskii reaction is widely studied both theoretically and experimently, since it plays an important role in description of chemical reactions. A simplified reaction based on the Noyes-Field model was suggested by Murray [Mu1]. By the law of mass action and an appropriate nondimensionlization, we have

$$\begin{cases} U_t = U_{xx} + U(1 - U - rV) + LrV, \\ V_t = V_{xx} - bVV - MV \end{cases} \tag{1.1}$$

where $L \approx M = O(10^{-4}), b = O(1), r$ can be varied from 5 to 50 approximately. Since $L \ll 1$ and $M \ll 1$, we may neglect the corresponding terms and obtain the following system

$$\begin{cases} U_t = DU_{xx} + U(1 - U - rV), \\ V_t = V_{xx} - bUV \end{cases} \tag{1.2}$$

where D, r and b are positive constants. Sometimes we take $D = 1$[Mu1]. There are many literatures concerning travelling wave solutions of (1.2) [L-Y, Mu2, T, W-X-Y, Y-W].

In this paper, we shall apply Painlevé analysis to (1.2). As we know, this approach was renewed by Ablowitz, Ramani, Segur [A-R-S] and Mcleod, Olver [M-O]. While Weiss, Tabor, Carnevale [W-T-C] applied this technique to partial differential equations. Originally it was used to test the integrability of nonlinear systems. But it can be also used to deduce Lax-Pair, Bäcklund transformation, Hirota bilinear transformation and other properties such as symmetry [S-E]). Furthermore people found special solutions of nonlinear partial differential equations by painlevé analysis. Recently, Conte, Musette [C-M], Cariello, Tabor [C-T], Hereman [H] and the authors [B-Z, C-G, Z-B] extended this approach to nonintegrable systems, and obtained many interesting solutions such as periodic solutions, travelling waves and coalescences of two waves.

We can also verify that (1.2) is a nonintegrable system. We shall use Painlevé expansion to derive its analytic solutions. Two kinds of expansions are considered, which imply two classes of solutions. Many explicit expressions of travelling wave solutions are obtained. We also find travelling wave solutions with arbitrary constant speed c, provided that certain conditions are fulfilled.

II. Painlevé Analysis, The First Expansion

The basic idea of painlevé anlysis is the expansions about solution singularity manifold and truncation technique. Let $\phi(x,t) = 0$ be the solution singularity manifold of (1.2), $\phi_x\phi_t \neq 0$, and

$$U = \sum_{j=0}^{\infty} U_j \phi^{j-\mu}, \qquad V = \sum_{j=0}^{\infty} V_j \phi^{j-\nu}.$$

By substituting the above expansions into (1.2) and analysing the leading parts, we find that

$$\mu = \nu = 2, \qquad \text{or} \qquad \mu = 2, \nu = 1.$$

Next, by truncation technique, let

$$U = \sum_{j=0}^{2} U_j \phi^{j-2}, \qquad V = \sum_{j=0}^{\nu} V_j \phi^{j-\nu}, \qquad \nu = 1 \text{ or } 2.$$

We substitute the above expansions into (1.2) to get the recursion relationships for U_j and V_j. If they are consistent, then we can derive the solutions of (1.2).

In this section, we consider the first expansion with $\mu = \nu = 2$. Firstly, let $bD \neq 3$ and

$$U = \frac{6}{b}u, \qquad V = \frac{6(bD-1)}{rb}v, \qquad t = \frac{1}{5(3-bD)}\tau. \tag{2.1}$$

Then (1.2) becomes

$$\begin{cases} 5(3-bD)u_\tau = Du_{xx} + u(1 - \frac{6}{b}u - \frac{6}{b}(bD-1)v), \\ 5(3-bD)v_\tau = v_{xx} - 6uv. \end{cases} \tag{2.2}$$

Let

$$u = \frac{u_0}{\phi^2} + \frac{u_1}{\phi} + u_2, \qquad v = \frac{v_0}{\phi^2} + \frac{v_1}{\phi} + v_2. \tag{2.3}$$

By putting (2.3) in (2.2) and comparing the coefficients of ϕ^j, we obtain that

$$\begin{cases} Du_0\phi_x^2 - \frac{1}{b}u_0^2 - \frac{1}{b}(bD-1)u_0v_0 = 0, \\ v_0\phi_x^2 - u_0v_0 = 0, \end{cases} \tag{2.4}$$

$$\begin{cases} -5(3-bD)u_0\phi_\tau = D(-2u_{0x}\phi_x - u_0\phi_{xx} + u_1\phi_x^2) - \dfrac{6}{b}u_0u_1 \\ \qquad -\dfrac{3}{b}(bD-1)(u_0v_1+u_1v_0), \\ -5(3-bD)v_0\phi_\tau = -2v_{0x}\phi_x - v_0\phi_{xx} + v_1\phi_x^2 - 3u_0v_1 - 3u_1v_0, \end{cases} \tag{2.5}$$

$$\begin{cases} 5(3-bD)(u_{0\tau}-u_1\phi_\tau) = D(u_{0xx} - 2u_{1x}\phi_x - u_1\phi_{xx}) + u_0 - \dfrac{6}{b}(u_1^2+2u_0u_2) \\ \qquad -\dfrac{6}{b}(bD-1)(u_0v_2+u_1v_1+u_2v_0), \\ 5(3-bD)(v_{0\tau}-v_1\phi_\tau) = v_{0xx} - 2v_{1x}\phi_x - v_1\phi_{xx} - 6(u_0v_2+u_1v_1+u_2v_0), \end{cases} \tag{2.6}$$

$$\begin{cases} 5(3-bD)u_{1\tau} = Du_{1xx} + u_1 - \dfrac{12}{b}u_1u_2 - \dfrac{6}{b}(bD-1)(u_1v_2+u_2v_1), \\ 5(3-bD)v_{1\tau} = v_{1xx} - 6(u_1v_2+u_2v_1), \end{cases} \tag{2.7}$$

and

$$\begin{cases} 5(3-bD)u_{2\tau} = Du_{2xx} + u_2 - \dfrac{6}{b}u_2^2 - \dfrac{6}{b}(bD-1)u_2v_2, \\ 5(3-bD)v_{2\tau} = v_{2xx} - 6u_2v_2. \end{cases} \tag{2.8}$$

We now derive the expressions of u_j and v_j. Firstly, (2.4) gives

$$u_0 = v_0 = \phi_x^2. \tag{2.9}$$

Next, by substituting (2.9) into (2.5), we get

$$\begin{cases} u_1 = -\phi_{xx} - (3bD-2b-3)\phi_\tau, \\ v_1 = -\phi_{xx} + (2bD-3b+3)\phi_\tau. \end{cases} \tag{2.10}$$

Finally (2.6), (2.9) and (2.10) lead to

$$\begin{cases} u_2 = -\dfrac{\phi_{xx}^2}{4\phi_x^2} + \dfrac{\phi_{xxx}}{3\phi_x} + (3bD-2b-3)\dfrac{\phi_{x\tau}}{\phi_x} \\ \qquad -(3bD-2b-3)\dfrac{\phi_{xx}\phi_\tau}{2\phi_x^2} + \dfrac{p\phi_\tau^2}{12\phi_x^2} + \dfrac{b}{12}, \\ v_2 = -\dfrac{\phi_{xx}^2}{4\phi_x^2} + \dfrac{\phi_{xxx}}{3\phi_x} - (2bD-3b+3)\dfrac{\phi_{x\tau}}{\phi_x} \\ \qquad +(2bD-3b+3)\dfrac{\phi_{xx}\phi_\tau}{2\phi_x^2} + \dfrac{q\phi_\tau^2}{12\phi_x^2} - \dfrac{b}{12} \end{cases} \tag{2.11}$$

where

$$p = 10b^3D^3 - 10b^3D - 79b^2D^2 + 72b^2D + 6b^2 + 78bD - 72b - 9,$$
$$q = -10b^3D^3 + 10b^3D + 131b^2D^2 - 198b^2D + 66b^2 - 12bD + 18b - 9.$$

On the other hand, u_j and v_j must fulfil (2.7) and (2.8). In other words, if ϕ satisfies four equations related to (2.7) and (2.8), then the recursion relationship for u_j and v_j are consistent. To do this, we introduce the Schwartzian derivative

$$S(\phi) = \frac{\phi_{xxx}}{\phi_x} - \frac{3}{2}(\frac{\phi_{xx}}{\phi_x})^2 \tag{2.12}$$

and

$$C(\phi) = \frac{-\phi_\tau}{\phi_x} \tag{2.13}$$

which has the dimension of velocity. Both $S(\phi)$ and $C(\phi)$ are invariant under the homographic transformation

$$\phi \to \frac{\alpha\phi + \beta}{\gamma\phi + \delta}, \qquad \alpha\delta - \beta\gamma \neq 0.$$

Besides let

$$Q = -\frac{\phi_{xx}}{2\phi_x}. \tag{2.14}$$

By using the above notations, (2.11) can be rewritten as

$$\begin{cases} u_2 = \frac{b}{12} + \frac{1}{3}S - (3bD - 2b - 3)C_x + \frac{p}{12}C^2 + (3bD - 2b - 3)CQ + Q^2, \\ v_2 = -\frac{b}{12} + \frac{1}{3}S + (2bD - 3b + 3)C_x + \frac{q}{12}C^2 - (2bD - 3b + 3)CQ + Q^2. \end{cases} \tag{2.15}$$

Furthermore, (2.7) and (2.8) lead to other four equations.

Obviously, (2.8) is the same as (2.2) and so it leads to the auto-Backlund transformation as follows

$$\begin{cases} u = [-\frac{\partial^2}{\partial x^2} - (3bD - 2b - 3)\frac{\partial}{\partial \tau}] \ln \phi + u_2, \\ v = [-\frac{\partial^2}{\partial x^2} + (2bD - 3b + 3)\frac{\partial}{\partial \tau}] \ln \phi + v_2. \end{cases} \tag{2.16}$$

Once we get a solution of (2.2), denoted by u^* and v^*, we may derive new solution by putting $u_2 = u^*$ and $v_0 = v^*$ in (2.16). Since (1.2) is nonintegrable, it may repeat several different solutions in turn.

In the case with $bD = 3$, we find that there is no solutions u_1 and v_1 if $D \neq 1$. Otherwise u_1 or v_1 is arbitrary. We do not consider it in this paper.

III. The First Class of Travelling Wave Solutions

In this section, we focus on travelling wave solutions. It means that S and C are constants. By computation, we get

$$C^2 = -\frac{b}{2(3bD-2b-3)(2b^2D^2+2b^2D-21bD+6b+9)}, \tag{3.1}$$

$$S = \frac{p}{2}C^2. \tag{3.2}$$

We also obtain

$$2b(2bD-3)(bD+b-6)(bD^2+bD+9D-15)=0 \tag{3.3}$$

which implies three consistent conditions for the existence of travelling wave solutions. We shall deal with these three cases in detail.

In the first case, $b = \frac{3}{2D}$. So (3.1) and (3.2) lead to

$$C_1^2 = \frac{D}{9(2-D)(3-4D)}, \qquad S_1 = -\frac{3-4D}{8D(2-D)}.$$

Thus for $D \in (0, \frac{3}{4}) \bigcup (2, +\infty)$,

$$C_1 = \frac{\varepsilon}{3}\sqrt{\frac{D}{(2-D)(3-4D)}}, \qquad \varepsilon = \pm 1.$$

Moreover

$$\phi = \frac{\overline{A}e^{k_1\xi_1} + \overline{B}e^{-k_1\xi_1}}{Ae^{k_1\xi_1} + Be^{-k_1\xi_1}}$$

where

$$k_1 = \frac{1}{4}\sqrt{\frac{3-4D}{D(2-D)}}, \quad \xi_1 = x - C_1\tau,$$

$A, B, \overline{A}$ and $\overline{B}$ are arbitrary constants satisfying

$$\overline{A}B - A\overline{B} = \frac{1}{2k_1}.$$

By (2.11), we obtain

$$\left\{\begin{aligned} u_2 &= \frac{1}{8D} + \frac{\varepsilon}{8D} sign(D-2) \cdot \frac{Ae^{k_1\xi_1} - Be^{-k_1\xi_1}}{Ae^{k_1\xi_1} + Be^{-k_1\xi_1}} \\ &\quad - \frac{AB(3-4D)}{4D(2-D)(Ae^{k_1\xi_1} + Be^{-k_1\xi_1})^2}, \\ v_2 &= -\frac{1}{8D} + \frac{\varepsilon(3-4D)}{8D|D-2|} \cdot \frac{Ae^{k_1\xi_1} - Be^{-k_1\xi_1}}{Ae^{k_1\xi_1} + Be^{-k_1\xi_1}} \\ &\quad + \frac{5(1-D)(A^2e^{2k_1\xi_1} + B^2e^{-2k_1\xi_1}) + 2AB(2-D)}{8D(2-D)(Ae^{k_1\xi_1} + Be^{-k_1\xi_1})^2}. \end{aligned}\right. \tag{3.4}$$

Next by (2.1),

$$\left\{\begin{aligned} U_2 &= \frac{A^2(2-D)e^{2k_1\eta_1} + AB(3D-1)}{(2-D)(Ae^{k_1\eta_1} + Be^{-k_1\eta_1})^2}, \\ V_2 &= \frac{B^2(3-4D)e^{-2k_1\eta_1}}{2r(2-D)(Ae^{k_1\eta_1} + Be^{-k_1\eta_1})^2}, \end{aligned}\right. \tag{3.5}$$

where

$$\eta_1 = x - \frac{5\varepsilon}{2}\sqrt{\frac{D}{(2-D)(3-4D)}}\,t.$$

and

$$\varepsilon = \begin{cases} -1, & \text{for } D \in (0, \frac{3}{4}), \\ 1, & \text{for } D \in (2, +\infty). \end{cases}$$

Also, (3.4) yields that

$$\left\{\begin{aligned} U_2 &= \frac{B^2(2-D)e^{-2k_1\eta_1} + AB(3D-1)}{(2-D)(Ae^{k_1\eta_1} + Be^{-k_1\eta_1})^2}, \\ V_2 &= \frac{A^2(3-4D)e^{-2k_1\eta_1}}{2r(2-D)(Ae^{k_1\eta_1} + Be^{-k_1\eta_1})^2} \end{aligned}\right. \tag{3.6}$$

with

$$\varepsilon = \begin{cases} 1, & \text{for } D \in (0, \frac{3}{4}), \\ -1, & \text{for } D \in (2, +\infty). \end{cases}$$

According to (2.8), the above expressions of U_2 and V_2 are also solutions of (1.2). New solutions may be obtained by auto-Bäcklend transformation. The derivation gives that

$$\left\{\begin{aligned} U &= -4D\left(\frac{\partial^2}{\partial x^2} - \frac{2-D}{5D}\frac{\partial}{\partial t}\right)\ln\phi + \overline{U}, \\ V &= -\frac{2D}{r}\left(\frac{\partial^2}{\partial x^2} + \frac{3-4D}{5D}\frac{\partial}{\partial t}\right)\ln\phi + \overline{V} \end{aligned}\right.$$

where $\overline{U}$ and $\overline{V}$ are solutions of (1.2), and ϕ satisfies (3.3).

In the second case, $b = \dfrac{6}{D+1}$. We have

$$C_2 = \frac{\varepsilon\sqrt{D+1}}{15(D-1)}, \qquad S_2 = -\frac{1}{2(D+1)}, \qquad \varepsilon = \pm 1.$$

Then we obtain

$$\begin{cases} U_2 = \dfrac{A^2 e^{2k_2\eta_2}}{(Ae^{k_2\eta_2} + Be^{-k_2\eta_2})^2}, & \varepsilon = 1, \\ V_2 = \dfrac{B^2(5D-1)e^{-2k_1\eta_2}}{r(D+1)(Ae^{k_2\eta_2} + Be^{-k_2\eta_2})^2}, & \varepsilon = 1, \end{cases}$$

and

$$\begin{cases} U_2 = \dfrac{B^2 e^{-2k_2\eta_2}}{(Ae^{k_2\eta_2} + Be^{-k_2\eta_2})^2}, & \varepsilon = -1, \\ V_2 = \dfrac{A^2(5D-1)e^{2k_1\eta_2}}{r(D+1)(Ae^{k_2\eta_2} + Be^{-k_2\eta_2})^2}, & \varepsilon = -1. \end{cases}$$

where A and B are arbitrary constants and

$$k_2 = \frac{1}{2\sqrt{D+1}}, \qquad \eta_2 = x + \frac{\varepsilon}{\sqrt{D+1}}t.$$

In the third case, $b = \frac{3(5-3D)}{D(D+1)}$. We have

$$C_3 = \frac{\varepsilon}{60(D-1)}\sqrt{\frac{D(D+1)(5-3D)}{(D-1)(2D-3)}}, \quad S_3 = -\frac{(2D-3)(5-3D)}{2D(D^2-1)}, \quad \varepsilon = \pm 1.$$

Then we obtain

$$\begin{cases} U_2 = \dfrac{A^2(D^2-1)e^{2k_3\eta_3} + AB(D+1)(5-3D)}{(D^2-1)(Ae^{k_3\eta_3} + Be^{-k_3\eta_3})^2} \\ V_2 = \dfrac{B^2(D-3)(7-5D)e^{-2k_3\eta_3} + AB(9-7D)(7-5D)}{r(D^2-1)(Ae^{k_3\eta_3} + Be^{-k_3\eta_3})^2} \end{cases}$$

where A and B are arbitrary constants,

$$k_3 = \frac{1}{2}\sqrt{\frac{(2D-3)(5-3D)}{D(D^2-1)}}, \qquad \eta_3 = x - \varepsilon\sqrt{\frac{D(5-3D)}{(2D-3)(D^2-1)}}t$$

and

$$\varepsilon = \begin{cases} 1, \text{for} \quad D \in (0,1), \\ -1, \text{for} \quad D \in (\frac{3}{2}, \frac{5}{3}). \end{cases}$$

Also we have

$$\begin{cases} U_2 = \dfrac{B^2(D^2-1)e^{-2k_3\eta_3} + AB(D+1)(5-3D)}{(D^2-1)(Ae^{k_3}\eta_3 + Be^{-k_3\eta_3})^2}, \\ V_2 = \dfrac{A^2(D-3)(7-5D)e^{2k_3\eta_3} + AB(9-7D)(7-5D)}{r(D^2-1)(Ae^{k_3\eta_3} + Be^{-k_3\eta_3})^2} \end{cases}$$

with

$$\varepsilon = \begin{cases} -1, \text{for} \quad D \in (0,1), \\ 1, \text{for} \quad D \in (\frac{3}{2}, \frac{5}{3}). \end{cases}$$

IV. The Second Painlevé Expansion

In this section, we consider the second class of solutions. Let $\mu = 2$ and $\nu = 1$ in the Painlevé expansion. By substituting it into (1.2) and comparing the coefficients of the leading term ϕ^{-3}, we find that this system is consistent if and only if $bD = \frac{1}{3}$. Now, let $b = \frac{1}{3D}$ and

$$U = 6Du, \quad V = \frac{D}{r}v, \quad t = \frac{1}{5D}\tau.$$

Then (1.2) changes into

$$\begin{cases} 5u_\tau = u_{xx} + \frac{1}{D}u - 6u^2 - uv, \\ 5Dv_\tau = v_{xx} - 2uv. \end{cases} \tag{4.1}$$

Let

$$u = \frac{u_0}{\phi^2} + \frac{u_1}{\phi} + u_2, \quad v = \frac{v_0}{\phi} + v_1. \tag{4.2}$$

By substituting (4.2) into (4.1), we get that

$$\begin{aligned}
&u_0\phi_x^2 - u_0^2 = 0, \\
&v_0\phi_x^2 - u_0v_0 = 0, \\
&10u_0\phi_\tau = 4u_{0x}\phi_x + 2u_0\phi_{xx} - 2u_1\phi_x^2 + 12u_0u_1 + u_0v_0, \\
&5Dv_0\phi_\tau = 2v_{0x}\phi_x + v_0\phi_{xx} + 2u_1v_0 + 2u_0v_1, \\
&5(u_{0\tau} - u_1\phi_\tau) = u_{0xx} - 2u_{1x}\phi_x - u_1\phi_{xx} + \frac{u_0}{D} - 6u_1^2 - 12u_0u_2 \\
&\qquad - u_0v_1 - u_1v_0, \\
&5Dv_{0\tau} = v_{0xx} - 2u_2v_0 - 2u_1v_1, \\
&5u_{1\tau} = u_{1xx} + \frac{u_1}{D} - 12u_1u_2 - u_2v_0 - u_1v_1
\end{aligned}$$

and

$$\begin{cases} 5Dv_{1\tau} = v_{1xx} - 2v_1u_2, \\ 5u_{2\tau} = u_{2xx} + \frac{u_2}{D} - 6u_2^2 - u_2v_1. \end{cases} \tag{4.3}$$

From the above equations, we know that $u_0 = \phi_x^2$, and that v_0 is an arbitrary function. It corresponds to the resonance point $j = 0$. For simplicity, let $v_0 = 10m\phi_x$, m being a constant. Then

$$u_1 = \phi_\tau - \phi_{xx} - m\phi_x,$$

$$v_1 = 5m(5D-2)\frac{\phi_\tau}{\phi_x} - 5m\frac{\phi_{xx}}{\phi_x} + 10m^2 \tag{4.4}$$

and

$$\begin{aligned} u_2 = &-\frac{\phi_\tau^2}{12\phi_x^2} - \frac{\phi_{x\tau}}{\phi_x} + \frac{m}{12}(7-25D)\frac{\phi_\tau}{\phi_x} + \frac{\phi_\tau\phi_{xx}}{2\phi_x^2} \\ &+ \frac{\phi_{xxx}}{3\phi_x} - \frac{\phi_{xx}^2}{4\phi_x^2} + \frac{m\phi_{xx}}{2\phi_x} + \frac{1}{12}(\frac{1}{D} - 6m^2). \end{aligned} \tag{4.5}$$

Obviously (4.3) is the same as (4.1), and thus it gives the Bäcklund transformation

$$\begin{cases} u = (-\frac{\partial^2}{\partial x^2} - m\frac{\partial}{\partial x} + \frac{\partial}{\partial \tau})\ln\phi + u_2, \\ v = 10m\frac{\partial}{\partial x}(\ln\phi) + v_1. \end{cases}$$

V. The Second Class of Travelling Wave Solutions.

In this section, we look for the second class of travelling wave solutions. By using the notations (2.12)—(2.14), the expressions (4.4) and (4.5) become

$$v_1 = 5m[(2-5D)C + 2Q + 2m] \tag{5.1}$$

and

$$u_2 = C_x - \frac{1}{12}C^2 + \frac{m}{12}(25D-7)C + \frac{1}{3}S - (C+m)Q + Q^2 + \frac{1}{12D} - \frac{m^2}{2}. \tag{5.2}$$

Also we get other four equations.

We now focus on the travelling waves. Let S and C be constants. Then we get

$$[C+m][(15D-6)C^2 + m(15D-12)C - 6m^2 + \frac{1}{2D}] = 0. \tag{5.3}$$

Clearly $C_4 = -m$ is one of the roots of (5.3). The corresponding value of S is as follows

$$S_4 = \frac{1}{2D}(1 - 25m^2D^2).$$

Therefore for $|m| > \frac{1}{5D}$,

$$\phi = \frac{\overline{A}e^{k_4\xi_4} + \overline{B}e^{-k_4\xi_4}}{Ae^{k_4\xi_4} + Be^{-k_4\xi_4}},$$

where

$$k_4 = \frac{1}{2}\sqrt{\frac{25m^2D^2 - 1}{D}}, \qquad \xi_4 = x + m\tau.$$

Hereafter, $A, B, \overline{A}$ and $\overline{B}$ are arbitrary constants provided

$$\overline{A}B - A\overline{B} = \frac{1}{2k_4}.$$

Furthermore

$$\begin{cases} U_2 = \frac{6AB(1-25m^2D^2)}{(Ae^{k_4\eta_4}+Be^{-k_4\eta_4})^2}, \\ V_1 = \frac{25D^2m^2}{r} + \frac{10mDk_4(Ae^{k_4\eta_4}-Be^{-k_4\eta_4})}{r(Ae^{k_4\eta_4}+Be^{-k_4\eta_4})}, \\ k_4 = \frac{1}{2}\sqrt{\frac{25m^2D^2-1}{D}}, \\ \eta_4 = x + 5Dmt. \end{cases} \tag{5.4}$$

The corresponding Bäcklund transformation is

$$\begin{cases} U = -6D(\frac{\partial^2}{\partial x^2} + m\frac{\partial}{\partial x} - \frac{6}{5}\frac{\partial}{\partial t})\ln\phi + \overline{U}, \\ V = \frac{10mD}{r}\frac{\partial}{\partial x}(\ln\phi) + \overline{V}. \end{cases}$$

It can be seen that (4.1) possesses chaotic behaviour.

We next consider the other two roots of (5.3). If $D \neq \frac{2}{5}$ and

$$|m| \geq \frac{1}{5D^2}\sqrt{\frac{2D(5D-2)}{3}},$$

then we have

$$C_5 = \frac{-m(5D-4) - \beta}{2(5D-2)},$$

and

$$C_6 = \frac{-m(5D-4) + \beta}{2(5D-2)},$$

where

$$\beta = \sqrt{25m^2D^2 - \frac{10}{3} + \frac{4}{3D}}.$$

Consequently

$$S_5 = \frac{1}{4(5D-2)^2}(\lambda - m\sigma\beta),$$
$$S_6 = \frac{1}{4(5D-2)^2}(\lambda + m\sigma\beta)$$

where

$$\lambda = -625m^2D^3 + 600m^2D^2 - 150m^2D + \frac{5}{3} - \frac{2}{3D},$$
$$\sigma = 125D^2 - 80D + 10.$$

For simplicity, we only consider a special case. It means that

$$m = \frac{1}{5D\sqrt{6D}}(1 - 5D).$$

In this case, we have that

$$C_5 = -\frac{1}{5D\sqrt{6D}}, \quad S_5 = -\frac{1}{12D}$$

and

$$\begin{aligned} C_6 &= \frac{25D^2 - 20D + 2}{5D(5D-2)\sqrt{6D}}, \\ S_6 &= -\frac{1}{12D^2(5D-2)^2}(625D^4 - 875D^3 + 425D^2 - 84D + 6). \end{aligned} \tag{5.5}$$

Furthermore for $C = C_5$ and $S = S_5$,

$$\begin{cases} U_2 = \left(\frac{Ae^{k_5\eta_5}}{Ae^{k_5\eta_5} + Be^{-k_5\eta_5}}\right)^2, \\ V_1 = \frac{B(5D-1)}{3rD}\frac{e^{-k_5\eta_5}}{(Ae^{k_5\eta_5} + Be^{-k_5\eta_5})}, \\ k_5 = \frac{1}{2\sqrt{6D}}, \\ \eta_5 = x + \frac{1}{\sqrt{6D}}t. \end{cases} \tag{5.6}$$

where A and B are arbitrary constants.

Finally we deal with the case with $C = C_6$ and $S = S_6$. We obtain

$$(25D^3 - 45D^2 + 20D - 2)(5D-1)(5D-3) = 0. \tag{5.7}$$

Obviously (5.7) is fulfilled for $D_1 = \frac{1}{5}$, and $D_2 = \frac{3}{5}$, and the roots of the equation

$$D^3 - \frac{9}{5}D^3 + \frac{4}{5}D - \frac{2}{25} = 0. \tag{5.8}$$

(5.8) possesses three real roots denoted by D_3, D_4 and D_5. It can be verified that for all $D_j, S_6^{(j)} < 0$ and so they lead to five kinds of travelling waves.

VI Discussion

It is shown in [W-X-Y] that (1.2) has travelling wave front solutions $U(x+ct)$ and $V(x+ct)$, satisfying

$$\begin{aligned} &U \to 0, \quad V \to 1, \quad \text{as } x \to -\infty, \\ &U \to 1, \quad V \to 0, \quad \text{as } x \to +\infty. \end{aligned}$$

provided that one of the following conditions are fufilled

$$\text{(i) } b = \frac{3}{2D}, \quad r = \frac{4D-3}{2(D-2)}, \quad \frac{1}{3} \le D < \frac{3}{4}, \tag{6.1}$$

$$\text{(ii) } b = \frac{6}{D+1}, \quad r = \frac{5D-1}{D+1}, \quad D > \frac{1}{5}, \tag{6.2}$$

$$\text{(iii) } b = \frac{3(5-3D)}{D(D+1)}, \quad r = \frac{(7-5D)(D-3)}{D^2-1}, \quad \frac{3}{2} < D < \frac{5}{3}, \tag{6.3}$$

$$\text{(iv) } b = \frac{1}{3D}, \quad r = \frac{5D-1}{3D}, \quad D > \frac{1}{5}. \tag{6.4}$$

Indeed if we take $r = \dfrac{4D-3}{2(D-2)}, \varepsilon = -1$ and $A = B = 1$ in (3.5), then we obtain

$$\begin{cases} U_2 = \dfrac{(e^{2k_1(x+ct)} + \frac{3D-1}{2-D})e^{2k_1(x+ct)}}{(e^{2k_1(x+ct)}+1)^2}, \\ V_2 = \dfrac{1}{(e^{2k_1(x+ct)}+1)^2}, \\ c = \dfrac{5}{2}\sqrt{\dfrac{D}{(3-4D)(2-D)}} \end{cases}$$

which is exactly the same solution as in [W-X-Y], when (6.1) holds. We can also refind all of other solutions in [W-X-Y]. Clearly this paper not only gives the same solutions as in [W-X-Y] with weaker restrictions on parameters of (1.2), but also provides other kinds of special solutions The technique used in this paper is also quite powerful for other nonlinear reaction-diffusion systems.

References

1. Ablowitz,M.J., Ramani,A. and Segur. H.A., *A connection between nonlinear evolution Equations and ordinary differential equations of P-type.I,*, J.Math. Phys., **21** (1980,), 715-721..
2. Ben-yu Guo and Zhi-xiong Chen, *Analytic solutions of the Fisher equation,*, J. Phys. A: Math, Gen., **24,** (1991,), 645-650..

3. Cariello,F. and Tabor, M., *Painlevé expansions for nonintegrable evolution equations,*, Physica D, **39,** (1989,), 77-94.
4. Chen Zhi-xiong and Guo Ben-yu, *Analytic solutions of Nagumo equation,*, IMA J. Appl. Math., **48,** (1992,), 107-115.
5. Conte,R. and Musette,M., *Painlevé analysis and Bäcklund transformation in the Kuramoto-Sivashinsky equation,*, J. Phys. A: Math, Gen., **22,** (1989,), 169-177.
6. Hereman,W., *Application of a Macsyma program for the Painlevé test to the Fitzhugh-Nagumo equation*, (unpublished).
7. Li,Z.Y. and Ye,Q.X., *Travelling wave front solutions for reaction-diffusion systems,*, Accepted by J.Diff. Equations.
8. McLeod, J.B. and Olver,P.J., *The connection between partial differential equations soluble by inverse scattering and ordinary differential equations of Painlevé type,*, SIAM J. Math. Anal., **14,** (1983,), 488-506.
9. Murray,J.D., *Mathematical Biology, Springer Verlag,*, Berin, (1989.).
10. Murray,J.D., *On travelling wave solutions in a model for B-Z reaction,*, J.Theor.Biol., **56,** (1976,), 329-353.
11. Steeb,W.H. and Euler, N., *Nonlinear Evolution Equations and Painlevé Test,*, World Scientific, Singapore, (1988.).
12. Troy,W.C., *The existence of travelling wave front solutions of a model of the Belousov-Zhabotinskii reaction,*, J.Diff. Equations, **36,** (1980,), 89-98.
13. Wang,M., Xiong,S. and Ye,Q., *Explicit wave front solutions of Noyes-Field systems for Belousov-Zhabotinskii reaction,*, To appear in J.Partial Differential Equation.
14. Weiss, J., Tabor, M. and Carnevale, G., *The Painlevé property for partial differential equations,*, J. Math. Phys., **24,** (1983,), 522-526.
15. Ye,Q.X. and Wang M,X., *Travelling wave front solutions of Noyes-Field system for Belousov-Zhabotinskii reaction, Nonlinear Analysis TMA* .
16. Zhi-Xiong Chen and Ben-yu Guo, *Analytic solutions of carrier flow equations via the painlevé analysis approach,*, J. Phys. A: Math, Gen., **22,** (1989,), 5187-5194.

Shanghai University of Science and Technology
Shanghai City, 201800, China

AMS/IP Studies in Advanced Mathematics
Volume **3**, 1997

The Global Attractors for the Landau-Lifshitz Equation of the Ferromagnetic Spin Chain on Compact Manifolds

Guo Boling*, Hong Min-chun** and Su Fengqiu*

1 Introduction

Let Ω be a domain in an n-dimensional Euclidean space R^n. The n-dimensional classical system for the isotropic Heisenberg Chain with spin-density $u^i(x,t)$(i=1,2,3) is described by the Hamiltonian density

$$H = \frac{\alpha_2}{2}|\nabla \mathbf{u}|^2 - H_0\mathbf{u},$$

where α_2 is the exchange constant, $\mathbf{u}$ the spin vector and H_0 an external magnetic field. The spin equation of motion with the Gilbert damping term (without the external magnetic field) has the form:

$$\partial_t \mathbf{u} = -\alpha_1 \mathbf{u}\times(\mathbf{u}\times\Delta\mathbf{u}) + \alpha_2 \mathbf{u}\times\Delta\mathbf{u}, \tag{1.1}$$

where "$\times$" denotes the vector cross product in R^3, $\mathbf{u} = (u^1, u^2, u^3)$: $\Omega\times[0,\infty) \to R^3$ is the spin vector and $\alpha_1 \geq 0$ is a Gilbert damping constant (see [1]). The system (1.1) is implied by the conservation of energy and magnitude of $\mathbf{u}$, and is a version which gives rise to a continuum spin wave theory. The magnitude of the spin is finite, i.e. $|\mathbf{u}|^2 = (u^1)^2 + (u^2)^2 + (u^3)^2 = 1$.

The continuous Heisenberg spin chain aroused considerable interest among physicists. The above precessional equation (1.1) of motion was first derived on phenomenological grounds by Landau-Lifshitz in [2]. The equation (1.1) bears a fundamental role in the understanding of nonequilibrium magnetism. A lot of work contributed to the study of the soliton for the Landau-Lifshitz equation of the one dimensional motion spin chain has been made by physicists and mathematicians [3-6]. But little is known for the higher dimensional motion about the Heisenberg spin chain in physics.

The Landau-Lifshitz equation ($\alpha_1 = 0$) is a nonlinear parabolic system with strong degeneration. The existence of global weak solution for the initial value problems and the initial-boundary value problems of this equation in n ($n \geq 1$) dimensions has been proved in [7-12]. The global smooth solution for the equation (1.1) ($\alpha_1 \geq 0$) in one dimension also has been obtained in [13].

In this paper, let (M, γ) and (N, g) be two compact Riemannian manifolds, M without boundary or has a boundary $\partial M \in C^2$, and N is S^2. Our aim is to prove the existence of the global attractors and give the estimates of their Hausdorff dimensions and fractal dimensions. For the following Landau-Lifshitz type equation of M into S^2

$$\partial_t \mathbf{u} = -\alpha_1 \mathbf{u} \times (\mathbf{u} \times \Delta_M \mathbf{u}) + \alpha_2 \mathbf{u} \times \Delta_M \mathbf{u} \tag{1.2}$$

with the initial-boundary conditions

$$\frac{\partial \mathbf{u}}{\partial \nu}|_{\partial M} = 0, \tag{1.3}$$

$$\mathbf{u}|_{t=0} = \mathbf{u}_0(x),\ |\mathbf{u}_0(x)|^2 = 1,\ x = (x_1, ..., x_n) \in M \tag{1.4}$$

or M without boundary with initial condition (1.4). Where the Laplace-Beltrami operator is expressed by

$$\Delta_M := \frac{1}{\sqrt{\gamma}} \frac{\partial}{\partial x^\beta}(\gamma^{\alpha\beta} \sqrt{\gamma} \frac{\partial}{\partial x^\alpha}) = \gamma^{\alpha\beta} \frac{\partial^2}{\partial x^\alpha \partial x^\beta} - \Gamma^k_{\alpha\beta} \frac{\partial}{\partial x^k},$$

ν denotes the exterior normal vector of ∂M. From [14], we know that, in the classical sense, $\mathbf{u}$ is a solution of problem (1.2) (1.4), if and only if $\mathbf{u}$ is a solution of the problem

$$\partial_t \mathbf{u} = \alpha_1 \frac{1}{\sqrt{\gamma}} \frac{\partial}{\partial x^\alpha}(\gamma^{\alpha\beta} \sqrt{\gamma} \frac{\partial \mathbf{u}}{\partial \mathbf{x}^\beta}) + \alpha_1 |\nabla \mathbf{u}|^2 \mathbf{u}$$

$$+ \alpha_2 \mathbf{u} \times \frac{1}{\sqrt{\gamma}} \frac{\partial}{\partial x^\beta}(\gamma^{\alpha\beta} \sqrt{\gamma} \frac{\partial \mathbf{u}}{\partial \mathbf{x}^\alpha}) \tag{1.5}$$

with the same initial condition (1.4), where M is a n-dimensional Riemannian manifold of the metric γ without boundary. We also know that in the classical sense, $\mathbf{u} : M \to S^2$ is a harmonic map, if and only if $\mathbf{u}$ satisfies (1.2), and $\partial_t \mathbf{u}(x, t) = 0$ for $t \geq 0$, while the heat flow for harmonic maps of M into S^2 is given by

$$\partial_t \mathbf{u} = \Delta_M \mathbf{u} + |\nabla \mathbf{u}|^2 \mathbf{u}. \tag{1.6}$$

This paper is organized as follows: In section 2, the uniformly a priori estimates for $t \geq 0$ to the solution of problem (1.2) (1.3) (1.4) are given. In section 3, under some conditions, the global smooth solution of problem (1.2) (1.4) in two dimensions is obtained, and the global weak solutions of problem (1.2) (1.4) and problem (1.2) (1.3) (1.4) in multi-dimensions are also obtained. In section 4, one prove the existence of bounded absorbing sets and global attractors for the mentioned problems, and the estimates of the upper bounds of Hausdorff and fractal dimensions for the global attractors are given. Finally, the lower bounds on dimensions of attractors of Landau-Lifshitz equation are obtained.

2 The uniform a priori estimates

Let (M, γ) be a compact Riemannian manifold with (or without) boundary, and ∇ denotes the conection (called covaiant derivative) corresponding to the metric γ,

$$|\nabla \mathbf{u}(x)|^2 = \sum_{\alpha\beta}\sum_{i}\gamma^{\alpha\beta}\frac{\partial u^i}{\partial x^\alpha}\frac{\partial u^i}{\partial x^\beta}. \tag{2.1}$$

For a real function φ belonging to $C^k(M)$ ($k{\geq}0$ is an integer), we define:

$$|\nabla^k\varphi|^2 = \nabla^{\alpha_1}.\nabla^{\alpha_2}...\nabla^{\alpha_k}.\nabla_{\alpha_1}.\nabla_{\alpha_2}...\nabla_{\alpha_k}\varphi.$$

In particular, $|\nabla^1\varphi| = |\nabla\varphi|$, $|\nabla^1\varphi|^2 = |\nabla\varphi|^2 = \nabla^\nu\varphi\nabla_\nu\varphi$. $\nabla^k\varphi$ will mean any k-th covariant derivative of φ.

Let us consider the vector space L^p_k of C^∞ function φ, such that $|\nabla^l\varphi|\in L^p(M)$, for all l with $0{\leq}l{\leq}k$, where k and l are integers and $p{\geq}1$ is a real number. The Sobolev space $W^k_p(M)$ is the completion of $L^p_k(M)$ with respect to the norm

$$\|\varphi\|_{W^k_p(M)} = \sum_{i=0}^{k}\|\nabla^i\varphi\|_p.$$

In particular, let $W^k_2(M) = H^k(M)$, $\|.\|_2 = \|.\|$.

Lemma 2.1. Suppose that $|\mathbf{u}_0(x)|^2 = 1$, then for the smooth solution of problem (1.2) (1.3) (1.4), we have

$$|\mathbf{u}(x,t)|^2 = 1,\ \forall\ (x,t){\in}M{\times}[0,\infty), \tag{2.2}$$

Lemma 2.2. Suppose that the condition of Lemma 2.1 is satisfied and assume that $\|\nabla\mathbf{u}_0(x)\| < \infty$, then for the smooth solution of problem (1.2) (1.3) (1.4), we have

$$\|\nabla\mathbf{u}(.,t)\|^2{\leq}\|\nabla\mathbf{u}_0(x)\|^2,$$

$$2\alpha_1\int_0^t\|\mathbf{u}{\times}\Delta_M\mathbf{u}\|^2dt{\leq}\|\nabla\mathbf{u}_0\|^2,\ \forall\ 0{\leq}t<\infty, \tag{2.3}$$

Lemma 2.3. (Sobolev's interpolation inequality on compact Riemannian manifold). M will be a compact Riemannian manifold with smooth boundary. Let q, r be real numbers, $1 \le q,\ r \le \infty$, and j, m integers $0 \le j < m$. Then there exists C, a constant depending only on n, m, j, q, r, and a, and on the manifold, such that for all $f \in W_r^m(M) \cap L_q(M)$,

$$\|\nabla^j f\|_p \le C \|f\|^a_{W_r^m(M)} \|f\|_q^{1-a}, \tag{2.4}$$

where

$$\frac{1}{p} = \frac{j}{n} + a\left(\frac{1}{r} - \frac{m}{n}\right) + (1-a)\frac{1}{q} \tag{2.5}$$

for all a in the interval $\frac{j}{m} \le a \le 1$, for which p is non-negative.

Lemma 2.4. Suppose that the conditions of Lemma 2.2 are satisfied. Assume that

$$\|\nabla \mathbf{u}_0(x)\| \le \lambda, \ (n=2), \tag{2.6}$$

where the constant λ is small, which will be determined in the proof of the lemma, then for the smooth solution of problem (1.2) (1.4), we have

$$\|\Delta_M \mathbf{u}(.,t)\| \le \frac{E_1}{t}, \ \forall \mathbf{x} = (x_1, ..., x_n), \ t > 0, \ n \le 2, \tag{2.7}$$

where the constant E_1 only depends on $\|\nabla \mathbf{u}_0(x)\|_{H^1(M)}$, and $0 < t \le T$, and manifold M without boundary.

Proof. Acting (1.2) by the operator Δ_M, then taking the inner product with $t\Delta_M \mathbf{u}$, we have

$$\begin{aligned}(t\Delta_M \mathbf{u},\ \Delta_M \mathbf{u}_t - \alpha_1 \Delta_M^2 \mathbf{u} - \alpha_1 \Delta_M(|\nabla \mathbf{u}|^2 \mathbf{u}) \\ -\alpha_2 \Delta_M(\mathbf{u} \times \Delta_M \mathbf{u})) = 0,\end{aligned} \tag{2.8}$$

where

$$t(\Delta_M \mathbf{u}, \Delta_M \mathbf{u}_t) = \frac{1}{2}\frac{d}{dt}(t\|\Delta_M \mathbf{u}\|^2) - \frac{1}{2}\|\Delta_M \mathbf{u}\|^2,$$

$$(\Delta_M \mathbf{u}, \alpha_1 \Delta_M^2 \mathbf{u}) = \alpha_1 \int \Delta_M \mathbf{u}. \frac{1}{\sqrt{\gamma}} \frac{\partial}{\partial x^\beta}\left(\gamma^{\alpha\beta}\sqrt{\gamma}\frac{\partial \Delta_M \mathbf{u}}{\partial x^\alpha}\right)\sqrt{\gamma}dx$$

$$= -\alpha_1 \int \gamma^{\alpha\beta}\sqrt{\gamma}\frac{\partial \Delta_M \mathbf{u}}{\partial x^\alpha}\frac{\partial \Delta_M \mathbf{u}}{\partial x^\beta}dx = -\alpha_1 \|\nabla \Delta_M \mathbf{u}\|^2,$$

$$\begin{aligned}|\int \Delta_M \mathbf{u}.\Delta_M(\mathbf{u}\times\Delta_M \mathbf{u})\sqrt{\gamma}dx| &= |\int \Delta_M \mathbf{u}.\frac{\partial}{\partial x^\beta}\left(\gamma^{\alpha\beta}\sqrt{\gamma}\frac{\partial}{\partial x^\alpha}(\mathbf{u}\times\Delta_M \mathbf{u})\right)dx| \\ &= |\int \gamma^{\alpha\beta}\frac{\partial \Delta_M \mathbf{u}}{\partial x^\beta}\frac{\partial(\mathbf{u}\times\Delta_M \mathbf{u})}{\partial x^\alpha}\sqrt{\gamma}dx| \\ &= |\int \gamma^{\alpha\beta}\frac{\partial \Delta_M \mathbf{u}}{\partial x^\beta}\left[\frac{\partial \mathbf{u}}{\partial \mathbf{x}^\alpha}\times\Delta_M \mathbf{u} + \mathbf{u}\times\frac{\partial \Delta_M \mathbf{u}}{\partial x^\alpha}\right]\sqrt{\gamma}dx| \\ &\le C_1 \|\nabla \Delta_M \mathbf{u}\| \|\nabla \mathbf{u}\|_\infty \|\Delta_M \mathbf{u}\|.\end{aligned} \tag{2.9}$$

In (2.9), we have used the fact:

$$\int \gamma^{\alpha\beta}\frac{\partial \Delta_M \mathbf{u}}{\partial x^\beta}.\left(\mathbf{u}\times\frac{\partial \Delta_M \mathbf{u}}{\partial x^\alpha}\right)\sqrt{\gamma}dx = 0. \tag{2.10}$$

Lemma 2.3 implies that

$$\|\nabla \mathbf{u}\|_\infty \le C_2\|\nabla^3\mathbf{u}\|^{\frac{1}{2}}\|\nabla \mathbf{u}\|^{\frac{1}{2}} + C_3,\ (n=2)$$

$$\|\Delta_M \mathbf{u}\| \le C_4\|\nabla^3\mathbf{u}\|^{\frac{1}{2}}\|\nabla \mathbf{u}\|^{\frac{1}{2}} + C_5, \tag{2.11}$$

where the constants C_3 and C_5 depend on $\|\nabla \mathbf{u}_0\|$.

Substituting (2.11) into (2.9), it follows

$$|\int_M \Delta_M \mathbf{u}.\Delta_M(\mathbf{u}\times\Delta_M\mathbf{u})dM| \le 2C_1C_2C_4C_6\|\nabla\mathbf{u}\|\|\nabla^3\mathbf{u}\|^2 + C_6', \tag{2.12}$$

Here

$$\|\nabla\Delta_M\mathbf{u}\| \le C_6\|\nabla^3\mathbf{u}\|,$$

and the constant C_6' depends on $\|\nabla\mathbf{u}_0\|$.

Now we are going to estimate the term $(\Delta_M(|\nabla\mathbf{u}|\mathbf{u}), {\Delta_M}^2\mathbf{u})$ in (2.8).

$$\begin{aligned}
&|(\Delta_M\mathbf{u}, \Delta_M(|\nabla\mathbf{u}|^2\mathbf{u}))| \\
&= |\int_M \Delta_M\mathbf{u}.\frac{\partial}{\partial x^\beta}\gamma^{\alpha\beta}\sqrt{\gamma}\frac{\partial|\nabla\mathbf{u}|^2\mathbf{u}}{\partial x^\alpha}dx| \\
&= |\int_M \gamma^{\alpha\beta}\sqrt{\gamma}\frac{\partial}{\partial x^\alpha}|\nabla\mathbf{u}|^2\mathbf{u}.\frac{\partial}{\partial x^\beta}\Delta_M\mathbf{u}dx| \\
&= |\int_M \gamma^{\alpha\beta}\sqrt{\gamma}((\frac{\partial}{\partial x^\alpha}|\nabla\mathbf{u}|^2)\mathbf{u} + |\nabla\mathbf{u}|^2\frac{\partial\mathbf{u}}{\partial \mathbf{x}^\alpha})\frac{\partial}{\partial x^\beta}\Delta_M\mathbf{u}dx| \\
&= |\int \gamma^{\alpha\beta}[2\gamma^{l\delta}(x)\frac{\partial^2\mathbf{u}}{\partial x^\alpha\partial x^l}\frac{\partial\mathbf{u}}{\partial x^\delta}\mathbf{u} + \gamma^{l\delta}(x)'\frac{\partial\mathbf{u}}{\partial x^l}\frac{\partial\mathbf{u}}{\partial x^\delta}\mathbf{u} \\
&\quad + |\nabla\mathbf{u}|^2\frac{\partial\mathbf{u}}{\partial x^\alpha}]\frac{\partial\Delta_M\mathbf{u}}{\partial x^\beta}dM| \\
&\le C_7[\|\nabla\mathbf{u}\|_\infty\|\mathbf{u}\|_\infty\|\nabla^2\mathbf{u}\|\|\nabla\Delta_M\mathbf{u}\| + \|\nabla\mathbf{u}\|^2\|\mathbf{u}\|_\infty\|\nabla\Delta_M\mathbf{u}\| \\
&\quad + \|\nabla\mathbf{u}\|_\infty^2\|\nabla\mathbf{u}\|\|\nabla\Delta_M\mathbf{u}\|] \\
&\le C_7[2C_2C_4\|\nabla\nabla\mathbf{u}\| + (2C_2^2+1)\|\nabla\mathbf{u}\|^2]\|\nabla^3\mathbf{u}\| + C_8,
\end{aligned} \tag{2.13}$$

where the constants C_7 and C_8 depend on $\|\nabla\mathbf{u}_0\|$ and $\sup_{x\in M}(|\gamma^{\alpha\beta}(x)|, |\gamma^{\alpha\beta'}|(x))$.

Hence from (2.8) (2.12) (2.13), we have

$$\begin{aligned}
&\frac{1}{2}\frac{d}{dt}t\|\Delta_M\mathbf{u}\|^2 - \frac{1}{2}\|\Delta_M\mathbf{u}\|^2 + \alpha_1 t\|\nabla\Delta_M\mathbf{u}\|^2 \\
&\le 2t[|\alpha_2|C_1C_2C_4C_6 + \alpha_1C_7C_2C_4 \\
&\quad + \alpha_1(C_2^2+1)C_7\|\nabla\mathbf{u}\|]\|\nabla\mathbf{u}\|\|\nabla^3\mathbf{u}\| + C_9.
\end{aligned} \tag{2.14}$$

Now we estimate the lower bound for $\|\nabla\Delta_M\mathbf{u}\|^2$.

By the Ricci formula

$$\Delta(\nabla^k f) = \nabla^k(\Delta f) + \sum_{i=0}^{k-1} S_{ki}(\nabla^{k-i} f), \tag{2.15}$$

where S_{ki} is a linear functional depending on the covariant derivative $\nabla^i R$ of curvature tensor, we get

$$\begin{aligned} \|\nabla\Delta_M\mathbf{u}\|_2 &= \|\Delta_M(\nabla\mathbf{u}) - S_{10}(\nabla\mathbf{u})\| \\ &\geq \|\Delta_M(\nabla\mathbf{u})\| - S_{10}\|\nabla\mathbf{u}\|. \end{aligned} \tag{2.16}$$

Since

$$\begin{aligned} &\|\Delta_M(\nabla\mathbf{u})\|^2 \\ &= \int_M \sum_j (\sum_i u_{jii})(\sum_k u_{jkk}) dM \\ &= \sum_j \int_M (\sum_i u_{jii})(\sum_k u_{jkk}) dM \\ &= -\sum_j \int_M (\sum_i u_{ji})(\sum_k u_{jkki}) dM \\ &= -\sum_j \int_M (\sum_i u_{ji})(\sum_k (u_{jkik} + u_{jl}R^l_{kki} + u_{lk}R^l_{jki}) dM \\ &= -\sum_j \int_M (\sum_i u_{ji})(\sum_k (u_{jikk} + u_l R^l_{jkik} \\ &\quad + u_{lk}R^l_{jki} + u_{jl}R^l_{kki} + u_{lk}R^i_{jki}) dM \\ &\geq \int_M |\nabla^3\mathbf{u}|^2 dM - C_{11}\int_M |\nabla\mathbf{u}|^2 dM. \end{aligned} \tag{2.17}$$

Hence from (2.14) (2.16) (2.17), we have

$$\begin{aligned} &\frac{d}{dt} t\|\Delta_M\mathbf{u}\|^2 - \|\Delta_M\mathbf{u}\|^2 + 2t[\alpha_1 - (2\alpha_2 C_1C_2C_4C_6 + 2\alpha_1 C_7C_2C_4 \\ &+ 2\alpha_1(C_2^2+1)C_7\|\nabla\mathbf{u}\| + C_{12}\|\nabla\mathbf{u}\|)]\|\nabla\mathbf{u}\|\|\nabla^3\mathbf{u}\|^2 \leq C_{13}. \end{aligned} \tag{2.18}$$

Choosing $\|\nabla\mathbf{u}_0\|$ suitably small, such that

$$\begin{aligned} &\alpha_1 - (2|\alpha_2|C_1C_2C_4C_6 + 2\alpha_1C_7C_2C_4 + 2\alpha_1(C_2^2+1)C_7\|\nabla\mathbf{u}_0\| \\ &+ C_{12}\|\nabla\mathbf{u}_0\|)\|\nabla\mathbf{u}_0\| > \frac{\alpha_1}{2} > 0. \end{aligned}$$

From (2.18), we get

$$t\|\Delta_M\mathbf{u}\|^2 - \int_0^t \|\Delta_M\mathbf{u}\|^2 dt + \frac{1}{2}\alpha\int_0^t t\|\Delta_M^3\mathbf{u}\|^2 dt$$
$$\leq C_{13}t. \tag{2.19}$$

In order to estimate the term $\int_0^t\|\Delta_M\mathbf{u}\|^2 dt$ in the left side of inequality (2.19), we need the following lemma.

Lemma 2.5. Under the conditions of Lemma 2.2, we have

$$\int_0^t \|\frac{\partial\mathbf{u}}{\partial\mathbf{t}}\|^2 dt \leq \frac{(\alpha_1^2+\alpha_2^2)}{\alpha_1}\|\nabla\mathbf{u}_0\|^2, \ \forall t\in R^+, \tag{2.20}$$

$$\int_0^t \|\Delta_M\mathbf{u}\|^2 dt \leq C_{14}, \ \ 0\leq t\leq T, \tag{2.21}$$

where the constant C_{14} depends on $\|\nabla\mathbf{u}_0\|$ and T.

Proof. Multiplying (1.5) by $\partial_t\mathbf{u}$ and integrating for x, $t\in M\times[0,t)$, we have

$$\int_0^t\int_M |\partial_t\mathbf{u}|^2 dM dt + \frac{\alpha_1}{2}\int_0^t \frac{d}{dt}\|\nabla\mathbf{u}\|^2 dt$$
$$-\alpha_2\int_0^t\int_M \partial_t\mathbf{u}.(\mathbf{u}\times\Delta_M\mathbf{u})dM dt = 0. \tag{2.22}$$

By the cross product of equation (1.5) with $\mathbf{u}$, we get

$$\begin{aligned}\mathbf{u}\times\partial_t\mathbf{u} &= \alpha_1(\mathbf{u}\times(\Delta_M\mathbf{u}) + \alpha_2(\mathbf{u}\times(\mathbf{u}\times\Delta_M\mathbf{u}))\\ &= \alpha_1(\mathbf{u}\times\Delta_M\mathbf{u}) - \alpha_2\Delta_M\mathbf{u} - \alpha_2|\nabla\mathbf{u}|^2\mathbf{u}.\end{aligned}$$

Because

$$-\Delta_M\mathbf{u} - |\nabla\mathbf{u}|^2\mathbf{u} = -\frac{1}{\alpha_1}\partial_t\mathbf{u} + \frac{\alpha_2}{\alpha_1}(\mathbf{u}\times\Delta_M\mathbf{u}).$$

This gives

$$\mathbf{u}\times\partial_t\mathbf{u} + \frac{\alpha_2}{\alpha_1}\partial_t\mathbf{u} = (\alpha_1 + \frac{\alpha_2^2}{\alpha_1})(\mathbf{u}\times\Delta_M\mathbf{u}).$$

Multiplying the above equation with $\partial_t\mathbf{u}$, we have

$$\partial_t\mathbf{u}.(\mathbf{u}\times\Delta_M\mathbf{u}) = \alpha_2(\alpha_1^2+\alpha_2^2)^{-1}|\partial_t\mathbf{u}|^2.$$

Then we have

$$\alpha_2\int_0^t\int_M \partial_t\mathbf{u}.(\mathbf{u}\times\Delta_M\mathbf{u})dM dt = \frac{\alpha_2^2}{\alpha_1^2+\alpha_2^2}\int_0^t\int_M |\partial_t\mathbf{u}|^2 dM dt.$$

From (2.21), it follows

$$\frac{\alpha_1^2}{\alpha_1^2+\alpha_2^2}\int_0^t\int_M |\partial_t\mathbf{u}|^2 dM dt + \frac{\alpha_1}{2}(\|\nabla\mathbf{u}(.,t)\|^2 - \|\nabla\mathbf{u}_0\|^2) = 0,$$

i.e.,

$$\int_0^t \int_M |\partial_t \mathbf{u}|^2 dM dt \leq \frac{\alpha_1^2 + \alpha_2^2}{2\alpha_1} \|\nabla \mathbf{u}_0\|^2, \quad \forall t \in R^+. \tag{2.23}$$

From (1.5), we have

$$\int_0^t \|\Delta_M \mathbf{u}\|^2 dt \leq C_{15}\Big(\int_0^t \|\mathbf{u}_t\|^2 dt + \int_0^t \|\mathbf{u}\times\Delta_M \mathbf{u}\|^2 dt + \int_0^t \int_M |\nabla \mathbf{u}|^4 dM dt\Big). \tag{2.24}$$

By the Sobolev's inequality (2.4), it follows

$$\int_M |\nabla \mathbf{u}|^4 dM \leq C_{16} \|\nabla^2 \mathbf{u}\|^2 \|\nabla \mathbf{u}\|^2 + C_{17}, \; 0 \leq t \leq T, \tag{2.25}$$

where the constants C_{17} depends on $\|\nabla \mathbf{u}_0\|$ and $0 \leq t \leq T$.

The definition of Laplace-Beltrami operator implies

$$C_{18}\|\nabla^2 \mathbf{u}\|^2 + C_{19}\|\nabla \mathbf{u}\|^2 \geq \|\Delta_M \mathbf{u}\|^2 \geq C_{18}\|\nabla^2 \mathbf{u}\|^2 - C_{19}\|\nabla \mathbf{u}\|^2, \tag{2.26}$$

where the constants C_{18} and C_{19} depend on $\sup_{x\in M}|\gamma^{\alpha\beta}(x)|$ and $\sup_{x\in M}|\Gamma^k_{\alpha\beta}|$ respectively. Therefore, from (2.23), we get

$$C_{18}\int_0^t \|\nabla^2 \mathbf{u}\|^2 dt - C_{15}C_{16}\int_0^t \|\nabla^2 \mathbf{u}\|^2 dt . \|\nabla \mathbf{u}_0\|^2 \leq C_{15}\Big(\int_0^t \Big\|\frac{\partial \mathbf{u}}{\partial \mathbf{t}}\Big\|^2 dt + \int_0^t \|\mathbf{u}\times\Delta_M \mathbf{u}\|^2 dt + C_{17}\Big) + C_{19}\|\nabla \mathbf{u}_0\|^2 t. \tag{2.27}$$

Choosing $\|\nabla \mathbf{u}_0\|$ suitably small, such that

$$C_{18} - C_{15}C_{16}\|\nabla \mathbf{u}_0\|^2 \geq \frac{C_{18}}{2}. \tag{2.28}$$

From (2.22) (2.3) (2.25) (2.26), we have

$$\int_0^t \|\Delta_M \mathbf{u}\|^2 dt \leq C_{14}, \quad 0 \leq t \leq T. \tag{2.29}$$

The lemma is proved.

From (2.28) and (2.19), we get

$$\|\Delta_M \mathbf{u}\|^2 \leq \frac{E_1}{t}, \forall\, x = (x_1, x_2, ..., x_n), \; 1 \leq n \leq 2, \; t > 0, \tag{2.30}$$

where the constant E_1 only depends on $\|\nabla \mathbf{u}_0\|$ and T.

Now we prove that the inequality (2.7) holds in the case n=1 without the restriction $\|\nabla \mathbf{u}_0\| \leq \lambda$. In fact, from Lemma 2.3, we have

$$\|\nabla \mathbf{u}\|_\infty \leq C_2 \|\nabla^3 \mathbf{u}\|^{\frac{1}{4}} \|\nabla \mathbf{u}\|^{\frac{3}{4}} + C_3,$$

$$\|\Delta_M \mathbf{u}\| \leq C_4 \|\nabla^3 \mathbf{u}\|^{\frac{1}{2}} \|\nabla \mathbf{u}\|^{\frac{1}{2}} + C_5, (n = 1),$$

$$\int_M |\nabla \mathbf{u}|^4 dM \leq C_{16} \|\nabla^2 \mathbf{u}\| \|\nabla \mathbf{u}\|^3 + C_{17}.$$

The conclusion of the inequality (2.7) is easy to be obtained.

3 The existence of global solution of problem (1.2) (1.4)

Now we consider the existence of global smooth solution for the initial value problem

$$\partial_t \mathbf{z} = -\alpha_1 \mathbf{z} \times (\mathbf{z} \times \Delta_M \mathbf{z}) + \alpha_2 \mathbf{z} \times \Delta_M \mathbf{z}, \tag{3.1}$$

$$\mathbf{z}|_{t=0} = \mathbf{z}_0(x),\ |\mathbf{z}_0(x)|^2 = 1,\ x = (x_1, x_2, ..., x_n) \in M,\ 1 \leq n \leq 2, \tag{3.2}$$

where M is a compact manifold without boundary.

By using a priori estimates and continuously extended theorem for local solution, we have

Theorem 3.1. Let M be a compact manifold without boundary, and suppose that the following conditions are satisfied:

(1) $\mathbf{z}_0(x) \in H^m(M),\ m \geq 2,\ |\mathbf{z}_0(x)|^2 = 1,$

$$x = (x_1, ..., x_n) \in M,\ 1 \leq n \leq 2,$$

(2) In the case of n=2,

$$\|\nabla \mathbf{z}_0(x)\| \leq \lambda, \tag{3.3}$$

where the constant λ is suitably small. Then there exists a unique global smooth solution $\mathbf{z}(x,t) : M \times [0, \infty) \to S^2$ of the initial value problem (3.1) (3.2)

$$\mathbf{z}(x,t) \in L^\infty(0, \infty; H^m(M)).$$

Under the constraint $|\mathbf{z}| = 1$, we can prove that the equation (3.1) is equivalent to the following equation

$$\frac{\alpha_1}{\alpha_1^2 + \alpha_2^2} \partial_t \mathbf{z} - \frac{\alpha_2}{\alpha_1^2 + \alpha_2^2} \mathbf{z} \times \partial_t \mathbf{z} = \Delta_M \mathbf{z} + |\nabla \mathbf{u}|^2 \mathbf{z}. \tag{3.4}$$

First, we consider the initial boundary value problem for equation (3.4) with the following initial boundary conditions

$$\mathbf{z}|_{t=0} = \mathbf{z}_0(x),\ |\mathbf{z}_0(x)| = 1,\ x \in M, \tag{3.5}$$

$$\frac{\partial \mathbf{z}}{\partial \nu}|_{\partial M} = 0,\ t{\geq}0, \tag{3.6}$$

where ν is the outward normal of the boundary $\partial M{\in}C^2$ of M.

By using penalized Galerkin method, we have

Theorem 3.2. Let $\alpha_1 > 0$ in (3.4), and for all $\mathbf{x} = (x_1, ..., x_m){\in}M, m{\geq}3$, $|\mathbf{z}_0(x)|^2 = 1$, $\mathbf{z}_0(x){\in}H^1(M)$, then there exists a global weak solution of the initial boundary value problem (3.4) (3.5) (3.6), $\mathbf{z}(x,t){\in}L^\infty(0,\infty;H^1(M))$ $\cap C^{(\frac{1}{2},\frac{1}{2})}(0,T;L_2(M))$ and $\partial_t\mathbf{z}{\in}L_2(0,\infty;L_2(M))$.

Next, we consider the following initial boundary value problem for the generalized Landau-Lifshitz equation with several variables

$$\mathbf{z}_t = -\alpha_1\mathbf{z}{\times}(\mathbf{z}{\times}\Delta_M\mathbf{z}) + \alpha_2\mathbf{z}{\times}\Delta_M\mathbf{z}$$

$$+\mathbf{f}(x,t,\mathbf{z}),\ x{\in}M,\ t{\in}R^+, \tag{3.7}$$

$$\frac{\partial \mathbf{z}}{\partial \nu}|_{\partial M} = 0,\ \forall\, t{\in}R^+, \tag{3.8}$$

$$\mathbf{z}|_{t=0} = \mathbf{z}_0(x),\ \ x = (x_1, ..., x_m){\in}M, \tag{3.9}$$

where $\mathbf{z} = (z_1, z_2, z_3)$ is a 3-dimensional vector valued unknown function, $\mathbf{f}(x,t,\mathbf{z})$ is a given 3-dimensional vector function in $x{\in}M$, $t{\in}R^+$, $\mathbf{z}{\in}R^3$.

Suppose that the following conditions are satisfied:

(1) The 3×3 Jacobi derivative matrix $\frac{\partial \mathbf{f}(x,t,\mathbf{z})}{\partial \mathbf{z}}$ is semi- bounded, that is, there exist a constant b such that for any 3-dimensional vector $\xi{\in}R^3$,

$$\xi.\frac{\partial \mathbf{f}}{\partial \mathbf{z}}\xi{\leq}b|\xi|^2, \tag{3.10}$$

where "," denotes the scalar product in the 3-dimensional Euclidean space.

(2) The system (3.7) is homogeneous, that is $\mathbf{f}(x,t,\mathbf{0}) = \mathbf{0}$. Furthermore, suppose that $\mathbf{f}(x,t,\mathbf{z})$ itself and its derivatives with respect to x satisfy the following inequalities respectively:

$$|\mathbf{f}(x,t,\mathbf{z})|{\leq}A|\mathbf{z}|^l + B,\ 2{\leq}l{\leq}2 + \frac{4}{m-2}\ (m{\geq}2), \tag{3.11}$$

$$|\nabla_x\mathbf{f}(x,t,\mathbf{z})|{\leq}A|\mathbf{z}|^{1+\frac{2}{m}} + B,\ t{\in}R^+, \tag{3.12}$$

where A and B are given nonnegative constants.

(3) The components of 3-dimensional initial vector valued function $\mathbf{z}_0(x)$ belong to $H^1(M)$.

By using the Galerkin method and the integral estimates, we can get

Theorem 3.3. Suppose that the assumptions (1) (2) and (3) are fulfilled. Then the homogeneous boundary value problem (3.7) (3.8) (3.9) has at least one global weak solution

$$\mathbf{z}(x,t){\in}L_\infty(0,T;H^1(M)){\cap}C^{(\frac{1}{2},\frac{1}{r^*})}(0,T;L^r(M)),$$

where

$$r = min\{\frac{2m}{l(m-2)}, \frac{m}{m-1}\},\ (m{\geq}2).\ \frac{1}{r} + \frac{1}{r^*} = 1.$$

4 The global attractors and its dimensions

In this section, we prove the existence of the global attractors for the mensioned problems and give the upper and lower estimates of their Hausdorff and fractal dimensions.

Theorem 4.1. Let $\mathbf{f}(x,t,\mathbf{z}) = \mathbf{f}(\mathbf{z})+\mathbf{g}(x)$, and suppose that the following conditions are satisfied:

(1) The 3×3 Jacobi derivative matrix $\frac{\partial \mathbf{f}(\mathbf{z})}{\partial \mathbf{z}}$ is semi-bounded, that is, there exists a negative constant b, such that

$$\xi.(\frac{\partial \mathbf{f}}{\partial \mathbf{z}}\xi)\leq b|\xi|^2, \tag{4.1}$$

(2) $\mathbf{f}(0) = \mathbf{0}$.

$$|\mathbf{f}(\mathbf{z})|\leq A|\mathbf{z}|^l + B, \quad 2\leq l\leq 2+\frac{4}{m-2}, \tag{4.2}$$

where A and B are given nonnegative constants.

(3) $\mathbf{z}_0(x)$, $\mathbf{g}(x)\in H^1(M)$.

Then the initial boundary value problem (3.7) (3.8) (3.9) has a global attractor A, which is weakly compact in $H^1(M)$, where

$$\mathrm{A} = \omega^w(B_1) = \bigcap_{s\geq 0}\bigcap_{t\geq s}\overline{\mathrm{S}(t)B_1}. \tag{4.3}$$

Here $\mathrm{S}(t)\mathbf{z}_0$ denotes the semigroup generated by problem (3.7) (3.8) (3.9),

$$B_1 = \{\mathbf{z}\in H^1(M), \|\mathbf{z}\|_{H^1(M)}\leq\rho_1\} \tag{4.4}$$

is a bounded absorbing set for $\mathrm{S}(t)$ in $H^1(M)$. Here ρ_1 is a positive constant. The closure are taken with respect to the weak topology of $H^1(M)$.

Proof. From Theorem 3.3, we know that the problem (3.7) (3.8) (3.9) generates a semi-group $\mathrm{S}(t)\mathbf{z}_0$. Taking the inner product (3.7) with $\mathbf{z}$, we have

$$\|\mathbf{z}(x,t)\|^2\leq exp(bt)\|\mathbf{z}_0(x)\|^2 + \frac{1}{b^2}(1-exp(bt))\|\mathbf{g}(x)\|^2.$$

Therefore

$$\begin{aligned}\|\mathbf{z}(x,t)\|^2 &= \|\mathrm{S}(t)\mathbf{z}_0(x)\|^2\\ &\leq\frac{2}{b^2}\|\mathbf{g}(x)\|^2 + \|\mathbf{z}_0(x)\|^2 = E_0\end{aligned} \tag{4.5}$$

for $t\geq T_0 = T_0(\|\mathbf{z}_0(x)\|, \|\mathbf{g}(x)\|)$.

Taking the inner product (3.7) with $\Delta_M\mathbf{z}$, we get

$$\|\nabla\mathbf{z}(.,t)\|^2\leq exp(bt)\|\nabla\mathbf{z}_0(x)\|^2 + \frac{1}{b^2}(1-exp(bt))\|\nabla\mathbf{g}\|^2.$$

Thus

$$\|\nabla\mathbf{z}(.,t)\|^2\leq\frac{2}{b^2}\|\nabla\mathbf{g}(x)\|^2 + \|\nabla\mathbf{z}_0\|^2 = E_1. \tag{4.6}$$

Hence $B_1 = \{\mathbf{z}\in H^1(M), \|\mathbf{z}\|_{H^1(M)} \le E_0 + E_1 = \rho_1\}$ is the bounded absorbing set of the semi-group operator $S(t)$, and B_1 is weakly compact in $H^1(M)$. From [18, Theorem 1.1], $A = \omega^w(B_1) = \bigcap_{s\ge 0}\bigcap_{t\ge s}\overline{S(t)B_1}$ is a global attractor. The set $\omega^w(B_1)$ is nonempty, weakly compact and invariant, attracts B_1 in weak sence.

Similarly, we have

Theorem 4.2. Assume that the conditions of Theorem 3.1 are satisfied. Then the initial value problem (3.1) (3.2) of Landau- Lifshitz equation with Gilbert term has an attractor A which is compact in $H^1(M)$, where

$$A = \omega(B^1) = \bigcap_{s\ge 0}\bigcap_{t\ge s}\overline{S(t)B_1}, \tag{4.7}$$

Here

$$B_1 = \{\mathbf{z}\in H^1(M), |\mathbf{z}| = 1, \|\mathbf{z}\|_{H^1(M)} \le \rho_1\} \tag{4.8}$$

is a bounded absorbing set for $S(t)$ in $H^1(M)$, ρ_1 is a positive constant.

Now we give the estimates of the upper and lower bounds of the Hausdorff and fractal dimensions for the attractors of mensioned problems.

We consider the linear variation corresponding to the problem (3.1) (3.2)

$$\mathbf{v}(t) + L(\mathbf{z}(t))\mathbf{v} = \mathbf{0}, \tag{4.9}$$

$$\mathbf{v}(0) = \mathbf{v}_0(x), \tag{4.10}$$

where

$$\begin{aligned} L(\mathbf{z}(t))\mathbf{v} &= -\alpha_1\Delta_M\mathbf{v} - \alpha_1|\nabla\mathbf{z}|^2\mathbf{v} - 2\alpha_1\nabla\mathbf{z}\mathbf{z}\nabla\mathbf{v} \\ &\quad -\alpha_2 A'(\mathbf{z})\Delta_M\mathbf{z}\mathbf{v} - \alpha_2 A(\mathbf{z})\Delta_M\mathbf{v}. \end{aligned} \tag{4.11}$$

$$\mathbf{z} = (z_1, z_2, z_3)^T,$$

$$A(\mathbf{z}) = \begin{pmatrix} 0 & -z_3 & z_2 \\ z_3 & 0 & -z_1 \\ -z_2 & z_1 & 0 \end{pmatrix}.$$

In order to estimate the upper bounds of the Hausdorff and fractal dimensions of attractors, according to [19], we need to estimate the lower bounds of $Tr(L(\mathbf{z}(t)).Q_J)$, where "Tr" denotes the trace of operator, and $Q_J(t)$ the orthogonal projection of space $L_2(M)$ to the spanning subspace generated by $V_1(t), V_2(t), ..., V_J(t)$, the solution of the linear equation (4.9) corresponding respectively to the initial data $\mathbf{V}_1(0) = \xi_1, ..., \mathbf{V}_J(0) = \xi_J$. From the following lemmas:

Lemma 4.1[21]. Let (M,g) be a Riemannian manifold of dimension n. For every p with

$$max(1, \frac{n}{2m}) < p \le 1 + \frac{n}{2m},$$

there exist two positive constants K(M) and $\chi(M)$, such that for every finite family $\{\varphi_1, ..., \varphi_m\}$ in $H^m(M)$, which is (sub-orthonormal) orthonomal in $L^2(M)$, we have

$$(\int_M \rho^{\frac{p}{p-1}} dM)^{\frac{2m(p-1)}{n}} \leq K(M) \sum_{j=1}^N \int_M |\nabla^m \varphi_j|^2 dM$$

$$+\chi(M) \int_M \rho dM. \tag{4.12}$$

Here $\rho = \sum_{j=1}^N |\varphi_j(x)|^2$, and constants K(M) and $\chi(M)$ depend on m, n, p and (M,g).

Lemma 4.2[16]. Let M be a compact Riemannian manifold without (or with) boundary, and $\{\lambda_j\}$ is the spectrum of M with respect to Laplace-Beltrami operator. Then the following inequalities hold

$$\lambda_k \geq \frac{\delta}{e} (\frac{k}{Vol(M)})^{\frac{2}{n}}, \tag{4.13}$$

where n=dim M and δ is the Sobolev constant of M, that is, for any $\mathbf{z} \in C^\infty(M)$

$$\int_M |\nabla \mathbf{z}|^2 dM \geq \delta (\int_M |\mathbf{z}|^{\frac{2n}{n-2}})^{\frac{n-2}{n}}. \tag{4.14}$$

We can get

Theorem 4.3. Under the conditions of Theorem 4.2, then the Hausdorff and fractal dimensions of the global attractor A of the problem (3.1) (3.2) are finite, and

$$d_H(\mathrm{A}) \leq J_0, \ d_F(\mathrm{A}) \leq 2J_0, \tag{4.15}$$

where J_0 is the smallest integer, that is such that

$$J_0 \geq C_0 \alpha_1^{\frac{-4n}{(4-n)(2+n)}}, \ (1 \leq n < 4), \tag{4.16}$$

where C_0 is a constant depends on M and the norms $\|\mathbf{z}\|_\infty$, $\|\nabla \mathbf{z}\|$ and $\|\nabla^2 \mathbf{z}\|_2$.

Similarly, we can get the estimates of the Hausdorff and fractal dimensions of the attractors for problem (3.7) (3.8) (3.9).

Finally, we consider the initial boundary value problem for the following Landau-Lifshitz equation

$$\mathbf{z}_t = -\alpha_1 (\mathbf{z} \times (\mathbf{z} \times \Delta_M \mathbf{z})) + \mathbf{z} \times J\mathbf{z}, \tag{4.17}$$

$$\mathbf{z}|_{t=0} = \mathbf{z}_0(x), \quad x \in M, \tag{4.18}$$

$$\frac{\partial \mathbf{z}}{\partial \nu}|_{\partial M} = 0, \quad t \geq 0, \tag{4.19}$$

where $J = diag(J_1, J_2, 0)$, $\alpha_1 > 0$. From [20], we can get the lower bounds on dimensions of the attractors for the initial boundary value problem (3.7) (3.8) (3.9), we have

Theorem 4.4. Assume that A is the global attractor of problem (4.17) (4.18) (4.19), and $J_1 J_2 < 0$. Then we have

$$dim\mathrm{A} \geq C\alpha_1^{-\frac{n}{2}},$$

where dimA is the Hausdorff or fractal dimension of A, C is a positive constant.

References

[1] M.Lakshmanan and K.Nakamura, Landau-Lifshitz equation of ferromagnetism: Exact treatment of the Gilbert Damping, Phys. Rew. Lett. 53(26), 1984, 2497- 2499.

[2] L.D.Landau and E.M.Lifshitz, On the theory of the dispersion of magnetic permeability in ferromagnetic bodies, Phys. Z. Sowej., 8, 1935, 153-, Reproduced in 1965, 101-114, collected papers of L.D.Landau (D.ter Haar, eds.), Pergamon, New york.

[3] H.C.Fogedby, Theoretical aspects of mainly low dimensional magnetic systems, Lecture Notes in Phys.(131), Springer-Verlag, Berlin, Heideberg, 1980.

[4] K.Nakamura and T.Sasada, Soliton and wave trains in ferromagnets, Phys. Lett., 48(A), 1974, 321-322.

[5] L.A.Takhtajan, Integration of the continuous Heisenberg spin chain through the inverse scattering method, Phys. Lett. 64(A), 1977, 235.

[6] J.Tjon and J.wright, Soliton in the continuous Heisenberg chain, Phys. Rev., 15B, 1977, 3470-3476.

[7] Zhou Yulin and Guo Boling, On the solvability of the initial value problem for the quasilinear degenerate parabolic system, Proceedings of DD-3 symposium, 1982, 713-732.

[8] Zhou Yulin and Guo Boling, Existence of weak solution for boundary problem of ferromagnetic chain, Scientia Sinica (A), 27(8), 1984, 799-811.

[9] Zhou Yulin and Guo Boling, Some boundary problems of the spin system and the system of ferromagnetic chain I: Nonlinear boundary problem, Acta Math. Scientia, 6(3), 1986, 321-337.

[10] Zhou Yulin and Guo Boling, Some boundary problems of the spin system and the system of ferromagnetic chain II: Mixed problems and others, Acta Math. Scientia, 7(2), 1987, 121-132.

[11] Zhou Yulin and Guo Boling, The weak solutions of homogeneous boundary value problem for the system of ferromagnetic chain with several variables, Scientia Sinica, 4(A), 1986, 337-349.

[12] Zhou Yulin, Sun Hesheng and Guo Boling, Geometrical extensions for systems of ferromagnetic chain, Advances in Math., China, 21(4), 1992, 497-501.

[13] Zhou Yulin, Guo Boling and Tan Shaobin, Existence and uniqueness of smooth solution for system of ferromagnetic chain, Science in China, 34(A), 1991, 257-266.

[14] Boling Guo and Min-Chun Hong, The Landau- Lifshitz equation of the ferromagnetic spin chain and harmonic maps, Calculus of variations and PDE, 1, 1993, 311-334.

[15] T.Aubin, Nonlinear Analysis on Manifolds, Monge-Amp ere Equations, Springer-Verlag, New York, Heidelberg, Berlin, 1982.

[16] R.Schoen and S.T.Yau, Differential Geom. Academic Press, Sinica, 1988.

[17] A.V.Babin and M.I.Visik, Attractors of partial differential equations and estimate of their dimension, Uspekhi Mat. Nank, 38, 1983, 133-187.

[18] R.Temam, Infinite Dimensional Dynamical in mechanics and Physics, Springer-Verlag, 1988.

[19] P.Constantin, C.Foias and R.Temam, Attractor representing tubulent flows, Mem. Am. Math. Sci., 53, No.3, 1985, 314.

[20] A.V.Babin and M.I.Vishik, Regular attractors of semi-groups and evolution equations, J. Math. Pures Appl., 62, 1983, 441-491.

* Institute of Applied Physics
and Computational Mathematics
P.O.Box 8009, Beijing, P.R.C

** Department of Mathematics, Zhejiang University
Hangzhou, P.R.C

AMS/IP Studies in Advanced Mathematics
Volume **3**, 1997

DISCONTINUOUS SOLUTIONS OF THE NAVIER-STOKES EQUATIONS FOR MULTIDIMENSIONAL, COMPRESSIBLE FLOW

DAVID HOFF*

We discuss some recent results concerning the global existence of weak solutions of the Navier–Stokes equations

$$\rho_t + \operatorname{div}(\rho u) = 0 \tag{1}$$

$$(\rho u^j)_t + \operatorname{div}(\rho u^j u) + P(\rho)_{x_j} = \varepsilon \Delta u^j + \lambda \operatorname{div} u_{x_j} \tag{2}$$

in two and three space dimensions when the Cauchy data (ρ_0, u_0) is small but possibly discontinuous. Here $t \geq 0$ is time, $x \in I\!R^n, n = 2, 3$, is the spatial coordinate, and $\rho(x,t), u(x,t) = \big(u^1(x,t), \ldots, u^n(x,t)\big)$, and $P = K\rho$ represent the fluid density, velocity, and pressure. $\varepsilon > 0$ and $\lambda \geq 0$ are viscosity constants, and div and Δ are the usual spatial divergence and Laplace operators. Specifically, we fix a positive constant density $\tilde{\rho}$, and we assume that $\rho_0 - \tilde{\rho}$ is small in L^∞, that $\rho_0 - \tilde{\rho}$ and u_0 are small in L^2 (when $n = 2$ the L^2 norm must be weighted slightly), and that u_0 is bounded in L^{2^n}. In particular, the initial data may be discontinuous across a hypersurface of $I\!R^n$. A more precise formulation of our results will be given below, and complete technical details may be found in [7].

Several significant mathematical and physical issues arise in the analysis of the Cauchy problem for such general data. One is the need to find a mechanism in the solution operator which enforces appropriate pointwise bounds on the density ρ in the absence of information concerning $\nabla\rho$ or even $\operatorname{Var}(\rho)$. More generally, it is both physically significant and mathematically necessary to understand in detail precisely which quantities are smoothed out in the flow, which are not, and how discontinuities evolve in the fluid. For example, we find that in the natural splitting of the velocity gradient, the antisymmetric part is relatively smooth, and the symmetric part is not. One of the important observations we make is that the trace of the symmetric part, plus a suitable multiple of the pressure, is in fact smooth. This quantity, which we call the "effective viscous flux" plays

---

* This research was supported in part by the NSF under Grant No. DMS-9201597.

a crucial role in the entire analysis. Related to this question of smoothing, we find that it is necessary to understand in a detailed way the various rates of regularization near the initial layer $t = 0$; these rates are coupled, and are largely determined by the dimension of the space via the relevant Sobolev inequalities. We shall discuss these issues in greater detail below, following the statement of our main theorem.

The questions addressed here are motivated by the fact that discontinuous solutions are fundamental both in the physical theory of nonequilibrium thermodynamics as well as in the mathematical study of inviscid models for compressible flow. It is natural, therefore, to seek a rigorous theory which accomodates these discontinuities. This question is largely settled for the case of one space dimension, even for nonisentropic, nonisothermal flows (see Hoff [2] – [5], Serre [10] – [11], and Shelukhin [13]), and for multidimensional, spherically symmetric flows which are isothermal, even when the initial data is both discontinuous and large (Hoff [6]). For general multidimensional flows, the most important work has been that of Matsumura and Nishida [9], who prove the global existence of solutions of the full Navier-Stokes equations when the initial data is close to a constant in $H^3(I\!R^n)$, hence in $C^1(I\!R^n)$. Recently, Lions [8] has announced existence results for the system (1) - (2) which are closely related to our own. He proves the global existence of solutions of (1) – (2) with *large*, discontinuous initial data for the isentropic case $P(\rho) = K\rho^\gamma$, where γ is sufficiently large, depending on n. Although the details are unavailable as of this writing, it appears that his techniques and analysis are substantially different from ours.

We now give a precise formulation of our results. First, we say that the pair (ρ, u) is a weak solution of (1) – (2) with Cauchy data (ρ_0, u_0) provided that $\rho, u, \nabla u \in L^1_{loc}(\{t \geq 0\})$ (∇ denotes the usual spatial gradient, and $\{t \geq 0\}$ is the set $I\!R^n \times [0, \infty)$), and for all test functions $\psi \in \mathcal{D}(I\!R^n \times (-\infty, \infty))$,

$$\int_{I\!R^n} \rho_0 \psi(\cdot, 0) dx + \int_0^\infty \int_{I\!R^n} (\rho\psi_t + \rho u \cdot \nabla\psi)\, dx dt = 0 \tag{3}$$

and

$$\begin{aligned} \int_{I\!R^n} \rho_0 u_0^j \psi(\cdot, 0) dx &+ \int_0^\infty \int_{I\!R^n} \left[\rho u^j u \cdot \nabla\psi + P(\rho)\psi_{x_j}\right] dx dt \\ &= -\int_0^\infty \int_{I\!R^n} \left[\varepsilon \nabla u^j \cdot \nabla\psi + \lambda(\operatorname{div} u)\psi_{x_j}\right] dx dt, \end{aligned} \tag{4}$$

$j = 1, \ldots, n$. (The solutions we obtain below will be somewhat more regular than is required here.)

We fix constants $\varepsilon > 0$ and $\lambda \geq 0$, and assume that $\lambda/\varepsilon < b_n$, where

$$b_n = \begin{cases} 1 + \sqrt{2}\,, & n = 2 \\ (1 + \sqrt{3})/3\,, & n = 3. \end{cases} \tag{5}$$

The case of an isotropic stress tensor $\lambda = \varepsilon/3$ is therefore included (see Batchelor [1], pg. 147). We must assume that the pressure is given by an isothermal equation of state, $P(\rho) = K\rho$, where K is a positive constant (but see the remarks below following the statement of our theorem). The initial data (ρ_0, u_0) will be measured in the norm given by

$$(6) \qquad C_0 = \|\rho_0 - \tilde{\rho}\|^2_{L^\infty(\mathbb{R}^n)} + \int_{\mathbb{R}^n} \left[(\rho_0(x) - \tilde{\rho})^2 + |u_0(x)|^2\right](1 + |x|^2)^\beta \, dx.$$

Here β is positive but arbitrarily small when $n = 2$, and $\beta = 0$ when $n = 3$. (This weighting of the L^2 norm serves to compensate for the growth of the fundamental solution of the Laplacian at infinity in two space dimensions.)

Given a velocity field $u(x,t)$ we define the corresponding vorticity matrix ω by

$$\omega^{j,k} = u^j_k - u^k_j$$

(subscripts denote differentiation with respect to x: $u^k_j = \dfrac{\partial u^k}{\partial x_j}$, etc.), and the material derivative $\dfrac{d}{dt} = \dot{}$, given by

$$\dot{w} = \frac{dw}{dt} = \frac{\partial w}{\partial t} + \nabla w \cdot u = w_t + w_j u^j$$

(summation over repeated indices is understood). Given a density-velocity pair (ρ, u), we define the corresponding effective viscous flux F by

$$(7) \qquad F = (\varepsilon + \lambda)\operatorname{div} u - P(\rho) + P(\tilde{\rho}).$$

As we shall see, F is one of the fundamental variables for the problem. Its significance will be discussed in detail below.

Our main existence result may then be stated as follows:

Theorem. *Let $n = 2$ or 3; fix ε, λ, $P = K\rho$, $\tilde{\rho}$, and β as above, and let $N > 0$ be given (N may be arbitrarily large). Then there are positive constants a and C, depending on K, $\tilde{\rho}$, β, N, and on upper bounds for ε, λ, and $(b_n\varepsilon - \lambda)^{-1}$ (see (5)), such that, given initial data (ρ_0, u_0) satisfying*

$$(8) \qquad C_0 \le a \quad \textit{and} \quad \|u_0\|_{L^{2n}} \le N,$$

the Cauchy problem for (1) – (2) with data (ρ_0, u_0) has a global weak solution (ρ, u) for which: u is bounded and Hölder continuous in each set $\mathbb{R}^n \times [\tau, \infty), \tau > 0$,

$$(9) \qquad u(\cdot, t) \in H^1(\mathbb{R}^n)\ ,\ t > 0,$$

$$(10) \qquad \omega(\cdot, t),\ F(\cdot, t) \in H^1(\mathbb{R}^n)\ ,\ a.a.\ \ t > 0,$$

(11) $\qquad F(\cdot,t)$ *is Hölder continuous, a.a.* $t>0$.

Also, $(\rho,u)\to(\tilde{\rho},0)$ *as* $t\to\infty$ *in the sense that, for all* $q\in(2,\infty)$.

(12) $$\lim_{T\to\infty}\left[\|\rho-\tilde{\rho}\|_{L^\infty(I\!R^n\times[T,\infty))}+\|u(\cdot,T)\|_{L^q(I\!R^n)}\right]=0.$$

The isothermal hypothesis $P=K\rho$ is required at a single point of the proof, in order to pass weak limits of approximate smooth densities through the function P. None of the a priori estimates required in the analysis depend on this assumption, however. If we assume only that P is C^1 with $P,P'>0$ in a neighborhood of $\tilde{\rho}$, then we can still make somewhat weaker, but nevertheless important statements. For this more general P we can assert that, if (ρ_0,u_0) satisfies (8), and if $\rho_0-\tilde{\rho},\ u_0\in H^3(I\!R^n)$, then the corresponding Cauchy problem for (1) – (2) has a global weak solution (ρ,u) satisfying all the conclusions of the theorem. And if (ρ_0,u_0) satisfies (8) only, then there is instead a triple $(\rho,u,\mathcal{P})$ which satisfies (3) – (4) with $P(\rho)$ replaced by $\mathcal{P}$. The conclusions of the theorem still hold, provided we take $F=(\varepsilon+\lambda)\mathrm{div}\,u-\mathcal{P}+P(\tilde{\rho})$; $\mathcal{P}$ and $P(\rho)$ are at the very least related by the fact that there is a sequence of smooth functions ρ^δ converging to ρ weakly in L^2_{loc} such that $P(\rho^\delta)$ converges weakly to $\mathcal{P}$.

We now give an overview of the analysis, paying particular attention to the issues raised at the beginning of this paper. We first discuss the nature of jump discontinuities in solutions of (1) – (2). This discussion will be completely heuristic, and is reflected only partially in the rigorous results (9) – (12). Nevertheless, it does provide significant insight into the nature of the problem, and it serves to motivate the overall direction of the rigorous analysis presented in [7]. Many of these observations concerning jump discontinuities are also discussed at the formal level by Serre in [12].

Suppose then that (ρ,u) is a smooth solution of (1) – (2), except possibly across a hypersurface S in $I\!R^n\times[0,\infty)$, and that ρ and u have one–sided limits at each point of S. Let (ν,ν_0) be the normal to S at such a point, where ν_0 denotes the scalar t–component and ν the vector x–component. Then (ρ,u) will be a distribution solution of the mass equation (1) across S if and only if, at each such point of S,

(13) $$\nu_0[\rho]+\nu\cdot[\rho u]=0.$$

Here $[\cdot]$ denotes the difference between the limit in a given quantity from one side of S and the limit from the other. Now, we expect that, due to the diffusion terms in (2), u should become smooth for $t>0$, so that $[u]=0$; no such smoothing is apparent in (1), however, so we expect that $[\rho]\neq0$. (These conjectures are amply justified by the one dimensional theory of [2] – [5].) (13) therefore becomes

(14) $$\nu_0+\nu\cdot u=0,$$

which indicates that the surface S consists of particle trajectories; in other words, discontinuities in the solution are convected with the flow.

Before computing the jump conditions associated with the momentum equation, we first rewrite this equation as follows. Observe that

$$\begin{aligned}\varepsilon\Delta u^j + \lambda \operatorname{div} u_j &= \varepsilon u^j_{kk} + \lambda u^k_{kj}\\ &= (\varepsilon+\lambda)u^k_{kj} + \varepsilon(u^j_k - u^k_j)_k\\ &= (\varepsilon+\lambda)\operatorname{div} u_j + \varepsilon\omega^{j,k}_k.\end{aligned}$$

Letting e_j be the standard basis vector in the x_j direction, we may then write (2) in the form

$$(15)\qquad \begin{aligned}(\rho u^j)_t + \operatorname{div}\Big(\rho u^j u + \big(P - (\varepsilon+\lambda)\operatorname{div} u\big)e_j\Big)\\ = \varepsilon\omega^{j,k}_k.\end{aligned}$$

Now, as is well known, the curl of the momentum equation (2) is an evolution equation for the vorticity ω in which the pressure does not appear; this differential equation is therefore diffusive, and with no singular driving term. We therefore expect that the vorticity ω should be smooth for positive time, at least smooth enough that $\nabla\omega$ contains no concentration measure on S. The term $\varepsilon\omega^{j,k}_k$ in (15) consequently should play no role in the computation of jump conditions. The momentum equation (15) will therefore be satisfied across S in the sense of distributions if and only if

$$\nu_0[\rho u^j] + \nu\cdot\big[\rho u^j u + \big(P - (\varepsilon+\lambda)\operatorname{div} u\big)e_j\big] = 0,$$

$j = 1, \dots, n$. The first two terms here cancel by (14) and the fact that $[u] = 0$. Since j is arbitrary, we then conclude that

$$[P(\rho) - (\varepsilon+\lambda)\operatorname{div} u] = 0.$$

Evidently, although u is continuous in $\{t > 0\}$, ∇u is not, nor is ρ; but the jumps in P and in $(\varepsilon+\lambda)\operatorname{div} u$ should exactly cancel, so that $P - (\varepsilon+\lambda)\operatorname{div} u$ becomes smooth in $\{t > 0\}$. This motivates our definition (7) of the effective viscous flux F (the constant $P(\tilde\rho)$ is added so that F will vanish at ∞). These conjectures concerning the smoothness of the variables u, ω, and F are made precise in the assertions (9) – (11) of the theorem.

Observe that, in the decomposition of the Jacobian matrix ∇u into the sum of its symmetric and antisymmetric parts, the antisymmetric part, which is the vorticity, is thus smooth. Concerning the symmetric part, however, we have but one piece of information: that its trace, minus a suitable multiple of the pressure, is also smooth. This information about ∇u, while incomplete, turns out to be sufficient to close a fairly involved sequence of energy estimates required in the analysis.

We shall discuss this last point in more detail. Standard energy estimates show that the L^2 norms of u and various of its derivatives can be bounded by initial data terms plus higher order terms, one of which is $\iint (1 \wedge t)^n |\nabla u|^4 \, dx dt$. It is therefore necessary to bound such a term by the aforementioned L^2 norms. Now, $W^{2,2} \to W^{1,4}$, so that an energy estimate for Δu would suffice. However, as we have observed, ∇u can be discontinuous across hypersurfaces of $\mathbb{R}^n$, so that $D_x^2 u$ cannot be locally integrable. On the other hand, we can relate Δu to quantities under better control, as follows. Applying again the fact that

$$\Delta u^j = \operatorname{div} u_j + \omega_k^{j,k},$$

we obtain from the definition of F that

$$(\varepsilon + \lambda)\Delta u^j = F_j + (\varepsilon + \lambda)\omega_k^{j,k} + (P - \tilde{P})_j. \tag{16}$$

Since we anticipate that $F(\cdot, t)$, $\omega(\cdot, t) \in H^1$ and $P(\cdot, t) - \tilde{P} \in L^2 \cap L^\infty$, the right side of (16) should be in $W^{-1,p}$ for various p (including $p = 4$). Applying a suitable Fourier multiplier theorem, we can then derive the required estimate for ∇u in L^4. This device succeeds by virtue of the decomposition (16), in which F and ω are sufficiently smooth and ρ is pointwise bounded.

The effective viscous flux F plays a more subtle role in the derivation of pointwise bounds for the density ρ. Before making this connection, we shall derive a differential equation for F. If we subtract u^j times (1) from (2) we obtain an equivalent form of the momentum equation:

$$\rho \dot{u}^j - F_j = \varepsilon \omega_k^{j,k}.$$

Upon taking $\dfrac{\partial}{\partial x_j}$ and summing over j, we find that the vorticity drops out, and that there results the elegantly simple equation

$$\Delta F = \operatorname{div} \rho \dot{u}. \tag{17}$$

This relation plays an essential role in the analysis; we regard it as one of the fundamental equations for this fluid mechanics problem.

To derive pointwise bounds for ρ, then, we first write the mass equation (1) in the form

$$\dot{\rho} = -\rho \operatorname{div} u.$$

The previous discussion suggests that we eliminate the term $\operatorname{div} u$ in favor of F. Applying the definition (7) of F and rearranging, we thus obtain that

$$\frac{d}{dt}(\rho - \tilde{\rho}) + (\varepsilon + \lambda)\rho(P - \tilde{P}) = -\rho F. \tag{18}$$

There are two important points to be made here. The first is that, because $P' > 0$, the term $\rho(P - \tilde{P})$ on the left has the "right sign", and so enforces a certain exponential decay. This exponential decay is crucial in deriving bounds for ρ which are independent of t as $t \to \infty$. This effect occurs precisely because we substituted for $\operatorname{div} u$ in terms of F. The driving term ρF on the right side of (18) can be controlled by means of energy estimates for $\dot{u}$ and $\nabla \dot{u}$, which serve to bound in $\dot{u}$ in L^p, and thereby F in L^∞, via the essential relation (17).

The other issue concerns the implications for the ode (18) of the degradation of the regularity estimates for F near $t = 0$. If we divide (18) by ρ and then integrate along a particle trajectory $x(t)$, we obtain an equation for $\log \rho/\tilde{\rho}$ which involves the term

$$\int_0^t F\big(x(s), s\big)\, ds. \tag{19}$$

This integral must be evaluated in L^∞, for small time. Our estimates for $\|F(\cdot, s)\|_\infty$ fail to show, however, that $\int_0^t \|F(\cdot, s)\|_\infty ds$ is finite (elementary computations for solutions of the heat equation suggest that it is not). On the other hand, we can argue from (17) that

$$F = \Gamma * \operatorname{div} \rho \dot{u} = \Gamma_j * \rho \dot{u}^j,$$

where Γ is the fundamental solution for the Laplace operator on $\mathbb{R}^n$. The integral (19) then becomes

$$\int_0^t \big(\Gamma_j * (\rho \dot{u}^j)(\cdot, s)\big)\big(x(s)\big) ds.$$

This integral is taken over a particle path $x(t)$, as is the material derivative $\dot{u}^j$. There is therefore a cancellation (integration by parts) which results in integrals involving lower order regularity, and therefore greater integrability in t. The integrand above is replaced essentially by a modulus of Hölder continuity for $u(\cdot, s)$ and by the norm $\|u(\cdot, s)\|_{L^{2n}}$. The first of these can be estimated in terms of the L^2 norms of u and its derivatives, and the second will be finite, by virtue of the second hypothesis in (8).

To prove the theorem, then, we first mollify the initial data and obtain smooth local solutions (ρ^δ, u^δ), depending upon a mollifying parameter δ. The required regularity information for u^δ, F^δ, and ω^δ is then obtained in a sequence of L^2 energy estimates which incorporate the correct smoothing rates near the initial layer $t = 0$. These bounds can be closed, modulo pointwise bounds for the densities ρ^δ, by exploiting the essential relation (17) in an appropriate way. Pointwise bounds for ρ, on the other hand, can be derived in turn from these same regularity estimates, as indicated above in (17)-(19). The smallness hypotheses then enable us to close this entire sequence of estimates, allowing for the extraction of a suitably converging sequence $(\rho^\delta, u^\delta) \to (\rho, u)$ as $\delta \to 0$. Complete technical details are given in [7].

Department of Mathematics
Indiana University
Bloomington, IN 47405 USA

References

[1] G.K. Batchelor, An Introduction to Fluid Dynamics, Cambridge Univ. Press, 1967.

[2] David Hoff, Construction of solutions for compressible, isentropic Navier-Stokes equations in one space dimension with nonsmooth initial data, Proc. Royal Soc. Edinburgh, Sect. A103 (1986), 301-315.

[3] David Hoff, Global existence for ID, compressible, isentropic Navier-Stokes equations with large initial data, Trans. AMS 303, no. 11 (1987), 169-181.

[4] David Hoff, Discontinuous solutions of the Navier-Stokes equations for compressible flow, Archive Rational. Mech. Ana. 114 (1991), 15-46.

[5] David Hoff, Global well-posedness of the Cauchy problem for nonisentropic gas dynamics with discontinuous initial data, J. of Differential Equations 95, No.1 (1992), 33-73.

[6] David Hoff, Spherically symmetric solutions of the Navier-Stokes equations for compressible, isothermal flow with large, discontinuous initial data, Indiana Univ. Math. J. 41 (1992), 1225-1302.

[7] David Hoff, Global solutions of the Navier-Stokes equations for multidimensional, compressible flow with discontinuous initial data, submitted to J. Diff. Eqns.

[8] P.L. Lions, Existence globale de solutions pour les équations de Navier-Stokes compressibles isentropiques, C.R. Acad. Sci. Paris 316 (1993), 1335-1340.

[9] Akitaka Matsumura and Takaaki Nishida, The initial value problem for the equations of motion of viscous and heat-conductive gases, J. Math. Kyoto Univ. 20 (1980), 67-104.

[10] Denis Serre, Solutions faibles globales des équations de Navier-Stokes pour un fluide compressible, C.R. Acad. Sc. Paris 303 (1986), 639-642.

[11] Denis Serre, Sur l'équation monodimensionnelle d'un fluide visqueux, compressible et conducteur de chaleur, C.R. Acad. Sc. Paris 303 (1986), 703-706.

[12] Denis Serre, Variations de grande amplitude pour la densite d'un fluide visqueux compressible, Physics D 48 (1991), 113-128.

[13] V.V. Shelukhin, On the structure of generalized solutions of the one dimensional equations of a polytropic viscous gas, Prikl. Matem. Mekhan. 48 (1984), 912-920.

AMS/IP Studies in Advanced Mathematics
Volume 3, 1997

Global existence and qualitative behavior of solutions for the system of compressible adiabatic flow with damping

L. Hsiao

Abstract

We consider a 3×3 system of quasilinear hyperbolic conservation laws with damping, which may be used to model the compressible adiabatic gas flow through a porous media. We have found that the damping dissipation is strong enough to preserve the smoothness of the initial data when it is of smallness. Namely, there exists a globally defined classical solution for the Cauchy problem if the initial data is smooth and of smallness.

Consider the system

$$\begin{cases} v_t - u_x = 0 \\ u_t + p(v,s)_x = -\alpha u, \alpha > 0 \\ s_t = 0 \end{cases} \tag{1}$$

which can be used to model the adiabatic gas flow through a porous media. Where v denotes the specific volume, u denotes the velocity, s denotes the entropy, p denotes the pressure with $p_v < 0$. This system is strictly hyperbolic with eigenvalues $\lambda_1 = -\sqrt{-p_v}, \lambda_2 \equiv 0, \lambda_3 = \sqrt{-p_v}$.

---

* Partially Supported by the National Natural Science Foundation of China

The system (1) can be written in a different form which is equivalent to (1) for smooth solutions, namely

$$\begin{cases} v_t - u_x = 0 \\ u_t + p(v,s)_x = -\alpha u, \quad \alpha > 0 \\ (e(v,s) + \frac{u^2}{2})_t + (pu)_x = -\alpha u^2 \end{cases} \tag{2}$$

where e denotes the specific internal energy for which $e_s \neq 0$ and $e_v + p = 0$ holds due to the second law of thermodynamics.

For the case $\alpha = 0$, it has been shown that there is no globally defined classical solution, in general, for a Cauchy problem. In fact, consider the following initial data

$$u(x,0) = \varepsilon u_0(x), \quad v = \overline{v} + \varepsilon v_0(x), \quad s = \varepsilon s_0(x)$$

where $\overline{v}$ is a positive constant and $(u_0(x), v_0(x), s_0(x)) \in C^1$ with a compact support, it was proved in [LI] and [LZG] that there exists a small $\varepsilon_0 > 0$ such that for any $\varepsilon \in (0, \varepsilon_0]$ the C^1 solution of this Cauchy problem blow-up at a finite time, the life span $\widetilde{T}(\varepsilon) = O(\varepsilon^{-1})$.

For the system (1) (or (2)) with damping term, however, the situation is different. Consider any initial data

$$u(x,0) = u_0(x), \quad v(x,0) = \overline{v} + v_0(x), \quad s(x,0) = \overline{s} + s_0(x) \tag{3}$$

where $\overline{v} > 0, \overline{v}$ and $\overline{s}$ are constants, $(u_0(x), v_0(x)) \in C^1$ with a compact support and $s_0(x) \in C^2$ with a compact support, it can be proved that there exists a globally defined classical solution for the Cauchy problem if the c^1-norm of $(u_0(x), v_0(x))$ and c^2-norm of $s_0(x)$ are small (see [HS]). This means that the damping dissipation is strong enough to preserve the smoothness of the initial data when it is of the smallness.

Another aspect is the influence to the large-time behavior, caused by the damping mechanism.

For the case of isentropic flow, namely, $s(x,t) \equiv$ constant, it has been proved that the solution of the Cauchy problem

$$\begin{cases} v_t - u_x = 0 \\ u_t + p(v)_x = -\alpha u, \alpha > 0, p'(v) < 0 \text{ for } v > 0 \\ v(x,0) = v_0(x), u(x,0) = u_0(x) \text{ with} \\ \lim_{x\to\mp\infty}(v_0(x), u_0(x)) = (v^\mp, u^\mp) \end{cases} \tag{4}$$

can be described by the solution of the following problem

$$\begin{cases} v_t = -\frac{1}{\alpha}p(v)_{xx} \\ u = -\frac{1}{\alpha}p(v)_x \\ v(x,0) = \tilde{v}_0(x) \text{ with } \lim_{x\to\mp\infty}\tilde{v}_0(x) = v^\mp \end{cases} \tag{5}$$

time-asymptotically. The system in (5) is obtained by approximating the momentum equation in (4) with Day's law. Moreover, the rate of $t^{-\frac{1}{2}}$ for the L_2-norm and L_∞-norm of the difference between these two solutions has been obtained [HL]. This shows that certain nonlinear diffusive phenomena occur for the solution of (4) which is caused by the damping mechanism.

For the system (1), the corresponding simplified system takes the form

$$\begin{cases} v_t = -\frac{1}{\alpha}p(v,s)_{xx} \\ u = -\frac{1}{\alpha}p(v,s)_x \\ s = s(x) = \bar{s} + s_0(x) \end{cases} \tag{6}$$

Consider the initial data

$$v(x,0) = \bar{v} + \tilde{v}_0(x), \quad \tilde{v}_0 \in C^1 \text{ with a compact support,} \tag{7}$$

it is expected that the large-time behavior of the solution of Cauchy problem for (1) (3) can be described by the solution for (6), (7). This is discussed in a comping paper.

We will concentrate in the result about the global existence which can be established by the L_2-estimates for the solutions and L_∞-estimates on the first derivatives. Denote $c = (-p_v(v,s))^{\frac{1}{2}}$. Let

$$\begin{cases} w = u - h(v,s), \\ z = u + h(v,s) \end{cases} \tag{8}$$

where $h(v,s)$ satisfies the equation

$$ch_v - p_v = 0 \tag{9}$$

then the system (1) can be written as

$$\begin{cases} w_t - cw_x = -\frac{\alpha}{2}(z+w) + (ch_s - p_s)s_x \\ z_t + cz_x = -\frac{\alpha}{2}(z+w) + (ch_s - p_s)s_x \\ s_t = 0 \end{cases} \tag{10}$$

which is equivalent to (1) for classical solutions.

For the system of gas dynamics, $p(v,s) = (\gamma-1)v^{-\gamma}e^s(1<\gamma<3)$, it is easy to find that

$$h(v,s) = 2\sqrt{\frac{\gamma}{\gamma-1}}v^{\frac{1-\gamma}{2}} \cdot e^{\frac{s}{2}}$$

and

$$ch_s - p_s = v^{-\gamma}e^s$$

Thus, the system (10) takes the form that

$$\begin{cases} D^- w = -\frac{\alpha}{2}(z+w) + f(z-w,s,s_x) \\ D^+ z = -\frac{\alpha}{2}(z+w) + f(z-w,s,s_x) \\ s(t,x) = s(0,x) \end{cases} \tag{11}$$

where $D^- = \partial_t - c\partial_x, D^+ = \partial_t + c\partial_x$,

$$f(z-w,s,s_x) = A(\gamma)\cdot(z-w)^{\frac{2\gamma}{\gamma-1}} \cdot e^{\frac{-s}{\gamma-1}} \cdot s_x,$$

$$c = c(z-w,s) = B(\gamma)\cdot e^{\frac{-s}{\gamma-1}}\cdot(z-w)^{\frac{\gamma+1}{\gamma-1}},$$

$$A(\gamma) = \left(\frac{\gamma-1}{16\gamma}\right)^{\frac{\gamma}{\gamma-1}},$$

$$B(\gamma) = \sqrt{\gamma(\gamma-1)}\cdot\left(\frac{\gamma-1}{16\gamma}\right)^{\frac{\gamma+1}{2(\gamma-1)}}.$$

For simplicity, we will discuss the system (1) for gas dynamics next.

Introduce $g(z-w,s,s_x) = g_1(z-w,s) + g_2(z-2,s)s_x$ in which g_1 and g_2 are chosen so that

$$g'_1 = -\frac{\alpha}{4}c^{-1/2}$$

$$g'_2 = -\frac{1}{4(\gamma-1)}c^{1/2}$$

where g'_i denotes $\frac{\partial g_i}{\partial(z-w)}(i=1,2)$. We may choose

$$g_1(z-w,s) = \frac{\alpha(\gamma-1)}{2(3-\gamma)}[B(\gamma)]^{-1/2}\cdot e^{\frac{s}{2(\gamma-1)}}\cdot(z-w)^{\frac{\gamma-3}{2(\gamma-1)}} \tag{12}$$

$$g_2(z-w,s) = -\frac{1}{2(3\gamma-1)}[B(\gamma)]^{1/2}\cdot e^{\frac{-s}{2(\gamma-1)}}\cdot(z-w)^{\frac{3\gamma-1}{2(\gamma-1)}} \tag{13}$$

Then, it can be shown that

$$D^+[c^{1/2}z_x+g] = -l[c^{1/2}z_x+g]^2 + m_F[c^{1/2}z_x+g] - G_F \tag{14}$$

$$D^-[c^{1/2}w_x+g] = -l[c^{1/2}w_x+g]^2 + m_B[c^{1/2}w_x+g] - G_B \tag{15}$$

where

$$l(z-w,s) = c^{-1/2}\cdot c'$$

$$m_F(z-w,s,s_x) = -\frac{\alpha}{2} + 2g\cdot c^{-1/2}c' + \frac{cs_x}{\gamma-1}$$

$$m_B(z-w,s,s_x) = -\frac{\alpha}{2} + 2g\cdot c^{-1/2}c' - \frac{cs_x}{\gamma-1}$$

$$G_F(z-w,s,s_x,s_{xx}) = n(z-w,s) + Q_{1F}(z-w,s)s_x + Q_{2F}(z-w,s)s_x^2 + Q_{3F}(z-w,s)s_{xx}$$

$$G_B(z-w,s,s_x,s_{xx}) = n(z-w,s) + Q_{1B}(z-w,s)s_x + Q_{2B}(z-w,s)s_x^2 + Q_{3B}(z-w,s)s_{xx}$$

$$n(z-w,s) = \frac{\alpha(\gamma-1)}{3-\gamma}g_1(z-w,s)$$

$$Q_{1F}(z-w,s) = 2g_1g_2c^{-1/2}c' - \frac{\alpha}{2}g_2 + \frac{c}{2(\gamma-1)}g_1$$

$$Q_{1B}(z-w,s) = 2g_1g_2c^{-1/2}c' - \frac{\alpha}{2}g_2 - \frac{c}{2(\gamma-1)}g_1$$

$$Q_{2F}(z-w,s) = -g_{2s}c + \frac{(z-w)}{4\gamma(\gamma-1)}\cdot c^{3/2} + g_2[g_2\cdot c^{-1/2}\cdot c' + \frac{c}{\gamma-1}]$$

$$Q_{2B}(z-w,s) = g_{2s}c + \frac{(z-w)}{4\gamma(\gamma-1)}\cdot c^{3/2} + g_2[g_2\cdot c^{-1/2}\cdot c' - \frac{c}{\gamma-1}]$$

$$Q_{3F}(z-w,s) = -g_2c - \frac{(z-w)}{4\gamma}\cdot c^{3/2}$$

$$Q_{3B}(z-w,s) = g_2c - \frac{(z-w)}{4\gamma}\cdot c^{3/2}$$

Furthermore, it can be shown that

$$m_F^2 - 4lG_F = \frac{\alpha^2}{4} + c_{1F}(z-w,s)s_x + c_{2F}(z-w,s)s_x^2 + c_{3F}(z-w,s)s_{xx}$$

where

$$c_{1F}(z-w,s) = \frac{2\alpha}{3-\gamma}c - \frac{2\alpha(\gamma+1)}{(3-\gamma)g_1}Q_{1F}$$

$$c_{2F}(z-w,s) = \frac{4c^2}{(3\gamma-1)^2} - \frac{2\alpha(\gamma+1)}{(3-\gamma)g_1}Q_{2F}$$

$$c_{3F}(z-w,s) = -\frac{2\alpha(\gamma+1)}{(3-\gamma)g_1}Q_{3F}$$

It is easy to show that $l > 0$ for $z - w > 0$. Moreover, $m_F^2 - 4lF_F > 0$ if $|s_x|$ and $|s_{xx}|$ are small.

Denote $c^{1/2}z_x + g = y$, (14) implies that

$$D^+[y] = -l(y-y_1)(y-y_2) \tag{16}$$

where

$$y_1 = \frac{(3-\gamma)}{\alpha(\gamma+1)}g_1\{m_F - \sqrt{m_F^2 - 4lG_F}\}$$

$$y_2 = \frac{(3-\gamma)}{\alpha(\gamma+1)}g_1\{m_F + \sqrt{m_F^2 - 4lG_F}\}$$

Thus, it follows that

$$0 < y_1(z-w,s,s_x,s_{xx}) < y_2(z-w,s,s_x,s_{xx})$$

if $z - w > 0$ and both $|s_x|$ and $|s_{xx}|$ are small. Similarly, denote $c^{1/2}w_x + g = z$, (15) implies that

$$D^-[z] = -l(z-z_1)(z-z_2) \tag{17}$$

where

$$z_1 = \frac{(3-\gamma)}{\alpha(\gamma+1)}g_1\{m_B - \sqrt{m_B^2 - 4lG_B}\}$$

$$z_2 = \frac{(3-\gamma)}{\alpha(\gamma+1)}g_1\{m_B + \sqrt{m_B^2 - 4lG_B}\}$$

$$m_B^2 - 4lG_B = \frac{\alpha^2}{4} + c_{1B}(z-w,s)s_x + c_{2B}(z-w,s)s_x^2 + c_{3B}(z-w,s)s_{xx}$$

$$c_{1B}(z-w,s) = -\frac{4\alpha\gamma}{(3-\gamma)(\gamma-1)}c - \frac{2\alpha(\gamma+1)}{(3-\gamma)g_1}\cdot Q_{1B}$$

$$c_{2B}(z-w,s) = \frac{16\gamma^2}{(3\gamma-1)^2(\gamma-1)^2}c^2 - \frac{2\alpha(\gamma+1)}{(3-\gamma)g_1}\cdot Q_{2B}$$

$$c_{3B}(z-w,s) = -\frac{2\alpha(\gamma+1)}{(3-\gamma)g_1}\cdot Q_{3B}$$

Thus, it follows that $0 < z_1(z-w, s, s_x, s_{xx}) < z_2(z-w, s, s_x, s_{xx})$ if $z - w > 0$ and both $|s_x|$ and $|s_{xx}|$ are small.

Moreover, it can be shown that

$$\begin{cases} y(0,x) > y_1(z-w, s, s_x, s_{xx})(0,x) \text{ and} \\ z(0,x) > z_1(z-w, s, s_x, s_{xx})(0,x) \end{cases} \tag{18}$$

if the c_1-norm of $(u_0(x), v_0(x))$ is small and the c_2-norm of $s_0(x)$ is small. (see [HS] for the detail).

In view of (16)–(18), it can be proved that

Lemma 1. Suppose that the solution of (1) (3) defined for $0 \leq t \leq T$ satisfies $(z, w, s \in D) = \{(z, w, s) : a_1 h(\overline{v}, \overline{s}) < z - w < a_2 h(\overline{v}, \overline{s}), |s - \overline{s}| < 1,$ where a_1 and a_2 are constants such that $0 < a_1 < 2 < a_2 < \infty\}$ Then, the following estimates hold if the c_1-norm of $(u_0(x), v_0(x))$ is sufficiently small and the c_2-norm of $s_0(x)$ is small enough.

For any $(t, x) \in [0, T] \times R$,

$$\min\left\{(c^{1/2} z_x + g)(0, x_3(0; t, x)), \inf_{\substack{x_3(\tau; t, x) \\ 0 \leq \tau \leq t}} y_2\right\}$$
$$\leq \quad (c^{1/2} z_x + g)(t, x) \leq \max\left\{(c^{1/2} z_x + g)(0, x_3(0; t, x)), \sup_{\substack{x_3(\tau; t, x) \\ 0 \leq \tau \leq t}} y_2\right\}$$
$$\min\left\{(c^{1/2} w_x + g)(0, x_1(0; t, x)), \inf_{\substack{x_1(\tau; t, x) \\ 0 \leq \tau \leq t}} z_2\right\}$$
$$\leq \quad (c^{1/2} w_x + g)(t, x) \leq \max\left\{(c^{1/2} w_x + g)(0, x_1(0; t, x)), \sup_{\substack{x_1(\tau; t, x) \\ 0 \leq \tau \leq t}} z_2\right\}$$

where $x = x_3(\tau; t, x)$ denotes the forward characteristic which passes through (t, x) at $(\tau = t)$ and meets $(0, x_3(0; t, x))$ at $\tau = 0, x = x_1(\tau; t, x)$ denotes the backward characteristic passing through (t, x) at $\tau = t$ and meeting $(0, x_1(0; t, x))$ at $\tau = 0$.

The proof is given in [HS].

In order to get the L_2-norm of the solution, we use the form (2) next. The equation $(2)_3$ can be written as

$$[\frac{1}{2}u^2 + e(v,s) - e(\overline{v},s) + \overline{p}v - \overline{pv}]_t + [(p - p(\overline{v},\overline{s}))u]_x = -\alpha u^2 \tag{19}$$

where $\overline{p} = p(\overline{v},\overline{s})$.

Integrate (19) over $[0,T] \times R$, it reads off that

Lemma 2. Suppose that the solution of (1) (3) is defined for $0 \leq t \leq T$, with $(z,w,s) \in D$, then it holds that

$$\frac{1}{2}\int_{-\infty}^{\infty} u^2(t,x)dx + \frac{a}{4}\int_{-\infty}^{\infty}(v-\overline{v})^2(t,x)dx \leq \int_{-M}^{M}[\frac{1}{2}u_0^2(x) + \frac{3\widehat{a}}{4}v_0^2(x) + (\frac{b^2}{a} + \frac{b^2}{\widehat{a}})s_0^2(x)]dx \tag{20}$$

where

$$a = \inf_{(z,w,s)\in D} e_{vv} > 0, \quad \widehat{a} = \sup_{(z,w,s)\in D} e_{vv} > 0,$$

$b = \sup_{(z,w,s)\in D} |e_v| > 0$, $[-M,M]$ is the compact support for $(u_0(x), v_0(x), s_0(x))$.

The global existence theorem follows then from these lemmas.

Theorem There exists a globally defined classical solution of (1) (3) if the c_1-norm of $(u_0(x), v_0(x))$ is small and the c_2-norm of $s_0(x)$ is small.

A sketch of the proof.

There exists a local solution of (1) (3) for $0 \leq t \leq t_0$ which satisfies $(z,w,s) \in D$ if $(u_0(x), v_0(x), s_0(x))$ is so small that

$$\begin{aligned} &(z(x,0), w(x,0), s(x,0)) \in D^* \\ = \ &\{(z,w,s) : A_1 h(\overline{v},\overline{s}) < z - w < A_2 h(\overline{v},\overline{s}), |s-\overline{s}| < 1, \end{aligned}$$

where A_1 and A_2 are constants such that $a_1 < A_1 < 2, 2 < A_2 < a_2\}$

Thus, Lemma 1 implies that

$$|u_x . v_x|_{L_\infty} \leq M^*$$

if the c_1-norm of $(u_0(x), v_0(x))$ is small and the c_2-norm of $s_0(x)$ is small, where M^* only depends on D. Therefore, Lemma 2 implies that $(z, w, s) \in D^*$, if

$$\int_{-M}^{M} [\frac{1}{2}u_0^2(x) + \frac{3\hat{a}}{4}v_0^2(x) + (\frac{1}{a} + \frac{1}{\hat{a}})b^2 \cdot s_0^2(x)]dx \leq \frac{1}{3M^*}[h(\overline{v}, \overline{s})]^3 \cdot (4\beta)^{-3},$$

where

$$\beta = \sup_D |h_v|$$

Then, there exists a positive τ_0 such that the solution of (1) (3) can be defined for $0 \leq t \leq t_0 + \tau_0$ where $(z, w, s) \in D$. Repeat the above argument, the theorem follows. The detail can be found in [HS].

References

[HL] L. Hsiao and Tai-ping Liu , Convergence to nonlinear diffusion waves for solutions of a system of hyperbolic conservation laws with damping, Commun. Math. Phys. 143, 599–605(1992).

[HS] L.Hsiao and D. Serre, Global existence of solutions for the system of compressible adiabatic flow through porous media. (to appear)

[LI] Tai-ping Liu, Development of singularities in the nonlinear waves for quasilinear hyperbolic partial differential equations, J. Diff. Eqns., 33(1979), 92–111.

[LZK] T.T. Li, Y. Zhou and D.X. Kong, Weak linear degeneracy and global classical solutions for first order quasilinear hyperbolic systems. (to appear)

Academia Sinica

Institute of mathematics

Beijing 100080, China

AMS/IP Studies in Advanced Mathematics
Volume **3**, 1997

Slow dynamics of linear waves in nonlinear systems of conservation laws

F. Hubert
D. Serre*
Unité de Maths. Pures & Appl.
ENS de Lyon-CNRS (UMR 128)
46, allée d'Italie
69364 Lyon Cedex 07
FRANCE

December 6, 1995

Abstract

Several models of continuum mechanics involve systems of conservation laws in the form

$$u_t + div\, f(u) = \epsilon P[u],$$

where $\epsilon > 0$ is a small parameter. The perturbation $\epsilon P[u]$ is usually a (possibly non linear) PDO and might represent the influence of viscosity, heat conduction, friction, etc...

For smoth data, this perturbation can often be ignored as long as we restrict our attention to bounded times, say $t << \epsilon^{-1}$. This paper is devoted to a qualitative description of the dynamics for times $t \sim \epsilon^{-1}$, assuming that the inviscid system $u_t + div\, f(u) = 0$ is hyperbolic and is endowed with both linear and nonlinear characteristic fields.

1 Introduction

For the sake of simplicity, one restricts our attention to one-dimensional systems of conservation laws :

(1)
$$u_t + f(u)_x = \epsilon P[u],\ t > 0,\ x \in \mathrm{R},$$

where $u(x,t)$ belongs to some open convex set $\mathcal{U}$ of R^n and $f : \mathcal{U} \to \mathrm{R}^n$ is a smooth vector field. In the perturbation, P is a (possibly nonlinear) partial differential operator in the space variable x with constant coefficients. One assumes that the unperturbed system

(2)
$$u_t + f(u)_x = 0,\ t > 0,\ x \in \mathrm{R},$$

*member of the Institut Universitaire de France

is strictly hyperbolic : the Jacobian matrix $Df(u)$ has simple and real eigenvalues $\lambda_1(u) < \cdots < \lambda_n(u)$.

We consider the Cauchy problem for (1) and (2) :

$$u(x,0) = a(x), \ x \in \mathrm{R}, \tag{3}$$

with a periodic initial data, say of unit period : $a(x+1) = a(x)$. Although the existence and uniqueness remain widely open questions for general unperturbed systems, one often is able to solve uniquely and globally in time the Cauchy problem for the perturbed system ($\epsilon > 0$) because ϵP induces either a smoothing effect or damping. We shall not discuss these aspects here because each system has to be studied with appropriate ideas or *a priori* estimates and with specific tools. Let us only give a few references concerning parabolic perturbations [1], semi-parabolic ones [2, 3], or damping terms [5].

We shall restrict our attention to *stable* perturbations in the sense that the solution u^ϵ of (1,3) is smooth and converges boundedly almost everywhere to a solution (in the distributional sense) u of (2, 3) as $\epsilon \to 0$ (we thus reject dispersive perturbations, like the term $\partial_x^3 u$ in the Korteweg-deVries equation, which give rise to large amplitude oscillations of arbitrarily large frequency). Stable perturbations will have a very weak influence on the solution u^ϵ for bounded times $t = \mathcal{O}(1)$ ($\mathcal{O}(1)$ refers to the asymptotic analysis as $\epsilon \to 0$). For instance, parabolic perturbations will only select the (physically) relevant shock waves among all possible discontinuities of (2), but we emphasize that many distinct such perturbations would yield to the same selection, so that u does not really depend on the choice of P in a rather wide class.

The perturbed solution u^ϵ behaves in a completely different way once we consider larger times, say of order ϵ^{-1} or even larger. The first reason is that the term $\epsilon P[u]$ has a non negligible effect on time range of order ϵ^{-1}. The second one is that such times are large times, regarding the inviscid part of the system, so that entropy dissipation, which is associated to nonlinear fields, tends to shrink the amplitude of the so-called *non-linear waves* to zero, according to an algebraic rate $t^{-\alpha}$. This point has been rigorously proved in simple cases by Glimm and Lax [6] and Temple [7], using precise estimates in the random choice method. Thus u^ϵ approach some (infinite-dimensional) manifold L which consists, roughly speaking, of all the global solutions involving only linear waves. When the system (2) has only one linear field (see below for a precise definition), then L is made of periodic travelling waves and can be completely described. We then are concerned with the slow dynamics induced by the perturbation on the manifold L.

For larger times, say $t >> \epsilon^{-1}$, one expect that the dissipative effect has damped everything, so that u^ϵ has gone to a constant state which solves (1) trivially. When the mass conservation occurs, this constant is the mean value of the initial data. Otherwise, if the perturbation is not differential (it looks like a friction term), this constant would be an equilibrium of the differential equation $u' = P(u)$.

In the sequel of this section, we shall give three simple examples in order to make the most important facts as clear as possible. The next section will be devoted to a formal asymptotic analysis for general perturbed systems. Then we close this paper with two important examples in continuum mechanics.

Let us finally mention that a deeper analysis was done since this paper was submitted. See [11, 12].

1.1 A linear case : the heat equation

One considers the heat equation with a small heat conductivity. Here one has $n = 1$, $f =: 0$ and $P[u] =: u_{xx}$:

$$u_t = \epsilon u_{xx}. \tag{4}$$

A rescaling shows that the solution u^ϵ of (4,3) depends only on the space variable and a slow-time variable $\tau = \epsilon t$, but not on the fast time t :

$$u^\epsilon(x,t) = u_0(x, \epsilon t), \tag{5}$$

where u_0 is the unique solution of

$$\begin{aligned} \partial_\tau u_0 &= \partial_x^2 u_0, \\ u_0(x,0) &= a(x). \end{aligned}$$

More generally, assuming that the Cauchy problem for

$$\left\{ \begin{aligned} \partial_\tau u_0 &= P[u_0], \\ u_0(x,0) &= a(x) \end{aligned} \right. \tag{6}$$

is well-posed, then the solution u^ϵ of (1,3) with $f \equiv 0$ is still given by (5), u_0 solving (6). Thus the behaviour of u^ϵ as the time is of order ϵ^{-1} depends on the perturbation operator, this one being given by physical considerations.

1.2 A non-linear case : the Hopf-Burgers's equation

We know consider

$$u_t + uu_x = \epsilon u_{xx}, \tag{7}$$

known as the Hopf-Burgers's equation. Explicit integration of the Cauchy problem can be performed using the Hopf-Cole transformation [10]. This has been used by Lax [8] in order to prove the convergence of u^ϵ to the entropy solution of $u_t + (\frac{u^2}{2})_x = 0$ as $\epsilon \to 0$.

Let us note first that the mean value of u^ϵ is conserved for all time, as it should be for general systems (1) as long as P has the form $\partial_x Q[u]$:

$$\int_y^{y+1} u^\epsilon(x,t)\, dx = \int_0^1 a(x)\, dx =: A. \tag{8}$$

The next observation is that the first derivative $z =: u_x^\epsilon$ satisfies a transport-diffusion equation of Ricatti type :

$$(\partial_t + u\partial_x - \epsilon\partial_x^2)z = -z^2.$$

A simple application of the maximum principle provides the Oleinik's inequality, which depends neither on ϵ nor on a :

$$\partial_x u^\epsilon \le \frac{1}{t}. \tag{9}$$

Both (8) and (9) imply

$$\sup_{x \in \mathbb{R}} | u^\epsilon(x,t) - A | \le \frac{1}{2t},$$

which is an $\mathcal{O}(\epsilon)$ for t of order ϵ^{-1}. In this example, the large time behaviour is very simple (we shall say "trivial"), the solution being approximately equal to a constant which is nothing but the mean value of the initial data. One does not distinguish here between the viscous and the inviscid ($\epsilon = 0$) cases.

1.3 The simplest non-trivial example

One considers now a system which has been introduced for studying elastic waves (see [9]). Here $n = 2$, $\mathcal{U} = \mathrm{R}^2$ and $f(u) =: \phi(r)u$ where $r =| u |$ stands for the euclidean norm of u :

(10) $$u_t + (\phi(r)u)_x = \epsilon(\nu(r)u_x)_x.$$

The description of the slow dynamics for general periodic data will be made later on. We restrict in this section our attention to the rotationally invariant solutions, that is those of the following form, with complex representation

$$u^\epsilon(x,t) = r^\epsilon(t)e^{2i\pi(x-X^\epsilon(t))}.$$

Such a formula provides an explicit solution of (10) whenever r^ϵ and X^ϵ solve the following ODEs :

$$\begin{cases} r^\epsilon_t &= -4\epsilon\pi^2\nu(r^\epsilon)r^\epsilon, \\ X^\epsilon_t &= \phi(r^\epsilon), \end{cases}$$

which reduces to

$$\begin{cases} r^\epsilon &= R(\epsilon t), \\ X^\epsilon &= \int_0^t \phi(R(\epsilon s))\, ds =: c(\epsilon t)t, \end{cases}$$

with

(11) $$c(\tau) = \tau^{-1}\int_0^\tau \phi(R(\sigma))\, d\sigma,$$

(12) $$R'(\tau) = -4\pi^2\nu(R)R,$$

(13) $$R(0) = r_*,$$

r_* being a value given by the Cauchy data.

When we consider the fast dynamics on a time interval of the form $| t - \epsilon^{-1}\tau |= \mathcal{O}(1)$ for a given value of τ, u^ϵ might be viewed as a travelling wave of amplitude $R(\tau)$ and velocity $c(\tau)$, both depending on the slow time. One notices that the velocity does not depend on the precise form of the perturbation, whereas the amplitude does, as in the scalar case.

2 The asymptotic analysis

2.1 Linear and non-linear waves

We recall well-known concepts due to Lax [8] about characteristic fields of the hyperbolic part (2). Let $1 \leq k \leq n$ be an integer and denote by r_k a smooth eigen-vector field of Df associated with its k-th eigenvalue. Then the k-th characteristic field is said to be *linearly degenerate* if $d\lambda_k \cdot r_k \equiv 0$ or *genuinely nonlinear* if $d\lambda_k \cdot r_k$ does not vanish. Most of practical examples are endowed with both types of characteristic fields. In the examples of the previous section, the first one has only a linear field, the second one has only a nonlinear field whereas the last one has a linear (with $\lambda = \phi(r)$) and a nonlinear ($\lambda = \phi + r\phi'$, provided $2\phi' + r\phi" \neq 0$) fields.

Linear and nonlinear fields in inviscid models give rise to completely different behaviours as the time goes to infinity. Genuine nonlinearity makes blow up the smooth solutions in finite time, beyond which shock waves develop. Schock waves are irreversible processes associated to entropy dissipation. This dissipation has a (quite weak) damping effect.

Conversely, linear degeneracy allows some non-constant periodic waves to survive forever without damping. Let us give a precise description of this claim. Let a characteristic field (λ, r) be linearly degenerate and let us choose an integral curve Γ of the eigenfield r. Then λ assumes a constant value $\sigma(\Gamma)$ along Γ and any bounded measurable function $g : \mathrm{R} \to \Gamma$ yields a weak solution of (2) in the form

$$u(x,t) = g(x - \sigma(\Gamma)t).$$

Moreover, this solution is admissible for the Lax's entropy criterion, since it satisfies in the distributional sense the equality

$$E(u)_t + F(u)_x = 0,$$

for any pair entropy-entropy flux (E, F). In particular, such solutions are reversible.

2.2 Asymptotic behaviour for the unperturbed system

Let us assume from now on that the hyperbolic part (2) has exactly **one** linearly degenerate field, denoted by (λ, r) because the value of the index k plays no role here, the $n-1$ other fields being genuinely nonlinear. Then a reasonable conjecture is the following.

Conjecture 2.1 *Any space-periodic entropy solution of (2) converges orbitally in the* sup*-norm to one of the travelling waves described above as time goes to* $+\infty$.

In other words, for a given periodic data a, the solution (we implicitly assume that a unique entropy solution exists globally in time) u converges towards a travelling wave $g(x - \sigma(\Gamma)t)$ for a suitable curve Γ and a suitable periodic profile $g : \mathrm{R}/\mathbb{Z} \to \Gamma$ in the sense

$$\lim_{t \to +\infty} \sup_{x \in \mathrm{R}} \mid u(x,t) - g(x - \sigma(\Gamma)t) \mid = 0.$$

A natural question is whether the curve Γ and the profile g could be computed from the data a, using exact conserved quantities. The answer seems to be negative in general because only n such conserved quantities are known, say the components of

$$\int_0^1 u(x)\, dx.$$

However, the third example of the previous section is more favourable because of some exact decoupling. First of all, the solution satisfies weakly the conservation law

$$E(u)_t + (\phi(r)E(u))_x = 0, \tag{14}$$

as long as E is a scalar function, homogeneous of degree 1. In particular, r is the entropy solution of the scalar conservation law of Burgers's type

$$r_t + (\phi(r)r)_x = 0,$$

so that, according to the Oleinik's inequality, $r(\cdot,t)$ converges uniformly to

$$r_* = \int_0^1 \mid a(x) \mid dx. \tag{15}$$

This formula shows that Γ is the circle of radius r_*, centered at the origin, since the integral curves of the linear eigen-field are such circles. Moreover, (14) allows us to compute the momenta of g. If $a(x) = R(x)e^{i\alpha(x)}$ and $g(x) = r_* e^{i\theta(x)}$, then

$$r_* \int_0^1 h(\theta(x))\, dx = \int_0^1 R(x) h(\alpha(x))\, dx,$$

for all smooth function h. More precisely, one can show that $\theta(x) = \alpha \circ \rho^{-1}(r_* x)$, where ρ is some antiderivative of R (thus is monotonous increasing). Unfortunately, one cannot compute θ more accurately, which means that θ is known only up to a shift.

2.3 The geometrical framework

Paying attention to times which are of order ϵ^{-1}, one assumes from now on that the solution lies in the vicinity of Γ, some integral curve of the linear eigen-field. Indeed, since we expect a slow dynamics on L induced by the perturbation, one will allow Γ to depend on the slow time variable $\tau =: \epsilon t$. In order to write analytically this asumption, one introduces (see [4]) new coordinates in the phase space $\mathcal{U}$. These coordinates $(v_1, \cdots, v_{n-1}, w)$ are such that the integral curves of the linear eigen-field are defined by collections of $n-1$ equations :

$$v_1(u) = \alpha_1,\ v_2(u) = \alpha_2, \cdots, v_{n-1}(u) = \alpha_{n-1}.$$

In other words, $v_1, \cdots, v_{n-1}$ are independent solutions of the differential equation $dv \cdot r = 0$.

With these new variables. the system (1) rewrites as

$$v_t + B(v,w)v_x = \epsilon\, p[v,w], \tag{16}$$

$$w_t + \mu(v)w_x + \eta(v,w)\cdot v_x = \epsilon\, q[v,w], \tag{17}$$

where B is an $(n-1)\times(n-1)$ valued matrix with real and distinct eigenvalues. The spectra of A and B differ only from $\mu(v) = \lambda(u)$, so that $B(v,w) - \mu(v)I_{n-1}$ is always invertible. The derivative w_x does not appear in the left-hand side of (16) because of the choice of the variables v, whereas the identity $\frac{\partial\mu}{\partial w} \equiv 0$ is equivalent to the linear degeneracy of the characteristic field (λ, r).

The asumption made at the begining of this paragraph says that v^ϵ is uniformly close to some slowly varying vector $v_0(\tau)$. Trying to mimic the result that we got in the non-trivial example, we look for solutions (v^ϵ, w^ϵ) admitting the following asymptotic development :

$$\begin{aligned} v^\epsilon(x,t) &= v_0(\epsilon t) + \epsilon v_1(x - c(\epsilon t)t, \epsilon t) + \mathcal{O}(\epsilon^2), \\ w^\epsilon(x,t) &= w_0(x - c(\epsilon t)t, \epsilon t) + \epsilon w_1(x - c(\epsilon t)t, \epsilon t) + \mathcal{O}(\epsilon^2). \end{aligned}$$

One has chosen a dependence on some travelling variable $x - c(\tau)t$ in the terms v_1, w_0, w_1 but not in v_0, which has to depend only on the slow time. We have seen that the velocity will not be equal (in general) to the naive value $\mu(v_0(\tau)) = \sigma(\Gamma(\tau))$. Finally, as quoted in the third example, we do not expect to determine completely w_0 ; we shall merely look for (v_0, w_0) *modulo* a shift, this one being an arbitrary function of the slow time (nevertheless, everything goes in a better shape when μ is a constant ; see the application to the gas dynamics).

2.4 The slow dynamics

We plug the asymptotic development in the system (16,17) and identify the coefficients of each power ϵ^j to zero. Doing that, we have to keep in mind that, being concerned by large times $t \sim \epsilon^{-1}$, the slow time τ will be of order ϵ^0. For instance, the leading term in the development of $\partial_t w^\epsilon$ will be $-(c(\tau) + \tau c'(\tau))\partial_x w_0$.

The lower order terms in the equations are thus the coefficients of ϵ^0 in (17) :

$$(\mu(v_0) - c - \tau c')\partial_x w_0 = 0.$$

Since we do not study trivial (say constant) linear waves, $\partial_x w_0$ does not vanish identically and we find the differential equation governing c :

$$\frac{d}{d\tau}(\tau c(\tau)) = \mu(v_0(\tau)). \tag{18}$$

The initial value of $\tau c(\tau)$ is clearly zero, so that we rewrite a formula analogous to (11) :

$$c(\tau) = \tau^{-1} \int_0^\tau \mu(v_0(s))\, ds.$$

We now turn to the ϵ^1-terms in (16). They write, using (18) :

$$\frac{dv_0}{d\tau} + (B(v_0, w_0) - \mu(v_0)I_{n-1})\partial_x v_1 = p[v_0, w_0], \tag{19}$$

since the derivatives in p act only on the space variable. We eliminate the term $\partial_x v_1$ by inverting $B - \mu$ and averaging the resulting equation on a period. Such averaged quantities are written in brackets :

$$< F >=: \int_0^1 F(x)\, dx.$$

One obtains

$$< (B(v_0, w_0) - \mu(v_0)I_{n-1})^{-1} > \frac{dv_0}{d\tau} =< (B(v_0, w_0) - \mu(v_0)I_{n-1})^{-1} p[v_0, w_0] > . \tag{20}$$

This equality is an evolution equation for v_0 as soon as the hyperbolic system (2) is *globally hyperbolic*, according to [4].

2.4.1 Remark

Let us consider the case of a parabolic conservative perturbation of the form $P[u] =: \partial_x(D(u)\partial_x u)$ where D is some diffusion tensor. Then $p = \frac{Dv}{Du}P$ and the right-hand side of (20) takes the form

$$\begin{aligned} < C(v_0, w_0)\partial_x(E(v_0, w_0)\partial_x v_0 &+ F(v_0, w_0)\partial_x w_0) > \\ &= - < \partial_x C(E\partial_x v_0 + F\partial_x w_0) > \\ &= - < \frac{\partial C}{\partial w} E\partial_x w_0 > \partial_x v_0 - < \frac{\partial C}{\partial w} F(\partial_x w_0)^2 > \\ &= < G(v_0, w_0)(\partial_x w_0)^2 >, \end{aligned}$$

for some scalar function G. We have make use of integration by parts and of the fact that $< h(w_0)\partial_x w_0 >= 0$ for any smooth function h.

We conclude from this calculus that the non-constancy of the profile w_0 induces a slow evolution of v^ϵ. This effect did not occur at the level of the unperturbed system and actually depends crucially on the form of the perturbation (hereabove, G linearly depends on the diffusion tensor D).

2.4.2 Going back to the general case

It remains to find an evolution equation for w_0 (at least *modulo* a shift), using the ϵ^1-terms in (17). Because of (18), it comes

$$\partial_\tau w_0 + (\frac{D\mu}{Dv}(v_0) \cdot v_1)\partial_x w_0 + \eta(v_0, w_0) \cdot \partial_x v_1 = q[v_0, w_0]. \tag{21}$$

We then compute $\partial_x v_1$ from (19,20) :

$$\begin{aligned} \partial_x v_1 = \ & (B(v_0, w_0) - \mu(v_0)I_{n-1})^{-1}\{p[v_0, w_0] \\ & - < (B - \mu I)^{-1} >^{-1} < (B - \mu I)^{-1} p[v_0, w_0] >\}. \end{aligned} \tag{22}$$

This determines v_1 up to the addition of some slowly varying vector-valued function $\tau \mapsto K$, so that the equation (22) does not seem to be well-written. However, as is quoted above, we only want to calculate w_0 up to some shift with respect to a slowly varying function. Thus the knowledge of $\partial_x v_1$ in (21) is exactly the one that we need, as we explain below.

Let us choose the special value V_1 of v_1 which is of zero mean. Then (21) is perfectly written and the Cauchy problem for (20,21) might be well-posed and define a unique solution (v_0, W_0). For any slowly varying function $\tau \mapsto m$, let us define $w_0(x,\tau) =: W_0(x - m(\tau), \tau)$. Then $\partial_\tau w_0 = \partial_\tau W_0 - m'\partial_x W_0$, so that (v_0, w_0) solves (20,21) with $v_1(x,\tau) = V_1(x - m(\tau), \tau) + K(\tau)$ as soon as $m' = \frac{D\mu}{Dv}(v_0) \cdot K$. Thus any choice of the mean value $K(\tau)$ of v_1 (such a choice completely determines the system (20,21)) gives the same solution (v_0, w_0) *modulo* a shift in space concerning w_0, this shift depending only on the slow time variable. Let us emphasize that this shift does change the involved average values ; in particular, the equation (20) does not depend on this shift.

2.4.3 Remark

When μ is constant (which is true for gas dynamics when we make use of lagrangian coordinates), then the above discussion concerning the shift is irrelevant since the equation (21) is perfectly written, so that we expect that w_0 is uniquely determined, whatever he choice of v_1 is.

3 The system of elastic waves

3.1 The asymptotic analysis

We go back to the system (10)

$$u_t + (\phi(r)u)_x = \epsilon(\nu(r)u_x)_x, \tag{23}$$

where we assume that the function ϕ' does not vanish.

The data will take values in the domain $\mathcal{U} =]0, +\infty[\times \mathrm{R}$, which is a positively invariant region for (23), due to the maximum principle. Then we may make use of the polar coordinates $u = re^{i\theta}$ with $r(x,t) > 0$ and $\mid \theta(x,t) \mid < \frac{\pi}{2}$. The system rewrites as

$$r_t + (\phi(r)r)_x = \epsilon\,(r_{xx} - r\theta_x^2), \tag{24}$$

$$\theta_t + \phi(r)\theta_x = \epsilon\,(\theta_{xx} + 2\frac{r_x\theta_x}{r}). \tag{25}$$

Let us emphasize that the inviscid part ($\epsilon = 0$) is not coupled (this gives a way to solve uniquely the Cauchy problem) whereas the viscous one is strongly coupled through the perturbation.

The radius r plays the role of the variable v and the angle θ stands for w. The travelling waves of the inviscid system are given by

$$r = cst, \;\; \theta(x,t) = \Theta(x - \phi(r)t).$$

Applying the ideas of the previous section, one considers solutions (r_0, θ_0) in (24,25) of the form[1]

$$\begin{aligned} r^\epsilon(x,t) &= r_0(\tau) + \epsilon r_1(x - c(\tau)t, \tau) + \mathcal{O}(\epsilon^2), \\ \theta^\epsilon(x,t) &= \theta_0(x - c(\tau)t, \tau) + \epsilon\theta_1(x - c(\tau)t, \tau) + \mathcal{O}(\epsilon^2). \end{aligned}$$

The linear field corresponds to the eigenvalue $\lambda = \phi(r)$, so that the equation (18) becomes

$$\frac{d}{d\tau}(\tau c(\tau)) = \phi(r_0(\tau)). \tag{26}$$

Next, the ϵ^1-terms in (24) give

$$\frac{dr_0}{d\tau} + r_0\phi'(r_0)\partial_x r_1 = -r_0(\partial_x\theta_0)^2,$$

so that the averaging gives both

$$\frac{dr_0}{d\tau} = -r_0 < (\partial_x\theta_0)^2 >, \tag{27}$$

and

$$\phi'(r_0)\partial_x r_1 = < (\partial_x\theta_0)^2 > - (\partial_x\theta_0)^2. \tag{28}$$

Last but not least, the ϵ^1-terms in (25) give

$$\partial_\tau\theta_0 + \phi'(r_0)r_1\partial_x\theta_0 = \partial_x^2\theta_0. \tag{29}$$

We observe that θ_0 obeys to a transport-diffusion equation, whereas r_0 is monotonous non-increasing. Let us also point out that the following conservation law holds as a

[1] added in proofs : since this paper was submitted, we were able to prove the relevance of this asymptotic development in the case of a viscosity $\nu(r) = r^{-2}$. We use a lagrangian change of variable and the maximum principal for a scalar balance law. See [11, 12].

consequence of the above equations (it is indeed a direct consequence of the conservation laws (23)) :

$$\partial_\tau(r_0 e^{i\theta_0}) + \partial_x(r_1\phi'(r_0)r_0 e^{i\theta_0}) = \partial_x^2(r_0 e^{i\theta_0}). \tag{30}$$

Finally, the system given by (26,27,28,29) has to be supplemented with a periodic initial data :

$$\begin{cases} r_0(0) & = & r_*, \\ \theta_0(0,x) & = & \alpha(x). \end{cases} \tag{31}$$

3.2 The well-posedness of the slow dynamics

We derive in this section the *a priori* estimates which are used in the proof of the following two results. The details will be given in a forthcoming paper.

Theorem 3.1 *Let r_* and $\alpha \in H^1(\mathrm{R}/\mathbb{Z})$ satisfy $|\ \alpha\ |_\infty < \frac{\pi}{2}$. Then there is a (possibly short) time $\tau_* > 0$ and a smooth solution (r_0,θ_0,r_1) of (27,28,29,31) defined on $\mathrm{R}/\mathbb{Z}\times]0,\tau_*[$.*

Theorem 3.2 *There is an absolute constant C_* such that the solution given above is globally defined whenever $TV(\alpha) \le C_*$. Moreover, (r_0,θ_0) converges exponentially fast (with respect to τ) in $L^2(\mathrm{R}/\mathbb{Z})$ towards a constant state $(\hat{r},\hat{\theta})$ given by*

$$\hat{r}e^{i\hat{\theta}} =: r_* < e^{i\alpha} > .$$

Let (r_0,r_1,θ_0) be a smooth solution of the system on some time interval $I =]0,\tau_1[$. Then (30) implies

$$< r_0 e^{i\theta_0} >= r_* < e^{i\alpha} >, \tag{32}$$

from which we conclude

$$r_0(\tau) \ge r_* \,|< e^{i\alpha} >| > 0. \tag{33}$$

Denoting $\partial_x\theta_0$ by z, (27) gives, thanks to (33) :

$$\int_I < z^2 > \,d\tau \le -\log|< e^{i\alpha} >|, \tag{34}$$

which is our first a priori estimate.. Then, differentiating (29), one obtain :

$$z_\tau + \phi'(r_0)r_1 z_x + (< z^2 > - z^2)z = z_{xx}. \tag{35}$$

Let us consider any smooth convex function j defined on the real line. We deduce from (35) and (28) :

$$< j(z) >_\tau + < z^2 >< zj' - j > + < z^2 j - z^3 j' >= - < j'' z_x^2 > \le 0. \tag{36}$$

Let us choose $j(z) = j_\eta(z) = \sqrt{\eta + z^2}$ hereabove, η being a small positive parameter. Pointing out that $0 < j(z) - zj'(z) \le \sqrt{\eta}$, one obtain

$$< j_\eta(z) >_\tau \le \sqrt{\eta} < z^2 > . \tag{37}$$

The right-hand side of (37) tends to zero in $L^1(I)$, thanks to (34), whereas $< j_\eta(z) >$ converges uniformly towards $<| z |>$. Passing to the limit in the sense of distributions in (37) gives

$$<| z |>_\tau \leq 0,$$

from which we receive our second estimate (this one and (34) do not depend on the length of I) :

(38) $$VT(\theta_0(\cdot, \tau)) =<| z |> \leq VT(\alpha).$$

We now choose $j(z) = z^2$ in (36) :

(39) $$< z^2 >_\tau + < z^2 >^2 +2 < z_x^2 >=< z^4 > .$$

The theorem 3.1 is obtain from (39), using the Sobolev-Gagliardo-Nirenberg inequality

$$< z^4 > \leq c_1 < z_x^2 >^{\frac{1}{2}} < z^2 >^{\frac{3}{2}} .$$

One obtain, thanks to a Young inequality :

(40) $$< z^2 >_\tau + < z_x^2 > + < z^2 >^2 \leq c_2 < z^2 >^3 .$$

The Gronwall's lemma gives then an estimate of τ_1 in terms of $< \alpha_x^2 >$, followed by an estimate of z_x in $L^2(\mathrm{R}/\mathbb{Z} \times I)$, and so on ...

The second theorem is proved using a different case of the Sobolev-Gagliardo-Nirenberg inequality :

(41) $$< z^4 > \leq c_3 < z_x^2 ><| z |>^2,$$

which gives, thanks to (38) :

$$< z^2 >_\tau + < z^2 >^2 \leq < z_x^2 > (c_3 VT(\alpha)^2 - 2).$$

The right-hand side will be non-positive whenever $VT(\alpha) \leq C_*$. Then the solution is global and satisfies an Oleinik's type inequality (!) :

$$< z^2 > \leq \frac{1}{\tau}.$$

But $<| z |> \leq < z^2 >^{\frac{1}{2}}$ (the Cauchy-Schwarz inequality), so that $<| z |> \to 0$ at infinity and, using again (40) and (41), we get

$$< z^2 >_\tau +b < z_x^2 > \leq 0,$$

for any constant $b < 2$, beyond some large enough time. This last inequality implies that $< z^2 >$ decreases exponentially to zero. Thus θ_0 converges uniformly towards some constant, as well as r_0. Passing to the limit in (32), one determines these constants by :

$$\hat{r} e^{i\hat{\theta}} = r_* < e^{i\alpha} > .$$

4 The gas dynamics

4.1 The model

We are concerned with a fluid described by its density ρ, pressure p, specific internal energy e and velocity u. These are related by an equation of state, say $p = p(v, e)$. Let y be the space variable in the physical space and t the time variable. Thanks to the law of mass conservation $\rho_t + (\rho u)_y = 0$, one may define the lagrangian mass variable x by the formula $dx =: \rho dy - \rho u\, dt$. In the lagrangian frame, the balance laws of mass, momentum and energy take the following form :

$$\begin{aligned} v_t &= u_x, && (42)\\ u_t + p_x &= \epsilon(\nu(v,e)u_x)_x, && (43)\\ (\frac{u^2}{2} + e)_t + (pu)_x &= \epsilon(\nu(v,e)uu_x)_x + \epsilon(k(v,e)T_x)_x. && (44)\end{aligned}$$

Hereabove, $v =: \rho^{-1}$ is the specific volume, $(v, e) \mapsto \nu$ is the newtonnian viscosity, $(v, e) \mapsto T$ is the temperature and $(v, e) \mapsto k$ is the heat conductivity. The temperature is such that the differential form $T^{-1}(de + pdv)$ is the differential of some scalar function S, called the specific entropy. These quantities satisfy the inequalities $\nu \geq 0$ and $p, v, e, T, k > 0$. One may consider the case $\nu \equiv 0$ if one admit that the viscous effect is negligible, compared to the heat diffusion, as is usually considered for a gas.

It will be easier to use p and S as independent thermodynamical variables instead of (v, e). We then rewrite the system as :

$$\begin{aligned} u_t + p_x &= \epsilon(\nu(S,p)u_x)_x, && (45)\\ p_t + c(S,p)^2 u_x &= \epsilon p_e(\nu u_x^2 + (kT_x)_x), && (46)\\ S_t &= \epsilon T^{-1}(\nu u_x^2 + (kT_x)_x), && (47)\end{aligned}$$

where $c(S, p)$ stands for the sound speed (this indeed does not have the dimension of a velocity, but is a length per mass), which is determined by the equation of state ; $c^2 = pp_e - p_v$.

The inviscid part ($\epsilon = 0$) of this system is strictly hyperbolic whenever $pp_e - p_v$ is positive. The wave speeds are $-c, 0, c$, where $\pm c$ are (usually) genuinely nonlinear. The eigenvalue $\lambda = 0$ is trivialy linearly degenerate. The variables v_1, v_2 associated to this linear field are the velocity u and the pressure p, whereas a good choice of the complement variable w is S, so that (45,46,47) is the appropriate form of the system in order to perform our asymptotic analysis.

4.2 The asymptotic analysis

In our expansion, only S may oscillate with amplitude $\mathcal{O}(1)$. Because $\lambda \equiv 0$, the velocity of the travelling waves vanishes identically (see(18)) and the asymptotic development takes the form

$$\begin{aligned} u^\epsilon(x,t) &= u_0(\epsilon t) + \epsilon u_1(x, \epsilon t) + \mathcal{O}(\epsilon^2),\\ p^\epsilon(x,t) &= p_0(\epsilon t) + \epsilon p_1(x, \epsilon t) + \mathcal{O}(\epsilon^2),\\ S^\epsilon(x,t) &= S_0(x, \epsilon t) + \epsilon S_1(x, \epsilon t) + \mathcal{O}(\epsilon^2).\end{aligned}$$

We now consider the ϵ^1-terms in the equations (45,46,47) :

$$\frac{du_0}{d\tau} = 0, \tag{48}$$

$$\frac{dp_0}{d\tau} = p_e^0 \partial_x (k^0(S_0) T_S^0(S_0) \partial_x S_0), \tag{49}$$

$$\frac{dS_0}{d\tau} = \frac{1}{T^0(S_0)} \partial_x (k^0(S_0) T_S^0(\dot{S_0}) \partial_x S_0). \tag{50}$$

Hereabove, we have make use of the notation $g^0 =: g(p_0, \cdot)$.

4.3 Remarks

References

[1] D. Hoff, J. Smoller
Solutions in the large for certain nonlinear parabolic systems. Ann. Inst. Henri Poincaré, vol. 2, 1985, p 213-235,
and
Global existence for systems of parabolic conservation laws in several space variables. J. Diff. Eq., vol 68, 1987, p 210-220.

[2] D. Hoff
Global existence for $1-d$, compressible , isentropic Navier-Stokes equations with large initial data. Trans. AMS, vol 303, 1987, p 169-181.

[3] D. Serre
Solutions faibles globales des équations de Navier-Stokes pour un fluide compressible. C. R. Acad. Sc. Paris, vol 303, 1986, p 639-642.

[4] D. Serre
Oscillations nonlinéaires de haute fréquence. Collège de France Seminar, vol XI, 1993.

[5] L. Hsiao, T.-P. Liu
Convergence to nonlinear diffusion waves for solutions of a system of hyperbolic conservation laws with damping. Comm. Math. Phys., vol 143, 1992, p 599-605.

[6] J. Glimm, P. D. Lax
Decay of solutions of systems of nonlinear hyperbolic conservation laws. Mem. AMS, vol 101, 1970.

[7] B. J. Temple
Decay with a rate for noncompactly supported solutions of conservation laws. Trans. AMS, vol 298, 1986, p 43-82.

[8] P. D. Lax
Hyperbolic systems of conservation laws, II Comm. Pure Appll. Maths, vol 10, 1957, p 537-566.

[9] B. L. Keyfitz, H. C. Kranzer
A system of non-strictly hyperbolic conservation laws arising in elasticity theory. Arch. Rat. Mech. Anal., vol 72, 1980, p 219-241.

[10] E. Hopf
The partial differential equation $u_t + uu_x = \mu u_{xx}$. Comm. Pure Appl. Maths., vol 3, 1950, p 201-230.

[11] F. Hubert
Dynamique lente-rapide pour des perturbations de systèmes de lois de conservation. Thèse de Doctorat, ENS de Lyon, novembre 1995.

[12] F. Hubert, D. Serre
Dynamique lente-rapide pour des perturbations de systèmes de lois de conservation. Comptes rendus Acad. Sci. Paris, 1995, note to appear. An article is in preparation.

AMS/IP Studies in Advanced Mathematics
Volume 3, 1997

A study of the global weak solution for the isentropic equations of a polytropic gas

A. JEFFREY AND Y.G. LU*[1]

Abstract

A compactness frame of the viscosity method for the isentropic equations of a polytropic gas in Lagrange coordinates is obtained by applying Lax entropy-entropy flux pairs and the theory of compensated compactness under some assumptions about estimates for the corresponding viscous solutions.

§1. INTRODUCTION

This paper considers the isentropic equations of a polytropic gas in Lagrange coordinates

$$\left.\begin{aligned} v_t - u_x &= 0 \\ u_t + (\frac{1}{\gamma} v^{-\gamma})_x &= 0 \end{aligned}\right\} \tag{1.1}$$

with initial data

$$(v,u)|_{t=0} = (v_0(0), u_0(x)), \tag{1.2}$$

where u is the velocity, v is the reciprocal of the density and $\gamma \in (1,3]$ is the adiabatic exponent. A convergence theorem for the method of artificial viscosity applied to the Cauchy problem (1.1), (1. 2) is established by using Lax entropy-entropy flux pairs and compensated compactness under some assumptions about estimates for the viscous solutions.

The Two eigenvalues of system (1.1) are

$$\lambda_1 = -v^{-(\gamma+1)/2}, \quad \lambda_2 = v^{-(\gamma+1)/2} \tag{1.3}$$

[1]Partially supported by grants from the National Youth Science Foundation of China

with the corresponding right eigenvectors

$$r_1 = \left[1, v^{-(\gamma+1)/2}\right], \quad r_2 = \left[1, -v^{-(\gamma+1)/2}\right].$$

Letting $d\lambda = (\partial\lambda/\partial v, \partial\lambda/\partial u)$ we see that

$$d\lambda_1 = \left[\frac{\gamma+1}{2} v^{-(\gamma+3)/2}, 0\right], \quad d\lambda_2 = \left[-\frac{\gamma+1}{2} v^{-(\gamma+3)/2}, 0\right]. \tag{1.4}$$

When viscosity terms are added to the right-hand sides of (1.1) we obtain

$$\left.\begin{aligned} v_t - u_x &= \varepsilon v_{xx} \\ u_t + (\frac{1}{\gamma} v^{-\gamma})_x &= \varepsilon u_{xx} \end{aligned}\right\} \tag{1.5}$$

By the theory of positively invariant regions given in [3], we know that

$$R = \{(v,u) : w(v,u) \leq const, \quad z(v,u) \geq const\}$$

are invariant regions (see Fig.1),

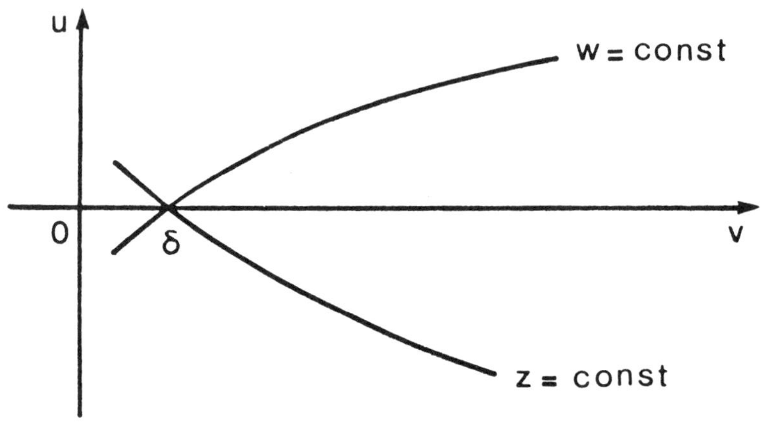

Fig.1

where w, z are the two Riemann invariants of (1.1) given by

$$w = u - \frac{2}{1-\gamma} v^{(1-\gamma)/2}, \quad z = u + \frac{2}{1-\gamma} v^{(1-\gamma)/2}.$$

From Fig. 1, we obtain the estimate for v from below, $v \geq \delta > 0$. If another estimate for v from above $v \leq C(T)$ could be obtained, then from (1.3) and (1.4), (1.1) would be a genuinely nonlinear and strictly hyperbolic system. In such circumstances, Di Perna's frame-work about compensated compactness described in [6] could then be easily used to establish the convergence of the viscous approximate solutions of (1.1). However, in general, the vacuum state, namely $v = +\infty$ appears. In paper [5], Di Perna considered the Cauchy problem for the isentropic equations of a polytropic gas in Eulerian coordinates and established the convergence of the corresponding viscous approximate solutions for the adiabatic exponent $\gamma = 1 + \frac{2}{2n+1} (n \geq 2)$. In later papers [1, 4], other authors extended the exponent γ to the interval $(1, \frac{5}{3}]$ and obtained the convergence of the Lax-Friedrichs scheme for such a system. All these results contained the vacuum state $\rho = 1/v = 0$ and were restricted by the initial data being confined to bounded functions. In this paper, we study the Cauchy problem for system (1.1) subject to the initial data (1.2) being bounded in $L^p (p > 2)$.

§2. Lax entropies

We recall that a pair (y, q) of real valued maps is an entropy– entropy flux pair of (1.1) if all smooth solutions satisfy

$$\nabla y(v,u) \nabla f(v,u) = \nabla q(v,u). \tag{2.1}$$

where

$$\nabla f = \begin{bmatrix} 0 & -1 \\ -v^{-(\gamma+1)} & 0 \end{bmatrix}.$$

The entropy equations (2.1) are equivalent to

$$y_{vv} = v^{-(\gamma+1)} y_{uu}. \tag{2.2}$$

We choose special entropy functions of the form $y = h(v)e^{ku}$, where k denotes a constant. Then from (2.2)

$$h''(v) - k^2 v^{-(\gamma+1)} h = 0.$$

Let $h(v) = a(v)\phi(s(v))$, where $a(v) = v^{(1+\gamma)/4}$ and $s = \frac{2k}{\gamma-1} v^{(1-\gamma)/2}$. Then ϕ satisfies a standard Fuchsian equation

$$\phi'' - (1 + \mu s^{-2})\phi = 0, \tag{2.3}$$

where $\mu = (4 - (1-\gamma)^2)/4(1-\gamma)^2$.

Appeal to the method of Frobenius gives a solution of equation (2 .3) of the following form

$$\phi_+(s) = s^{j_+} \sum_{n=0}^{\infty} c_{2n} s^{2n} = s^{j_+} g(s),$$

where $j_+ = \frac{1+\sqrt{1+4\mu}}{2} = \frac{\gamma+1}{2(\gamma-1)}$ is a root of the equation $j(j-1) = \mu$, and

$$c_{2n} = \frac{c_{2(n-1)}}{(2n+j_+)(2n+j_+-1)-\mu}, \quad n \geq 1$$

with c_0 an arbitrary positive constant. Let another independent solution ϕ_- of (2.3) satisfy $\phi_- = \phi_+$, then solving for ϕ gives

$$\phi'' + \frac{2\phi_+'}{\phi_+}\phi = 0,$$

so that $\phi' = \phi_+^{-2}$. Thus

$$\phi = \int \frac{1}{s^{2j}+g^2(s)}ds = \frac{1-\gamma}{2}s^{2/(1-\gamma)}g^{-2}(s) + (1-\gamma)\int s^{2/(1-\gamma)}g^{-3}(s)g'(s)ds.$$

We make the special choice

$$\phi_- = s^{\frac{\gamma-3}{2(\gamma-1)}}g^{-1}(s) + 2s^{\frac{\gamma+1}{2(\gamma-1)}}g(s)\int_C^S \tau^{\frac{2}{1-\gamma}}g^{-3}(\tau)g'(\tau)d\tau.$$

Since

$$\frac{\gamma-3}{2(\gamma-1)} < \frac{\gamma-1}{2(\gamma-1)} + \frac{2}{1-\gamma} + 2 = \frac{5\gamma-7}{2(\gamma-1)},$$

we may choose c to be a small positive constant such that $\phi_- > 0$ for $s > 0$.

Using (2.1), we have

$$q_v = -v^{-(\gamma+1)}y_u, \quad q_u = -y_v.$$

Then two progressing waves of system (1.1) are provided by

$$\left.\begin{aligned} y_+ &= a(v)\phi_+e^{ku}, \\ q_+ &= -\frac{e^{ku}}{k}(a(v)\phi_+)_v = y_+[\frac{\phi_+'}{\phi_+}\lambda_2 - \frac{1}{4kv}(1+\gamma)] \end{aligned}\right\} \tag{2.4}$$

and

$$\left.\begin{aligned} y_- &= a(v)\phi_-e^{-ku}, \\ q_- &= \frac{e^{-ku}}{k}(a(v)\phi_+)_v = y_-[-\frac{\phi_-'}{\phi_-}\lambda_2 + \frac{1}{4kv}(1+r)] \end{aligned}\right\} \tag{2.5}$$

Since ϕ_+, ϕ_- are two independent solutions of the Fuchsian equation (2. 3) , they behave exponentially at infinity like the solutions to the equation $\phi'' - \phi = 0$. More precisely, according to the comparison theorem given in [7], ϕ_+, ϕ_- satisfy

$$\left.\begin{array}{ll}\phi_+ e^{-S} = 1 + O(s^{-2}), & \phi'_+\phi_+^{-1} = 1 + O(s^{-2}), \\ \phi_- e^{S} = 1 + O(s^{-2}), & \phi'_-\phi_-^{-1} = -1 + O(s^{-2}),\end{array}\right\} \tag{2.6}$$

as s approaches infinity. Then it follows that

$$y_+ = a(v)\phi_+ e^{kw} e^{-s} = e^{kw}\left[a(v) + O\left[\frac{v^{(5\gamma-3)/4}}{k^2}\right]\right] = e^{kw}\left[a(v) + O\left[\frac{1}{k^2}\right]\right]$$

on compact subsets of $v \geq \delta$, and we have the uniform ratio

$$\frac{q_+}{y_+} = \lambda_2 + O\left[\frac{v^{(\gamma-3)/2}}{k^2}\right] - \frac{1}{4kv}(1+\gamma) = \lambda_2 - \frac{1+\gamma}{4v}\cdot\frac{1}{k} + O\left[\frac{1}{k^2}\right]$$

on $v \geq \delta$. The entropy

$$y_- = a(v)\phi_- e^{-kw} e^{s} = e^{-kw}\left[a(v) + O\left[\frac{v^{(5\gamma-3)/4}}{k^2}\right]\right] = e^{kw}\left[a(v) + O\left[\frac{1}{k^2}\right]\right]$$

on compact subsets of $v \geq \delta$, and the corresponding flux

$$q_- = y_-\left[\lambda_2 + \frac{1+\gamma}{4v}\cdot\frac{1}{k} + O\left[\frac{1}{k^2}\right]\right]$$

is uniform on $v \geq \delta$.

§3. H_{loc}^{-1} COMPACTNESS CONDITIONS

In this section we further analyse the entropy-entropy pairs (y_+, q_+) and (y_-, q_+), and obtain the following results.

Theorem 3.1: Suppose that $(v^\varepsilon(x,t), u^\varepsilon(x,t)$ is the sequence of viscosity solutions determined by the Cauchy problem (1.5), (1.2). Then we have

$$\partial_t y_+(v^\varepsilon(x,t), u^\varepsilon(x,t)) + \partial_x q_+(v^\varepsilon(x,t), u^\varepsilon(x,t)) \tag{3.1}$$

is relatively compact in $H_{loc}^{-1}(R \times R^+)$. Moreover,

$$\partial_t y_-(v^\varepsilon(x,t), u^\varepsilon(x,t)) + \partial_x q_-(v^\varepsilon(x,t), u^\varepsilon(x,t)) \tag{3.2}$$

is relatively compact in $H_{loc}^{-1}(R \times R)$ provided the viscosity solutions have the a priori estimates

$$v^\varepsilon \in L_{loc}^p(R \times R^+) \quad (p > 2), \quad \varepsilon(v^\varepsilon)^{-1}(v_x^\varepsilon)^2 \in L_{loc}^1(R \times R^+). \tag{3.3}$$

Remark 3.2: According to the results in [12], under the corresponding conditions on the initial data we have

$$\int_{-\infty}^{\infty} y(v^\varepsilon, u^\varepsilon)|_{t=T} dx + \varepsilon \int_0^T \int_{-\infty}^{\infty} [(u^\varepsilon)_x]^2 + (v^\varepsilon)^{-(\gamma+1)}[(v^\varepsilon)_x]^2 dx dt$$

$$\leq \int_{-\infty}^{\infty} y(v^\varepsilon, u^\varepsilon|_{t=0} dx, \tag{3.4}$$

where y is a non-negative entropy of (1.1),

$$y(v,u) = \int_\delta^v \frac{1}{\gamma}(\delta^{-\gamma} - s^{-\gamma} ds + \frac{1}{2}u^2.$$

Thus we obtain the following estimates

$$v^\varepsilon \in L^1_{loc}(R \times R^+),$$

$$\sqrt{\varepsilon}(u^\varepsilon)_x \in L^2_{loc}(R \times R^+), \sqrt{\varepsilon}(v^\varepsilon)^{-(\gamma+1)/2}(v^\varepsilon)_x \in L^2_{loc}(R \times R^+). \tag{3.5}$$

If it was possible to obtain the estimate $v \leq c(T)$, then assumption (3.3) would be easily determined from (3.5) . However, in general, the estimate $v \leq c(T)$ is not true.

Proof of Theorem 3.1 : From the theory of invariant regions given in [4], we have $|u^\varepsilon| \leq M$ for $1 < \gamma \leq 3$. Then

$$|y_+|_{L^\infty} \leq M(k), \quad |q_+|_{L^\infty} \leq M(k). \tag{3.6}$$

A simple calculation gives

$$(y_+)_t + (q_+)_x = \varepsilon(y_+)_{xx} - \varepsilon\nabla^2 y_+(v^\varepsilon_x, u^\varepsilon_x),$$

and

$$\begin{aligned}|\varepsilon\nabla^2 y_+(v^\varepsilon_x, u^\varepsilon_x)|_{L^1_{loc}} &\leq \varepsilon|k^2 y_+(u^\varepsilon_x)^2 + O(((v^\varepsilon)^{-\gamma}u^\varepsilon_x v^\varepsilon_x) + O\left((v^\varepsilon)^{-(\gamma+1)}(v^\varepsilon_x)^2\right)|_{L^1_{loc}} \\ &= \varepsilon|O\left((u^\varepsilon_x)^2\right) + O\left((v^\varepsilon)^{-2\gamma}(v^\varepsilon_x)^2\right) + O\left((v^\varepsilon)^{-(\gamma+1)}(v^\varepsilon_x)^2\right)|_{L^1_{loc}} \\ &= O(1),\end{aligned}$$

resulting from (3.5). Thus $\varepsilon\nabla^2 y_+(v^\varepsilon_x, u^\varepsilon_x)$ is bounded in $C_0^*(\Omega)$, and then is compact in $W^{-1,q_0}(\Omega)$ $(1 < q_0 < 2)$ for any bounded open region Ω. Moreover

$$|\sqrt{\varepsilon}(y_+)_x|_{L^2_{loc}} = |\sqrt{\varepsilon}O(u^\varepsilon_x) + \sqrt{\varepsilon}O\left[(v^\varepsilon)^{-\gamma}v^\varepsilon_x\right]|_{L^2_{loc}} = O(1),$$

so $\varepsilon(y_+)_{xx}$ is compact in $W^{-1,2}(\Omega)$. From the boundedness of y_+, q_+, we have $(y_+)_t + (q_+)_x$ is bounded in $W^{-1,q}(\Omega)$ for any $q > 2$. The proof of (3.1) follows by using Theorem 3 in [4].

To prove (3.2) , we use the fact that

$$|\varepsilon\nabla^2 y_-(v_x^\varepsilon, u_x^\varepsilon)| = \varepsilon[O(v^\varepsilon(u_x^\varepsilon)^2) + O(u_x^\varepsilon \cdot v_x^\varepsilon) + O((v^\varepsilon)^{-\gamma}(v_x^\varepsilon)^2)], \tag{3.7}$$

then

$$\varepsilon v^\varepsilon(u_x^\varepsilon)^2 \in L^1_{lov}(R \times R^+) \tag{3.8}$$

follows from (3.7) and the assumption (3.3). Thus $\varepsilon\nabla^2 y_-(v_x^\varepsilon, u_x^\varepsilon)$ is bounded in $C_0^*(\Omega)$, and furthermore is compact in $W^{-1,q_0}(\Omega)$ $(1 < q_0, 2)$. Since

$$(y_-)_x = O(v^\varepsilon u_x^\varepsilon) + O(v_x^\varepsilon),$$

from (3.3) we have

$$|\varepsilon(y_-)_x\phi_x|_{L^1_{loc}} \leq M\sqrt{\varepsilon}\Big[|\sqrt{\varepsilon}(v^\varepsilon)^{\frac{1}{2}}u_x^\varepsilon|_{L^2_{loc}} \cdot |(v^\varepsilon)^{\frac{1}{2}}\phi_x|_{L^2_{loc}}$$

$$+|\sqrt{\varepsilon}(v^\varepsilon)^{-\frac{1}{2}}v_x^\varepsilon|_{L^2_{loc}} \cdot |(v^\varepsilon)^{\frac{1}{2}}\phi_x|_{L^2_{loc}}\Big] \to 0$$

as $\varepsilon \downarrow 0$ for any $\phi \in W_0^{2,p}(\Omega)$. Thus $\varepsilon(y_-))_{xx}$ is compact in $W^{-1,2}(\Omega)$. By assumption (3.3) , $(y_-)_t + (q_-)_x$ is bounded in $W^{-1,p}(\Omega)$ $(p > 2)$. Therefore we derive (3.2) from the Theorem 3 in [4], and Theorem 3.1 is proved.

§4. Reduction of Young measures

In this section we use the entropy-entropy flux pairs of Lax type constructed in the last two sections to prove that the positive measures $\theta_{(x,t)}$ given in Lemmas 4.1, 4.2 are indeed Dirac measures.

Lemma 4.1 Suppose that w^ε is bounded in $L^p_{loc}(R\times R^+, R^2)$, where $1 \leq p < \infty$. Then there exists a subsequence w^{ε_n} of w^ε and a family of positive measures $\theta_{(x,t)} \in M(R^2), (x,t) \in R \times R^+$, and $M(R^2)$ is the dual space of $C_0(R^2)$, such that for any bounded set $\Omega \subset R \times R^+$,

$$f(w^{\varepsilon_n}) \rightharpoonup < f(\lambda), \theta_{(x,t)} > \quad in \quad L^1(\Omega),$$

whenever $f \in C(R^2)$ satisfies

$$\lim_{|\lambda|\to\infty} \frac{f(\lambda)}{|\lambda|^p} = 0$$

This Lemma comes directly from [8]. The conclusion in the following Lemma can be derived from Lemma 4.1 and the well-known Div-Curl Lemma.

Lemma 4.2 There exist a subsequence $(v^{\varepsilon_n}, u^{\varepsilon_n})$ of $(v^\varepsilon, u^\varepsilon)$ and a family of positive measures $\theta_{(x,t)} \in M(R^2)$ such that

$$< \theta_{(x,t)}, y_\pm q - y q_\pm >=< \theta_{(x,t)}, y_\pm >< \theta_{(x,t)}, q > - < \theta_{(x,t)}, y >< \theta_{(x,t)}, q_\pm > \tag{4.1}$$

for $(x, t) \in R \times R^+$ and any entropy-entropy flux pair (y, q) satisfying the conditions that

$$y(v^{\varepsilon_n}, u^{\varepsilon_n})_t + q(v^{\varepsilon_n}, u^{\varepsilon_n})_x \quad is \quad compact \quad in \quad W_{loc}^{-1,2}(\Omega)$$

and

Since $v^\varepsilon \in L^p_{loc}(R \times R^+)$, there exits a set of measure zero $\Omega_1 \subset \Omega$ such that $S = \{(v, u), supp\theta_{(x,t)} \backslash \{v = +\infty\}\}$ is a non-empty set for any point $(x, t) \in \Omega \backslash \Omega_1$. In what follows we only consider the points in $\Omega \backslash \Omega_1$.

Let Q denote the smallest characteristic rectangle

$$Q = \{(v, u) : w_- \leq w \leq w_+, \quad z_- \in z \leq z_+, \quad z \leq w\}.$$

As was done in [5], we can easily prove that the point $(w, z) = (w_+, z_+)$ is contained in the support set of $\theta_{(x,t)}$. We define two sequences of measures $\theta_{+k}, \theta_{-k} \in M(R^2)$ as follows:

$$< \theta_{+k}, h >= \frac{< \theta_{(x,t)}, hy_+ >}{< \theta_{(x,t)}, y_+ >}, \qquad < \theta_{-k}, h >= \frac{< \theta_{(x,t)}, hy_- >}{< \theta_{(x,t)}, y_- >}$$

for any $(x, t) \in \Omega \backslash \Omega_1$ and $h \in C(R^2)$ satisfying $|h(v, u)| = O(|v|)$. Since $||\theta_{\pm k}|| = 1$, there is a subsequence of $\{k\}$, also denoted by $\{k\}$, such that

$$\theta_{+k} \stackrel{\star}{\rightharpoonup} \theta_+ \in M(R^2), \quad \theta_{-k} \stackrel{\star}{\rightharpoonup} \theta_- \in M(R^2) \quad as \quad k \to \infty$$

We can easily prove that (see [6,8,9])

$$supp\{\theta_+\} \subset \{(w, z) \in Q, w = w^+\}, supp\{\theta_-\} \subset \{(w, z) \in Q, w = w^-\}$$

and

$$< \theta_+, \lambda_2 y - q >=< \theta_-, \lambda_2 y - q > \tag{4.2}$$

follows from (4.1). Especially, if we choose (y_+, q_+) as (y, q) in (4.2) , then

$$< \theta_+, \left[\frac{1+\gamma}{4v} \cdot \frac{1}{k} + O\left(\frac{1}{k^2}\right)\right]_{y_+} =< \theta_-, \left[\frac{1+\gamma}{4v} \cdot \frac{1}{k} + O\left(\frac{1}{k^2}\right)\right]_{y_+} > . \tag{4.3}$$

Since the point $(w,z)=(w_+,z_+)$ is contained in the set supp θ_+, we can obtain the contradictory equality from (4.3)

$$\begin{aligned}\frac{c_1e^{kw_+}}{k}\leq &< \theta_+, \left[\frac{1+\gamma}{4v}\frac{1}{k}+O\left(\frac{1}{k^2}\right)\right]_{y+} > \\ &=< \theta_-, \left[\frac{1+\gamma}{4v}\cdot\frac{1}{k}+O\left(\frac{1}{k^2}\right)\right]_{y+} > \\ &\leq \frac{c_2e^{kw_-}}{k},\end{aligned}$$

for some positive constants c_1, c_2. So w_- must equal w_+. Similarly z_- equals z_+. Therefore $\theta_{(x,t)}$ is a Dirac measure for almost all points $(x,t)\in\Omega$.

References

1. G.Q. Chen, Covergence of the Lax- Friedrichs scheme for isentropic gas dynamics, III, *Acta Math. Sci.* 1 (1986), 75-120
2. G.Q. Chen, & Y.G. Lu, A study on the application of compensated compactness theory, *Chin, Sci.* Bull, 34 (1989) , 1: 15-19.
3. K.N. Chueh, C.C Conley & J.A. Smoller, Positively invariant regions for systems of nonlinear diffusion equations, *Indiana Univ. Math. J.* 26 (1977), 372-411.
4. X.X. Ding , G.Q. Chen & P.Z. Luo, Convergence of the Lax- Friedrichs scheme for isentropic gas dynamics, I,II, *Acta Math. Sci.*, 4 (1985) , 415-472.
5. R.J. Di Perna, Convergence of the viscosity method for isentropic gas dynamics, *Comm. Math. Phys.*, 91 (1983), 1-30.
6. R.J. Di Perna, Convergence of approximate solutions to conservation laws, *Arch. Rational Mech. Anal.* , 82 (1983) , 27-70.
7. E. Kamake, *Differentialgleichungen*, (1971).
8. P. X. Lin, Young measures and an application of compensated compactness to one-dimensional nonlinear elastodynamics, *Trans. Amer. Math. Soc.* , 329 (1992) , 1: 377-413.
9. Y.G. Lu, Convergence of the viscosity method for nonstrictly hyperbolic conservation laws, *Comm. in Math. Phys.*, 150, (1992) ,59-64.
10. Y.G. Lu, Cauchy problem for an extended model of combustion, *Proceedings of the Royal Society of Edinburgh,* 120A (1992), 349-360.
11. Y.G. Lu, Global continuous solution of isentropic gas dynamics, *Proceedings of the Royal Society of Edinburgh,* 123A (1993).
12. J.A. Smoller, Shock waves and reaction- diffusion equations, Springer, New York, (1983).

A. Jeffrey
Department of Engineering Mathematics
University of Newcastle upon Tyne
U.K.

Y.G. Lu
Wuhan Institute of Mathematical Sciences
Academia Sinica
P.O. Box 71007, Wuhan 430071
People's Republic of China.

AMS/IP Studies in Advanced Mathematics
Volume **3**, 1997

SMOOTH SHOCK PROFILES IN VISCOELASTICITY WITH MEMORY

SHUICHI KAWASHIMA & HARUMI HATTORI

Abstract

We survey our recent results on the smooth monotone shock profiles for viscoelaticity with fading memory. We discuss the existence, uniqueness and asymptotic stability without assuming the strict convexity of the stress functions.

1. Introduction

The aim of this note is to survey our recent results in [5,3], which are on the existence, uniqueness and asymptotic stability of smooth shock profiles for the equation of viscoelasticity with fading memory.

We consider a system of conservation equations describing a one-dimensional motion of a viscoelastic material with fading memory:

$$\begin{aligned} &v_t - u_x = 0, \\ &u_t - \left\{\sigma(v) + \int_{-\infty}^{t} a'(t-\tau)\eta(v)(\tau)d\tau\right\}_x = 0, \end{aligned} \tag{1.1}$$

where the unknown functions v and u denote the strain and the velocity, respectively. Throughout this paper we make the following assumptions on $a(t), \sigma(v)$ and $\eta(v)$:

Assumption 1.1. *(i) $a(t)$ is a function in $W^{3,1}(0,\infty)$, with the following sign conditions:*

$$a(t) > 0, \quad a'(t) < 0 \quad \textit{for } t \geq 0. \tag{1.2}$$

(ii) $\sigma(v)$ and $\eta(v)$ are smooth functions such that

$$\sigma'(v) > 0, \quad \eta'(v) > 0, \quad \chi'(v) > 0 \tag{1.3}$$

for any v under consideration, where

$$\chi(v) \equiv \sigma(v) - a(0)\eta(v). \tag{1.4}$$

Usually, $\sigma(v)$ and $\chi(v)$ are called the instantaneous stress and the equilibrium stress, respectively.

We want to discuss the existence, uniqueness and asymptotic stability of shock profiles in a situation that allows $\chi(v)$ to have a inflection point: In particular, we include the case where

$$\chi''(v) \gtrless 0 \quad \text{for } v \gtrless 0, \tag{1.5}$$

or

$$\chi''(v) \lessgtr 0 \quad \text{for } v \gtrless 0. \tag{1.6}$$

A class of shock profiles considered here is defined as follows:

Definition 1.1. *Let $(v_\pm, u_\pm)$ be two different constant states and let s be a constant. Then a function $(V, U)(x - st)$ is called a shock profile for (1.1) which connects (v_-, u_-) and (v_+, u_+) if the following two properties hold:*
(i) $(V, U)(\xi)$ is smooth and strictly monotone as a function of $\xi \in I\!R$, and satisfies

$$(V, U)(\xi) \to (v_\pm, u_\pm) \qquad \text{as } \xi \to \pm\infty. \tag{1.7}$$

(ii) $(v, u) = (V, U)(x - st)$ solves (1.1).

In the definition above, s is called the shock speed. In our analysis, when $s \gtrless 0$, the rate of convergence of $V(\xi) \to v_\pm$ as $\xi \to \pm\infty$ becomes important. We characterize this by the following property: When $s \gtrless 0$,

$$\omega_\pm(\tau) \equiv \lim_{\xi \to \pm\infty} \frac{V(\xi + s\tau) - v_\pm}{V(\xi) - v_\pm} > 0 \tag{1.8}$$

(that is, the limit $\omega_\pm(\tau)$ exists and is strictly positive) for each $\tau \geq 0$. We note that for $s > 0$, (1.8) holds true if $V(\xi) - v_+$ behaves like $e^{-\lambda\xi}$ or ξ^{-k} but not if $V(\xi) - v_+$ is like $e^{-\lambda\xi^2}$, where λ and k are positive numbers.

The content of this paper is as follows. In Section 2 we give the necessary and sufficient conditions for the existence of shock profiles for (1.1) satisfying (1.8). The uniqueness of shock profiles is discussed in Section 3. The discussions in these two sections include the degenerate case where $\chi'(v_\pm) = s^2$ for $s \gtrless 0$, which was not covered in the previous literatures ([2,7]). We note that such a degenerate case can occur only if $\chi(v)$ has an inflection point. The final section proves the asymptotic stability of shock profiles in some restricted situation.

2. Existence

Let us consider a shock profiles for (1.1) which connects (v_-, u_-) and (v_+, u_+) at the speed s. We want to show that the necessary and sufficient conditions for the existence of such a shock profile consist of the following three conditions. The first is the *Rankine-Hugoniot condition*:

$$\begin{aligned}&-s(v_+ - v_-) - (u_+ - u_-) = 0,\\&-s(u_+ - u_-) - (\chi(v_+) - \chi(v_-)) = 0.\end{aligned} \tag{2.1}$$

The second is the *entropy condition*: For $s \gtrless 0$,

$$s^2 \lessgtr \frac{\chi(v) - \chi(v_-)}{v - v_-} \quad \text{for } v \in (v_+, v_-). \tag{2.2}$$

Note that under (2.1) this entropy condition is equivalent to the following: For $s \gtrless 0$,

$$s^2 \gtrless \frac{\chi(v) - \chi(v_+)}{v - v_+} \quad \text{for } v \in (v_+, v_-). \tag{2.2$'$}$$

The third is the *non-resonance condition*:

$$s^2 < \sigma'(v) \quad \text{for } v \in [v_+, v_-]. \tag{2.3}$$

Here and in what follows, (v_+, v_-) means the open interval between v_+ and v_-, and it does not necessarily imply $v_+ < v_-$; the closed interval $[v_+, v_-]$ is understood similarly.

The following first theorem states that the above three conditions are necessary for the existence of a shock profile satisfying (1.8).

Theorem 2.1. *(Necessary conditions for the existence) Suppose Assumption 1.1 is satisfied. Suppose that (1.1) admits a shock profile $(V, U)(x - st)$ connecting (v_-, u_-) and (v_+, u_+). Then $(v_\pm, u_\pm)$ and s must satisfy (2.1), (2.2) and the following condition:*

$$s^2 < \sigma'(v) \quad \textit{for } v \in (v_+, v_-). \tag{2.4}$$

Furthermore, if the shock profile satisfies (1.8), then (2.3) (which is stronger than (2.4)) is satisfied.

An outline of the proof is as follows. Let $(V, U)(x - st)$ be a shock profile connecting (v_-, u_-) and (v_+, u_+). Then we see that

$$\begin{aligned}&-sV - U = A_1,\\&-sU - \left\{\sigma(V) + \int_0^\infty a'(\tau)\eta(V)(\xi + s\tau)d\tau\right\} = A_2,\end{aligned} \tag{2.5}$$

where A_1 and A_2 are constants of integration. Letting $\xi \to \pm\infty$ in (2.5), we have $A_1 = -sv_\pm - u_\pm$ and $A_2 = -su_\pm - \chi(v_\pm)$, which implies the Rankine-Hugoniot condition (2.1). We eliminate U from (2.5), obtaining

$$-s^2V + \sigma(V) = -\int_0^\infty a'(\tau)\eta(V)(\xi + s\tau)d\tau + A, \tag{2.6}$$

where $A = sA_1 - A_2 = -s^2 v_\pm + \chi(v_\pm)$. The integration of the memory term in (2.6) yields

$$(2.7) \qquad -s^2 V + \chi(V) - A = s \int_0^\infty a(\tau)(\eta'(V)V')(\xi + s\tau)d\tau,$$

while the differentiation of (2.6) with respect to ξ gives

$$(2.8) \qquad (-s^2 + \sigma'(V))V' = -\int_0^\infty a'(\tau)(\eta'(V)V')(\xi + s\tau)d\tau.$$

The necessity of the entropy condition (2.2) and the weaker non-resonance condition (2.4) can be shown easily by making use of (2.7) and (2.8), respectively. Moreover, it follows from (2.2) and (2.2)$'$ that $s^2 \le \chi'(v_\mp)$ for $s \gtrless 0$, which implies $s^2 < \sigma'(v_\mp)$ for $s \gtrless 0$. Consequently, in order to prove the necessity of the non-resonance condition (2.3), it suffices to verify

$$(2.9) \qquad s^2 < \sigma'(v_\pm) \quad \text{for } s \gtrless 0.$$

This inequality is crucial in the proof of Theorem 2.1 and can be verified by using (1.8) as follows. Let $s \gtrless 0$. We expand (2.6) about $v_\pm$ and divide the resulting equation by $V(\xi) - v_\pm$. This yields

$$(2.10) \qquad \begin{aligned} &-s^2 + \sigma'(v_\pm) + O(|V(\xi) - v_\pm|) \\ &= -\int_0^\infty a'(\tau)\left\{\eta'(v_\pm)\frac{V(\xi + s\tau) - v_\pm}{V(\xi) - v_\pm} + O(|V(\xi + s\tau) - v_\pm|)\right\} d\tau \end{aligned}$$

as $\xi \to \pm\infty$. Taking the limit as $\xi \to \pm\infty$, we obtain

$$(2.11) \qquad -s^2 + \sigma'(v_\pm) = -\int_0^\infty a'(\tau)\eta'(v_\pm)\omega_\pm(\tau)d\tau,$$

which together with (1.8) proves (2.9). The proof of Theorem 2.1 is complete.

Next we discuss the sufficiency of the conditions (2.1), (2.2) and (2.3). We observe that (2.2) and (2.2)$'$ imply $\chi'(v_\pm) \le s^2$ for $s \gtrless 0$. So we shall treat the following two cases:

Case A: $\chi'(v_\pm) < s^2$ for $s \gtrless 0$.

Case B: $\chi'(v_\pm) = s^2$ and $\chi''(v_\pm) \ne 0$ for $s \gtrless 0$.

Case A is a non-degenerate case. If $\chi(v)$ is strictly convex (or concave), we have only Case A. Case B is one of the simplest degenerate cases and can occur only if $\chi(v)$ has an inflection point. We do not deal with the cases where $\chi(v)$ has higher degeneracy.

Our second theorem concerning the sufficiency of the conditions (2.1), (2.2) and (2.3) is now stated as follows.

Theorem 2.2. *(Sufficient conditions for the existence) Suppose Assumption 1.1 is satisfied. Assume that* $(v_\pm, u_\pm)$ *and* s *satisfy (2.1), (2.2) and (2.3), and consider Case A or Case B; in Case B we require that* $(1+t)^3 a'(t) \in L^1(0,\infty)$. *Then, in either case, there exists a shock profile* $(V, U)(x - st)$ *for (1.1) which connects* (v_-, u_-) *and* (v_+, u_+), *and satisfies (1.8).*

To prove this theorem, we must solve the integral equation (2.6) under the restriction (1.7). This can be done by applying the method of Greenberg-Hasting [2], and the solution $V(\xi)$ is constructed as a limit of a suitable successive approximation sequence $\{V_n(\xi)\}$. Our proof in Case A is essentially the same as that in Greenberg-Hasting [2] where they discussed the case where $\chi(v)$ is strictly concave. On the other hand, in Case B, we need some technical modifications. Though we omit the details of our proof, we describe the behavior of the solution $V(\xi)$ as $\xi \to \pm\infty$ for $s \gtrless 0$, in order to show the difference between Case A and Case B.

Let $s > 0$. Then we have: In Case A,

$$V(\xi) = v_+ + \delta e^{-\lambda\xi} + O(e^{-2\lambda\xi}), \quad \xi \to +\infty, \tag{2.12}$$

and in Case B,

$$V(\xi) = v_+ + \alpha\xi^{-1} + \beta\xi^{-2}\log\xi + O(\xi^{-3}|\log\xi|^2), \quad \xi \to +\infty, \tag{2.13}$$

Here in (2.12), $\delta = v_- - v_+$, and λ is a constant uniquely determined by

$$- s^2 + \sigma'(v_+) + \eta'(v_+)\int_0^\infty a'(\tau)e^{-\lambda s\tau}d\tau = 0. \tag{2.14}$$

We see $\lambda > 0$ because we have $\chi'(v_+) < s^2 < \sigma'(v_+)$ in Case A with $s > 0$. The coefficients α and β in (2.13) are uniquely determined by

$$\frac{1}{2}\chi''(v_+)\alpha + sa_1\eta'(v_+) = 0, \tag{2.15}$$

$$- sa_1\eta'(v_+)\beta + \frac{1}{6}\chi'''(v_+)\alpha^3 + sa_1\eta''(v_+)\alpha^2 - s^2a_2\eta'(v_+)\alpha = 0, \tag{2.16}$$

respectively, where

$$a_j = -\int_0^\infty a'(\tau)\tau^j d\tau > 0 \quad j = 1, 2. \tag{2.17}$$

We note that $\alpha \gtrless 0$ for $v_+ \lessgtr v_-$; in fact, in Case B with $s > 0$, we have $\chi''(v_+) \lessgtr 0$ for $v_+ \lessgtr v_-$ and this determines the sign of α.

The behavior of $V(\xi)$ as $\xi \to -\infty$ for $s < 0$ is analogous and we omit it. We note that (2.12) in Case A is essentially due to [2,7] but (2.13) in Case B is not contained in the previous literatures.

Finally in this section, we restrict ourselves to the case where $\chi(v)$ satisfies (1.5) or (1.6). We see that if $\chi(v)$ satisfies (1.6), then only Case A occurs. Also, if $\chi(v)$ satisfies (1.5), only Case A or Case B is possible and the higher degeneracy can not occur. Therefore, combining theorems 2.1 and 2.2, we have:

Corollary 2.3. *(Necessary and sufficient conditions for the existence) Suppose that Assumption 1.1 is satisfied and that $\chi(v)$ satisfies (1.5) or (1.6); in the case where (1.5) holds, we require that $(1+t)^3a'(t) \in L^1(0,\infty)$. Then there exists a shock profile $(V,U)(x - st)$ for (1.1) which connects (v_-, u_-) and (v_+, u_+), and satisfies (1.8) if and only if $(v_\pm, u_\pm)$ and s satisfy (2.1), (2.2) and (2.3).*

3. Uniqueness

We discuss the uniqueness of shock profiles only for $s > 0$. The case $s < 0$ can be treated similarly. The following two theorems state the uniqueness in Case A and Case B, respectively.

Theorem 3.1A. *(Uniqueness in Case A) Suppose Assumption 1.1 is satisfied. Let $s > 0$ and assume (2.1) and (2.3). Consider Case A and suppose that the problem (2.6), (1.7) admits two smooth and strictly monotone solutions $V_1(\xi)$ and $V_2(\xi)$. If*

$$V_1(\xi) - V_2(\xi) = O(e^{-k\xi}) \quad as\ \xi \to +\infty \tag{3.1}$$

for some $k > \lambda$, then $V_1 \equiv V_2$, where $\lambda > 0$ is the constant determined in (2.14). In particular, the solution of the form (2.12) (which is constructed in Theorem 2.2) is unique.

Theorem 3.1B. *(Uniqueness in Case B) Suppose Assumption 1.1 is satisfied. Let $s > 0$ and assume (2.1), (2.2) and (2.3). We require $(1+t)^2 a'(t) \in L^1(0,\infty)$ and consider Case B. Suppose that the problem (2.6), (1.7) admits two smooth and strictly monotone solutions $V_1(\xi)$ and $V_2(\xi)$ which are of the form*

$$V(\xi) = v_+ + \alpha\xi^{-1} + o(\xi^{-1}) \quad as\ \xi \to +\infty, \tag{3.2}$$

where α is determined in (2.15). If

$$V_1(\xi) - V_2(\xi) = O(\xi^{-m}) \quad as\ \xi \to +\infty \tag{3.3}$$

for some $m > 2$, then $V_1 \equiv V_2$. In particular, the solution of the form (2.13) (which is constructed in Theorem 2.2) is unique.

Theorem 3.1A is essentially due to Greenberg-Hasting [2]. Here we give an outline of the proof of Theorem 3.1B.

Let $V_1(\xi)$ and $V_2(\xi)$ be two solutions in Case B. Then the difference satisfies

$$(V_1 - V_2)(\xi) = -\int_0^\infty a'(\tau)\frac{H(\xi + s\tau)}{-s^2 + S(\xi)}(V_1 - V_2)(\xi + s\tau)d\tau, \tag{3.4}$$

where

$$S(\xi) = \int_0^1 \sigma'(V_2(\xi) + \theta(V_1 - V_2)(\xi))d\theta,$$

$$H(\xi) = \int_0^1 \eta'(V_2(\xi) + \theta(V_1 - V_2)(\xi))d\theta.$$

Taking (3.3) into account, we introduce

$$N_k(\xi) = \sup_{y\geq\xi} y^k|(V_1 - V_2)(y)|, \tag{3.5}$$

where $2 < k \le m$, and estimate (3.4) by using $N_k(\xi)$. Then we see that

$$\xi^k |(V_1 - V_2)(\xi)| \le D(\xi) N_k(\xi), \tag{3.6}$$

where

$$D(\xi) = -\int_0^\infty a'(\tau) \frac{H(\xi + s\tau)}{-s^2 + S(\xi)} (1 + s\tau\xi^{-1})^{-k} d\tau. \tag{3.7}$$

We consider the asymptotic behavior of $D(\xi)$ as $\xi \to +\infty$. A careful computation, using (3.2) and (3.3), shows that

$$D(\xi) = 1 - (k-2)\frac{sa_1}{a(0)}\xi^{-1} + o(\xi^{-1}) \quad \text{as } \xi \to +\infty, \tag{3.8}$$

where a_1 is given in (2.17). We fix k such that $2 < k < m$. Then it follows from (3.6) and (3.8) that $N_k(R) = 0$ for a sufficiently large R. Consequently, we have $V_1(\xi) = V_2(\xi)$ for $\xi \ge R$. Once this is known, we can easily deduce from (3.4) that $V_1 \equiv V_2$. The proof of Theorem 3.1B is complete.

4. Asymptotic stability

In this section we restrict ourselves to the case where $\chi(v)$ satisfies (1.5), and discuss the asymptotic stability of shock profiles; as pointed out in [6], when $\chi(v)$ satisfies (1.6), the stability analysis is open even for the equation of viscoelasticity of rate type.

Let us consider a shock profile $(V, U)(x - st)$ which connects (v_-, u_-) and (v_+, u_+), and satisfies (1.8); then $(v_\pm, u_\pm)$ and s satisfy (2.1), (2.2) and (2.3) by Theorem 2.1. Without loss of generality we may assume that $s > 0$ and $v_+ < v_-$. Also, since we are interested in the case where (v_+, v_-) contains the inflection point $v = 0$ of $\chi(v)$, we assume that $v_+ < 0 < v_-$. Moreover, for technical reason, we only treat Case A. Thus we are to discuss the stability in the following situation:

$$s > 0, \quad v_+ < 0 < v_-, \quad \chi'(v_+) < s^2. \tag{4.1}$$

As stated later, our stability analysis requires the smallness conditions not only on the strength $|v_+ - v_-|$ of the shock profile and on the initial disturbance but also on $|v_+|$ relative to v_-. This last condition will be given in terms of the following quantity:

$$H(v_+, v_-) = \sup_{v \in [v_+, 0]} \frac{|\chi''(v)|}{(s^2 - \chi'(v))^2} + \sup_{v \in [0, v_-]} |\chi''(v)|. \tag{4.2}$$

We state some properties of our shock profile. Obviously, $V(\xi)$ is strictly decreasing, i.e., $V'(\xi) < 0$ for $\xi \in \mathbb{R}$. Moreover, when $|v_+ - v_-|$ is suitably small, say $|v_+ - v_-| \le \delta_0$, we have

$$|\partial_\xi^k V(\xi)| \le C|v_+ - v_-|, \quad \xi \in \mathbb{R}, \tag{4.3}$$

for each $k \geq 1$, where C is a constant. This is proved as follows. We integrate the memory term in (2.8), obtaining

$$(-s^2 + \sigma'(V))V' = \frac{1}{s}\int_0^\infty a''(\tau)\{\eta(V)(\xi + s\tau) - \eta(V)(\xi)\}d\tau.$$

This equality and its differentiation with respect to ξ give the estimate (4.3). We note here that (4.3) is weaker than the usual estimate $|\partial_\xi^k V(\xi)| \leq C|v_+ - v_-|^2$, $\xi \in \mathbb{R}$, for the equation of viscoelasticity of rate type.

Now we consider the initial-history value problem for (1.1) in a neighborhood of the shock profile $(V, U)(x - st)$ mentioned above. The initial-history data are given as

$$(v, u)(0, x) = (v_0, u_0)(x), \quad x \in \mathbb{R}, \tag{4.4}$$

$$(v, u)(t, x) = (V, U)(x - st), \quad t < 0, \quad x \in \mathbb{R}. \tag{4.5}$$

As in the previous works, we assume that the initial disturbance $(v_0, u_0)(x) - (V, U)(x)$ is integrable over $\mathbb{R}$, and that the integral

$$(\phi_0, \psi_0)(x) = \int_{-\infty}^{x} \{(v_0, u_0)(y) - (V, U)(y)\}dy \tag{4.6}$$

belongs to $H^3(\mathbb{R})$. This implies the zero integral condition

$$\int_{-\infty}^{\infty} \{(v_0, u_0)(x) - (V, U)(x)\}dx = 0.$$

Our stability result is then stated as follows.

Theorem 4.1. *(Stability) Suppose that Assumption 1.1 is satisfied and that $a(t)$ is strongly positive definite and $\chi(v)$ satisfies (1.5). Let $(V, U)(x - st)$ be a shock profile connecting (v_-, u_-) and (v_+, u_+), and let $(\phi_0, \psi_0) \in H^3(\mathbb{R})$. Consider the case where (4.1) holds. Then there are positive constants δ_1, δ_2 and ϵ_1 such that if*

$$|v_+ - v_-| \leq \delta_1, \quad |v_+ - v_-| \cdot H(v_+, v_-) \leq \delta_2, \quad \|(\phi_0, \psi_0)\|_3 \leq \epsilon_1, \tag{4.7}$$

then the initial-history value problem (1.1), (4.4), (4.5) has a unique global solution $(v, u)(t, x)$ satisfying

$$(v, u)(t, x) - (V, U)(x - st) \in C^0([0, \infty); H^2(\mathbb{R})) \cap C^1([0, \infty); H^1(\mathbb{R})). \tag{4.8}$$

Furthermore, the solution approaches the shock profile as $t \to \infty$, that is

$$\sup_{x \in \mathbb{R}} |(v, u)(t, x) - (V, U)(x - st)| \to 0 \quad \text{as} \quad t \to \infty. \tag{4.9}$$

Remark 4.1. (i) For the definition of strongly positive definite kernel and its basic properties, the reader is referred to [4].

(ii) In the simpler case where $0 < v_+ < v_-$ so that (v_+, v_-) does not contain the inflection point $v = 0$ of $\chi(v)$, we can prove the same stability result without assuming the smallness of $|v_+ - v_-| \cdot H(v_+, v_-)$.

As in the previous papers [8,1,6], our solution is given in the form

$$(v,u)(t,x) = (V,U)(x-st) + (\phi,\psi)_x(t,x), \tag{4.10}$$

where $(\phi,\psi)(t,x)$ is constructed as a solution to the following "integrated" problem:

$$\begin{aligned} &\phi_t - \psi_x = 0, \\ &\psi_t - \left\{\sigma(V+\phi_x) - \sigma(V) + \int_0^t a'(t-\tau)(\eta(V+\phi_x) - \eta(V))(\tau)d\tau\right\} = 0, \end{aligned} \tag{4.11}$$

$$(\phi,\psi)(0,x) = (\phi_0,\psi_0)(x). \tag{4.12}$$

Notice that in (4.11) the memory term is reduced to the integral over $[0,t]$ and this is due to (4.5).

For the proof of Theorem 4.1, we need to show the global solvability of the problem (4.11), (4.12), together with the decay law

$$\sup_{x\in \mathbb{R}} |(\phi,\psi)_x(t,x)| \to 0 \quad \text{as } t \to \infty. \tag{4.13}$$

However, as easily seen, (4.11) is a fully nonlinear (not quasilinear) hyperbolic system with integral term, and therefore the standard argument does not give us an existence result even local in time. To overcome this difficulty, we consider the system satisfied by the derivative $(p,q) \equiv (\phi,\psi)_x$:

$$\begin{aligned} &p_t - q_x = 0, \\ &q_t - \sigma'(V+p)p_x = (\sigma'(V+p) - \sigma'(V))V_x \\ &\qquad\qquad + \left\{\int_0^t a'(t-\tau)(\eta(V+p) - \eta(V))(\tau)d\tau\right\}_x. \end{aligned} \tag{4.14}$$

This is a quasilinear symmetric hyperbolic system because the memory term can be regarded as a lower order term. Therefore, the standard method applies to (4.14) and we can get a unique local solution $(p,q)(t,x)$ of (4.14) with the initial data $(p,q)(0,x) = (\phi_0,\psi_0)_x(x)$. We now put

$$\begin{aligned} \phi(t,x) &= \phi_0(x) + \int_0^t q(\tau,x)d\tau, \\ \psi(t,x) &= \psi_0(x) + \int_0^t \Big\{(\sigma(V+p) - \sigma(V))(\tau,x) \\ &\qquad + \int_0^\tau a'(\tau-s)(\eta(V+p) - \eta(V))(s,x)ds\Big\}d\tau. \end{aligned} \tag{4.15}$$

Then we see that this $(\phi,\psi)(t,x)$ becomes a solution to the original problem (4.11), (4.12). Thus we have proved the following local existence result.

Proposition 4.2. *(Local existence) Suppose Assumption 1.1 is satisfied. Let $(\phi_0, \psi_0) \in H^3(\mathbb{R})$. Then there exists a positive constant T_0 depending only on $\|(\phi_0, \psi_0)\|_3$ such that the "integrated" problem (4.11), (4.12) admits a unique local solution satisfying*

$$(\phi, \psi) \in C^0([0, T_0]; H^3(\mathbb{R})) \cap C^1([0, T_0]; H^2(\mathbb{R})). \tag{4.16}$$

For the global solvability of the problem (4.11), (4.12), it is now sufficient to prove the following a priori estimate of solutions.

Proposition 4.3. *(A priori estimate) Suppose that Assumption 1.1 is satisfied and that $a(t)$ is strongly positive definite and $\chi(v)$ satisfies (1.5). Consider the case where (4.1) holds. Let $(\phi, \psi)(t, x)$ be a solution to the "integrated" problem (4.11), (4.12) such that*

$$(\phi, \psi) \in C^0([0, T]; H^3(\mathbb{R})) \cap C^1([0, T]; H^2(\mathbb{R})) \tag{4.17}$$

for some $T > 0$. Then there are constants δ_1, δ_2 and ϵ_1^* (independent of T) such that if

$$\begin{aligned} &|v_+ - v_-| \le \delta_1, \quad |v_+ - v_-| \cdot H(v_+, v_-) \le \delta_2, \\ &\sup_{0 \le t \le T} \{\|(\phi, \psi)(t)\|_3 + \|\partial_t(\phi, \psi)(t)\|_2\} \le \epsilon_1^*, \end{aligned} \tag{4.18}$$

then we have

$$\begin{aligned} &\|(\phi, \psi)(t)\|_3^2 + \|\partial_t(\phi, \psi)(t)\|_2^2 + \int_0^t \{\|\partial_x(\phi, \psi)(\tau)\|_2^2 + \|\partial_t(\phi, \psi)(\tau)\|_2^2 d\tau\} \\ &\le C\|(\phi_0, \psi_0)\|_3^2 \end{aligned} \tag{4.19}$$

for $t \in [0, T]$, where C is a constant independent of T.

This a priori estimate can be proved by an energy method which combines two different techniques employed in [4] and [6], respectively. In fact, in the proof, we basically apply the energy method in [4] which makes use of some properties of the strongly positive definite kernel $a(t)$, but in the derivation of the first energy estimate we introduce the weight function $w(V)$ used in [6], where $V(x - st)$ is the shock profile, and compute the energy density

$$w(V)\phi^2 + w(V)\chi'(V)\psi^2.$$

Here $w(v)$ is given explicitly as

$$w(v) = \begin{cases} (s^2 - \chi'(0))^2/(s^2 - \chi'(v))^2, & v \in [v_+, 0], \\ \chi'(0)/\chi'(v), & v \in [0, v_-]. \end{cases} \tag{4.20}$$

Though we omit the computation, we note that the quantity $H(v_+, v_-)$ in (4.2) is closely related to the above choice of the weight function.

If Propositions 4.2 and 4.3 are proved, then we can apply the standard continuation argument and obtain a unique global solution $(\phi, \psi)(t, x)$ to the "integrated" problem (4.11), (4.12). This solution verifies the uniform estimate (4.19) for all $t \geq 0$, from which we can deduce the decay law (4.13). Now the $(v, u)(t, x)$ given by the formula (4.10) becomes a desired global solution to the original problem (1.1), (4.4), (4.5), and therefore the proof of Theorem 4.1 is complete.

References

[1] J. Goodman, Nonlinear asymptotic stability of viscous shock profiles for conservation laws, Arch. Rat. Mech. Anal., 95(1986), 325-344.

[2] J. M. Greenberg and S. Hasting, Progressive waves in materials exhibiting long range memory, SIAM J. Appl. Math., 30(1976), 31-41.

[3] H. Hattori and S. Kawashima, Nonlinear stability of travelling wave solutions for viscoelastic materials with memory, to appear.

[4] S. Kawashima, Global solutions to the equation of viscoelasticity with fading memory, J. Diff. Equations, 101(1993), 388-420.

[5] S. Kawashima and H. Hattori, Existence of shock profiles for viscoelastic materials with memory, to appear.

[6] S. Kawashima and A. Matsumura, Stability of shock profiles in viscoelasticity with non-convex constitutive relations, to appear.

[7] T.-P. Liu, Nonlinear waves for viscoelasticity with fading memory, J. Diff. Equations, 76(1988), 26-46.

[8] A. Matsumura and K. Nishihara, On the stability of traveling wave solutions of a one-dimensional model system for compressible viscous gas, Japan J. Appl. Math., 2(1985), 17-25.

Department of Applied Science
Faculty of Engineering 36
Kyushu University
Fukuoka 812, Japan

and

Department of Mathematics
West Virginia University
Morgantown, WV 26506, USA

AMS/IP Studies in Advanced Mathematics
Volume **3**, 1997

Weak linear degeneracy and global existence of classical solutions for quasilinear hyperbolic systems

Li Ta-tsien, Zhou Yi and Kong De-xing

Abstract—By introducing the concept of the weak linear degeneracy, one gets the global existence and the sharp estimate of the lifespan of C^1 solutions to the Cauchy problem for general first order quasilinear hyperbolic systems with initial data small in C^1 norm and with compact support.

1 Introduction

Consider the Cauchy problem for general quasilinear strictly hyperbolic systems of first order:

$$\frac{\partial u}{\partial t} + A(u)\frac{\partial u}{\partial x} = 0, \tag{1.1}$$

$$t = 0: \quad u = \phi(x), \tag{1.2}$$

where $u = (u_1, \cdots, u_n)^T$ is the unknown vector function of (t, x), $A(u) = (a_{ij}(u))$ is an $n \times n$ matrix with suitably smooth elements $a_{ij}(u)$ $(i, j = 1, \cdots, n)$, and $\phi(x) = (\phi_1(x), \cdots, \phi_n(x))^T$ is a C^1 vector function with compact support:

$$\text{supp } \phi \subseteq [\alpha_0, \beta_0]. \tag{1.3}$$

By strict hyperbolicity, on the domain under consideration $A(u)$ has n distinct real eigenvalues

$$\lambda_1(u) < \lambda_2(u) < \cdots < \lambda_n(u). \tag{1.4}$$

Let $l_i(u) = (l_{i1}(u), \cdots, l_{in}(u))$ and $r_i(u) = (r_{i1}(u), \cdots, r_{in}(u))^T$ be the left and the right eigenvectors respectively, corresponding to $\lambda_i(u)$ $(i = 1, \cdots, n)$. All $\lambda_i(u)$, $l_{ij}(u)$ and $r_{ij}(u)$ $(i, j = 1, \cdots, n)$ have the same regularity as $a_{ij}(u)$ $(i, j = 1, \cdots, n)$.

Without loss of generality, we may suppose that

$$A(0) = \operatorname{diag}\{\lambda_1(0), \cdots, \lambda_n(0)\} \tag{1.5}$$

and

$$l_i(u)r_j(u) \equiv \delta_{ij} \qquad (i, j = 1, \cdots, n), \tag{1.6}$$

$$r_i^T(u)r_i(u) \equiv 1 \qquad (i = 1, \cdots, n), \tag{1.7}$$

where δ_{ij} stands for the Kronecker's symbol.

Generally speaking, our aim is to study the following two kinds of problems:

(1) Under what conditions does Cauchy problem (1.1)-(1.2) admit a unique global C^1 solution $u = u(t, x)$ on $t \geq 0$?

(2) Under what conditions does the C^1 solution $u = u(t, x)$ to Cauchy problem (1.1)-(1.2) blow up in a finite time? In this case we want to get a sharp estimate on the lifespan of classical solution.□

These problems have been completely solved in the case $n = 1$ or 2. For the general quasilinear hyperbolic system composed of $n(> 2)$ equations, however, up to now we only know a few of results.

Suppose that in a neighbourhood of $u = 0$, system (1.1) is not only strictly hyperbolic but also genuinely nonlinear in the sense of P.D.Lax:

$$\nabla\lambda_i(u)r_i(u) \neq 0 \qquad (i = 1, \cdots, n), \tag{1.8}$$

F.John [1] proved that if $A(u)$ and $\phi(x)$ are C^2 functions and

$$\theta \triangleq (\beta_0 - \alpha_0)^2 \cdot \sup_{x \in \mathbb{R}} |\phi''(x)| > 0 \tag{1.9}$$

is small enough, then the first order derivatives of C^2 solution $u = u(t, x)$ to Cauchy problem (1.1)-(1.2) must blow up in a finite time.

T.P.Liu [2] generalized F.John's result to the case that a part of characteristics is genuinely nonlinear, while the other part is linearly degenerate in a neighbourhood of $u = 0$ in the sense of P.D.Lax, i.e., for the corresponding indices i we have

$$\nabla\lambda_i(u)r_i(u) \equiv 0. \tag{1.10}$$

In this situation he showed that for a quite large class of initial data, the first order derivatives of the solution still blow up in a finite time.

L.Hörmander [3] reproved F.John's result.

Since both the genuinely nonlinear characteristic and the linearly degenerate characteristic are only two extreme cases, a general consideration is needed for the general quasilinear hyperbolic system. In this lecture, stimulated by the null condition in the study of the global existence of classical solutions to the Cauchy problem for nonlinear wave equations with small initial data (see S.Klainerman [4], D.Christodoulou [5]), we shall introduce the concept of weak linear degeneracy and present a quite complete result on the global existence and the lifespan of C^1 solution to Cauchy problem (1.1)-(1.2) for small initial data with compact support.□

2 Main results

First of all, we give the following

Definition 1 The i-th characteristic $\lambda_i(u)$ is called to be *weakly linearly degenerate*, if

$$\nabla\lambda_i(u)r_i(u) \equiv 0, \qquad \forall\ |u| \text{ small} \tag{2.1}$$

along the i-th characteristic trajectory $u = u^{(i)}(s)$ passing through the origin $u = 0$, defined by

$$\begin{cases} \dfrac{du}{ds} = r_i(u), \\ s = 0 : u = 0, \end{cases} \tag{2.2}$$

namely, if

$$\lambda_i(u^{(i)}(s)) \equiv \lambda_i(0), \qquad \forall\ |s| \text{ small.} \tag{2.3}$$

System (1.1) is weakly linearly degenerate, if all characteristics are *weakly linearly degenerate.*□

Obviously, if in a neighbourhood of $u = 0$, $\lambda_i(u)$ is linearly degenerate (resp. genuinely nonlinear) in the sense of P.D.Lax, then $\lambda_i(u)$ is (resp. is not) weakly linearly degenerate.

Theorem 1 *Suppose that in a neighbourhood of $u = 0$, $A(u) \in C^2$, and system (1.1) is strictly hyperbolic and weakly linearly degenerate. Suppose furthermore that $\phi(x)$ is a C^1 vector function such that (1.3) holds. Let*

$$\theta \triangleq (\beta_0 - \alpha_0) \cdot \sup_{x\in\mathbb{R}} |\phi'(x)|. \tag{2.4}$$

Then there exists $\theta_0 > 0$ so small that for any given $\theta \in [0, \theta_0]$, Cauchy problem (1.1)-(1.2) admits a unique global C^1 solution $u = u(t,x)$ on $t \geq 0$.□

As a direct consequence, we get

Corollary 1 *If system (1.1) is linearly degenerate in a neighbourhood of* $u = 0$, *then the conclusion of Theorem 1 holds.*□

We now consider the opposite case that system (1.1) is not weakly linearly degenerate. Suppose that $a_{ij}(u)$ $(i, j = 1, \cdots, n)$ are sufficiently smooth, we will show that for a quite large class of initial data, the first order derivatives of the C^1 solution to Cauchy problem (1.1)-(1.2) must blow up in a finite time and we will give a sharp estimate on the lifespan of the C^1 solution.

Since system (1.1) is not weakly linearly degenerate, there exists a nonempty set $J \subseteq \{1, 2, \cdots, n\}$ such that $\lambda_i(u)$ is not weakly linearly degenerate if and only if $i \in J$.

Noting (2.3), for any fixed $i \in J$, suppose that there exists an integer $\alpha_i \geq 0$ such that

$$\left.\frac{d^l \lambda_i(u^{(i)}(s))}{ds^l}\right|_{s=0} = 0 \ (l = 1, \cdots, \alpha_i) \quad \text{but} \quad \left.\frac{d^{\alpha_i+1} \lambda_i(u^{(i)}(s))}{ds^{\alpha_i+1}}\right|_{s=0} \neq 0, \quad (2.5)$$

where $u = u^{(i)}(s)$ is defined by (2.2).

Theorem 2 *Suppose that in a neighbourhood of* $u = 0$, $A(u)$ *is a sufficient smooth* $n \times n$ *matrix with (1.5), and system (1.1) is strictly hyperbolic but not weakly linearly degenerate. Suppose furthermore that* $\phi(x) = \varepsilon\psi(x)$, *where* $\varepsilon > 0$ *is a small parameter and* $\psi(x)$ *is a* C^1 *vector function with compact support:*

$$supp\ \psi \subseteq [\alpha_0, \beta_0]. \quad (2.6)$$

Let

$$\alpha = \min\{\alpha_i, i \in J\} \quad (2.7)$$

and

$$J_1 = \{i \mid i \in J, \alpha_i = \alpha\}, \quad (2.8)$$

where for each $i \in J$, $\alpha_i \geq 0$ *is defined by (2.5). If*

$$\sum_{i \in J_1} (\psi_i(x))^2 \not\equiv 0, \quad (2.9)$$

then there exists $\varepsilon_0 > 0$ *so small that for any fixed* ε *with* $0 < \varepsilon \leq \varepsilon_0$, *the first order derivatives of the* C^1 *solution* $u = u(t, x)$ *to Cauchy problem (1.1)-(1.2) must blow up in a finite time and there exist two positive constants c and C independent of* ε, *such that the lifespan* $\widetilde{T}(\varepsilon)$ *of the* C^1 *solution satisfies*

$$c\varepsilon^{-(1+\alpha)} \leq \widetilde{T}(\varepsilon) \leq C\varepsilon^{-(1+\alpha)}, \quad (2.10)$$

denoted by

$$\widetilde{T}(\varepsilon) \approx \varepsilon^{-(1+\alpha)}.□ \quad (2.11)$$

Corollary 2 *If system (1.1) is genuinely nonlinear in a neighbourhood of $u = 0$, then, under the assumption that $a_{ij}(u)$ $(i, j = 1, \cdots, n)$ are C^2 functions of u, for any nontrivial C^1 initial data with compact support $\phi(x) = \varepsilon\psi(x)$ with $\varepsilon > 0$ small enough, the first order derivatives of the C^1 solution to Cauchy problem (1.1)-(1.2) must blow up in a finite time and the lifespan*

$$\widetilde{T}(\varepsilon) \approx \varepsilon^{-1}.\square \tag{2.12}$$

Corollary 2 gives F.John's result as well as the sharp estimate on the lifespan only under the hypothesis of C^1 regularity on $\phi(x)$.

Corollary 3 *Suppose that in a neighbourhood of $u = 0$, a part of characteristics is genuinely nonlinear, while the other part is linearly degenerate. Under hypothesis (1.5), if a part of initial data corresponding to the genuinely nonlinear characteristics is nontrivial, then for $\varepsilon > 0$ small enough, the conclusion of Corollary 2 is still valid.*$\square$

This corollary gives T.P.Liu's result and the sharp estimate on the lifespan, however, his hypothesis that "linear waves do not generate nonlinear waves" is eliminated.

Some applications of Theorems 1 and 2 have been given for the system of one-dimensional gas dynamics and for the system of one-dimensional magneto-hydrodynamics etc.$\square$

3 Idea of the proof

In order to prove Theorem 1 and Theorem 2, we shall use the following

Lemma 1 *Suppose that $A(u) \in C^k$ (k is an integer ≥ 1). There exists an invertible C^{k+1} transformation $u = u(\widetilde{u})$ $(u(0) = 0)$ such that in $\widetilde{u}$-space, for each $i = 1, \cdots, n$, the i-th characteristic trajectory passing through $\widetilde{u} = 0$ coincides the $\widetilde{u}_i$-axis at least for $|\widetilde{u}_i|$ small, namely,*

$$\widetilde{r}_i(\widetilde{u}_i e_i) \equiv e_i, \quad \forall\ |\widetilde{u}_i| \text{ small}, \tag{3.1}$$

where $\widetilde{r}_i(\widetilde{u})$ denotes the corresponding i-th right eigenvector in $\widetilde{u}$-space and

$$e_i = (0, \cdots, 0, \overset{(i)}{1}, 0, \cdots, 0)^T.\square \tag{3.2}$$

Remark 1 Lemma 1 is still valid if (1.1) is only a hyperbolic (not strictly hyperbolic!) system.$\square$

We refer to the transformation given in Lemma 1 as the *normalized transformation*, and the corresponding unknown variables $\widetilde{u} = (\widetilde{u}_1, \cdots, \widetilde{u}_n)$ are called the *normalized variables* or *normalized coordinates*.

The system in normalized coordinates can be regarded as a standard form of system (1.1), which is characterized by (3.1) and then

$$\widetilde{A}(0) = \mathrm{diag}\{\lambda_1(0), \cdots, \lambda_n(0)\}. \tag{3.3}$$

Without loss of generality, in the proof of Theorems 1 and 2, we may suppose that $u = (u_1, \cdots, u_n)$ are the normalized coordinates. Thus, we have (1.5) and

$$r_i(u_i e_i) \equiv e_i, \quad \forall\, |u_i| \text{ small} \quad (i = 1, \cdots, n); \tag{3.4}$$

moreover, if $\lambda_i(u)$ is weakly linearly degenerate, we simply have

$$\lambda_i(u_i e_i) \equiv \lambda_i(0), \quad \forall\, |u_i| \text{ small}. \tag{3.5}$$

Let

$$v_i = l_i(u)u \quad (i = 1, \cdots, n), \tag{3.6}$$

$$w_i = l_i(u)\frac{\partial u}{\partial x} \quad (i = 1, \cdots, n), \tag{3.7}$$

where $l_i(u)$ denotes the i-th left eigenvector. According to F.John's formula on the decomposition of waves (see F.John [1]), $v = (v_1, \cdots, v_n)$ and $w = (w_1, \cdots, w_n)$ satisfy the following inhomogeneous system of diagonal form:

$$\frac{\partial v_i}{\partial t} + \lambda_i(u)\frac{\partial v_i}{\partial x} = \sum_{j,k=1}^{n} \beta_{ijk}(u) v_j w_k \quad (i = 1, \cdots, n), \tag{3.8}$$

$$\frac{\partial w_i}{\partial t} + \lambda_i(u)\frac{\partial w_i}{\partial x} = \sum_{j,k=1}^{n} \gamma_{ijk}(u) w_j w_k \quad (i = 1, \cdots, n) \tag{3.9}$$

with

$$\beta_{iji}(u) \equiv 0, \quad \forall\, j, \tag{3.10}$$

$$\gamma_{ijj}(u) \equiv 0, \quad \forall\, j \neq i, \tag{3.11}$$

$$\gamma_{iii}(u) = -\nabla\lambda_i(u) r_i(u), \quad \forall\, i \tag{3.12}$$

and, since $u = (u_1, \cdots, u_n)$ are supposed to be the normalized coordinates,

$$\beta_{ijj}(u_j e_j) = 0, \quad \forall\, |u_j| \text{ small}, \ \forall\, j, \tag{3.13}$$

and if $\lambda_i(u)$ is weakly linearly degenerate,

$$\gamma_{iii}(u_i e_i) \equiv 0, \quad \forall\, |u_i| \text{ small}. \tag{3.14}$$

Thus, a modification of the method given in T.P.Liu [2] can be used to obtain all results mentioned above.

4 Motivation

Definition 2 System (1.1) satisfies the *null condition* if each small plane wave solution $u = u(s)$ $(u(0) = 0)$, where $s = ax+bt$ (a, b constants), to the linearized system

$$u_t + A(0)u_x = 0 \tag{4.1}$$

is always a solution to system (1.1).□

We can easily get the following

Lemma 2 *Under hypothesis (1.5), system (1.1) satisfies the null condition if and only if*

$$\lambda_i(u_i e_i) \equiv \lambda_i(0), \quad \forall\, |u_i| \text{ small} \quad (i = 1, \cdots, n) \tag{4.2}$$

and

$$r_i(u_i e_i) \equiv e_i, \quad \forall\, |u_i| \text{ small} \quad (i = 1, \cdots, n), \tag{4.3}$$

namely, $u = (u_1, \cdots, u_n)$ are the normalized coordinates and system (1.1) is weakly linearly degenerate.□

Definition 3 System (1.1) satisfies the *generalized null condition* if there exists a suitably smooth invertible transformation $u = u(\widetilde{u})$ $(u(0) = 0)$ such that the system for $\widetilde{u}$ satisfies the null condition.□

The transformation given in Definition 3 must be a normalized one and $\widetilde{u} = (\widetilde{u}_1, \cdots, \widetilde{u}_n)$ are the corresponding normalized coordinates. Thus, system (1.1) satisfies the generalized null condition if and only if the corresponding system in normalized coordinates satisfies the null condition.

Since for each $i = 1, \cdots, n$, $\nabla\lambda_i(u)r_i(u)$ is a scalar under any smooth invertible transformation $u = u(\widetilde{u})$, system (1.1) satisfies the generalized null condition if and only if for any fixed $i = 1, \cdots, n$, (2.1) hold along the i-th characteristic trajectory $u = u^{(i)}(s)$ passing through the origin $u = 0$, i.e., we have

Lemma 3 *System (1.1) satisfies the generalized null condition if and only if the system is weakly linearly degenerate.*□

5 Remarks

1. If the initial data have no compact support but possess certain decay properties as $|x| \to +\infty$, for example, if

$$\theta \triangleq \sup_{x\in \mathbb{R}} \{(1 + |x|)^{1+\delta}(|\phi(x)| + |\phi'(x)|)\} \quad (\delta > 0 \text{ constant}) \tag{5.1}$$

is small enough, we can obtain the same result as in Theorems 1 and 2.□

2. If system (1.1) is hyperbolic but not strictly hyperbolic, say

$$\lambda_1(0) = \lambda_2(0) < \lambda_3(0) < \cdots < \lambda_n(0), \tag{5.2}$$

for $i = 1, 2$, the corresponding definition of weak linear degeneracy should be modified as follows: in the normalized coordinates

$$\lambda_i(u_1e_1 + u_2e_2) \equiv \lambda_1(0) = \lambda_2(0), \quad \forall\, |u_1|, |u_2| \text{ small} \quad (i = 1, 2). \tag{5.3}$$

Thus, we can still get the same conclusion as in Theorem 1.□

3. If system (1.1) is hyperbolic and the multiplicity of characteristics is constant, say

$$\lambda_1(u) \equiv \lambda_2(u) \equiv \cdots \equiv \lambda_p(u) < \lambda_{p+1}(u) < \cdots < \lambda_n(u) \quad (1 < p \leq n), \tag{5.4}$$

we still have the same conclusion as in Theorems 1 and 2.□

4. For the inhomogeneous system

$$\frac{\partial u}{\partial t} + A(u)\frac{\partial u}{\partial x} = F(u) \tag{5.5}$$

with (1.5) and

$$F(0) = 0, \quad \nabla F(0) = 0, \tag{5.6}$$

the null condition implies (4.2)-(4.3) and

$$F(u_ie_i) \equiv 0, \quad \forall\, |u_i| \text{ small} \quad (i = 1, \cdots, n). \tag{5.7}$$

Hence, the generalized null condition asks that all characteristics $\lambda_i(u)$ $(i = 1, \cdots, n)$ are weakly linearly degenerate and F must be zero along all characteristic trajectories passing through the origin for small $|u|$. If system (5.5) satisfies the generalized null condition, then we still have the global existence of C^1 solution to the Cauchy problem, provided that the C^1 norm of the initial data is small enough.□

References

[1] F.John, Formation of singularities in one-dimensional nonlinear wave propagation, *Comm. Pure Appl. Math.*, **27** (1974), 377-405.

[2] Liu Tai-ping, Development of singularities in the nonlinear waves for quasi-linear hyperbolic partial differential equations, *J. of Diff. Equations*, **33** (1979), 92-111.

[3] L.Hörmander, *The life span of classical solutions of nonlinear hyperbolic equations*, Institute Mittag-Leffler, Report No.5, 1985.

[4] S.Klainerman, The null condition and global existence to nonlinear wave equations, *Lectures in Applied Mathematics*, **23** (1986), 293-326.

[5] D.Christodoulou, Global solution of nonlinear hyperbolic equations for small initial data, *Comm. Pure Appl. Math.*, **39** (1986), 267-282.

[6] Li Ta-tsien, Zhou Yi and Kong De-xing, Dégénérescence linéaire faible et solutions régulières globales pour les systèmes hyperboliques quasi linéaires d'ordre 1, C. R. Acad. Sci. Paris, **t.317, Série I** (1993), 571-574.

[7] Li Ta-tsien, *Global classical solutions for quasilinear hyperbolic systems*, to be published in Recherches en Mathématiques Appliquées, MASSON, 1994.

Institute of Mathematics
Fudan University
Shanghai 200433
P.R.China

AMS/IP Studies in Advanced Mathematics
Volume 3, 1997

SURVEY OF VACUUM THEORY AND CONJECTURE*

LONGWEI LIN

1. Introduction

It is well–known that, by compensated compactness theory, DiPerna (see [3,4]), Ding,Chen and Luo[2] solved the system of isentropic gas dynamics equations in Eulerian coordinates. Their framework essentially depends on the boundness of invariant regions of the systems. we now consider the systems of isentropic gas dynamics equations in Lagrangian coordinates, whose invariant regions are unbounded. DiPerna showed that[3] in order to solve this system it suffices to prove the uniform boundness of the difference approximations of the Lax–Friedrichs' scheme or the solutions of the systems with small viscous terms. However the solutions of Riemann problems to this system may contain vacuum. One thereby tends to believe that if the initial data is not far from the vacuum, then the solution will contain vacuum at a later time. It implies that the approximations are not uniformly bounded.

When the vacuum appears, the system is not strictly hyperbolic and waves behave in singular way causing serious analysis difficulties[2,18]. Thus it is valuable to study how the vacuum can be attained in the solution. We conjecture that the aforementioned solution of Riemann problem is the only circumstance where the vacuum can occur.

CONJECTURE 1. *If the isentropic gas flow is bounded and bounded away from vacuum at the initial instant $t = 0+$, then it is also bounded and bounded away from vacuum in any finite time, even though the density may tend to zero as time tends to infinity.*

*This work is supported in part by the Foundation of Zhongshan University Advanced Research Centre.

If the flow at the initial instant $t = 0+$ consists of rarefaction waves, the conjecture has been proved[8,9,11–14,16]. Precisely, we consider the Cauchy problem for the system of isentropic gas dynamics equations in Lagrangian coordinates

$$v_t - u_x = 0, \quad u_t + p(v)_x = 0, \quad (x,t) \in R \times R_+ \tag{E}$$

$$(u(x,0), v(x,0)) = (u_0(x), v_0(x)), \qquad x \in R \tag{I}$$

where the pressure $p = p(v) > 0$ is a smooth function of the specific volume $v > 0$, and u is the velocity of the flow. The eigenvalues of the Jacobian of (E) are $\lambda_1(v) = -\lambda_2(v) = -\sqrt{-p'(v)}$. We assume that $p'(v) < 0,\ p''(v) > 0$ as $v > 0$, i.e, the system (E) is genuinely nonlinear in Lax's sense and strictly hyperbolic. Moreover, we assuame that

$$\int_1^{\infty} \sqrt{-p'(v)}dv < \infty \tag{C}$$

Condition (C), which is dicated by physics, implies that the solution to the Riemann problem of (E) may contain vacuum $v = \infty$. In vacuum states, $\lambda_1(\infty) = \lambda_2(\infty) = 0$, the system (E) is no longer strictly hyperbolic. The Riemann invariants are taken as $r = u + \int_1^v \sqrt{-p'(z)}dz$, $s = u - \int_1^v \sqrt{-p'(z)}dz$.

THEOREM 1.1. *Let $u_0(x)$, $v_0(x)$ be bounded $u_* \le u_0(x) \le u^*$, $0 < v_* \le v_0(x) \le v^*$, where u_*, u^*, v_* and v^* are constants, If*

$$0 \le r_0(x_2) - r_0(x_1), \ 0 \le s_0(x_2) - s_0(x_1), \quad as\ x_1 < x_2, \tag{1.1}$$

$$r_0(x_2) - r_0(x_1) \le L(x_2 - x_1), \ s_0(x_2) - s_0(x_1) \le L(x_2 - x_1), \quad as\ x_1 < x_2 \tag{1.2}$$

where $r_0(x) = r(u_0(x), v_0(x))$, $s_0(x) = s(u_0(x), v_0(x))$, $L > 0$, then there is global continuous solution $(u(x,t), v(x,t))$ of (E), (I), and

$$\sup_{0\le t\le T} v(x,t) \le \sup_x v_0(x) + LT \tag{1.3}$$

This theorem has been proved by several ways[8,9,11–14,16], it will be described briefly in section 2.

We now discuss the case where the initial functions are piecewise Lipschitz continuous. In such case, there are centered rarefaction waves issued from the initial discontinuity points.

THEOREM 1.2. *Let $u_0(x), v_0(x)$ be bounded, $0 < v_* \leq v_0(x) \leq v^*$. If $r_0(x)$ and $s_0(x)$ satisfy (1.1) and piecewise Lipschitz continuous (cp. (1.2)), furthermore,*

$$\sup_x \{r_0(x+0) - s_0(x-0)\} < 2\Phi(\infty), \tag{1.4}$$

then there is a global continuous solution $(u(x,t), v(x,t))$ of (E), (I).

P.S.Lin proved that[16] the condition piecewise Lipschitz continuous can be released. But the condition (1.4) is necessary, it implies that there is no vacuum at the initial instant $t = 0+$

It is well-known that v increase with respect to time t if and only if the rarefaction waves of different kinds collide. Thus, at first glance, Conjecture 1 seems to have been proved by Theorem 1.2. However, a new centered rarefaction wave will arise as a shock is overtaken by other shock of the same kind. The new rarefaction wayes will collide the rarefaction waves of the different kind and v will then increase. In order to achieve the proof of Conjecture 1, the main task is to study the collisions between two new centered rarefaction waves. The study of the collisions between a new centered rarefaction wave and an original one (issued from initial line $t = 0$) will be simpler.

In section 3, we will show that if we restrict our attention on the collisions of new centered rarefaction waves, our goal can't be reached. However, there must be a shock of another kind at the center of a new centered rarefaction wave, since a new centered rarefaction wave arises if and only if a shock is overtaken by other shock of the same kind. On account of this fact, we have the following conjecture.

CONJECTURE 2. *Let $u_0(x), v_0(x)$ be bounded, $0 < v_* \leq v_0(x) \leq v^*$. if*

$$r_0(x_2) - r_0(x_1) \leq 0, \quad s_0(x_2) - s_0(x_1) \leq 0, \quad as\ x_1 < x_2, \tag{1.5}$$

and there is a global generalized solution $(u(x,t), v(x,t))$ of (E), (I), then

$$\sup_{x,t} v(x,t) = \sup_x v_0(x) < \infty \tag{1.6}$$

(cp. Theorem 1.1)

In paper [15], we illustrate Conjecture 2 in detail, where $p(v) = v^{-\gamma}$, $1 < \gamma \leq 3$, It will be described briefly in Section 3.

If Conjecture 2 is verified, then the following Conjecture 3 will be true, and Conjecture 1 will then follow.

CONJECTURE 3. *Let $u_0(x), v_0(x)$ be bounded, $0 < v_* \leq v_0(x) \leq v^*$, if*

$$r_0(x_2) - r_0(x_1) \leq L(x_2 - x_1), \quad s_0(x_2) - s_0(x_1) \leq L(x_2 - x_1), \quad as\ x_1 < x_2 \tag{1.7}$$

where $L \geq 0$, and there is a global generalized solution $(u(x,t), v(x,t))$ of $(E), (I)$, then

$$\sup_{0 \leq t \leq T} v(x,t) \leq \sup_x v_0(x) + LT \tag{1.8}$$

NOTE 1.3. *When $L = 0$, Conjecture 3 is deduced to Conjecture 2. Cancelling (1.1), Theorem 1.1 turns out to be Conjecture 3.*

If this Conjecture is proved, the existence theorem will follow. We omit the precise statement on Conjecture 1.

2. Initial Flow Consists of Rarefaction Waves

We now discuss Theorem 1.1. It is well-known that (1.1) is a necessary and sufficient condition for the existence of a global bounded continuous solution of $(E), (I)$.

Notice that this assertion was proved only under the following additional condition

$$\{(r, s) \mid r_0(-\infty) \leq r \leq r_0(\infty),\ s_0(-\infty) \leq s \leq s_0(\infty)\} \subseteq \Omega \tag{2.1}$$

where, $\Omega = \{(r, s) \mid r - s \leq 2\int_1^\infty \sqrt{-p'(z)}dz\}$, or other no essentially different additional conditions. This condition assures that the solution is globally bounded. When it fails, the methods used are not effective. For instance, if one attempts to solve $(E), (I)$ by the standard application of the Glimm's scheme, he may end up with a difference approximation in which vacuum appears. However, condition (2.1) is not necessarily justfied by physics and mathematics. What is the matter ? It is that the random sequence in the Glimm's scheme should be equidistributed [9,11,12,17], in order that the difference approximations approach to the exact solution. In the standard application of the Glimm's scheme, however, the estimates are established for all sequence and not only for the equidistributed ones. It causes that the vacuum may appear in the difference approxomations. The author gave[9,11,12] an application of Glimm's scheme in which the estimates are established only for equidistributed ones. (T.P.Liu has used this technique to prove existence in his earlier well-known paper[17]). By this technique, condition (2.1) can be released.

THEOREM 2.1. *Let $u_0(x), v_0(x)$ be bounded, $0 < v_* \leq v_0(x) \leq v^*$, Lipschitz continuous, $r_0(x), s_0(x)$ are monotonic non-decreasing. Suppose that the random sequence is uniformly eqidistributed on the interval $(-1, 1)$. For any given $T > 0$, if the mesh length in t-direccion is sufficiently small, then the Glimm's approximations of $(E), (I)$ are uniformly bounded with respect to h in the strip $0 \leq t \leq T$. Of course, we assume that Courant-Friedrich-Lewy condition is satisfied.*

The proof of Theorem 2.1 appeared in [9], and Theorem 1.1 is a corollary of Theorem 2.1. Roughly speaking, the reason for the nonoccurrence of the vacuum is that the characteristic speed is small near the vacuum, thereby slowing down the interaction of waves. Our main task is to estimate the rate of waves interactions. Notice that the nonoccurrence of the vacuum is not always true for any 2×2 genuinely nonlinear in Lax's sense strictly hyperbolic system. We will discuss this topic in section 4.

If the intial functions are piecewise Lipschitz contiuous, there are centered rarefaction waves issued from the initial discontinuity points. The estimates are much more complicated than that in Theorem 2.1[11,12], it is due to the random sequence is uniformly equidistributed. This restriction causes serious technical difficulties to study the solutions with shocks by Glimm's scheme.

We then used various method to prove Theorem 1.1 in order to find a better way to verify Conjecture 1. We have used Polygonal approxim— ations[8], Lax-Friedrichs scheme[13], Viscosity method[14], and Upwind scheme[16] (by P.S.Lin). We now only describe Lax-Friedrichs scheme briefly.

Riemann invariants are very useful for studing system (E), but Lax-Friedrichs' scheme is too complicated if it is rewritten in terms of Riemann invariants. However, it can be rewritten in terms of the difference in x-direction of Riemann invariants.

Let the upper half plane $t \geq 0$ be covered by a grid $t = kh$, $x = nl$, $n = 0, \pm 1, \pm 2, \cdots$, $k = 1, 2, \cdots$ where h and l are the mesh length in t-direction and x-direction respectively. The Lax-Friedrichs'scheme for system (E) is

$$\begin{cases} u_n^{k+1} = \frac{1}{2}(u_{n+1}^k + u_{n-1}^k) - \frac{\delta}{2}(p_{n+1}^k - p_{n-1}^k) \\ v_n^{k+1} = \frac{1}{2}(v_{n+1}^k + v_{n-1}^k) + \frac{\delta}{2}(u_{n+1}^k - u_{n-1}^k) \end{cases}$$

defined on the grid[20]. Let

$$\Delta r_n^k = \Delta u_n^k + \lambda_n^k \Delta v_n^k, \quad \Delta s_n^k = \Delta u_n^k - \lambda_n^k \Delta v_n^k$$

$$\lambda_n^k = \sqrt{-\frac{\Delta p_n^k}{\Delta v_n^k}}$$

where $\Delta F_n^k =: \frac{1}{2}(F_{n+1}^k - F_{n-1}^k)$. By straight calculation, we have

$$\begin{cases} \Delta r_n^{k+1} = A_{n+1}^k B_{n+1,n}^k \Delta r_{n+1}^k + a_{n+1}^k b_{n+1,n}^k \Delta s_{n+1}^k \\ \qquad + a_{n-1}^k B_{n-1,n}^k \Delta r_{n-1}^k + A_{n-1}^k b_{n-1,n}^k \Delta s_{n-1}^k \\ \Delta s_n^{k+1} = A_{n+1}^k b_{n+1,n}^k \Delta r_{n+1}^k + a_{n+1}^k B_{n+1,n}^k \Delta s_{n+1}^k \\ \qquad + a_{n-1}^k b_{n-1,n}^k \Delta r_{n-1}^k + A_{n-1}^k B_{n-1,n}^k \Delta s_{n-1}^k \end{cases} \tag{2.4}$$

where

$$a_{n\pm1}^k = \frac{1}{2}(1 - \delta\lambda_{n\pm1}^k), \quad A_{n\pm1}^k = \frac{1}{2}(1 + \delta\lambda_{n\pm1}^k)$$

$$b_{n\pm1,n}^k = \frac{1}{2}(1 - \frac{\lambda_n^{k+1}}{\lambda_{n\pm1}^k}), \quad B_{n\pm1,n}^k = \frac{1}{2}(1 + \frac{\lambda_n^{k+1}}{\lambda_{n\pm1}^k})$$

(2.4) can be rewritten into following form

$$\Delta r_n^{k+1} = A_{n+1}^k \Delta r_{n+1}^k + a_{n-1}^k \Delta r_{n-1}^k + P_n^k$$

$$\Delta s_n^{k+1} = a_{n+1}^k \Delta s_{n+1}^k + A_{n-1}^k \Delta s_{n-1}^k - P_n^k$$

where

$$P_n^k = b_{n-1,n}^k \Delta Y_{n-1}^k + b_{n+1,n}^k \Delta Z_{n+1}^k$$

$$\Delta Y_{n\pm1}^k = A_{n\pm1}^k \Delta s_{n\pm1}^k - a_{n\pm1}^k \Delta r_{n\pm1}^k$$

$$\Delta Z_{n\pm1}^k = A_{n\pm1}^k \Delta r_{n\pm1}^k - a_{n\pm1}^k \Delta s_{n\pm1}^k$$

THEOREM 2.2[13]. *Let the initial data smooth, bounded, $0 < v_* \le v_0(x) \le v^*$, $u_* \le u_0(x) \le u^*$, where v_*, v^*, u_* and u^* are constants. If*

$$0 \le u_0'(x) + \lambda(v_0(x))v_0'(x) \le L, \ 0 \le u_0'(x) - \lambda(v_0(x))v_0'(x) \le L$$

then the Lax-friedrichs'difference approximations of (E), (I) are uniformly bounded with respect to the mesh length in the strip $(-\infty, \infty) \times [0, T]$.

The sketch of the proof by the assumptions, it is easy to see

$$\begin{cases} -G_1 l^2 \le \Delta r_n^0, \ \Delta s_n^0 \le Ll + G_1 l^2, & n = odd, \\ |\Delta^2 r_n^0|, \ |\Delta^2 s_n^0| \le G_2 l^2, & n = even, \end{cases}$$

where G_1, G_2 are generic constants. For any small $\varepsilon > 0$, let

$$D_\varepsilon = \{(u,v) \mid u_* - \varepsilon \leq u \leq u^* + \varepsilon, \quad (1-\varepsilon)v_* \leq v \leq LT + (1+\varepsilon)v^*\}$$

Our aim is to verify $(u_n^k, v_n^k) \in D_\varepsilon$ as $l > 0$ is small. To do this , it suffices to prove

$$-H_1 l^2 \leq m_1^j \leq M_1^j \leq Ll + H_1 l^2, \tag{2.5}$$

$$M_2^j \leq H_2 l^2 \tag{2.6}$$

as $l > 0$ is small, where $m_1^j = \inf_n (\Delta r_n^j, \Delta s_n^j)$, $M_1^j = \sup_n (\Delta r_n^j, \Delta s_n^j)$, $M_2^j = \sup_n (|\Delta^2 r_n^j|, |\Delta^2 s_n^j|)$, $0 \leq jh < T$ and H_1, H_2 are postive generic constants. Moreover, in order to prove (2.5),(2.6), it suffices to verify

$$\min\{0, (1+\beta l)m_1^{j-1}\} - \beta l^3 \leq m_1^j \leq M_1^j \leq M_1^{j-1} + \beta l^3, \tag{2.7}$$

$$M_2^j \leq (1+\alpha l)M_2^{j-1} + \alpha l^3, \tag{2.8}$$

where α, β are positive generic constants. (2.5)–(2.8) can be verified by induction, in which we want to prove $P_n^k = O(l^3)$.

3. Initial Flow Consists of Compressive Waves

In paper [15], we illustrate Conjecture 2 in detail, we now describe it briefly.

Suppose that several 2-shocks (1-shocks) interact at a point $B(A)$. They combine to form a 2-shock $S_2(B)$ (a 1-shock $S_1(A)$) and reflect a new 1-centered rarefaction wave $R_1(B)$ (2- centered rarefaction wave $R_2(A)$). To our end, it is no harm to assume that the points A and B lie on a horizontal line, and A is on the left, and the strengths of $R_2(A)$ and $R_1(B)$ ($S_1(A)$ and $S_2(B)$) are equal. Thus we consider Cauchy problem $(E), (I)$ with following initial data

$$(u_0(x), v_0(x)) = \begin{cases} (u_1, v_1), & \text{as } x < -b, \\ (u_2, v_2), & \text{as } |x| \leq b, \\ (u_3, v_3), & \text{as } x > b, \end{cases} \tag{3.1}$$

$$(r_0(x), s_0(x)) = \begin{cases} (r_1, s_1), & \text{as } x < -b, \\ (r_2, s_2), & \text{as } |x| \leq b, \\ (r_3, s_3), & \text{as } x > b, \end{cases} \tag{3.1}$$

where $b > 0$ is a constant, $r_i = r(u_i, v_i)$, $s_i = s(u_i, v_i)$, $i = 1,2,3$, $r_1 \geq r_2 \geq r_3$, $s_1 \geq s_2 \geq s_3$.

In order to simplify our consideration, we now give a note. Suppose $p(v) = v^{-\gamma}$, $1 < \gamma \leq 3$, then the Riemann invariants $r = u + \Phi(v)$, $s = u - \Phi(v)$, where $\Phi(v) = \frac{\sqrt{\gamma}}{\varepsilon}(1 - v^{-\varepsilon}), 0 < \varepsilon = \frac{\gamma-1}{2} \leq 1$. Following [19,20], the equation of 1-shock curves S_1 can be written in the form

$$S_1 : h^{-1}(r_0 - r) = KF_+(\alpha),\ h^{-1}(s_0 - s) = KF_-(\alpha),\ \alpha = \rho\rho_0^{-1} > 1 \quad (3.2)$$

where (r_0, s_0) is the state on the left, and (r, s) on the right, $F\pm(\alpha) =: \sqrt{(1-\alpha^{-1})(\alpha^{\gamma}-1)} \pm \frac{\sqrt{\gamma}}{\varepsilon}(\alpha^{\varepsilon} - 1)$, $K = \varepsilon(2r)^{-\frac{1}{2}}$, and h is the distance from (r_0, s_0) to the vacuum line $\rho = 0$ which equation in (r, s) plane is $r - s = \frac{2\sqrt{\gamma}}{\varepsilon}$. Similarly, the equations of 2-shock inverse curves s_2^{-1} can be written in

$$S_2^{-1} : h^{-1}(s_0 - s) = KF_+(\alpha),\ h^{-1}(r_0 - r) = KF_-(\alpha),\ \alpha = \rho\rho_0^{-1} > 1 \quad (3.3)$$

where (r_0, s_0) is the state on the right, and (r, s) on the left.

NOTE 3.1. *The expressions on the right side of (3.2), (3.3) contain only one variable α. They are independent of u_0 and v_0.*

We now go back to (3.1). let

$$r_2 = r_3,\ s_1 = s_2,\ r_1 - r_2 = s_2 - s_3 > 0 \quad (3.4)$$

We will show that this is a typical case to illustrate Conjecture 2 (see Note 3.2). By Note 3.1, it is no harm to put any value on u_2 and $v_2 > 0$ for our purpose. Thus we set $u_2 = 0$, i.e, $r_2 + s_2 = 0$ and $v_2 = a > 0$. Our problem only depends on $\alpha = \frac{v_1}{a} = \frac{v_3}{a} > 1$. To sum up, we put

$$(u_0(x), v_0(x)) = \begin{cases} (u_1, \alpha a), & \text{as } x < -b, \\ (0, a), & \text{as } |x| \leq b, \\ (-u_1, \alpha a), & \text{as } x > b, \end{cases} \quad (3.5)$$

where $u_1 = \Phi(\alpha a) - \Phi(a)$, i.e.

$$(r_0(x), s_0(x)) = \begin{cases} (r_1, s_2), & \text{as } x < -b, \\ (-s_2, s_2), & \text{as } |x| \leq b, \\ (-s_2, -r_1), & \text{as } x > b, \end{cases} \quad (3.5)$$

where $s_2 = -\Phi(a)$, $r_1 = u_1 + \Phi(\alpha a) = 2\phi(\alpha a) - \Phi(a)$. We now consider Cauchy problem (E),(3.5). There is a 2-shock $S_2(b)$ and a 1-centered rarefation wave $R_1(b)$ issued from $x = b$; a 1-shock $S_1(-b)$ and a 2-centered rarefation wave $R_2(-b)$ issued from $x = -b$. If we restrict our attention to the collision between $R_2(-b)$ and $R_1(b)$, by straight calculation, while they are weak, i.e, $\alpha - 1 > 0$ is small, they will penetrate each other completely, and the maxium v value is less than αa; while they are strong, i.e,$\alpha - 1 > 0$ is large, they will not penetrate each other wholly but partly, and v value is upper unbounded. In former case, the Conjecture is trivally true. In later case, if we attend not only $R_2(-b)$ and $R_1(b)$, but also $S_1(-b)$ and $S_2(b)$, the part of $R_1(b)$ penetrating $R_2(-b)$ must overtake $S_1(-b)$, and then a new 2-compressive flow (it can be seen as a family of new 2-shocks) arises, it must overtake and weaken $R_2(-b)$, and then new flows arise, but they are weak. The main interections are the aforementioned collision and overtaking. It is very complicated to estimate the rate of the wave interctions by theoretical analysis, but it can be done numerical analysis. In [15], we solve Cauchy problem (3.1), (3.4) by Lax-Friedrichs' scheme, where $p(v) = v^{-2}$, $a = 10^{-1}$ and put $\alpha = 10, 100, 300, 1000$. The numerical solutions illustrate that the Conjecture is true. In a rough precision scale $v(nl, kh)$ is convex upward with respect to n, where $v(nl, kh)$ is the value of the numerical solution, l and h are the mesh length in x-direction and t-direction respectively, $n = 0, \pm 1, \cdots$, k are large positive integer.

NOTE 3.2. *We now show that (3.4) is a typical case to illustrate Conjecture 2. If there are n 1-shocks interacting at the point A, and they are equal in strength, the left state and right state of the point A are denoted by (r_1, s_1) and (r_2, s_2) respectively, if $r_1 - r_2 > 0$ is a constant, i.e, $r_1 - r_2$ is independent of n, then $s_1 - s_2 > 0$ monotonical tends to zero as $n \to \infty$, Thus (3.4) is the limit case, $R_2(-b)$ is strongest. This is the worst case to Conjecture 2.*

4. Nonlinear Hyperbolic Conservation Laws

In paper [10], we showed that the nonoccurrence of vacuum does not generally hold for 2×2 nonlinear strictly hyperbolic conservation laws. Even Theorem 1.1 can't be extended in general. For nonlinear hyperbolic conservation laws, so-called "vacuum state" means that, in the state, the speed of the characteristics of two families coincides with each other.

Example 4.1 Cosider the system

$$u_t - \sigma(v)_x = 0, \quad v_t - u_x = 0, \tag{4.1}$$

where $\sigma(v) = v^\gamma$, $v > 0$, $\gamma > 1$, thus $\sigma'(v) = \gamma v^{\gamma-1} > 0$, $\sigma''(v) = \gamma(\gamma-1)v^{\gamma-2} > 0$. The eigenvalues of the Jacobian matrix of (4.1) are $\lambda_1(v) = -\lambda_2(v) = -\sqrt{\gamma}v^{\frac{\gamma-1}{2}}$. The system is strictiy hyperbolic in the region $D_1 =: \{(u,v) \mid v > 0,\ u \in R\}, v = 0$ is the "vacuum state". The Riemann invariants are taken as $z = -u - \Phi(v)$, $w = -u + \Phi(v)$ where $\Phi(v) =: \int_0^v \mu(s)ds = Av^{\frac{\gamma+1}{2}}, A =: 2\sqrt{\gamma}/(\gamma+1), (\lambda =: \lambda_1, \mu =: \lambda_2)$. The Riemann invariants give a smooth bijective mapping from D_1 to the region $D =: \{(z,w) \mid w - z > 0\}$. And $\mu_w = \lambda_z = -\mu_z = -\lambda_w = \frac{\gamma-1}{4v} > 0$ in D. The system is genuinely nonlinear in Lax's sense. The initial data is taken as

$$u_0(x) = \begin{cases} LN, & \text{as } x < -N, \\ -Lx, & \text{as } |x| \leq N, \\ -LN, & \text{as } x > N, \end{cases}$$

$$v_0(x) = v_0, as\ x \in R$$

where L, N and v_0 are positive constants, then

$$(z_0(x), w_0(x)) = \begin{cases} (-LN - \Phi_0, -LN + \Phi_0), & \text{as } x < -N, \\ (Lx - \Phi_0, Lx + \Phi_0), & \text{as } |x| \leq N, \\ (LN - \Phi_0, -LN + \Phi_0), & \text{as } x > N, \end{cases}$$

where $\Phi_0 = \Phi(v(x,0)) =: Av_0^{\frac{\gamma+1}{2}} > 0$, $z_0'(x) = w_0'(x) = L > 0$, as $|x| \leq N$, $z_0'(x) = w_0'(x) = 0$ as $|x| > N$. We now consider the determinate region, denoted by B_N, of the interval $(-N, N)$ on the initial line $t = 0$. It is easy to check

$$u(x,t) = -Lx,\ v(x,t) = v_0 - Lt, \quad (x,t) \in B_N$$

is a solution in B_N; and the left (right) side of B_N is the λ_2-characteristic (λ_1-characteristic) issued from the initial point $(x,t) = (-N,0)$, $((x,t) = (N,0))$, which is denoted by $x_2^-(t)$ $(x_1^+(t))$, $x_2^-(t) = -N + L^{-1}(\Phi_0 - \Phi(t))$, $x_1^+(t) = N - L^{-1}(\Phi_0 - \Phi(t))$, when $N > L^{-1}\Phi_0$ the upper side of B_N is horizontal segment: $b_0 = \{(x,t) \mid |x| \leq N - L^{-1}\Phi_0,\ t = L^{-1}v_0\}$. On b_0, $v = 0$, $\lambda_1 = \lambda_2 = 0$. "vacuum" occur on b_0.

Remark 4.2 Example 4.1 implies that the solution to the Cauchy problem of 2×2 nonlinear conservation laws, in general, may go out the strictly hyperbolic region of the system in a finite time, here the initial data is given in the strictly hyperbolic region and bounded away from the boundary of the region. In other words, in general, we can't confine our attention to solve the Cauchy problem of 2×2 nonlinear conservation laws in the strictly hyperbolic region of the system. However, the following Theorem 4.3 shows that under the assumptions (A) and (B), we can expect that the (generalized) solution will stay in the strictly hyperbolic region of the system in any finite time, if the initial data is given in it. In other words, we can expect that the Cauchy problem 2×2 nonlinear conservation laws may be solved in the strictly hyperbolic region of the conservation laws under the assumptions (A) and (B).

We are concerned with the Cauchy problem of

$$u_t + f(u,v)_x = 0, \quad v_t + g(u,v)_x = 0, \quad (x,t) \in R \times R_+ \tag{4.2}$$

$$u(x,0) = u_0(x), \quad v(x,0) = v_0(x), \quad x \in R \tag{4.3}$$

here $f(u,v), g(u,v) \in C^2(D_0), D_0$ is a region on (u,v) plane, the system (4.2) is strictly hyperbolic in a subregion $D_1 \subset D_0$, the Jacobian matrix of the system (4.2)has two real and distinct eigenvalues $\lambda_1 < \lambda_2$ in D_1. Let $G =: \lambda_2 - \lambda_1$, then $0 < G < \infty$ in D_1. Riemann invariants of (4.2) $z = z(u,v)$, $w = w(u,v)$ give a bijective smooth mapping from D_1 onto a region D on (z,w) plane. Thus $0 < G < \infty$ in D. The Riemann invariants diagonalize principle part of the system (4.2) as

$$z_t + \lambda_1(z,w) z_x = 0, \quad w_t + \lambda_2(z,w) z_x = 0, \quad (x,t) \in R \times R_+ \tag{4.4}$$

and reduce the initial datd (4.3) to

$$z(x,0) = z_0(x), \quad w(x,0) = w_0(x), \quad x \in R \tag{4.5}$$

We assume that

$$0 \le \lambda_z(z,w) < \infty, \ 0 \le \mu_w(z,w) < \infty \ \ in \ D \tag{4.6}$$

The initial data is smooth and satisfies the condition

$$I \subset\subset D \tag{4.7}$$

where the bounded set $I =: \{(z,w) \mid z = z_0(x), \ w = w_0(x), \ x \in R\}$.

THEOREM 4.3 Under the aforementioned assumptions and the additional coditions

$$\mu_w + \mu_z \ge 0, \ \lambda_w + \lambda_z \ge 0 \ \ in \ D \tag{A}$$

$$|G_z| < \infty, \quad |G_w| < \infty, \quad in\ \overline{D} \tag{B}$$

if

$$0 \le z_0'(x) \le N, \quad 0 \le w_0'(x) \le N, \quad x \in R \tag{4.8}$$

where the constant N can be arbitrarily large, then the Cauchy problem (4.2), (4.3) has a unique global bounded classical solution $z = z(x,t)$, $w = w(x,t)$

$$(z(x,t), w(x,t)) \subset Q, (x,t) \in R \times R_+$$

$$(z(x,t), w(x,t)) \subset\subset D, \ x \in R, \quad 0 \le t \le T$$

where $T > 0$ can be arbitrarily large, and Q is a bounded rectangular $Q = \{(z,w) \mid z_* \le z \le z^*, \ w_* \le w \le z^*\}$, $z_* = \inf_x z_0(x)$, $z^* = \sup_x z_0(x)$, $w_* = \inf_x w_0(x)$, $w^* = \sup_x w_0(x)$

The proof see [10]. It is easy to check the assumptions of Theorem 4.2 hold for the systems of isentropic gas dynamics equations in Eulerian coordinates and in Lagrangian coordinates under the ordinary assumptions to the state equation $p(v)$[1].

Example 4.1 shows that once the assumption (B) is violated "vacuum" may occur at a later time. Another example (see [10]) shows that once the assumption (A) is violated, no matter how narrowly, "vacuum" also may occur at a later time, and wave behaves in a really singular way.

REFERENCE

[1]. R.Courant and K.O.Friedrichs, "Supersonic flow and shock waves," Wiley-Interscience, New York, 1948.

[2]. X.X.Ding, G.Q.Chen, P.Z.Luo, *Convergence of the Lax-Friedrichs scheme for isentropic gas dynamics (1);(2)*, Acta.Math.Sci. **5** (1985), 415-432;433-472.

G.Q.Chen, *Convergence of the Lax-Friedrichs scheme for isentropic gas dynamics (3)*, Acta.Math.Sci. **6** (1986), 75-120.

[3]. R.DiPerna, *Convergence of approximate solutions to conservation laws*, Arch.Rational Mech.Anal. **82** (1983), 27-70.

[4]. R.DiPerna, *Convergence of the viscosity method for isentropic gas dynamics*, Comm.Math.Phys. **91** (1983), 1-30.

[5]. P.D.Lax, *Hyperbolic systems of conservation laws and the mathematical theory of shock waves*, Conf.Board Math.Sci.11, SIMA (1973).

[6]. D.T.Lee(D.Q.Li), *Global regularity and formation of singularities of solutions to first order quasilinear hyperbolic system*, Proceeding of Royal Society of Edingburgh **87A** (1981), 255-261.

[7]. L.W.Lin, *On the global existence of the continuous solutions of the reducible quasilinear hperbolic systems*, Acta of Jilin university **4** (1963), 83-96.

[8]. L.W.Lin, *On the vacuum states for the equations of isentropic gas dynamics,*, J.Math.Anal.Appl. **121** (1987), 406-425.

[9]. L.W.Lin, *Vacuum states and quidistribution of the random sequence for Glimm's scheme*, J.Math.Anal.Appl. **124** (1987), 117-126.

[10]. L.W.Lin, *On the occurrence of "vacuum states" for* 2×2 *quasilinear hyperbolic conservation laws (to appear).*

[11]. L.W.Lin and F.P.Hsieh, *Vacuum state and equidistribution of the random sequence for Glimm's scheme (Continuation)*, J.Part.Diff.Equ. **2** 1 (1989), 9-14..

[12]. L.W.Lin and F.P.Hsieh, *Characteristic method in certain macros—copic sense for Glimm's scheme*, Comm.Part.Diff.Equ. **15** (1990), 1179 – 1197.

[13]. L.W.Lin and T.Yang, *The convergence of the Lax-Friedrichs'scheme for isentropic gas dynamics in Lagrangian coordinates*, Comm.Part. Diff.Equ. **16** (8,9) (1991), 1441-1460..

[14]. L.W.Lin and T.Yang, *Convergence of the viscosity method for the systems of isentropic gas dynamics in Lagrangian coordinates*, J.Diff. Equ. **102** (2) (1993), 330-341.

[15]. L.W.Lin and others, *Conjecture on the nonoccurrence of the vacuum of isentropic gas flows with shocks (to appear).*

[16]. P.S.Lin, *The construction of global continuous solutions for the isentropic gas dynamics equations*, Acta.Math.Sci. **7** (1987), 177-186.

[17]. T.P.Liu, *The deterministic version of the Glimm's scheme*, Comm. Math.Phys. **57** (1977), 135-148.

[18]. T.P.Liu and J.Smoller, *On the vacuum state for the isentropic gas dynamics*, Adv. Appl.Math. **1** (1980)), 345-359.

[19]. T.Nishida and J.A.Smoller, *Solutions in the large for some nonlinear hyperbolic conservation laws*, Comm.Pure.Appl.Math. **26** (1973), 193-200.

[20]. J.Smoller, "Shock waves and reaction-diffusion equations," Springer-Verlag, New York Inc, 1983.

Macau University,
Macau

AMS/IP Studies in Advanced Mathematics
Volume **3**, 1997

THE INITIAL VALUE PROBLEM OF A SYSTEM OF CONSERVATION LAWS

PEIZHU LUO[1]

It is very important to study the theory of the multidimensional hyperbolic conservation laws. A typical model is the following initial value problem

$$\begin{cases} \frac{\partial u_i}{\partial t} + \sum_{j=1}^n \frac{\partial}{\partial x_j}(u_i u_j) = 0, \quad i = 1, \cdots n, & (1) \\ u_i(x,0) = \emptyset_i(x). & (2) \end{cases}$$

In 2-dimensional case, Tong Zhang and others investigated its Riemann problem. But for general initial value problem up to now there is no result.

A system closely related with (1) is the following

$$\frac{\partial u_i}{\partial t} + \sum_{j=1}^n \frac{\partial}{\partial x_j}(u_i u_j) = \mu \Delta u_i, \quad i = 1, \cdots, n, \tag{3}$$

which can be used as a simple model of the multidimensional Navier-Stocks equations. We hope that the solution of (3) approaches that of (1) as $\mu \to 0$. But then firstly we meet the problem whether the global solution of (3), (2) exists or not. If there exists one, does it approach the solution of (1), (2) as $\mu \to 0$? All of the problems are difficult.

In this paper we shall solve the problem (3), (2) with small initial data and we shall discuss the multistep difference method to the problem (1), (2) in some special cases.

When we discuss the problem (3), (2), we want to consider the decay property of the solutions at infinity. So we consider the Lebesgue class D_p^l with power weight.

We define

$$D_p^1 \equiv \{u | (1 + |x|)^l u \in L_p\},$$

[1]Partially supported by grants from National Science Foundation

$$\|u\|_{D_p^l} = \|u\|_{p,l} = \|u\|_p + \||x|^l u\|_p.$$

Besides these, we want to use the weighted Young inequality. Generalized Young inequality. If $l < n(1-\frac{1}{p}), h < n(1-\frac{1}{q}), h+l \geq 0, \frac{1}{r} = \frac{1}{p'} + \frac{1}{q'}$, then

$$\||x|^{-h} f\|_{q'} \leq c\||x|^l g\|_p \||x|^{-h-j} G\|_r, (p, r \geq 1), \tag{4}$$

where f denotes the convolution of g and G,

$$f(x) = \int G(x-y)g(y)dy$$

where $G = G(|x|) = |x|^\mu k(|x|)$, and $k(|x|)$ is non-decreasing. We get the following theorem [1].

Theorem. If $\emptyset_i \in D_q^{k\alpha}$, and $\|\emptyset\|_{q,k\alpha}$ are sufficuently small, then the problem (3), (2) has a global solution $u_i(x,t)$. Moreover, when $t > 0$,
$u \in C([0,\infty), D_q^{k\alpha}), t^{\frac{n}{4q}} u \in C([0,\infty), D_{2q}^{\alpha})$,

$$\|u\|_{q,k\alpha} \leq C, \quad C \quad independent \quad of \quad t,$$

$q = \frac{n}{1-\mu\alpha}, q \geq n, \alpha$ satisfies
i). If $0 \leq k \leq 1$, then

$$1 - \frac{n}{q} \leq \alpha < n - 1$$

ii) if $k > 1$, then $k < \frac{2(n-1)}{n(1-\frac{1}{q})} \leq 2$,

$$\frac{1-\frac{n}{q}}{2-k} \leq \alpha < \frac{n-1}{k}.$$

In $\xi 3$, we discuss the multistep method of (1). This means we want to discuss the system of equations

$$\begin{cases} \frac{\partial u}{\partial t} + \frac{\partial u^2}{\partial x} = 0, \\ \frac{\partial v}{\partial t} + \frac{\partial uv}{\partial x} = 0, \\ u(x,0) = u_0(x), v(x,0) = v_0(x). \end{cases}$$

We use the self-adjusting hybrid scheme and get a better value of a parameter in this scheme[2],[3].

Reference

[1] Ding Xiaxi, Luo Peizhu, Yan Guogheng, Suplinear conservation law with viscosity, Acta Math. Sci., 10 (1990), 85-99.

[2] Harten, A., G. Zwas, Self-adjusting hybrid schemes for shock computation, J. of Comp. Phys., 9 (1972), 568-583.
[3] Wilde, D.J., Oplimum Seeking Method, 1964

Institute of Applied Mathematics
Academia Sinica
Beijing 100080, P. R. China

AMS/IP Studies in Advanced Mathematics
Volume 3, 1997

ENTROPY SOLUTIONS FOR A HYDRODYNAMIC MODEL FOR SEMICONDUCTORS

P.A. MARCATI and R. NATALINI

Abstract. We investigate the Cauchy problem for a hydrodynamic model for semiconductors. An existence theorem of global weak solutions with large initial data is obtained by using the fractional step Lax–Fiedrichs scheme and Godounov scheme.

We are interested in the existence of weak solutions for the one–dimensional Euler–Poisson (or hydrodynamic) model for semiconductors. Consider the following one–dimensional system (see, e.g., [11])

$$\begin{cases} n_t + (nu)_x = 0 \ , \\ (nu)_t + (nu^2 + p(n))_x = nE - \dfrac{nu}{\tau} \ , \\ E_x = n - b(x) \ , \end{cases} \tag{1}$$

in the region $\Pi := \mathbb{R} \times [0, \infty)$. The initial conditions are given by

$$n(x,0) = n_0(x), \qquad u(x,0) = u_0(x) \tag{2}$$

for all $x \in \mathbb{R}$. Here $n \geq 0$ denotes the electron density, u the (average) particle velocity and E the (negative) electric field, which is generated by the Coulomb force of the particles. The function $b = b(x) \geq 0$ stands for the density of fixed, positively charged background ions, $p = p(n)$ is the pressure–density relation; in particular we shall use here $p(n) := \frac{1}{\gamma}\, n^\gamma$, where $1 < \gamma \leq 5/3$. Also $J := nu$ will denote the electron current density

Here we will discuss a construction of the weak solutions by using the approximations given by the Lax–Friedrichs and Godounov schemes, combined with the fractional step method.

The main tools, we will use to study this Cauchy problem, are the fractional step Lax–Friedrichs and Godounov schemes, which have been used in [5] for the system of isentropic gas–dynamics.

Concerning the hydrodinamic model for the theory of semiconductors we refer to [1], [6] and [10] and the standard reference book [11].

However in the present case we will use essentially the mathematical theory of quasilinear hyperbolic systems [7]. In particular for the method of compensated compactness we refer to the original papers of Tartar [12] and Di Perna [3].

Finally we notice that even the theory of non homogeneous hyperbolic systems of quasilinear equation is not yet fully developed, but some interesting contributions can be found in [13], [2] where various kind of local source terms have been analyzed for some special systems. More details and proofs will be found in [8].

Consider the locally bounded measurable functions $n(x,t)$, $J(x,t)$, $E(x,t)$, where E is absolutely continuous in x, a.e. in t; the triple (n,J,E) is a weak solution to (1)–(2), if and only if, for all smooth test functions ϕ, ψ, χ with compact support in Π, one has

$$\begin{aligned}\iint \{n(x,t)\phi_t(x,t) + J(x,t)\phi_x(x,t)\phi(x,t)\}\,\mathrm{d}x\,\mathrm{d}t \\ + \int n_0(x)\phi(x,0)\,\mathrm{d}x = 0\ ,\end{aligned} \tag{3}$$

$$\begin{aligned}\iint \Big\{ J(x,t)\psi_t(x,t) + \left(\frac{J^2(x,t)}{n(x,t)} + p(n(x,t))\right)\psi_x(x,t) \\ + [n(x,t)E(x,t) - \frac{J}{\tau}]\psi(x,t)\Big\}\,\mathrm{d}x\,\mathrm{d}t \\ + \int n_0(x)u_0(x)\psi(x,0)\,\mathrm{d}x\ ;\end{aligned} \tag{4}$$

moreover

$$\iint \{E(x,t)\chi_x(x,t) + (n(x,t) - b(x))\chi(x,t)\}\,\mathrm{d}x\,\mathrm{d}t = 0\ . \tag{5}$$

We have also to supplement (3)–(5) with a condition at $-\infty$ for the electric field; for some fixed constant E^- this is

$$\lim_{x\to-\infty} E(x,t) = E^-, \qquad \text{for a.e. } t > 0\ . \tag{6}$$

The standard entropy inequalities are also introduced. For our system a natural example of entropy–entropy flux pair, in the case $p(n) = \frac{1}{\gamma}n^\gamma$, is given by the mechanical energy function

$$\eta^* := \frac{J^2}{2n} + \frac{n^\gamma}{\gamma(\gamma-1)}\ ,$$

and the mechanical entropy–flux function

$$q^* := \frac{J^3}{2n^2} + \frac{Jn^{\gamma-1}}{(\gamma-1)}\ ,$$

Let us recall that in our case the characteristic speeds are given by

(7) $$\lambda_1 := \frac{J}{n} - c(n)\ , \quad \lambda_2 := \frac{J}{n} + c(n)\ ,$$

where $c(n) := \sqrt{p'(n)} = n^\theta$, $\theta := \frac{\gamma-1}{2}$, is the sound–speed; the Riemann invariants are then given by

(8) $$w := \frac{J}{n} + \frac{n^\theta}{\theta}\ , \quad z := \frac{J}{n} - \frac{n^\theta}{\theta}\ .$$

Hence, along smooth solutions the system (1) could be diagonalized by using the functions w, z as unknown

(9) $$\begin{cases} w_t + \lambda_2 w_x = E - \frac{w+z}{2\tau}\ , \\ z_t + \lambda_1 z_x = E - \frac{w+z}{2\tau}\ , \\ E_x = \left(\frac{\theta}{2}(w-z)\right)^{\frac{1}{\theta}} - b(x)\ . \end{cases}$$

Next we define our difference scheme, the fractional step Lax–Friedrichs scheme and the Godounov scheme, which have been proved to converge for the isentropic gas dynamics [3], [4] and for the nonhomogeneous isentropic gas dynamics [5].

Let us define the schemes in the following way; first we give a partition of the strip Π into horizontal layers

$$S_k := \{(x,t)|\, kh \le t < (k+1)h\}\ ,$$

for some fixed time step $\Delta t := h > 0$ and for any integer $k \ge 0$. If $I_k := \{j;\ k+j$ is even$\}$, let us set for any $j \in I_k$,

$$Q_{k,j} := S_k \cap \{(x,t)|\,(j-1)l < x < (j+1)l\}$$

for some space step $\Delta x := l > 0$, such that the ratio $\frac{h}{l}$ satisfies the usual CFL condition.

The schemes will be defined as customary for the three points monotone schemes; namely, first we shall define the schemes at $t = 0$ and then we shall construct the approximate solution on $Q_{0,j}$ for any $j \in I_0$. Later we shall continue the solution on the strip S_{k+1}, by using the solution previously defined on S_k.

Set

$$V_0^l(x) := \begin{cases} V_0((j-1)l)\ , & x < jl\ , \\ V_0((j+1)l)\ , & x > jl\ , \end{cases}$$

for $x \in ((j-1)l, (j+1)l]$, where $V_0 := (n_0, J_0)$. Now we define $V_R^l := (n_R^l, J_R^l)$ be the piecewise smooth solutions to the Riemann problem for the homogeneous gas–dynamics on $Q_{0,j}$ with initial data V_0^l (see [4]). We set also

$$E_R^l(x,t) := \int_{-\infty}^{x} [n_R^l(y,t) - b(y)]\, dy + E^-\ ,$$

E^- given by (6). The notation $V_R^{j,k} := V_R^l(jl, kh)$ will be used in what follows. Finally we define, for $(x,t) \in Q_{0,j}$,

$$(10) \qquad \begin{cases} n^l(x,t) := n_R^l(x,t) \,, \\ J^l(x,t) := J_R^l(x,t) + t\,[n_R^l(x,t)E_R^l(x,t) - \frac{J_R^l(x,t)}{\tau}] \,, \end{cases}$$

and

$$(11) \qquad E^l(x,t) := \int_{-\infty}^{x} [n^l(y,t) - b(y)]\, dy + E^- \,.$$

So $V^l := (n^l, J^l)$ is completely defined in the strip S_0. Now we define V^l in the strip S_{k+1}, provided that we know V^l in the strip S_k.

For any $j \in I_{k+1}$, define

$$(12) \qquad V^{j,k+1} := \frac{1}{2l} \int_{(j-1)l}^{(j+1)l} V^l(x, (k+1)h - 0)\, dx \,,$$

for the Lax–Friedrichs scheme, and

$$(13) \qquad V^{j,k+1} := \frac{1}{l} \int_{(j-1/2)l}^{(j+1/2)/l} V^l(x, (k+1)h - 0)\, dx \,,$$

for any integer j and $k \geq 0$ for the Godounov scheme. Then, as before, we denote by V_R^l the Riemann solutions for the homogeneous gas–dynamics in $Q_{k+1,j}$, for $j \in I_{k+1}$, by taking as initial datum at $t = (k+1)h$

$$V_{k+1}^l(x) := \begin{cases} V^{j-1,k+1} & x < jl \,, \\ V^{j+1,k+1} & x > jl \,, \end{cases}$$

for $x \in ((j-1)l, (l+1)l)$. The approximate (homogeneous) electric field E_R^l is defined as in the strip S_0. Then we set

$$(14) \qquad \begin{cases} n^l(x,t) := n_R^l(x,t) \,, \\ J^l(x,t) := J_R^l(x,t) + (t-(k+1)h)[n_R^l(x,t)E_R^l(x,t) - \frac{J_R^l(x,t)}{\tau}] \,, \end{cases}$$

for all $(x,t) \in S_{k+1}$. The approximate electric field E^l is given again by (11). The Godounov solutions will be defined in a similar way, the only difference consisting in the Riemann Problems used to define n_R^l and J_R^l; they are solved on any rectangle $(jl, (j+1)l) \times (kh, (k+1)h)$, by using elementary waves centered in $\left((j+\frac{1}{2})\, l \,,\, kh\right)$.

The classical CFL condition has to be imposed in order to guarantee the stability of the previous schemes. Namely, in the Lax–Friedrichs case, we have to require

$$\sup |\lambda_i(n^l, J^l)| \le \frac{l}{2h} .$$

Then, if n^l, J^l are uniformely bounded, we take the CFL number $\frac{h}{l}$ small enough to achieve the condition.

We shall begin by considering some sharp uniform bounds on the approximate solutions. The main idea consists in dealing with the system in the diagonal form given by the Riemann invariants state variables, and to reduce the analysis to standard dynamical systems methods.

Theorem 1. *Assume that $n_0(x)$, $u_0(x)$ are compactly supported initial data which satisfy*

$$0 \le n_0(x) \le C_1 , \tag{15}$$

$$|u_0(x)| = |J_0(x)/n_0(x)| \le C_2 , \tag{16}$$

a.e. $x \in \mathbb{R}$. *Assume that $b \in L^1(\mathbb{R})$. Moreover, let E^- be some given constant and define $E_0 = E_0(x)$ by the following relation*

$$E_0(x) := \int_{-\infty}^{x} (n_0(y) - b(y))\,\mathrm{d}y + E^- . \tag{17}$$

Then there exists some constants $D_0, D_1, D_2 > 0$ and $l^* > 0$ such that

$$0 \le n^l(x,t) \le (\theta(D_0 + tD_1))^{\frac{1}{\theta}} , \tag{18}$$

$$|u^l(x,t)| = |J^l(x,t)/n^l(x,t)| \le D_0 + tD_1 , \tag{19}$$

$$|E^l(x,t)| \le D_2 , \tag{20}$$

for all $(x,t) \in \Pi$ and for any $l \in (0, l^*)$.

Now it is possible to obtain our convergence result:

Theorem 2. *Under the assumptions of the Theorem 1, there is a convergent subsequence $(n^{l_\nu}, J^{l_\nu}, E^{l_n u})$ and some bounded measurable functions (n, J, E) in Π, such that*

$$(n^{l_\nu}, J^{l_\nu}) \to (n, J)$$

for a.e. $(x,t) \in \Pi$, and

$$E^{l_\nu} \to E$$

uniformly on the compact subsets of Π.

The proof of this result is essentially based on the compensated compactness arguments [3], [12]. The main step in the proof is the study of the behaviour of the Young measures generate by the sequence $\{V^l\}$. In the case of the non-homogeneous isentropic gas–dynamics this study was done in [5]. The the basic argument is given by the following proposition.

Proposition 1. *Under the assumptions of the Theorem 1, the sequence*

$$\{\Lambda^l\} := \partial_t \eta(V^l) + \partial_x q(V^l)$$

is relatively compact in H^{-1}_{loc}, *for any entropy–entropy flux pair* (η, q).

To prove the convergence of the approximate sequence $\{V^l, E^l\}$ to the solution of the system (1.1) we have only to show that the previous schemes are consistent.

Theorem 3. *Assume that the initial data* $(n_0(x), J_0(x))$ *are compactly supported and satisfy (15) and (16). Assume also that* $b \in L^1(\mathbb{R})$ *and* $E_0 = E_0(x)$ *is a function defined by the relation (17) for some given constant* E^-. *Then, for all* γ, $1 < \gamma \leq 5/3$, *let* (n, J, E) *be the limit functions given by the Theorem 2. If we define* $u(x,t) := \dfrac{J(x,t)}{n(x,t)}$, *then the measurable functions* (n, u, E) *are a bounded entropy solution of the problem (1)–(2) in* Π.

In a more recent paper [9] we study the relaxation problem for the system (1) when the momentum relaxation time τ tends to zero.

We introduce the scaled variables $\mathcal{N}^\tau(x,s) = n\left(x, \frac{s}{\tau}\right)$, $\mathcal{J}^\tau(x,s) = \frac{1}{\tau} J\left(x, \frac{s}{\tau}\right)$ and $\mathcal{E}^\tau(x,s) = E\left(x, \frac{s}{\tau}\right)$. We can prove ([9], Theorem 5.1) that, as τ tends to zero, $\mathcal{N}^\tau$ converges to a density profile $\mathcal{N}(s,x)$, strongly in L^p_{loc}, $p < +\infty$, $\mathcal{J}^\tau$ converges weakly in L^2_{loc} to a current profile $\mathcal{J}$ and $\mathcal{E}^\tau$ converges uniformly to a limit electric field $\mathcal{E}$. Moreover the limit profiles $(\mathcal{N}, \mathcal{E})$ satisfy the classical drift–diffusion equation (see [11])

$$\begin{cases} \mathcal{N}_S + (\mathcal{N}\mathcal{E} - p(\mathcal{N})_x)_x = 0 \\ \mathcal{E}_x = \mathcal{N} - b(x) , \end{cases} \tag{21}$$

and the limit current is ricovered from the usual current relation

$$\mathcal{J} = \mathcal{N}\mathcal{E} - p(\mathcal{N})_x .$$

Let us observe that the result concerning the Cauchy problem for the system (1) obtained in [8] need to be improved because the "a priori" L^∞ and energy estimates were not uniform with respect to the relaxation time τ.

Therefore to overcome such a difficulty we introduce here a more sophisticated version of the fractional step method, which provides the required bounds. Then we can pass into the limit by using again the methods of compensated compactness .

REFERENCES

[1] U.M. ASCHER, P.A. MARKOWICH, P. PIETRA, C. SCHMEISER, "A phase plane analysis of transonic solutions for the hydrodynamic semiconductor model", *Math. Mod. Math. Appl. Sci.*, **1** (3) (1991), 347–376.

[2] C.M. Dafermos and L. Hsiao, "Hyperbolic systems of balance law with inhomogeneity and dissipation", *Indiana U. Math. J.*, **31** (1982), 471–491.
[3] R. Di Perna, "Convergence of the viscosity method for isentropic gas dynamics", *Comm. Math. Phys.*, **91** (1983), 1–30.
[4] X. Ding, G.–Q. Chen, P. Luo, "Convergence of the Lax–Friedrichs scheme for isentropic gas dynamics (I) (II)", *Acta Math. Sci.*, **5** (1985), 483–500, 501–540.
[5] X. Ding, G.–Q. Chen, P. Luo, "Convergence of the fractional step Lax–Friedrichs scheme and Godounov scheme for isentropic gas dynamics", *Comm. Math. Phys.*, **121** (1989), 63–84.
[6] I.M. Gamba, "Stationary transonic solutions of a one–dimensional hydrodynamic model for semiconductors", *Comm. P.D.E.*, **17 (3&4)** (1992), 553–577.
[7] P.D. Lax, "Hyperbolic systems of conservation laws and the mathematical theory of shock waves", *Conf. Board Math. Sci.*, **11**, SIAM (1973).
[8] P.A. Marcati and R. Natalini, "Weak solutions to a hydrodynamical model for semiconductors: the Cauchy problem", *Quaderno IAC*, **2** (1993); to appear in Proc. Roy. Soc. Edinburgh.
[9] P.A. Marcati and R. Natalini, "Weak solutions to a hydrodynamical model for semiconductors and relaxation to the drift– diffusion–equation", *Quaderno IAC*, **27** (1993).
[10] P.A. Markowich and P. Pietra,"A Non–Isentropic Euler-Poisson Model for a Collisionless Plasma", *Preprint Istituto di Analisi Numerica del CNR, PAVIA*, **794** (1991).
[11] P.A. Markowich, C. Ringhofer, C. Schmeiser, *Semiconductor Equations*, Springer–Verlag, Wien–New York (1990).
[12] L. Tartar, "Compensated compactness and applications to partial differential equations", Research notes in mathematics, nonlinear analysis and mechanics: Heriot–Watt Symposium, Vol. **4**, ed. R. J. Knops, New York, Pitman Press (1979), 136–212.
[13] L. Ying and C. Wang, "Global solutions of the Cauchy problem for a nonhomogeneous quasilinear hyperbolic system", *Comm. Pure Appl. Math.*, **33** (1980), 579–597.

Dipartimento di Matematica Pura e Applicata,
Università degli Studi dell'Aquila,
I–67000 L'Aquila, Italia

Istituto per le Applicazioni del Calcolo "M. Picone", C.N.R.,
Viale del Policlinico 137,
I–00161 Roma, Italia

AMS/IP Studies in Advanced Mathematics
Volume 3, 1997

A SURVEY ON STABILITY OF VISCOUS SHOCK PROFILES WITHOUT GENUINE NONLINEARITY

AKITAKA MATSUMURA

ABSTRACT. We survey some recent developments on the stability of viscous shock profiles for the system of viscous conservation law without genuine nonlinearity.

1. INTRODUCTION

This talk is a brief survey of our recent works mainly Kawashima-Matsumura [5] and Matsumura-Nishihara [9] on the stability of viscous shock profiles for the system of viscous conservation law without genuine nonlinearity.

We first consider the Cauchy problem for the scalar viscous conservation law:

$$u_t + f(u)_x = \mu u_{xx}, \quad x \in R, \ t > 0, \tag{1.1}$$

$$u(0,x) = u_0(x), \quad x \in R, \tag{1.2}$$

where f is a smooth function of u, μ is a positive constant and the initial data $u_0(x)$ asymptotically tends toward the constant states $u_\pm$ as $x \to \pm\infty$. For given constants $u_\pm$ and s (shock speed), we call a smooth traveling wave solution $u = U(\xi)$ $(\xi = x - st)$ of (1.1) satisfying $U(\pm\infty) = u_\pm$ as a *viscous shock profile* which connects u_- and u_+. In order for Eq(1.1) to admit a viscous shock profile, it is well known that $u_\pm$ and s must satisfy the Rankine-Hugoniot condition

$$-s(u_+ - u_-) + f(u_+) - f(u_-) = 0 \tag{1.3}$$

and the generalized shock condition

$$h(u) \equiv -s(u - u_\pm) + f(u) - f(u_\pm) \begin{cases} < 0 & \text{if} \quad u_+ < u < u_- \\ > 0 & \text{if} \quad u_- < u < u_+ \end{cases} \tag{1.4}$$

which is known as Oleinik's shock condition. It is noted that if f has a convex nonlinearity (genuine nonlinearity), *i.e.*,

$$f''(u) > 0 \quad \text{for all } u \text{ under consideration}, \tag{1.5}$$

the Oleinik's shock condition (1.4) is equivalent to

$$f'(u_+) < s < f'(u_-) \tag{1.6}$$

which is known as Lax's shock condition.

In the case f has a convex nonlinearity, the stability theorems on the viscous shock profiles have been investigated by Il'in-Oleinik [2], Sattinger [12], Nishihara [11], Kawashima-Matsumura [4] and etc. The first stability results were obtained by Il'in-Oleinik[2] based on the maximum principle. Subsequently, another proof based on the spectral analysis was given by Sattinger [12]. They also obtained the time decay rate of asymptotics such that if the initial disturbance decays exponentially with respect to x, the solution tends toward the viscous shock profile at an exponential decay rate. In a particular case $f = u^2/2$, Nishihara [11] studied the case the initial disturbance has an algebraic decay rate with respect to x, and then obtained a corresponding optimal algebraic time decay rate by using the exact representation of the solution. For general convex function f, Kawashima-Matsumura [4] investigated the algebraic time decay rate in L^2-framework for small initial disturbances by an elementary energy method.

Recently, in the absence of $f'' > 0$, the stability theorems have been studied by Kawashima-Matsumura [5], Jones-Gardner-Kapitula [3], Mei [10] and Matsumura-Nishihara [9]. Kawashima-Matsumura [5] started with the case

$$f''(u) \gtrless 0 \quad \text{for} \quad u \gtrless 0, \tag{1.7}$$

by an elementary energy method. We note that in the case (1.7) the Oleinik's shock condition (1.4) is equivalent to

$$f'(u_+) \leq s < f'(u_-). \tag{1.8}$$

Under the Lax's shock condition $f'(u_+) < s < f'(u_+)$, they showed the stability and Mei [10] studied the time decay rate by an elementary energy method. Mei [10] also proved the stability for the first time in the case (1.7) with the degenerate shock condition $f'(u_+) = s$. Jones-Gardner-Kapitula [3] treated much more general case (1.4) than (1.7) under the Lax's shock condition (1.6), and showed both the stability and the time decay rate by the spectral analysis. However, their decay rate seems to be less sufficient than that in [4] and [10]. Just recently, Matsumura-Nishihara [9] showed the stability and the time decay rate only under the Oleinik's shock condition (1.4) by an elementary energy method. Even under the same conditions as in [3,4,10], our decay rate is sharper than the previous results in [4,10], and seems to be almost optimal from a view point of the optimal results in Nishihara [11]. All proofs in [4,9] are based on an elementary weighted energy method where suitable selections of weight functions play an essential role.

On the other hand, the stability theory in system case has started independently by Matsumura-Nishihara [8] and Goodman [1] based on an energy method. After these two papers, the stability of shock profiles was proved for several interesting systems both in viscous gas dynamics and in more general frameworks [4,6,7,13]; the simplest example among such systems is the equation of barotorpic viscous gas considered in [8]:

$$\begin{aligned} &v_t - u_x = 0, \\ &u_t + p(v)_x = \mu(u_x/v)_x, \end{aligned} \tag{1.9}$$

where $\mu > 0$ is a constant, and $p'(v) < 0$ and $p''(v) > 0$ for $v > 0$. The stability proofs in all these papers are, however, making use of the genuine nonlinearity of the system (for example, the convexity of $p(v)$ in (1.9)) and hence are not applicable straightforwardly to non-genuine (non-convex) nonlinearity case. Recently, Kawashima-Matsumura [5] showed a stability theory of viscous shock profiles for some regularized conservation law systems without genuine nonlinearity. A typical example of such systems is the following one of viscoelasticity:

$$\begin{aligned} &v_t - u_x = 0, \\ &u_t - \sigma(v)_x = \mu\, u_{xx}, \end{aligned} \tag{1.10}$$

where $\mu > 0$ is a constant, and $\sigma(v)$ is usually a non-convex function like $\sigma(v) = v + v^3$. More specifically, we assume that $\sigma'(v) > 0$ for all v and $\sigma''(v) \gtrless 0$ for $v \gtrless 0$. For such a non-convex $\sigma(v)$, we note that the system (1.10) is neither genuinely nonlinear nor linearly degenerate around $v = 0$. Under some additional conditions, we can prove the stability of a viscous shock profile around $v = 0$ by applying an energy method. Our energy method is a modification of the previous one in [4,8] where we made use a relatively simple weight function which is constructed by the shock profile itself. This time we manipulate such a weight function as to switch its form depending on the sign of $\sigma''(V)$ where V denotes the shock profile. To expose this technical idea clearly, we shall explain it in details for the scalar equation (1.1) with a non-convex $f(u)$. Our stability analysis is also applicable to the equation of van der Waals viscous fluid, i.e., the system (1.9) with $p(v) = R\theta/(v-b) - a/v^2$, where R, a and b are positive constants. Under suitable conditions, we can prove the stability of the shock profile around the inflection point of $p(v)$.

Notations. We denote generic positive constants by C without confusions. We also denote $f(x) \sim g(x)$ as $x \to a$ when $C^{-1}g < f < Cg$ in a neighborhood of a. For function spaces, L^2 denotes the space of square integrable functions on R with the norm

$$\| f \| = \left(\int_R |f(x)|^2 dx\right)^{1/2}.$$

Here and after the integrand R is abrèviated. $H^l (l \geq 0)$ denotes the usual l-th order Sobolev space with the norm

$$\| f \|_l = \left(\sum_{j=0}^{l} \| \partial_x^j f \|^2\right)^{1/2}.$$

For the weight function w , L^2_w denotes the space of measurable functions f satisfying $\sqrt{w} f \in L^2$ with the norm

$$|f|_w = \left(\int w(x)|f(x)|^2 dx\right)^{1/2}.$$

When $w(x) = \langle x \rangle^\alpha = (1+x^2)^{\alpha/2}$, we write $L^2_w = L^2_\alpha$ and $|\cdot|_w = |\cdot|_\alpha$ without confusions. Moreover when w is replaced by $\langle x \rangle^\alpha w$, we denote that space by $L^2_{\alpha,w}$

with the norm

$$|f|_{\alpha,w} = \Big(\int \langle x\rangle^{\alpha} w(x)|f(x)|^2 dx\Big)^{1/2}.$$

When $C^{-1} \le w(x) \le C$, we note that $L^2 = H^0 = L_0^2 = L_w^2$ with $\|\cdot\| = \|\cdot\|_0 = |\cdot|_0 \sim |\cdot|_w$ and that $L^2_{\alpha,w} = L^2_\alpha$ with $|\cdot|_{\alpha,w} \sim |\cdot|_\alpha$.

2. Scalar viscous conservation law

In this section, we show the existence and stability of viscous shock profiles for a scalar viscous conservation law (1.1) with the inital data (1.2). As the nonlinearity of f, we first assume the condition (1.7)

$$f''(u) \gtrless 0 \quad \text{for} \quad u \gtrless 0$$

to explain the idea in [5]. It is proved that if the viscous shock profile of (1.1) exists, $U(\xi)(\xi = x - st)$ must satisfy $-sU' + f(U)' = \mu U''$ and hence $u_\pm$ and s must satisfy the Rankine-Hugoniot condition (1.3) and the generalized shock condition (1.8)

$$f'(u_+) \le s < f'(u_-).$$

Then $U(\xi)$ must satisfy the ordinary differential equation

$$\mu U' = -sU + f(U) - a \equiv h(U), \tag{2.1}$$

where a is an integral constant given by $a = -su_\pm + f(u_\pm)$. Moreover, we can show that the conditions (1.3) and (1.8) are necessary and sufficient for the existence of U , and $U(\xi)$ is unique up to a shift in ξ and is a monotone function of ξ. In particular, when $u_+ \lessgtr u_-$, we have

$$u_+ \lessgtr U(\xi) \lessgtr u_-, \quad U'(\xi) \gtrless 0 \quad \text{for } \xi \in R. \tag{2.2}$$

Next we investigate the stability. We assume that $u_\pm$ and s satisfy the Rankine-Hugoniot condition (1.3) and the Lax's shock condition (1.6) (We here exclude the degenerate case $f'(u_+) = s$). Let $U(x - st)$ be a viscous shock profile connecting u_- and u_+, and let us define x_0 by

$$\int_{-\infty}^{\infty} (u_0(x) - U(x))\, dx = x_0(u_+ - u_-), \tag{2.3}$$

where $u_0(x)$ is the initial data in (1.2); this formula determines x_0 uniquely, provided that $u_0(x) - U(x)$ is integrable over R. The shifted $U(x - st + x_0)$ is also a viscous shock profile connecting u_- and u_+ and satisfies

$$\int_{-\infty}^{\infty} (u_0(x) - U(x + x_0))\, dx = 0. \tag{2.4}$$

This shifted viscous shock profile $U(x - st + x_0)$ is proved to be stable as $t \to \infty$, if the following $\phi_0(x)$ is suitable small:

$$\phi_0(x) \equiv \int_{-\infty}^{x} (u_0(y) - U(y + x_0))\, dy. \tag{2.5}$$

Theorem 2.1([4]). *Suppose that (1.7) holds and that $u_\pm$ and s satisfy the Rankine-Hugoniot condition (1.3) and the Lax's shock condition (1.6). When $u_\pm$ have opposite signs (i.e., $u_+u_- < 0$), we also require that*

$$f'''(u) \geq 0 \quad \text{for all } u \text{ between } u_+ \text{ and } 0. \tag{2.6}$$

Let $U(x-st)$ be a shock profile connecting u_- and u_+ and suppose that $u_0(x) - U(x)$ is integrable over R. We define x_0 and $\phi_0(x)$ by (2.4) and (2.5), respectively, and assume that $\phi_0 \in H^2$. Then there is a constant $\delta_0 > 0$ such that if $\|\phi_0\|_2 \leq \delta_0$, then the problem (1.1), (1.2) has a unique global solution $u(t,x)$ satisfying

$$u - U \in C^0([0,\infty); H^1) \cap L^2([0,\infty); H^2),$$

and the solution verifies

$$\sup_{x\in R} |u(t,x) - U(x - st + x_0)| \to 0 \quad \text{as } t \to \infty.$$

To prove the stability result, as in the previous papers [1,4,8], we seek the solution of (1.1), (1.2) in the form

$$u(t,x) = U(x - st + x_0) + \phi_x(t,x).$$

Here $\phi(t,x)$ is to be determined as a solution to the "integrated" problem:

$$\phi_t + f(U + \phi_x) - f(U) = \mu\phi_{xx}, \tag{2.7}$$

$$\phi(0,x) = \phi_0(x), \tag{2.8}$$

where $U = U(x - st + x_0)$, and $\phi_0(x)$ is the function in (2.5). To show the global solvability of this integrated problem, we need to derive a suitable *a priori* estimate of solutions, which is as follows.

Proposition 2.2. (*A priori* estimate) *We assume the same conditions as in Theorem 2.1. Let $\phi(t,x)$ be a solution of (2.7), (2.8) satisfying*

$$\phi \in C^0([0,T]; H^2) \cap C^1([0,T]; L^2), \ \phi_x \in L^2([0,T]; H^2)$$

for some $T > 0$, and put

$$N(t) = \sup_{0\leq\tau\leq t} \|\phi(\tau)\|_2.$$

Then there are constants $\delta_1 > 0$ and $C_1 > 0$ (both are independent of T) such that if $N(T) \leq \delta_1$, then

$$\|\phi(t)\|_2^2 + \int_0^t \|\phi_x(\tau)\|_2^2 \, d\tau \leq C_1\|\phi_0\|_2^2, \quad t \in [0,T].$$

Once we have the Proposition 2.2, the proof of the Theorem 2.1 is on the same line of the previous papers[4,8]. A key to the proof of this *a priori* estimate is to show the following first energy estimate. We omit the estimates of the heigher derivatives.

Lemma 2.3. *For any fixed number $\overline{\delta} > 0$ (for example, $\overline{\delta} = 1$), we assume that $N(T) \leq \overline{\delta}$. Then it holds that*

$$\begin{aligned} \|\phi(t)\|^2 + \int_0^t \|\phi_x(\tau)\|^2 d\tau + \int_0^t \int_{-\infty}^{\infty} |f'(U)_\xi| |\phi|^2 \, dx d\tau \\ \leq C\|\phi_0\|^2 + CN(t) \int_0^t \|\phi_x(\tau)\|^2 \, d\tau, \quad t \in [0,T], \end{aligned} \tag{2.9}$$

where $U = U(\xi)$ with $\xi = x - st + x_0$, and C is a positive constant depending on $\overline{\delta}$ but not on T.

Proof of Lemma 2.3. First, we rewrite (2.7) in the form

$$\phi_t + f'(U)\phi_x - \mu\phi_{xx} = F, \tag{2.10}$$

where $U = U(\xi)$ with $\xi = x - st + x_0$ and

$$F = -\{f(U + \phi_x) - f(U) - f'(U)\phi_x\}.$$

Now we introduce a weight function $w(u) > 0$ belonging to $C^1[u_+, u_-]$ and multiply (2.10) by $w(U)\phi$ to get:

$$\begin{aligned} (\tfrac{1}{2}w(U)\phi^2)_t + (\tfrac{1}{2}w(U)f'(U)\phi^2 - \mu w(U)\phi\phi_x)_x \\ - \tfrac{1}{2}w(U)z_\epsilon(U)U_\xi\phi^2 + (1-\epsilon)\mu w(U)\phi_x^2 \leq w(U)\phi F, \end{aligned} \tag{2.11}$$

for any $\epsilon \in (0,1)$ where

$$z_\epsilon(u) = f''(u) - (s - f'(u))\frac{w'(u)}{w(u)} + \frac{h(u)}{2\epsilon}\left(\frac{w'(u)}{w(u)}\right)^2. \tag{2.12}$$

Here we have used (2.1), i.e., $\mu U_\xi = h(U)$ and the Schwarz inequality. Since our interest is in the study of non-convex nonlinearity case, we restrict our attention to the case where $u_+ < 0 < u_-$. when $u_+ < u_-$, from (2.2), we have $u_+ < U < u_-$ and $U_\xi < 0$ for $\xi \in R$. Therefore, in order to show (2.9), it is sufficient to show

$$z_\epsilon(u) \geq c|f''(u)| \quad \text{for } u \in [u_+, u_-]. \tag{2.13}$$

where c is a constant independent of T. Inspired by the form (2.12), we choose $w(u)$ as

$$w(u) = \begin{cases} 1, & u \in [0, u_-], \\ c_0(s - f'(u))^{-2}, & u \in [u_+, 0], \end{cases}$$

where $c_0 = (s - f'(0))^2$. Our $w(u)$ verifies $w(u) \geq 1$ and is in $C^1[u_+, u_-]$. Then,by the assumptions on f it is not difficult to see the desired estimate (2.13) holds for a suitable ϵ. Therefore, the proof of Lemma 2.3 is completed.

Remark 2.1. We can also replace the conditions (1.7),(2.6) respectively by

$$f''(u) \lesseqgtr 0 \quad \text{for } u \gtreqless 0 \text{ under consideration,} \tag{2.14}$$

$$f'''(u) \leq 0 \quad \text{for all } u \text{ under consideration.} \tag{2.15}$$

In fact, if u solves (1.1) with $f(u)$ satisfying (2.14) and (2.15), then the change of independent variable, $y = -x$, transforms (1.1) into

$$u_t + \tilde{f}(u)_y = \mu u_{yy},$$

where $\tilde{f}(u) = -f(u)$, and this $\tilde{f}(u)$ verifies (1.7) and (2.6). Simple examples of $f(u)$ satisfying either (1.7), (2.6) or (2.14), (2.15) are

$$f(u) = \alpha u + \beta u^{2m+1}, \tag{2.16}$$

$$f(u) = \beta u(1+u^2)^{-1/2}, \tag{2.17}$$

where α, β are constants with $\beta \neq 0$, and $m \geq 1$ is an integer. For instance, if $\beta > 0$, then the $f(u)$ in (2.16) verifies (1.7), (2.6) while the $f(u)$ in (2.17) satisfies (2.14), (2.15) in $|u| < 1/2$.

Rmark 2.2 Mei [10] studied the time decay rate by the same arguments as in [4] and obtained that if ϕ_0 furthermore belongs to L^2_α for some $\alpha > 0$, then it holds $\sup_R |u - U| \leq Ct^{-[\alpha]/2}$. So, if α is a positive integer, this decay rate seems to be optimal since Nishihara [11] showed the decay rate $t^{-\alpha/2}$ is optimal in the case $f = u^2/2$. Mei [10] also studied the stability in the degenerate case $f(u_+) = s$. He introduced a new unknown function $\Phi = \mu^{1/2}|h(U(\xi))|^{-1/2}\phi$ and then the equation (2.7) is rewritten in the form

$$\Phi_t + s\Phi_x + Z(U)\Phi - \mu\Phi_{xx} = \mu^{1/2}|h(U(\xi))|^{-1/2}F,$$

where

$$Z(U) = \frac{1}{4\mu}(f'(U) - s)^2 - \frac{1}{2\mu}f''(U)h(U).$$

Since it is easy to see $Z(U) \geq 0$ even for $f'(u_+) = s$, the corresponding *a priori* estimate for Φ to Proposition 2.2 is expected to hold provided $\Phi(0, x)$ is suitably small in H^2.

Just recently, in stead of the condition (1.7), Matsumura-Nishihara [9] investigated the stability of the viscous shock profile only under the Oleinik's shock condition (1.4)

$$h(u) \equiv -s(u - u_\pm) + f(u) - f(u_\pm) \begin{cases} < 0 & \text{if} \quad u_+ < u < u_- \\ > 0 & \text{if} \quad u_- < u < u_+. \end{cases}$$

In this case, as the properties of the viscous shock profiles, we have the following by the standard arguments on the ordinary differential equations.

Lemma 2.4. *Assume (1.3), (1.4) and*

$$|h(U)| \sim |U - u_\pm|^{1+k_\pm}, \quad U \to u_\pm$$

for some $k_\pm \geq 0$. *Then there exists a viscous shock profile* $U(\xi)$ *of (1.1) with* $U(\pm\infty) = u_\pm$, *unique up to a shift, which is determined by the ordinary differential equation (2.1). Moreover, it holds as* $\xi \to \pm\infty$

$$|U(\xi) - u_\pm| \sim \exp(-c|\xi|) \quad \textit{if} \quad f'(u_+) < s < f'(u_-),$$

$$|U(\xi) - u_+| \sim |\xi|^{-1/k_+} \quad \textit{if} \quad s = f'(u_+)$$

and

$$|U(\xi) - u_-| \sim |\xi|^{-1/k_-} \quad \textit{if} \quad s = f'(u_-).$$

Then our new stability theorem is

Theorem 2.5([9]). *Suppose that* $u_\pm$ *and* s *satisfy the Rankine-Hugoniot condition (1.3) and the Oleinik's shock condition (1.4). Let* $U(x - st)$ *be a shock profile connecting* u_- *and* u_+ *and suppose that* $u_0(x) - U(x)$ *is integrable over* R. *We define* x_0 *and* $\phi_0(x)$ *by (2.4) and (2.5), respectively, and assume that* $\phi_0 \in H^2$. *Then the followings hold.*
(i) When $f'(u_+) < s < f'(u_-)$, *there exists a positive constant* ε_1 *such that if* $\| \phi_0 \|_2 < \varepsilon_1$, *then the Cauchy problem (1.1), (1.2) has a unique global solution* $u(t, x)$ *satisfying*

$$u - U \in C^0([0, \infty); H^1) \cap L^2(0, \infty; H^2)$$

and moreover

$$\sup_R |u(t, x) - U(x - st)| \to 0 \quad \textit{as} \quad t \to \infty.$$

(ii) When $s = f'(u_+) < f'(u_-)$, *there exists a positive constant* ε_1 *such that if* $\| \phi_0 \|_2 + |\phi_0|_{\langle\xi\rangle_+} < \varepsilon_1$ *then the Cauchy problem (1.1), (1.2) has a unique global solution* $u(t, x)$ *satisfying*

$$u - U \in C^0([0, \infty); H^1) \cap L^2([0, \infty); H^2 \cap L^2_{\langle\xi\rangle_+})$$

and moreover

$$\sup_R |u(t, x) - U(x - st)| \to 0 \quad \textit{as} \quad t \to \infty,$$

where $\langle\xi\rangle_+ = \begin{cases} \sqrt{1 + \xi^2} & \xi \geq 0 \\ 1 & \xi < 0. \end{cases}$
(iii) When $f'(u_+) < s = f'(u_-)$ *or* $s = f'(u_+) = f'(u_-)$, *then* $L^2_{\langle\xi\rangle_+}$ *in (ii) should be replaced by* $L^2_{\langle\xi\rangle_-}$ *or* $L^2_{\langle\xi\rangle} = L^2_1$, *respectively. Here* $\langle\xi\rangle_- = \begin{cases} \sqrt{1 + \xi^2} & \xi < 0 \\ 1 & \xi \geq 0. \end{cases}$

Key of the proof of Theorem 2.5. As in the previous papers, the essential part of the proof is to get the corresponding first energy estimate (2.9) in Lemma 2.3. For simplicity, we assume $u_+ < u_-$ and $h(U) < 0$ for $U \in (u_+, u_-)$. In this case, we may select a positive weight as

$$w = w(U) = \frac{(U - u_+)(U - u_-)}{h(U)}. \tag{2.18}$$

Since $w(U) \sim$ *positive const.* for the case $f'(u_+) < s < f'(u_-)$, $L^2_{\alpha,w(U)} = L^2_\alpha$. While if $s = f'(u_+) < f'(u_-)$, then $w(U) \sim |U-u|^{-k_+}$ as $U \to u_+$ and $w(U(\xi)) \sim \langle\xi\rangle$ as $\xi \to +\infty$, and hence $L^2_{w(U)} = L^2_{\langle\xi\rangle_+}$. Then, multiplying (2.10) by $w(U(\xi))\phi$ we have

$$(\frac{1}{2}w(U)\phi^2)_t + (\dots)_x + \mu w(U)\phi_x^2 - \frac{1}{2}(wh)''(U)U_x\phi^2 = w(U)\phi F.$$

Here we have used $\mu U_\xi = h(U)$. Since we have taken the weight w as (2.18), it is not hard to obtain

$$\frac{1}{2}|\phi(t)|^2_{w(U)} + \int_0^t (\| \sqrt{-U_x}\phi(\tau) \|^2 + \mu|\phi_x(\tau)|^2_{w(U)})d\tau$$

$$\leq \frac{1}{2}|\phi_0|^2_{w(U)} + \int_0^t \int w(U)\phi F dx d\tau$$

whitch implies the basic first energy estimate. We omit the details.

Although we omit the details, we can also obtain the following theorems concerning the time decay rate (for more details, see [9]).

Theorem 2.6 (Rate of asymptotic speed for $f'(u_+) < s < f'(u_-)$). *Let u be a solution obtained in Theorem 2.5(i) and let ϕ_0 lie in L^2_α for some $\alpha > 0$. If α is an integer, then it holds*

$$\sup_R |u(t,x) - U(x-st)| \leq C(1+t)^{-\alpha/2}(\| u_0 - U \|_1 + |\phi_0|_\alpha),$$

while if α is not an integer, then

$$\sup_R |u(t,x) - U(x-st)| \leq C_\varepsilon(1+t)^{-\alpha/2+\varepsilon}(\| u_0 - U \|_1 + |\phi_0|_\alpha)$$

for any constant $\varepsilon > 0$ and some constant C_ε such that $C_\varepsilon \to \infty$ as $\varepsilon \to 0$.

Remark 2.3. For general case (1.4) with (1.6), Jones et al.[3] have shown $\sup_R |u - U| \leq Ct^{-[\alpha]/4}$ and, Kawashima and Matsumura [4] for convex case (1.5) and Mei [10] for the case (1.7) with (1.6) have shown $\sup_R |u-U| \leq Ct^{-[\alpha]/2}$. So, Theorems 2.5 and 2.6 improve their results. Further, for $f = u^2/2$ Nishihara [11] showed, using an explicit formula, that if $|\phi_0(x)| \leq C|x|^{-\alpha/2}$ as $|x| \to \infty$, then $\sup_R |u - U| \leq Ct^{-\alpha/2}$ and this decay rate is optimal in general. Hence the decay rate in Theorem 2.6 seems to be almost optimal in L^2-space framework.

Theorem 2.7(Rate of asymptotic speed for $f'(u_+) = s < f'(u_-)$). *Let u be a solution obtained in Theorem 2.5(ii) and $f''(u_+) = \dots = f^{(n)}(u_+) = 0$ and $f^{(n+1)}(u_+) \neq 0$ for $n \geq 1$. Then if $\phi_0 \in L^2_{\alpha,\langle\xi\rangle_+}$ $(0 < \alpha < 2/n)$, it holds for any $\varepsilon > 0$*

$$\sup_R |u(t,x) - U(x-st)| \leq C_\varepsilon(1+t)^{-\alpha/4+\varepsilon}(\| u_0 - U \|_1 + |\phi_0|_{\alpha,\langle\xi\rangle_+}).$$

Remark 2.4. Our rate seems to be not so sharp. But, the rate in this case is not known as far as we know. More contributions will be expected.

3. Equation of viscoelasticity

In this section we study the viscous shock profiles for the equation of viscoelasticity with a non-convex constitutive relation:

$$\begin{aligned} v_t - u_x &= 0, \\ u_t - \sigma(v)_x &= \mu u_{xx}, \end{aligned} \tag{3.1}$$

$$(v,u)(0,x) = (v_0,u_0)(x). \tag{3.2}$$

Here v denotes the deformation gradient, u the velocity, $\sigma(v)$ the stress, and μ is the coefficient of viscosity; $\mu > 0$ is a constant and $\sigma(v)$ is a smooth function satisfying the following sign conditions:

$$\sigma'(v) > 0 \text{ for all } v \text{ under consideration}, \tag{3.3}$$

$$\sigma''(v) \lesseqgtr 0 \text{ for } v \lesseqgtr 0 \text{ under consideration}, \tag{3.4}$$

so that $\sigma(v)$ has a point of inflection at $v = 0$. We see that the system (3.1) with $\mu = 0$ is strictly hyperbolic, with the characteristic roots

$$\lambda = \pm\lambda(v), \quad \text{where } \lambda(v) = \sqrt{\sigma'(v)}, \tag{3.5}$$

and the corresponding right eigenvectors $r_\pm(v)$. Moreover, we find that both characteristic fields are neither genuinely nonlinear nor linearly degenerate around $v = 0$. Let $v_\pm, u_\pm$ and s be constants such that $(v_+,u_+) \neq (v_-,u_-)$. As in the previous section, we can show that a viscous shock profile $(V,U)(x - st)$ connecting (v_-,u_-) and (v_+,u_+) exists if and only if $(v_\pm,u_\pm)$ and s satisfy the Rankine-Hugoniot condition for (3.1)

$$\begin{aligned} -s(v_+ - v_-) - (u_+ - u_-) &= 0, \\ -s(u_+ - u_-) - (\sigma(v_+) - \sigma(v_-)) &= 0. \end{aligned} \tag{3.6}$$

and the generalized shock condition for (3.4)

$$-\lambda(v_+) < s \le -\lambda(v_-) \quad \text{or} \quad \lambda(v_+) \le s < \lambda(v_-). \tag{3.7}$$

From now on, we restrict our attention to the case $s > 0$ and consider a shock profile $(V,U)(x - st)$ connecting (v_-,u_-) and (v_+,u_+), where $(v_\pm,u_\pm)$ and s satisfy the Rankine-Hugoniot condition (3.6) and the following shock condition:

$$\lambda(v_+) < s < \lambda(v_-). \tag{3.8}$$

(The degenerate case $\lambda(v_+) = s$ is excluded here.)

To show the stability of such a shock profile, we need to formulate conditions on the initial data $(v_0,u_0)(x)$. We write $w_0 = (v_0,u_0)$, $W = (V,U)$ and $w_\pm =$

$(v_\pm, u_\pm)$, where these vectors are understood as row vectors. We assume the integrability of $w_0(x) - W(x)$ over R and express that integral in the form

$$\int_{-\infty}^{\infty} (w_0(x) - W(x))\, dx = x_0(w_+ - w_-) + \beta r_-(v_-), \tag{3.9}$$

where $r_-(v_-)$ is the right eigenvector evaluated at $v = v_-$; the coefficients x_0 and β are uniquely determined because the vectors $w_+ - w_-$ and $r_-(v_-)$ are linearly independent. We now assume that $\beta = 0$ holds eventually in (3.9). Then (3.9) becomes

$$\int_{-\infty}^{\infty} (w_0(x) - W(x + x_0))\, dx = 0. \tag{3.10}$$

Then we define $\zeta_0 = (\phi_0, \psi_0)$ by

$$\zeta_0(x) = \int_{-\infty}^{x} (w_0(y) - W(y + x_0))\, dy. \tag{3.11}$$

Now we state our result concerning the stability of viscous shock profiles without the proof (for details, see[4]).

Theorem 3.2([4]). *Suppose that (3.3) and (3.4) hold and that $(v_\pm, u_\pm)$ and s satisfy the Rankine-Hugoniot condition (3.6) and the shock condition (3.8). Moreover, when $v_\pm$ have opposite signs, we assume*

$$\sigma'''(v) \geq 0 \quad \textit{for all } v \textit{ between } v_+ \textit{ and } 0. \tag{3.12}$$

Let $(V, U)(x - st)$ be a shock profile connecting (v_-, u_-) and (v_+, u_+), and suppose that $(v_0, u_0)(x) - (V, U)(x)$ is integrable over R and that (3.17) holds for some x_0 and $\beta = 0$. In addition we assume that the $(\phi_0, \psi_0)(x)$ defined by (3.11) belongs to H^2. Then there is a constant $\delta_0 > 0$ such that if $\|(\phi_0, \psi_0)\|_2 + |v_+ - v_-| \leq \delta_0$, then the problem (3.1), (3.2) has a unique global solution $(v, u)(t, x)$ satisfying

$$\begin{aligned} v - V &\in C^0([0,\infty); H^1) \cap C^1([0,\infty); L^2), \\ u - U &\in C^0([0,\infty); H^1) \cap L^2([0,\infty); H^2), \end{aligned} \tag{3.13}$$

where $(V, U) = (V, U)(x - st + x_0)$ is the shifted shock profile. Furthermore, the solution verifies

$$\sup_{x \in R} |(v, u)(t, x) - (V, U)(x - st + x_0)| \to 0 \quad \textit{as } t \to \infty.$$

Remark 3.1. In (3.9) we assumed that $\beta = 0$. The case $\beta \neq 0$ was treated in [6,7] but the stability proofs seem to contain a serious mistake. Recently, Szepessy and Zhouping [13] announced to succeed in treating the case $\beta \neq 0$ for general systems with genuine nonlinearity. So it sounds interesting to combine our arguments together with that in [13].

Remark 3.2. It seems interesting to discuss the case where $\sigma(v)$ satisfies (3.3) together with the following set of sign conditions:

$$\sigma''(v) \lesseqgtr 0 \quad \text{for } v \gtreqless 0 \text{ under consideration,} \tag{3.14}$$

$$\sigma'''(v) \leq 0 \quad \text{for all } v \text{ under consideration.} \tag{3.15}$$

However, this remains an open question, since we could not find a recipe for reducing that case to the former case discussed in the previous paragraphs. Consequently, an important example $\sigma(v) = v(1+v^2)^{-1/2}$ is excluded from our stability theory. This forms a striking contrast to the case of a single equation.

Remark 3.3. It also seems interesting to consider the degenerate case $\lambda(v_+) = s$ we excluded. We may expect the similar arguments as in the scalar case should be done.

4. Equation of van der Waals viscous fluid

As an application of the stability theory developed in the previous section, we discuss the equation of van der Waals viscous fluid:

$$\begin{aligned} &v_t - u_x = 0, \\ &u_t + p(v)_x = \mu(u_x/v)_x, \end{aligned} \tag{4.1}$$

where v denotes the specific volume, u the velocity, $p(v)$ the pressure, and μ is the coefficient of viscosity; $\mu > 0$ is a constant and $p(v)$ is given explicitly as

$$p(v) = \frac{R\theta}{v-b} - \frac{a}{v^2} \quad \text{for } v > b, \tag{4.2}$$

where $R > 0$ is the gas constant, $\theta > 0$ the absolute temperature (assumed to be constant), and a and b are positive constants.

We are interested in the case where $p(v)$ is strictly decreasing and has just two points of inflection in $v > b$. This is the case when

$$m_1 a < bR\theta < m_2 a, \tag{4.3}$$

where $m_1 = (2/3)^3$ and $m_2 = (3/4)^4$. In fact, if (4.3) holds, then

$$p'(v) < 0 \quad \text{for } v > b. \tag{4.4}$$

In addition, there are constants $\underline{v}$ and $\overline{v}$ with $3b < \underline{v} < 4b < \overline{v}$ such that $p''(\underline{v}) = p''(\overline{v}) = 0$, $p''(v) < 0$ on $(\underline{v}, \overline{v})$, and $p'''(v) > 0$ otherwise. Therefore, $p(v)$ is strictly decreasing and has two point of inflection at $v = \underline{v}$ and $\overline{v}$. Furthermore, there are v_1 and v_2 with $\underline{v} < v_1 < \overline{v} < v_2$ and with $v_2 > 5b$ such that $p'''(v_1) = p'''(v_2) = 0$, $p'''(v) > 0$ on (v_1, v_2), and $p'''(v) < 0$ otherwise. Consequently, we have: On the interval $b < v < v_1$,

$$p''(v) \lesseqgtr 0 \quad \text{for } v \gtreqless \underline{v}, \quad \text{and} \quad p'''(v) \leq 0 \quad \text{for all } v, \tag{4.5}$$

and on the interval $v_1 < v < v_2$,

$$p''(v) \gtreqless 0 \quad \text{for } v \gtreqless \overline{v}, \quad \text{and} \quad p'''(v) \geq 0 \quad \text{for all } v. \tag{4.6}$$

We can show that in the region $b < v < v_1$, (4.1) is reduced to the equation of the form (3.1) with a $\sigma(v)$ satisfying (3.3), (3.4) and (3.12). To this end, we change the dependent variable v as $v = \underline{v} + \tilde{v}$ and transform (4.1) into

$$\begin{aligned} &\tilde{v}_t - u_x = 0, \\ &u_t - \sigma(\tilde{v}_x) = (\tilde{\mu}(\tilde{v})u_x)_x, \end{aligned} \tag{4.7}$$

where $\sigma(\tilde{v}) = -p(\underline{v} + \tilde{v})$ and $\tilde{\mu}(\tilde{v}) = \mu/(\underline{v} + \tilde{v})$. For this transformed equation, thanks to (4.4) and (4.5), we have:

$$\sigma'(\tilde{v}) > 0, \quad \sigma''(\tilde{v}) \gtrless 0 \quad (\text{for } \tilde{v} \gtrless 0), \quad \sigma'''(\tilde{v}) \geq 0 \tag{4.8}$$

for $\tilde{v}$ in a neighborhood of $\tilde{v} = 0$, which just correspond to (3.3), (3.4) and (3.12), respectively. Therefore, the stability theory developed in Section 3 applies to (4.7); the dependence of $\tilde{\mu}(\tilde{v})$ upon $\tilde{v}$ does not cause any serious difficulty in the application (in this respect, see [4,8]). Consequently, shock profiles for (4.1) in a neighborhood of $v = \underline{v}$ are proved to be stable as $t \to \infty$, provided that initial disturbances are suitably small and of integral zero. Next, we consider (4.1) in the region $v_1 < v < v_2$. The change of the dependent variable $v = \overline{v} + \tilde{v}$, transforms (4.1) into (4.7) with $\sigma(\tilde{v}) = -p(\overline{v} + \tilde{v})$ and $\tilde{\mu}(\tilde{v}) = \mu/(\overline{v} + \tilde{v})$. In this case, from (4.4) and (4.6) we have:

$$\sigma'(\tilde{v}) > 0, \quad \sigma''(\tilde{v}) \lesseqgtr 0 \quad (\text{for } \tilde{v} \gtrless 0), \quad \sigma'''(\tilde{v}) \leq 0 \tag{4.9}$$

for $\tilde{v}$ in a neighborhood of $\tilde{v} = 0$. Unfortunately, this corresponds to the case where (3.3), (3.14) and (3.15) hold, so that our stability theory is not applicable to this case.

REFERENCES

1. GOODMAN, J., *Nonlinear asymptotic stability of viscous shock profiles for conservation laws*, Arch. Rat. Mech. Anal.**95**, pp. 325-344 (1986).
2. IL'IN, A. M. AND OLEINIK, O. A., *Asymptotic behavior of the solutions of the Cauchy problem for certain quasilinear equations for large time*, (Russian), Mat. Sbornik **51**, pp. 191-216 (1960).
3. JONES, C., GARDNER, R. AND KAPITULA, T., *Stability of travelling waves for non-convex scalar viscous conservation laws*, Commun. Pure Appl. Math. **46**, pp. 505-526 (1993).
4. KAWASHIMA, S. AND MATSUMURA, A., *Asymptotic stability of traveling wave solutions of systems for one-dimensional gas motion*, Commun. Math. Phys. **101**, pp. 97-127 (1985).
5. KAWASHIMA, S. AND MATSUMURA, A., *Stability of shock profiles in viscoelasticity with non-convex constitutive relations*, (to appear).
6. LIU, T. -P., *Nonlinear stability of shock waves for viscous conservation laws*, Mem. Amer. Math. Soc. **56**, no. 328 (1985).
7. LIU, T. -P., *Shock waves for compressible Navier-Stokes equations are stable*, Comm. Pure Appl. Math. **39**, pp. 565-594 (1986).
8. MATSUMURA, A. AND NISHIHARA, K., *On the stability of traveling wave solutions of a one-dimensional model system for compressible viscous gas*, Japan J. Appl. Math. **2**, pp. 17-25 (1985).

9. MATSUMURA, A. AND NISHIHARA, K., *Asymptotic stability of traveling waves for scalar viscous conservation laws with non-convex nonlinearity*, (to appear).
10. MEI, M., *Stability of shock profiles for non-convex scalar viscous conservation laws with non-degenerate and degenerate shock conditions*, (to appear).
11. NISHIHARA, K., *A note on the stability of traveling wave solutions of Burgers' equation*, Jpn. J. Appl. Math. **2**, pp. 27-35 (1985).
12. SATTINGER, D. H., *On the stability of waves of nonlinear parabolic systems*, Adv. Math. **22**, pp. 312-355 (1976).
13. SZEPESSY, A. AND ZHOUPING, X., *Nonlinear Stability of Viscous Shock Waves*, (to appear).
14. WEINBERGER, H. F., *Long-time behavior for a regularized scalar conservation law in the absence of genuine nonlinearity*, Ann. Inst. Henri Poincare **7**, pp. 407-425 (1990).

DEPARTMENT OF MATHEMATICS, OSAKA UNIVERSITY, TOYONAKA 560, JAPAN
E-mail address: akitaka@math.sci.osaka-u.ac.jp

AMS/IP Studies in Advanced Mathematics
Volume **3**, 1997

ASYMPTOTIC CONFIGURATION OF STATIONARY INTERFACIAL PATTERNS FOR REACTION DIFFUSION SYSTEMS

Yasumasa NISHIURA † Hiromasa SUZUKI ††

Abstract

Asymptotic configuration of interfacial patterns of reaction diffusion systems is considered when the interfacial thickness tends to zero. Under several hypotheses derived by formal asymptotic analysis, it is shown that there are no smooth limiting configuration of interface for generic domains. This partially explains the fact that interfacial patterns become fine and complicated as in the micro-phase separation of block copolymer in this limit.

1. Introduction

Morphology of *final* patterns in phase transition are usually simple ones: only one phase dominates the whole domain (non-conserved) or it is decomposed into simple subdomains (conserved) after coarsening process. This is due to the tendency to minimize the area of interface. However, if there is a microscopic constraint to the system, the final pattern becomes much richer and has in general a variety of morphologies from lamellar to labyrinthine patterns. Block copolymer is one of such materials where two monomers (say, A and B) are connected at some point (constraint), and this is responsible for the formation of very fine and complicated structures depending on the ratio of composite monomers in the process of micro-phase separation [3][2][13]. Locally each monomer moves in a random way and tends to segregate each other (bistability), however connectivity does not allow them to form a large domain consisting of only one monomer (nonlocality). Ohta and Kawasaki [9] proposed the following model system to describe such a phenomenon.

$$
(1.1)\qquad \begin{cases} u_t = \varepsilon^2 \Delta u + f(u,v) & \\ & \text{in } \Omega, \\ 0 = D\Delta v + u & \\ \dfrac{\partial u}{\partial n} = 0 = \dfrac{\partial v}{\partial n} & \text{on } \partial\Omega. \end{cases}
$$

where u is the *order parameter* indicating A-rich or B-rich phase, v represents the nonlocal effect due to connectivity, $\varepsilon(\ll 1)$ corresponds to the interfacial thickness and $D(\gg 1)$ is proportional to the square of the polymerization index (namely, the length of block copolymer) which is usually quite large, and $f(u,v)$ is a cubic nonlinearity (typically of the form $u - u^3 - v$). It is anticipated that many other phenomena could be described by similar models to (1.1), since the basic mechanism creating a variety of patterns is due to the competition between local dynamics and nonlocal effect. In fact similar patterns are observed in liquid crystal, magnetic thin film, and so on. The arguments in this note is valid to slightly more general system:

$$(1.2)\qquad \begin{cases} u_t = \varepsilon^2 \Delta u + f(u,v) & \\ & \text{in } \Omega, \\ \delta v_t = D\Delta v + g(u,v) & \\ \dfrac{\partial u}{\partial n} = 0 = \dfrac{\partial v}{\partial n} & \text{on } \partial\Omega. \end{cases}$$

where δ is a nonnegative constant. Although the precise assumptions for (f,g) are delegated to [7], they are qualitatively the same as (1.1). A naive approach to find nontrivial patterns of (1.2) is to consider the limiting case either $\varepsilon \downarrow 0$ or $D \uparrow \infty$. For the latter case it is known (see [5][4]) that the resulting equations become a scalar equation with a constraint of integral type and it is unlikely to have a stable complicated pattern for such a system, since there are no stable multi-layered solutions even in 1D case [6]. On the other hand we know very little about the former case in higher space dimensions, since it has been regarded to be extremely difficult to find the first approximate stationary solutions in the limit of $\varepsilon \downarrow 0$. Especiallly we are interested in the behavior of the asymptotic configuration of the interface Γ^ε (see (2.1)). The aim of this note is to answer (at least partially) the following question.

Does (1.2) *has an ε-family of stationary layered solutions with smooth interface Γ^ε up to $\varepsilon = 0$?*

The answer is obviously affirmative, since we know planar and spherical layered solutions (see [12][10]). However those domains have very *special* geometries, i.e., rectangles and spheres and it is not a priori clear that such smooth interfaces persist up to $\varepsilon = 0$ for *generic* domains. It turns out that the answer is negative for generic ones under several hypotheses derived by the formal asymptotic analysis. Here we only consider the case where Γ^ε is a simple closed curve inside of Ω. $(U^\varepsilon, V^\varepsilon)$ is called an *ε-family of matched asymptotic solutions* to (1.2), if it has an matched asymptotic expansion mentioned in Section 2.

MAIN THEOREM
(*a*) *(Disk Symmetry) Suppose that* (1.2) *with $n = 2$ has an ε-family of matched asymptotic solutions with simple closed smooth interfaces Γ^ε (see* (2.1)*) up to $\varepsilon = 0$ and that the matching condition $(MC)_1$ holds (see Lemmas* 3.4.1 *and* 3.4.2*), then Γ^0 must be a circle.*
(*b*) *(Non-existence) Moreover under Hypothesis* 4.1 *in Section* 4, (*a*) *implies that*

the reduced problem (2.26) *has no solutions for generic domains* Ω, *and hence there does not exist associated* ε*-family of matched asymptotic solutions.*

REMARK 1.1 *The hypothesis in Main Theorem* (*a*) *can be weakened, in fact, it suffices to assume the existence of principal orders of outer and inner expansions. The extension to* $\{n \geq 3\}$*-case can be done without difficulty. More precisely, see the forthcoming paper [8].*

The above non-existence result is not a dissapoiting result and, in fact, it suggests an important thing about the behavior of the interface as $\varepsilon \downarrow 0$. Namely, if some stationary pattern of (1.2) exists up to $\varepsilon = 0$, but does not have a smooth limiting interface, then the configuration of the interface must become fine and complicated as $\varepsilon \downarrow 0$. In order to understand the morphology of the complicated patterns, it seems necessary to apply an appropriate rescaling to blow up the degenerate situation since there are no well-defined asymptotic limit of interfaces in the original framework. The study in this direction is now currently being done and will be reported elsewhere. The outline of this note is as follows. In Section 2 we shall display the expansion of ε-family of matched asymptotic solutions. In section 3 we shall derive $(v_0)_y$ = *constant* along the interface, which is a byproduct of the analysis of the formal matched asymptotic expansions of the eigenvalue problem of Allen-Cahn operator associated with the original system (1.2). Based on these observations we shall prove the Main Theorem in Section 4. The key ingredient is the Serrin's result [11] (and its generalizations) for the over-determined Poisson equation.

2.Matched asymptotic expansion of singularly perturbed stationary solutions

In this section, we summarize the hypothesis for the stationary solutions which have interior transition layers, especially, a precise definition of matched asymptotic expansion of them is presented. We assume that there exist an ε-family of smooth stationay solutions $(U^\varepsilon(x), V^\varepsilon(x))$ to (1.2) with interior transition layers such that the interface Γ^ε:

$$\Gamma^\varepsilon \equiv \{x \in \Omega | U^\varepsilon(x) = \frac{1}{2}(h_+(v^*) + h_-(v^*))\} \tag{2.1}$$

is a smooth simple closed curve in $\mathbf{R}^2$ and have a definite limit Γ_0 with same properties as $\varepsilon \downarrow 0$, where $u_\pm$ are two stable branches of $f(u, v) = 0$ (see [7]). Let Ω_0^+ be the region surrounded by Γ_0 and $\Omega_0^- \equiv \Omega \backslash \bar{\Omega}_0^+$. In some tubular neighbourhood $T(\Gamma_0)$ of Γ_0, local coordinate system (s, y) is defined and for $x \in T(\Gamma_0)$

$$x = \Gamma_0(s(x)) + y(x)\nu(s(x)) \tag{2.2}$$

holds, where $s(x)$ is the parameter measuring the arclength along Γ_0 to the point on Γ_0 closest to x, $\nu(s)$ is outward unit normal vector at s, and $y(x)$ is signed distance from x to Γ_0 that is positive if $x \in \Omega_0^-$. We use the notation $u(s,y)$ for the representation of $u(x)$ by the local coordinate system. Using this local coordinate system, Γ^ε can be expanded into

$$\Gamma^\varepsilon = \Gamma_0(s) + \gamma(s,\varepsilon)\nu(s), \quad \gamma(s,\varepsilon) = \sum_{k=1}^{m} \gamma_k(s)\varepsilon^k + \varepsilon^m \hat{\gamma}_{m+1}(s,\varepsilon). \tag{2.3}$$

$(U^\varepsilon(x), V^\varepsilon(x))$ is called an *ε-family of matched asymptotic solutions* when it has the following expansion (2.4) (matched asymptotic expansion (MAE) procedure). Roughly speaking, $(U^\varepsilon(x), V^\varepsilon(x))$ is expanded separately in two regions $\Omega_\varepsilon^\pm$ and they are matched smoothly at Γ^ε. More precisely we have

$$\begin{cases} U_\pm^\varepsilon(x) = \displaystyle\sum_{k=0}^{m} u_k^\pm(x)\varepsilon^k + \Phi_m^\pm(x,\varepsilon) + \varepsilon^m R^\pm(x,\varepsilon) \\ V_\pm^\varepsilon(x) = \displaystyle\sum_{k=0}^{m} v_k^\pm(x)\varepsilon^k + \varepsilon^2\Psi_m^\pm(x,\varepsilon) + \varepsilon^m S^\pm(x,\varepsilon) \end{cases} \qquad x \in \Omega_\varepsilon^\pm \tag{2.4}$$

where

$$\Phi_m^\pm(x,\varepsilon) = \begin{cases} \omega\left(\frac{y(x)-\gamma(s,\varepsilon)}{d}\right) \displaystyle\sum_{n=0}^{m} \phi_n^\pm\left(s(x), \frac{y(x)-\gamma(s,\varepsilon)}{\varepsilon}\right)\varepsilon^k, \\ \qquad\qquad |y(x)-\gamma(s,\varepsilon)| \le d, \\ 0, \qquad\qquad |y(x)-\gamma(s,\varepsilon)| > d, \end{cases} \tag{2.5}$$

$$\Psi_m^\pm(x,\varepsilon) = \begin{cases} \omega\left(\frac{y(x)-\gamma(s,\varepsilon)}{d}\right) \displaystyle\sum_{k=0}^{m-2} \psi_k^\pm\left(s(x), \frac{y(x)-\gamma(s,\varepsilon)}{\varepsilon}\right)\varepsilon^k, \\ \qquad\qquad |y(x)-\gamma(s,\varepsilon)| \le d, \\ 0, \qquad\qquad |y(x)-\gamma(s,\varepsilon)| > d, \end{cases} \tag{2.6}$$

$\omega(\tau) \in C^\infty(\mathbf{R})$ is a cut off function such that

$$\omega(\tau) = 1 \text{ for } |\tau| \le \frac{1}{2}, \omega(\tau) = 0 \text{ for } |\tau| \ge 1, 0 \le \omega(\tau) \le 1, \tag{2.7}$$

$d > 0$ is some small constant, and $R(x,\varepsilon)$ and $S(x,\varepsilon)$ are remainders. $\phi_k^\pm$ and $\psi_k^\pm$ are functions of s and ξ, and ξ is stretched variable $\xi \equiv (y - \gamma(s,\varepsilon))/\varepsilon$. The coefficients $u_k^\pm$, $v_k^\pm$, $\phi_k^\pm$, and $\psi_k^\pm$ satisfy some equations and relations. We can obtain these equations by making outer and inner expansions and equating the same powers of ε^k. The resulting relations are, what we call, *matching conditions* between inner and outer expansions and *C^1-matching conditions* between

$(U_+^\varepsilon, V_+^\varepsilon)$ and $(U_-^\varepsilon, V_-^\varepsilon)$. Let $\beta_\varepsilon(s) = v^* + \sum_{k=1}^m \beta_k(s)\varepsilon^k + \varepsilon^m \hat{\beta}_{m+1}(s,\varepsilon)$ be the value of $V^\varepsilon(x)$ on Γ^ε, Ω_ε^+ be the region surrounded by Γ^ε, and $\Omega_\varepsilon^- \equiv \Omega\backslash\bar{\Omega}_\varepsilon^+$.

We display the equations and the relations up to order $O(\varepsilon)$. Higher order terms are given by the same procedures.

$O(\varepsilon^0)$:

$$\begin{cases} u_0^\pm = h_\pm(v_0^\pm) \\ D\Delta v_0^\pm + g(h_\pm(v_0^\pm), v_0^\pm) = 0 \end{cases} \qquad x \in \Omega_0^\pm \tag{2.8}$$

$$v_0^\pm(s,0) = v^*, \quad \frac{\partial v_0^-}{\partial n} = 0 \text{ on } \partial\Omega, \tag{2.9}$$

$$(v_0^+)_y(s,0) = (v_0^-)_y(s,0) \tag{2.10}$$

$$\begin{aligned} &\phi_0^\pm(s,0) = \tfrac{1}{2}(h_+(v^*) + h_-(v^*)) - u_0^\pm(s,0), \quad \phi_0^\pm(s,\mp\infty) = 0 \\ &\psi_0^\pm(s,\mp\infty) = 0 = \dot{\psi}_0^\pm(s,\mp\infty) \end{aligned} \tag{2.11}$$

$$\dot{\phi}_0^+(s,0) = \dot{\phi}_0^-(s,0) \tag{2.12}$$

$$\begin{cases} \ddot{\phi}_0^\pm + f(h_\pm(v^*) + \phi_0^\pm, v^*) = 0 \\ D\ddot{\psi}_0^\pm = g(h_\pm(v^*), v^*) - g(h_\pm(v^*) + \phi_0^\pm, v^*) \end{cases} \qquad \xi \in I^\pm, \quad 0 \le s \le \ell \tag{2.13}$$

$O(\varepsilon^1)$:

$$\begin{cases} f_u^{0\pm} u_1^\pm + f_v^{0\pm} v_1^\pm = 0 \\ D\Delta v_1^\pm + g_u^{0\pm} u_1^\pm + g_v^{0\pm} v_1^\pm = 0 \end{cases} \qquad x \in \Omega_0^\pm \tag{2.14}$$

$$v_1^\pm(s,0) = \beta_1(s) - (v_0^\pm)_y(s,0)\gamma_1(s), \quad \frac{\partial v_1^-}{\partial n} = 0 \text{ on } \partial\Omega, \tag{2.15}$$

$$\begin{aligned} &(v_0^+)_{yy}(s,0)\gamma_1(s) + (v_1^+)_y(s,0) + \dot{\psi}_0^+(s,0) = \\ &\qquad (v_0^-)_{yy}(s,0)\gamma_1(s) + (v_1^-)_y(s,0) + \dot{\psi}_0^-(s,0) \end{aligned} \tag{2.16}$$

$$\begin{cases} \ddot{\phi}_1^\pm + \tilde{f}_u^{0\pm}\phi_1^\pm = -D_1\phi_0^\pm - \tilde{F}^{1\pm} \\ D\ddot{\psi}_1^\pm = -DD_1\phi_0^\pm - E^{1\pm}(s,\xi) - \tilde{G}^{1\pm}, \end{cases} \qquad \xi \in I^\pm, \quad 0 \le s \le \ell \tag{2.17}$$

$$(2.18)\quad \begin{aligned} &\phi_1^{\pm}(s,0) = -(u_0^{\pm})_y(s,0)\gamma_1(s) - u_1^{\pm}(s,0), \quad \phi_1^{\pm}(s,\mp\infty) = 0, \\ &\psi_1^{\pm}(s,\mp\infty) = 0 = \dot{\psi}_1^{\pm}(s,\mp\infty) \end{aligned}$$

$$(2.19)\qquad (u_0^{+})_y(s,0) + \dot{\phi}_1^{+}(s,0) = (u_0^{-})_y(s,0) + \dot{\phi}_1^{-}(s,0),$$

where $\cdot = \frac{\partial}{\partial \xi}$, $I^+ \equiv (-\infty, 0)$, $I^- \equiv (0,\infty)$, ℓ is the total arclength of Γ_0, $f_u^{0\pm} \equiv \frac{\partial}{\partial u} f(u_0^{\pm}, v_0^{\pm})$, $\tilde{f}_u^{0\pm} \equiv \frac{\partial}{\partial u} f(h_{\pm}(v^*) + \phi_0^{\pm}, v^*)$, and D_k $(k = 0, 1, \cdots m)$ are the coefficients of the expansion of Laplacian Δ in the local coordinate system (s,ξ);

$$(2.20)\qquad \Delta_x \equiv \frac{1}{\varepsilon^2}\sum_{k=0}^{m} \varepsilon^k D_k.$$

Here D_k is at most second order differential operator in s and ξ. For example,

$$(2.21)\quad \begin{aligned} &D_0 \equiv \frac{\partial^2}{\partial \xi^2}, \qquad D_1 \equiv -\kappa(s)\frac{\partial}{\partial \xi}, \qquad D_2 \equiv \Lambda_0 - \kappa(s)^2(\xi + \gamma_1(s))\frac{\partial}{\partial \xi}, \\ &\Lambda_0 \equiv \frac{\partial^2}{\partial s^2} - 2\gamma_1' \frac{\partial^2}{\partial \xi \partial s} - \gamma_1'' \frac{\partial}{\partial \xi} + (\gamma_1')^2 \frac{\partial^2}{\partial \xi^2}, \end{aligned}$$

where $' = \frac{\partial}{\partial s}$ and $\kappa(s)$ is curvature of $\Gamma_0(s)$. $\tilde{F}^{k\pm}$, $\tilde{G}^{k\pm}$, and $E^{k\pm}$ are defined by

$$(2.22)\quad \begin{aligned} \tilde{F}^{k\pm}(s,\xi) \equiv \frac{1}{k!}\frac{d^k}{d\varepsilon^k} f\Bigg(&\sum_{i=0}^{m} u_i^{\pm}(s, \varepsilon\xi + \gamma(s,\varepsilon))\varepsilon^i + \sum_{i=0}^{m} \phi_i^{\pm}(s,\xi)\varepsilon^i, \\ &\sum_{i=0}^{m} v_i^{\pm}(s, \varepsilon\xi + \gamma(s,\varepsilon))\varepsilon^i + \varepsilon^2 \sum_{i=0}^{m-2} \psi_i^{\pm}(s,\xi)\varepsilon^i \Bigg)\Bigg|_{\varepsilon=0} \end{aligned}$$

$$(2.23)\quad \begin{aligned} \tilde{G}^{k\pm}(s,\xi) \equiv \frac{1}{k!}\frac{d^k}{d\varepsilon^k} g\Bigg(&\sum_{i=0}^{m} u_i^{\pm}(s, \varepsilon\xi + \gamma(s,\varepsilon))\varepsilon^i + \sum_{i=0}^{m} \phi_i^{\pm}(s,\xi)\varepsilon^i, \\ &\sum_{i=0}^{m} v_i^{\pm}(s, \varepsilon\xi + \gamma(s,\varepsilon))\varepsilon^i + \varepsilon^2 \sum_{i=0}^{m-2} \psi_i^{\pm}(s,\xi)\varepsilon^i \Bigg)\Bigg|_{\varepsilon=0} \end{aligned}$$

$$(2.24)\qquad E^{k\pm}(s,\xi) \equiv \frac{1}{k!}\frac{d^k}{d\varepsilon^k}\left[D\Delta\left(\sum_{i=0}^{m} v_i^{\pm}\varepsilon^i\right)(s, \varepsilon\xi + \gamma(s,\varepsilon))\right]\Bigg|_{\varepsilon=0}$$

Here we present the precise form of $\tilde{F}^{1\pm}$ for later use in Section 3;

$$(2.25)\quad \begin{aligned} \tilde{F}^{1\pm} \equiv &\{\tilde{f}_u^{0\pm}(u_0^{\pm})_y(s,0) + \tilde{f}_v^{0\pm}(v_0^{\pm})_y(s,0)\}(\gamma_1(s) + \xi) \\ &- \tilde{f}_u^{0\pm} u_1^{\pm}(s,0) - \tilde{f}_v^{0\pm} v_1^{\pm}(s,0). \end{aligned}$$

We call the second equation of (2.8) with (2.9) and (2.10) the *reduced problem*, that is

$$(2.26)\qquad \begin{cases} D\Delta v_0^{\pm} + g(h_{\pm}(v_0^{\pm}), v_0^{\pm}) = 0, & \text{in } \Omega_0^{\pm} \\ v_0^{\pm}(s,0) = v^*, \quad \dfrac{\partial v_0^-}{\partial n} = 0 \text{ on } \partial\Omega, \\ (v_0^+)_y(s,0) = (v_0^-)_y(s,0) \end{cases}$$

3. Asymptotic formula for critical eigenvalues of Allen-Cahn operator associated with Activator-Inhibitor systems

In this section, we consider the following eigenvalue problem

$$(3.1)\qquad \begin{cases} L^{\varepsilon} w = \lambda w & \text{in } \Omega, \\ \dfrac{\partial w}{\partial n} = 0 & \text{on } \partial\Omega. \end{cases}$$

where

$$(3.2)\qquad L^{\varepsilon} \equiv \varepsilon^2 \Delta + f_u^{\varepsilon},$$

$f_u^{\varepsilon} = \frac{\partial}{\partial u} f(U^{\varepsilon}, V^{\varepsilon})$, and $(U^{\varepsilon}, V^{\varepsilon})$ is a stationary solution of (1.2) having the expansion as in Section 2. The information on the spectral behavior of (3.1) is basic for the study of the stability of $(U^{\varepsilon}, V^{\varepsilon})$. Especially the behavior of critical eigenvalues (i.e., those which tend to zero as $\varepsilon \downarrow 0$) play the key role to investigate the spectrum of the linearized problem for the full system. However in this short note, we only focus on the asymptotic expansions of eigen-pairs of (3.1) and their matching conditions, which gives us an important result Proposition 3.4 to prove the Main theorem.

Our approach is the matched asymptotic expansion method. We divide (3.1) into two problems as follows;

$$(3.3)_-\qquad \begin{cases} \varepsilon^2 \Delta w^- + f_u^{\varepsilon} w^- = \lambda^{\varepsilon} w^- & \text{in } \Omega_{\varepsilon}^-, \\ \dfrac{\partial w^-}{\partial n} = 0 \quad \text{on } \partial\Omega, \quad w^- = \Theta^{\varepsilon} & \text{on } \Gamma^{\varepsilon}. \end{cases}$$

$$(3.3)_+\qquad \begin{cases} \varepsilon^2 \Delta w^+ + f_u^{\varepsilon} w^+ = \lambda^{\varepsilon} w^+ & \text{in } \Omega_{\varepsilon}^+, \\ w^+ = \Theta^{\varepsilon} \quad \text{on } \Gamma^{\varepsilon}, \end{cases}$$

where

$$(3.4)\qquad \Theta^{\varepsilon} \equiv \sum_{k=0}^{m} \varepsilon^k \Theta_k(s), \qquad \lambda^{\varepsilon} \equiv \sum_{k=1}^{m} \varepsilon^k \lambda_k.$$

$\Omega_\varepsilon^\pm$ and Γ^ε are defined in Section 2. Θ^ε and λ^ε are determined by C^1-matching condition of w_-^ε and w_+^ε on Γ^ε.

OUTER EXPANSION Let

$$w^\pm = \sum_{k=0}^m \varepsilon^k w_k^\pm(x) \quad \text{and} \quad f_u^{\varepsilon\pm} = \sum_{k=0}^m \varepsilon^k F_u^{k\pm} \tag{3.5}$$

where

$$F_u^{k\pm} \equiv \frac{1}{k!}\frac{d^k}{d\varepsilon^k} f_u\left(\sum_{i=0}^m \varepsilon^i u_i^\pm(x), \sum_{i=0}^m \varepsilon^i v_i^\pm(x)\right)\Bigg|_{\varepsilon=0}. \tag{3.6}$$

Substituting (3.5) and (3.6) into $(3.3)_\pm$, we have

$$\sum_{k=2}^m \varepsilon^k \Delta w_{k-2}^\pm + \sum_{k=0}^m \left(\varepsilon^k \sum_{i+j=k} F_u^{i\pm} w_j^\pm\right) = \sum_{k=1}^m \left(\varepsilon^k \sum_{i+j=k, i\geq 1} \lambda_i w_j^\pm\right).$$

Equating like power of ε^k, we obtain the following equations:

$$k=0: \quad F_u^{0\pm} w_0^\pm = 0,$$

$$k=1: \quad F_u^{0\pm} w_1^\pm + F_u^{1\pm} w_0^\pm = \lambda_1 w_0^\pm,$$

$$2 \leq k \leq m: \quad \Delta w_{k-2}^\pm + \sum_{i+j=k} F_u^{i\pm} w_j^\pm = \sum_{i+j=k, i\geq 1} \lambda_i w_j^\pm,$$

By using induction arguments, we see $w_k^\pm \equiv 0 \quad (0 \leq k \leq m)$.

INNER EXPANSION We introduce the stretched variable ξ in the neighbourhood of Γ^ε as in Section 2. Substituting

$$w^\pm = \sum_{k=0}^m \varepsilon^k \zeta_k^\pm(s,\xi) \quad \text{and} \quad f_u^{\varepsilon\pm} = \sum_{k=0}^m \tilde{F}_u^{k\pm}\varepsilon^k$$

into $(3.3)_\pm$

$$\sum_{k=0}^m \left(\varepsilon^k \sum_{i+j=k} D_i \zeta_j^\pm\right) + \sum_{k=0}^m \left(\varepsilon^k \sum_{i+j=k} \tilde{F}_u^{i\pm} \zeta_j^\pm\right) = \sum_{k=1}^m \left(\varepsilon^k \sum_{i+j=k, i\geq 1} \lambda_i \zeta_j^\pm\right),$$

where

$$\tilde{F}_u^{k\pm} \equiv \frac{1}{k!}\frac{d^k}{d\varepsilon^k} f_u\left(\sum_{i=0}^m \varepsilon^i u_i^\pm(s, \varepsilon\xi + \gamma(s,\varepsilon)) + \sum_{i=0}^m \varepsilon^i \phi_i^\pm(s,\xi),\right.$$

$$\left.\sum_{i=0}^m \varepsilon^i v_i^\pm(s, \varepsilon\xi + \gamma(s,\varepsilon)) + \sum_{i=2}^m \varepsilon^i \psi_i^\pm(s,\xi)\right)\Bigg|_{\varepsilon=0}.$$

Equating like powers of ε^k, we have the following problems for $\zeta_k^\pm$.

$$
(3.7)\qquad \begin{cases} \ddot{\zeta}_0^\pm + \tilde{F}_u^{0\pm}\zeta_0^\pm = 0, & \xi \in I^\pm, \quad 0 \le s \le \ell \\ \zeta_0^\pm(s, \pm\infty) = 0, \qquad \zeta_0^\pm(s,0) = \Theta_0(s), & \end{cases}
$$

$$
(3.8)\qquad \begin{cases} \ddot{\zeta}_k^\pm + \tilde{F}_u^0\zeta_k^\pm = \sum_{i=1}^{k} R_i^\pm \zeta_{k-i}^\pm, & \xi \in I^\pm, \quad 0 \le s \le \ell \\ \zeta_k^\pm(s, \pm\infty) = 0, \qquad \zeta_k^\pm(s,0) = \Theta_k(s), & \end{cases}
$$

where

$$
(3.9)\qquad R_i^\pm(s,\xi) \equiv \lambda_i - D_i - \tilde{F}_u^{i\pm}.
$$

By the similar argument in Fife [1], we can see that the inhomogeneous terms of (3.8) and their derivatives of any order with respect to s and ξ decay exponentially as $|\xi| \to \infty$. Then noting that $\dot{\phi}_0^\pm$ (> 0) are fundamental solutions of (3.7) and (3.8), the solutions of them are given by

$$
(3.10)\qquad \zeta_0^\pm = \frac{\dot{\phi}_0(\xi)}{\dot{\phi}_0(0)}\Theta_0(s)
$$

$$
(3.11)\qquad \begin{aligned} \zeta_k^\pm(s,\xi) &= \tfrac{\dot{\phi}_0(\xi)}{\dot{\phi}_0(0)}\Theta_k(s) + \dot{\phi}_0(\xi)\textstyle\int_0^\xi (\dot{\phi}_0(t))^{-2} \\ &\times \textstyle\int_{\mp\infty}^t \left(\sum_{i=1}^k \dot{\phi}_0(\tau) R_i^\pm(s,\tau)\zeta_{k-i}^\pm(s,\tau)\right) d\tau dt, \quad (1 \le k \le m). \end{aligned}
$$

Finally we define the formal approximation to the solution $W_m^\pm(x)$ of $(3.3)_\pm$. We take a sufficiently small $\varepsilon_0 > 0$ and let $\kappa_\varepsilon(s)$ be the curvature of Γ^ε. Define $d > 0$ by

$$
\max_{0 \le s \le \ell, 0 \le \varepsilon \le \varepsilon_0} |\kappa_\varepsilon(s)| \le \frac{1}{2d}
$$

and $W_m^\pm(x)$ on $\Omega_\varepsilon^\pm$ by

$$
(3.12)\qquad W_m^\pm(x) = \begin{cases} \omega\left(\frac{y(x)-\gamma(s;\varepsilon)}{d}\right) \sum_{k=0}^{m} \varepsilon^k \zeta_k^\pm\left(s, \frac{y(x)-\gamma(s;\varepsilon)}{\varepsilon}\right), & |y(x) - \gamma(s;\varepsilon)| \le d, \\ 0, & |y(x) - \gamma(s;\varepsilon)| > d, \end{cases}
$$

LEMMA 3.1. *There exist $K > 0$ independent of ε, such that for all $x \in \Omega^{\pm}$ and sufficiently small ε,*

$$|(\varepsilon^2 \Delta + f_u^\varepsilon - \lambda^\varepsilon)[W_m^\pm]| \leq K\varepsilon^{m+1}$$

holds.

In order to determine Θ^ε and λ^ε, $W_m^\pm$ must satisfy the C^1-matching conditions

$$\varepsilon(W_m^+)_y(s,\gamma(s,\varepsilon)) - \varepsilon(W_m^-)_y(s,\gamma(s,\varepsilon)) = O(\varepsilon^{m+1}), \tag{3.13}$$

which can be rewritten as

(3.14)

$$\varepsilon(W_m^+)_y(s,\gamma(s,\varepsilon)) - \varepsilon(W_m^-)_y(s,\gamma(s,\varepsilon)) = \sum_{k=0}^{m} \varepsilon^k \left\{\dot{\zeta}_k^+(s,0) - \dot{\zeta}_k^-(s,0)\right\} + O(\varepsilon^{m+1}).$$

Noting (2.12), we see that $\dot{\zeta}_0^+(s,0) - \dot{\zeta}_0^-(s,0) = 0$ is already satisfied. For (3.14), we see that

LEMMA 3.2. *Each C^1-matching condition*

$$\dot{\zeta}_k^+(s,0) - \dot{\zeta}_k^-(s,0) = 0 \qquad (1 \leq k \leq m) \tag{3.15}$$

is equivalent to the formal solvability condition for (3.8);

$$(MC)_k \qquad \left\langle \sum_{i=1}^{k} \hat{R}_i(s,\xi)\hat{\zeta}_{k-i}, \phi_0(\xi) \right\rangle_\xi = 0,$$

where $\hat{R}_i(s,\xi)$ and $\hat{\zeta}_i(s,\xi)$ are

$$\hat{R}_i(s,\xi) = \begin{cases} R_i^+(s,\xi) & \xi \in (-\infty,0) \\ R_i^-(s,\xi) & \xi \in [0,\infty) \end{cases}$$

and

$$\hat{\zeta}_i(s,\xi) = \begin{cases} \zeta_i^+(s,\xi) & \xi \in (-\infty,0) \\ \zeta_i^-(s,\xi) & \xi \in [0,\infty) \end{cases}$$

respectively, and $\langle \cdot,\cdot \rangle_\xi$ denotes L^2-inner product with respect to ξ.

PROOF. Differentiating (3.11) with respect to ξ at $\xi = 0$, we have

$$\dot{\zeta}_k^\pm(0,\xi) = \frac{\ddot{\phi}_0(\xi)}{\dot{\phi}_0(0)}\Theta_k(s) + \frac{1}{\dot{\phi}_0(0)} \sum_{i=1}^{k} \int_{\mp\infty}^{0} \dot{\phi}_0(\tau) R_i^\pm(s,\tau)\zeta_{k-i}^\pm(s,\tau)d\tau dt.$$

Then,

$$0 = \dot{\zeta}_k^+(s,0) - \dot{\zeta}_k^-(s,0) = \tfrac{1}{\dot{\phi}_0(0)}\left[\int_{-\infty}^0 \sum_{i=1}^k \dot{\phi}_0(\tau) R_i^+(s,\tau)\zeta_{k-i}^+(s,\tau)d\tau dt\right.$$

$$\left. - \int_{+\infty}^0 \sum_{i=1}^k \dot{\phi}_0(\tau) R_i^-(s,\tau)\zeta_{k-i}^-(s,\tau)d\tau dt\right] = \left\langle \sum_{i=1}^k \hat{R}_i(s,\xi)\hat{\zeta}_{k-i}, \dot{\phi}_0(\xi)\right\rangle_\xi .$$

In the following of this section, we study how Θ^ε and λ^ε are determined. When $k = 1$, we have the following result.

LEMMA 3.3. *λ_1 is determined by $(MC)_1$ and is given by*

$$\lambda_1 = (v_0)_y(s,0)\int_{h_-(\beta^*)}^{h_+(\beta^*)} f_v(u,\beta^*)du \Big/ \int_{-\infty}^{+\infty} (\dot{\phi}_0(t))^2 dt \tag{3.16}$$

PROOF. $\zeta_1^\pm$ satisfies

$$\ddot{\zeta}_1^\pm + \tilde{F}_u^{0\pm}\zeta_1^\pm = \frac{\Theta_0}{\dot{\phi}_0(0)}[\lambda_1\dot{\phi}_0^\pm + H^\pm], \qquad H^\pm \equiv \kappa\ddot{\phi}_0 - \tilde{F}_u^{1\pm}\dot{\phi}_0$$

and $(MC)_1$ is rewritten as

$$\dot{\zeta}_1^+(s,0) - \dot{\zeta}_1^-(s,0) = \frac{\Theta_0(s)}{(\dot{\phi}_0(0))^2}\left[\lambda_1 \int_{-\infty}^{+\infty} (\dot{\phi}_0(t))^2 dt\right.$$

$$\left. + \int_{-\infty}^0 H^+\dot{\phi}_0(t)dt + \int_0^{+\infty} H^-\dot{\phi}_0(t)dt\right] = 0.$$

We see that $r^\pm \equiv \dot{\phi}_1$ satisfy the next equation;

$$\ddot{r}^\pm + \tilde{F}_u^{0\pm} r^\pm = \Omega^\pm$$

$$\Omega^\pm(s,\xi) \equiv H^\pm(s,\xi) - \{(u_0^\pm)_y(s,0)\tilde{f}_u^{0\pm} + (v_0^\pm)_y(s,0)\tilde{f}_v^{0\pm}\}.$$

This can be solved as

$$r^\pm(s,\xi) = \frac{r^\pm(0)}{\dot{\phi}_0(0)}\dot{\phi}_0^\pm(\xi) + \dot{\phi}_0(\xi)\int_0^\xi (\dot{\phi}_0(t))^{-2}\int_{\mp\infty}^t \Omega^\pm(s,\tau)\dot{\phi}_0(\tau)d\tau dt. \tag{3.17}$$

Differentiating (3.17) with respect to ξ at $\xi = 0$, we have

$$\dot{r}^\pm(s,0) = \frac{r^\pm(0)}{\dot{\phi}_0(0)}\ddot{\phi}_0^\pm(0) + \frac{1}{\dot{\phi}_0(0)}\int_{\mp\infty}^0 \Omega^\pm(s,\tau)\dot{\phi}_0(\tau)d\tau. \tag{3.18}$$

Using

$$\int_{\mp\infty}^{0} \tilde{f}_u^{0\pm}\dot{\phi}_0(\tau)d\tau = \ddot{\phi}_0^{\pm}(0), \qquad \int_{\mp\infty}^{0} \tilde{f}_v^{0\pm}\dot{\phi}_0(\tau)d\tau = \int_{h_\pm(\beta_0^*)}^{\alpha} f_v(u,\beta_0^*)du$$

and after some computation, (3.18) can be rewritten as

$$\int_{\mp\infty}^{0} H^{\pm}(s,\tau)\dot{\phi}_0(\tau)d\tau = \ddot{\phi}_1^{\pm}(s,0)\dot{\phi}_0(0) - \ddot{\phi}_0(0)\{\dot{\phi}_1^{\pm}(s,0) + (u_0^{\pm})_y(s,0)\}$$

$$+(v_0^{\pm})_y(s,0)\int_{h_\pm(\beta_0^*)}^{\alpha} f_v(u,\beta_0^*)du,$$

where $\alpha = (h_+(\beta_0^*) + h_-(\beta_0^*))/2$. Moreover, noting that $\ddot{\phi}_1^-(s,0) = \ddot{\phi}_1^+(s,0)$, $v_0^{\pm}$ are C^1-matched on Γ_0 (see (2.10)) (so we omit the superscript $\pm$ of $(v_0)_y$) and (2.19), we have

$$\dot{\zeta}_1^+(s,0)-\dot{\zeta}_1^-(s,0) = \frac{\Theta_0(s)}{(\dot{\phi}_0(0))^2}\left[\lambda_1\int_{-\infty}^{+\infty}(\dot{\phi}_0(t))^2dt + (v_0)_y(s,0)\int_{h_+(\beta_0^*)}^{h_-(\beta_0^*)} f_v(u,\beta_0^*)du\right],$$

which implies (3.16).

Noting that the inhomogeneous terms of the equation ζ_1^+ and ζ_1^- are continuous at $\xi = 0$, once $\lambda_1$is determined, we can regard $\hat{\zeta}_1$ as C^2-solution on $\mathbf{R}$. So we omit ^ and write ζ_1 as

$$\zeta_1(s,\xi) = \frac{\dot{\phi}_0(\xi)}{\dot{\phi}_0(0)}\Theta_1(s) + \frac{\dot{\phi}_0(\xi)}{\dot{\phi}_0(0)}\left[\int_0^{\xi}(\dot{\phi}_0(t))^{-2}\int_{-\infty}^{t}\dot{\phi}_0(\tau)(R_1(s,\tau)\dot{\phi}_0(\tau))d\tau dt\right]\Theta_0(s)$$

As a corollary of Lemma 3.3, we have an important consequence.

PROPOSITION 3.4.

$$(v_0)_y \equiv \mathit{constant} \quad \mathit{on}\ \Gamma_0$$

Next, for $(MC)_2$ we have

LEMMA 3.5. *$(MC)_2$ is equivalent to the following problem;*

$$(3.19)\qquad \begin{cases} \dfrac{d^2}{ds^2}\Theta_0 + P(s)\Theta_0 = \lambda_2\Theta_0, & s \in (0,\ell), \\[2ex] \Theta_0(0) = \Theta_0(\ell), & \dfrac{d\Theta_0}{ds}(0) = \dfrac{d\Theta_0}{ds}(\ell). \end{cases}$$

where, $P(s)$ is a smooth bounded function of s.

PROOF. First note that $(MC)_2$ is rewritten as

$$(3.20)\qquad 0 = \langle R_1\zeta_1 + R_2\zeta_0, \dot{\phi}_0\rangle_\xi = \langle R_1\zeta_1, \dot{\phi}_0\rangle_\xi + \langle R_2\zeta_0, \dot{\phi}_0\rangle_\xi$$

Substituting the representation of ζ_1 into the first term of (3.20), we have

$$\langle R_1\zeta_1,\dot\phi_0\rangle_\xi = \frac{1}{\dot\phi_0(0)}\Big[\Theta_1\langle R_1\dot\phi_0,\dot\phi_0\rangle_\xi + \Theta_0\int_{-\infty}^{+\infty}\dot\phi_0(\xi)(R_1(s,\xi)\dot\phi_0(\xi))$$

$$\times\int_0^\xi(\dot\phi_0(\xi))^{-2}\int_{-\infty}^{t}(R_1(s,\tau)\dot\phi_0(\tau))d\tau dt d\xi\Big] = \mathcal{G}(\Theta_0),$$

where

$$\mathcal{G}(\Theta_0) \equiv \frac{\Theta_0}{\dot\phi_0(0)}\int_{-\infty}^{+\infty}\dot\phi_0(\xi)(R_1(s,\xi)\dot\phi_0(\xi))\int_0^\xi(\dot\phi_0(t))^{-2}$$

$$\times\int_{-\infty}^{t}\dot\phi_0(\xi)(R_1(s,\tau)\dot\phi_0(\tau))d\tau dt d\xi.$$

On the other hand, recalling (2.21), (3.9) and (3.10), the second term of (3.20) is computed as follows

$$\begin{aligned}\langle R_2\zeta_0,\dot\phi_0\rangle_\xi &= \frac{\Theta_0}{\dot\phi_0(0)}\int_{-\infty}^{+\infty}(\lambda_2-\tilde F_u^2)(\dot\phi_0(\xi))^2d\xi - \frac{1}{\dot\phi_0(0)}\int_{-\infty}^{+\infty}\Big[\dot\phi_0(\xi)\partial_{ss}\Theta_0\\ &\quad -2\gamma_1'\ddot\phi_0(\xi)\partial_s\Theta_0 - \gamma_1''\dot\phi_0(\xi)\Theta_0 + (\gamma_1')^2(\phi_0)_{\xi\xi\xi}(\xi)\Theta_0 - \kappa^2\xi\ddot\phi_0(\xi)\Theta_0\\ &\quad -\kappa^2\gamma_1\ddot\phi_0(\xi)\Theta_0\Big]\dot\phi_0(\xi)d\xi\\ &= \frac{1}{\dot\phi_0(0)}\Big[\lambda_2\Theta_0\int_{-\infty}^{+\infty}(\dot\phi_0(\xi))^2d\xi - \partial_{ss}\Theta_0\int_{-\infty}^{+\infty}(\dot\phi_0(\xi))^2d\xi\\ &\quad -\Theta_0\Big\{\int_{-\infty}^{+\infty}\tilde F_u^2(\dot\phi_0(\xi))^2d\xi - (\gamma_1')^2\int_{-\infty}^{+\infty}(\ddot\phi_0(\xi))^2d\xi\\ &\quad +(\kappa(s))^2\int_{-\infty}^{+\infty}\xi\dot\phi_0(\xi)\ddot\phi_0(\xi)d\xi\Big\}\Big] \equiv \mathcal{F}(\Theta_0).\end{aligned}$$

Hence $(MC)_2$ is equivalent to

$$\mathcal{F}(\Theta_0)+\mathcal{G}(\Theta_0)=0,$$

and this yields (3.19).

REMARK 3.6. *(3.19) is a Sturm-Liouville eigenvalue problem with periodic boundary condition, and the existence and asymptotic behavior of the eigenvalues and their eigenfunctions are well-studied. We denote those eigenpairs by* $\{(\hat\lambda_2^{(n)},\hat\Theta_0^{(n)})\}_{n=1}^\infty$.

By Proposition 3.5, λ_2 and Θ_0 are determined. Generally λ_k and Θ_{k-2} ($k\ge 3$) are determined by $(MC)_k$, that is

PROPOSITION 3.7. *For each $n\ge 1$, $(MC)_k$ ($k\ge 3$) is equivalent to the following problem*

$$\frac{d^2}{ds^2}\Theta_{k-2}+P(s)\Theta_{k-2}-\lambda_2^{(n)}\Theta_{k-2}=\Psi_k(s),\qquad 0\le s\le \ell \tag{3.21}$$

with Periodic boundary condition, where $P(s)$ is the same one appeared in Proposition 3.5 and $\Psi_k(s)$ is a smooth bounded function of s depending on $\lambda_k, \lambda_{k-1}, \cdots, \hat{\lambda}_2^{(n)}, \lambda_1, \Theta_{k-3}, \cdots, \hat{\Theta}_0^{(n)}$, and s. Then λ_k and Θ_{k-2} are determined once $\lambda_{k-1}, \cdots, \lambda_1$ and $\Theta_{k-3}, \cdots, \hat{\Theta}_0^{(n)}$ are known. In fact, λ_k is uniquely determined by the solvability condition

$$\langle \Psi_k, \hat{\Theta}_0^{(n)} \rangle_s = 0, \tag{3.22}$$

where $\langle \cdot, \cdot \rangle_s$ denotes L^2-inner product with respect to s, and then, Θ_{k-2} is also uniquely determined by (3.21) with $\langle \Theta_{k-2}, \hat{\Theta}_0^{(n)} \rangle_s = 0$.

PROOF. We prove by using induction arguments. Assume that λ_j and Θ_{j-2} $(2 \leq j \leq k)$ are known, and we determine λ_{k+1} and Θ_{k-1}. $(MC)_{k+1}$ is rewritten as

$$\begin{aligned} 0 &= \left\langle \sum_{i=1}^{k+1} R_i(s,\xi)\zeta_{k+1-i}, \dot{\phi}_0(\xi) \right\rangle_\xi \\ &= \left\langle R_1\zeta_k, \dot{\phi}_0 \right\rangle_\xi + \left\langle R_2\zeta_{k-1}, \dot{\phi}_0 \right\rangle_\xi + \left\langle \sum_{i=3}^{k+1} R_i\zeta_{k+1-i}, \dot{\phi}_0 \right\rangle_\xi . \end{aligned} \tag{3.23}$$

Then note that the third term of (3.23) does not depend on Θ_k and Θ_{k-1}, and λ_{k+1} is only involved in the third term and the coefficient of λ_{k+1} is given by $\dfrac{\dot{\phi}_0(\xi)}{\dot{\phi}_0(0)}\Theta_0(s)$. Noting that the coefficients of Θ_k and Θ_{k-1} in the representation of ζ_k are $\dfrac{\dot{\phi}_0(\xi)}{\dot{\phi}_0(0)}$ and $\dfrac{\dot{\phi}_0(\xi)}{\dot{\phi}_0(0)}\displaystyle\int_0^\xi (\dot{\phi}_0(t))^{-2}\int_{-\infty}^t \dot{\phi}_0(\tau)(R_1(s,\tau)\dot{\phi}_0(\tau))d\tau dt$, respectively, the first term of (3.23) is computed as follows:

$$\begin{aligned} \langle R_1\zeta_k, \dot{\phi}_0\rangle_\xi &= \frac{1}{\dot{\phi}_0(0)}\Big[\Theta_k\langle R_1\dot{\phi}_0, \dot{\phi}_0\rangle_\xi + \Theta_{k-1}\int_{-\infty}^{+\infty}(\dot{\phi}_0(\xi))^2\int_0^\xi (\dot{\phi}_0(t))^{-2} \\ &\times \textstyle\int_{-\infty}^t \dot{\phi}_0(\tau)(R_1(s,\tau)\dot{\phi}_0(\tau))d\tau dt d\xi\Big] + \Sigma_{k-1}(s) = \mathcal{G}(\Theta_{k-1}) + \Sigma_{k-1}(s). \end{aligned} \tag{3.24}$$

Here, $\mathcal{G}(\cdot)$ is the same one defined in Proposition 3.5 and $\Sigma_{k-1}(s)$ is a known function depending on λ_j $(1 \leq j \leq k)$, Θ_i $(0 \leq i \leq k-2)$, its derivatives, and so on. On the other hand, noting that ζ_{k-1} is can be represented by

$$\zeta_{k-1}(s,\xi) = \frac{\dot{\phi}_0(\xi)}{\dot{\phi}_0(0)}\Theta_{k-1}(s) + Z_{k-1}(s,\xi),$$

($Z_{k-1}(s,\xi)$ is a known function independent of Θ_{k-1} and λ_{k+1}) we have

$$\left\langle R_2\zeta_{k-1}, \dot{\phi}_0 \right\rangle_\xi = \mathcal{F}(\Theta_{k-1}) + \left\langle R_2 Z_{k-1}, \dot{\phi}_0 \right\rangle_\xi . \tag{3.25}$$

Using (3.24) and (3.25), we see that (3.23) is equivalent to (3.21). The remainder of the statement is obvious from Riesz-Schauder theory.

4 Over-determined reduced elliptic problem and the proof of Main Theorem

Formal asymptotic analysis in the previous sections tells us that the normal derivative of the C^1-matched solution v_0 at Γ_0 is constant along the interface (Prop. 3.4). On the other hand, v_0 satisfies the reduced problem (2.26) subject $v_0 = v^*$ on Γ_0. This is apparently an *over-determined* problem and only special Γ_0 and v_0 are allowable. In fact, the first part of the Main Theorem in Section 1 is clear from Theorem 2 of J. Serrin [11]. To show the second part, the following hypothesis is necessary.

HYPOTHESIS 4.1. *The nonlinear elliptic problem*

$$D\Delta v_0 + g(h_-(v_0), v_0) = 0 \qquad \text{in } \mathbf{R}^2 \backslash \Omega_0^+,$$

$$v_0 = v^* \quad \text{and} \quad (v_0)_y \equiv \text{constant} \qquad \text{on } \Gamma_0 = \partial\Omega_0^+$$

has only axisymmetric solutions.

Under this hypothesis it is clear that the Neumann boundary condition on $\partial\Omega$ of the reduced problem (2.26) cannot be satisfied for generic domains. This completes the proof of the second part of the Main Theorem.

References

[1] P.C.Fife, Semilinear elliptic boundary value problems with small parameters, Arch. Rational Mech. Anal., **52** (1973), 205-232.

[2] H.Hasegawa, H.Tanaka, K.Yamasaki, and T.Hashimoto, Macromolecules, **20**, 1651 (1987).

[3] T.Hashimoto, H.Tanaka, H.Hasegawa, (M.Nagasawa Ed.) *Molecular Conformation and Dynamics of Macromolecules in Condensed Systems*, Elsevier (1988).

[4] D.Hilhorst, Y.Nishiura, and M.Mimura, A free boundary problem arising in some reacting-diffusing system, Proc. Roy. Soc. Edinburgh, **118A** (1991), 355-378.

[5] Y.Nishiura, Global structure of bifurcating solutions of some reaction-diffusion systems, SIAM J. Math. Anal., **13** (1982), 555-593.

[6] Y.Nishiura, Coexistence of infinitely many stable solutions to reaction diffusion systems in the singular limit, Dynamics reported, Springer-Verlag, in press.

[7] Y.Nishiura and H.Fujii, Stability of singularly perturbed solutions to systems of reaction diffusion equations, SIAM J. Math. Anal., **18** (1987), 1726-1770.

[8] Y.Nishiura and H.Suzuki, in preparation.

[9] T.Ohta and K.Kawasaki, Macromolecules, **19**, 261 (1986).

[10] T.Ohta, M.Mimura, and R.Kobayashi, Higher-dimensional localized pattern in excitable media, Physics **34** D (1989), 115-144.

[11] J.Serrin A symmetry problem in potential theory, Arch. Rational Mech. Anal., **43** (1971), 304-318.

[12] M.Taniguchi and Y.Nishiura, Instability of planar interfaces in reaction diffusion systems, SIAM J. Math. Anal., in press.

[13] E.L.Thomas, D.B.Alward, D.J.Kinning, D.C.Martin, D.L.Handlin, and Jr.,L.J.Fetters, Macromolecules, **19**, 2197 (1986).

†Division of Mathematics and Informatics, Faculty of Integrated Arts and Sciences, Hiroshima University, Higashi-Hiroshima 724, Japan

††Department of Mathematics, Faculty of Science, Hiroshima University, Higashi-Hiroshima 724, Japan

AMS/IP Studies in Advanced Mathematics
Volume **3**, 1997

A symplectic difference scheme for the PDEs*

Qin Mengzhao

Abstract. Feng Kang Introduced in [1] the concept of symplectic scheme of the Hamiltonian equation, and used the generating function Method to construct the symplectic scheme with arbitrarily order of accuracy in the finite dimensional case. The author in paper [3] use the method in the infinite dimensional case following [1, 2], that is using generating function methods to construct the difference schemes of arbitrary order of accuracy for PDEs which can be written as Hamiltonian system in the Banach space.

§1 Preliminaries

In the 19th century Hamilton proposed another form by using momentum $p = M\dot{q}$ instead of velocity. Then $T = \frac{1}{2}(p, M^{-1}p)$, total energy $H = T + V$ is represented by a function of p and q, namely,

$$H(p,q) = T(p) + V(q) = \frac{1}{2}(p, M^{-1}p) + V(q), \tag{0}$$

and furthermore for general $H(p,q)$ not only separable form (0), we have

$$\begin{cases} \dot{p} = -\dfrac{\partial H}{\partial q}, \\ \dot{q} = \dfrac{\partial H}{\partial p}, \end{cases} \tag{1}$$

which is called the Hamiltonian system in canonical form. It is seen that (1) is a first order differential equation system with variable $(p_1, \cdots, p_n, q_1, \cdots, q_n)$ in $2n$-dimensional phase space or symplectic space and it is in a form with remarkable simplicity and symmetry property.

A basic but unwritten rule for the research on numerical methods is that the properties of the original problem should be preserved as much as possible under discretization. According to the above guideline, numerical methods for Hamiltonian systems which serve the essential properties of the system are proposed. So, Hamiltonian algorithm should be originated from within the work of symplectic geometry. This is our basic approach for research.

---

*) The Project is Supported by National Natural Science Foundation of China

We illustrate concisely symplectic geometry by analogy with Euclidean geometry. An Euclidean structure in Euclidean space R^n is determined by a bilinear, symmetric and nondegenerate inner product

$$\langle x, y\rangle = \langle x, Iy\rangle, \quad I = I_n \quad \text{is identity.} \tag{2}$$

When $x \neq 0, \langle x, x\rangle$ is positive, and so, we can define the length $\|x\| = \sqrt{\langle x, x\rangle} > 0$. Linear operators which keep inner product (i.e. length) or satisfy $A'A = I$ form a group $O(n)$, i.e. orthogonal group. It is a classical Lie group and its Lie algebra $o(n)$ consists of anti-symmetric operators, i.e. $A' + A = A'I + IA = 0$, or of infinitesimal orthogonal operators.

Symplectic geometry is the geometry of phase space; a symplectic structure in symplectic space, i.e. phase space R^{2n} is determined by bilinear, anti-symmetric and non-degenerate inner product-symplectic inner product:

$$[x, y] = \langle x, Jy\rangle, \quad J = J_{2n} = \begin{pmatrix} 0 & +I_n \\ -I_n & 0 \end{pmatrix}. \tag{3}$$

When $n = 1$,

$$[x, y] = \begin{vmatrix} x_1 & y_1 \\ x_2 & y_2 \end{vmatrix} \tag{4}$$

which is the area of the parallelogram formed by vectors x, y. In general, symplectic inner product is area measure. Since symplectic inner product is anti-symmetric, $[x, x] = 0$ for any x, and so it s impossible to define length by symplectic structure. This distinguishes symplectic geometry from Euclidean geometry. Linear operators which keep symplectic inner product, i.e. $A'JA = J$, from a group $Sp(2n)$, i.e. symplectic group. It is also a classical Lie group, and its Lie algebra consists of infinitesimal symplectic operators B, satisfying $B'J + JB = 0$, which is denoted by $Sp(2n)$. In odd dimensions anti-symmetric matrix always degenerates, and thus, the dimensions of symplectic space should be even, so should the dimensions of phase space.

In symplectic geometry, a symplectic operator (or, in classical language, canonical transformation) is a diffeomeorphism g of R^{2n} onto itself with jacobi matrix g_z everywhere symplectic. Naturally, this transformation plays an important role in symplectic geometry. Let us come back to the Hamiltonian dynamics. If a pair of n-dimensional vectors are presented as a $2n$-dimensional vector $z = (p, q)$, then canonical Hamiltonian Eq. (1) can be written as

$$\frac{dz}{dt} = J^{-1}\nabla H(z), \tag{5}$$

and symplectic transformation keeps the canonical form of Hamiltonian Eq. (5). The fundamental theorem on Hamiltonian formalism says that the solution $z(t)$ of the canonical system (5) of energy H with initial value $z(t_0)$ can be generated by a one-parameter group $g^{t,t_0} = g_H^{t,t_0}$, called the phase flow of H, of smylectic operators of R^{2n} (locally in t) such that

$$z(t) = g_H^{t,t_0} z(0),$$

i.e. g_H^{t,t_0} transforms the state at time t_0 to the state at time t. Therefore the evolutions of Hamiltonian dynamical systems are exclusively canonical transformations. This is the fundamental property of Hamiltonian formalism. When the system H is time-independent then $g_H^{t,t_0} = g_H^{t-t_0}$, i.e. the phase flow depends only on the difference of parameters $t - t_0$. For simplicity, we denote it by $g^t = g_H^{t,0}$.

Suppose B is a reflexive Banach space and B^* its dual. E^n is an Euclidean space and n its dimension. The generalized coordinate in the Banach space is function $q(r, t) : E^n \times R \rightarrow$

$R, \forall t \in R$. We have $q(r,t) \in R$. B corresponds to the configuration space. We introduce $p(r,t) \in B^*$, generalized momentum space. $B \times B^*$ is the phase space.

Let H be a functional in Hamiltonian mechanics. We have the Hamiltonian equation in $B \times B^*$:

$$\begin{aligned} \frac{dp}{dt} &= -\frac{\delta H}{\delta q}(q,p,t), \\ \frac{dq}{dt} &= \frac{\delta H}{\delta p}(q,p,t), \end{aligned} \tag{6}$$

where $\frac{\delta}{\delta q}, \frac{\delta}{\delta p}$ are functional derivatives.

We denote $z = (q,p) \in B \times B^*$, then we get formally the form of (6)

$$\frac{dz}{dt} = \begin{bmatrix} 0 & I \\ -I & 0 \end{bmatrix} H_z.$$

We define the operator J as

$$J: \quad B \times B^* \to (B \times B^*)^* = B^* \times B$$

$$J(q,p) = (-p,q)$$

then it is easy to prove that $J \in GL((B \times B^*), (B \times B^*)^*)$.

$$J = \begin{bmatrix} 0 & -I_{B^*} \\ I_B & 0 \end{bmatrix} \qquad J^{-1} = \begin{bmatrix} 0 & I_B \\ -I_{B^*} & 0 \end{bmatrix},$$

$$I_B : B \to B \text{ identity operator,}$$

$$I_{B^*} : B^* \to B^* \text{ identity operator,}$$

we write them briefly as I.

Then the H-system has the form

$$\frac{dz}{dt} = J^{-1} H_z, \tag{7}$$

where $H_z = (\frac{\delta H}{\delta q}, \frac{\delta H}{\delta p})$, with canonical 2-form:

$$\omega(\alpha_1, \alpha_2) = \langle J\alpha_1, \alpha_2 \rangle$$

where $\alpha_1, \alpha_2 \in B \times B^*$, and $\langle , \rangle$ is the dual product.

§2 Generating Functional and Difference Schemes

Suppose $f : B \times B^* \to (B \times B^*)^*$

$$\widehat{w} = f(w)$$

We call f a gradient transformation, or potential operator, if there is $\phi : B \times B^* \to R$ such that:

$$\widehat{w} = \phi_w(w)$$

Let $g : B \times B^* \to B \times B^*$ be a canonical transformation

$$z \to \widehat{z} = g(z)$$

$T \in GL((B \times B^*) \times (B \times B^*), (B \times B^*)^* \times (B \times B^*))$

$$T = \begin{bmatrix} A & B \\ C & D \end{bmatrix}$$

in which $A, B \in GL(B \times B^*, (B \times B^*)^*), C, D \in GL(B \times B^*, B \times B^*)$.

Lemma (Vainberg) 1. If the following conditions are satisfied:

1. F is an operator: $E \to E^*, E$ is a reflexive Banach space.

2. F has Gateaux differential $DF(x,h)$ in U which belongs to E and is defined by $|x - x_0| < r$.

3. Functional $\langle DF(x,h_1), h_2\rangle$ is continuous in U. Then F is a potential operator iff $\langle DF(x,h_1,h_2)\rangle$ is symmetric for $\forall x \in U$:

$$\langle DF(x,h_1),h_2\rangle = \langle DF(x,h_2),h_1\rangle, \forall h_1, h_2 \in E.$$

Theorem 2: T carries every symplectic transformation into gradient transformation, if:

$$T^* \begin{bmatrix} 0 & I \\ -I & 0 \end{bmatrix} T = \mu \begin{bmatrix} -J & 0 \\ 0 & J \end{bmatrix}$$

$$\begin{aligned} &\text{i.e.} \quad A_1 = -\mu^{-1} J C^*, \quad B_1 = \mu^{-1} J A^*, \\ &C_1 = \mu^{-1} J D^*, \quad D_1 = -\mu^{-1} J B^*, \end{aligned} \tag{8}$$

$$T = \begin{bmatrix} A & B \\ C & D \end{bmatrix}, \quad T^{-1} = \begin{bmatrix} A_1 & B_1 \\ C_1 & D_1 \end{bmatrix}.$$

With the aid of th above lemma, like the finite dimensional case, we have

Lemma 3: Let T be defined as above. $z \to \widehat{z} = g(z,t)$ is a canonical transformation. $z(t)$ is the solution of (7).

$$M(z,t) = g_z(z,t).$$

1) $g(*,0)$ is a linear canonical transformation,

$$M(z,0) = M_0 \text{ is independent of } z.$$

2) $g^{-1}(z,0)g(*,t)$ is a canonical transformation dependent on t, and carries $z(t)$ into $z(0)$.

3) $CM_0 + D$ is nonsingular,

then there is a gradient transformation dependent on $t : w \to \widehat{w} = f(w,t)$ and a generating functional $\phi(w,t)$ for sufficiently small $|t|$, such that:

(1) $[\widehat{w} - f(w,t)]_{\widehat{w}=A\widehat{z}+Bz, w=C\widehat{z}+Dz} = 0$

(2) $\phi_t(w,t) = -\mu H(C_1\phi_w(w,t) + D_1 w) \mid_{w=C\widehat{z}+Dz}$ (9)

Like the finite dimensional case, we can determine by recursion all possible time-dependent generating functional for Hamiltonians $H(z)$ analytic in z.

$$
\begin{aligned}
&\phi(w,t) = \sum_{k=0}^{\infty} \phi^{(k)}(w),\\
&\phi^{(0)}(w,t) = \frac{1}{2}\langle w, N_0 \cdot w\rangle,\\
&N_0 = (AM_0 + B)(CM_0 + D)^{-1},\\
&\phi^{(1)}(w) = -\mu H(E_0 w), \quad E_0 = (CM_0 + D)^{-1},\\
k \geq 1: \quad &\phi^{(k+1)}(w)\\
= \quad &-\frac{1}{k+1}\sum_{m=1} \frac{\mu}{m!} \sum_{k_1+k_2+\cdots+k_m=k} H^{(m)}(E_0 w)\\
&\times (C_1 \phi_w^{(k_1)}) \cdots (C_1 \phi_w^{(k_m)}(w)).
\end{aligned} \tag{10}
$$

Remark . All conclusion concerning construction of S-scheme via generating functional is true when the J-symplectic form is replaced by K-symplectic form in finite or infinite dimensional case.

Remark. When we take $B = R^n, q = (q_1, \cdots, q_n) \in R^n, p = (p_1, \cdots, p_n) \in (R^n)^* = R^n$, equations (6) become the usual canonical Hamiltonian equations. All the conclusion of the finite dimensional case can be regarded as a special case of infinite dimensional case.

Remark. Using above method, we get a semi-discrete S-scheme (involving time discretization only), in other words, we obtained an ∞-dim symplectic transformation in space $B \times B^*$. In order to obtain a fully discrete S-scheme in $R^n \times R^n$, we must approximate spatial derivative by central difference.

Another way is a method of lines: discretizing spatial derivatives, then we get a system of hamiltonian ODE's. For fully discrete S-scheme, we use methods of constructing S-scheme for finite dimensional H-system [1].

Wave equation. Let

$$B = H^1, \quad P = H^1 \times H^{-1}.$$

In $H^2 \times H^1$, a subspace of P, we have the Hamilton functional:

$$
\begin{aligned}
H(\phi, \dot{\phi}) = &\int_{R^2} \frac{1}{2}\{\dot{\phi}^2 + (\nabla\phi)^2 + m^2\phi^2\} + F\{\phi\} dx,\\
dH = \ & i_{X_H}\omega,\\
X_H(\phi, \dot{\phi}) \equiv \ & (\dot{\phi}, \Delta\phi - m^2\phi - F'(\phi)).
\end{aligned}
$$

So, setting $u = \phi$ and $v = \dot{\phi}$ we have

$$\frac{dv}{dt} = -\frac{\delta H}{\delta u},$$

$$\frac{du}{dt} = \frac{\delta H}{\delta v}.$$

For B being the reflexive Banach space, we choose

$$T=\begin{pmatrix} J & -J \\ \frac{1}{2}I & \frac{1}{2}I \end{pmatrix}$$

briefly for $m=0$ and F=constant. We construct a difference scheme which is of fourth order accuracy.

$$\begin{aligned} Z^{n+1} &= Z^n+\tau JH_z\left(\frac{Z^{n+1}+Z^n}{2}\right) \\ &\quad -\frac{2\tau^3}{3!4}J^{-1}H_{zz}J^{-1}H_{zz}J^{-1}H_z\left(\frac{Z^{n+1}+Z^n}{2}\right), \end{aligned}$$

For compressible flow.

$$H=C\rho^r+(\sigma\nabla\mu+\rho\nabla\phi)^2/2\rho,$$

where $(\rho,\phi)(\sigma,\mu)$ is conjugate variable,

$$\sigma=-\rho\lambda.$$

For flow field: $u=\lambda\nabla\mu-\nabla\phi, \lambda, \mu$ are vortex labels. The equation has the form

$$\begin{aligned} \frac{\partial\rho}{\partial t} &= -\frac{\partial H}{\partial\phi}, \\ \frac{\partial\phi}{\partial t} &= \frac{\partial H}{\partial\rho}, \\ \frac{\partial\mu}{\partial t} &= \frac{\partial H}{\partial\sigma}, \\ \frac{\partial\sigma}{\partial t} &= -\frac{\partial H}{\partial\mu}. \end{aligned}$$

Denote $p=(\rho,\sigma), q=(\phi,\mu)$, We have $\frac{\partial}{\partial t}\begin{pmatrix} p \\ q \end{pmatrix}=J^{-1}\begin{pmatrix} \frac{\partial H}{\partial p} \\ \frac{\partial H}{\partial q} \end{pmatrix}$.

We choose the generating functional of first kind with second-order accuracy. Then

$$\begin{aligned} p^{n+1} &= p^n-\tau H_q(p^n,q^{n+1/2}). \\ q^{n+1+1/2} &= q^{n+1/2}+\tau H_p(p^{n+1},q^{n+1/2}). \end{aligned}$$

References

[1] Feng Kang: Difference scheme for Hamilton formalism and symplectic geometry, J. Comp. Math., 4:3(1986), 279–289.

[2] Feng Kang and Qin Meng-zhao: The symplectic methods for the computation of Hamiltonian equations, Proc. of 1st Chinese Conf. on Numerical Methods of PDE's, March 1986, Shanghai, Lecture Notes in Mathematics, No. 1297, Ed. Zhu You-lan and Gu Ben-yu, Spring, Berlin, pp. 1–37(1987).

[3] Li Chun-wang and Qin Meng-zhao: A symplectic difference scheme for the infinite dimensional Hamiltonian system, J. Comput. Math. 6(2)(1988).

(State Key Laboratory of Scientific and Engineering Computing, Computing Center, Academia Sinica, Beijing 100080)

AMS/IP Studies in Advanced Mathematics
Volume 3, 1997

THE ESSENCE OF GLIMM'S SCHEME

STEVEN SCHOCHET

ABSTRACT. Convergence and consistency are proven for a variant of Glimm's scheme that combines the esthetic features of both the original and the front-tracking versions: The approximation is an exact solution between collision times, and those times are determined by the actual collisions of waves. This allows both the scheme and the proof to be reduced to their essential elements. The scheme is obtained by modifying the front-tracking scheme of Bressan so as to avoid the use of nonphysical waves.

1. THE ELEMENTS OF GLIMM SCHEMES

Roughly speaking, weak solutions to a system of conservation laws having small BV initial data are constant outside of a finite number of shocks, rarefactions, and contact discontinuities, which interact when they meet to produce the waves of the same three types that solve the Riemann problem for the jump over the interacting waves.

This idyllic picture is inaccurate in three ways: First, the initial data need not be piece-wise constant with a finite number of jumps. Second, even if a finite number of waves are present initially, the interactions of those waves can produce infinitely many new waves in a finite time. Third, interactions involving a rarefaction wave actually produce something more complicated than the above elementary waves that make up the solution to Riemann problems.

Nevertheless, Glimm's scheme [Gli] proves the existence of a weak solution for which the above rough picture is approximately correct: The scheme approximates solutions of systems of conservations laws by a finite number of waves that interact approximately according to the solution of the Riemann problem. In order to avoid the above three difficulties,

The approximate initial data must contain only a finite number of waves, (1.1)

and the way that waves interact must be altered slightly in order to ensure that

Only a finite number of new waves are produced in a finite time (1.2)

1991 *Mathematics Subject Classification.* 35L65.
Key words and phrases. conservation laws, weak solutions, Glimm's scheme, front tracking.

and that

$$\text{New waves are also (at least approximately) elementary waves.} \tag{1.3}$$

(Actually, on account of the finite propagation speed of the system it suffices for the number of waves to be locally finite.) These alterations must be chosen so that the approximate solutions will approximately satisfy the conservation law

$$J(u,\phi) \equiv \iint_{\substack{0<t<\infty \\ -\infty<x<\infty}} \{\phi_t u + \phi_x f(u)\}\, dx\, dt + \int_{-\infty<x<\infty} \phi(0,x) u_0\, dx = 0$$

for all smooth functions ϕ having compact support.

In order to prove the existence of solutions to conservation laws, it suffices to show that the approximate solutions u_{app} converge to some limit, that $J(u_{\text{app}}, \phi)$ converges to zero, and that J of the limit is the limit of the Js. For these purposes it suffices to prove boundedness in BV and L^∞:

$$\|u_{\text{app}}(t,\cdot)\|_{BV} \le c, \qquad \|u_{\text{app}}\|_{L^\infty} \le c, \tag{1.4}$$

plus approximate continuity in time:

$$\|u_{\text{app}}(t_2,\cdot) - u_{\text{app}}(t_1,\cdot)\|_{L^1_{\text{loc}}} \le c|t_2 - t_1| + o(1), \tag{1.5}$$

and approximate consistency:

$$J(u_{\text{app}}, \phi) = o(1) \text{ for each test function } \phi. \tag{1.6}$$

These suffice because an extension of Helly's theorem [Gli] shows that estimates (1.4,1.5) imply convergence of a subsequence in L^1_{loc}, which is a strong enough sense of convergence to obtain the continuity of J under the limit.

Thus, Glimm's scheme must be defined so as to satisfy conditions (1.1,1.2,1.3) and estimates (1.4,1.5,1.6). Now BV initial data can be approximated arbitrarily closely in L^1_{loc} by piece-wise constant data having a finite number of jumps just by setting the approximate data equal to the actual data at a sufficient number of sufficiently closely spaced points. This shows that replacing u_0 in J by its approximation makes a $o(1)$ change in J, and hence that it suffices to satisfy (1.2,1.3,1.4,1.5,1.6) under the assumption that the initial data obeys condition (1.1).

Futhermore, the BV bound in (1.4) follows from the fact that each pair of waves can interact only once, the L^∞ bound is a consequence of the BV bound, and the time estimate (1.5) follows from the BV bound plus the finite propagation speed of the system.

Hence, the only flexible point in Glimm's scheme is the method of wave interaction. This must be close enough to "replacement by the solution of the Riemann problem upon collision" to be approximately consistent (estimate (1.6)), yet different enough to satisfy (1.2,1.3).

In Glimm's original scheme [Gli], the outgoing waves of an interaction are exactly the solution of the Riemann problem, but the timing of interactions is determined by a superimposed space-time grid rather than actual collisions of waves. At each time $n\Delta t$, all waves that are in an interval of length $2\Delta x$ are moved to the spatial grid point contained in that interval, forcing them to interact; the center of these interaction intervals are offset from the corresponding spatial grid points by an amount $a_n\Delta x$. Moving the waves introduces an inconsistency, i.e makes J nonzero. However, the inconsistencies from different time steps mostly cancel, thereby making the total inconsistency tend to zero with the grid size, provided that the sequence a_n is chosen to be random [Gli], or more precisely, equidistributed [Liu]. Also, provided that $\frac{\Delta x}{\Delta t}$ is greater than the largest possible wave speed (the CFL condition), no waves can meet between mesh times, which ensures that only a finite number of waves are created in a finite time, and that the approximation is an exact solution of the conservation laws between mesh times.

Recently, alternative interaction methods [Bre], [Ris] have been proven to satisfy the above conditions and estimates, generalizing earlier results for systems of one [Daf] and two [DiP] equations. These methods do not use a grid, and so determine the time and position of interactions by the actual meeting of waves, but in order to satisfy (1.2,1.3) they employ nonphysical waves, i.e. discontinuities that are not by themselves solutions of the conservation law. These nonphysical waves are of two types: approximate rarefactions consisting of a finite sequence of entropy-violating approximate shock waves, introduced in order to avoid the complications of actual rarefaction interactions, and arbitrary small nonphysical waves introduced in order to limit the number of new waves produced. Since nonphysical waves are used, the approximate solution is not an exact solution of the conservation law between collision times, i.e. there is a nonzero contribution to J at all times, not just at collisions. However, the integral over time of the absolute value of the inconsistency is small, so the fact that J tends to zero does not depend on delicate cancellations.

A variant will be presented here that combines the simplifying features of both types of Glimm's scheme. Like the original scheme, the approximate solutions are exact solutions between interaction times. Like the alternative schemes, no grid is used, the interaction times are determined by actual wave collisions, and the total of all inconsistencies is small in absolute value. Hence the scheme here most closely ressembles the rough picture described above, and the deviations from that picture are related directly to the need to satisfy conditions (1.1, 1.2,1.3).

The Glimm scheme presented here is obtained by modifying that of [Bre] to avoid the use of nonphysical waves. Following [Wen], approximate rarefaction waves are replaced by actual rarefaction waves that are constrained to interact in discrete units. This still ensures that the outgoing solution at interactions will consist only of elementary waves. It was shown in [Wen] that such a replacement is feasible for systems whose interactions do not increase the number of waves present; it will be shown here to be feasible in the general case. The remaining nonphysical waves of [Bre] will simply be eliminated; this is equivalent to taking their artificial speed in [Bre] to be infinite. Although the consistency proof in [Bre] uses the fact that the speed of these waves is finite, it will be modified here to allow their speed to be infinite, which eliminates them by sending them instantly to spatial infinity.

As mentioned above, the BV bound (1.4) is a consequence of the fact that

$$\text{Each pair of waves interacts but once.} \tag{1.7}$$

Following [Gli], all the works cited obtain the BV bound by showing that in each interaction the increase in the total variation is less than the decrease of a quadratic interaction potential, said decrease being due to the above fact. In order to emphasize that the BV bound is a consequence of (1.7), it will be obtained here directly from (1.7) by calculating the increase in the total variation that that principal allows.

2. A Glimm Scheme

The scheme depends upon three parameters: Δx, the discretization length of the initial data; δ, the discretization size of rarefaction waves; and n, the number of generations of new waves allowed to be produced. As discussed in the previous section, the contribution to J caused by approximating the initial data tends to zero with Δx.

The approximate solution will be composed of rarefactions, shock waves, and contact discontinuities. As shown in [Lax], given any sufficiently close states u_L and u_R, there is a unique set of such waves solving the Riemann problem for the jump $u_L \to u_R$, i.e. satisfying the conservation law with initial data $u_0 = u_L$ for $x < x_0$ and $u_0 = u_R$ for $x > x_0$, and obeying the Lax entropy inequalities. This set of waves consists of one "j-wave" from each "family" j, i.e. one wave whose speed approximately equals the jth eigenvalue of f_u. This result holds for systems all of whose families are either genuinely nonlinear or linearly degenerate [Lax].

To each wave is associated a strength μ whose absolute value measures the size of the wave [Lax]; the sum of these absolute values over all waves gives a metric equivalent to the BV norm [Gli]. As in [Wen], any rarefaction wave having strength μ greater than δ will be divided upon creation into $O(\frac{\mu}{\delta})$ pieces of strength no more than δ, each of which will be considered to be a distinct "discrete" rarefaction wave. Since pieces of a rarefaction wave are noninteracting rarefaction waves, this division makes no difference until the rarefaction interacts with another wave. When another wave collides with a discrete rarefaction wave, the latter will be made to contract instantaneously to the point of collision. For example, if the right end of the rarefaction is hit at time t then at time $t + 0$ the the solution in the region occupied by the rarefaction will be set to the left state of the rarefaction. This pushes the entire change in u occurring in the rarefaction to the point of collision, making the discrete rarefaction interact with the colliding wave as a unit. If waves collide with both sides of a rarefaction simultaneously the rarefaction may be contracted from an arbitrary point towards the two collision points.

The contribution to J from the instantaneous change in u over the region of the rarefaction is $\int \phi(t)\{u(t+0) - u(t-0)\}\,dx$ [Gli]. Now, rarefactions created at time t_1 have the form [Lax]

$$u = \begin{cases} u(\lambda_1) & \frac{x-x_1}{t-t_1} \le \lambda_1 \\ u(\frac{x-x_1}{t-t_1}) & \lambda_1 < \frac{x-x_1}{t-t_1} < \lambda_2 \\ u(\lambda_2) & \frac{x-x_1}{t-t_1} \ge \lambda_2. \end{cases}$$

Thus, if the rarefaction contracts at time t_2 then the contribution to J is bounded by $\|\phi\|_{L^\infty}$ times

$$\int_{x_1+\lambda_1(t_2-t_1)}^{x_1+\lambda_2(t_2-t_1)} |u(\lambda_1) - u(\tfrac{x-x_1}{t_2-t_1})|\,dx$$
$$= (t_2 - t_1)\int_{\lambda_1}^{\lambda_2} |u(\lambda_1) - u(\lambda)|\,d\lambda \le c(t_2 - t_1)\mu^2 \le c\delta\mu\,(t_2 - t_1).$$

Since at any time the sum of the strengths of all rarefaction waves existing that will contract in the future is bounded by the BV norm of u, for test functions ϕ that vanish for $t > T$ the total contribution to J from rarefaction contractions is bounded by $\delta T\|u\|_{BV}$, which tends to zero with δ.

In order to show that each pair of waves can only interact once, it is necessary to maintain the identity of waves through collisions. Each outgoing wave from a collision is considered to be the continuation of all incoming waves of the same family and type (shock, rarefaction, contact). If no such waves were present then the outgoing wave is a new wave. If a wave has no continuation then it ends at the collision. When two interacting waves are from different families then the one that is is from the higher family leaves to the right of the one from the lower family and forever travels faster, so the two can never meet again. Two interacting waves from the same family leave the interaction either as a single wave, which cannot interact with itself, or as a set of discrete rarefaction waves of the same family, which never interact among themselves. Hence (1.7) indeed holds.

Each wave will also be assigned a "generation", so that the finiteness of the set of waves can be achieved by not creating waves of generation n. The waves initially present, i.e. those produced as the solutions of the Riemann problems for the jumps in the approximate initial data, are the first generation. A new wave leaving an interaction is assigned the generation one more than the second smallest generation of all waves entering the interaction. For example, if the incoming waves have generations $\{1,2\}$, $\{2,2\}$, or $\{1,2,4\}$ then new outgoing waves are of generation 3. A continuing wave leaving an interaction is assigned the lowest generation of the waves it continues, or the generation it would have as a new wave, whichever is lower. Essentially the same scheme for assigning generations was used in [Bre].

Since new waves of each generation are produced only by the interactions of previous generations, the finiteness of all previous generations, together with the facts that waves only interact once and that each interaction produces a finite number of new waves, implies that each given generation contains a finite number of waves. Since only waves of generation less than n are produced, the total number of waves is finite, i.e. condition (1.2) holds.

Whenever two or more waves collide, the outgoing waves of the Riemann problem are calculated, after taking into account the contraction of rarefaction waves described above. Any outgoing waves of generation n are then eliminated by changing the next wave to the right to be a wave of the same family, type, and strength, but with left state equal to the left state of the wave that is being eliminated. This replaces the wave to be eliminated with a nonphysical wave lying to the right of wave that was shifted. Each such nonphysical wave is then eliminated in turn in the same manner, until the wave to be eliminated is the rightmost wave of all waves in existence at the time of collision. The unwanted wave is then eliminated completely by making the solution everywhere to the right of the left side of the wave equal the state on the left side of the wave. Since all outgoing waves are exact elementary waves, the approximation is an exact solution between collision times.

Because the right state of a wave of a given type and family is a C^1 function of its left state and strength, moving the wave to be eliminated past a wave of strength ν multiples its total variation $1 + O(|\nu|)$. All such moves together therefore multiply its strength by $e^{O(\|u\|_{TV})}$. For a test function ϕ having compact support in $|x| < M$, the contribution to J from eliminating a wave of strength μ is therefore bounded by $cM\|\phi\|_{L^\infty} e^{c\|u\|_{TV}}|\mu|$. Thus,

The total contibution to J from eliminating n^{th} generation waves is bounded by a constant times the total strength of those waves. (2.1)

When a set μ of zero or more waves plus a single wave ν_j are forced to interact together to produce the outgoing waves $\varepsilon(\mu, \nu_j)$, then by the mean-value theorem $\varepsilon_k(\mu, \nu_j) - \varepsilon_k(\mu, 0) = \frac{\partial \varepsilon_k}{\partial \nu_j}(\mu, z_j)\nu_j$. But $\varepsilon_k(0, z_j) = \delta_{jk} z_j$, so $\frac{\partial \varepsilon_k}{\partial \nu_j}(\mu, z_j) = \delta_{jk} + O(|\mu|)$. Hence, if μ by itself solves some Riemann problem so that $\varepsilon(\mu, 0) = \mu$, then

$$|\varepsilon_k - (\mu_k + \delta_{jk}\nu_j)| \leq c|\nu_j| \sum_i |\mu_i|. \tag{2.2}$$

This proof of (2.2) is adapted from the proof in [You] of Glimm's [Gli] more refined interaction estimate, which, as for all wave-tracking schemes, will not be needed here. Whenever two or more waves interact we imagine that the two waves on the left interact first, then the outgoing set of waves interacts with the next wave, etc., until all waves have interacted. Rarefaction waves are not split into pieces and waves of generation n are not eliminated until the interaction is complete. In this way, we can calculate the effect of multiple interactions as if they were a finite sequence of interactions satisfying (2.2). By induction we see from (2.2) that when waves $\nu^{(i)}$ interact then the outgoing j-wave ε_j satisfies

$$|\varepsilon_j - \sum_{\substack{\text{incoming} \\ j\text{-waves}}} \nu^{(i)}| \le \sum_{p\ge 2} c^{p-1} \sum_{i_1<\cdots<i_p} |\nu^{(i_1)}|\cdots|\nu(i_p)| \le \frac{c\sum_{i<k}|\nu^{(i)}||\nu^{(k)})|}{1-c\sum_r |\nu^{(r)}|} \quad (2.3)$$

provided that $1 - c\sum_r |\nu^{(r)}| > 0$.

Let A be the sum of strengths of the waves present at time zero, and let $B < 1/c$ be a bound for the maximum total strengths of all waves ever existing. When waves of the same family and type interact and combine, this maximum is calculated by considering all possibilities. For example, if two waves collide and combine then their maximum total strength is the greater of the maximum of the sum of their strengths at any time while they are separate or of the maximum of their combined strength at any time while they are together. By (1.7,2.3) we see that the total increase in the strength of all original waves plus the total strength of all new waves is bounded by $cB^2/(1-cB)$, so the total strength of all waves is bounded by $A + cB^2/(1-cB)$. More precisely, the latter bound holds through the collision following the last one at which the bound B holds. For A sufficiently small, there exists a $B < 1/c$ such that $A + cB^2/(1-cB) < B$. Hence both bounds are valid for all time, i.e. the total variation bound in (1.4) holds.

Since waves of second and higher generations are not present initially but are only created by interactions of all waves, and revert to the first generation whenever they combine with a wave of first generation, the maximum total strength of second and higher generation waves is bounded by $c_1\|u\|_{BV}^2$, where $c_1 = c/(1-c\|u\|_{BV})$. Since waves of generation j and higher are created only by interactions in which all but one wave are from generation $j-1$ and higher, and revert to a lower generation whenever they combine with waves of a lower generation or interact with two or more waves of generation lower than $j-1$, we find by induction that the maximum total strength of the waves of generation j and higher is bounded by $(c_1\|u\|_{BV})^{j-1}\|u\|_{BV}$. This bound was first proven in [Bre]. Since the above smallness condition $\|u_0\|_{BV} + c_1\|u\|_{BV}^2 < \|u\|_{BV}$ implies that $c_1\|u\|_{BV} < 1$, the maximum total strength of waves of generation n is $\theta^{n-1}\|u\|_{BV}$ for some $\theta < 1$. Since this tends to zero as $n \to \infty$, (2.1) shows that the contribution to J from eliminating the waves of generation n shrinks to zero as n tends to infinity. This completes the proof of the consistency estimate (1.6).

Since the leftmost state of the approximation remains fixed for all time, the BV bound immediately yields an L^∞ bound, which completes the proof of (1.4).

Furthermore,

$$\|u(t_2) - u(t_1)\|_{L^1_{\text{loc}}}$$
$$\leq \sum \left[\|u(\tau_{j+1} - 0) - u(\tau_j + 0)\|_{L^1_{\text{loc}}} + \|u(\tau_j + 0) - u(\tau_j - 0)\|_{L^1_{\text{loc}}} \right],$$

where the times τ_j include t_1, t_2, and all interaction times between them. Each wave μ present between times τ_j and τ_{j+1} contributes at most $c|\mu|(\tau_{j+1} - \tau_j)$ to $\|u(\tau_{j+1} - 0) - u(\tau_j + 0)\|_{L^1_{\text{loc}}}$, so the sum of all such terms over all j is bounded by $c\|u\|_{BV}(\tau_2 - \tau_1)$. Finally, the proof of consistency showed precisely that the sum over all interactions of $\|u(\tau_j + 0) - u(\tau_j - 0)\|_{L^1_{\text{loc}}}$ is $o(1)$, so (1.5) holds.

References

[Bre] A. Bressan, *Global solutions of systems of conservation laws by wave-front tracking*, J. Math. Anal. Appl. **170** (1992), 414-432.

[Daf] C. Dafermos, *Polygonal approximations of the solutions of the initial value problem for a conservation law*, J. Math. Anal. Appl. **38** (1972), 33-41.

[DiP] R. DiPerna, *Global existence of solutions to nonlinear hyperbolic systems of conservation laws*, J. Diff. Eq. **20** (1976), 187-212.

[Gli] J. Glimm, *Solutions in the large for nonlinear hyperbolic systems of equations*, Commun. Pure Appl. Math. **18** (1965), 697-715.

[Lax] P. Lax, *Hyperbolic systems of conservation laws II*, Commun. Pure Appl. Math. **10** (1957), 537-566.

[Liu] T.-P. Liu, *The deterministic version of the Glimm scheme*, Commun. Math. Phys. **57** (1977), 135-148.

[Ris] N. Risebro, *A front-tracking alternative to the random choice method*, Proc. AMS **117** (1993), 1125-1139.

[Wen] B. Wendroff, *An analysis of front tracking for chromotography*, Acta Appl. Math. **30** (1993), 265-285.

[You] R. Young, *A third order Glimm scheme*, Commun. Pure Appl. Math. (to appear).

School of Mathematical Sciences, Raymond and Beverly Sackler Faculty of Exact Sciences, Tel Aviv University, Ramat Aviv 69978 Israel

E-mail address: steve@math.tau.ac.il

AMS/IP Studies in Advanced Mathematics
Volume **3**, 1997

NONLINEAR GALERKIN METHOD USING LEGENDRE POLYNOMIALS

Jie Shen† & Roger Temam‡

Abstract. A nonlinear Galerkin method using Legendre polynomials is presented for solving linear elliptic and nonlinear dissipative evolution equations. The essential idea is to decompose the approximation space into a long-wavelength part and a short-wavelength part which are orthogonal with respect to the principal elliptic operator of the equation.

1. Introduction. The exchange of energy between the long- and short-wavelength components of a flow is an important aspect of nonlinear phenomena. Although the short-wavelength component usually carries only a small fraction of the total energy, its effect over a long time integration, through the nonlinear interaction with the long-wavelength component, is not negligible and sometimes essential (cf. [5], [4]). However, it is computationally inefficient to allocate as much computing resources to compute the short-wavelength component of the flow carrying little energy as we do with the long-wavelength component of the flow which carries most of the energy. The nonlinear Galerkin method introduced in [8] was a first attempt to address this issue from a computational point of view. It has been applied to many different space discretizations (cf. [9], [2], [10]) and has proven to be computationally efficient (cf. [3], [7]).

We consider in this article the application of the nonlinear Galerkin method to spectral discretizations using Legendre polynomials. *Our objective is to derive simplified versions of the classical algorithms which produce better conditioned systems and a reduction in computing time without affecting the discretization error of the scheme under consideration.*

This work was supported in part by NSF grants DMS-9205300 and DMS-9024769 and DOE grant DE-FG02-92ER25120.

We now describe the nonlinear Galerkin Method in an abstract setting. Let H be a given Hilbert space with a scalar product $(\cdot,\cdot)$ and a norm $\|\cdot\|$. We shall consider a class of nonlinear evolution equations of the form:

$$\frac{du}{dt} + \nu Au + Cu + B(u) = f,\ t > 0, \tag{1.1}$$

with $u(0) = u_0$. The operator A is a strictly positive linear unbounded self-adjoint operator with domain $D(A)$ dense in H. Hence, A^s $(s \in \mathbb{R})$ is well defined and we set $V = D(A^{\frac{1}{2}})$. B is a nonlinear operator such that $B(u) = B(u,u)$, where $B(\cdot,\cdot)$ is a bilinear form from $V \times V$ into V', C is a linear operator from V into H and there exists $c_1 > 0$ such that

$$((A+C)v, v) \geq c_1 \|v\|_V^2\ ,\ \forall\, v \in V.$$

The variational formulation for (1.1) is to find $u(t) \in V$ such that

$$(\frac{du}{dt}, v) + \nu(Au, v) + (Cu, v) + (B(u), v) = (f, v)\ ,\ \forall\, v \in V \tag{1.2}$$

with $u(0) = u_0$. Let V_h, where h is the discretization parameter, be a finite dimensional approximation of V. The classical Galerkin method for (1.2) in V_h is to find $u_h(t) \in V_h$ such that

$$(\frac{du_h}{dt}, v) + \nu(Au_h, v) + (Cu_h, v) + (B(u_h), v) = (f, v)\ ,\ \forall\, v \in V_h, \tag{1.3}$$

with $u_h(0) = u_{0h}$, where u_{0h} is some projection of u_0 in V onto V_h.

The nonlinear Galerkin method stems from the recently developed theory on Inertial Manifolds and Approximate Inertial Manifolds. It is intended to provide an approximate interaction law between the long- and short-wavelength components. Essential to the method is a decomposition of the space $V_h = Y_{h_c} \oplus Z_h$, where $h < h_c \to 0$, Y_{h_c} consists of the long-wavelength elements and Z_h consists of the short-wavelength elements. With this decomposition, we can set $u_h = y + z$ for $y \in Y_{h_c}$ and $z \in Z_h$, and rewrite (1.3) as follows:

$\forall\, v \in Y_{h_c}$ and $w \in Z_h$,

$$\frac{d}{dt}(y+z, v) + \nu(A(y+z), v) + (C(y+z), v) + (B(y+z), v) = (f, v), \tag{1.4}$$

$$\frac{d}{dt}(y+z, w) + \nu(A(y+z), w) + (C(y+z), w) + (B(y+z), w) = (f, w), \tag{1.5}$$

together with $y(0) = u_{0h_c}$, $z(0) = u_{0h} - u_{0h_c}$. Let us note that the short-wavelength component z usually carries only a small part of the total energy Hence, we can drop out some higher order (in z) terms without affecting the accuracy. If in addition the decomposition is L^2-orthogonal, i.e.

$$(A) \qquad (y, z) = 0, \ \forall y \in Y_{h_c}, z \in Z_h,$$

we can then approximate the system (1.4)-(1.5) by the following nonlinear Galerkin scheme:

Find $y(t) \in Y_{h_c}$, $z(t) \in Z_h$ such that $\forall v \in Y_{h_c}$, $w \in Z_h$,

$$(1.6) \quad \frac{d}{dt}(y, v) + \nu(A(y+z), v) + (C(y+z), v) + (B(y+z), v) = (f, v),$$

$$(1.7) \quad \gamma\frac{d}{dt}(z, w) + \nu(A(y+z), w) + (C(y+z), w) + (B(y), w) = (f, w),$$

with $y(0) = u_{0h_c}$; γ is either 1 or 0; $z(0) = u_{0h} - u_{0h_c}$ if $\gamma = 1$.

We use $\gamma = 0$ in case $\frac{dz}{dt}$ is small compared to other terms.[1] Otherwise we take $\gamma = 1$.

Let us mention that the wavelet-like incremental unknowns method [2] and the two-grid finite element method [10] provide decompositions which are L^2-orthogonal.

The above system can be interpreted in two different aspects:

(i) Equation (1.7) defines a short-wavelength correction z in terms of the long-wavelength component y. Hence, (1.6)-(1.7) can be viewed as *a corrected Galerkin scheme in* Y_{h_c}.

(ii) Since (1.6)-(1.7) was obtained by neglecting some small terms in (1.4)-(1.5), we can also view (1.6)-(1.7) as *a simplified Galerkin scheme in* V_h.

The first interpretation leads us to expect that (1.6)-(1.7) gives a better approximation than the Galerkin scheme in Y_{h_c}, while the second interpretation leads us to expect that (1.6)-(1.7) provides the same accuracy as the Galerkin scheme in V_h.

The system (1.6)-(1.7) can be further simplified if the decomposition is also A-orthogonal, i.e.

$$(B) \qquad (Ay, z) = 0, \ \forall y \in Y_{h_c}, z \in Z_h.$$

[1] This fact is rigorously established for some special decompositions (cf. [4], [10]).

In this case, we can approximate (1.4)-(1.5) by the following Nonlinear Galerkin scheme:

Find $y \in Y_{h_c}$, $z \in Z_h$ such that $\forall v \in Y_{h_c}$, $z \in Z_h$,

$$\frac{d}{dt}(y,v)+\nu(Ay,v)+(Cy,v)+(B(y+z),v)=(f,v), \tag{1.8}$$

$$\gamma\frac{d}{dt}(z,w)+\nu(Az,w)+(Cz,w)+(B(y),w)=(f,w), \tag{1.9}$$

with $y(0)=u_{0h_c}$; γ is either 1 or 0; $z(0)=u_{0h}-u_{0h_c}$ if $\gamma=1$.

The above scheme is certainly computationally more efficient than the scheme (1.6)-(1.7). The advantage of the nonlinear Galerkin scheme (1.8)-(1.9) over the classical Galerkin scheme (1.3) is two-fold: Firstly, the system (1.8)-(1.9) is greatly simplified compared to the Galerkin scheme (1.3). Secondly, the system (1.8)-(1.9) significantly reduces the stiffness of the system (1.3), since the stability or CFL (Courant-Friedrichs-Levy) condition for a full or partially explicit time discretization scheme for (1.6)-(1.7) only depends on the long-wavelength components (see for instance [12], [15]). As a result, larger time steps can be used which would greatly reduce the computing time, especially when we have to integrate over a long time interval.

However, the only available decomposition satisfying both (A) and (B) is based on the eigenfunctions of A. Unfortunately, this decomposition is only practical for problems with periodic boundary conditions. In order to apply scheme (1.8)-(1.9) to more general problems, we propose to construct decompositions which are A-orthogonal and "nearly" L^2-orthogonal (in a sense to be specified later), and we shall show that with this type of decompositions (1.8)-(1.9) is still as accurate as (1.3). In the next section, we construct several decompositions of this type with respect to different elliptic operators. Then in Section 3, we present an error estimate for the nonlinear Galerkin scheme.

2. Long- and short-wavelength decompositions. The following notations will be used hereafter. Let us set $I=(-1,1)$, and denote by $H^s(I)$, $H_0^s(I)$ the usual Sobolev spaces with the norm $\|\cdot\|_s$. Let $L_n(x)$ be the nth order Legendre polynomial. We are interested in using a spectral discretization with Legendre polynomials. In analogy with the Fourier series, we consider a high (resp. low) order (or mode) Legendre polynomial as a short (resp. long) wavelength element. The main difficulty is that the Legendre polynomials do not satisfy the boundary conditions associated with the operator A.

For any $d \geq 1$ such that dm is an integer, we set

$$\begin{aligned} S_{dm} &= \text{span}\{L_0(x),\ L_1(x),\ \ldots,\ L_{dm}(x)\}, \\ V_{dm} &= \{v \in S_{dm} : v(\pm 1) = 0\}, \\ W_{dm} &= \{v \in S_{dm} : v(\pm 1) = v'(\pm 1) = 0\}. \end{aligned}$$

S_{dm}, V_{dm} and W_{dm} are respectively finite dimensional approximations of $L^2(I)$, $H_0^1(I)$ and $H_0^2(I)$. We denote by π_{dm} the orthogonal projector from $L^2(I)$ onto S_{dm}.

2.1. 1-D case: decomposition of V_{dm} with respect to $A = -\frac{d^2}{dx^2}$.

Let us write

$$\phi_k(x) = L_k(x) - L_{k+2}(x).$$

It is an easy matter to verify that

$$V_{dm} = \text{span}\{\phi_0(x), \phi_1(x), \cdots, \phi_{dm-2}(x)\}. \tag{2.1}$$

Furthermore, using properties of the Legendre polynomials, we can show [11] that

$$\int_I \phi_k'(x)\phi_j'(x)dx = 0\ ,\ \forall\, k \neq j, \tag{2.2}$$

$$\int_I \phi_k(x)\phi_j(x)dx = 0\ ,\ \forall\, j \neq k-2, k, k+2. \tag{2.3}$$

Hence, a natural decomposition of V_{dm} is: $V_{dm} = V_m \oplus Z_{dm}$ where

$$Z_{dm} = \text{span}\{\phi_{m-1}(x), \phi_m(x), \cdots, \phi_{dm-2}(x)\}.$$

We observe immediately that $(Ay, z) = 0\ ,\ \forall\, y \in V_m, z \in Z_{dm}$, and

$$(y, z) = (y, (I - \pi_{m-2})z)\ ,\ \forall\, y \in V_m, z \in Z_{dm}. \tag{2.4}$$

In other words, V_m and Z_{dm} are A-orthogonal and nearly L^2-orthogonal.

We note that the scheme (1.8)-(1.9) can also be applied to a linear steady problem. Let us consider for instance the second order linear steady equation:

$$\alpha u - u_{xx} = f \text{ in } I;\ u(\pm 1) = 0, \tag{2.5}$$

where α is a constant such that the operator $\alpha I - D_{xx}$ is positive in $H_0^1(I)$. The classical Galerkin approximation of (2.5) in V_{dm} is: Find $u_{dm} \in V_{dm}$ such that

$$\alpha(u_{dm}, v) + (u'_{dm}, v') = (f, v)\ ,\ \forall\, v \in V_{dm}. \tag{2.6}$$

It can be easily shown [1] that

$$\|u_{dm} - u\|_1 \leq C(s)(dm)^{1-s}\|f\|_{s-2}\ ,\ \forall\, s \geq 1.$$

The scheme (1.8)-(1.9) applied to (2.6) is:

Find $y_m \in V_m$ and $z_{dm} \in Z_{dm}$ such that

$$\alpha(y_m, v) + ({y_m}', v') = (f, v)\ ,\ \forall\, v \in V_m, \tag{2.7}$$

$$\alpha(z_{dm}, w) + ({z_{dm}}', w') = (f, w)\ ,\ \forall\, w \in Z_{dm}. \tag{2.8}$$

We note that $y_m + z_{dm} = u_{dm}$ if and only if $\alpha = 0$. However, we can prove the following results.

Lemma 2.1.

$$\|y_m + z_{dm} - u\|_1 \leq C(s)\left(|\alpha| m^{-1-s} + (dm)^{1-s}\right)\|f\|_{s-2}\ ,\ \forall\, s \geq 1.$$

Remark 2.1. *We have approximately decomposed a large system (2.6) into two small systems (2.7)-(2.8) with the same accuracy as long as $|\alpha| d^{s-1} \lessapprox m^2$. We note that relation (2.4) is of crucial importance for this type of results.*

A proof of Lemma 2.1 with $\alpha = 1$ is given in [13]. However the same proof carries over to the general case.

2.2. 1-D case: decomposition of W_{dm} with respect to $A = \frac{d^4}{dx^4}$. Let us denote

$$\psi_k(x) = L_k(x) - \frac{2(2k+5)}{2k+7} L_{k+2}(x) + \frac{2k+3}{2k+7} L_{k+4}(x). \tag{2.9}$$

It is shown in [11] that

$$W_{dm} = \text{span}\{\psi_0(x), \psi_1(x), \cdots, \psi_{dm-4}(x)\},$$

and

$$(2.10)\qquad (\psi_i''(x), \psi_j''(x)) = 0\ ,\ \forall\, i \neq j.$$

Hence a natural decomposition of W_{dm} is: $W_{dm} = W_m \oplus \tilde{Z}_{dm}$ where

$$\tilde{Z}_{dm} = \mathrm{span}\{\phi_{m-3}(x), \phi_{m-2}(x), \cdots, \phi_{dm-4}(x)\}.$$

Thanks to (2.9) and (2.10), we derive that $(Ay, z) = 0$ and

$$(2.11)\qquad (y, z) = (y, (I - \pi_{m-4})z)\ ,\ \forall\, y \in W_m, z \in \tilde{Z}_{dm}.$$

Let us now apply the decomposition to the fourth order linear steady equation:

$$(2.12)\qquad \alpha u - \beta u_{xx} + u_{xxxx} = f \text{ in } I,\ u(\pm 1) = u_x(\pm 1) = 0,$$

where α and β are two constants such that $\alpha I - \beta D_{xx} + D_{xxxx}$ is positive in $H_0^2(I)$. The Galerkin approximation of (2.12) in W_{dm} is:

Find $u_{dm} \in W_{dm}$ such that

$$(2.13)\qquad \alpha(u_{dm}, v) + \beta(u'_{dm}, v') + (u''_{dm}, v'') = (f, v)\ ,\ \forall\, v \in W_{dm}.$$

and it can be shown [1] that

$$\|u_{dm} - u\|_2 \leq C(s)(dm)^{2-s}\|f\|_{s-4}\ ,\ \forall\, s \geq 2.$$

The scheme (1.8)-(1.9) applied to (2.12) is:

Find $y_m \in W_m$ and $z_{dm} \in \tilde{Z}_{dm}$ such that

$$(2.14)\qquad \alpha(y_m, v) + \beta(y'_m, v') + (y_m'', v'') = (f, v)\ ,\ \forall\, v \in W_m,$$

$$(2.15)\qquad \alpha(z_{dm}, w) + \beta(z'_{dm}, w') + (z_{dm}'', w'') = (f, w)\ ,\ \forall\, w \in \tilde{Z}_{dm}.$$

Equations (2.14)-(2.15) can be efficiently solved as indicated in [11]. The following lemma indicates that (2.14)-(2.15) is as accurate as (2.13) if $|\alpha| d^{s-2} \lessapprox m^4$ and $|\beta| d^{s-2} \lessapprox m^2$.

Lemma 2.2.

$$\|u - y_m - z_{dm}\|_2 \le C(s)(|\alpha|m^{-2-s} + |\beta|m^{-s} + (dm)^{2-s})\|f\|_{s-2}\ ,\ \forall\, s \ge 2.$$

Sketch of the Proof. The proof is quite similar to the proof of Lemma 2.1 given in [13]. Let $\pi_{dm}^{2,0}$ be the orthogonal projector from $H_0^2(I)$ onto W_{dm}, i.e. $((u - \pi_{dm}^{2,0}u)'', v'') = 0$ for all $v \in W_{dm}$. It can be shown [1] that

$$\|u - \pi_{dm}^{2,0}u\|_\mu \le C(s)(dm)^{\mu-s}\|u\|_s\ ,\ \forall\, 0 \le \mu \le 2,\ s \ge 2. \tag{2.16}$$

We derive from (2.12) that $\forall\, v \in W_m$, and $w \in \tilde{Z}_{dm}$,

$$\alpha(u, v + w) + \beta(u', v' + w') + ((\pi_{dm}^{2,0}u)'', v'' + w'') = (f, v + w). \tag{2.17}$$

Let us denote $\xi = \pi_m^{2,0}u - y_m$ and $\eta = (\pi_{dm}^{2,0} - \pi_m^{2,0})u - z_{dm}$. We note that $\eta \in \tilde{Z}_{dm}$. Indeed, by definition of $\pi_m^{2,0}$ we find that

$$(\pi_{dm}^{2,0}u - \pi_m^{2,0}u, v'''') = 0\ ,\ \forall\, v \in W_m.$$

Since the map $\frac{d^4}{dx^4}$: $W_m \to S_{m-4}$ is surjective, we infer that $(\pi_{dm}^{2,0}u - \pi_m^{2,0}u, v) = 0\ ,\ \forall\, v \in S_{m-4}$. Therefore, $\pi_{dm}^{2,0}u - \pi_m^{2,0}u \in \tilde{Z}_{dm}$, which implies that $\eta \in \tilde{Z}_{dm}$.

Now subtracting (2.14) and (2.15) from (2.17), we obtain $\forall\, v \in V_m$ and $w \in Z_{dm}$,

$$\begin{aligned} &\alpha(u - y_m, v) + \alpha(u - z_{dm}, w) + \beta((u - y_m)', v') \\ &\quad + \beta((u - z_{dm})', w') + ((\pi_{dm}^{2,0}u - y_m - z_{dm})'', (v + w)'') = 0. \end{aligned} \tag{2.18}$$

Since

$$\begin{aligned} \alpha(u - y_m, v) &= \alpha(u - \pi_m^{2,0}u, v) + \alpha(\xi, v), \\ \alpha(u - z_{dm}, w) &= \alpha(u - (\pi_{dm}^{2,0} - \pi_m^{2,0})u, w) + \alpha(\eta, w), \\ \beta((u - y_m)', v') &= \beta((u - \pi_m^{2,0}u)', v') + \beta(\xi', v'), \\ \beta((u - z_{dm})', w') &= \beta((u - (\pi_{dm}^{2,0} - \pi_m^{2,0})u)', w') + \beta(\eta', w'), \end{aligned}$$

setting $v = \xi$ and $w = \eta$ in (2.18), we find that

$$\begin{aligned} &\alpha(\|\xi\|^2 + \|\eta\|^2) + \beta(|\xi|_1^2 + |\eta|_1^2) + (|\xi|_2^2 + |\eta|_2^2) \\ &\quad = -\,\alpha(u - \pi_m^{2,0}u, \xi) - \alpha(u - (\pi_{dm}^{2,0} - \pi_m^{2,0})u, \eta) \\ &\qquad - \beta((u - \pi_m^{2,0}u)', \xi') - \beta((u - (\pi_{dm}^{2,0} - \pi_m^{2,0})u)', \eta'), \end{aligned} \tag{2.19}$$

where $|\cdot|_i$, $(i = 1, 2)$ is the semi-norm defined by $|v|_i = \sqrt{(D^i v, D^i v)}$.

The Lemma can then be established by bounding properly the four terms on the right-hand side, using various inequalities of the projection operators π_m and $\pi_m^{2,0}$. $\square$

2.3. 2-D periodic-nonperiodic case. It is in general very difficult, except in the pure periodic case, to construct a decomposition which is A-orthogonal in a multi-dimensional domain. We shall make a first attempt to construct such a decomposition for $A = -\Delta$ subject to the homogeneous Dirichlet boundary condition in one direction and to the periodic boundary condition in the other direction.

Let us write

$$H_p^n(0,2\pi) = \left\{ v = \sum_{k=-\infty}^{+\infty} a_k e^{ikx} \; : \; \bar{a}_j = a_{-j}; \right.$$
$$\left. \sum_{k=-\infty}^{+\infty} |k|^l |a_k| < +\infty \, , \; \forall \, l = 0, 1, \cdots, n \right\},$$

$$P_{dm} = \left\{ v = \sum_{|k| \le \frac{dm}{2}} a_k e^{ikx} \; : \bar{a}_j = a_{-j} \right\}.$$

Let $\Omega = [0, 2\pi] \times I$ and $\boldsymbol{m} = (m_1, m_2)$. We then set

$$\boldsymbol{H} = H_p^0(0,2\pi) \times L^2(I), \; \boldsymbol{V} = H_p^1(0,2\pi) \times H_0^1(I),$$

$$\boldsymbol{S}_{dm} = P_{dm_1} \times S_{dm_2}, \; \boldsymbol{V}_{dm} = P_{dm_1} \times V_{dm_2}.$$

We denote by Π_{dm} the orthogonal projector in $\boldsymbol{H}$ onto $\boldsymbol{S}_{dm}$.

We are looking for a decomposition $\boldsymbol{V}_{dm} = \boldsymbol{Y}_{\boldsymbol{m}} \oplus \boldsymbol{Z}_{dm}$ which is A-orthogonal and nearly L^2-orthogonal. In general we can fix either $\boldsymbol{Y}_{\boldsymbol{m}}$ or $\boldsymbol{Z}_{dm}$ and try to find its complement in $\boldsymbol{V}_{dm}$ with the desired properties. Since it is important that $\boldsymbol{Z}_{dm}$ only contains high-mode elements, we set

$$\boldsymbol{Z}_{dm} = \left\{ v = \sum_{\substack{\frac{m_1}{2} < |k| \le dm_1, \text{ or} \\ m_2+1 \le j \le m_2-2}} a_{kj} e^{ikx} \phi_j(y) : \bar{a}_{kj} = a_{-kj} \right\}.$$

We note that $z = (I - \Pi_{\boldsymbol{m}})z \, , \; \forall \, z \in \boldsymbol{Z}_{dm}$.

Thanks to (2.2) and (2.3), we observe immediately that for $|k| \le \frac{m_1}{2}$ and $0 \le j \le m_2 - 2$, we have

$$(Ae^{ikx}\phi_j(y), w) = k^2(e^{ikx}\phi_j(y), w) - (e^{ikx}\phi_j''(y), w) = 0 \, , \; \forall \, w \in \boldsymbol{Z}_{dm}.$$

Therefore, $e^{ikx}\phi_j(y) \in \boldsymbol{Y}_{\boldsymbol{m}}$ for $|k| \le \frac{m_1}{2}$ and $0 \le j \le m_2 - 2$.

In order to obtain a complete basis for $\boldsymbol{Y_m}$, we only need to construct $\gamma_{k,j}(y)$ for $1 \le k \le \frac{m_1}{2}$ and $j = m_2 - 1, m_2$ such that

$$(Ae^{ikx}\gamma_{k,j}(y), w) = 0 \ , \ \forall\, w \in Z_{dm}, \quad \begin{cases} 1 \le k \le \frac{m_1}{2} \\ j = m_2 - 1, m_2 \end{cases}.$$

In fact, the above condition reduces to:

$$\begin{aligned} &k^2 \int_I \gamma_{k,j}(y)\phi_l(y)dy + \int_I \gamma'_{k,j}(y)\phi'_l(y)dy = 0, \\ &l = m_2 + 1, \cdots, dm_2 - 2, \quad \begin{cases} 1 \le k \le \frac{m_1}{2} \\ j = m_2 - 1, m_2 \end{cases}. \end{aligned} \tag{2.20}$$

These relations lead us to search $\gamma_{k,j}(y)$ in the following form:

$$\gamma_{k,j}(y) = \phi_j(y) + \sum_{l=m_2+1}^{m_2-2} \gamma_l^{(k,j)}\phi_l(y), \quad \begin{cases} 1 \le k \le \frac{m_1}{2} \\ j = m_2 - 1, m_2 \end{cases}.$$

Again thanks to (2.2) and (2.3), for each pair (k, j), relations (2.20) become a symmetric positive definite tridiagonal system which uniquely determines $\{\gamma_l^{(k,j)}\}_{l=m_2+1,\cdots,dm_2-2}$.

To summarize, we write

$$\begin{aligned} &\gamma_{k,j}(y) = \phi_j(y), \text{ if } j = 0, 1, \cdots, m_2 - 2 \text{ or } k = 0, \\ &\gamma_{-k,j}(y) = \gamma_{k,j}(y), \text{ if } k = 1, \cdots, \frac{m_1}{2}, \text{ and } j = m_2 - 1, m_2. \end{aligned}$$

Then setting

$$\boldsymbol{Y_m} = \left\{ v = \sum_{k=-\frac{m_1}{2}}^{\frac{m_1}{2}} \sum_{j=0}^{m_2} a_{kj} e^{ikx}\gamma_{k,j}(y) : \bar{a}_{kj} = a_{-kj} \right\},$$

we have $\boldsymbol{V_{dm}} = \boldsymbol{Y_m} \oplus \boldsymbol{Z_{dm}}$ by construction and

$$(Ay, z) = 0, \ (y, z) = (y, (I - \Pi_m)z) \ , \ \forall\, y \in \boldsymbol{Y_m}, \ z \in \boldsymbol{Z_{dm}}.$$

Remark 2.2. *We can apply in particular the above decomposition to the Helmholtz equation: Find* $y_m \in \boldsymbol{Y}_m$, $z_{dm} \in \boldsymbol{Z}_{dm}$ *such that*

$$\alpha(y_m, v) + (\nabla y_m, \nabla v) = (f, v) \ , \ \forall \, v \in \boldsymbol{Y}_m,$$
$$\alpha(z_m, w) + (\nabla z_m, \nabla v) = (f, w) \ , \ \forall \, v \in \boldsymbol{Z}_{dm}.$$

As in the 1-D case, it can be shown that the above scheme is as accurate as the Galerkin scheme in $\boldsymbol{V}_{dm}$ *provided* $|\alpha| d^{s-1} \lessapprox m_2^2$, *where* s *is the largest number such that* $f \in H^{s-2}(\Omega)$.

The procedure can be directly applied to construct a proper decomposition with respect to the biharmonic operator. However, the extension to the two-dimensional case with Dirichlet boundary condition in both directions is more involved and will be considered in a future work [14].

3. Error estimates for the nonlinear Galerkin method. Let us consider the abstract nonlinear evolution equation (1.1). We set $b(u, v, w) = (B(u, v), w)$ and assume that the trilinear form b satisfies

$$b(u, v, v) = 0 \ , \ \forall \, u, v \in V, \tag{3.1}$$

$$b(u, v, w) \leq C \|u\|_V \|v\|_V \|w\| \ , \ \forall \, u, v, w \in V. \tag{3.2}$$

It is standard to show that the initial value problem (1.2) is well posed in the following sense: *for any* $u_0 \in V$ *and* $f \in L^2(0, T; H)$, *there exists a unique solution* u *of* (1.2) *such that* $u \in C([0, T]; V)$.

We can perform error analyses for the nonlinear Galerkin method based on all the decompositions presented in the last section. To fix the idea, we shall restrict ourselves to the decomposition in Section 2.1. In other words, we consider a 1-D problem with $A = -D_{xx}$, $H = L^2(I)$ and $V = H_0^1(I)$. A particular example is the 1-D Burgers' equation, for which we set

$$C = 0, \ B(u) = u u_x, \ B(u, v) = \frac{2}{3} u v_x + \frac{1}{3} v u_x.$$

One can readily check that the assumptions (3.1)-(3.2) are satisfied.

Using the notations in Section 2.1, the Galerkin approximation in V_{dm} to (1.2) in the above case is:

Find $u_{dm} \in V_{dm}$ such that $\forall \, v \in V_{dm}$,

$$\frac{d}{dt}(u_{dm}, v) + \nu(u'_{dm}, v') + (C u_{dm}, v) + b(u_{dm}, u_{dm}, v) = (f, v), \tag{3.3}$$

with $u_{dm}(\cdot,0) = \pi_{dm}u_0(\cdot)$.

We propose to approximate (3.3) by the following nonlinear Galerkin scheme:

Find $y_m \in V_m$ and $z_{dm} \in Z_{dm}$ such that $\forall\, v \in V_m$ and $w \in Z_{dm}$,

$$(3.4)\quad \frac{d}{dt}(y_m, v)+\nu(y_m', v')+(Cy_m, v)+b(y_m+z_{dm}, y_m+z_{dm}, v) = (f, v),$$

$$(3.5)\quad \gamma\frac{d}{dt}(z_{dm}, w)+\nu(z_{dm}', w')+(Cz_{dm}, w)+b(y_m+z_{dm}, y_m, w) = (f, w),$$

with $y_m(x,0) = u_{0m}(x)$; γ is either 1 or 0; $z_{dm}(x,0) = u_{0dm}(x) - u_{0m}(x)$ if $\gamma = 1$; u_{0m} (resp. u_{0dm}) is the projection of u_0 in V onto V_m (resp. V_{dm}).

We can prove the following error estimates for the above scheme.

Theorem 4.1. *Under the assumptions (3.1), (3.2) and for m sufficiently large, we have*

(3.6)

$$\Big(\int_0^T \|(u - y_m - z_{dm})(\rho)\|_1^2 d\rho\Big)^{\frac{1}{2}} \le$$

$$K\left((dm)^{1-s}\|u\|_{L^2(0,T;H^s)} + \gamma(dm)^{1-\alpha}\|u_t\|_{L^2(0,T;H^{\alpha-2})}\right)$$
$$+K\left(m^{-s}\|u\|_{L^2(0,T;H^s)} + (1-\gamma)m^{1-\alpha}\|u_t\|_{L^2(0,T;H^{\alpha-2})}\right),$$

where K is a constant independent of d and m.

Remark 3.1.

(a) *The first two terms on the right hand side of the error estimate in (3.6) are inherited from the classical Galerkin approximation (3.3), while the last two terms are due to the nonlinear Galerkin treatment. We suggest to use $\gamma = 1$ if f is time dependent and $\gamma = 0$ if f is time independent. In the latter case, we can show (see [6]) that u is analytic in time with value in V. Using Cauchy's formula for analytic functions, we can prove $\|u_t\|_s \sim \|u\|_s$. Hence, by taking $\alpha = s+2$ in (3.6), the last term can be absorbed into the first term. Therefore, in both cases we find that as long as $d^{s-1} \lessapprox m$, (3.4)-(3.5) is as accurate as the classical Galerkin scheme (3.3) in V_{dm}.*

(b) *It is also possible to derive error estimates in $L^\infty(0,T;H^1(\Omega))$ norm. We shall not pursue in this direction since the main purpose of our analysis is to determine a quantitative guideline for the proper choice of d.*

(c) *A proof of the Theorem in the special case $C = 0$ is given in [13]. The case $C \neq 0$ can be proved similarly. Moreover, the same technique can be used to obtain error estimates for nonlinear Galerkin schemes using the decomposition in Sections 2.2 and 2.3 (cf. [14]).*

(d) *We note that compared to* (1.9), *there is an additional term $b(z_{dm}, y_m, w)$ in* (3.5). *This term is added due to a technical difficulty. It is believed that the same results would hold without this additional term.*

REFERENCES

[1] C. BERNARDI AND Y. MADAY, *Approximations Spectrales de Problèmes aux Limites Elliptiques*, Springer-Verlag, Paris, 1992.

[2] M. CHEN AND R. TEMAM, *Nonlinear Galerkin Method in the Finite Difference Case and Wavelet-like Incremental Unknowns*, Numer. Math., 64 (1993), pp. 271–294 .

[3] T. DUBOIS, F. JAUBERTEAU, and R. TEMAM, *Solution of the incompressible Navier-Stokes equations by the nonlinear Galerkin method*, to appear in J. Scientific Computing.

[4] C. FOIAS, O. MANLEY, and R. TEMAM, *Modeling of the interaction of small and large eddies in two dimensional turbulent flows*, Math. Model. and Num. Anal., 22 (1988).

[5] C. FOIAS, G. R. SELL, and R. TEMAM, *Inertial manifolds for nonlinear evolutionary equations*, J. Diff. Eqn., 73 (1988), pp. 308–353.

[6] C. FOIAS AND R. TEMAM, *Some analytic and geometric properties of the evolution Navier-Stokes equations*, J. Math. Pure et Appl., 58 (1979), pp. 339–368.

[7] F. JAUBERTEAU, C. ROSIER, and R. TEMAM, *The nonlinear Galerkin method in computational fluid dynamics*, Appl. Numer. Math., 6 (1989/90), pp. 361–370.

[8] M. MARION AND R. TEMAM, *Nonlinear Galerkin Methods*, Siam. J. Numer. Anal., 26 (1989), pp. 1139–1157.

[9] ———, *Nonlinear Galerkin Methods: the finite element case*, Numer. Math., 57 (1990), pp. 205–226.

[10] M. MARION AND J. C. XU, *Error estimates on a new nonlinear Galerkin method based on two-grid finite elements*, Submitted to Siam J. Numer. Anal.

[11] J. SHEN, *Efficient spectral-Galerkin method I. Direct solvers for second and fourth order equations by using Legendre polynomials*, To appear in SIAM J. Sci. Comput.

[12] ———, *Long time stability and convergence for fully discrete nonlinear Galerkin methods*, Appl. Anal., 38 (1990), pp. 201–229.

[13] J. Shen and R. Temam, *Nonlinear Galerkin methods using Chebyshev or Legendre polynomials I. One dimensional case*, To appear in Siam J. Numer. Anal .

[14] ———, *Nonlinear Galerkin methods using Chebyshev or Legendre polynomials II. multi-dimensional cases*, in preparation.

[15] R. Temam, *Stability analysis of the nonlinear Galerkin method*, Math. Comp., 57 (1991), pp. 477–505.

†Department of Mathematics, Penn State University
E-mail address: shen_j@math.psu.edu

‡The Institute for Scientific Computing and Applied Mathematics, Indiana University

AMS/IP Studies in Advanced Mathematics
Volume 3, 1997

MULTI-DIMENSIONAL SHOCK-WAVES FOR RELATIVISTIC FLUIDS

Joel Smoller[1] *Blake Temple*[2]

Abstract

We discuss the issues in our recent paper [14] in which we study multi-dimensional shock waves via a study of the relativistic Euler equations which appear as a subsystem of the Einstein gravitational field equations of general relativity. In particular, we discuss the general theory of shock waves in this setting, and we apply this theory to explicitly construct spherically symmetric shock-wave solutions of the Einstein equations that generalizes the famous Oppenheimer-Snyder model for gravitational collapse in stars, thereby solving the problem first posed by Oppenheimer and Snyder in 1939 of generalizing their model to the case when the pressure is non-zero. This solution opens up intriguing possibilities in cosmology, but in terms of the classical theory of shock waves, there are remarkable simplifying features of the Einstein equations over the classical Euler equations that enable us to explicitly construct shock-wave solutions within the framework of general relativity. Our solution models an explosion which has no known classical analog. From the point of view of the theory of general relativity, it is interesting that the O-S solution is a contact discontinuity, and hence is time reversible, while in contrast, our shock-wave solution in which $p \neq 0$ is an irreversible solution of the Einstein gravitational field equations in which the irreversibility, loss of information, and increase of entropy in the fluids puts irreversibility into the dynamics of the gravitational field.

[1]Department of Mathematics, University of Michigan, Ann Arbor, MI 48109; Supported in part by NSF Applied Mathematics Grant Number DMS-92-03972 and the Institute of Theoretical Dynamics, UC-Davis.

[2]Department of Mathematics, University of California, Davis, Davis CA 95616; Supported in part by NSF Applied Mathematics Grant Number DMS-92-06631, the Institute of Theoretical Dynamics, UC-Davis.

1 Introduction

The modern theory of shock waves and conservation laws grew out of a study of the classical Euler equations for 3-dimensional compressible flow in gas dynamics. The Euler equations can be written in the form

$$\begin{aligned} \rho_t + div\{\rho u\} &= 0, \\ (\rho u)_t + div\{\rho u^i u^j + \delta^i_j p\} &= 0, \end{aligned} \tag{1.1}$$

where ρ denotes the density, $\mathbf{u} = (u^1, \cdots, u^3)$ the velocity, and p the pressure of the fluid, and *div* denotes the divergence taken with respect to the spatial coordinates $\mathbf{x} = (x^1, x^2, x^3)$. We write the equations (1.1) in the simple conservation form

$$DivT = 0, \tag{1.2}$$

where *Div* denotes the divergence taken (on each row of T) in the combined variables (t, x^1, x^2, x^3), and T denotes the 4×4 matrix of entries in $(\rho, \mathbf{u}, p)$ that makes (1.1) agree with (1.2); the components of T in a given coordinate system are denoted by T_{ij}, $i, j = 0, ..., 3$. It is well known that shock waves form in solutions of (1.1), and these model the steep fronts that propagate in the underlying problem. Shock waves represent the fundamental physics in solutions of (1.1), but it is well known that the presence of shocks introduces many difficulties for both computing and analyzing solutions. This is because at a shock-wave, the fluid variables in T become discontinuous, and thus any scheme for approximating the derivatives in $DivT = 0$ must involve differencing discontinuous functions, and these produce delta function singularities in the derivatives. This is what leads to the well known Gibbs type oscillations that typically appear in numerical calculations of shock waves by finite difference schemes. Although there is a well developed (though highly incomplete!) theory of shock waves when (1.1) is restricted to one space and one time dimension, it is fair to say that a general mathematical theory for shock waves in more than one space dimension is essentially non-existent. The geometry of shock-wave propagation in several space dimensions can be exceedingly complicated, and most of the work in this field has involved the numerical simulation of special solutions-there being few general principles emerging from such studies. Our idea here is to approach the theory of shock waves for (1.1) through the study of shock-wave propagation in the Einstein field equations of general relativity. At first it appears that the Einstein equations are much more complicated than the Euler equations since they include the effects of gravity, but in fact the Einstein equations have some remarkable simplifying features that have for the most part been overlooked in the subject of shock waves. One simplifying feature is that the Einstein equations are based on the highly developed theory of Riemannian geometry, a theory that allows one to work in arbitrary coordinate systems $(t, \mathbf{x})$. This can be used to advantage because geometrical complications can appear simple in special coordinate systems that change from point to point. Indeed, in our generalized Oppenheimer-Snyder construction, we obtain explicit formulas for the shock surface and shock speed, and we can interpret these physically in terms of a global conservation of mass principle. However, due to the covariance (coordinate freedom) of the Einstein equations, we succeed in this without ever having to construct a single coordinate system in which the solutions on each side of the shock are simultaneously expressed. Thus the Einstein equations are a natural setting in which the powerful techniques of Riemannian geometry can be brought to bear upon the problem of constructing multi-dimensional shock waves. But there is a more subtle and more remarkable simplification that occurs in the Einstein equations over the Euler equations, and this simplification is one of the miracles that led Einstein to discover the field equations in the first place. The idea here is that to get the Einstein equations from the Euler equations, one introduces the gravitational metric potentials g_{ij} that give the properties of the underlying spacetime. The Einstein equations are then equations in the fluid variables $(\rho, \mathbf{u}, p)$, together with the gravitational potentials g_{ij}, and these equations take the form

$$G = \kappa T, \tag{1.3}$$

where G is the Einstein curvature tensor, (the simplest 4×4 matrix of entries constructed from second order derivatives of g_{ij} that is automatically divergence free, and has the same transformation properties as T under arbitrary $(t, \mathbf{x})$ coordinate transformations), T is the relativistic version of the T in (1.2),

$$T_{ij} = (p + \rho c^2)u_i u_j + p g_{ij}, \tag{1.4}$$

and $\kappa = 8\pi/c^2$, where c denotes the speed of light. The great subtlety of (1.3) is that the relativistic Euler equations $Div T = 0$ follow *as a consequence of* (1.3), and needn't be imposed on solutions! This is because G is constructed from the Riemann curvature tensor so as to be *identically* divergence free as a consequence of the well known Bianchi identities of geometry. Said differently, since (1.3) does not involve derivatives of T, using (1.3) one can solve shock-wave problems without ever taking a derivative of the fluid variables in T, the variables $(\rho, \mathbf{u}, p)$ that become discontinuous and non-differentiable at the shock[1]. Now because G involves second derivatives of the metric potentials g_{ij}, it is natural to ask whether the difficulties in solving $DivT = 0$ when T becomes discontinuous at shock waves is replaced by corresponding difficulties in the Einstein equations due to the formation of discontinuities in the first derivatives of the metric variables g_{ij}. The results we present on the general theory of shock waves in Section 3 of [14] shows that this is *not* the case. (These results are based on the work of Israel, [3].) Indeed, we show that the Einstein tensor G, and hence the stress tensor T itself, are free of delta function sources at a shock *in every coordinate system* if and only if there exists some coordinate system in a neighborhood of the shock in which the metric potentials are continuously differentiable functions of these coordinates, with Lipschitz continuous second derivatives, (i.e., $C^{1,1}$ functions)[2]. Thus, in numerically simulating the second derivatives that appear in the Einstein tensor G when we solve $G = \kappa T$, we need only difference Lipschitz continuous functions at a shock-wave, while in numerically simulating $DivT = 0$, we must difference discontinuous functions, which produces delta function singularities in the derivatives of the terms that appear in each component of T, together with the disturbing Gibbs type oscillations

[1]The gravitational potentials g_{ij} play a role similar to the vector potential A in the theory of electro-magnetism. In the latter, choosing $F = dA$ has the effect of making the Maxwell equations $dF = 0$ hold automatically, in the same way that choosing $G = \kappa T$ has the effect of making the Euler equations $DivT = 0$ hold automatically. More precisely, $divG = 0$ is a geometric identity, independent of the Einstein equations, and holds as a consequence of the Bianchi identities, and thus $DivT = 0$ holds as a consequence of the identity $divG = 0$ once the Einstein equations $G = \kappa T$ are solved.

[2]The point here is that in an arbitrary coordinate system in which the metric components g_{ij} are Lipschitz continuous functions of the spacetime coordinates $(t, \mathbf{x})$ across a shock-wave, the terms that are differenced in G can have delta function singularities in the second derivatives, but these cancel out in each component of G, if and only if there exists a $C^{1,1}$ transformation of the spacetime coordinates such that, in the new coordinates, the metric potentials are $C^{1,1}$ functions across the shock. This leads to the conclusion that delta function singularities can never appear in the second derivatives of *any* of the terms that are differenced in G when these derivatives are calculated in these special coordinates. This is a covariant version of a result first given by Israel, [3]

that typically destroy the accuracy of numerical computations.

The advantage of having the Euler equations $DivT = 0$ as identities in the Einstein equations is also essential in our construction of the Oppenheimer-Snyder shock-wave solutions in sections 4 and 5 of [14]. In this setting, the presence of the metric potentials g_{ij} allows us to solve the problem by first matching *the metric* across the shock, and this gives us directly an explicit formula for the shock position without requiring that we solve the implicit jump conditions (the weak form of $DivT = 0$) for the fluids across the shock. Moreover, after the matching is done, we have shown in [14] that the jump conditions reduce in complexity from two to one nontrivial constraint. This reduction of the jump conditions after the metrics are matched, also represents a constraint on the weak solutions of $DivT = 0$, which follows because $DivT = 0$ is an identity on solutions of $G = \kappa T$.

To summarize, in [14] we use these simplifying features of (1.3) over (1.2) to construct an explicit spherically symmetric shock-wave solution of (1.3) that solves the longstanding open problem of generalizing the Oppenheimer-Snyder model of gravitational collapse to the case of non-zero pressure,[9]. We know of no classsical analog of our solutions, but said in plain terms, we have constructed a solution which models a spherically symmetric explosion described by an expanding shock-wave that is driven by the expansion behind the shock into a fixed ambient fluid. In this construction we heavily exploit the coordinate freedom of the Einstein equations and use the fact that it is easier to match the metric potentials g_{ij} across a shock than it is to satisfy the Rankine-Hugoniot jump conditions directly.

Our solution of the Oppenheimer-Snyder problem also opens up intriguing possibilities in cosmology which we discuss below, but in terms of the classical theory of shock waves, we plan to investigate whether the classical limit $c \to \infty$, (c denotes the speed of light), will yield new descriptions of classical shock waves as well.

2 Statement of Results

We now state the results in [14], which sets out a general theory of shock waves, (based on work of Israel, [3]), as well as a generalization of the Oppenheimer-Snyder (O-S) solution. We refer the reader to our paper for details. Our work here is best introduced in the first paragraph of our paper:

In their classic 1939 paper, Oppenheimer and Snyder [9] introduced the first mathematical model for gravitational collapse of stars based on spherically symmetric solutions of the Einstein gravitational field equations. In this pioneering paper, Oppenheimer and Snyder gave the first rigorous results describing gravitational collapse of stellar objects, and the remarkable conclusion of this work was that "black holes" could form from gravitational collapse in massive stars. In his comprehensive article on the history of the subject of gravitation, ([4], page 226, paragraph 4), Israel references the Oppenheimer-Snyder paper as having "strong claims to be considered the most daring and uncannily prophetic paper ever published in the field". Indeed, the paper appeared a quarter of a century before the process of gravitational collapse was widely accepted as the explanation for a variety of astronomical events. The Oppenheimer-Snyder paper also provided the first example in which a solution of the Einstein equations having interesting dynamics, was constructed by using the covariance of the equations to match two simpler solutions across an interface. However, it is well known that the Oppenheimer-Snyder model requires the simplifying assumption that the pressure be identically zero. In this paper we obtain a generalization of the Oppenheimer-Snyder model describing gravitational collapse which extends their model to the case when the pressure is non-zero. Our idea is to treat the case $p \neq 0$ *by replacing the boundary surface of the star in the Oppenheimer-Snyder model by a shock-wave interface across which mass and momentum are transported. In the limit* $p = 0$ *we obtain the Oppenheimer-Snyder solution, and in this limit we observe that the interface reduces to what is referred to as a "contact discontinuity" in the mathematical theory of shock waves, a degenerate discontinuous solution in which neither mass nor momentum crosses the interface, [1, 5, 12].*

Our mathematical procedure is to simultaneously find a (t, r) coordinate transformation and a shock surface $r = r(t)$ such that the Robertson-Walker (R-W) metric and the Interior Schwarzschild (I-S) metric, (two well known exact solutions of the Einstein equations $G = kT$ that are spherically symmetric in radial coordinate r, [16]), match in a Lipschitz continuous fashion across the shock surface. The (R-W) metric describes a uniformly expanding solution of the Einstein equations that is used in cosmology as a model for the universe as a whole. The R-W metric is determined by one unknown function $R(t)$, the *cosmological scale factor*, and in the universe model, the function $R(t)$ alone determines the redshift factor for electromagnetic radiation emmision from distant stars. The I-S metric is a time independent spherically symmetric solution of $G = \kappa T$, which has been used to model the interior of a stable star. Both metrics are determined by a system of ordinary differential equations that close when a particular equation of state $p = p(\rho)$ is imposed, (c.f [14], page 33). In our dynamically matched solution, we imagine the R-W metric as an exploding inner core (of a star or the universe as a whole), and the boundary of this inner core is a shock surface that is driven into the static I-S solution, which we imagine as the outer layers of a star, or the outer regions of the universe. Since the solution is time reversible for any equation of state, we can reverse the solution and equivalently obtain a shock-wave that collapses in on itself. We anticipate that the expanding shock is stable, while the collapsing shock solution is actually the analog of an unstable rarefaction shock, (in the sense of Lax [5]), except in the limit case $p = 0$ treated by Oppenheimer and Snyder. In the case $p = 0$, the shock interface reduces to a contact discontinuity, and in this degenerate case both the expanding and contracting solutions should be stable, (at least to radial perturbation). In this case, Oppenheimer and Snyder showed that the contracting shock-wave collpses to a "black hole", and they made the now well known paradoxical observation that the collapse to the black hole takes an infinite time as measured by the observer at infinity, but takes only a finite proper time as measured by the observer sitting on the shock-wave itself.

In [14] we construct a dynamical shock-wave solution of the Einstein equations that consists of the R-W metric on the inside connected to the I-S solution on the outside, with a shock-wave interface connecting the two, and we construct such a solution assuming arbitrary equations of state $p = p(\rho)$ and $\bar{p} = \bar{p}(\bar{\rho})$ in the R-W and I-S solutions, respectively. Our matching procedure is complicated by the fact that the coordinate systems for the R-W and I-S metrics must be treated differently on the inside and outside of the shock separately, (see [14]). An explicit formula for the shock position in terms of the jump in density emerges from the analysis, and this allows us to identify a *global* conservation of energy principle, despite the fact that, in general, conservation is only locally valid in Einsteins theory due to the effects of the curvature in the underlying metric g. For our general result here, the metrics match in a Lipschitz continuous fashion across the shock interface. There then arises the interesting issue of whether the weak, local form of conservation holds across this shock-wave. In Section 3 of our paper

[14], we present a self-contained treatment of the general theory of shock waves for the Einstein equations that re-organizes and extends previous results of Israel. (See also [8] and references therein where this topic is referred to by the heading *Junction Conditions*). We consider the general problem of allowing the metric to be only Lipschitz continuous across a hypersurface Σ in spacetime. In this case, the metric loses continuity in the first derivative. It turns out that the issue of conservation across an arbitrary Lipschitz continuous shock-wave is intimately related to when there exist delta function sources in G on the surface, and also intimately related to when there exists a $C^{1,1}$ coordinate transformation defined in a neighborhood of the shock that removes the jump in the derivatives of the metric potentials g_{ij} across the shock. We present a complete resolution of these issues in Theorem 3, Section 3 of our paper, which can be stated as follows:

Theorem 1 *Let Σ denote a smooth, 3-dimensional shock surface in spacetime with spacelike normal vector* $\mathbf{n}$. *Assume that the components g_{ij} of the gravitational metric g are smooth on either side of Σ, (continuous up to the boundary on either side separately), and Lipschitz continuous across Σ in some fixed coordinate system. Then the following statements are all equivalent:*
(i) $[K]=0$ at each point of Σ.
(ii) The curvature tensors R^i_{jkl} and G_{ij}, viewed as second order operators on the metric components g_{ij}, produce no delta function sources on Σ.
(iii) There exist locally Lorentzian coordinate frames at each point $P\in\Sigma$.
(iv) For each point $P\in\Sigma$ there exists a $C^{1,1}$ coordinate transformation defined in a neighborhood of P, such that, in the new coordinates, (which can be taken to be the Gaussian normal coordinates for the surface), the metric components are $C^{1,1}$ functions of these coordinates.

Moreover, if any one of these equivalencies hold, then the Rankine-Hugoniot jump conditions

$$[G]^\sigma_i n_\sigma = 0, \tag{2.5}$$

(which express the weak form of conservation of energy and momentum across Σ when $G=\kappa T$), hold at each point on Σ, a covariant statement.

Here $[K]$ denotes the jump in the second fundamental form (extrinsic curvature) K across Σ, (this being determined by the metric separately on each side of Σ because g_{ij} is only Lipschitz continuous across Σ), and by $C^{1,1}$ we mean that the first derivatives are Lipschitz continuous. Theorem 1 should be credited mostly to Israel, who obtained these results in Gaussian normal coordinates. Our contribution was to identify the covariance class of $C^{1,1}$ transformations, and to thereby obtain precise coordinate independent statements for (ii)-(iv). As a consequence of this, we obtain the result that the Ricci scalar curvature R *never* has delta function sources at a Lipschitz continuous matching of the metrics, as well as the following theorem that partially validates the statement that shock wave singularities in the source free Einstein equations $R_{ij} = 0$ or $G_{ij} = 0$ can only appear as coordinate anomalies, and can be transformed away by coordinate transformation:

Theorem 2 *If a smooth shock surface Σ forms in weak solutions of $R_{\alpha\beta} = 0$ or $G_{\alpha\beta} = 0$ posed in some given coordinate system y, such that the y-components $g_{\alpha\beta}$ of the metric tensor g are Lipschitz continuous across Σ, and C^k functions of y on either side of Σ (continuous up to the boundary on either side separately), then there exists a regular $C^{1,1}$ coordinate transformation taking $y \to x$, such that, the components g_{ij} of g in x-coordinates are actually C^k functions of x in a neighborhood of each point on the surface Σ.*

Note that when there are delta function sources in G on a surface Σ, the surface should be interpreted as a surface layer (because $G = \kappa T$), and not a shock wave, [3, 8]. In [14] we show that for spherically symmetric solutions, the weak form of conservation of energy and momentum (2.5) implies the absence of surface layers, so long as the areas of the spheres of symmetry match smoothly at Σ. We use this result in our construction of the shock waves that extend the Oppenheimer-Snyder model to the case of non-zero pressure.

We now discuss how we explicitly construct spherically symmetric shock-wave solutions of the Einstein equations (1.3) for a perfect fluid (1.4) by matching the Robertson-Walker (R-W) metric,

$$(R-W) \qquad ds^2 = -dt^2 + R^2(t)\left\{\frac{1}{1-kr^2}dr^2 + r^2 d\Omega^2\right\}. \tag{2.6}$$

and Interior Schwarzschild (I-S) metric,

$$(I-S) \qquad d\bar{s}^2 = -B(\bar{r})d\bar{t}^2 + \left(1 - \frac{2M(\bar{r})}{\bar{r}}\right) d\bar{r}^2 + \bar{r}^2 d\Omega^2, \tag{2.7}$$

at shock wave interfaces across which the metrics are Lipschitz continuous, and the conditions (i)-(iv) of Theorem 1 above hold. Here, the quantity $d\Omega^2 = d\theta^2 + sin^2\theta d\varphi^2$ denotes the standard metric on the 2-sphere. The R-W metric is homogeneous at each fixed t, and expands or contracts in time according to the Cosmological Scale Factor $R(t)$. The fluid is assumed to be co-moving with the metric, c.f. [16], and the Einstein equations (1.3) reduce to a system of two ODE's in two unknowns when an equation of state $p = p(\rho)$ is specified. The R-W metric has been accepted as a model for the universe as a whole. In this interpretation, $R(t)$ determines the red-shift factors for far away objects, and the existence of a singularity in the R-W metric in backwards time has been interpreted as the original *big bang*. The I-S metric is a time independent solution of (1.3) that is used to model the interior of a star. In this case the stress tensor is again taken to be that of a perfect fluid that is co-moving with the metric, and the functions $B(\bar{r})$ and $M(\bar{r})$, (the total mass inside radius $\bar{r}$), are also determined by an autonomous system of two equations in two unknowns when an equation of state $\bar{p} = \bar{p}(\bar{\rho})$ is chosen, and the Einstein equations (1.3) are imposed.

The idea in the matching is to define a coordinate transformation mapping $(t, r) \to (\bar{t}, \bar{r})$ such that, under this identification of coordinates, the metrics R-W and I-S agree Lipschitz continuously on a 3-dimensional shock surface. We set $\bar{r} = Rr$ in order that the sphere's of symmetry match smoothly at the surface, and with this it remains to define $\bar{t} = \bar{t}(t, r)$. The following general theorem states that a Lipschitz continuous matching of metrics can be achieved for arbitrary R-W and I-S metrics,[14]

Theorem 3 *Let $B(\bar{r}), M(\bar{r}), \bar{\rho}(\bar{r}), \bar{p}(\bar{r})$ denote any I-S solution of the Einstein equations (1.3), and let $R(t), \rho(t), p(t)$ denote any R-W solution of (1.3). Then, (under non-degeneracy assumptions, [14]), there exists a coordinate mapping $(t, r) \to (\bar{t}, \bar{r})$ of the form $\bar{r} = Rr$ and $\bar{t} = \bar{t}(t, r)$ such that, under this identification, the metrics agree and are Lipschitz continuous at the shock surface*

$$M(\bar{r}) = \frac{4\pi}{3}\rho(t)\bar{r}^3. \tag{2.8}$$

The equation (2.8) defines the shock surface $\bar{r} = \bar{r}(t)$ implicitly so long as $\bar{\rho}'(\bar{r}) < 0$.

Note that for this theorem, arbitrary equations of state $p = p(\rho)$ and $\bar{p} = \bar{p}(\bar{\rho})$ can be assigned. The identity (2.8) can be interpreted as a global conservation of mass principle for such matchings: in words, (2.8) says that the total mass that the I-S solution would see inside the shock interface $\bar{r} = \bar{r}(t)$ if it were continued on into the origin $\bar{r} = 0$, equals the total mass inside a sphere of radius $\bar{r} = \bar{r}(t)$ and constant density $\rho(t)$.

A further constraint on the metrics must be imposed to insure that the Rankine-Hugoniot jump conditions (2.5) for conservation hold across a shock. By the general theory, it suffices to match the second fundamental forms $[K] = 0$ across the surface, and this alone will insure that the surface is not a surface layer, and that conditions of Theorem 1 will then all apply at the shock surface. In [14] we show that in the presence of spherical symmetry, $[K] = 0$ is equivalent to the single condition

$$[T^i_j]n_i n^j = 0. \tag{2.9}$$

Using this, we derive a system of two autonomous ordinary differential equations in the shock position $r(t)$ and the cosmological scale factor $R(t)$ that determine the R-W metrics that will match any given I-S metric at the shock surface (2.8) such that energy and momentum are conserved across the shock. In order to impose the extra constraint (2.9), we must allow the R-W pressure $p(t)$ to be determined by the equations, rather than by an equation of state. This is summarized in the following theorem,[14]:

Theorem 4 *Let $B(\bar{r}), M(\bar{r}), \bar{\rho}(\bar{r}), \bar{p}(\bar{r})$ denote any fixed I-S solution of the Einstein equations. Then, (under non-degeneracy assumptions), the R-W metric $R(t), \rho(t), p(t)$ will satisfy conservation across the shock surface (2.8), (under the identification $\bar{r} = Rr, \bar{t} = \bar{t}(t, r)$), if and only if $(r(t), R(t))$ solve the system of ODE's*

$$\dot{R}^2 = \frac{8\pi G}{3}\rho R^2 - k, \tag{2.10}$$

$$\alpha \dot{r}^2 + \beta \dot{r} + \gamma = 0, \tag{2.11}$$

where all of the functions $\rho, \alpha, \beta, \gamma$ appearing in (2.10) and (2.11) can be expressed in terms of the unknowns $(r(t), R(t))$ through the shock surface equation (2.8) and the identity $\bar{r} = Rr$. The R-W pressure is then given by

$$p = -\frac{\frac{d}{dt}(\rho R^3)}{3R^2\dot{R}}. \tag{2.12}$$

Thus system (2.10), (2.11) gives the conservation constraint in terms of a first order *autonomous* ODE in these unknowns, and the solutions $(r(t), R(t))$ of this ODE determine the R-W metrics that will match a given I-S metric across a shock surface, such that the additional condition of conservation is maintained across the shock. The solutions that solve this ODE reduce to the Oppenheimer-Snyder solutions when M is constant and the I-S metric reduces to the empty space Schwarzschild metric. In the O-S case, the solution of (2.10), (2.11) reproduces $p \equiv 0$. Note that for a given I-S solution, we can in principle construct a shock wave at radius $\bar{r} = Rr$ with arbitrary R-W pressure p assigned at that radius by specifying the appropriate initial conditions for r and R.

3 Astrophysics and Cosmology

We now discuss possible applications of our Oppenheimer-Snyder type shock waves to astrophysics and cosmology. First of all, our solutions can be taken as a dynamical model for an evolving star which is either collapsing or exploding within the core, but which shows no visible signs of this collapse on the surface of the stationary I-S solution which sits on the outside. The possibility of such a dynamical solution raises interesting questions concerning limiting radii of stable stars as first discussed by Oppenheimer-Volkov and Chandrasekhar ([10]). In particular, when the I-S solution is taken as a model for the interior of a star, there is a theorem that states that for essentially any possible equation of state within the star, the I-S metric must have an infinite pressure at the center $r = 0$ whenever the entire mass of the star is compressed within 9/8 of it's Schwarzschild radius (c.f. [15], page 231). This implies that there are no stable configuations of stars that are sufficiently compressed, as it would require an infinite pressure at the center to hold up such a star once it falls within 9/8 of it's Schwarzschild radius. This has interesting cosmological consequences. In particular, a calculation shows that the maximum redshift factor[4] for light emitted from the surface of a stable star can be no more than 2, (see [15]). This becomes all the more interesting because of the discovery of quasars, (relatively bright stellar

[4]The redshift factor is the ratio given by the frequency of light received by an observer in a stationary solution, divided by the frequencey of light emitted by the source.

objects which are point sources of electro-magnetic radiation which are red-shifted all the way into the radio wave frequencies), which have a red shift factor much larger than 2. Now redshifting of the light emitted by a stellar object can arise in one of two ways: either the object is far away, so that the expansion of the universe moves the object away at a high velocity, or else the light is climbing out of a large gravitational potential. Since the stability theorem appears to rule out the latter, it has been conjectured that quasars are far away, and this leads to the incredible conclusion that the quasars must be fantastically large objects in order to explain their relatively bright emissions. But our model applies to highly compressed I-S solutions; specifically, to solutions whose total mass $M(r)$ lies well within the stability limit of 9/8 of the associated Schwarzschild radius. Since our procedure enables us to replace the inner core of any I-S solution with a dynamical expanding smooth solution, we can apply this to I-S solutions having a singular infinite pressure at $\bar{r} = 0$, and thereby obtain a solution with no singularity at $\bar{r} = 0$. Thus the conclusion we can make based upon our model is that a shock-wave in the core can generate the enormous pressures necessary to hold up the outer layers of the star even when the outer layers are well within 9/8 of the Schwarzschild radius. (For comparison, the Schwarzschild radius of the sun is about 3 kilometers.) Thus our *dynamical* shock-wave solution allows for the possibility of a stellar object that appears to us as a point source of light in the heavens that is emitting highly redshifted frequencies of radiation, and, according to our model, the surface of such a star would show no signs of the dynamics within the interior. Moreover, since (if the calculation of Oppenheimer-Snyder carries over to the case of non-zero pressure), the collapse of the shock-wave may well take an infinite time as measured by an observer far away from the star, our dynamical solution allows for the possibility that such an object would be stable for long periods of time. Thus our dynamical model supports the possibility that certain objects emitting highly red-shifted frequencies of electromagnetic radiation could really be dynamical, quasi-stable objects that are relatively nearby, thus resolving the paradox of their brightness.

Another intriguing possibility which our Oppenheimer-Snyder type shocks lead us to, is the possibility of a new model for the universe at large. The widely held view is that the universe as a whole can be described by an expanding solution of the Einstein equations which can be approximately modeled by the Roberson-Walker metric. Under these assumptions, the cosmological scale factor $R(t)$ in the R-W metric is the only physical parameter to be determined, and this is determined by the average mass-energy density contained within the universe at large. The Hubble constant gives a measure of the expansion in this model, and all of the information relating stellar distances to red-shift factors is based on the assumption of this model. One of the great open questions in physics involves the determination of the average energy density of the universe, and this is complicated by an inability to determine the amount of *dark matter*, or un-illuminated matter in the universe. Our shock-wave solution that connects the R-W metric on the inside to the I-S metric on the outside, opens the possibility that there is a shock-wave at the edge of the universe. It is interesting to pursue the possibility that the 2.7 degree Kelvin background radiation observed throughout the universe represents the black-body radiation from the boundary of such a shock. Assuming this, we can guess the energy density on the I-S side of the shock, and thereby, from our model, we should be able to determine the average energy density in the universe from certain theoretically observable parameters in the model, such as the distance to the shock-wave and the speed of the shock.[5] Even if this model is not exactly correct, these relations which are determined by the model may have an interest that transcends the model itself.

These possible physical applications make the problem of determining the large time dynamics of our Oppenheimer-Snyder shocks all that much more interesting.

References

[1] R. Courant and K.O. Friedrichs, *Supersonic Flow and Shock Waves*, John Wiley & Sons, New York, 1948.

[2] J. Glimm, *Solutions in the large for nonlinear hyperbolic systems of equations*, Comm. Pure Appl. Math., **18**(1965), pp. 697-715.

[3] W. Israel, *Singular hypersurfaces and thin shells in general relativity*, IL Nuovo Cimento, Vol. XLIV B, N. 1, 1966.

[5]Similar ideas were suggested to us by D. Christodoulou, to whom we are indebted for sharing his insights.

[4] W. Israel, *Dark stars: the evolution of an idea*, in: *300 years of gravitation*, edited by S. W. Hawking and W. Israel, Cambridge University Press, 1987, pp. 199–276.

[5] P.D. Lax, *Hyperbolic systems of conservation laws, II*, Comm. Pure Appl. Math., **10**(1957), PP. 537–566.

[6] G.C. McVittie, *Gravitational collapse to a small volume*, Astro. Phys. Jour., **140**(1964), pp. 401–416.

[7] C. Misner and D. Sharp, *Relativistic equations for adiabatic, spherically symmetric gravitational collapse*, Phys. Rev., **26**(1964), p. 571–576.

[8] C. Misner, K. Thorne and J. Wheeler, *Gravitation*, W.H. Freeman and Co., New York, 1973.

[9] J.R. Oppenheimer and J.R. Snyder,*On continued gravitational contraction*, Phys. Rev., **56**(1939), p. 455–459.

[10] J.R. Oppenheimer and G.M. Volkoff, *On massive neutron cores*, Phys. Rev., **55**(1939), pp. 374–381.

[11] R. Penrose, *The Emperor's New Mind*, Oxford University Press, 1989.

[12] J. Smoller, *Shock Waves and Reaction-Diffusion Equations*, Springer Verlag, 1983.

[13] J. Smoller and B. Temple *Global solutions of the relativistic Euler equations*, Commun. Math. Phys. 156,67-97 (1993).

[14] J. Smoller and B. Temple *Shock-wave solutions of the Einstein equations: the Oppenheimer-Snyder model of gravitational collapse extended to the case of non-zero pressure*, (preprint).

[15] H. Stephani, *General Relativity, An Introduction to the Theory of the Gravitational Field*, Cambridge University Press, 2nd ed., 1990.

[16] S. Weinberg, *Gravitation and Cosmology: Principles and Applications of the General Theory of Relativity*, John Wiley & Sons, New York, 1972.

AMS/IP Studies in Advanced Mathematics
Volume 3, 1997

Nonlinear system of equations of mixed type and nonlinear complex equation of mixed type

Sun Hesheng

Abstract

In this paper a nonlinear system of equations of mixed type and a nonlinear complex equation of mixed type are considered. The existence of H^1 strong solutions for the modified Tricomi problem is proved.

1 Nonlinear system of equations of mixed type

In domain $\mathcal{D} \subset R^2$ we consider the second order nonlinear system of partial differential equations

$$\begin{aligned} Lu \equiv A_{11}u_{xx} - 2A_{12}u_{xy} + A_{22}u_{yy} + B_1 u_x + B_2 u_y \\ +Cu - \operatorname{grad}F(u) = g \end{aligned} \tag{1}$$

where $u(x,y) = (u_1,\ u_2,\ \cdots,\ u_N)$ and $g(x,y) = (g_1,\ g_2,\ \cdots,\ g_N)$ are N-dimensional vector functions; A_{ij}, B_j $(i,\ j = 1,\ 2)$ and C are $N \times N$ symmetric matrices, whose elements a_{kl}^{ij}, b_{kl}^{j} and c_{kl} $(i,\ j = 1,\ 2;\ k,\ l = 1,\ \cdots,\ N)$ are functions of x and y, a_{kl}^{ij}, b_{kl}^{j}, $c_{kl} \in C^1$; F is a positive nonlinear scalar function of vector u.

Assume that the coefficients in (1) satisfy the condition

$$A_{ij} = \left(a_{kl}^{ij}\right) = a^{ij}(x,y)I,\quad i,\ j = 1,\ 2 \tag{2}$$

where I is a $N \times N$ unit matrix, $a^{ij}(x,y) = a^{ji}(x,y)$ $(i,\ j = 1,\ 2)$ are scalar functions of x and y, and satisfy the condition

$$\Delta \equiv \left(a^{12}\right)^2 - a^{11}a^{22} \begin{cases} < 0, & y > 0 \\ = 0, & y = 0 \\ > 0, & y < 0 \end{cases} \quad \text{in } \mathcal{D} \tag{3}$$

Then, (1) is a second order system of equations of mixed type, is elliptic in $\mathcal{D}^+ = \mathcal{D} \cap \{y > 0\}$ and hyperbolic in $\mathcal{D}^- = \mathcal{D} \cap \{y < 0\}$.

Suppose again that

$$a^{11} > 0 \quad \text{and} \quad a_x^{11} = a_y^{12} \tag{4}$$

and let y_0 be a point on the y-axis such that $a^{12}(x, y_0) = 0$.

We consider the system (1) in such a domain $\mathcal{D}$, containing a part of x-axis inside:

Γ_0 is the outer boundary of $\mathcal{D}$ in the part $\mathcal{D}^+$, which is connected with the x-axis at points A and B; Γ_+ and Γ_- are the outer boundaries of $\mathcal{D}$ in the part $\mathcal{D}^-$, which are two common characteristic lines of system (1), issuing from the points A and B respectively and are defined by the following equations:

$$\Gamma_+ : \ dy + \frac{a^{12} + \sqrt{\Delta}}{a^{11}} dx = 0, \quad \Gamma_- : \ dy + \frac{a^{12} - \sqrt{\Delta}}{a^{11}} dx = 0. \tag{5}$$

Assume that the boundary curves Γ_0, Γ_+ and Γ_- satisfy the following conditions:

$$n_2|_{\Gamma_0} \geq 0, \qquad n_2|_{\Gamma_+ \cup \Gamma_-} < 0 \tag{6}$$

where n_1 and n_2 are two components of the unit vector $\vec{n}$ of the outward normal to the boundary curve $\partial\mathcal{D}$, $n_1 = \frac{dy}{ds}$, $n_2 = -\frac{dx}{ds}$.

Then, under the conditions (3), (4) and (6) we may consider the modified Tricomi problem:

$$u = 0 \qquad \text{on } \Gamma_0 \tag{7}$$

for the system (1) in the domain $\mathcal{D} = \mathcal{D}^+ \cup \mathcal{D}^-$.

Let

$$L_0 u \equiv a^{11} I u_{xx} - 2a^{12} I u_{xy} + a^{22} I u_{yy} + B_1 u_x + B_2 u_y + Cu = f \tag{8}$$

Make the double integral

$$J = \int\int_{\mathcal{D}} q u_y \cdot L_0 u dx dy \tag{9}$$

here

$$q = \frac{1}{a^{11}} e^{\epsilon \int_{y_0}^{y} a^{11} dy} \tag{10}$$

where $\epsilon > 0$ is a sufficiently small constant. After integral by parts, we get

$$\begin{aligned} J = \ & \int\int_{\mathcal{D}} \Big\{ \tfrac{\epsilon}{2}\big(\sqrt{a^{11}} u_x - \tfrac{a^{12}}{\sqrt{a^{11}}} u_y\big) \cdot \big(\sqrt{a^{11}} u_x - \tfrac{a^{12}}{\sqrt{a^{11}}} u_y\big) \\ & + \Big[\Big(\tfrac{a^{22}}{a^{11}}\Big)_x - \tfrac{1}{2}\Big(\tfrac{a^{12}}{a^{11}}\Big)_y + \tfrac{1}{a^{11}} B_2 + \epsilon \tfrac{(a^{12})^2 - a^{11}a^{22}}{2a^{11}} \Big] u_y \cdot u_y \\ & + \tfrac{1}{a^{11}} B_1 u_x \cdot u_y - \tfrac{1}{2}\Big(\Big(\tfrac{C}{a^{11}}\Big)_y + \epsilon C \Big) u \cdot u \Big\} e^{\epsilon \int_{y_0}^{y} a^{11} dy} dx dy \\ & + \int_{\partial\mathcal{D}} \Big\{ \tfrac{1}{2} \Big(\tfrac{a^{22}}{a^{11}} u_y \cdot u_y - u_x \cdot u_x + \tfrac{1}{a^{11}} Cu \cdot u \Big) n_2 \\ & + \Big(u_x \cdot u_y - \tfrac{a^{12}}{a^{11}} u_y \cdot u_y \Big) n_1 \Big\} e^{\epsilon \int_{y_0}^{y} a^{11} dy} ds \\ & = I_1 + I_2 \end{aligned} \tag{11}$$

In consideration of the boundary condition, $u_x = u_n n_1,\ u_y = u_n n_2\ \left(u_n = \frac{du}{dn}\right)$, we have

$$\begin{aligned} I_2 = \ & \int_{\Gamma_0} \frac{1}{2a^{11}} \left(a^{22}n_2^2 - 2a^{12}n_1n_2 + a^{11}n_1^2\right) n_2 u_n \cdot u_n e^{\epsilon \int_{y_0}^{y} a^{11} dy} ds \\ & - \int_{\Gamma_+} \frac{1}{2} \left\{ \left(u_x - \frac{a^{12}+\sqrt{\Delta}}{a^{11}} u_y\right) \cdot \left(u_x - \frac{a^{12}+\sqrt{\Delta}}{a^{11}} u_y\right) \right. \\ & \left. + \frac{\Delta}{(a^{11})^2} u_y \cdot u_y - \frac{1}{a^{11}} Cu \cdot u \right\} n_2 e^{\epsilon \int_{y_0}^{y} a^{11} dy} ds \\ & - \int_{\Gamma_-} \frac{1}{2} \left\{ \left(u_x - \frac{a^{12}-\sqrt{\Delta}}{a^{11}} u_y\right) \cdot \left(u_x - \frac{a^{12}-\sqrt{\Delta}}{a^{11}} u_y\right) \right. \\ & \left. + \frac{\Delta}{(a^{11})^2} u_y \cdot u_y - \frac{1}{a^{11}} Cu \cdot u \right\} n_2 e^{\epsilon \int_{y_0}^{y} a^{11} dy} ds \end{aligned} \tag{12}$$

Now we assume that the following conditions are satisfied:

$$\begin{cases} (i) & \left(\frac{a^{22}}{a^{11}}\right)_x I - \frac{1}{2}\left(\frac{a^{12}}{a^{11}}\right)_y I + \frac{1}{a^{11}} B_2 \\ & \text{is a positive definite matrix in } \mathcal{D} \\ (ii) & B_1 = 0 \text{ is a zero matrix in } \mathcal{D} \\ (iii) & C \text{ and } \left(\frac{C}{a^{11}}\right)_y \text{ are non-positive definite matrices in } \mathcal{D} \end{cases} \tag{13}$$

Then, in consideration of the condition (6), we have only need to take ϵ sufficiently small, we derive from (11) and (12) immediately

$$J \geq C\|u\|^2_{H^1(\mathcal{D})} \tag{14}$$

Lemma 1. Under the conditions (4), (6) and (13), the soluiton of the problem (8)(7) satisfies the inequality

$$\|L_0 u\|_{L_2(\mathcal{D})} \geq C\|u\|_{H^1(\mathcal{D})} \tag{15}$$

Proof. Since $J \leq C\|u\|_{H^1(\mathcal{D})}\|L_0 u\|_{L_2(\mathcal{D})}$, from (14) it follows (15) immediately.

Now we consider the adjoint system of equations for the system of equations (8). In consideration of the conditions (4) and (13)(ii), we have following system

$$\begin{aligned} L_0^* v \equiv\ & a^{11} I v_{xx} - 2a^{12} I v_{xy} + a^{22} I v_{yy} + \left(\left(2a^{22}_y - 2a^{12}_x\right) I - B_2\right) v_y \\ & + \left(\left(a^{22}_{yy} - a^{12}_{xy}\right) I - B_{2y} + C\right) v = f^* \end{aligned} \tag{16}$$

and the adjoint boundary condition

$$v = 0 \qquad \text{on } \Gamma_0 \cup \Gamma_+ \cup \Gamma_- \tag{17}$$

where L_0^* is the adjoint operator of L_0.

Lemma 2. The solution to the adjoint problem (16)(17) satisfies inequality

$$\|L_0^* v\|_{H^{-1}(\mathcal{D})} \geq C\|v\|_{L_2(\mathcal{D})} \tag{18}$$

Proof. For any function $v \in C^2$ and $v|_{\partial D} = 0$, we consider the problem of first order system of differential equations:

$$\begin{cases} qu_y = v & \text{in } D \\ u = 0 & \text{on } \Gamma_0 \end{cases} \tag{19}$$

where the function q is taken from (10). Under the assumptions (4) and (6), obviously this problem has a unique solution $u \in C^2$ in D. Then, we have inequality:

$$\|L_0^* v\|_{H^{-1}} \|u\|_{H^1} \geq (L_0^* v,\ u)_0 = (v,\ L_0 u)_0$$

$$= (qu_y,\ L_0 u)_0 \geq C\|u\|_{H^1}^2 \geq C'\|u\|_{H^1}\|v\|_{L_2(D)}$$

From this, it follows the inequality (18).

From this dual inequality it follows the existence of weak solution to the problem (8)(7). Then, we have

Theorem 1. Assume that the conditions (4), (6) and (13) are satisfied. Then, the H^1 weak solution to the linear problem (8)(7) exists. Moreover, the strong solution is unique.

Since the intersecting points A and B of Γ_0 with Γ_+ and Γ_- in our problem are all regular points, at such points the weak and strong solutions are coincide. Hence, by applying the argument about the coincidence of strong and weak solutions in [1], it follows immediately

Theorem 2. Assume that the conditions (4), (6) and (13) are satisfied. Then, the H^1 strong solution to the linear problem (8)(7) uniquely exists. Moreover, the strong solution coincides with the weak one.

Now we consider the nonlinear problem (1)(7). In order to prove the existence of solution of the nonlinear problem (1)(7), we assume that the nonlinear vector valued function $F(u)$ satisfies the following conditions:

$$\begin{cases} F(0) = 0, \qquad F(u) \geq C_1|u|^{\rho+2} \\ |\mathrm{grad}F(u)| \leq C_2 + C_3|u|^{\rho+1}, \qquad \rho > 0,\ C_i > 0,\ i = 1,\ 2,\ 3 \end{cases} \tag{20}$$

Lemma 3. Assume that the coefficients in system (1) satisfy the conditions in Theorem 1, the conditions in (20) and the condition

$$a_y^{11} \leq 0 \tag{21}$$

Then, under the selection of function q in (10), we have

$$\int\int_D qu_y \cdot (-\mathrm{grad}F(u))dxdy \geq C\|u\|_{L_{\rho+2}(D)}^{\rho+2} \tag{22}$$

Proof. Since

$$-\tfrac{1}{a^{11}} e^{\epsilon\int_{y_0}^{y} a^{11}dy} u_y \cdot \mathrm{grad}F(u)$$

$$= -\left(\tfrac{1}{a^{11}} e^{\epsilon\int_{y_0}^{y} a^{11}dy} F(u)\right)_y + \left(-\tfrac{a_y^{11}}{(a^{11})^2} + \epsilon\right) e^{\epsilon\int_{y_0}^{y} a^{11}dy} F(u)$$

Hence, in consideration of the conditions (4) and (6), we have

$$\begin{aligned}
& -\int\int_{\mathcal{D}} \frac{1}{a^{11}} e^{\epsilon\int_{y_0}^{y} a^{11}dy} u_y \cdot \mathrm{grad}F(u)dxdy \\
= & \int\int_{\mathcal{D}} \left(-\frac{a^{11}_y}{(a^{11})^2}+\epsilon\right) e^{\epsilon\int_{y_0}^{y} a^{11}dy} F(u)dxdy \\
& -\int_{\partial\mathcal{D}} \frac{1}{a^{11}} e^{\epsilon\int_{y_0}^{y} a^{11}dy} F(u)n_2 ds \\
= & \int\int_{\mathcal{D}} \left(-\frac{a^{11}_y}{(a^{11})^2}+\epsilon\right) e^{\epsilon\int_{y_0}^{y} a^{11}dy} F(u)dxdy \\
& -\int_{\Gamma_+\cup\Gamma_-} \frac{1}{a^{11}} e^{\epsilon\int_{y_0}^{y} a^{11}dy} F(u)n_2 ds \\
\geq & \; C\|u\|^{\rho+2}_{L_{\rho+2}(\mathcal{D})}
\end{aligned}$$

By this lemma we may take the functional space $G \equiv L_{2\rho+2}(\mathcal{D})$ as the base space and apply the Leray-Schauder's fixed point principle.

For every N-dimensional vector $w(x, y) \in G$, we construct a vector function u, satisfying the following linear system

$$L_0 u = \tau \mathrm{grad}F(w) + g \tag{23}$$

with a parameter τ $(0 \leq \tau \leq 1)$. By the assumption in (20) the right hand side of the above equality (23) belongs to $L_2(\mathcal{D})$. Hence, from Theorem 2 the strong solution $u \in Z \equiv H^1(\mathcal{D})$ to the linear problem (23)(7) uniquely exists.

Then, using the same discussion about the nonlinear problem in [2] and applying the Leray-Schauder's fixed point principle we get

Theorem 3. Assume that the conditions (4)(6)(13)(20) and (21) are satisfied. Then, there exists at least one H^1 strong solution to the nonlinear problem (1)(7).

2 Nonlinear complex equation of mixed type

In domain Ω in the complex plane z $(= x + iy)$ consider the second order nonlinear complex equation

$$\begin{aligned}
& w_{z\bar{z}} + q_1(z)w_{zz} + q_2(z)w_{\bar{z}\bar{z}} + q_3(z)\bar{w}_{zz} + q_4(z)\bar{w}_{\bar{z}\bar{z}} \\
& +r_1(z)w_z + r_2(z)w_{\bar{z}} + r_3(z)\bar{w}_z + r_4(z)\bar{w}_{\bar{z}} \\
& +s_1(z)w + s_2(z)\bar{w} - |w|^\rho w = \Phi(z)
\end{aligned} \tag{24}$$

where $w = u(x, y) + iv(x, y)$ is a complex function, $\bar{w}$ is the complex conjugate of w; $q_j(z)$, $r_j(z)$ $(j = 1, 2, 3, 4)$, $s_k(z)$ $(k = 1, 2)$, $\Phi(z)$ are all complex functions of z; $\rho > 0$ is any constant; $\frac{\partial}{\partial \bar{z}} = \frac{1}{2}\left(\frac{\partial}{\partial x} + i\frac{\partial}{\partial y}\right)$, $\frac{\partial}{\partial z} = \frac{1}{2}\left(\frac{\partial}{\partial x} - i\frac{\partial}{\partial y}\right)$.

It is well known that if $q_j(z)$ $(j = 1, 2, 3, 4)$ satisfy condition

$$\sum_{j=1}^{4} |q_j(z)| < 1 \quad \text{in } \Omega \tag{25}$$

then the equation (24) is of elliptic type, this case had been investigated by many authors. If the condition (25) is not satisfied, then the case becomes more complicated, up to now there is no any work about this.

Let

$$\begin{cases} q_j(z) = a_j(x,y) + ib_j(x,y), & j = 1,\ 2,\ 3,\ 4 \\ r_j(z) = c_j(x,y) + id_j(x,y), & j = 1,\ 2,\ 3,\ 4 \\ s_k(z) = e_k(x,y) + ih_k(x,y), & k = 1,\ 2 \\ \Phi(z) = f(x,y) + ig(x,y) \end{cases} \tag{26}$$

Separating the real and imaginary parts of (24) we obtain the system of equations:

$$\begin{aligned} &(1 + a_1 + a_2 + a_3 + a_4)u_{xx} + (1 - a_1 - a_2 - a_3 - a_4)u_{yy} \\ &+2(b_1 - b_2 + b_3 - b_4)u_{xy} + (-b_1 - b_2 + b_3 + b_4)v_{xx} \\ &+(b_1 + b_2 - b_3 - b_4)v_{yy} + 2(a_1 - a_2 - a_3 + a_4)v_{xy} \\ &+2(c_1 + c_2 + c_3 + c_4)u_x + 2(d_1 - d_2 + d_3 - d_4)u_y \\ &+2(-d_1 - d_2 + d_3 + d_4)v_x + 2(c_1 - c_2 - c_3 + c_4)v_y \\ &+4(e_1 + e_2)u + 4(h_2 - h_1)v - 4|u^2 + v^2|^{\frac{\rho}{2}}u = 4f \end{aligned} \tag{27a}$$

$$\begin{aligned} &(b_1 + b_2 + b_3 + b_4)u_{xx} - (b_1 + b_2 + b_3 + b_4)u_{yy} \\ &+2(-a_1 + a_2 - a_3 + a_4)u_{xy} + (1 + a_1 + a_2 - a_3 - a_4)v_{xx} \\ &+(1 - a_1 - a_2 + a_3 + a_4)v_{yy} + 2(b_1 - b_2 - b_3 + b_4)v_{xy} \\ &+2(d_1 + d_2 + d_3 + d_4)u_x + 2(-c_1 + c_2 - c_3 + c_4)u_y \\ &+2(c_1 + c_2 - c_3 - c_4)v_x + 2(d_1 - d_2 - d_3 + d_4)v_y \\ &+4(h_1 + h_2)u + 4(e_1 - e_2)v - 4|u^2 + v^2|^{\frac{\rho}{2}}v = 4g \end{aligned} \tag{27b}$$

Let

$$U = \begin{pmatrix} u(x,y) \\ v(x,y) \end{pmatrix}, \quad G = \begin{pmatrix} f \\ g \end{pmatrix} \tag{28}$$

the system of equations (27a)(27b) may be written as

$$\begin{aligned} &A_{11}U_{xx} - 2A_{12}U_{xy} + A_{22}U_{yy} + 2B_1U_x + 2B_2U_y \\ &+4CU - 4|U|^{\rho}U = 4G \end{aligned} \tag{29}$$

where

$$A_{11} = \begin{pmatrix} 1 + a_1 + a_2 + a_3 + a_4 & -b_1 - b_2 + b_3 + b_4 \\ b_1 + b_2 + b_3 + b_4 & 1 + a_1 + a_2 - a_3 - a_4 \end{pmatrix}$$

$$A_{12} = \begin{pmatrix} -b_1 + b_2 - b_3 + b_4 & -a_1 + a_2 + a_3 - a_4 \\ a_1 - a_2 + a_3 - a_4 & -b_1 + b_2 + b_3 - b_4 \end{pmatrix}$$

$$A_{22} = \begin{pmatrix} 1 - a_1 - a_2 - a_3 - a_4 & b_1 + b_2 - b_3 - b_4 \\ -b_1 - b_2 - b_3 - b_4 & 1 - a_1 - a_2 + a_3 + a_4 \end{pmatrix} \tag{30}$$

$$B_1 = \begin{pmatrix} c_1 + c_2 + c_3 + c_4 & -d_1 - d_2 + d_3 + d_4 \\ d_1 + d_2 + d_3 + d_4 & c_1 + c_2 - c_3 - c_4 \end{pmatrix}$$

$$B_2 = \begin{pmatrix} d_1 - d_2 + d_3 - d_4 & c_1 - c_2 - c_3 + c_4 \\ -c_1 + c_2 - c_3 + c_4 & d_1 - d_2 - d_3 + d_4 \end{pmatrix}$$

$$C = \begin{pmatrix} e_1 + e_2 & h_2 - h_1 \\ h_1 + h_2 & e_1 - e_2 \end{pmatrix}$$

At first we assume that

$$\begin{cases} (i) & q_1(z) = \bar{q}_2(z), \quad q_3(z) = q_4(z) = 0 \\ (ii) & q_2(z) \text{ is analytic and belongs to } C^1 \end{cases} \tag{31}$$

i.e.

$$\begin{cases} (i) & a_1 = a_2, \ b_1 = -b_2, \ a_3 = a_4 = b_3 = b_4 = 0 \\ (ii) & a_{2x} = b_{2y}, \ a_{2y} = -b_{2x} \end{cases} \tag{32}$$

In this case (29) may be written as

$$(\frac{1}{2} + a_1)IU_{xx} + 2b_1 IU_{xy} + (\frac{1}{2} - a_1)IU_{yy} + B_1 U_x$$
$$+B_2 U_y + 2CU - 2|U|^\rho U = 2G \tag{33}$$

The discriminant of this system is $\Delta \equiv a_1^2 + b_1^2 - \frac{1}{4}$, from this we may see that, if the coefficients of complex equation (24) satisfy the conditions in (31) and the following condition

$$|q_1| \begin{cases} < \frac{1}{2}, & \text{Im} z > 0 \\ = 0, & \text{Im} z = 0 \\ > \frac{1}{2}, & \text{Im} z < 0 \end{cases} \tag{34}$$

(evidently, it not satisfies the condition (25)), then the real system (33) is a system of equations of mixed type, i.e. the complex equation (24) is a complex equation of mixed type.

For the system (33) satisfying the condition (34) we may consider two cases:

$$(i) \quad \frac{1}{2} + a_1 > 0 \tag{35}$$

and

$$(ii) \quad \frac{1}{2} - a_1 > 0 \tag{36}$$

For the case (ii) we shall discuss it in another paper (see [3]).

Now for the case (i), in a domain D containing a segment of x-axis inside we consider the real system (33). The outer boundary of the elliptic domain $D^+ \subset \{(x,y)|\ |q_1| < \frac{1}{2},\ y > 0\}$ is Γ_0; the degenerate line is $\overline{AB} \subset \{(x,y)|\ |q_1| = \frac{1}{2},\ y = 0\}$; the outer boundaries of the hyperbolic domain $D^- \subset \{(x,y)|\ |q_1| > \frac{1}{2},\ y < 0\}$ are Γ_+ and Γ_-, being two common characteristic lines of (33), issuing from points A and B respectively. Assume that the boundaries Γ_0, Γ_+ and Γ_- of D satisfy the conditions in (6). Then, we may apply the results in above section. Assume that

$$\begin{cases} (i) & c_1 = -c_2,\ c_3 = -c_4,\ d_1 = -d_2,\ d_3 = -d_4 \quad \text{in } D \\ (ii) & \left(\frac{\frac{1}{2}-a_1}{\frac{1}{2}+a_1}\right)_x I - \frac{1}{2}\left(\frac{-b_1}{\frac{1}{2}+a_1}\right)_y I \\ & + \frac{2}{\frac{1}{2}+a_1}\begin{pmatrix} d_1+d_3 & c_1-c_3 \\ -c_1-c_3 & d_1-d_3 \end{pmatrix} \\ & \text{is a positive definite matrix in } D \\ (iii) & \begin{pmatrix} e_1+e_2 & h_2-h_1 \\ h_1+h_2 & e_1-e_2 \end{pmatrix} \text{ and} \\ & \left[\frac{1}{\frac{1}{2}+a_1}\begin{pmatrix} e_1+e_2 & h_2-h_1 \\ h_1+h_2 & e_1-e_2 \end{pmatrix}\right]_y \\ & \text{are non-positive definite matrices in } D \\ (iv) & a_{1y} \le 0 \quad \text{in } D \end{cases} \tag{37}$$

Then, from Theorem 3 we have

Theorem 4. Assume that the conditions (32) (34) (35) (37) and (6) are satisfied. Then, the H^1 strong solution to the modified Tricomi problem for the nonlinear system (33) exists.

Now, returning to the complex equation (24), we have

Theorem 5. Assume that the coefficients of (24) in Ω satisfy the conditions (31),

(34) and the following conditions:

$$\begin{cases} (i) & \mathrm{Re}q_1 > -\frac{1}{2}, \quad q_2(z) \text{ is analytic and belongs to } C^1; \\ (ii) & r_1(z) = -r_2(z),\ r_3(z) = -r_4(z); \\ (iii) & \left(\frac{\frac{1}{2}-\mathrm{Re}q_1}{\frac{1}{2}+\mathrm{Re}q_1}\right)_x I - \frac{1}{2}\left(\frac{-\mathrm{Im}q_1}{\frac{1}{2}+\mathrm{Re}q_1}\right)_y I \\ & +\frac{2}{\frac{1}{2}+\mathrm{Re}q_1}\begin{pmatrix} \mathrm{Im}(r_1+r_3) & \mathrm{Re}(r_1-r_3) \\ -\mathrm{Re}(r_1+r_3) & \mathrm{Im}(r_1-r_3)\end{pmatrix} \\ & \text{is a positive definite matrix in } D; \\ (iv) & \mathrm{Re}s_1 \le \mathrm{Re}s_2 \le 0,\ (\mathrm{Re}s_1)^2 - |s_2|^2 \ge 0; \\ (v) & \mathrm{Re}s_{1y} \le \mathrm{Re}s_{2y} \le 0,\ (\mathrm{Re}s_{1y})^2 - |s_{2y}|^2 \ge 0; \\ (vi) & \frac{\partial}{\partial y}(\mathrm{Re}q_1) \le 0; \\ (vii) & n_2|_{\Gamma_0} \ge 0, \quad n_2|_{\Gamma_+\cup\Gamma_-} \le 0 \end{cases} \tag{38}$$

Then, the H^1 strong solution to the modified Tricomi problem

$$w(z) = 0 \qquad \text{on } \Gamma_0 \tag{39}$$

for the nonlinear complex equation (24) exists.

In particular, in the case $r_1(z) \equiv r_3(z) = 0$ the condition (38)(iii) may be simplified as

$$(iii') \ \frac{\partial}{\partial x}(|q_1|^2 + 5\mathrm{Re}q_1) \le -\delta < 0 \tag{40}$$

where $\delta > 0$ is an arbitrary constant.

References

[1] Karatoprakliev G. D., On the theory of boundary value problem for the equations of mixed type in multidimensional domain, Differencial'nye Uravnenija 13, 1(1977), 64-75. (in Russian)

[2] Sun Hesheng, Tricomi problem for nonlinear equation of mixed type, Science in China (series A) 35, 1(1992), 14-20.

[3] Sun Hesheng, Second order nonlinear system of equations of mixed type, Preprint IAPCM#93-07 (to appear).

Institute of Applied Physics and
Computational Mathematics
Beijing P. O. Box 8009
100088, China

AMS/IP Studies in Advanced Mathematics
Volume **3**, 1997

ON THE STABILITY OF BROADWELL SHOCKS

ANDERS SZEPESSY

1. Introduction

We study the asymptotic behavior of perturbations of traveling shock waves of the Broadwell model,

$$
\begin{aligned}
&(\frac{\partial}{\partial t} + \frac{\partial}{\partial x}) f_+ = f_0^2 - f_+ f_- \qquad , x \in \boldsymbol{R}^1, t > 0,\\
&\frac{\partial}{\partial t} f_0 = -\frac{1}{2}(f_0^2 - f_+ f_-),\\
&(\frac{\partial}{\partial t} - \frac{\partial}{\partial x}) f_- = f_0^2 - f_+ f_- \quad ,
\end{aligned}
\tag{1.1a}
$$

with initial data

$$
(f_+, f_0, f_-)(\cdot, 0) = f_I(\cdot) \ , \tag{1.1b}
$$

where $f = (f_+, f_0, f_-)$ represents the density of particles moving with speeds $(1, 0, -1)$ in the x-direction. Using the fluid dynamical variables

$$
\rho = f_+ + 4f_0 + f_- \quad , \quad m = f_+ - f_- \quad , \quad z = f_+ + f_- \quad ,
$$

where ρ is the density and m the momentum, the equation (1.1a) takes the form

$$
\begin{aligned}
\rho_t + m_x &= 0,\\
m_t + z_x &= 0,\\
z_t + m_x &= \frac{1}{8}((\rho - z)^2 - 4(z^2 - m^2)) \ .
\end{aligned}
\tag{1.2}
$$

This model of a discrete kinetic gas was first introduced by Broadwell [2] to study kinetic shock waves. Caflisch and Papanicolaou shows in [5] that the fluid dynamical limit of the model in many aspects has the same qualitative behavior as the Navier-Stokes equations related to the Boltzmann equation. Global existence of solutions to (1.1) has been proved by Tartar [16]. A recent review on mathematical results for the Broadwell model is given in [13] and [12]. Here we are interested in stability results, with respect to small perturbations of initial data, that are valid for long time.

The asymptotic behavior has been studied for some special cases of initial data. These initial data are perturbations of Riemann data for the related Euler equation

$$\rho_t + m_x = 0 \ ,$$
$$m_t + \hat{z}(\rho, m)_x = 0 \ ,$$
$$(\rho, m)(x, 0) = \begin{cases} (\rho_+, m_+) & x > 0 \ , \\ (\rho_-, m_-) & x < 0 \ , \end{cases}$$

where

$$z = \hat{z}(\rho, m) = \rho\sigma(u) \ ,$$
$$u \equiv m/\rho \quad , \quad \sigma(u) \equiv (2\sqrt{1+3u^2} - 1)/3 \ .$$

The Euler equation is obtained by assuming that $f(x,t) = (\rho, m, z)(x,t)$ is in equilibrium, i.e. that the collision term satisfies

$$Q(f) \equiv \frac{1}{8}((\rho - z)^2 - 4(z^2 - m^2)) = 0 \ , \quad \text{for all} \quad (x, t) \ ,$$

use this to eliminate $z = \hat{z}(\rho, m)$ and then drop the third equation in (1.2). Here $\rho_\pm, m_\pm$ and $z_\pm = \hat{z}(\rho_\pm, m_\pm)$ are constant states in equilibrium.

The cases that have been studied include perturbations of a constant equilibrium state [7] (where the asymptotic behavior is described by diffusion waves), the vacuum state [1], rarefaction waves [11] and weak shock waves. Perturbations of data that in the fluid dynamical limit corresponds to a shock wave solution was first studied by Kawashima and Matsumura [8] and then by Caflish and Liu [4]. Kawashima and Matsumura proved that the Broadwell solution tends asymptotically in time to a smooth traveling shock wave of (1.1) provided the initial perturbation is small and its fluid dynamical moments are zero. Caflish and Liu generalized this to certain nonlocal perturbations $w_0(x)$ with prescribed decay for $x \to \pm\infty$ of the form $\lim_{x\to\pm\infty} |x|^{3/2} w_0(x) = c$, where c depends on the initial excess mass $\int_{-\infty}^{\infty} w_0(x)dx$. In this case the asymptotic behavior is described by a translated shock wave and a diffusion wave in the opposite field. The translation and mass of the diffusion wave is given by the fluid dynamical moments of the initial perturbation.

Here we extend the work [4] to consider general localized small perturbations of shock waves. We begin by making the Ansatz

$$f = f_S + f_\theta + f_\eta + f_e \ , \tag{1.3}$$

where f_S is the shock wave, f_θ the diffusion wave, f_η a linear coupled viscous wave explained in more detail in Section 2 and f_e is the remaining error. In the region outside the shock layer, f_S is almost an equilibrium state and the asymptotics behaves locally as a perturbation of a constant equilibrium state. Therefore the Chapman-Enskog expansion is relevant and the perturbation is described by a diffusion wave f_θ as in [7].

The behavior in the shock region is related to the corresponding case for viscous conservation laws. In the shock region, the perturbation f_θ will be small and if in addition the fluid dynamical moments vanish, then also f_η is small and the energy estimate of [8] works. For general localized perturbations with non-vanishing moments the shock wave is translated, cf. [4], and the perturbation f_η in the shock region is not well described by the Chapman-Enskog expansion as was first observed for the special non-local perturbations in [4]. The principle of the Chapman-Enskog expansion is that it assumes that higher order derivates in all directions decay faster, which is true for the diffusion wave f_θ, but not for the perturbation f_η in the shock region. In the case of viscous conservation laws we know this from [14], [6]. The perturbation is compressed and moves with the shock and higher order derivatives of the perturbation f_η do not decay faster in general. Instead the derivative of f_η in the direction of the shock is small. Therefore, for general localized perturbations the asymptotic behavior is different from the diffusion waves and a new expansion is necessary. Our expansion is inspired by the work [9] of Liu on relaxation problems, where the fluid dynamical limit is a scalar equation. Our expansion will be based on the fact that the derivatives in the direction of the shock is small. As compared to the Navier–Stokes equations, obtained from the Chapman-Enskog expansion [5], the new expansion results in an equation of the same form for f_η, where the viscosity matrix has changed in the shock region. This fact is not only an artifact of the technique used in the proof, but is a real phenomenon, since the estimates of Theorem 1 and Remark 1 shows that $f_S + f_\theta + f_\eta$ is a more accurate description of the asymptotic behavior than f_S or $f_S + f_\theta$.

In [15], we prove

Theorem 1.*Assume that $\rho_\pm, m_\pm$ generate a shock wave f_S for* (1.1) *with speed s. There is a translation, δ, uniquely determined by the initial data and there are positive constants $c_i, i = 1, ..., 4$ so that if $f(\cdot, 0) - f_S \in H^1(\boldsymbol{R})$ and*

$$\epsilon \equiv |\rho_+ - \rho_-| + |m_+ - m_-| \le c_1 ,$$

$$\Big|\int_{-\infty}^{\infty} \big(f(x,0) - f_S(x-\delta)\big)\, dx\Big| \le c_2 \epsilon ,$$

$$\int_{-\infty}^{\infty} (1 + (x-\delta)^2)\, |f(x,0) - f_S(x-\delta)|^2\, dx \le c_3 ,$$

$$\int_{-\infty}^{\infty} |\frac{d}{dx} f(x,0) - \frac{d}{dx} f_S(x-\delta)|^2\, dx \le c_4 ,$$

then there exists a unique global (in time) smooth solution, $f(x,t)$, to problem (1.1) *satisfying*

$$f(x,t) - f_S(x - st - \sigma t) \in C([0,+\infty); H^1(\boldsymbol{R})) ,$$

$$\lim_{t\to\infty} ||f(x,t) - f_S(x - st - \delta)||_{L^p(\boldsymbol{R})} = 0 \ , p \ge 2 \ , \text{ and}$$

$$\lim_{t\to\infty}\int_t^\infty\int_{-\infty}^\infty |f(x,\tau)-(f_S(x-s\tau-\delta)+f_\theta(x,\tau)+f_\eta(x,\tau)|^2\,dxd\tau=0\ ,$$

where f_θ is the diffusion wave and f_η the linear viscous wave described in Section 2.

Remark 1. *By the estimates of f_θ, f_η in Lemma 1, we have in general, cf.* [6]

$$\lim_{\tau\to\infty}\int_\tau^\infty\int_{-\infty}^\infty |f_\theta+f_\eta|^2 dx\,dt=\infty,$$

$$\lim_{\tau\to\infty}\int_\tau^\infty\int_{|x-st|<1/\epsilon} |f_\eta|^2 dx\,dt=\infty,$$

$$\lim_{\tau\to\infty}\int_\tau^\infty\int_{|x-\lambda_1^- t|<t^{-1/2}} |f_\eta|^2 dx\,dt=\infty,$$

so without f_η , the last statement in Theorem 1 could not hold. Moreover f_η depends on the viscosity resulting from the new expansion.

The proof of Theorem 1 is divided into two main steps, consisting of pointwise estimates of f_η and energy estimates of the error term f_e. In Section 2, we study the first step and derive the equations for the diffusion wave f_θ and the linear coupled viscous wave f_η, state two lemmas of their estimates and sketch the proof. Compared to the analogous linear coupled viscous wave in the proof of stability of viscous shock waves [14], the present wave satisfies an equation where the viscosity matrix is only semi-definite. In Section 3, we modify the energy estimate of [8] to be applicable to the present case. This is accomplished by introducing carefully chosen weight functions.

We denote by $\lambda_i, r_i, i=1,2$ the eigenvalues and right eigenvectors of the Jacobian of $(m,\hat{z}(\rho,m))$ and write $F^-\equiv\lim_{x\to-\infty}F(x,\cdot)$, $F_{\eta,x}\equiv(F_\eta)_x$, $F_{\eta,t}\equiv(F_\eta)_t$ for any function F.

2. The Ansatz

Let us briefly describe the Ansatz (1.3) $f=f_S+f_\theta+f_\eta+f_e$. The shock wave $f_S=(\rho_S,m_S,z_S)$ is a traveling wave solution of (1.1a) with equilibrium boundary conditions at $x=\pm\infty$ which satisfy the Rankine-Hugoniot conditions and the Lax entropy condition. The shock wave can be expressed in explicit form with the hyperbolic tangent function, [3][4]. Let us for definiteness assume that it is a forward shock, i.e. $s>0$.

The diffusion wave f_θ is obtained from the Navier-Stokes equations of (1.1) which are given by the Chapman-Enskog expansion, see [4]. This expansion is obtained by inserting the improved Ansatz $z=\hat{z}+z^*$ in (1.2) and neglecting second order terms in z^* and derivatives of z^*. The result is

$$z^*=-2(1-\sigma(u))(1+3u^2)^{-3/2}u_x\ . \tag{2.1}$$

The explicit form

$$f_\theta(x,t)=\theta(x,t)r_1^-, \tag{2.2}$$

of the diffusion wave is then found by solving a convected Burgers equation in the diagonal part of the Navier-Stokes equation in the coordinate system (r_1^-, r_2^-). The qualitative behavior of f_θ is similar to a heat kernel convected with characteristic speed λ_1^- [4], and it solves (1.2) by adding an error $E^\theta = (E^\theta_{\rho,x}, E^\theta_{m,x}, E^\theta_z)$ to the right hand side, where

$$\begin{aligned} &E^\theta_\rho, E^\theta_m = O(\theta^2)r_1^- + O(\theta^3) + R, \\ (2.3)\qquad &E^\theta_z = R, \\ &R \equiv O(\theta)|\theta_x| + O(\theta_{xx}) + O(\theta_{xt}) + O((\theta_x)^2) + O(\theta)|f_S - f_S^-| \end{aligned}$$

Following [4], the mass of f_θ and the translation δ of the shock is determined by the fluid dynamical moments $\int_{-\infty}^{\infty}(\rho - \rho_S)(x,0)dx$ and $\int_{-\infty}^{\infty}(m - m_S)(x,0)dx$

The linear coupled viscous wave f_η will be different from the hyperbolic wave used in [4]. Linearizing the the collision term around $f_S + f_\theta$ we have

$$(2.4)\qquad Q(f_S + f_\theta + f_\eta) - Q(f_S + f_\theta) = A\rho_\eta + Mm_\eta - Bz_\eta + O(|f_\eta|^2)$$

where

$$A \equiv (\rho_S + \rho_\theta - (z_S + z_\theta))/4 \ , B \equiv (\rho_S + \rho_\theta + 3(z_S + z_\theta))/4 \ , M \equiv m_S + m_\theta \ .$$

Let

$$\bar\lambda(x,t) \equiv \begin{cases} \lambda_1^- & \text{if} \quad x \le (2\lambda_1^-/3 + s/3)t, \\ s & \text{if} \quad x \ge (\lambda_1^-/3 + 2s/3)t, \\ \lambda_1^- \frac{(2s+\lambda_1^-)t - x}{(s-\lambda_1^-)t} + s\frac{x - (s+2\lambda_1^-)t}{(s-\lambda_1^-)t} & \text{otherwise,} \end{cases}$$

and

$$D_{\bar\lambda} \equiv \partial_t + \bar\lambda\partial_x.$$

Then, assuming that

$$\begin{aligned} \rho_{\eta,t} + m_{\eta,x} &= -E^\theta_{\rho,x} \ , \\ m_{\eta,t} + z_{\eta,x} &= -E^\theta_{m,x}, \end{aligned}$$

we have

$$(2.5)\qquad \begin{aligned} z_{\eta,t} &= -\bar\lambda z_{\eta,x} + D_{\bar\lambda} z_\eta = D_{\bar\lambda} z_\eta + \bar\lambda m_{\eta,t} + \bar\lambda E^\theta_{m,x} \\ &= D_{\bar\lambda} z_\eta + \bar\lambda D_{\bar\lambda} m_{\eta,t} - \bar\lambda^2 m_{\eta,x} + \bar\lambda E^\theta_{m,x} \ . \end{aligned}$$

Therefore, by defining

$$z_\eta = \frac{A}{B}\rho_\eta + \frac{M}{B}m_\eta - \frac{1 - \bar\lambda^2}{B}m_{\eta,x}$$

and

$$\begin{aligned} &\rho_{\eta,t} + m_{\eta,x} = -E^{\theta}_{\rho,x}\ , \\ (2.6)\qquad &m_{\eta,t} + (\frac{A}{B}\rho_\eta + \frac{M}{B}m_\eta - \frac{1-\bar{\lambda}^2}{B}m_{\eta,x})_x = -E^{\theta}_{m,x}\ , \\ &(\rho_\eta, m_\eta)(\cdot,0) = 0\ , \end{aligned}$$

we get by (2.4) and (2.5)

$$\begin{aligned} (2.7)\qquad &z_{\eta,t} + m_{\eta,x} = Q(f_S + f_\theta + f_\eta) - Q(f_S + f_\theta) + O(|f_\eta|^2) \\ &+ D_{\bar{\lambda}} z_\eta + \bar{\lambda} D_{\bar{\lambda}} m_{\eta,t} + O(\theta)|\theta_x| + O(\theta)|f_S - f_{\bar{S}}|\ . \end{aligned}$$

Let us rewrite (2.6) in the form

$$(2.8)\qquad \eta_t + (F'(f_S)\eta)_x - (\tilde{B}\eta_x)_x = E_x\ .$$

Compared to the analogous case studied in [14], the difference is that the diffusion matrix $\tilde{B}$ is now only semi-definite. We also need more refined estimates of $D_{\bar{\lambda}}\eta$ here, because of (2.7). The linear viscous wave η in [14] was estimated by using scalar dual waves. We will modify this to treat (2.8). The semi definiteness introduces a distribution of zero order in the dual waves , which decays exponentially in time. Estimates showing this was presented in this conference and [10] by Tai-Ping Liu and Yanni Zeng, treating perturbations of constant states in viscous conservation laws.

In our case, the modification is follows. For short time $T - C < t < T$ the linear coupled equation is solved exactly with suitable frozen coefficients. Then at time $t = T - C$, the dual wave is divided into two parts, using the explicit expression. One part, ϕ, is smooth and has $||\phi_{xxx}||_{L^1(R)} << 1$. The other part is a zero order distribution. The smooth part is then given as initial data for certain coupled dual waves, as given below, which remain smooth and localized along the characteristic directions. The non smooth part decays exponentially in time and the singularity is convected along the particle path x =constant. For repeated time intervals of length C this procedure is continued.

To evolve the smooth part, some coupling from the diffusion matrix is used. Let the dual equation be

$$-\Psi_t - \Lambda\Psi_x - (D\Psi_x)_x = 0$$

where Λ and D are F' and $\tilde{B}$, respectively, in the coordinate system $(r_1(f_S(x,t)), r_2(f_S(x,t)))$. Assume that $w \sim \Psi_2$ solves the 2-component of the diagonal part in the coordinate system (r_1, r_2). Then the constants γ, β can be chosen such that $\Psi_1(x,t) = \gamma w_x(x, \beta t)$ is a good approximation of the coupling [15].

Based on this and a similar technique to the one used in [14], we have the following estimate, now taking into account also the exponential decay in x of η for $x > st$, [15].

Lemma 1. *The solution f_η of (2.6) satisfies $f_\eta(x,t) \le S(x,t)$, where*

$$S(x,t) \equiv \begin{cases} C\log t/(t^5(1+x)^2) & \text{if} \quad x \le \lambda_1^- t - 5\sqrt{t}\log(t), \\ C\log(t)/t^{3/4} & \text{if} \quad |\frac{x}{\lambda_1^-} - t| < 5\sqrt{t}\log(t), \\ C\log(t)/t^{1/2} & \text{if} \quad |\frac{x}{s} - t| < 5\log(t)/\epsilon, \\ \frac{C\log(t)}{t^{1/2}(\min(|\frac{x}{\lambda_1^-}-t|,|\frac{x}{s}-t|)+t^{1/2})^{1/2}} & \text{otherwise,} \end{cases}$$

To estimate $D_{\bar\lambda}\eta$ we use the difference quotient

$$D_{\bar\lambda}\eta(x,t) = \lim_{s\to 0}(\eta(x+\bar\lambda s, t+s) - \eta(x,t))/s\ .$$

The main idea is then to apply this limit using the pointwise estimates of η, and to integrate by parts to obtain

$$D_{\bar\lambda}\eta = -\int_{T/2}^{T}\int_{R}\Psi D_{\bar\lambda}E\,dx\,dt + \int_{T/2}^{T}\int_{R}E D_{\bar\lambda}\Psi\,dx\,dt$$
$$-\int_{R}E\Psi(x,T/2)dx + \text{ small error terms}\ .$$

We have

Lemma 2. *The solution f_η of (2.6) satisfies*

$$D_{\lambda_1^-}f_\eta(x,t) \le S(x,t)/(1+|x-st|)\ ,$$
$$D_s f_\eta(x,t) \le S(x,t)/(1+|x-\lambda_1^- t|)\ .$$

3. The Energy estimate

Combining (1.1),(1.3),(2.3),(2.6) and (2.7) implies that the error $f_e = (\rho_e, m_e, z_e)$ satisfies

$$(3.1)\qquad \begin{aligned} &\rho_{e,t} + m_{e,x} = 0\ ,\\ &m_{e,t} + z_{e,x} = 0\ ,\\ &z_{e,t} + m_{e,x} = Q(f_s + f_\theta + f_\eta + f_e)\\ &\quad -[Q(f_s) + (Q(f_s+f_\theta) - Q(f_S)) + Q'(f_S+f_\theta)f_\eta] + R_1\\ &\quad = Q'(f_S+f_\theta)f_e + R_1 + O(|f_e|^2)\\ &\quad = A\rho_e + Mm_e - Bz_e + R_1 + O(|f_e|^2), \end{aligned}$$

where

$$R_1 = O(\theta_x^2) + O(\theta_{xt}) + O(\theta_{xx}) + O(\theta)|\theta_x| + O(\theta)|f_S - f_S^-| + O(|f_\eta|^2) + O(D_{\bar\lambda}f_\eta)\ .$$

By the construction we have for $t \geq 0$

$$\int_{-\infty}^{\infty} \rho_e(x,t)dx = 0,$$
$$\int_{-\infty}^{\infty} m_e(x,t)dx = 0.$$

Following [8] and [4], let

$$\Phi(x,t) = \int_{-\infty}^{x} \rho_e(y,t)dy,$$
$$\Psi(x,t) = \int_{-\infty}^{x} m_e(y,t)dy \ ,$$

and eliminate z_e in (3.1) to get

$$\begin{aligned} L_1(\Phi,\Psi) &\equiv \Phi_t + \Psi_x = 0 \ , \\ L_2(\Phi,\Psi) &\equiv \Psi_{tt} - \Psi_{xx} + A\Phi x + M\Psi_x + B\Psi_t = -\Gamma \ , \\ \Gamma &\equiv R_1 + O(|f_e|^2) \ . \end{aligned}$$

In [8], Kawashima and Matsumura prove an energy estimate for f_e in the case of zero total mass perturbation in ρ, m. Then, $f_\theta = f_\eta = 0$ and $D_sQ' = D_sQ'(f_S) = 0$. In our case $D_sQ' = D_sQ'(f_S + f_\theta)$ is small but not negligible. We shall now describe how the energy estimate of [8] can be modified using appropriate weights to handle the case of $D_sQ' \neq 0$.

The first estimate in [8] is obtained from

$$-\Psi_x L_1 + A^{-1}\Psi_t L_2 = -A^{-1}\Psi_t\Gamma$$

and needs no modification. The second estimate need a modification and with λ_a, λ_b being two positive weight functions, depending only on f_θ and satisfying $|\lambda_a - \lambda_b| = O(\theta)$, we shall use

$$-\lambda_b A^{-1}\Psi\Gamma = \lambda_a \Phi L_1 + \lambda_b A^{-1}\Psi L_2 \ .$$

Integrating over $\boldsymbol{R} \times (0,T)$ we have

$$\begin{aligned} &-\int_0^T \int_{-\infty}^{\infty} \lambda_b A^{-1}\Psi\Gamma dxdt = \int_{t=T} [\lambda_a\Phi^2 + \lambda_b A^{-1}\Psi\Psi_t \\ &+ \lambda_b A^{-1}B\Psi^2/2 + (\lambda_b A^{-1})_t\Psi^2/2]dx \\ &+ \int_0^T \int_{-\infty}^{\infty} [(\lambda_a - \lambda_b)\Phi\Psi_x - \lambda_{b,x}\Phi\Psi - ((\lambda_b A^{-1}M)_x + (\lambda_b A^{-1}B)_t)\Psi^2/2 \\ &- \lambda_{a,t}\Phi^2/2 + \lambda_b A^{-1}(\Psi_t)^2 + (\lambda_b A^{-1})_{tt}\Psi^2/2 + \lambda_b A^{-1}(\Psi_x)^2 - (\lambda_b A^{-1})_{xx}\Psi^2/2]dxdt \ . \end{aligned}$$

Using $\lambda_1 + \lambda_2 = M/B$ and $\lambda_1\lambda_2 = -A/B$ we have

$$\lambda_{b,x}\Phi\Psi - ((\lambda_b A^{-1}M)_x + (\lambda_b A^{-1}B)_t)\Psi^2/2 - \lambda_{a,t}\Phi^2/2$$
$$= -\lambda_{b,x}\Phi\Psi + \lambda_1\lambda_{a,x}\Phi^2/2 - ((\lambda_b\frac{\lambda_1+\lambda_2}{|\lambda_1|\lambda_2})_x - \lambda_1(\frac{\lambda_b}{|\lambda_1|\lambda_2})_x)\Psi^2/2 - D_{\lambda_1}(\lambda_a)\Phi^2/2$$
$$- D_{\lambda_1}(\frac{\lambda_b}{|\lambda_1|\lambda_2})\Psi^2/2 = -\lambda_{b,x}(|\lambda_1|^{1/2}\Phi - |\lambda_1|^{-1/2}\Psi)^2 - (\lambda_a - \lambda_b)_x(|\lambda_1|\Phi)^2/2$$
$$- \lambda_b\lambda_{1,x}(|\lambda_1|^{-1} + \lambda_2^{-1})(|\lambda_1|^{-1/2}\Psi)^2/2 - D_{\lambda_1}(\lambda_a)\Phi^2/2 - D_{\lambda_1}(\frac{\lambda_b}{|\lambda_1|\lambda_2})\Psi^2/2 .$$

Here the terms

$$||D_{\lambda_1}\lambda_a||_{L^\infty(|x-\lambda_1 t|<5t^{1/2}log(t))} + ||D_{\lambda_1}(\frac{\lambda_b}{|\lambda_1|\lambda_2})||_{L^\infty(|x-\lambda_1 t|<5t^{1/2}log(t))},$$

as well as

$$||(\partial_t^2 + \partial_x^2)(A^{-1}\lambda_b)||_{L^\infty(|x-\lambda_1 t|<5t^{1/2}log(t))}$$

are integrable in time, so they can be neglected. Let

$$S_x \equiv \lambda_b(|\lambda_1|^{-1} + \lambda_2^{-1})\lambda_2'\theta_x, \ \lambda_{b,x} = \beta|S_x|, \ \lambda_{a,x} = \alpha|S_x| . \tag{3.2}$$

where for some constants $\beta_\pm, \alpha_\pm$

$$\beta(x,t) = \begin{cases} \beta_+ & \text{if} \quad \theta_x > 0 \ , \\ \beta_- & \text{if} \quad \theta_x < 0 \ , \end{cases}$$
$$\alpha(x,t) = \begin{cases} \alpha_+ & \text{if} \quad \theta_x > 0 \ , \\ \alpha_- & \text{if} \quad \theta_x < 0 \ . \end{cases}$$

It turns out that $\beta_\pm, \alpha_\pm$ can be determined so that

$$I \equiv -\beta(|\lambda_1|^{1/2}\Phi - |\lambda_1|^{-1/2}\Psi)^2 - (\alpha - \beta)(|\lambda_1|^{1/2}\Phi)^2 + sign(S_x)(|\lambda_1|^{-1/2}\Psi)^2$$

becomes positive definite with the constraint

$$\alpha_+ + \beta_+ = \alpha_- + \beta_- \ ,$$

which implies

$$\lim_{x\to\pm\infty}(\lambda_a(x,t) - \lambda_b(x,t)) = 0. \tag{3.3}$$

By (3.2) and (3.3) we have

$$\lambda_a - \lambda_b = O(\theta),$$

so that with a small modification of λ_a, λ_b near $|x - \lambda_1^- t| << \sqrt{t}$ there holds

$$(\lambda_a - \lambda_b)\Phi\Psi_x << |S_x|I + c(\Psi_x)^2.$$

Using λ_a, λ_b as above, the energy estimate of [8] proves the estimates of the error f_e in Theorem 1.

References

[1] Beale J.I., Large-time behavior of the Broadwell model of a discrete velocity gas, Comm. Math. Phys., 102 (1985), 217-235.

[2] Broadwell J.E., Shock structure in a simple discrete velocity gas, Phys. Fluids, 7 (1964), 1243-1247.

[3] Caflisch R.E., Navier-Stokes and Boltzmann shock profiles for a model of a gas dynamics, Comm. Pure Appl. Math., 32 (1979), 521-554.

[4] Caflisch R.E. & Liu T.-P., Stability of shock waves for the Broadwell equations, Comm. Math. Phys., 114 (1988), 103-130.

[5] Caflisch R.E. & Papanicolaou G.C., The fluid-dynamical limit of a nonlinear model Boltzmann equation, Comm. Pure Appl. Math., 32 (1979), 589-619.

[6] Goodman J., Szepessy A. & Zumbrun K., A remark on the stability of viscous shock waves, to appear in SIAM J. Math. Anal.

[7] Kawashima S., Large-time behavior of solutions of the discrete Boltzmann equation, Comm. Math. Phys., 109 (1987), 563-589.

[8] Kawashima S. & Matsumura A., Asymptotic stability of travelling wave solutions of systems of one-dimensional gas motion, Comm. Math. Phys, 101 (1985), 97-127.

[9] Liu T.-P., Hyperbolic conservation laws with relaxation, Comm. Math. Phys, 108 (1987), 153-175.

[10] Liu T.-P., Hyperbolic conservation laws, seminar, NYU, April 1993.

[11] Matsumura A., Asymptotics towards rarefaction wave of solutions to the Broadwell Modell of a discrete velocity gas, Japan J. Appl. Math., 4 (1987), 489-502.

[12] Monaco R. & Preziosi L., Fluid Dynamical Applications of the Discrete Boltzmann Equation, World Scientific Publishing, 1991.

[13] Platkowski T. & Illner R., Discrete velocity models of the Boltzmann equation: a survey on the mathematical aspects of the theory, SIAM Rev., 30 (1988), 213-255.

[14] Szepessy A. & Xin Z., Nonlinear stability of viscous shock waves, Arch. Rational Mech. Anal., 122 (1993), 53-103.

[15] Szepessy A. & Xin Z., Stability of Broadwell Shocks, manuscript.

[16] Tartar L., Existence globale pour un systeme hyperbolic semilinearie de la theorie cinetique des gaz, Semin. Goulaouic-Schwarz, No. 1 (1975/76).

NADA
Royal Institute of Technology
S-100 44 Stockholm
Sweden

AMS/IP Studies in Advanced Mathematics
Volume **3**, 1997

RIEMANN PROBLEM FOR HYPERBOLIC COMBUSTION

Dechun TAN[1]

§1. Gas dynamical combustion. Combustion is a complicated process. Usually it is a high speed flow coupled with chemical reaction. It would both reveal all of the phenomena in fluid mechanics and include heat releasing. For gas dynamical combustion, it will produce shock, detonation and deflagration waves. Detonation is a flame with higher speed. And deflagration is a flame with lower speed. Early in 1880's, these waves were observed by french chemists. Chapman and Jouget proposed the hyperbolic theory about the waves. But it was in 1940's Courant and Friedrichs explained their substantial differences:

Detonation (DT), like a shock wave, is supersonic with respect to the flow in front of it and subsonic with respect to the flow behind it. The deflagration (DF) is subsonic with respect to the flow at its both sides. Here supersonic means $v/c > 1$ and subsonic $v/c < 1$. v is the velocity of the wave and c is the sound speed of the flow. These DT and DF are in fact corresponding to the SDT (strong detonation) and WDF (weak deflagaration) in the updated papers.

Our intention is to study the propagation of the combustion waves. So we neglect other properties, such as chaos, spiral, turbulence and the forth. The model is about 1-D exothermic reaction of gases in an infinitely long tube. For this model the governing equation can be obtained from the conservation laws of mass, momentum and energy. In Lagrangian coordinates, it can be written as

$$\begin{cases} \tau_t - u_x = 0, \\ u_t + p_x = 0, \\ E_t + (pu)_x = 0 \end{cases}$$

where u is the velocity, p is the pressure, $1/\tau$ is the density and $E = \frac{1}{2}u^2 + e + q_0 z$ is the specific total energy, z is the percentage of the unburnt gases. e is the internal energy and q_0 is the chemical binding energy of the reactant. Combustion is viewed as the liberating of the binding energy.

In addition to the above conservation laws, a reaction equation is also needed. According to the situation of combustion, there are three seperate experimental laws about the reaction rate:

[1]This work is supported by Italian Consiglio Nazionale delle Ricerche under the grant 211-01-26 and partially by NSFC

1). Chapman-Jouget: $k = +\infty$, the reaction region is infinitely narrow. When the reactant particle enters this region, its binding energy is released instantaneously.

2). Zeldovich-v. Neumann-Döring (ZND): $k = k_0\phi(T)$. Here T is the temperature, k_0 is a positive constant. $\phi(T)$ is the Heavisida function or its smoothing.

$$\phi(T) = \begin{cases} 1, & T > T_i, \\ 0, & T \leq T_i \end{cases}$$

T_i is regarded as the ignition temperature.

3). Arrhenius: $k = k_0 exp(-E_0/T)$, T is the temperature, E_0 is the activation energy.

Related to the above three laws we have three different reaction equations:

1). CJ-model:

$$Z(x,t) = \begin{cases} z(x,0), & if \ \sup_{0\leq y\leq t} T(x,y) \leq T_i \\ 0, & otherwise \end{cases}$$

2). ZND-model:

$$z_t = -k\phi(T)z$$

3). Arrhenius:

$$z_t = -kexp(-E_0/T)z$$

Here first we consider the CJ-model:

$$\begin{cases} \tau_t - u_x = 0, \\ u_t + p_x = 0, \\ E_t + (pu)_x = 0, \\ z(x,t) = \begin{cases} z(x,0), & if \ \sup_{0\leq y\leq t} T(x,y) \leq T_i, \\ 0, & otherwise. \end{cases} \end{cases} \tag{1.1}$$

The Riemann problem is to find the solution of (1.1) which satisfies

$$(u,p,\tau,z)(x,0) = (u^{\pm}, p^{\pm}, \tau^{\pm}, z^{\pm}), \qquad as \ \pm x > 0, \tag{1.2}$$

where $(u^{\pm}, p^{\pm}, \tau^{\pm}, z^{\pm})$ are constant vectors. The typical case is that $z^+ = 1$, $z^- = 0$ and $T^+ = p^+\tau^+/R < T_i$, $T^- = p^-\tau^-/R > T_i$.

[C] consider the Riemann problem of this typical Riemann initial data. The solution may involve detonation, deflagration and streamline between the burnt and unburnt gases. He solves the Riemann problem under the assumption that the solution may only involve one detonation. [TCL] assume a deflagration velocity law and construct the Riemann solutions involving DT and DF. The one which involves a deflagration may fail to exist since the temperature at the back bank of it may fail to be higher than the ignition point. Giving up the DF velocity law of [TCL] and assuming that the temperature at the front bank of DF is the ignition point, [ZZ] obtain the complete pointwise entropy solutions. For the typical initial data, the Riemann problem may admit three solutions as depicted in Fig. 1.

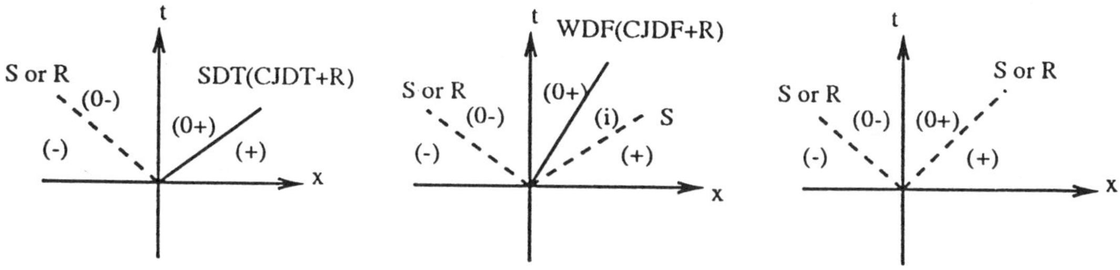

Fig. 1. Multiple pointwise entropy solutions to (1.1) with typical Riemann initial data.

The first two solutions are combustion ones and the third is a noncombustion solution. For the general Riemann initial data, such as $z^+ = z^- = 1$ and T^+, $T^- < T_i$, the problem may admit nine pointwise entropy solutions.

As we see in the figure of the deflagration solution, ahead of the DF wave, there is a shock wave which promote the temperature to the ignition point. In the three solutions, $x = 0$ is a streamline s.t. $[u] = [p] = 0$, $[\tau] \neq 0$.

Since the problem admits multiple solutions, one may ask: Is there any one which is not a physical solution? That is, the solutions you construct are for $k = +\infty$, whether they can be approached by the solutions to the model of finite reaction rate. Similar to [D1] [DD], [TZ1] study the following selfsimilar ZND-model:

$$\begin{cases} \tau_t - u_x = 0, \\ u_t + p_x = 0, \\ E_t + (pu)_x = 0, \\ z_t = -\frac{k}{t}\phi(T)z, \\ (u, p, \tau, z)(x, 0) = (u^\pm, p^\pm, \tau^\pm, z^\pm), \qquad as \ \pm x > 0. \end{cases} \tag{1.3}$$

For this problem, we prove that

(1). When k is large enough, (1.3) may possess three solutions, denoted by U_1^k, U_2^k and U_3^k. They structures of them are similar to those of the detonation, the deflagration and the noncombustion wave solutions depicted in Fig.1.

(2). When $k \to +\infty$, the following limits hold

$$\lim_{k \to +\infty} U_j^k(x,t) = U_j^0(x,t) \qquad a.e. \qquad j = 1, 2, 3,$$

where $U_j^0(x,t)$ represent the solutions of (1.1) depicted in Fig.1.

Here we would like to mention that in [ZZ] they also suggest a group of global entropy conditions so as to pick up the unique DT solution. However as we can see in [ZZ], whenever we have a DT solution, we always have a noncombustion wave solution $U_3^0(x,t)$ which is of course a Lax-entropy solution to the corresponding Riemann problem of the ZND-model for arbitary reaction rate $k > 0$. That is, if the global entropy solution is the limit of the solution to the ZND-model as $k \to +\infty$, the limit must be nonunique, which implies that the pointwise entropy condition (which allows multiple solutions) is more reasonable than the global entropy. And hence there comes the following conjecture:

Conjecture 1. For the pointwise entropy solutions depicted in Fig. 1., there always exists $\{k_n\}$: $k_n \to +\infty$ s.t. the following Riemann problem

$$\begin{cases} \tau_t - u_x = 0, \\ u_t + p_x = 0, \\ E_t + (pu)_x = 0, \\ z_t = -k\phi(T)z, \\ (u,p,\tau,z)(x,0) = (u^\pm, p^\pm, \tau^\pm, z^\pm), \qquad as\ \pm x > 0, \end{cases} \tag{1.4}$$

admits three solutions $\bar{U}_1^k(x,t)$, $\bar{U}_2^k(x,t)$ and $\bar{U}_3^k(x,t)$ and

$$\lim_{k \to +\infty} \bar{U}_j^k(x,t) = U_j^0(x,t) \qquad a.e. \qquad j = 1,2,3,$$

§2. Scalar Combustion Model. In this section, in stead of (1.1), here we use a more simple model to show how to approach the Riemann solutions. The model is as follows

$$\begin{cases} (u+qz)_t + f(u)_x = 0 \\ z(x,t) = \begin{cases} z(x,0), & if\ \sup_{0 \le y \le t} u(x,y) \le 0 \\ 0, & otherwise \end{cases} \end{cases} \tag{2.1}$$

where u is to represent a lumped variable the same as in [M]. $f \in C^2(R)$ satisfies $f''(u) > \delta > 0$ for all u under consideration. For the Riemann problem of (2.1), we clarify the following facts:

1). The discontinuity satisfies Rankine-Hugoniot condition

$$\sigma = [f]/[u+qz]$$

2). Combustion wave is defined as an discontinuity along which $[z] \neq 0$ and acocrding to the relation of σ, λ_ℓ and λ_r, we have six kinds of combustion waves: SDT, CJDT, WDT, WDF, CJDF, SDF (see [Z] for details)

3). Pointwise entropy conditions: SDT, CJDT, CJDF are admissible. The front bank of DF is the ignition temperature; noncombustion discontinuity satisfies Lax-entropy condition, unless it is a contact discontinuity.

4). Global entropy conditions: when we have multiple pointwise entropy solutions, we choose the one which is of minimum number of combustion waves as the admissible solution (refer to [LZ]).

5). Generalized entropy conditions: SDT, CJDT, WDT, WDF and CJDF are all admissible. The temperature u at the front bank of DF may be lower than (or equal) the ignition point $u_i = 0$; noncombustion wave satisfies the Lax-entropy condition (refer to [ST], [TZ2]). [LZ] directly get the global entropy solution for the Riemann problem of (2.1). [YT]'s results show that the global entropy solution is the limit of the Riemann solution to

$$\begin{cases} (u+qz)_t + f(u)_x = 0, \\ z_t = -k\phi(u)z, \end{cases} \tag{2.2}$$

as $k \to +\infty$. [JYZ] show that all the pointwise entropy solutions (the number of which may be four for some given Riemann initial data) are just the limits of the Riemann solutions of

$$\begin{cases} (u+qz)_t + f(u)_x = 0, \\ z_t = -\frac{k}{t}\phi(u)z, \end{cases} \tag{2.3}$$

as $k \to +\infty$. As we see that, under certain conditions, for instance, $f'(u) > 0,\ f''(u) > \delta > 0$, the limit solutions of (2.2) and (2.3) for $k \to +\infty$ are just the same. The multisolution phenomena occur only if f fails to fulfill the condition $f'(u) > 0$. [F] study the Riemann problem of

$$\begin{cases} (u+qz)_t + f(u)_x = \epsilon t u_{xx}, \\ z_t = -\frac{k}{t}\phi(u)z. \end{cases} \tag{2.4}$$

He obatins the existence for both the global solutions for $\epsilon,\ k > 0$ and the limit of $\lim_{\epsilon k = B, \epsilon \to 0^+, k \to +\infty} U^{\epsilon,k}(x/t)$ under the assumption that $f'(u) > 0,\ f''(u) > \delta > 0$. Under the same restrictions, [TZ2] prove that all of the WDT solutions (or all of the generalized entropy solutions) of (2.1), which form an one-parameter family, are the limits of the solution $U^{\epsilon,k}(x/t)$ of (2.4) as $\epsilon \to 0^+, k \to +\infty$. [ST] extend the results in [F] and [TZ2] to the case of WDF (under the assumption $f''(u) > \delta > 0$). And prove that all of the WDF solutions (which also form an one-parameter family) are the limits of the corresponding Riemann solutions of (2.4) as $\epsilon \to 0^+, k \to +\infty$.

For the scalar conservation laws of $f''(u) > 0$, [D] [DD]'s results imply that the vanishing viscosity limits of the Riemann solutions for the viscosity and the artificial viscosity models are the same. As stated above when $\epsilon = 0$, under the assumption $f'(u) > 0,\ f''(u) > \delta > 0$ the limit of the solution to (2.2) equals that of solution of (2.3). So we have the following

Conjecture 2. Assume $f'(u) > 0,\ f''(u) > \delta > 0$. Consider the Riemann problem of

$$\begin{cases} (u+qz)_t + f(u)_x = \epsilon u_{xx}, \\ z_t = -k\phi(u)z. \end{cases} \tag{2.5}$$

Then all of the WDT solutions for the Riemann problem of (2.1) are the limits of the solution of (2.5) as $\epsilon \to 0^+, k \to +\infty$

References

[C] A.J. Chorin, *Random choice methods with application to reacting gas flow*, J. Comput. Phys. 25(1977), 253-272

[CF] R. Courant and K.O. Friedrichs, *Supersonic Flow and Shock Waves*, Interscience, New York, 1948

[D] C.M. Dafermos, Solution of the Riemann problem for a class of hyperbolic systems of conservation laws by viscosity method, Arch. Rat. Mech. Anal. 52: 1-9

[DD] C.M. Dafermos and R.J. DiPerna, The Riemann problem for certain classes of hyperbolic systems of consrevation laws, J.D.E. 20: 90-114 (1976)

[F] E.Z. Fu, *Riemann problem for system of combustion model*, Master Thesis of Peking University, 1987

[JYZ] W. Jäger, H.M. Yang, T. Zhang, *Multisolution phenomena in combustion*, to appear

[LZ] T.P. Liu, T. Zhang, *A scalar combustion model*, Arch. Rat. Mech. Anal. 114 (1991), 297-312

[M] A. Majda, *A qualitative model for dynamic combustion*, SIAM. J. Appl. Math. vol. 41 (1981), p70-91

[ST] W.C. Sheng, D. Tan, *Weak deflagration wave solutions to the simplest combustion model*, to appear in J. Differential Eqs.

[TCL] Z.H. Teng, A.J. Chorin, T.P. Liu, *Riemann problem for reacting gas with application to transition*, SIAM J. Appl. Math. 42(1982), 964-981

[TY] Z. H. Teng, L.A. Ying, *The Riemann problem for hyperbolic model of combustion*, Chinese Anal. 6A(1985) 13-22

[TZ1] D. Tan, T. Zhang, *Riemann problem for the selfsimilar ZND-model in gas gynamical combustion*, J. Differential Eqs. vol. 95(1992), p331-369

[TZ2] D. Tan, T. Zhang, *Weak detonation wave solutions to the simplest combustion model*, to appear in Acta Mathematica Scientia

[YT] L.A. Ying, Z.H. Teng, *Riemann problem for a reaction and convection hyperbolic system*, Appr. Theor. Appl. 1(1984) pp95-122

[Z] T. Zhang, *The Riemann problem for combustion*, in Contemporary Math. 100 (1989), 114-124

[ZZ] T. Zhang, Y.X. Zheng, *Riemann problem for gasdynamic combustion*, J. Differential Eqs. vol. 77(1989) pp203-230

Current Address:
c/o Dr. Roberto Natalini
Istituto per le Applicazioni del Calcolo "M. Picone"
Viale de Policlinico, 137
00161 Roma, Italy

AMS/IP Studies in Advanced Mathematics
Volume **3**, 1997

OPTIMAL ERROR ESTIMATES FOR VISCOSITY METHOD AND MONOTONE SCHEME TO RIEMANN PROBLEM

TENG ZHEN-HUAN*

ABSTRACT. In this paper we derive optimal error bounds for viscosity method and monotone difference scheme to discontinuous solution of Riemann problem with both nonlinear fluxes and linear fluxes and show that the convergence rates for problem with nonlinear fluxes are much faster than that with linear fluxes. A rigorous analysis shows that both viscosity method and monotone scheme to approach the Riemann problem with nonlinear fluxes have error bounds of $O(\epsilon)$ and $O(\Delta x)$ in L^1 norm, respectively, where ϵ and Δx are their corresponding viscosity coefficient and discrete mesh length. Numerical experiments for Lax-Friedrichs scheme, Godunov scheme and Engquist-Osher scheme are presented and numerical results justify the theoretical analysis.

Key words. error estimate, viscosity method, monotone difference scheme, conservation laws

AMS(MOS) subject classifications. primary 65M25; secondary 76D05

§1 Introduction

It is widely believed that some approximate methods, such as monotone difference schemes, to approach discontinuous solutions for conservation laws with nonlinear fluxes perform better than that with linear fluxes (see Harten [4]). This property has important implications in numerical calculations ([4]). In this paper we justify this observation and derive optimal L^1 error estimates for both viscosity method and monotone difference scheme to the following Riemann problem:

$$\frac{\partial u}{\partial t} + \frac{\partial f(u)}{\partial x} = 0 \tag{1.1a}$$

$$u\Big|_{t=0} = u_0(x) \tag{1.1b}$$

with

$$u_0(x) = \begin{cases} u_l, & x < 0 \\ u_r, & x > 0, \end{cases} \tag{1.2}$$

* Supported in part by the National Natural Science Foundation of China and the Science Found of the Education Commission of China .

where u_l and u_r are given constants and the flux $f(u)$ with u_l and u_r satisfies Oleinik's condition E (see [12])

$$\frac{f(u)-f(u_l)}{u-u_l} > D > \frac{f(u)-f(u_r)}{u-u_r} \tag{1.3a}$$

for all u strictly between u_l and u_r, and satisfies Lax geometric condition ([10])

$$f'(u_l) > D > f'(u_r), \tag{1.3b}$$

where

$$D = \frac{f(u_l)-f(u_r)}{u_l-u_r} \tag{1.3c}$$

is the shock speed. It is known that under the entropy condition (1.3) the Riemann solution of (1.1) and (1.2) is a discontinuous solution

$$u(x,t) = \begin{cases} u_l, & x < Dt \\ u_r, & x > Dt, \end{cases} \tag{1.4}$$

where D is the speed of the discontinuity.

Remark If the flux $f(u)$ is strictly convex $f''(u) > 0$, then (1.3) is equivalent to $u_l > u_r$.

In this paper we will prove the following main results. For convenience only, we will take $u_r < u_l$ in what follows.

Theorem 1. *Let the entropy conditions (1.3a) and (1.3b) be satisfied. If v_ϵ and $w_{\Delta x}$ are solutions of viscosity method and monotone scheme, defined by (1.7) and (1.8) below respectively, to the Riemann problem (1.1) and (1.2), then the following error bounds are fulfilled for any $t > 0$*

$$\|v_\epsilon(\cdot,t) - u(\cdot,t)\|_{L^1(\mathbb{R})} \le C\epsilon \tag{1.5}$$

and

$$\|w_{\Delta x}(\cdot,t) - u(\cdot,t)\|_{L^1(\mathbb{R})} \le C\Delta x, \tag{1.6}$$

where u is the Riemann solution of (1.1) and (1.2) given by (1.4), ϵ and Δx are the viscosity coefficient and the discrete mesh length of the two approximate methods respectively and C are constants independent of ϵ, Δx and t.

We now give a more detailed description of the viscosity method and the monotone difference scheme. The viscosity method approaches the scalar conservation laws (1.1) by solving the parabolic equation

$$\frac{\partial v_\epsilon}{\partial t} + \frac{\partial f(v_\epsilon)}{\partial x} = \epsilon\frac{\partial^2 v_\epsilon}{\partial x^2} \tag{1.7a}$$

$$v_\epsilon\big|_{t=0} = u_0(x) \tag{1.7b}$$

and the entropy solution u of (1.1) can be constructed as the limit of solutions v_ϵ as $\epsilon \to +0$.

The monotone scheme is a finite-difference approach to (1.1). A $(p+q+1)$-point conservative finite-difference scheme

$$\begin{aligned} w_j^{n+1} &= H(w_{j-p}^n, w_{j-p+1}^n, \dots, w_{j+q}^n) \\ &= w_j^n - \lambda[\bar{f}(w_{j-p+1}^n, \dots, w_{j+q}^n) - \bar{f}(w_{j-p}^n, \dots, w_{j+q-1}^n)], \end{aligned} \tag{1.8a}$$

is said to be monotone if H is a monotone non-decreasing function of each of its arguments, and is said to be consistent with the scalar conservation laws (1.1) if the numerical flux $\bar{f}$ satisfies

$$\bar{f}(w, \dots, w) = f(w), \tag{1.8b}$$

where $\lambda = \Delta t/\Delta x = \text{const.}$, p and q are given nonnegative integers and

$$w_j^0 = T_{\Delta x}(u_0)(x_j) = \frac{1}{\Delta x}\int_{x_j-\Delta x/2}^{x_j+\Delta x/2} u_0(x)\,dx, \quad x_j = j\Delta x. \tag{1.8c}$$

Extend the lattice function w_j^n to continuous values of x and t by setting

$$w_{\Delta x}(x,t) = w_j^n \quad \text{for} \quad (x,t) \in [x_{j-1/2}, x_{j+1/2}) \times [t_n, t_{n+1}), \tag{1.8d}$$

where $x_{j+1/2} = (j+1/2)\Delta x$, $t_n = n\Delta t$, $n \in \mathbb{N}$ and $j \in \mathbb{Z}$.

Stability, convergence and error estimate for the monotone difference schemes can be found in [1] [9] and [13].

In order to compare the error bounds for nonlinear flux problem given in Theorem 1 with that for linear flux one, we consider the linear advection equation

$$\begin{aligned} \frac{\partial u}{\partial t} + a\frac{\partial u}{\partial x} &= 0 \qquad (a = \text{const.}) \\ u|_{t=0} &= u_0(x). \end{aligned} \tag{1.9}$$

Thus the viscosity solution v_ϵ for the linear advection equation is defined as the solution of the linear parabolic equation

$$\begin{aligned} \frac{\partial v_\epsilon}{\partial t} + a\frac{\partial v_\epsilon}{\partial x} &= \epsilon\frac{\partial^2 v_\epsilon}{\partial x^2} \\ v_\epsilon\big|_{t=0} &= u_0(x), \end{aligned} \tag{1.10}$$

and the liner $(p+q+1)$-point monotone difference scheme for (1.9) is of the form

$$w_j^{n+1} = \sum_{s=-p}^{q} a_s w_{j+s}^n \tag{1.11a}$$

where

$$a_s \geq 0 \qquad \text{for} \quad s = -p, \dots, q. \tag{1.11b}$$

The consistent condition (1.8b) implies that

$$\sum_{s=-p}^{q} a_s = 1 \tag{1.11c}$$

and

$$\sum_{s=-p}^{q} s a_s = -\lambda a. \tag{1.11d}$$

In the following Theorem 2 we will assume that the difference scheme (1.11a) is not a pure translation, i.e. there at least two coefficients among $\{a_s\}_{-p}^{q}$ are not equal to zero. Now we present an error estimate theorem for the linear advection equation.

Theorem 2. *If v_ϵ and $w_{\Delta x}$ are solutions of viscosity method and monotone scheme, defined by (1.10) and (1.11) respectively, to the Riemann problem (1.9) and (1.2), then the following error bounds are fulfilled for any $t > 0$*

$$c\sqrt{t}\sqrt{\epsilon} \leq \|v_\epsilon(\cdot, t) - u(\cdot, t)\|_{L^1(\mathbb{R})} \leq C\sqrt{t}\sqrt{\epsilon} \tag{1.12}$$

and

$$c\sqrt{t}\sqrt{\Delta x} \leq \|w_{\Delta x}(\cdot, t) - u(\cdot, t)\|_{L^1(\mathbb{R})} \leq C\sqrt{t}\sqrt{\Delta x}, \tag{1.13}$$

where u is the Riemann solution of (1.9) and (1.2) given by

$$u(x,t) = \begin{cases} u_l, & x < at \\ u_r, & x > at, \end{cases}$$

ϵ and Δx are the viscosity coefficient and the discrete mesh length of the two approximate methods respectively and c, C are positive constants independent of ϵ, Δx and t.

The proof of (1.12) follows easily from an explicit expression of (1.10) and (1.2)

$$v_\epsilon(x,t) = u_r + (u_l - u_r)\,\mathrm{erfc}\left(\frac{x - at}{2\sqrt{\epsilon t}}\right)$$

and the asymptotic expansion of erfc(z), and the estimation of (1.13) is rigorously established by Tang and Teng in [15].

In comparison of (1.5) and (1.6) with (1.12) and (1.13) we find that the performance of the two approximate methods for the nonlinear flux problem is indeed better than that for the linear flux one. Not only the order of the convergence rates for nonlinear fluxes increase from half order to one order, but also the coefficients C in (1.5) and (1.6) are independent of t while the coefficients $c\sqrt{t}$ and $C\sqrt{t}$ in (1.12) and (1.13) go to infinity as $t \to \infty$.

Theorem 1 also justifies an observation by Lucier in [11] that Godunov' method is of $O(\Delta x)$ accurate for (1.1) and (1.2) with flux $f(u) = u^2/2$. This is because monotone schemes include most of the popular first order schemes such as Lax-Friedrichs scheme, Godunov scheme [3] and Engquist-Osher scheme [2].

Several authors provide convergence rates for various approximate methods to discontinuous solutions of conservation laws (see, e.g. [5, 8, 9, 11, 13-18]). Although their methods are different, the convergence rates in L^1 norm obtained are half order in most of the cases. Therefore, Theorem 1 having one order convergence rate may provide a clue on how to break the half order barrier to the error estimate.

The rest of this paper is organized as follows: in Section 2 we will introduce stability lemmas and traveling wave lemmas for viscosity method and monotone scheme and then prove the main results, Theorem 1, and in section 3 we will present some numerical experiments to verify the theoretical analysis.

§2 Proof of Theorem 1

First we establish stability lemmas and travelling wave lemmas for both the viscosity method and the monotone difference scheme. These two kinds of lemmas play center role in proving Theorem 1.

Lemma 2.1 (Existence). *If $u_0(x)$ belongs to $L^\infty(\mathbb{R})$, then (1.7) has a unique solution $v_\epsilon(x,t)$ which satisfies:*

(I) $v_\epsilon(x,t)$ has continuous derivatives involved in (1.7) for $t > 0$,

(II) $v_\epsilon(x,t)$ is uniformly bounded by

$$\|v_\epsilon(\cdot,t)\|_{L^\infty(\mathbb{R})} \le \|u_0(\cdot)\|_{L^\infty(\mathbb{R})}, \quad \forall t > 0, \tag{2.1}$$

(III) $v_\epsilon(x,t)$ assumes the initial value $u_0(x)$ in the following sense

$$\lim_{t\to+0} \|v_\epsilon(\cdot,t) - u_0(\cdot)\|_{L^1([-R,R])} = 0, \quad \forall R > 0. \tag{2.2}$$

The proof of this lemma can be found in [7]. Next we introduce stability results for viscosity solutions of (1.7).

Lemma 2.2 (Stability). *If $v_\epsilon^{(i)}(x,t)$ for $i = 1,2$ are solutions of (1.7) with initial date $u_0^{(i)}(x) \in L^\infty(\mathbb{R})$ and $u_0^{(1)}(x) - u_0^{(2)}(x)$ belongs $L^1(\mathbb{R})$, then*

$$\|v_\epsilon^{(1)}(\cdot,t) - v_\epsilon^{(2)}(\cdot,t)\|_{L^1(\mathbb{R})} \le \|u_0^{(1)}(\cdot) - u_0^{(2)}(\cdot)\|_{L^1(\mathbb{R})}. \tag{2.3}$$

The proof is made by Lax in [10].

A travelling wave solution is solution of (1.7a) in the form $v_\epsilon(x,t) = V_\epsilon(x - St)$ where the wave speed S is a constant. We have the following existence lemma for the travelling wave solution.

Lemma 2.3 (Travelling wave). *Let the assumption (1.3) be satisfied, then, for each $u_m \in (u_r, u_l)$, there is a unique travelling wave solution $V_\epsilon(x - Dt)$ of (1.7a) taking on the values $V_\epsilon(0) = u_m$, $V_\epsilon(-\infty) = u_l$ and $V_\epsilon(\infty) = u_r$, which has the following properties:*

(I) *$V_\epsilon(x - Dt) = V((x - Dt)/\epsilon)$, where $V(\xi)$ is defined implicitly by*

$$\xi(V) = \int_{u_m}^{V} \Phi(u)^{-1}\, du, \tag{2.4a}$$

where

$$\Phi(u) = f(u) - f(u_r) - D(u - u_r); \tag{2.4b}$$

(II) *$u_l > V_\epsilon(\xi) > u_r$ for all $\xi \in \mathbb{R}$;*

(III) *$V(\xi)$ ($V_\epsilon(\xi)$ as well) is a decreasing function for $\xi \in \mathbb{R}$ and approaches u_l and u_r exponentially in ξ as $\xi \to \mp\infty$;*

(IV) *The following estimates hold:*

$$A \equiv \|u_0(\cdot) - V(\cdot)\|_{L^1(\mathbb{R})} < \infty \tag{2.5}$$

and

$$\|u_0(\cdot) - V_\epsilon(\cdot)\|_{L^1(\mathbb{R})} = A\epsilon, \tag{2.6}$$

where $u_0(x)$ is given by (1.2).

Proof. The expression (2.4) can be easily derived by substituting $V_\epsilon(x - Dt) = V((x-Dt)/\epsilon)$, with $V_\epsilon(0) = u_m$, into (1.7a) and, in view of the Oleinik's condition E (1.3a), or equivalent $\Phi(u) < 0$ for $u \in (u_r, u_l)$, and $\Phi(u_l) = \Phi(u_r) = 0$, the inverse function of (2.4a) also satisfies $V_\epsilon(-\infty) = u_l$, and $V_\epsilon(\infty) = u_r$.

Property (II) follows immediately from (2.4a) and $V'(\xi) = \Phi(V) < 0$.

In order to prove (III), we show that there exist $0 < \alpha < \beta$ such that

$$\beta(u - u_l)(u - u_r) < \Phi(u) < \alpha(u - u_l)(u - u_r), \quad \forall u \in (u_r, u_l). \tag{2.7}$$

This is due to the entropy conditions (1.3a) and (1.3b), or equivalent $\Phi(u) < 0$ for $u \in (u_r, u_l)$ and $\Phi'(u_r) < 0$ and $\Phi'(u_l) > 0$ (see Fig. 1).

Thus, (2.7) and (2.4) imply that

$$\alpha^{-1}\bar{\xi}(V) < \xi(V) < \beta^{-1}\bar{\xi}(V), \tag{2.8}$$

where

$$\begin{aligned}\bar{\xi}(V) &= \int_{u_m}^{V} \frac{1}{(u - u_l)(u - u_r)}\, du \\ &= \frac{1}{u_j - u_r} \ln\left[\left(\frac{V - u_l}{V - u_r}\right)\left(\frac{u_m - u_r}{u_m - u_l}\right)\right].\end{aligned}$$

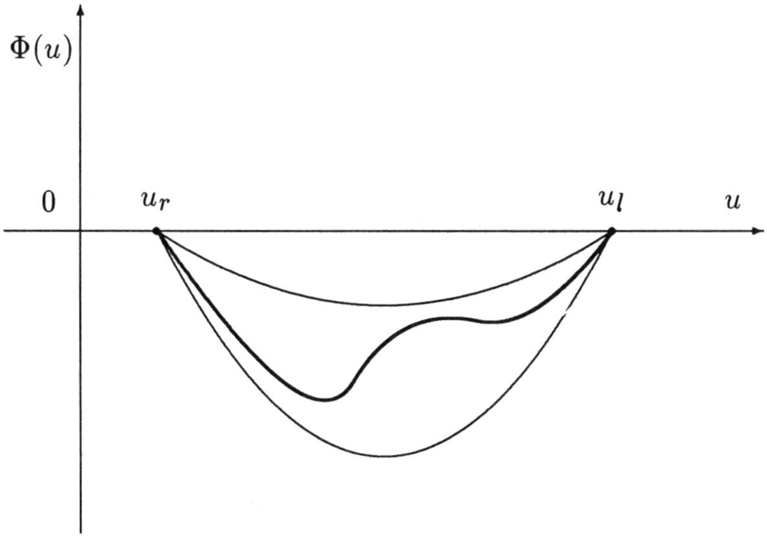

Figure 1. Entropy conditions (1.3a) and (1.3b). Dark line is $\Phi(u)$, light lines are $\alpha(u-u_l)(u-u_r)$ and $\beta(u-u_l)(u-u_r)$.

Since all three functions in (2.8) are decreasing functions for $V \in (u_r, u_l)$, their inverse functions also satisfy a similar inequality in the same order

$$\bar{V}(\alpha\xi) < V(\xi) < \bar{V}(\beta\xi), \quad \xi \in \mathbb{R}, \tag{2.9}$$

where

$$\begin{aligned}\bar{V}(\xi) = \left(u_l + u_r\left(\frac{u_l-u_r}{u_m-u_r}\right)\exp\{\xi(u_l-u_r)\}\right) \\ \times\left(1+\left(\frac{u_l-u_r}{u_m-u_r}\right)\exp\{\xi(u_l-u_r)\}\right)^{-1}.\end{aligned} \tag{2.10}$$

It is easy to see from (2.10) that $\bar{V}(\xi)$ approaches u_l and u_r exponentially in ξ as $\xi \to \mp\infty$, and so does $V(\xi)$. Hence (III) is proved.

We can see that (2.5) is an immediate conclusion from the property (III). Substituting $V_\epsilon(\xi) = V(\xi/\epsilon)$ and $u_0(\xi) = u_0(\xi/\epsilon)$ (because of the special form of (1.2)) into the left hand integration of (2.6) and using integration by substitution, we obtain the equality (2.6). We have now completed the proof of Lemma 2.3. □

Remark From the proof we see that the Lax geometric condition (1.3b) is necessary for proving Property (III). This condition is also needed for proving a similar property in the following Lemma 2.5.

Similarly, we can also establish the stability lemma and the travelling wave lemma for the monotone scheme.

Lemma 2.4 (Stability). *If $w^{(i)}_{\Delta x}(x,t)$ for $i = 1,2$ are solutions of (1.8) with initial date $u^{(i)}_0(x) \in L^\infty(\mathbb{R})$ and $u^{(1)}_0(x) - u^{(2)}_0(x)$ belongs $L^1(\mathbb{R})$, then*

$$\|w^{(1)}_{\Delta x}(\cdot,t) - w^{(2)}_{\Delta x}(\cdot,t)\|_{L^1(\mathbb{R})} \le \|u^{(1)}_0(\cdot) - u^{(2)}_0(\cdot)\|_{L^1(\mathbb{R})}. \tag{2.11}$$

The proof can be found in [1].

In what follows we introduce the travelling wave solution to the monotone scheme (1.8a) by following Jennings' work [6]. A travelling wave solution of (1.8a) moving with speed D satisfies the difference equation

$$W_{z-D\lambda} = H(W_{z-p}, W_{z-p+1}, \cdots, W_{z+q}). \tag{2.12}$$

In this paper we assume that D is defined by (1.3c).

The minimal domain on which (2.12) makes sense consists of functions defined on the linear span over the integers of $\eta = D\lambda$ and 1. Call the closure of the set $\mathcal{L}_\eta$. If η is rational, $\mathcal{L}_\eta$ is discrete, if η is irrational, then $\mathcal{L}_\eta$ is the entire real line. $L^1(\mathcal{L}_\eta)$ is the space of absolutely integrable functions on $\mathcal{L}_\eta$ with the usual sense.

Lemma 2.5 (Travelling wave). *Let the assumption (1.3) be satisfied and $|\eta| < 1$, then, for each $u_m \in (u_r, u_l)$, there is a travelling wave solution $W_{z-\eta}$ of (2.12) continuous on $\mathcal{L}_\eta$ taking on the values $W_0 = u_m$, $W_{-\infty} = u_l$ and $W_\infty = u_r$, which is a monotone nonincreasing function of z and approaches u_l and u_r exponentially in z as $z \to \mp\infty$.*

The lemma for strictly monotone scheme is due to Jennings [6] and it is generalized by Engquist and Osher [2] to any weakly monotone scheme of the form (1.8a) having the property that for any $\epsilon > 0$ sufficiently small there exists a strictly monotone scheme $H^\epsilon(w_{j-p}, \cdots, w_{j+q})$ such that $H^\epsilon(w_{j-p}, \cdots, w_{j+q}) \to H(w_{j-p}, \cdots, w_{j+q})$ as $\epsilon \to 0$.

For rational η, the set $\mathcal{L}_\eta$ is discrete. So it is of the form $\{z_j\}^\infty_{-\infty}$ with $z_j < z_{j+1}$, $\forall j \in \mathbb{Z}$. we extend the discrete travelling wave solution W_{z_j} ($j \in \mathbb{Z}$) to continuous values of ξ ($\xi \in \mathbb{R}$) by setting

$$W(\xi) = W_{z_j}, \quad \xi \in [z_j, z_{j+1}), \tag{2.13a}$$

where $j \in \mathbb{Z}$.

For irrational η we identify $W(\xi)$ with W_ξ i.e.,

$$W(\xi) = W_\xi, \quad \xi \in \mathbb{R}. \tag{2.13b}$$

For $W(\xi)$ we have the following lemma.

Lemma 2.6. *Under the conditions of Lemma 2.5 we have*

$$B \equiv \|W(\cdot) - u_0(\cdot)\|_{L^1(\mathbb{R})} < \infty, \tag{2.14}$$

where $u_0(x)$ is given by (1.3).

Proof. Lemma 2.5 indicates that $W(\xi)$ approaches u_l and u_r exponentially in ξ as $\xi \to \mp\infty$. Therefore the integration in (2.14) converges and it is bounded above. □

Now we turn to Theorem 1 and the rest of this section is devoted to proving it.

Proof of Theorem 1. We first to prove the estimate (1.5). From (1.4) and (1.2) we see that the Riemann solution of (1.1) and (1.2) can be expressed as

$$u(x,t) = u_0(x - Dt), \tag{2.15}$$

where $u_0(x)$ is given by (1.2). Substituting (2.15) into the right hand side of (1.5) and using the triangular inequality we obtain

$$\begin{aligned}
&\|v_\epsilon(\cdot,t) - u(\cdot,t)\|_{L^1(\mathbb{R})}\\
&\quad = \|v_\epsilon(\cdot,t) - u_0(\cdot - Dt)\|_{L^1(\mathbb{R})}\\
&\quad \le \|v_\epsilon(\cdot,t) - V_\epsilon(\cdot - Dt)\|_{L^1(\mathbb{R})} + \|V_\epsilon(\cdot - Dt) - u_0(\cdot - Dt)\|_{L^1(\mathbb{R})}\\
&\quad = \|v_\epsilon(\cdot,t) - V_\epsilon(\cdot - Dt)\|_{L^1(\mathbb{R})} + \|V_\epsilon(\cdot) - u_0(\cdot)\|_{L^1(\mathbb{R})},
\end{aligned} \tag{2.16}$$

where $V_\epsilon(x - Dt)$ is the travelling wave solution of (1.7a) given in Lemma 2.3. Since $V_\epsilon(x - Dt)$ is also a solution of (1.7a), so (2.3) implies that

$$\|v_\epsilon(\cdot,t) - V_\epsilon(\cdot - Dt)\|_{L^1(\mathbb{R})} \le \|u_0(\cdot) - V_\epsilon(\cdot)\|_{L^1(\mathbb{R})}.$$

Therefore substituting this into (2.16) yields

$$\|v_\epsilon(\cdot,t) - u(\cdot,t)\|_{L^1(\mathbb{R})} \le 2\,\|u_0(\cdot) - V_\epsilon(\cdot)\|_{L^1(\mathbb{R})},$$

and applying (2.6) to the above inequality gives the desired (1.5) with a constant $C = 2A$, where A is defined by (2.5) and is obviously independent of ϵ and t. This has verified (1.5).

Next we turn to (1.6). Let

$$W_{\Delta x}(x,t) = W\left(\frac{x - Dt_n}{\Delta x}\right), \tag{2.17}$$

where $W(\xi)$ is defined by (2.13) and $t_n = n\Delta t$ with $n = [t/\Delta t]$. Since $W(\xi)$ is the travelling wave solution of (1.8a), it satisfies

$$W(\xi - D\lambda) = H(W(\xi - p), \cdots, W(\xi + q)), \quad \xi \in \mathbb{R}.$$

From this we see that $W_{\Delta x}(x,t)$ is also a solution of (1.8a), therefore (2.11) implies that

$$\begin{aligned}
\|w_{\Delta x}(\cdot,t) - W_{\Delta x}(\cdot,t)\|_{L^1(\mathbb{R})} &\le \|w_{\Delta x}(\cdot,0) - W_{\Delta x}(\cdot,0)\|_{L^1(\mathbb{R})}\\
&= \|u_0(\cdot) - W(\cdot/\Delta x)\|_{L^1(\mathbb{R})}\\
&= \|u_0(\cdot) - W(\cdot)\|_{L^1(\mathbb{R})}\Delta x = B\Delta x,
\end{aligned} \tag{2.18}$$

where $u_0(x)$ is given by (1.2) and the last two equalities follow from $u_0(x) = u_0(x/\Delta x)$ and (2.14) respectively. By using (2.18), (2.17) and by using a similar argument as in (2.16) we have

$$\begin{aligned} &\|w_{\Delta x}(\cdot,t) - u(\cdot,t)\|_{L^1(\mathbb{R})} \\ &\qquad \leq \|w_{\Delta x}(\cdot,t) - W_{\Delta x}(\cdot,t)\|_{L^1(\mathbb{R})} + \|W_{\Delta x}(\cdot,t) - u_0(\cdot - Dt)\|_{L^1(\mathbb{R})} \\ &\qquad \leq B\Delta x + \|W((\cdot - Dt)/\Delta x) - u_0(\cdot - Dt)\|_{L^1(\mathbb{R})} = 2B\Delta x. \end{aligned}$$

So we arrive at the conclusion (1.6) with a constant $C = 2B$, where B is defined by (2.14) and is independent of Δx and t. This has finished the proof of Theorem 1. □

§3 Numerical Experiments

In this section we will present some numerical experiments to verify the theoretical analysis, i.e., the convergence rates of monotone schemes for the Riemann problem with nonlinear fluxes are of one order, but the rates for problem with linear fluxes are of only half order.

The Riemann problem we consider is of the form

$$\frac{\partial u}{\partial t} + \frac{\partial f(u)}{\partial u} = 0$$
$$u|_{t=0} = \begin{cases} u_l, & x < 0 \\ u_r, & x > 0 \end{cases} \tag{3.1a}$$

with

$$u_l = 2, \quad u_r = -1. \tag{3.1b}$$

We take

$$f(u) = \frac{u^2}{2} \tag{3.2}$$

for the nonlinear fluxes problem and

$$f(u) = \frac{1}{2}u \tag{3.3}$$

for the linear fluxes problem. It is easy to show that both problem (3.1), (3.2) and problem (3.1), (3.3) have the same discontinuous solution

$$u(x,t) = \begin{cases} 2, & x < \frac{1}{2}t \\ -1, & x > \frac{1}{2}t. \end{cases} \tag{3.4}$$

The monotone difference schemes we consider here are Lax-Friedrichs scheme, Godunov scheme and Engquist-Osher scheme, which are defined below.

$$w_j^{n+1} = \frac{1}{2}(w_{j-1}^n + w_{j+1}^n) - \frac{\lambda}{2}(f(w_{j+1}^n) - f(w_{j-1}^n)), \qquad \text{(Lax-Friedrichs)}$$

$$w_j^{n+1} = w_j^n - \lambda(f(U(0; w_j^n, w_{j+1}^n)) - f(U(0; w_{j-1}^n, w_j^n))) \qquad \text{(Godunov)}$$

and

$$w_j^{n+1} = w_j^n - \lambda(f_-(w_{j+1}^n) - f_-(w_j^n) + f_+(w_j^n) - f_+(w_{j-1}^n)), \qquad \text{(Engquist-Osher)}$$

where $U(x/t; u_l, u_r)$ is the Riemann solution of (3.1a) and, for convex $f(u)$ with $f'(\bar{u}) = 0$, $f_-(w) = f(\min(w, \bar{u}))$ and $f_+(w) = f(\max(w, \bar{u}))$. A simple analysis reveals that all of the three schemes are monotone schemes provided that the C.F.L. condition

$$\lambda \max_{|u| \leq \sup |w_j^0|} |f'(u)| \leq 1$$

holds.

For linear flux $f(u) = au$ with $a \geq 0$, the Lax-Friedrichs takes the form

$$w_j^{n+1} = \frac{1}{2}(w_{j-1}^n + w_{j+1}^n) - \frac{\lambda}{2} a(w_{j+1}^n - w_{j-1}^n), \qquad \text{(Lax-Friedrichs)}$$

while Both Engquist-Osher scheme and Godunov scheme reduce to the conventional upwind scheme

$$w_j^{n+1} = w_j^n - \lambda a(w_j^n - w_{j-1}^n). \qquad \text{(Upwind)}$$

Let us recall some of the notations defined earlier in this paper.

$$\lambda = \frac{\Delta t}{\Delta x} = \text{constant}; \quad w_j^n \approx u(x_j, t_n), \quad x_j = j\Delta x, \quad t_n = n\Delta t.$$

Numerical results for Riemann problems with nonlinear and linear fluxes are given in Table 1 and Table 2, respectively, where the errors mean $\|u - w_{\Delta x}\|_{L^1}$ at $t = 1$ with different steps Δx and $\lambda = 1/4$. The numerical results clearly indicate that for all of the three monotone difference schemes the convergence rates for nonlinear problem (3.1), (3.2) are one order, but the rates for linear problem (3.1), (3.3) are only half order. This complies well with the theoretical analysis of Theorem 1 and Theorem 2.

Table 1. L^1 errors and convergence rates for Riemann problem (3.1) and (3.2)

$\lambda = 1/4$	Lax-Friedrichs		Godunov		Engquist-Osher	
Δx	L^1-error	L^1-rate	L^1-error	L^1-rate	L^1-error	L^1-rate
2^{-1}	1.507616		0.0173977		0.0441255	
2^{-2}	0.9057875	0.7350247	0.0089001	0.9670025	0.0225604	0.9678159
2^{-3}	0.4976002	0.8641856	0.0044543	0.9985997	0.0112905	0.9986827
2^{-4}	0.2550494	0.9642104	0.0022272	0.9999565	0.0056492	1.0000000
2^{-5}	0.1277514	0.9974374	0.0011136	0.9999946	0.0028226	1.0000000
2^{-6}	0.0638763	0.9999852	0.0005568	1.0000000	0.0014113	1.0000000
2^{-7}	0.0319381	1.0000000	0.0002784	1.0000000	0.0007056	1.0000000
2^{-8}	0.0159690	1.0000000	0.0001392	1.0000000	0.0003528	1.0000000
2^{-9}	0.0079845	1.0000000	0.0000690	1.0000000	0.0001764	1.0000000

Table 2. L^1 errors and convergence rates for Riemann problem (3.1) and (3.3)

$\lambda = 1/4$	Lax-Friedrichs		Upwind	
Δx	L^1-error	L^1-rate	L^1-error	L^1-rate
2^{-1}	2.7253904		0.6358240	
2^{-2}	2.0394516	0.4182816	0.5168118	0.2977627
2^{-3}	1.5059696	0.4374886	0.4089656	0.3376593
2^{-4}	1.0987356	0.4548485	0.3152825	0.3753347
2^{-5}	0.7945614	0.4676135	0.2377254	0.4073488
2^{-6}	0.5709159	0.4768806	0.1761459	0.4325255
2^{-7}	0.4083289	0.4835464	0.1288193	0.4514214
2^{-8}	0.2910830	0.4883013	0.0933134	0.4651922
2^{-9}	0.2070248	0.4916266	0.0671389	0.4749352

REFERENCES

[1] M. Crandall and A. Majda, Monotone difference approximations for scalar conservation laws, *Math. Comp.* **34,** 1980, pp. 1-21.

[2] B. Engquist and S. Osher, One-side difference approximations for nonlinear conservation laws, *Math. Comp.* **36,** 1981, pp. 321-351.

[3] S.K. Godunov, Finite difference methods for numerical computations of discontinuous solutions of the equations of fluid dynamics, *Mat. Sb.*, **47**, 1959, pp. 271-295 (Russian).

[4] A. Harten, The artificial compression method for computation of shocks and contact discontinuities, *Comm. Pure Appl. Math.*, **30**, 1977, pp. 611-638.

[5] D. Hoff and J. Smoller, Error bounds for Glimm difference approximations for scalar conservation laws, *Trans. Amer. Math. Soc.*, **289**, 1985, pp. 611-642.

[6] G. Jennings, Discrete shocks, *Comm. Pure Appl. Math.*, **27**, pp. 25-37.

[7] S.N. Kruzkov, First order quasi-linear equations with several variables, *Math. USSR Sb.*, **10**, 1970, pp. 217-243.

[8] N.N. Kuznetsov, Accuracy of some approximate methods for computing the weak solutions of a first-order quasi-linear equation, *USSR Comput. Math. and Math. Phys.*, **16** 1976, pp.105-119.

[9] N.N. Kuznetsov, On stable methods for solving non-linear first order partial differential equations in the class of discontinuous functions, *Royal Irish Academy Conference on Numerical Analysis: Topics in Numerical Analysis III* (J.J.H. Miller, ed), 1977, pp. 183-197.

[10] P.D. Lax, Hyperbolic Systems of Conservation Laws and the Mathematical Theory of Shock Waves, *SIAM Regional Conference Series in Applied Mathematics, SIAM, Philadelphia*, 1973.

[11] B.J. Lucier, Error bounds for the methods of Glimm, Godunov and LeVeque, *SIAM J. Numer. Anal.*, **22**, 1985, pp. 1074-1081.

[12] O.A. Oleinik, Uniqueness and stability of the generalized solution of the Cauchy problem for a quasilinear equation, *Uspekhi Mat. Nauk* **2(86)**, 1959, pp. 165-170; English translation in *Amer. Math. Soc. Trans.*, **2(33)**, pp. 285-290.

[13] R. Sanders, On convergence of monotone finite difference schemes with variable spatial differencing, *Math. Comp.*, **40**, 1983, pp. 91-106.

[14] E. Tadmor, Local error estimates for discontinuous solutions of nonlinear hyperbolic equations, *SIAM J. Numer. Anal.*, **28**, 1991, pp. 891-906.

[15] T. Tang and Z.H. Teng, A lower error bound of monotone difference schemes for discontinuous solutions, 1993, to appear.

[16] T. Tang and Z.H. Teng, Error bounds for fractional step methods for conservation laws with source terms, 1992, to appear in *SIAM J. Numer. Anal.*

[17] Z.H. Teng, On the accuracy of fractional step methods for conservation laws in two-dimensions, *SIAM J. Numer. Anal.*, **31**, 1994.

[18] A. Tveito and R. Winther, An error estimate for a finite difference scheme approximating a hyperbolic system of conservation laws, *SIAM J. Numer. Anal.*, **30**, 1993, pp. 401-424.

DEPARTMENT OF MATHEMATICS, PEKING UNIVERSITY, BEIJING 100871, P.R. CHINA

E-mail address: tengzh%bepc2@scs.slac.stanford.edu

AMS/IP Studies in Advanced Mathematics
Volume 3, 1997

A NEW FRAMEWORK OF NONLINEAR HYDRODYNAMICS [1, 2]

Wu, Jianghang

ABSTRACT

In this framework a transformation for viscous incompressible flows in arbitrary bounded domain is provided by solving an eigenvalue problem of a symmetric positive definite matrix for generating a set of orthonormal solenoidal vector bases of velocity and a Helmholtz problem for generating a set of scalar bases such as density, temperature and concentration, which can transform the Navier-Stokes, Boussinesq and various scalar transport equations into a system of canonical nonlinear ordinary differential equations. The mathematical transformation for viscous homogenous and stratified and heat convective fluid flow is presented, and the numerical method is proposed to generate the bases and solve the canonical nonlinear differential equations. The numerical test of the new framework and the exploration of some directions in nonlinear hydrodynamics show that the new framework is of particular advantage and can be used to solve a wide range of problems in nonlinear hydrodynamics.

I INTRODUCTION

Recent striking advances in NONLINEAR HYDRODYNAMICS such as discovery of coherent structures in turbulence (both spatially organized large scale structures and disorganized motions, generally at smaller scales in existence); the development of hydrodynamics stability theory, which relates stability with bifurcation for the solutions of the Navier-Stokes equations; the formation of chaotic fluid dynamics, which will be one

1. The numerical computation is made by Zhu, Huayang

2. This research is supported by grant from the National Natural Science Foundation of China

of the most stimulating developments of the subject in our time; the theoretical proof and experimental confirmation of generic finiteness of the solutions of the Navier-Stokes equations, which implies a finite number of degrees of freedom is sufficient to resolve the structure of the attractors of infinite-dimensional flows; and so forth, challenge us to find new mathematical frameworks, analytical tools and numerical algorithms for solving the NONLINEAR HYDRODYNAMICS PROBLEMS in PRACTICAL APPLICATIONS. The original framework of Navier-Stokes equations for studies of these modern problems suffers from two aspects: (1) represented in form of the partial differential equations coupling the velocity with the pressure; (2) lack of the pressure control equations and boundary condition on the wall due to instantaneous spread of pressure information over all domain where the incompressible fluid occurs, and the available low-dimensional models and dynamical systems approaches[1][2], either too crude or applicable only to the simple geometric domain, are less connected with a real fluids system. This paper presents a new framework, GENERALIZED for complicated flows with an arbitrary boundary. In this framework a problem considered is decomposed into three operators: (1) solving a set of boundary problems of coupled Poisson's equations for construction of mathematical bases of the vector potential $\vec{\psi}$ and vorticity $\vec{\omega}$ and a Helmholtz problem for scalar bases such as density of ρ or temperature T or concentration C of stratified fluids, (2) an eigenvalue problem of a symmetric positive definite matrix for generating a complete set of orthonormal solenoidal vector bases of velocity $\vec{v}$ and, finally, (3) a system of canonical nonlinear ordinary differential equations, that is of great advantage to study the variety of possible motions of the viscous imcompressible fluids, involving the bifurcation, chaos and turbulence.

II FUNCTION REPRESENTATIONS OF VELOCITY, VORTICITY, DENSITY, TEMPERATURE AND CONCENTRATION OF HOMOGENEOUS AND STRATIFIED FLUIDS

2.1 Mathematical Bases for Vector Potential and Vorticity

Choosing properly a complete solenoidal vector set $\{\vec{h}_i(\vec{x})\}$, $i = 1, 2, \ldots$, for each $\vec{h}_i(\vec{x})$ one can find a couple of solution $\vec{\psi}_i(\vec{x})$ and $\vec{\omega}_i(\vec{x})$, satisfying the following equations:

$$\nabla^2\vec{\psi}_i(\vec{x}) + \vec{\omega}_i(\vec{x}) = 0, \quad \vec{x} \in \Omega \tag{2.1.a}$$

$$\nabla^2\vec{\omega}_i(\vec{x}) = \vec{h}_i(\vec{x}), \quad \vec{x} \in \Omega \tag{2.1.b}$$

subject to the boundary conditions:

$$\vec{\psi}_i(\vec{x}) = 0, \quad \vec{x} \in \partial\Omega \tag{2.1.c}$$

$$\frac{\partial\vec{\psi}_i}{\partial n}(\vec{x}) = 0, \quad \vec{x} \in \partial\Omega \tag{2.1.d}$$

As it will be shown later, $\{\vec{\psi}_i(\vec{x})\}$ and $\{\vec{\omega}_i(\vec{x})\}$, $i = 1, 2, \ldots$, can be taken as bases of the vector potential $\vec{\psi}$ and vorticity $\vec{\omega}$.

2.2 Mathematical Bases for Scalars Density, Temperature and Concentration of Stratified Fluids

As well known, the Helmholtz problem

$$\nabla^2\chi(\vec{x}) = -\sigma\chi(\vec{x}), \quad \vec{x} \in \Omega \tag{2.2.a}$$

$$\chi(\vec{x}) = 0, \quad \vec{x} \in \Gamma_E \tag{2.2.b}$$

$$\frac{\partial\chi}{\partial n}(\vec{x}) = 0, \quad \vec{x} \in \Gamma_N \tag{2.2.c}$$

has the solutions $\{\chi_i(\vec{x})\}, i = 1, 2, \ldots$, which can be taken as the orthonormal bases for scalars such as the density, temperature and concentration of stratified fluids.

The problems (2.1) and (2.2)can be solved by the finite-element, spectral-element and boundary element methods over arbitrary bounded domain Ω.

2.3 Construction of Orthonormal Bases of Velocity

Introducing

$$\vec{v}_i(\vec{x}) = \nabla \times \vec{\psi}_i(\vec{x}), \quad i = 1, 2, \ldots \tag{2.3}$$

for any positive function $\rho(\vec{x}) \in L^2(\Omega)$, one can obtain a symmetrical positive matrix:

$$K_{ij} = \int_\Omega \rho(\vec{x})\vec{v}_i(\vec{x}) \cdot \vec{v}_j(\vec{x}) d\Omega \tag{2.4}$$

that has simple positive eigenvalues $\lambda_k > 0$ and corresponding orthonormal independent eigenvectors:

$$\alpha^k = (\alpha_1^k, \alpha_2^k, \ldots), \quad k = 1, 2, \ldots \tag{2.5}$$

Obviously, the vector functions:

$$\vec{\varsigma}_k(\vec{x}) = \sum_i \alpha_i^k \vec{v}_i(\vec{x})/\sqrt{\lambda_k}, \quad k = 1, 2, \ldots \tag{2.6}$$

have the properties:

$$\nabla \cdot \vec{\varsigma}_k(\vec{x}) = 0, \quad \vec{x} \in \Omega \tag{2.7}$$

$$\vec{\varsigma}_k(\vec{x}) = 0, \quad \vec{x} \in \partial\Omega \tag{2.8}$$

$$\int_\Omega \vec{\varsigma}_k(\vec{x}) \cdot \nabla\Pi(\vec{x}) d\Omega = 0, \quad for \quad \forall\Pi(\vec{x}) \in W^{1,2}(\Omega) \tag{2.9}$$

$$\int_\Omega \rho(\vec{x})\vec{\varsigma}_k(\vec{x}) \cdot \vec{\varsigma}_l(\vec{x}) d\Omega = \delta_{kl} \tag{2.10}$$

2.4 Representation of Velocity, Vorticity and Density

If the considered system of viscous flows has the boundary conditions:

$$\vec{v}(\vec{x},t) = \vec{v}_\Gamma(\vec{x},t) = \textit{prescribed vector function}, \quad \vec{x} \in \partial\Omega \tag{2.11}$$

$$\rho(\vec{x},t) = \rho_0(z) = \textit{undisturbed density distribution}, \quad \vec{x} \in \Gamma_E \tag{2.12.a}$$

$$\frac{\partial \rho}{\partial n}(\vec{x},t) = 0, \quad \vec{x} \in \Gamma_N \tag{2.12.b}$$

where $\partial\Omega = \Gamma_E \bigcup \Gamma_N$. The temporally or spatially varying boundary values $\vec{v}_\Gamma(\vec{x},t)$ may be accommodated by setting:

$$\vec{v}(\vec{x},t) = \vec{v}_B(\vec{x},t) + \vec{U}(\vec{x},t) \tag{2.13}$$

where $\vec{v}_B(\vec{x},t)$ is chosen properly to satisfy:

$$\nabla \cdot \vec{v}_B(\vec{x},t) = 0, \quad \vec{x} \in \Omega, \ t \geq 0 \tag{2.14}$$

$$\vec{v}_B(\vec{x},t) = \vec{v}_\Gamma(\vec{x},t), \quad \vec{x} \in \partial\Omega, \ t \geq 0 \tag{2.15}$$

Thus, the velocity $\vec{v}(\vec{x},t)$ would taken expansion:

$$\begin{aligned}\vec{v}(\vec{x},t) &= \vec{v}_B(\vec{x},t) + \sum_k a_k(t)\vec{\varsigma}_k(\vec{x}) \\ &= \vec{v}_B(\vec{x},t) + \sum_k a_k(t) \sum_i \alpha_i^k \nabla \times \vec{\psi}_i(\vec{x})/\sqrt{\lambda_k}\end{aligned} \tag{2.16}$$

It is easy to show:

$$\nabla \cdot \vec{\psi}(\vec{x}) = 0, \quad \vec{x} \in \Omega \tag{2.17}$$

$$\nabla \times \nabla \times \vec{\psi}_i(\vec{x}) = \nabla\nabla \cdot \vec{\psi}_i(\vec{x}) - \nabla^2 \vec{\psi}_i(\vec{x}) = \vec{\omega}_i(\vec{x}), \quad \vec{x} \in \Omega \tag{2.18}$$

So, the vorticity can be expanded as:

$$\vec{\omega}(\vec{x},t) = \vec{\omega}_0(\vec{x},t) + \sum_k a_k(t) \sum_i \alpha_i^k \vec{\omega}_i(\vec{x})/\sqrt{\lambda_k}, \quad \vec{x} \in \Omega \tag{2.19}$$

where $\vec{\omega}_0(\vec{x},t) = \nabla \times \vec{v}_B(\vec{x},t)$.

Similarly, the density $\rho(\vec{x},t)$ may take expansion:

$$\rho(\vec{x},t) = \rho_0(z) + \rho'(\vec{x},t) \quad \vec{x} \in \Omega \tag{2.20.a}$$

$$\rho'(\vec{x},t) = \rho'_B(\vec{x}) + \sum_j b_j(t)\chi_j(\vec{x}) \quad \vec{x} \in \Omega \tag{2.20.b}$$

where $\rho'_B(\vec{x})$ is chosen properly to satisfy:

$$\rho'_B(\vec{x}) = 0, \quad \vec{x} \in \Gamma_E \tag{2.21.a}$$

$$\frac{\partial \rho'_B}{\partial n}(\vec{x}) = -\frac{d\rho_0(z)}{dz}\vec{e}_z \cdot \vec{n}, \quad \vec{x} \in \Gamma_N \tag{2.21.b}$$

III GALERKIN REPRESENTATION OF NAVIER-STOKES EQUATIONS

The viscous flow of homogeneous incompressible fluids may be described by the Navier-Stokes equations in the following dimensionless form:

$$\nabla \cdot \vec{v} = 0, \quad \vec{x} \in \Omega \tag{3.1.a}$$

$$\frac{\partial \vec{v}}{\partial t} + \vec{v} \cdot \nabla \vec{v} = -\nabla \Pi + \frac{1}{Re} \nabla^2 \vec{v}, \quad \vec{x} \in \Omega \tag{3.1.b}$$

For this flow $\rho(\vec{x}) = 1$ will take place in the matrix (2.4) and relation (2.10). Substituting the expansion of velocity (2.16) into the equations (3.1), one can obtain the Galerkin representation of Navier-Stokes equations:

$$\dot{a}_k(t) + \sum_l \sum_m B_{klm} a_l(t) a_m(t) + \sum_l C_{kl} a_l(t)$$

$$+ \frac{1}{Re} \sum_l D_{kl} a_l(t) + E_k = 0 \tag{3.2.a}$$

where

$$B_{klm} = \int_\Omega \vec{\varsigma}_k \cdot (\vec{\varsigma}_l \cdot \nabla \vec{\varsigma}_m) d\Omega = -B_{mlk} \tag{3.2.b}$$

$$C_{kl} = \int_\Omega \vec{\varsigma}_k \cdot (\vec{\varsigma}_l \cdot \nabla \vec{v}_B + \vec{v}_B \cdot \nabla \vec{\varsigma}_l) d\Omega \tag{3.2.c}$$

$$D_{kl} = \int_\Omega \nabla \vec{\varsigma}_k : \nabla \vec{\varsigma}_l d\Omega \tag{3.2.d}$$

$$E_k = \int_\Omega \vec{\varsigma}_k \cdot \frac{\partial \vec{v}_B}{\partial t} d\Omega + \int_\Omega \vec{\varsigma}_k \cdot (\vec{v}_B \cdot \nabla \vec{v}_B) d\Omega + \frac{1}{Re} \int_\Omega \nabla \vec{\varsigma}_k : \nabla \vec{v}_B d\Omega \tag{3.2.e}$$

$$\vec{\varsigma}_k = \sum_i \alpha_i^k \nabla \times \vec{\psi}_i(\vec{x}) / \sqrt{\lambda_k} \tag{3.2.f}$$

IV GALERKIN REPRESENTATION OF BOUSSINESQ EQUATIONS

The stratified flows resulting in the fluid, which will nearly always be regarded as incompressible, can be described by the following equations:

$$\nabla \cdot \vec{v} = 0, \quad \vec{x} \in \Omega \tag{4.1.a}$$

$$\rho(\frac{\partial \vec{v}}{\partial t} + \vec{v} \cdot \nabla \vec{v}) = -\nabla p + \rho \vec{g} + \mu \nabla^2 \vec{v}, \quad \vec{x} \in \Omega \tag{4.1.b}$$

$$\frac{\partial \rho}{\partial t} + \vec{v} \cdot \nabla \rho = K_\rho \nabla^2 \rho, \quad \vec{x} \in \Omega \tag{4.1.c}$$

where ρ is the density of stratified fluids, K_ρ is the diffusivity.

If p and ρ are expanded about the values $p_0(z)$ and $\rho_0(z)$ in a reference state of hydrostatic equilibrium for which $\nabla p_0(z) = \rho_0(z)\vec{g}$ (i.e. one sets $p(\vec{x},t) = p_0(z) + p'(\vec{x},t)$ and $\rho(\vec{x},t) = \rho_0(z) + \rho'(\vec{x},t)$, the Navier-Stokes equations (4.1) in the Boussinesq approximation (i.e. neglecting variations of density in so far as they affect inertia, but retaining them in the buoyancy terms) can by written as:

$$\nabla \cdot \vec{v} = 0, \quad \vec{x} \in \Omega \tag{4.2.a}$$

$$\rho_0(\frac{\partial \vec{v}}{\partial t} + \vec{v} \cdot \nabla \vec{v}) = -\nabla p' + \rho' \vec{g} + \mu \nabla^2 \vec{v}, \quad \vec{x} \in \Omega \tag{4.2.b}$$

$$\frac{\partial \rho'}{\partial t} + \vec{v} \cdot \nabla \rho' + \frac{d\rho_0(z)}{dz} \vec{v} \cdot \vec{e}_z = K_\rho \nabla^2 \rho' + K_\rho \frac{d^2 \rho_0(z)}{dz^2}, \quad \vec{x} \in \Omega \tag{4.2.c}$$

If the non-dimensional variables based on a length scale H_0, time scale $\frac{H_0}{U_0}$, velocity scale U_0, density scale ρ_1, pressure scale $\rho_1 U_0^2$ and buoyancy scale $\rho_1 U_0^2 / g H_0$ are used, the Boussinesq equations (4.2) will take the following dimensionless form:

$$\nabla \cdot \vec{v} = 0, \quad \vec{x} \in \Omega \tag{4.3.a}$$

$$\rho_0(\frac{\partial \vec{v}}{\partial t} + \vec{v} \cdot \nabla \vec{v}) = -\nabla q - r\vec{e}_z + \frac{1}{Re} \nabla^2 \vec{v}, \quad \vec{x} \in \Omega \tag{4.3.b}$$

$$\frac{\partial r}{\partial t} + \vec{v} \cdot \nabla r = \rho_0(z) N^2(z) \vec{v} \cdot \vec{e}_z + \frac{1}{PrRe} \nabla^2 r + \frac{1}{Fr^2 PrRe} \frac{d^2 \rho_0(z)}{dz^2}, \quad \vec{x} \in \Omega \tag{4.3.c}$$

where all variables are non-dimensional, such as $q = p'/\rho_1 U_0^2$, $r = \rho' g H_0 / \rho_1 U_0^2$ and
$N(z) = [-\frac{1}{\rho_0(z)} \frac{d\rho_0(z)}{dz} \frac{gH_0}{U_0^2}]^{1/2}$ is the non-dimensional Brunt- $\ddot{V}$aisälä frequency,
z is the non-dimensional vertical coordinate,
$R_e = \frac{\rho_1 H_0 U_0}{\mu} = \frac{H_0 U_0}{\nu}$ is the Reynolds number,
$P_r = \nu / K_\rho$ is the Prandtl number,
$F_r = U_0 / \sqrt{gH_0}$ is the Froude number.
Substituting the expansions:

$$\vec{v}(\vec{x},t) = \vec{v}_B(\vec{x},t) + \sum_k a_k(t) \vec{\varsigma}_k(\vec{x}) \tag{4.4.a}$$

$$r(\vec{x},t) = r_B(\vec{x},t) + \sum_j b_j(t) \chi_j(\vec{x}) \tag{4.4.b}$$

into the equations (4.3), one can obtain the Galerkin representation of them:

$$\dot{a}_k(t) + \sum_l \sum_m B_{klm} a_l(t) a_m(t) + \sum_l C_{kl} a_l(t) + \frac{1}{Re} \sum_l D_{kl} a_l(t) + \sum_j F_{kj} b_j(t) + E_k = 0 \tag{4.5.a}$$

$$\dot{b}_i(t) + \sum_l \sum_j P_{ilj} a_l(t) b_j(t) + \sum_j Q_{ij} b_j(t) + \frac{1}{PrRe} \sum_j R_{ij} b_j(t) + \sum_l T_{il} a_l(t) + S_i = 0 \tag{4.5.b}$$

where

$$B_{klm} = \int_\Omega \rho_0 \vec{\varsigma}_k \cdot (\vec{\varsigma}_l \cdot \nabla \vec{\varsigma}_m) d\Omega, \quad C_{kl} = \int_\Omega \rho_0 \vec{\varsigma}_k \cdot [\vec{\varsigma}_l \cdot \nabla \vec{v}_B + \vec{v}_B \cdot \nabla \vec{\varsigma}_l] d\Omega$$

$$D_{kl} = \int_\Omega \nabla\vec{\varsigma}_k : \nabla\vec{\varsigma}_l d\Omega, \qquad F_{kj} = \int_\Omega \vec{\varsigma}_k \cdot \vec{e}_z \chi_j d\Omega$$

$$E_k = \int_\Omega \rho_0\vec{\varsigma}_k \cdot \frac{\partial \vec{v}_B}{\partial t} + \int_\Omega \rho_0\vec{\varsigma}_k \cdot (\vec{v}_B \cdot \nabla\vec{v}_B)d\Omega + \frac{1}{Re}\int_\Omega \nabla\vec{\varsigma}_k : \nabla\vec{v}_B d\Omega + \int_\Omega \vec{\varsigma}_k \cdot \vec{e}_z r_B d\Omega$$

$$P_{ilj} = \int_\Omega \chi_i\vec{\varsigma}_l \cdot \nabla\chi_j d\Omega, \qquad Q_{ij} = \int_\Omega \chi_i\vec{v}_B \cdot \nabla\chi_j d\Omega, \qquad R_{ij} = \int_\Omega \nabla\chi_i \cdot \nabla\chi_j d\Omega$$

$$T_{il} = -\int_\Omega \chi_i\rho_0 N^2\vec{\varsigma}_l \cdot \vec{e}_z d\Omega + \int_\Omega \chi_i\vec{\varsigma}_l \cdot \nabla r_B d\Omega$$

$$S_i = \int_\Omega \chi_i\vec{v}_B \cdot \nabla r_B d\Omega - \int_\Omega \chi_i\rho_0 N^2\vec{v}_B \cdot \vec{e}_z d\Omega$$

$$-\frac{1}{PrRe}\int_\Omega \chi_i[\nabla^2 r_B + \frac{1}{Fr^2}\frac{d^2\rho_0}{dz^2}]d\Omega$$

For system of stratified fluids subject to the boundary conditions (2.11)(2.12), the $r_B(\vec{x})$ must be chosen to satisfy the following conditions:

$$r_B(\vec{x}) = 0, \quad \vec{x} \in \Gamma_E \tag{4.6.a}$$

$$\frac{\partial r_B}{\partial n}(\vec{x}) = -\frac{1}{Fr^2}\frac{d\rho_0(z)}{dz}\vec{n} \cdot \vec{e}_z, \quad \vec{x} \in \Gamma_N \tag{4.6.b}$$

where $\partial\Omega = \Gamma_E \bigcup \Gamma_N$. The equations (3.2) and (4.5) provide a transformation of the viscous incompressible flow of the homogeneous and stratified fluid, respectively, from the physical space in which it occurs into a phase space in which behavior of the Fourier coefficients is portrayed.

The initial values $\{a_k(0)\}$ and $\{b_i(0)\}$ are determined by:

$$a_k(0) = \int_\Omega \rho\vec{\varsigma}_k \cdot [\vec{v}(\vec{x},0) - \vec{v}_B(\vec{x},0)]d\Omega \tag{4.7.a}$$

$$b_i(0) = \int_\Omega \chi_i[r(\vec{x},0) - r_B(\vec{x})]d\Omega \tag{4.7.b}$$

V NUMERICAL METHODS

In two-dimensional case:

$$\vec{h}_i(\vec{x}) = (0,0,h_i)$$
$$\vec{\psi}_i(\vec{x}) = (0,0,\psi_i)$$
$$\vec{\omega}_i(\vec{x}) = (0,0,\omega_i)$$
$$\vec{x} = (x,y,0).$$

then, (2.1) can be simplified as the following forms:

$$\nabla^2\psi(\vec{x}) + \omega(\vec{x}) = 0, \quad \vec{x} \in \Omega \tag{5.1.a}$$

$$\nabla^2\omega(\vec{x}) = h(\vec{x}), \quad \vec{x} \in \Omega \tag{5.1.b}$$

$$\psi(\vec{x}) = 0, \quad \vec{x} \in \partial\Omega \tag{5.1.c}$$

$$\frac{\partial\psi}{\partial n}(\vec{x}) = 0, \quad \vec{x} \in \partial\Omega \tag{5.1.d}$$

where the subscript i is omitted.

Taking into account computing the coefficients of the spectral equations (3.2), it is more convenient to use spectral or finite element method whose detailed formulations will be given here seperately.

For any bounded domain Ω , we can divide the domain Ω into numerous finite elements $\{\Omega^{(e)}\}, e = 1, 2, \ldots, E$, *i.e.* $\Omega = \bigcup_{e=1}^{E} \Omega^{(e)}$. Let $\varphi_i(\vec{x})$ denote the global interpolation function, we can take the interpolation function $\varphi_i(\vec{x})$ as the generator function $h_i(\vec{x})$. We select the test function $\varphi_l(\vec{x}) \in H_0^1(\Omega)$ for (5.1.a), and $\varphi_l(\vec{x}) \in H^1(\Omega)$ for (5.1.b), then the variational equations of Galerkin spectral element for (5.1) can be written as:

$$\int_\Omega \omega\varphi_l d\Omega - \int_\Omega \nabla\psi \cdot \nabla\varphi_l d\Omega = 0, \quad l = 1, 2, \ldots, N_i \tag{5.2.a}$$

$$\int_\Omega \nabla\omega \cdot \nabla\varphi_l d\Omega = -\int_\Omega h\varphi_l d\Omega \quad l = 1, 2, \ldots, N \tag{5.2.b}$$

where N_i is the total number of internal nodes, N is the total number of all nodes.

After obtaining the solutions $\psi(\vec{x})$, we use the following variational form to get the generalized solution of $\vec{v}(\vec{x})$:

$$\int_\Omega \vec{v}\varphi_l d\Omega = \int_\Omega \nabla \times (\psi\vec{e}_z)\varphi_l d\Omega \tag{5.3}$$

where $\varphi_l \in H_0^1(\Omega)$, $l = 1, 2, \ldots, N_i$.

Representing the global interpolation through the local ones:

$$\varphi_i = \bigcup_{e=1}^{E} \sum_{\alpha=1}^{Ne} N_\alpha^{(e)} \triangle_{\alpha i}^{(e)} \tag{5.4.a}$$

$$(\omega, \psi, u, v) = \sum_{j=1}^{N} \bigcup_{e=1}^{E} \sum_{\beta=1}^{Ne} N_\beta^{(e)} \triangle_{\beta j}^{(e)} (\omega, \psi, u, v)_j \tag{5.4.b}$$

we can rewrite (5.2.a) and (5.2.b) as follows:

$$\sum_{j=1}^{N} G_{ij}\omega_j = -\int_\Omega h\varphi_i d\Omega, \quad i = 1, 2, \ldots, N_i \tag{5.5.a}$$

$$\sum_{j=1}^{N} M_{ij}\omega_j - \sum_{j=1}^{N_i} G_{ij}\psi_j = 0, \quad i = 1, 2, \ldots, N \tag{5.5.b}$$

$$\sum_{j=1}^{N} M_{ij}u_j = \sum_{j}^{N_i} D_{ij}\psi_j, \quad i = 1, 2, \ldots, N \tag{5.6.a}$$

$$\sum_{j=1}^{N} M_{ij} v_j = \sum_{j}^{N_i} C_{ij} \psi_j, \quad i = 1, 2, \ldots, N \tag{5.6.b}$$

where

$$M_{ij} = \sum_{e=1}^{E} \sum_{\alpha=1}^{Ne} \sum_{\beta=1}^{Ne} M_{\alpha\beta}^{(e)} \triangle_{\alpha i}^{(e)} \triangle_{\beta j}^{(e)} \tag{5.7.a}$$

$$G_{ij} = \sum_{e=1}^{E} \sum_{\alpha=1}^{Ne} \sum_{\beta=1}^{Ne} G_{\alpha\beta}^{(e)} \triangle_{\alpha i}^{(e)} \triangle_{\beta j}^{(e)} \tag{5.7.b}$$

$$C_{ij} = \sum_{e=1}^{E} \sum_{\alpha=1}^{Ne} \sum_{\beta=1}^{Ne} C_{\alpha\beta}^{(e)} \triangle_{\alpha i}^{(e)} \triangle_{\beta j}^{(e)} \tag{5.7.c}$$

$$D_{ij} = \sum_{e=1}^{E} \sum_{\alpha=1}^{Ne} \sum_{\beta=1}^{Ne} D_{\alpha\beta}^{(e)} \triangle_{\alpha i}^{(e)} \triangle_{\beta j}^{(e)} \tag{5.7.d}$$

$$M_{\alpha\beta}^{(e)} = \int_{\Omega^{(e)}} N_\alpha^{(e)} N_\beta^{(e)} d\Omega^{(e)} \tag{5.7.e}$$

$$G_{\alpha\beta}^{(e)} = \int_{\Omega^{(e)}} \nabla N_\alpha^{(e)} \cdot \nabla N_\beta^{(e)} d\Omega^{(e)} \tag{5.7.f}$$

$$C_{\alpha\beta}^{(e)} = \int_{\Omega^{(e)}} N_\alpha^{(e)} \frac{\partial N_\beta^{(e)}}{\partial x} d\Omega^{(e)} \tag{5.7.g}$$

$$D_{\alpha\beta}^{(e)} = \int_{\Omega^{(e)}} N_\alpha^{(e)} \frac{\partial N_\beta^{(e)}}{\partial y} d\Omega^{(e)} \tag{5.7.h}$$

and E is the total number of element, N_e is the number of nodes in the e-th element, $N_\alpha^{(e)}, N_\beta^{(e)}$ are the local interpolation function over the e-th element, $\triangle_{\alpha i}^{(e)}, \triangle_{\beta j}^{(e)}$ are the Boole's matrices.

With the available $\{\vec{v}_i(\vec{x})\}$, the symmetric positive definite matrix K_{ij} can be obained as follows,

$$K_{ij} = \sum_{m=1}^{N_i} \sum_{n=1}^{N_i} M_{ij} (\vec{v}_{im} \cdot \vec{v}_{jn}) \tag{5.8}$$

then the eigenvalue problem of symmetric positive definite matrix K_{ij} can be solved by QR algorithm.

5.1 Isoparametric Spectral Element Formulation

According to the isoparametric spectral element method, the approximate solution is presented as:

$$(x, y, \omega, \psi, u, v) = \bigcup_{e=1}^{E} \sum_{\beta=1}^{N_e} (x, y, \omega, \psi, u, v)_\beta^{(e)} N_\beta^{(e)}(\xi, \eta) \tag{5.9}$$

where

$$N_\alpha^{(e)}(\xi,\eta) = h_p(\xi)h_q(\eta);$$
$$N_\beta^{(e)}(\xi,\eta) = h_r(\xi)h_s(\eta);$$
$$\alpha = (q-1)N_1 + p, \quad \beta = (s-1)N_1 + r$$
$$p, r = 1, 2, \ldots, N_1, \quad q, s = 1, 2, \ldots, N_2$$

are the local interpolation function. $h_j (j = p, q, r, s)$ is the interpolation polynomials. (5.7) can be integrated over domain $\Omega^{(e)}$ by Gauss-Lobatto integral formula as follows,

$$M_{\alpha\beta}^{(e)} = \frac{J^{(e)}(\xi_p,\eta_q)\delta_{pr}\delta_{qs}}{w(\xi_p)w(\eta_q)} W_p W_q \tag{5.10.a}$$

$$G_{\alpha\beta}^{(e)} = \sum_{f=1}^{N_1}\sum_{g=1}^{N_2} \frac{W_f W_g}{w(\xi_f)w(\eta_g)J^{(e)}(\xi_f,\eta_g)} \tag{5.10.b}$$

$$\{[\frac{dh_p(\xi_f)}{d\xi}\delta_{qg}\sum_{l=1}^{N_2}\frac{dh_l(\eta_g)}{d\eta}y_{(l-1)N_1+f}^{(e)} - \delta_{pf}\frac{dh_q(\eta_g)}{d\eta}\sum_{k=1}^{N_1}\frac{dh_k(\xi_f)}{d\eta}y_{(g-1)N_1+k}^{(e)}]$$

$$[\frac{dh_r(\xi_f)}{d\xi}\delta_{sg}\sum_{l=1}^{N_2}\frac{dh_s(\eta_g)}{d\eta}y_{(l-1)N_1+f}^{(e)} - \delta_{rf}\frac{dh_s(\eta_g)}{d\eta}\sum_{k=1}^{N_1}\frac{dh_k(\xi_f)}{d\eta}y_{(g-1)N_1+k}^{(e)}]$$

$$+[\delta_{pf}\frac{dh_q(\eta_g)}{d\eta}\sum_{l=1}^{N_1}\frac{dh_k(\xi_f)}{d\xi}x_{(g-1)N_1+k}^{(e)} - \frac{dh_p(\xi_f)}{d\xi}\delta_{gq}\sum_{l=1}^{N_2}\frac{dh_l(\eta_g)}{d\eta}x_{(l-1)N1+f}^{(e)}]$$

$$[\delta_{rf}\frac{dh_s(\eta_g)}{d\eta}\sum_{l=1}^{N_1}\frac{dh_k(\xi_f)}{d\xi}x_{(g-1)N_1+k}^{(e)} - \frac{dh_r(\xi_f)}{d\xi}\delta_{gq}\sum_{l=1}^{N_2}\frac{dh_l(\eta_g)}{d\eta}x_{(l-1)N1+f}^{(e)}]\} \tag{5.10.c}$$

$$C_{\alpha\beta}^{(e)} = \int_{\Omega^{(e)}} N_\alpha^{(e)}\frac{\partial N_\beta^{(e)}}{\partial x}d\Omega^{(e)} = \frac{W_p W_q}{w(\xi_p)w(\eta_q)}$$

$$[\frac{dh_r(\xi_p)}{d\xi}\delta_{qs}\sum_{l=1}^{N_2}\frac{dh_l(\eta_q)}{d\eta}y_{(l-1)N1+p}^{(e)} - \delta_{pr}\frac{dh_s(\eta_q)}{d\eta}\sum_{k=1}^{N_1}\frac{dh_k(\xi_p)}{d\xi}y_{(q-1)N1+k}^{(e)}] \tag{5.10.d}$$

$$D_{\alpha\beta}^{(e)} = \int_{\Omega^{(e)}} N_\alpha^{(e)}\frac{\partial N_\beta^{(e)}}{\partial y}d\Omega^{(e)} = \frac{W_p W_q}{w(\xi_p)w(\eta_q)}$$

$$[\delta_{pr}\frac{dh_s(\eta_q)}{d\eta}\sum_{l=1}^{N_1}\frac{dh_k(\xi)}{d\xi}x_{(q-1)N1+k}^{(e)} - \frac{dh_r(\xi_p)}{d\xi}\delta_{qs}\sum_{l=1}^{N_2}\frac{dh_l(\eta_q)}{d\eta}x_{(l-1)N1+p}^{(e)}] \tag{5.10.e}$$

where

$$J^{(e)}(\xi_f,\eta_g) = \frac{\partial(x^{(e)},y^{(e)})}{\partial(\xi,\eta)}|_{(\xi_f,\eta_g)}$$

$$= \sum_{k=1}^{N_1}\frac{dh_k(\xi_f)}{d\xi}x_{(g-1)N_1+k}^{(e)}\sum_{l=1}^{N_2}\frac{dh_l(\eta_g)}{d\eta}y_{(l-1)N1+f}^{(e)}$$

$$-\sum_{k=1}^{N_1}\frac{dh_k(\xi_f)}{d\xi}y^{(e)}_{(g-1)N_1+k}\sum_{l=1}^{N_2}\frac{dh_l(\eta_g)}{d\eta}x^{(e)}_{(l-1)N1+f} \tag{5.10.f}$$

is the Jacobian matrix of transformation.

The coefficients in the spectral equations of the homogeneous imcompressible Navier-Stokes equations can be obtained by using the Gauss-Lobatto integral scheme as follows,

$$B_{klm}=\int_\Omega \vec{\varsigma}_k\cdot(\vec{\varsigma}_l\cdot\nabla\vec{\varsigma}_m)d\Omega=\sum_{i=1}^{N_i}\sum_{j=1}^{N_i}(C_{ij}\varsigma_{lix}+D_{ij}\varsigma_{liy})(\vec{\varsigma}_{ki}\cdot\vec{\varsigma}_{mj}) \tag{5.11.a}$$

$$C_{kl}=\int_\Omega \vec{\varsigma}_k\cdot(\vec{\varsigma}_l\cdot\nabla\vec{v}_B+\vec{v}_B\cdot\nabla\vec{\varsigma}_l)d\Omega$$

$$=\sum_{i=1}^{N}\sum_{j=1}^{N}\{(C_{ij}v_{Bix}+D_{ij}v_{Biy})(\vec{\varsigma}_{ki}\cdot\vec{\varsigma}_{lj})+(C_{ij}\varsigma_{lix}+D_{ij}\varsigma_{liy})(\vec{\varsigma}_{ki}\cdot\vec{v}_{Bj})\} \tag{5.11.b}$$

$$D_{kl}=\int_\Omega\nabla\vec{\varsigma}_k:\nabla\vec{\varsigma}_l d\Omega=\sum_{i=1}^{N_i}\sum_{j=1}^{N_i}G_{ij}(\vec{\varsigma}_{ki}\cdot\vec{\varsigma}_{lj}) \tag{5.11.c}$$

$$E_k=\int_\Omega\vec{\varsigma}_k\cdot\frac{\partial\vec{v}_B}{\partial t}d\Omega+\int_\Omega\vec{\varsigma}_k\cdot(\vec{v}_B\cdot\nabla\vec{v}_B)d\Omega+\frac{1}{Re}\int_\Omega\nabla\vec{\varsigma}_k:\nabla\vec{v}_B d\Omega$$

$$=\sum_{i=1}^{N}\sum_{j=1}^{N}M_{ij}(\vec{\varsigma}_{ki}\frac{\partial\vec{v}_{Bj}}{\partial t})+\sum_{i=1}^{N}\sum_{j=1}^{N}(C_{ij}v_{Bix}+D_{ij}v_{Biy})(\vec{\varsigma}_{ki}\cdot\vec{v}_{Bj})+\frac{1}{R_e}\sum_{i=1}^{N}\sum_{j=1}^{N}G_{ij}(\vec{\varsigma}_{ki}\cdot\vec{v}_{Bj}) \tag{5.11.d}$$

Similarly, we can obtain the coefficients in the spectral equations of Boussinesq equations which are omitted there because of the limits of space of this paper.

Here the main contributions of the error in the spatial discretization come from the fact that the discrete equations are obtained in a restricted subspace of the space $H_0(\Omega)$. Another errors may be introduced by inexact representation of the integrals involved and the insufficiency of the quadrature rules used. The choice of collocation points is, of course, a key factor in achieving high-order accuracy. In the early formulation of the method the Chebyshev-Gauss-Lobatto points were used as local and physical collocation points, mainly due to closed form analytical expressions for the quadratures involved in the computation and the possibility of using fast Fourier transformations. However, in recent implemetations, Legendre-Gauss-Lobatto points have been found to be a better choice with all quadratures exactly performed numerically using Gauss-Lobatto integration schemes.

Legendre-Gauss-Lobatto interpolants [4] are

$$h_j(\varsigma)=-\frac{1}{N(N-1)L_{N-1}(\varsigma_j)}\frac{(1-\varsigma^2)L'_{N-1}(\varsigma)}{\varsigma-\varsigma_j}$$

which satisfies

$$h_j(\varsigma_l)=\begin{cases}0 & \text{if } l\neq j\\ 1 & \text{if } l=j,\end{cases}$$

and the Legendre-Gauss-Lobatto collocation derivatives are

$$\frac{dh_j(\varsigma_l)}{d\varsigma} = \begin{cases} \frac{1}{\varsigma_l - \varsigma_j} \frac{L_{N-1}(\varsigma_l)}{L_{N-1}(\varsigma_j)} & \text{if } l \neq j \\ -\frac{1}{4}N(N-1) & \text{if } l = j = 1 \\ \frac{1}{4}N(N-1) & \text{if } l = j = N \\ 0 & \text{if } l = j \neq 1, N \end{cases}$$

where $l, j = 1, 2, \ldots, N$, and $\varsigma_j, j = 1, 2, \ldots, N$. are Legendre-Gauss-Lobatto collocation points as follows,

$$\varsigma_1 = -1, \ \varsigma_N = 1 \ \varsigma_j = \text{root of } L'_{N-1}(\varsigma) = 0, \ j = 2, 3, \ldots, N-1$$

For Legendre polynomials $w(\varsigma) = 1$, and the weights to the Legendre-Gauss-Lobatto collocation points are

$$W_j = \frac{2}{N(N-1)[L_{N-1}(\varsigma_j)]^2}, \quad j = 1, 2, \ldots, N$$

The fourth-order Runge-Kutta method is employed here to solve (3.2).

5.2 Finite Element Formulation

It is well known that the triangle element of finite element method can be very easy to adapt the complex geometric boundary. This paper employs the linear interpolant triangle element to generate the numerical bases and the spectral equations.

For the linear interpolant triangle element, the approxmate solution can be presented as:

$$(x, y, \omega, \psi, u, v) = \bigcup_{e=1}^{E} \sum_{\beta=1}^{N_e} (x, y, \omega, \psi, u, v)_\beta^{(e)} N_\beta^{(e)} \tag{5.12}$$

where

$$N_\beta^{(e)} = L_\beta, \quad , \beta = 1, 2, 3$$

and $L_\beta, \beta = 1, 2, 3$ is the area coordinate of the triangle element and satisfies the following relation,

$$L_1 + L_2 + L_3 = 1$$

Assuming that α, β, γ are the local nodes number, and denoting

$$a_\alpha = x_\beta^{(e)} y_\gamma^{(e)} - x_\gamma^{(e)} y_\beta^{(e)}$$

$$b_\alpha = y_\beta^{(e)} - y_\gamma^{(e)}$$

$$c_\alpha = x_\gamma^{(e)} - x_\beta^{(e)}$$

then the area of the triangle element can be written as:

$$\triangle^{(e)} = \frac{1}{2}(b_\alpha c_\beta - b_\beta c_\alpha)$$

and (5.7) can be integated as:

$$M_{\alpha\beta}^{(e)} = \begin{cases} \triangle^{(e)}/6 & \text{if } \alpha = \beta \\ \triangle^{(e)}/12 & \text{if } \alpha \neq \beta \end{cases} \tag{5.13.a}$$

$$G_{\alpha\beta}^{(e)} = \frac{(b_\alpha b_\beta + c_\alpha c_\beta)}{4\triangle^{(e)}} \tag{5.13.b}$$

$$C_{\alpha\beta}^{(e)} = \frac{b_\beta}{6} \tag{5.13.c}$$

$$D_{\alpha\beta}^{(e)} = \frac{c_\beta}{6} \tag{5.13.d}$$

The coefficients in the spectral equations of the homogeneous incompressible Navier-Stokes equations and the Boussinseq equations can also be obtained by using the finite element method. This paper only takes the homogeneous imcompressible Navier-Stokes equations as example in the following form,

$$B_{klm} = \int_\Omega \vec{\varsigma}_k \cdot (\vec{\varsigma}_l \cdot \nabla \vec{\varsigma}_m) d\Omega$$

$$= \sum_{e=1}^{E} \sum_{r,s,t=1}^{N} \sum_{\alpha,\beta,\gamma=1}^{3} (E_{\alpha\beta\gamma}^{(e)} \varsigma_{lsx} + F_{\alpha\beta\gamma}^{(e)} \varsigma_{lsy})(\vec{\varsigma}_{kr} \cdot \vec{\varsigma}_{mt}) \triangle_{r\alpha}^{(e)} \triangle_{s\beta}^{(e)} \triangle_{t\gamma}^{(e)} \tag{5.14.a}$$

$$C_{kl} = \int_\Omega \vec{\varsigma}_k \cdot (\vec{\varsigma}_l \cdot \nabla \vec{v}_B + \vec{v}_B \cdot \nabla \vec{\varsigma}_l) d\Omega$$

$$= \sum_{e=1}^{E} \sum_{r,s,t=1}^{N} \sum_{\alpha,\beta,\gamma=1}^{3} \{ E_{\alpha\beta\gamma}^{(e)} [\varsigma_{lsx} (\vec{\varsigma}_{kr} \cdot \vec{v}_{Bt}) + v_{Bsx} (\vec{\varsigma}_{rk} \cdot \vec{\varsigma}_{lt})]$$

$$+ F_{\alpha\beta\gamma}^{(e)} [\varsigma_{lsy} (\vec{\varsigma}_{kr} \cdot \vec{v}_{Bt}) + v_{Bsy} (\vec{\varsigma}_{kr} \cdot \vec{\varsigma}_{lt})] \} \triangle_{r\alpha}^{(e)} \triangle_{s\beta}^{(e)} \triangle_{t\gamma}^{(e)} \tag{5.14.b}$$

$$D_{kl} = \int_\Omega \nabla \vec{\varsigma}_k : \nabla \vec{\varsigma}_l d\Omega = \sum_{e=1}^{E} \sum_{r,s=1}^{N} \sum_{\alpha,\beta=1}^{3} G_{\alpha\beta}^{(e)} (\vec{\varsigma}_{kr} \cdot \vec{\varsigma}_{ls}) \triangle_{r\alpha}^{(e)} \triangle_{s\beta}^{(e)} \tag{5.14.c}$$

$$E_k = \int_\Omega \vec{\varsigma}_k \cdot \frac{\partial \vec{v}_B}{\partial t} d\Omega + \int_\Omega \vec{\varsigma}_k \cdot (\vec{v}_B \cdot \nabla \vec{v}_B) d\Omega + \frac{1}{Re} \int_\Omega \nabla \vec{\varsigma}_k : \nabla \vec{v}_B d\Omega$$

$$= \sum_{e=1}^{E} \sum_{r,s=1}^{N} \sum_{\alpha,\beta=1}^{3} M_{\alpha\beta}^{(e)} (\vec{\varsigma}_{kr} \cdot \frac{\partial \vec{v}_{Bs}}{\partial t}) \triangle_{r\alpha}^{(e)} \triangle_{s\beta}^{(e)}$$

$$+ \sum_{e=1}^{E} \sum_{r,s,t=1}^{N} \sum_{\alpha,\beta,\gamma=1}^{3} (E_{\alpha\beta\gamma}^{(e)} v_{Bsx} + F_{\alpha\beta\gamma}^{(e)} v_{Bsy})(\vec{\varsigma}_{kr} \cdot \vec{v}_{Bt}) \triangle_{r\alpha}^{(e)} \triangle_{s\beta}^{(e)} \triangle_{t\gamma}^{(e)}$$

$$+ \frac{1}{Re} \sum_{e=1}^{E} \sum_{r,s=1}^{N} \sum_{\alpha,\beta=1}^{3} G_{\alpha\beta} (\vec{\varsigma}_{kr} \cdot \vec{v}_{Bs}) \triangle_{r\alpha}^{(e)} \triangle_{s\beta}^{(e)} \tag{5.14.d}$$

where

$$E_{\alpha\beta\gamma}^{(e)} = \int_{\Omega^{(e)}} N_\alpha^{(e)} N_\beta^{(e)} \frac{\partial N_\gamma^{(e)}}{\partial x} d\Omega = \begin{cases} b_\gamma/12, & \text{if } \alpha = \beta \\ b_\gamma/24, & \text{if } \alpha \neq \beta \end{cases} \tag{5.15.a}$$

$$F^{(e)}_{\alpha\beta\gamma} = \int_{\Omega^{(e)}} N^{(e)}_\alpha N^{(e)}_\beta \frac{\partial N^{(e)}_\gamma}{\partial y} d\Omega = \begin{cases} c_\gamma/12, & \text{if } \alpha = \beta \\ c_\gamma/24, & \text{if } \alpha \neq \beta \end{cases} \tag{5.15.b}$$

VI NUMERICAL RESULTS

6.1 Flow over a backward-facing step

In order to examine the applicability and efficiency of the new framework for the numerical studies of incompressible viscous flow, Wu, Jianghang and Zhu, Huayang[5] employ the isoparametric spectral element technique to generate the mathematical bases of the streamfunction ψ and the vorticity ω in an arbitrary geometry, and to solve the Stokes equations for the non-zero velocity on the boundary and to form the symmetrical matrix K of the eigenvalue problem. The QR algorithm is used for the solution of the matrix eigenvalue problem and the 4th-order Runge-Kutta method for the system of the ordinary differential equations there. The numerical results of flow inside drived square cavity at Re=0–1000[5] show that both the new framework of nonlinear hydrodynamics and the numerical method have good efficiency and accurary. In this paper, this same numerical method is used to compute flow over a backward-facing step.

The flow over a backward-facing step in a channel provides an excellent test case for the accuracy of numerical method because of the dependence of the reattachment length x_r on the Reynolds number. In this paper we report our two-dimensional computations of laminar flow over backward-facing step. At the inflow boundary, located at the step, and the outflow boundary, a parabolic profile is prescribed. All results presented are obtained using 17×15 Legendre collocation points. Fig. 1. show the computed results of streamlines, contours of constant vorticity for Re=100, and 200, which are in good agreement with the results presented by J. Kim and P. Moin[6].

6.2 Flow in the thermally driven cavity

The natural convective flow can be described by the incompressible Navier-Stokes and energy equations. Assuming constant thermophysical properties, except for variation of density with temperature in the buoyancy force(i.e. Boussinesq approximation is valid), the governing equations can be written in dimensionless form as (Wu Jianghang and Han Qinshu[7]):

$$\nabla \cdot \vec{v} = 0 \tag{6.1.a}$$

$$\frac{\partial \vec{v}}{\partial t} + \vec{v} \cdot \nabla \vec{v} = -\nabla P + \frac{1}{Re}\nabla^2 \vec{v} + \frac{Ra}{Re^2 Pr}\theta \vec{e}_g \tag{6.1.b}$$

$$\frac{\partial \theta}{\partial t} + \vec{v} \cdot \theta = \frac{1}{RePr}\nabla^2 \theta \tag{6.1.c}$$

where $\vec{v}, \theta, \vec{e}_g, Ra, Pr$ and Re are velocity, $\frac{T-T_0}{T_w-T0}$, unit vector in the gravitational direction, Rayleigh, Prandtl and Reynolds number, respectively. If the characteristic velocity U^* is taken place by $\sqrt{\gamma g D \triangle T}$, then the non-dimensional governing equations of the thermal convection preblom above can be rewritten as follows,

$$\nabla \cdot \vec{v} = 0 \tag{6.2.a}$$

$$\frac{\partial \vec{v}}{\partial t} + \vec{v} \cdot \nabla \vec{v} = -\nabla P + \sqrt{\frac{Pr}{Ra}} \nabla^2 \vec{v} + \theta \vec{e}_g \tag{6.2.b}$$

$$\frac{\partial \theta}{\partial t} + \vec{v} \cdot \theta = \sqrt{\frac{1}{PrRa}} \nabla^2 \theta \tag{6.2.c}$$

We can similarily obtain the spectral equations of the natural thermal convection considered above, which is omitted here because of the limit space of this paper, on the new framework.

The problem under consideration is a heated closed square cavity of dimensions 1×1 in x and y directions.

No-slip conditions are imposed at the boundaries, i.e. the velocity components are set to zero at the boundaries. Two cases of temperature boundary conditions are considered here. The first case is that the temperature of the left wall is kept at higher value compared with the right wall, and temperature gradient normal to the boundaries of the bottom and upper wall is set to zero. The second case is the Rayleigh-Benard convection in two dimensional space, i.e. the left and right wall of the cavity is taken as insulated boundaries, and temperature of the bottom wall is more higher than that of the upper wall. Triangle element with 15×15 node points and 392 elements and linear interpolation function is used here.

Fig. 3. (a-c) show the contours of streamfunction, vorticity and temperature for the first case when $Ra = 5 \times 10^3, 10^4, 10^5$ and $Pr = 1.0$. Fig. 3. (d) are the corresponding local Nusselt number at the left and right wall which shows that heat transfer increases as Ra number increases, and Fig. 3. (e) is the velocity profiles of u and v at x=0.5.

Fig. 4. (a-c) are the contours of streamfunction, vorticity and temperature for the first case when $Pr = 0.1, 10$, and $Ra = 10^4$, and its corresponding local Nusselt number at the left and right wall is shown in Fig. 4.(d), and Fig. 4. (e) is the velocity profiles of u and v at x=0.5. The results is similar to that of J. Y. Choo and D. H. Schultz[8].

Fig. 5. and Fig. 6. show the results for the second case of Rayleigh-Benard convection with the same parameters as that in the first case.

6.3 Heat transfer in a driven cavity

Convection motion driven by buoyancy and surface tension plays an important role in the growth of crystals and alloys, especially in small scale and/or low-gravity hydrodynamics. Sufficient understanding of these physical phenomena is necessary to improve the property of the product. Recently, considerable interset has been developed concerning the interaction of buoyancy and surface tension driven convection. In particular this problem is interesting to numerical analysts because of non-linearity of the governing equations and the high singularity appearing on the free surface near the corners. The focus of this paper in this subsection is aimed at investigating the intercoupling relation between buoyancy and surface tension by using the new framework which solved by the finite element method. Fig. 6. (a-c) are the contours of the temperature and streamline for $Ra = 10^3, 10^4, 10^5$ and Re=50, Pr=1.0, Fig. 6. (d) is the local Nusselt number on left and right wall, the velocity profiles of u and v at x=0.5 is given in Fig. 6. (e). This results have the same properities as that presented by B. P. Leonard[10], J. Y. Choo and D. H. Schultz[8] and B. Ramaswamy and T. C. Jue[9].

In conclusion, numerical results above show that the new mathematical framework of nonlinear hydrodynamics can efficiently exploit the hydrodynamic property. Since the pressure term vanishes in the spectral space spanned by the special numerical or analytic bases of velocity and temperature, the trouble to handle the pressure is avoided. No limit to the bases exists, it is expected that the new mathematical framework can be applied to a wide range of problems.

REFERENCES

1. H. N. Shirer (1987), Nonlinear hydrodynamic modeling: a mathematical introduction, Lecture Notes in Physics 271, Springer-Verlag.
2. A. E. Perry (1989), The utility of dynamical systems approaches, Session Three in Lecture Notes in Physics 357, in Whither Turbulence? Turbulence at the Crossroads, Proceedings, Ithaca, New York, Ed. by J. L. Lumley, Springer-Verlag, 1990, 195-309.
3. Wu, Jianghang (1993), The formulations and applications of a new framework for nonlinear hydrodynamics mathematical transformation for viscous homogeneous and stratified fluid flows, Research Reports in Department of Mechanics, Peking University, 1993.
4. C. Canuto, M. Y. Hussaini, A. Quarteroni and T. A. Zang, Spectral Methods in Fluid Dynamics, Springer-Verlag, 1988.
5. Wu, Jianghang and Zhu, Huayang (1993) A decomposition operator—spectral element method for solution of the Navier-Stokes equations of incompressible fluids. In Proceedings of the 5th International Symposium on Computational Fluid Dynamics, Sendai, Japan, Ed. by H. Daiguji, August 31-September 3, 1993.
6. J. Kim and P. Moin (1985) Application of a fractional-step method to incompressible Navier-Stokes equations. J. Comput. Phys. 59(308-323).
7. Wu Jinghang and Han Qinshu (1988) Theory, Method and Application of Computational Fluid Mechanics(in Chinese), Academic Press(Beijing).
8. J. Y. Choo and D. H. Schultz (1992) A stable high-order method for the heated cavity problem, Inter. J. Numer. Methods Fluids 15(1313-1332).
9. B. Ramaswamy and T.C. Jue (1992) Some recent trends and developments in finite element analysis for incompressible thermal flows. Inter. J. Numer. Methods Eng. 35(671-707).
10. B. P. Leonard (1983) A convectively stable, third-order accurate finite-difference method for steady two-dimensional flow and heat transfer. in Numerical Properities and Methodologies in Heat Transfer, Proceedings of the Second National Symposium, Ed. by T.M. Shih, Hemisphere Publishing Corporation, 211-226.

Department of Mechanics
Peking University
Beijing, 100871
P. R. China

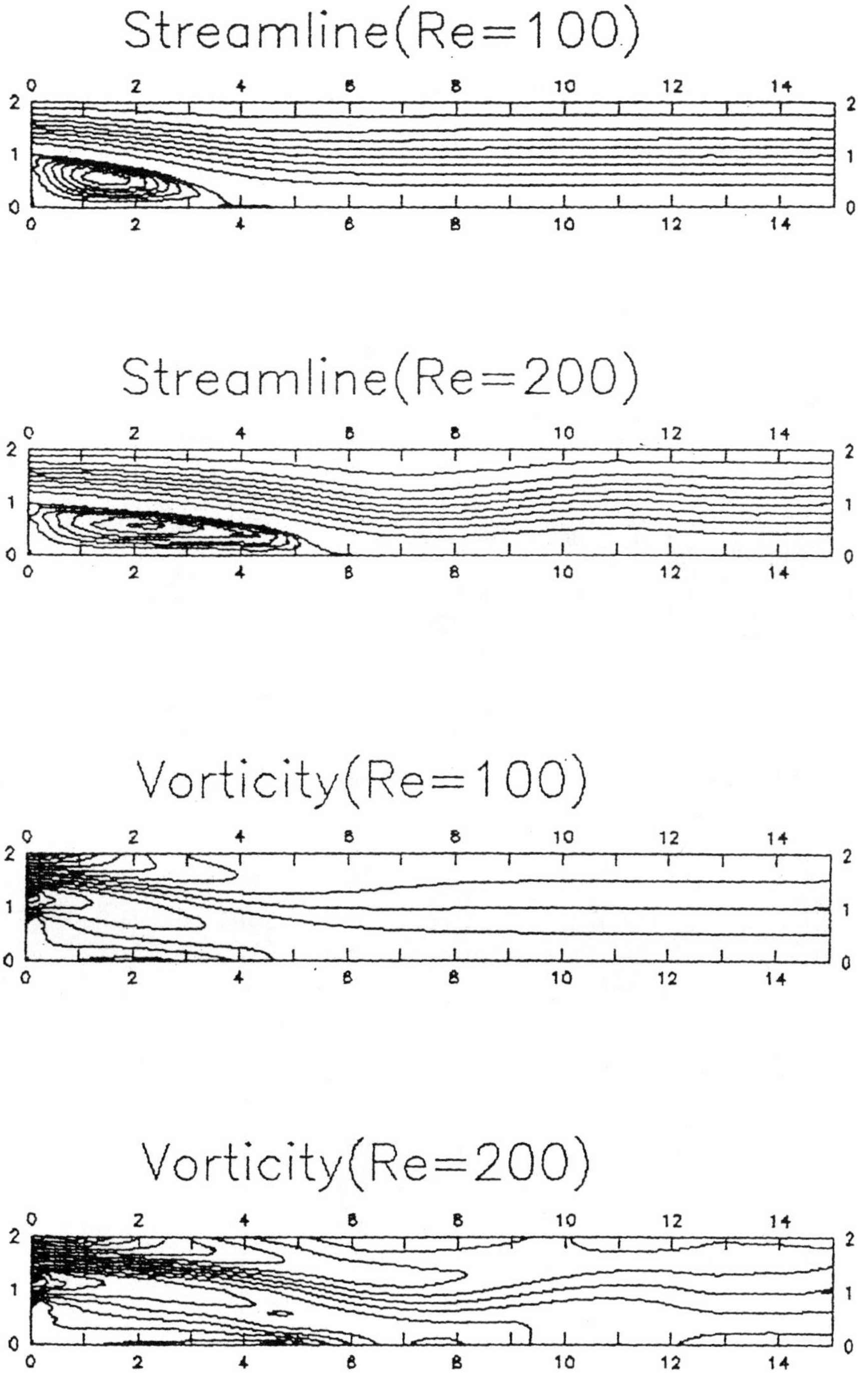

Fig. 1. Steady-state streamlines and contours of constant vorticity for flow over a backward-facing step

(2.a)
(2.b)
(2.c)
(2.d)
(2.e)

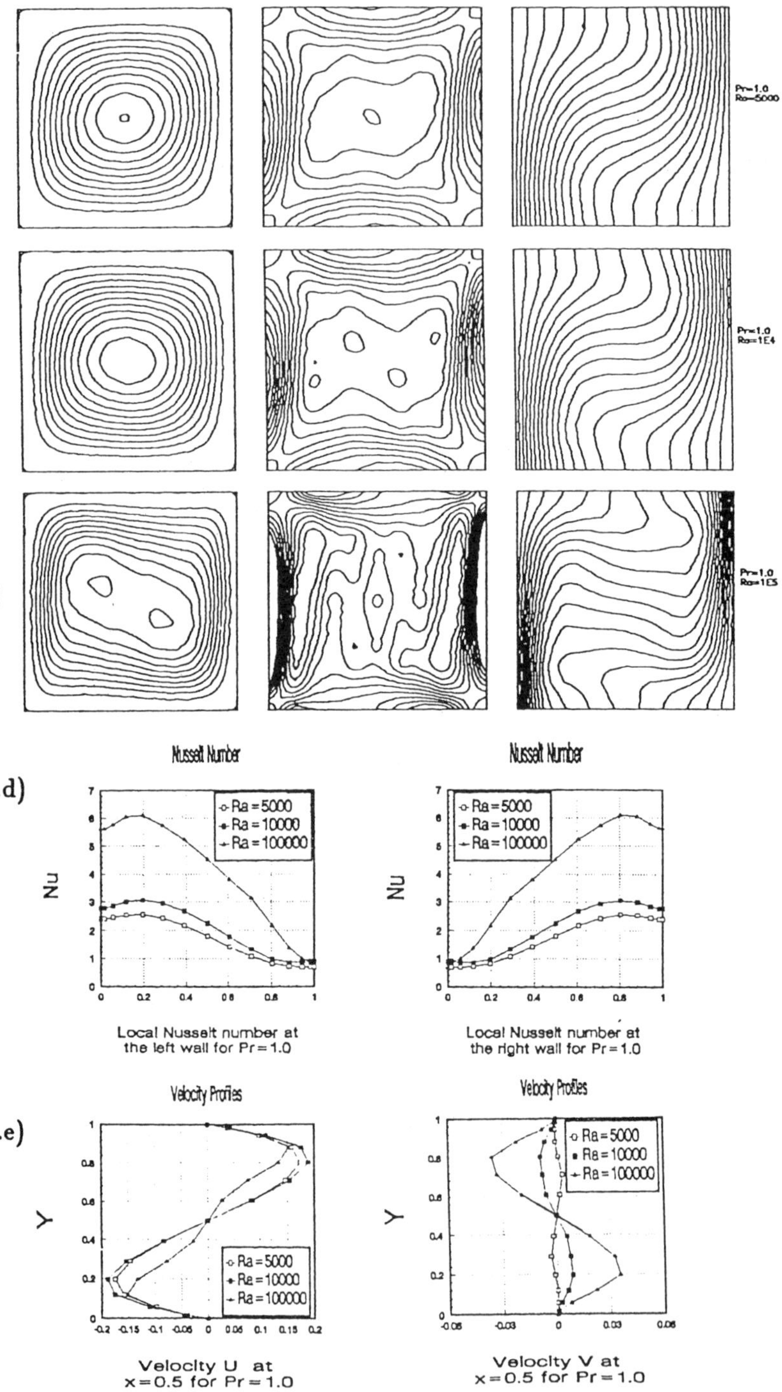

Fig. 2. Flow in the thermally driven cavity of the first case: heated from the left wall for $Ra = 5 \times 10^3, 10^4, 10^5$ and Pr=1.0. (a-c) Streamlines (left) and contours of constant vorticity (middle) and temperature (right), (d) local Nusselt number on hot wall(left) and cold wall (right) (e) velocity profiles u (left) and v (right) at x=0.5

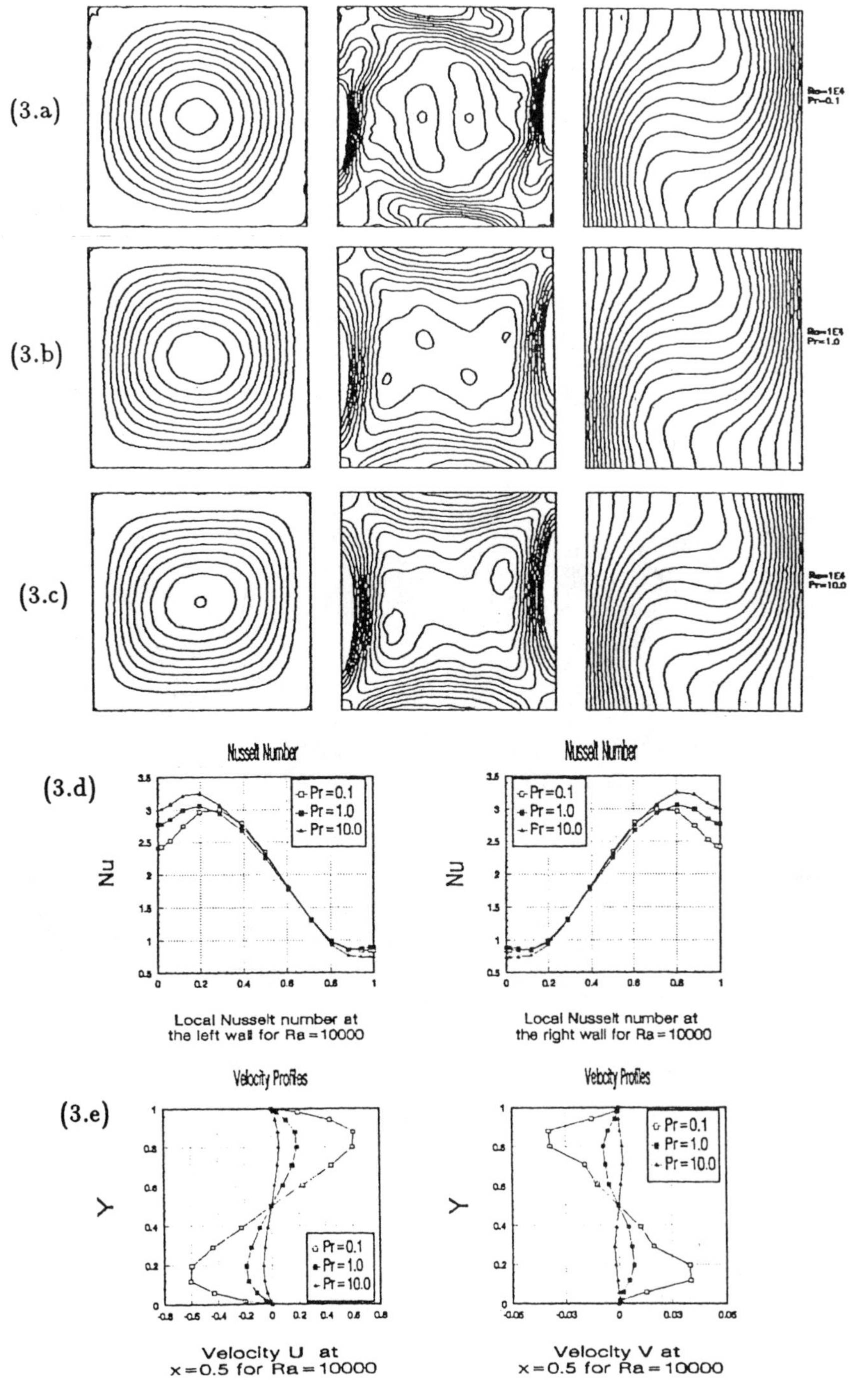

Fig. 3. Flow in the thermally driven cavity of the first case: heated from the left wall for $Pr = 0.1, 1, 10$ and $Ra = 10^4$. (a-c) Streamlines (left) and contours of constant vorticity (middle) and temperature (right), (d) local Nusselt number on hot wall (left) and cold wall (right), (e) velocity profiles u (left) and v (right) at x=0.5

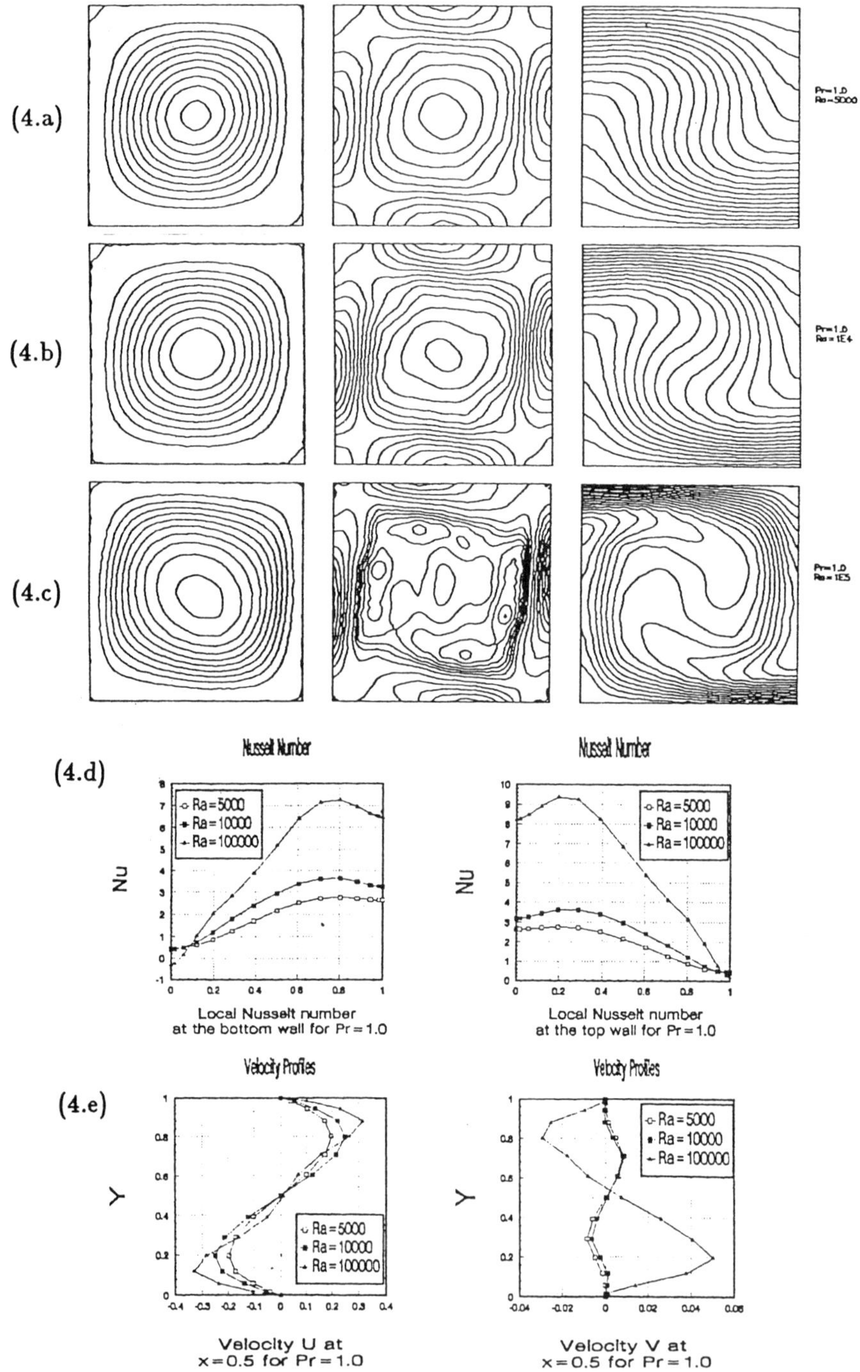

Fig. 4. Flow in the thermally driven cavity of the second case: heated from the bottom wall for $Ra = 5 \times 10^3, 10^4, 10^5$ and Pr=1.0. (a-c) Streamlines (left) and contours of constant vorticity (middle) and temperature (right), (d) local Nusselt number on hot wall (left) and cold wall (right), (e) velocity profiles u (left) and v (right) at x=0.5

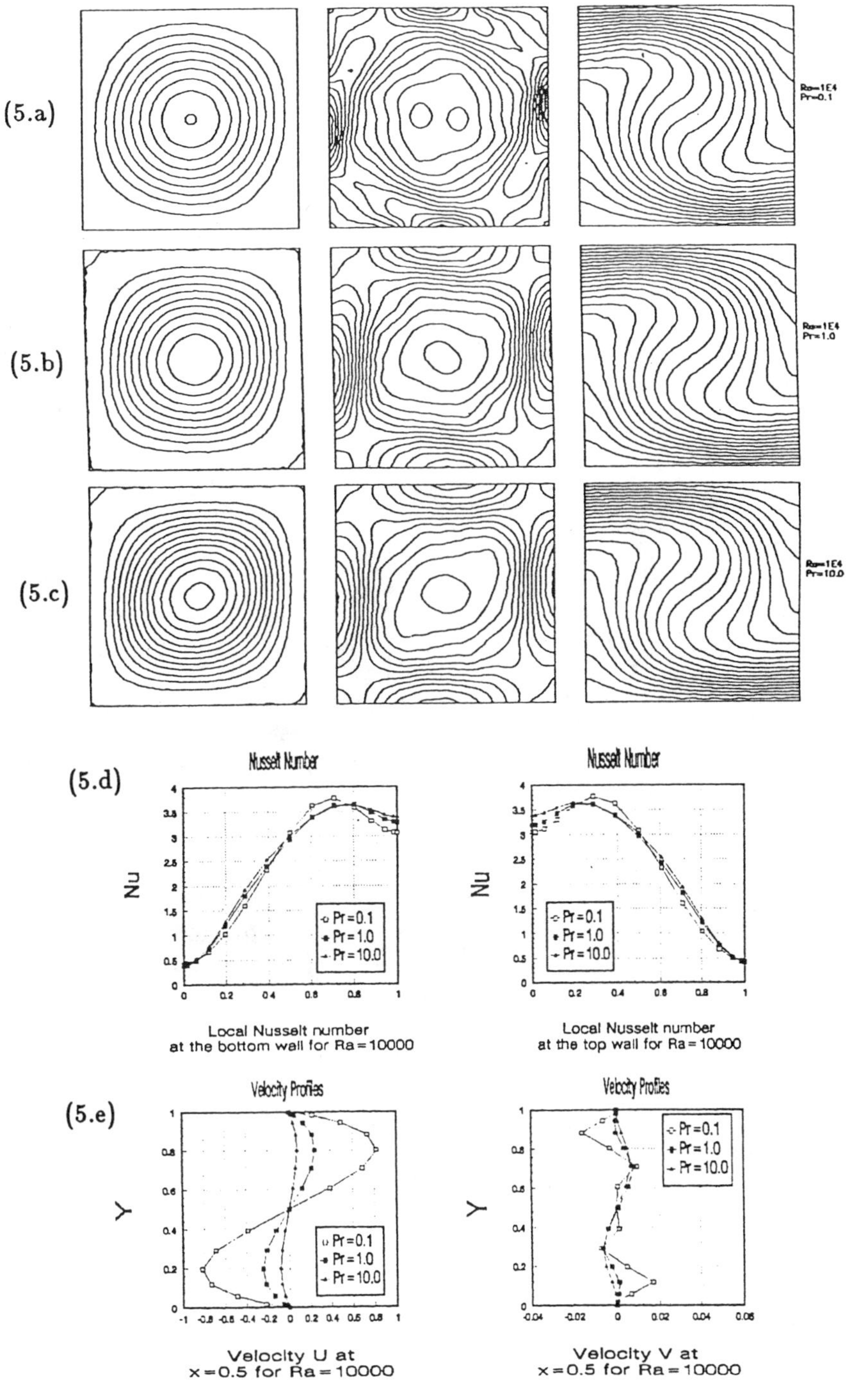

Fig. 5. Flow in the thermally driven cavity of the second case: heated from the bottom wall for $Pr = 0.1, 1, 10$ and $Ra = 10^4$. (a-c) Streamlines (left) and contours of constant vorticity (middle) and temperature (right), (d) local Nusselt number on hot wall (left) and cold wall (right), (e) velocity profiles u (left) and v (right) at x=0.5

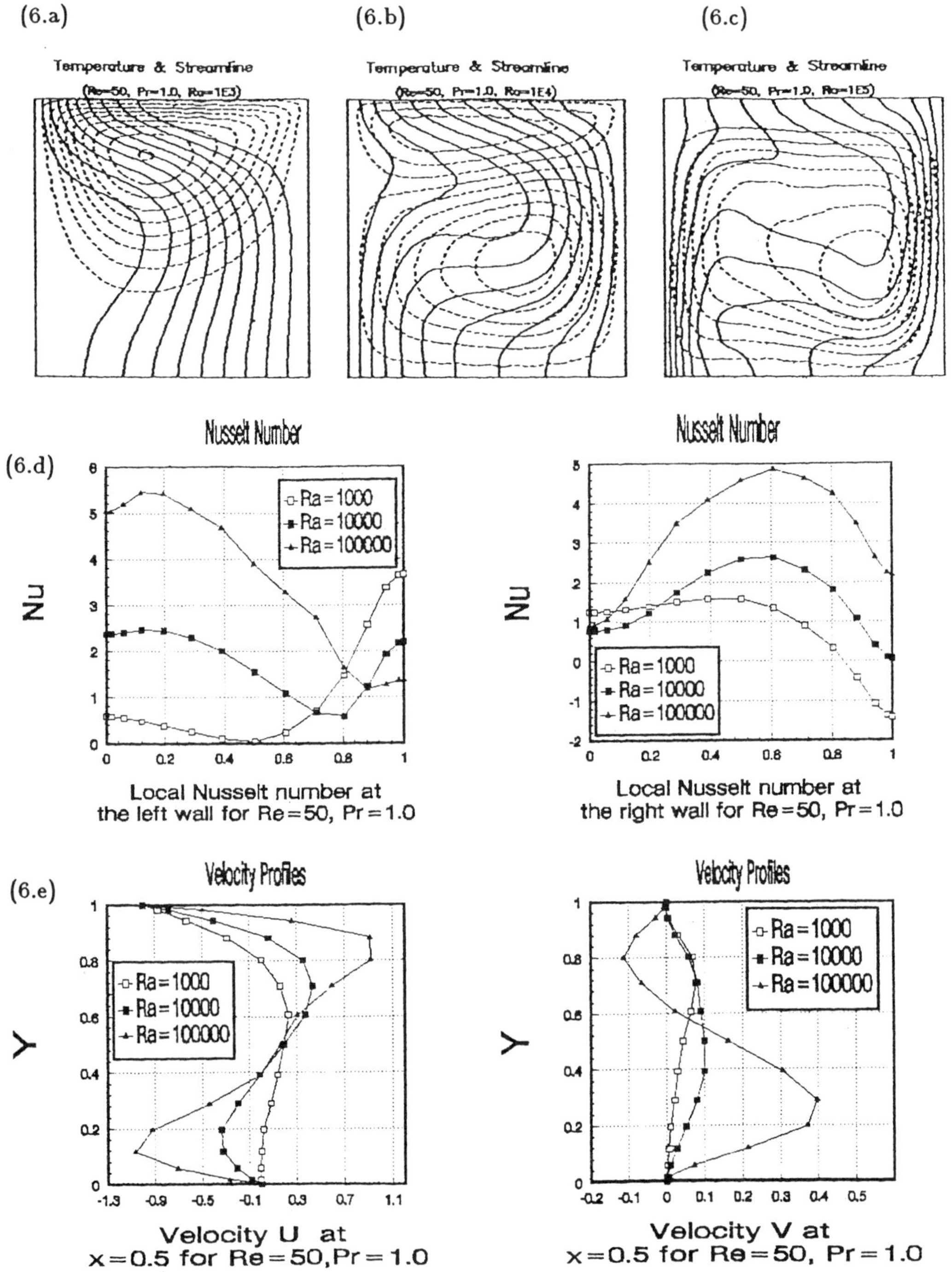

Fig. 6. Heat transfer in a driven cavity for $Ra = 10^3, 10^4, 10^5$ and Re=50, Pr=1.0. (a-c) Streamlines (dash line) and contours of temperature (solid line), (d) local Nusselt number on hot wall (left) and cold wall (right), (e) velocity profiles u (left) and v (right) at x=0.5

AMS/IP Studies in Advanced Mathematics
Volume **3**, 1997

Shock and Boundary Layer Behavior of Solutions to the Broadwell Equation in the Fluid Dynamic Limit

ZHOUPING XIN*[1]

Courant Institute of Mathematical Sciences
New York University
New York, NY 10012

and

School of Mathematics
Institute for Advanced Study
Princeton, NJ 08540

Abstract

This note concerns the shock and boundary layer behavior of solutions to the one-dimensional Broadwell model of the nonlinear Boltzmann equation. First we show that the fluid dynamical approximation is valid for the Broadwell model even if there are shocks in the fluid flow. The convergence and the detailed structure near shocks of the Broadwell solutions in the limit of small mean free path are given. We then study the behavior of the Broadwell solutions with boundary conditions which are either purely diffusive or diffusive-reflexive. It is found that even for such a simple model, there are boundary layers due to purely kinetic effects which cannot be detected by the corresponding Navier-Stokes system. It is found numerically that a compressive boundary layer is not always stable and

[1]Research was supported in part by a Sloan Foundation Fellowship, NSF Grants DMS-93-03887 and DMS 9304580, and DOE Grant DE-FG02-88ER-25053.
E-mail: xinz @ cims.nyu.edu or riemann @ math.ias.edu.

it may detach from the boundary and move into the interior of the gas as a shock layer. It is shown that both expansive and compressive boundary layers(before detachment) are nonlinearly stable and the layer effects are localized. The second part is a joint work with J.-G. Liu.

1. Introduction

The Boltzman equation of kinetic theory gives a statistical description of a gas of interacting particles. One of the important properties of this equation is its asymptotic equivalence to the Euler or Navier-Stokes equation of the compressible fluid dynamics, in the limit of the small mean free path [7, 8]. It is expected physically that away from initial layers, shock layers and boundary layers, the Boltzman solutions should relax to their equilibrium states (the local Maxwellian states) in the limit of a small mean free path, and the gases should be governed by the microscopic equations - the fluid equations. This is predicted formally by the normal solution method based on the Hilbert and Chapman-Enskog expansions. However, even this formal theory is not complete since it is necessary to provide suitable initial data, boundary condition, and matching conditions across shocks to carry out the Hibert expansion (or Chapman-Enskog expansion), i.e., one has to solve the three so-called connection problems across the layers within which the Hilbert expansion fails: to relate is a given initial distribution function to the Hilbert (or Chapman-Enskog) solution which takes over after an initial transient (initial layer problem); to find the correct matching conditions for the two Hilbert solutions prevailing on each side of a shock layer (shock layer problem); and to relate a given boundary condition on the microscopic distribution function to the Hilbert solution which holds outside the boundary layers. Since shock waves are basic nonlinear wave patterns for the compressible fluids, and boundary layers describe the interaction of molecules of the gas with the molecules of the solid body, to which one can trace the origin of the drag exerted by the gas on the body and the heat transfer between the gas and the solid boundary, it is important to gain deeper understanding of the asymptotic layer behavior of the Boltzmann solutions near shocks and near boundaries. This is a challenging problem both mathematically and physically due to the complex structures of the shock and boundary layers, and the nonlinear interaction of these layers. Thus, in order to study the fluid dynamic limit of the Boltzmann equation, it seems to be a crucial step to understand the geometric structure of these layers and their interactions, and to establish the asymptotic stability of the shock layers and boundary layers.

There is an extensive literature on the fluid-dynamic limit problem of the general Boltzmann equation or some models of the Boltzmann equation [3, 6, 8, 10, 12-15]. A rather complete theory exists for the initial layer problems. In fact, it has been shown by Caflisch [3] that when there exists a smooth solution to the nonlinear Euler equation, then one can construct corresponding solutions to the Boltzmann equations, and the fluid dynamic approximation is valid. A similiar theory with detailed initial layer structure has been obtained by Caflisch and Papanicolou for the Broadwell model of the nonlinear Boltzmann equation [6]. Moreover, the problems of shock layers and boundary layers remain largely open and far from being understood. In this direction, we note that the existence of Boltzmann shock profiles has been proved by Caflisch and Nicolaenko [5] for many important cases. However, structure of these Boltzmann profiles still needs to be further understood. The shock layer problems have been solved recently by the author only for the Broadwell equation (see Section 3 and [14]). Recently, the fluid-dynamic limit to the Riemann solutions of similiarity solutions to a modified Broadwell equation has been studied by Slemrod and Tzavaras [13]. As for the boundary layer problem, a qualitative theory exists for some models of steady Boltzmann equations [1], but very little is known for the unsteady problems. One of the main purposes of this note is to report some results of J. Liu and the author on the boundary layer problem for the Broadwell model (see Section 4 and [11]).

The main difficulties in analyzing the fluid dynamic limit problem are due to the complexity of the nonlocal collision operator in the Boltzmann equation which makes it difficult to study the structures of the layer problems associated with the formal matched asymptotic analysis. Even in the case that the structures of these layers are relatively easy to study for some models of the Boltzmann equations, the convergence cannot be obtained easily due to the fact that the fluid dynamic limits are highly singular, and the dissipative mechanisms are much weaker than those for the Navier-Stokes equations. We will consider a much simpler model, the Broadwell model, in which the velocity of the particle is discretized and the collision operator is localized, which enables one to find the shock profiles and boundary layer structure explicitly. The detailed geometrical structures of the layers make it possible to design careful energy method to study the stability of the solution. In Section 3, we will present a result on the shock layer problem which shows roughly speaking that, if the corresponding Euler system has a piecewise smooth solution with finitely many noninteracting shocks, then there exist solutions to the Broadwell equations such that the Broadwell solutions converge to the fluid dynamical solutions uniformly away from the shock. The investigation on boundary layer problem will be reported in Section 4. J. G. Liu and I consider both the analogues of the Maxwell's diffusive and diffusive-reflexive boundary conditions. We classify the boundary layers as either compressive or expansive in terms of the associated characteristic fields. It is found that even for such a simple model, there exists boundary layers due to purely kinetic effects which cannot be detected by the Chapman-Enskog expansion on the viscous level. Furthermore, by numerical experiment, we found that the compressive boundary layer may not be always stable, and it can detach from the boundary and moves into the interior of the gas as a shock layer. It is proved that both expansive and compressive boundary layers are nonlinearly stable before detachment and the layer effects are localized.

2. The Broadwell Model

the Broadwell model describes a gas as composed of particles of only six speeds with a binary collision law and spatial variation in only one direction, which was proposed by Broadwell [2]. In one space dimension, the model takes the following form.

$$\begin{cases} \partial_t f^+ + \partial_x f^+ = \dfrac{1}{\epsilon}(f^{02} - f^+ f^-), \\ \partial_t f^0 = \dfrac{1}{2\epsilon}(f^+ f^- - f^{02}), \\ \partial_t f^- - \partial_x f^- = \dfrac{1}{\epsilon}(f^{02} - f^+ f^-), \end{cases} \tag{2.1}$$

where ϵ is the mean free path, f^+, f^0 and f^- denote the mass densities of gas particles with speed $1, 0$, and -1, respectively. In terms of the hydrodynamical density ρ and momentum m, i.e.

$$\rho = f^+ + 4f^0 + f^-, \quad m = f^+ - f^-$$

and an additional quantity

$$z = f^+ - f^-$$

one can rewrite (2.1) as

$$\begin{cases} \partial_t \rho + \partial_x m = 0 \\ \partial_t m + \partial_x z = 0 \\ \partial_t z + \partial_x m = \frac{1}{8\epsilon}\{(\rho - z)^2 - 4(z^2 - m^2)\}. \end{cases} \tag{2.2}$$

The state (ρ, m, z) is said to be a local Maxwellian (or in equilibrium) if

$$\rho > 0, \left|\frac{m}{\rho}\right| < 1, and \quad z = \sigma\left(\frac{m}{\rho}\right),$$

where $\sigma(u) = \frac{2}{3}\sqrt{1+3u^2} - \frac{1}{3}$ and $u = \frac{m}{\rho}$ is the macroscopic fluid velocity. By assuming that the state is in equilibrium, one obtains the following model Euler Equations [6],

$$\left\{ \begin{array}{c} \partial_t \rho + \partial_x(\rho u) = 0 \\ \partial_t(\rho u) + \partial_x(\rho\sigma(u)) = 0. \end{array} \right. \tag{2.3}$$

This system is strictly hyperbolic and genuinely nonlinear with characteristic speeds $\lambda_1(u)$ and λ_2 satisfying

$$-1 < \lambda_1(u) < 0 < \lambda_2(u) < 1, if |u| < 1.$$

It is shown by Caflisch that system (2.3) shares many important properties of the isoentropic gas dynamics when the macroscopic speed of the gas is relatively small compared with the microscopic speed of the gas particles. We will consider both Cauchy problem and initial boundary value problem for (2.1) with suitably small mean free path. We start with shock layer problem.

3. Shock Layer Problem

In this section, we study the fluid dynamical limit of the Broadwell equations in the presence of shcoks. The main assumption is that the fluid flow is piecewise smooth with finitely many noninteracting entropy satisfying shocks. In this case, we can solve the shock layer problem, i.e., deriving the necessary boundary conditions for the Hilbert solutions prevailing on each side of the shock layers, and justifying the fluid dynamic approximation rigorously. To simplify the presentation, we focus on the case that the fluid flow contains only a single curved shock. More precisely, we assume that the Euler eqution (2.3) has a distributional solution $(\rho_0, m_0)(x,t)$ on $\Re \times [0,T]$ with $0 < T < +\infty$, such that
i) The initial data of the fluid solutions $(\rho_0, m_0)(x,t)$ satisfy

$$\rho_0(x,0) > 0, \quad |m_0(x,0)| < c\rho_0(x,0), x \in \Re, \tag{3.1}$$

ii) The fluid solution $(\rho_0, m_0)(x,t)$ contains a shock moving along a smooth curve $x = s(t)$ such that the fluid state $(\rho_0, m_0)(x,t)$ and its derivatives are uniformly smooth away from the shock but up to the shock, and crossing the shock $x = s(t)$, it satisfies the Rankine-Hugoniot conditions

$$\left\{ \begin{array}{c} \rho_0^l(u_0^l - \dot{s}) = \rho_0^r(u_0^r - \dot{s}) \\ \rho_0^l(\sigma(u_0^l) - \dot{s}u_0^l) = \rho_0^r(\sigma(u_0^r) - \dot{s}u_0^r) \end{array} \right. \quad 0 < t \leq T, \tag{3.2}$$

and the Lax entropy conditions

$$\lambda_2^r < \dot{s} < \lambda_2^l \quad or \quad \lambda_1^r < \dot{s} < \lambda_1^l.$$

We set $z_0(x,t) = \rho_0(x,t)\sigma(u_0(s,t))$, then $(\rho_0, m_0, z_0)(x,t)$ is in equilibrium state away from the shock $x = s(t)$. The following theorem asserts that there are solutions of the Broadwell equation that are close to this local Maxwellian[14].

Theorem 1(Shock Layer). Let $(\rho_0, m_0)(x,t)$ be a solution to the Euler equation (2.3) as described above. Then

i) the Hilbert expansion, which is valid away from the shock $x = s(t)$ and contains (ρ_0, m_0, z_0) as leading order term, can be carried out up to any order;

ii) a formal solution, which solves the Broadwell equation(2.1) asymptotically with any order of accuracy, can be constructed by matching the Hilbert solutions with the shock layer solutions;

iii) there exist suitably small positive constant δ_0 and ϵ_0 such that, as long as $(\rho_0, m_0, z_0)(x,t)$ lies in the δ_0–neighborhood of a state $(\bar{\rho}, \bar{m}, \bar{\rho}\sigma(\bar{m}/\bar{\rho}))$, and $0 < \epsilon < \epsilon_0$, a smooth solution $(\rho^\epsilon, m^\epsilon, z^\epsilon)(x,t)$ to the Broadwell equation (2.2) exists on $\Re \times [0,T]$ with

$$(\rho^\epsilon, m^\epsilon, z^\epsilon) - (\rho_0, m_0, z_0) \in C'([0,T], H^2).$$

Moreover, for any positive integer n, one can construct a smooth bounded function $(\rho, m, z)(x,t,\epsilon,n)$ such that

(3.3)
$$\sup_{0\leq 0\leq T} \|(\rho^\epsilon, m^\epsilon, z^\epsilon)(\cdot,t) - (\rho_0, m_0, z_0)(\cdot,t) - \epsilon(\rho, m, z)(\cdot,t,\epsilon,n)\|_{L^\infty(\Re)} \leq C_n \epsilon^{n+1}.$$

In particular, we obtain that

$$(3.4) \qquad \sup_{\substack{|x-s(t)|\geq h \\ 0\leq t\leq T}} |(\rho^\epsilon - \rho_0, m^\epsilon - m_0, z^\epsilon - z_0)(x,t)| \leq C_h \epsilon \quad for \quad any \quad h > 0,$$

where the constant C_h does not depend on ϵ

We remark that the asymptotic structure of the Broadwell solution $(\rho^\epsilon, m^\epsilon, z^\epsilon)$ in Theorem 1 can be described explicitly up to any order of accuracy by (3.3). We will sketch the proof of Theorem 1 here, details can be found in [14]. The proof has two main parts. First, we need to construct a smooth vector valued function $V_n^\epsilon(x,t) = (\rho_n^\epsilon, m_n^\epsilon, z_n^\epsilon)(x,t)$ with properties that

i) $V_n^\epsilon(x,t)$ is a smooth approximate solution to the Broadwell equation (2.2) with an error of order ϵ^{n+2};

ii) $V_n^\epsilon(x,t)$ approximates the given fluid flow $(\rho_0, m_0, z_0)(x,t)$ away from the shock with an error of order ϵ;
iii) up to leading order, in place of the shock discontinuity $V_n^\epsilon(x,t)$ has a smooth time dependent Broadwell shock profile of width ϵ
Such an approximate solution can be constructed by the method of matched asymptotic analysis as follows. The outer solution, which approximates the fluid flow $(\rho_0, m_0, z_0)(x,t)$ away from the shock, can be found by using the Hilbert expansions and solving a series of associated mixed problems for hyperbolic partial differential equations, while the inner solutions, which approximate the shock discontinuity near the shock, are found by using shock layer expansions which solve the systems of ordinary differential equations with prescribed boundary condition at infinities. The required boundary conditions for inner and outer solutions are obtained from matching them together. To this end, it is necessary to solve the differential equations for various order of outer and inner solutions simultaneously. The errors in the matching zone due to the matching are diffused out by introducing an appropriate linear hyperbolic wave. The approximate solution is constructed in such a way so that hydrodynamical mass and momentum are conserved which plays an important role in the following stability analysis. It should be noted that the key ingredients in the construction of the approximate solutions are the detailed structure of the Broadwell shock profile, Lax entropy condition and the observation that the shock location changes in the relaxation process. The second part of the proof is essentially a nonlinear stability analysis. We rigorously prove that there exists an exact solution to Broadwell equations in the neighborhood of the approximate solution constructed above, and the solutions converge to the given fluid flow. The main difficulty is that the linearized Broadwell operator has singularity of order $0(\frac{1}{\epsilon})$ due to the Broadwell shock profiles. We can overcome this difficulty by using appropriate scalings and applying an energy method which makes full use of the geometry of the time dependent shock profile for the Broadwell equation as in [9, 4, 14].

4. Boundary Layer Problem

We now turn to the layer behavior of solutions to the Broadwell equation near the boundary. We consier both diffusive and diffusive-reflesive boundary conditions. The appropriate boundary conditions for the corresponding model Euler equations is formulated so that the initial boundary value problem for the Euler equation is well posed and its solution can be realized as the limit of the corresponding Broadwell solution in the limit of small mean free path. This is achieved by matching the fluid solution with the boundary layer solution through conservation laws (see below). This matched asymptotic analysis produces an approximate solution for the Broadwell equation with detailed layer structures near the boundary. Then the existence of exact Broadwell solution and its convergence to the fuid solution away from the boundary are reduced to nonlinear stability analysis.

4.1. Initial- boundary value problems for the Broadwell and Euler equations

Consider the Broadwell equation (2.1) on the region

$$\Omega_T = \{(x,t), s(t) \le x < +\infty, 0 \le t \le T\} \tag{4.1}$$

with the moving boundary given by

$$x = -\alpha t \equiv s(t). \tag{4.2}$$

For simplicity, we assume $0 < \alpha < 1$. We remark that the cases $\alpha = 0$ and $\alpha = 1$ correspond to the uniform characteristic boundary conditions for the Broadwell equations, in which there are no strong boundary layers so that the fluid dynamic approximations can be justified easily. The initial data for (2.1) are given by

$$(f^+, f^0, f^-)(x, t=0) = (f_0^+, f_0^0, f_0^-)(x). \tag{4.3}$$

Two types of boundary conditions will be considered. One is a purely diffusive boundary conditin

$$f^+(s(t),t) = f_b^+(t), f^0(s(t),t) = f_b^-(t). \tag{4.4}$$

The other type is a diffusive-reflexive boundary condition

$$f^+(s(t),t) = a(t)f^-(s(t),t), 4f^0(s(t),t) = b(t)f^-(s(t),t), \tag{4.5}$$

where $a(t) > 0$ and $b(t) > 0$. In particular, the purely reflexive boundary condition corresponds to taking a and b in (4.5) such that

$$a + \alpha(1 + a + b) = 1$$

In this case, the mass flux is conserved on the boundary, i.e.,

$$(1 - \alpha)f^- = (1 + \alpha)f^+ + 4\alpha f^0.$$

The gas near the boundary in general is not in equilibrium state. In order to understand the leading order behavior of the kinetic boundary layer, one can use the stretched variable $\varsigma = \frac{x+\alpha t}{\epsilon}$ and look for the solution to (2.1) of the form $\mathbf{f}(\varsigma, t) = \mathbf{f}(\frac{x+\alpha t}{\epsilon}, t)$. Then up to leading order, the solution is governed by the following system of ordinary differential equations regarding t as parameter,

$$
\tag{4.6}
\begin{cases}
(\alpha+1)\dfrac{df^+}{d\varsigma} = f^0 f^0 - f^+ f^- \\
-2\alpha\dfrac{df^0}{d\varsigma} = f^0 f^0 - f^+ f^-, \\
(\alpha-1)\dfrac{df^-}{d\varsigma} = f^0 f^0 - f^+ f^-
\end{cases}
$$

The boundary data for (4.6) at $\varsigma = 0$ in the case of (4.4) is given by

$$
\tag{4.7}
f^+(0) = f_b^+, f^0(0) = b_b^0,
$$

while for (4.5) the boundary condition for (4.6) takes the form

$$
\tag{4.8}
f^+(0) - \alpha f^-(0) = 0, 4f^0(0) - bf^-(0) = 0.
$$

The state at $\varsigma = +\infty$ is in the fluid region and hence taken to be a local Maxwellian in both cases,

$$
\tag{4.9}
\mathbf{f}_\infty = (f_\infty^+, f_\infty^0, f_\infty^-), f_\infty^+ f_\infty^- = (f_\infty^0)^2.
$$

The system (4.6) can be integrated explicitly. As a consequence, one can obtain the appropriate boundary condition for the model Euler equation (2.3) as follows. We start with the case associated with the diffusive boundary condition (4.4). It follows from (4.6) that there exist functions $C_1(t)$ and $C_2(t)$ independent of ς such that

$$
\tag{4.10}
(\alpha+1)f^+ + 2\alpha f^0 = C_1(t), (\alpha-1)f^- + 2\alpha f^0 = C_2(t).
$$

Then the boundary condition (4.7) yields

$$
\tag{4.11}
C_1(t) = (\alpha+1)f_b^+ + 2\alpha f_b^0.
$$

On the other hand, rewriting (4.9) in terms of fluid moments leads to

$$
C_1(t) = \frac{\rho_b}{2}((\alpha+1)(u_b + \sigma(u_b)) + \alpha(1 - \sigma(u_b))).
$$

Setting

$$B(\rho,u)(t) \equiv \frac{1}{2}\rho(\sigma(u)+(\alpha+1)u+\alpha|_{(s(t),t)},$$

we arrive at the desirable boundary condition for the model Euler equation (2.3) as

$$B(\rho,u)(t) = (\alpha+1)f_b^+(t)+2\alpha f_b^0(t). \tag{4.12}$$

Next, we derive the boundary condition for (2.1) corresponding to the diffusive-reflexive boundary condition (4.5). Instead of (4. 11) , one obtains from (4.7) and (4.10) that

$$\frac{2(\alpha+1)a+\alpha b}{2(\alpha-1)+\alpha b} = \frac{C_1}{C_2}. \tag{4.13}$$

Evaluating (4.10) at $\varsigma = +\infty$ and using (4.9), one gets

$$C_1 = (\alpha+1)f_\infty^+ + 2\alpha f_\infty^0 = \frac{\rho_b}{2}(\alpha+u_b+\sigma(u_b)+\alpha u_b),$$
$$C_2 = (\alpha-1)f_\infty^- + 2\alpha f_\infty^0 = \frac{\rho_b}{2}(\alpha+u_b-\sigma(u_b)-\alpha u_b).$$

It follows from (4.13) and (4.14) that

$$(\alpha+u_b)(1+a-\alpha(1-a)) = (\sigma(u_b)+\alpha u_b)(1-a-\alpha(1+a+b)). \tag{4.15}$$

Let $u_b(t)$ be the solution of (4.15). Then the boundary condition for the Euler equation (2.3) should be taken as

$$u(s(t),t) = u_b(t). \tag{4.16}$$

In particular, for the purely diffusive boundary condition, one has that $u(s(t),t) = u_b(t) \equiv -\alpha$.

The initial data for the Euler equation (2.3) is given by

$$(\rho,m)(x,0) = (\rho^{in},m^{in})(x) \tag{4.17}$$

where to isolate the effects of the boundary layers, we assume that $(\rho^{in},m^{in},\rho^{in}\sigma^{in}(m^{in}/\rho^{in}))$ is a local Maxwellian.

It can be shown [11] that both initial boundary problems (2.3), (4.12), (4.17) and (2.3), (4.16), (4.17) are well posed at least locally in time. Denote the solution by $(\rho^0(x,t),m^0(x,t))$ which is assumed to be smooth up to $T<+\infty$.

4.2. Classification of the Boundary Layers

To determine the structure of the boundary layer, we can now integrate (4.6) with appropriate boundary condition. Let $(\rho_b, u_b)(t) = (\rho^0, \rho^0 m^0)(s(t), t)$. Define

$$f_\infty^0 = \frac{\rho_b}{4}(1 - \sigma(u_b)), f_{-\infty}^0 = f_\infty^0 + \frac{2\alpha}{3\alpha^2 + 1}(u_b + \alpha)\rho_b.$$

Then it follows from (4.6) and (4.10) that

$$\frac{df^0}{d\varsigma} = -C_\alpha(f^0 - f_\infty^0)(f^0 - f_{-\infty}^0) \tag{4.18}$$

where

$$C_\alpha = \frac{3\alpha^2 + 1}{2\alpha(1 - \alpha^2)}.$$

Solving (4.18) gives

$$f^0(\varsigma) - f_\infty^0 = \frac{(f_b^0 - f_\infty^0)(f_\infty^0 - f_{-\infty}^0)e^{-C_\alpha(f_\infty^0 - f_{-\infty}^0)\varsigma}}{f_b^0 - f_{-\infty}^0 - (f_b^0 - f_\infty^0)e^{-C_\alpha(f_\infty^0 - f_{-\infty}^0)\varsigma}}. \tag{4.19}$$

The corresponding formulas form f^+ and f^- are obtained from (4.10) and (4.19). The kinetic boundary layers can be classified as either *compressive* or *expansive* according to $\frac{d\lambda_1(u)}{d\varsigma} < 0$ or $\frac{d\lambda_1(u)}{d\varsigma} \geq 0$ along the boundary layer profile, respectively . It can be shown that there are four different cases depending on the speed of the wall and that of the fluid at the wall.

	$\frac{df^0}{d\xi} < 0$	$\frac{df^0}{d\xi} > 0$
compressive layer $\frac{d\lambda_1}{d\xi} < 0$	$u_b < -\alpha$	$u_b > -\alpha$
expansive layer $\frac{d\lambda_1}{d\xi} \geq 0$	$u_b \geq -\alpha$	$u_b \leq -\alpha$

We now observe the following facts:

1. Boundary layers, if they exist, have width of order mean free path, ϵ, and converge to the end Maxwellian state exponentially fast.

2. Boundary layers, corresponding to $u_b \leq -\alpha$, are purely due to the kinetic effects, which cannot be detected by the Chapman- Enskog expansions. Indeed, it can be shown that viscous boundary layers exist only when $u_b > -\alpha$

3. The kinetic boundary layers may not be always stable. In fact, by careful numerical experiments, we found that the compressive boundary layers may detach from the boundary and move into the gas as fluid shocks, which we will call the ***boundary enhanced shocks***, For details, see [11].

4.3 The Fluid-Dynamic Limit

We now show that the fluid-dynamic approximation is still valid away from the boundary before the detachment of the boundary layers, provided that the underlying fluid flow is smooth. Under some minor conditions (see [11]), our convergence can be stated as follows.

Theorem 2 (J. Liu and Z. Xin). Assume that the boundary layer is either compressive or expansive. Let $\mathbf{f}^0(x,t)$ be the microscopic density distribution associated with the local Maxwellian $(\rho^0, m^0)(x,t)$. Then there exists an $\epsilon_0 > 0$ such that for each $0 < \epsilon \leq \epsilon_0$, the initial-boundary value problem (2.1),(4.3) (4.4) or (2.1), (4.3), (4.5) has a unique smooth solution $\mathbf{f}^\epsilon(x,t)$ satisfying

$$\mathbf{f}^\epsilon(x+\alpha t, t) - \mathbf{f}^0(x+\alpha t, t) \in L^\infty([0,T], H^1(\Re_+)) \cap C([0,T] : L^2(\Re_+)),$$

$$\frac{d\mathbf{f}^\epsilon}{dt}(x+\alpha t, t) \in L^\infty([0,T] : L^2(\Re_+)).$$

Furthermore, for any integer $n > 0$, we can construct a bounded function $\mathbf{g}(x,t,\epsilon,n)$ such that

$$\sup_{0\leq t\leq T} \|\mathbf{f}^\epsilon(\cdot+\alpha t, t) - \mathbf{f}^0(\cdot+\alpha t, t) - \epsilon\mathbf{g}(\cdot+\alpha t, t, \epsilon, n)\|_{L^\infty(\Re_+)} \leq C_n \epsilon^{n+1}.$$

In particular, we have that for any $\delta > 0, \exists C_\delta > 0$ such that

$$\sup_{\substack{0\leq t\leq T \\ \delta \leq x+\alpha t}} \|\mathbf{f}^\epsilon(\cdot+\alpha t, t) - \mathbf{f}^0(\cdot+\alpha t, t)\|_{L^\infty(\Re)} \leq C_\delta \epsilon.$$

We note that the major difficulty in the proof of Theorem 2 lies in the stiffness of the approximate solution near the boundary. One needs to apply different energy methods depending on the classification of the underlying boundary layers. The basic idea is that the "better" approximate solution can reduce the nonlinearity in the problem, and the definiteness of the normalized linear Broadwell operator and the structure of the underlying boundary layers enable one to overcome the difficulty of stiffness near the boundary. For the detailed analysis, we refer to [11].

5. Concluding Remarks

To isolate the effects due to layers, we have avoided the interesting and important problem of interactions among initial, boundary, and shock layers in the Theorems 1 and 2. Rich wave phenomena should be expected in the study of layer interactions. The transition criteria from compressive boundary layers to shock layers needs to be further analyzed. All these are under investigation.

A thorough study on this problems for the Broadwell model should shed light on the more complicated and important problems such as reflection of a shock from a solid boundary, and interactions of the same and different kinds of layers in the rarefied gas dynamics. It is hoped that these studies can eventually lead to a resolution of the problem of the fluid- dynamic limit problem for the general Boltzmann equations.

References

[1] C.bardos, R. Caflisch and B. Nicolaenko, The Milne and Kramers problems for the Boltzmann equation of a hard sphere gas, Comm. Pure Appl. Math. 39 (1986).

[2] J. E. Broadwell, Shock struture in a simple discrete velocity gas, Phy. Fluids 7 (1964), 1243-1247.

[3] R. E. Caflisch, The fluid dynamic limit of the nonlineare Boltzmann equation, Comm. Pure Appl. Math. 33 (1980), 651-666.

[4] R. E. Caflisch and Tai-Ping Liu, Stability of shock waves for the Broadwell equations, Commun. Math. Phys. 114 (1980), 103-130.

[5] R. E. Caflisch and B. Nicolaenko, Shock profile solutions of the Boltzmann equation, Commun. Math. Phys. 86 (1982), 161-194.

[6] R. E. Caflisch and G. C. Papanicolaou, The fluid dynamical limit of a nonlinear model Boltzmann equation, Comm. Pure Appl. Math. 32 (1979), 589-616.

[7] C. Cercignani, The Boltzmann Equation and Its Applicatons, Springer - verlag, New York, 1988.

[8] H. Grad, Singular and nonuniform limits of solutions of the Boltzmann equations, SIAM-AMS Proceedings I, Transport Theory (1969) , 296-308.

[9] S. Kawashima and A. Matsumura, Asymptotic stability of traveling wave solutions pof a system of one-dimensional gas motion, Commun. Math . Phys. 101 (1985), 92-127.

[10] S. Kawashima, A. Matsumura and T. Nishida, On the fluid approximation to the Boltzmann at the level of the Navier- Stokes equation, Commun. Math. Phys. 70 (1979), 97-124.

[11] J.-G. Liu and Zhouping Xin, Boundary layer behavior in the fluid -dynamic limit for a nonlinear model Boltzmann equation, preprint, 1993

[12] T. Platkowski and R. Illner, Discrete velocity models of the Boltzmann equation: a survey of the mathematical aspects of the theory, SIAM Rev. 30 (1988), 231-255.

[13] M. Slemrod and A. E. Tzavaras, Self-similar fluid dynamic limits for the Broadwell system, preprint (1993).

[14] Zhouping Xin, The fluid-dynamic limit of the Broadwell model of the non-linear Boltzmann equation in the presence of shocks, Comm. Pure Appl . Math. 44 (1991), 679-713.

[15] Zhouping Xin, The fluid dynamic limit of the Broadwell equation, SIAM Volume on *Viscous and Numerical Approximation of shock waves*, ed. by M. Shearer (1991), 237-252.

AMS/IP Studies in Advanced Mathematics
Volume **3**, 1997

A NEW PROOF OF EXISTENCE OF THE WAVE FRONT SOLUTIONS FOR A KIND OF REACTION-DIFFUSION SYSTEM*

ZHOU LI AND YA I. KANEL

Abstract

In this paper we give a new proof of existence of wave front solutions for Fisher's equation. The method provided here come from the comparison existence theorem on unbounded domain and the construction of solutions in the neighborhood of positive infinity which can be expressed by seriers. Then we inverstigate the competition-diffusion system and obtain the existence of wave front solutions.

1. Scaler case

The travelling wave solutions of the Fisher's equation

$$u_t = u_{xx} + f(u), \quad f(u) = u(1-u) \tag{1.1}$$

were studied by many authors. Here we will give a new method. First to consider

$$u = u(\xi), \qquad \xi = x - ct \tag{1.2}$$

as a solution of (1.1), substituting it into (1.1), then $u(\xi)$ satisfies

$$u''(\xi) + cu' + u(1-u) = 0 \tag{1.3}$$

with boundary conditions at $\xi = \pm\infty$

$$u(+\infty) = 1, \qquad u(-\infty) = 0 \tag{1.4}$$

*The project supported by National Natural Science Foundation Of China

Theorem. *Suppose $c < -2$, then (1.3)-(1.4) has a solution, and this solution is a increasing function of ξ.*

The proof of the theorem is divided some parts.

(1) Local solution in the neighborhood of $\xi = +\infty$

Set $w = 1 - u$, then we get from (1.3) that

$$w'' + cw' - w(1-w) = 0, \tag{1.5}$$

linearilizing (1.5) and setting $w_1 = w'$ we get the following system

$$\begin{aligned} w' &= \quad w_1, \\ w_1' &= w - cw_1, \end{aligned} \tag{1.6}$$

There are two characteristic roots for the matrix corresponding to (1.6), these are

$$\lambda_1 = \frac{-c - \sqrt{c^2+4}}{2} < 0 < \frac{-c + \sqrt{c^2+4}}{2}.$$

we set $s = e^{\lambda_1 \xi}, w(s) = w(\xi)$, noting $\frac{d}{d\xi} = \lambda_1 s \frac{d}{ds}$, $\frac{d^2}{d\xi^2} = \lambda_1^2 s^2 \frac{d^2}{ds^2} + \lambda_1^2 s \frac{d}{ds}$, we get

$$\lambda_1^2 s^2 w'' + sw' - w(1-w) = 0$$

we solve this equation by looking for power series $w(s) = \sum_{i=1}^{\infty} a_i s^i$, we get a_1 is an arbitrary constant, $a_2 = -a_1^2/(1+2\lambda_1^2)$, recursively we get whole $a_i, i \geq 3$.

We can obtain some properties from the expression $w(\xi)$ which are useful below, here $u_L(\xi) = 1 - w(\xi)$.

(a)$\lim_{\xi \to \infty} w(\xi) = \lim_{s \to 0} w(s) = 0$.

(b)$w(s) < 1$ in the right neighborhood of $s = 0$, hence $0 < u_L(\xi)$, $1 - u_L(\xi) < 1$ in the neighborhood of $\xi = +\infty$.

(c) $\lim_{\xi \to \infty} w'(\xi) = 0$.

Therefore we can choose a ξ_0 such that

$$\frac{1}{2} < u_L(\xi_0) = u^*$$

and

$$0 < u_L'(\xi_0) < \frac{1}{2} \cdot \frac{-c - \sqrt{c^2 - 4}}{2}, \quad c < -2. \tag{1.8}$$

(2) Semiinfinite B.V.P.

Here we consider

$$u'' + cu' + u(1-u) = 0, \quad -\infty < \xi < \xi_0 \tag{1.9}$$

$$u(-\infty) = 0, \qquad u(\xi_0) = u^*. \tag{1.10}$$

Proposition. *There exists a solution of (1.9)-(1.10) for $c < -2$.*

Proof: Choose

$$k > -c, k' \in \left(\frac{-c - \sqrt{c^2 - 4}}{2}, \quad \frac{-c + \sqrt{c^2 - 4}}{2} \right) \tag{1.11}$$

and set

$$\overline{u}_s = u^* e^{k'(\xi - \xi_0)} \qquad \underline{u}_s = u^* e^{k(\xi - \xi_0)}, \tag{1.12}$$

since $0 < k' < k$, we have

$$\overline{u}_s(\xi) > \underline{u}_s(\xi) \qquad in \quad (-\infty, \xi_0).$$

According to (1.11) we get

$$\overline{u}_s'' + c\overline{u}_s' + \overline{u}_s(1 - \overline{u}_s) = u^* e^{k'(\xi - \xi_0)}(k'^2 + ck' + 1 - \overline{u}_s) \leq 0 \tag{1.13}$$

and

$$\underline{u}_s'' + c\underline{u}_s' + \underline{u}_s(1 - \underline{u}_s) = u^* e^{k(\xi - \xi_0)}(k^2 + ck) + \underline{u}_s(1 - \underline{u}_s) \geq 0, \tag{1.14}$$

the boundary comparison hold obviously, hence $\overline{u}$ and $\underline{u}$ are upper and lower solutions of the problem (1.9)-(1.10) respectively. By means of the comparison existence theorem [6] we get the existence of solution and we denote this solution by u_s.

Remark 1 Because of $u_s(\xi_0) = \overline{u}_s(\xi_0) = \underline{u}_s(\xi_0)$ and $\underline{u}_s(\xi) \leq u_s(\xi) \leq \overline{u}_s(\xi)$ so $u^* k' \leq u_s'(\xi_0) \leq u^* k$.

Remark 2 $u_s(\xi)$ is a monotone increasing function Of ξ. If it is not true, then there exists a ξ^* and ξ^* is a minimum point of $u_s(\xi)$ so $u_s'(\xi^*) = 0$, but $u_s(\xi^*) \geq \underline{u}_s(\xi^*) > 0$, therefore

$$u_s''(\xi^*) + cu_s'(\xi^*) + u_s(\xi^*)(1 - u_s(\xi^*)) > 0,$$

this is a contradiction.

The proof of theorem.

Go back to the problem (1.3)-(1.4), we define

$$\tilde{u}(\xi) = \begin{cases} u_s(\xi), & (-\infty, \xi_0) \\ u_L(\xi), & [\xi_0, \infty) \end{cases} \tag{1.16}$$

It follows from the remark 1 and (1.8) that $\tilde{u}'(\xi_0 - 0) > \tilde{u}'(\xi_0 + 0)$, so $\tilde{u}$ satisfies the require of general upper solution, then we choose $\overline{u} = \tilde{u}$ and $\underline{u} = 0$ as upper and lower solutions for the problem (1.3)-(1.4) respectively, and we get the existence of the problem (1.3)-(1.4). Similarly we can infer the monotonic of solution, so the theorem is proved.

2. Competition-diffusion system

Here we consider the competition-diffusion system

$$\begin{cases} u_t &= u_{xx} + u(a_1 - b_1 u - c_1 v) \\ v_t &= v_{xx} + v(a_2 - b_2 u - c_2 v) \end{cases} \tag{2.1}$$

the existence of a travelling front solution is proved. It is different with the references [2],[3] we have used alternative method to get the existence which has different equilibriums at $\xi = \pm\infty$.

Suppose that

$$\frac{c_1}{c_2} > \frac{a_1}{a_2} > \frac{b_1}{b_2} \tag{2.2}$$

and introducing travelling wave variable $\xi = x - ct$, we transform the problem looking for travelling wave solution into the following problem for differential equations

$$\begin{cases} u''(\xi) + cu'(\xi) + u(a_1 - b_1 u - c_1 v) &= 0, \\ v''(\xi) + cv'(\xi) + v(a_2 - b_2 u - c_2 v) &= 0. \end{cases} \tag{2.3}$$

we choose the equilibriums at $\xi = \pm\infty$ as follows: $(a_1/b_1, 0)$ and (u^*, v^*), where $u^* = (a_1 c_2 - a_2 c_1)/(b_1 c_2 - b_2 c_1)$, $v^* = (b_1 a_2 - b_2 a_1)/(b_1 c_2 - b_2 c_1)$. therefore at $\xi = \pm\infty$ u and v satisfy

$$u(-\infty) = u^*, \qquad v(-\infty) = v^*, \tag{2.4}$$

$$u(\infty) = \frac{a_1}{b_1}, \qquad v(\infty) = 0. \tag{2.5}$$

THEOREM. *For any $c < -c_0$ there exists a solution of the problem (2.3),(2.4) and (2.5) moreover $u(\xi)$ is a nondecreasing function and $v(\xi)$ is a nonincreasing function where $c_0 = 2\sqrt{c^*}$, $c^* = \max\left(\frac{a_1 c_1 b_2}{b_1 c_2}, \frac{c_1 b_2 v^*}{b_1}\right)$.*

The proof of above Theorem is divided several steps.

(I) By the method of [1] we can find the local solution $u_L(\xi)$ and $v_L(\xi)$ of the problem (2.3) and (2.5) in the neighborhood of $\xi = \pm\infty$. Let $w = \frac{a_1}{b_1} - u$ and we linearize the system (2.3) near $w = 0$ and $v = 0$. Then we obtain

$$\begin{cases} w'' + cw' - a_1 w + \frac{a_1 c_1}{b_1} v &= 0 \\ v'' + cv' + (a_2 - \frac{b_2 a_1}{b_1}) v &= 0 \end{cases} \tag{2.6}$$

or

$$\begin{aligned} \frac{dw}{d\xi} &= w_1 \\ \frac{dw_1}{d\xi} &= a_1 w - cw_1 - \frac{a_1 c_1}{b_1} v \\ \frac{dv}{d\xi} &= v_1 \\ \frac{dv_1}{d\xi} &= -(a_2 - \frac{b_2 a_1}{b_1}) v - cv_1 \end{aligned} \tag{2.7}$$

the characteristic equation of this system is

$$(\lambda^2 + c\lambda - a_1)[\lambda^2 + c\lambda - (\frac{a_1 b_2}{b_1} - a_2)] = 0.$$

This equation has two negative real roots:

$$\lambda_1 = \frac{1}{2}\left(-c - \sqrt{c^2 + 4a_1}\right), \qquad \lambda_2 = \frac{1}{2}\left(-c - \sqrt{c^2 + 4(\frac{a_1 b_2}{b_1} - a_2)}\right).$$

We consider the case that $|\lambda_1| > |\lambda_2|$. Denoting

$$\overline{D} = \frac{b_1}{a_1 c_1}(a_1 - \frac{a_1 b_2}{b_1} + a_2), \quad \kappa = \frac{\lambda_1}{\lambda_2}, \quad \sigma = \frac{d_1}{d_2^{\kappa}},$$

where d_2 is positive and d_1 and d_2 are constans which can be chosen suitably. Introducing the following transform

$$v = \overline{D}z, \quad s = d_2 e^{\lambda_2 \xi} \tag{2.8}$$

then the equations (2.3) become

$$\begin{aligned} &\lambda_2^2 s^2 w'' + (\frac{a_1 b_2}{b_1} - a_2) s w' - a_1 w \\ &\qquad + \frac{c_1 a_1}{b_1}\overline{D}z + w(b_1 w - c_1 \overline{D}z) = 0, \end{aligned} \tag{2.9}$$

$$\begin{aligned} &\lambda_2^2 s^2 z'' + (\frac{a_1 b_2}{b_1} - a_2) s z' \\ &\qquad - (\frac{a_1 b_2}{b_1} - a_2) z + z(b_2 w - c_2 \overline{D}z) = 0. \end{aligned} \tag{2.10}$$

Using the double series method of [1] to solve above system we have

$$\begin{aligned} w(s) &= s + w_{02}s^2 + \cdots + \sigma s^k(1 + w_{11}s + \cdots) + \cdots, \\ z(s) &= s + z_{02}s^2 + \cdots + \sigma s^k(z_{11}s + \cdots) + \cdots \end{aligned} \tag{2.11}$$

we can recursively determine these coefficients w_{ij}, z_{ij},for example,

$$\begin{aligned} z_{02} &= \frac{-b_2 + c_2\overline{D}}{2\lambda_2^2 + \frac{a_1 b_2}{b_1} - a_2}, \\ w_{02} &= \frac{-b_1 + c_1\overline{D} - \frac{c_1 a_1}{b_1}\overline{D}z_{02}}{2\lambda_2^2 + 2(\frac{b_2 a_1}{b_1} - a_2) - a_1} \\ z_{11} &= \frac{-b_2}{(\kappa(\kappa+1)\lambda_2^2 + \kappa(\frac{a_1 b_2}{b_1} - a_2)} \\ w_{11} &= \frac{-2b_1 - c_1\overline{D} - \frac{c_1 a_1}{b_1}\overline{D}z_{11}}{(\kappa+1)\kappa\lambda_2^2 + (\kappa+1)(\frac{a_1 b_2}{b_1} - a_2) - a_1} \end{aligned}$$

Hence we get $u_L(\xi)$ and $v_L(\xi)$ which defined in a neighborhood of $\xi = \infty$:

$$\begin{aligned} u_L(\xi) &= \frac{a_1}{b_1} - [d_2 e^{\lambda_2 \xi} + w_{02}(d_2 e^{\lambda_2 \xi})^2 + \cdots \\ &\quad + d_1 e^{\lambda_1 \xi}(1 + w_{11} d_2 e^{\lambda_2 \xi} + \cdots) + \cdots], \\ v_L(\xi) &= \overline{D}[d_2 e^{\lambda_2 \xi} + z_{02}(d_2 e^{\lambda_2 \xi})^2 + \cdots \\ &\quad + d_1 e^{\lambda_1 \xi}(z_{11} d_2 e^{\lambda_2 \xi} + \cdots) + \cdots], \end{aligned} \tag{2.12}$$

By the series expressions of $u_L(\xi)$ and $v_L(\xi)$ we obtain the following conclusions:
(a)

$$\lim_{\xi \to \infty} u_L(\xi) = \frac{a_1}{b_1}$$

$$\lim_{\xi \to \infty} (v_L(\xi), u'_L(\xi), v'_L(\xi)) = (0,0,0),$$

$$\lim_{\xi \to \infty} \frac{\frac{a_1}{b_1} - u_L}{v_L} = \overline{D}^{-1};$$

(b) For sufficiently large ξ, we have

$$\begin{aligned} &u'_L(\xi) > 0, \qquad v'_L(\xi) < 0, \\ &u^* < u_L(\xi) < \frac{a_1}{b_1}, \qquad 0 < v_L(\xi) < v^*, \\ &a_1 - b_1 u_L(\xi) - c_1 v_L(\xi) > 0. \end{aligned} \tag{2.13}$$

So we can choose ξ_0 such that

$$\frac{a_1}{b_1} > u_L(\xi_0) = u_1 > \frac{1}{2}(\frac{a_1}{b_1} + u^*), \quad 0 < v_L(\xi_0) = v_1 < \frac{v^*}{2}, \tag{2.14}$$

$$0 < u'_L(\xi_0) < \frac{1}{2}(\frac{a_1}{b_1} - u^*)k', \quad 0 > v'_L(\xi_0) > -\frac{v^* k'}{2} \tag{2.15}$$

where

$$0 < k' \in (\frac{-c - \sqrt{c^2 - 4c^*}}{2}, \quad \frac{-c + \sqrt{c^2 - 4c^*}}{2}). \tag{2.16}$$

(II) We now consider a semi-infinite boundary value problem.

We consider the boundary value problem for system (2.3) on $(-\infty, \xi_0)$ with the boundary value condition (2.4) and the condition

$$u(\xi_0) = u_1, \qquad v(\xi_0) = v_1 \tag{2.17}$$

where u_1 and v_1 are defined in (2.14).

Proposition 1. *Let $\underline{u}_1(\xi), \overline{u}_1(\xi), \underline{v}_1(\xi)$ and $\overline{v}_1(\xi)$ are in $C^2(-\infty, \xi_0) \cap C(-\infty, \xi_0]$, $0 \le \underline{u}_1(\xi) \le \overline{u}_1(\xi), o \le \underline{v}_1(\xi) \le \overline{v}_1(\xi)$ hold in $(-\infty, \xi_0)$, and they satisfy the following* *differential inequalities*

$$\begin{aligned}&\underline{u}_1'' + c\underline{u}_1' + \underline{u}_1(a_1 - b_1\underline{u}_1 - c_1\overline{v}_1)\\ &\quad \geq 0 \geq \overline{v}_1'' + c\overline{v}_1' + \overline{v}_1(a_2 - b_2\underline{u}_1 - c_2\overline{v}_1),\\ &\underline{v}_1'' + c\underline{v}_1' + \underline{v}_1(a_2 - b_2\overline{u}_1 - c_2\underline{v}_1)\\ &\quad \geq 0 \geq \overline{u}_1'' + c\overline{u}_1' + \overline{u}_1(a_1 - b_1\overline{u}_1 - c_1\underline{v}_1),\\ &\quad (-\infty < \xi < \xi_0)\end{aligned} \tag{2.18}$$

and the boundary value comparison relations

$$\underline{u}_1(\xi_0) \leq u_1 \leq \overline{u}_1(\xi_0), \quad \underline{v}_1(\xi_0) \leq v_1 \leq \overline{v}_1(\xi_0) \tag{2.19}$$

and the limit relations

$$\lim_{\xi \to -\infty} \underline{u}_1(\xi) = \lim_{\xi \to -\infty} \overline{u}_1(\xi) = u^*, \lim_{\xi \to -\infty} \underline{v}_1(\xi) = \lim_{\xi \to -\infty} \overline{v}_1(\xi) = v^*. \tag{2.20}$$

Then the solution of the problem (2.3),(2.4) and (2.17) exists.

Proof: First we consider the system (2.3),the condition (2.17) and the condition

$$u(-R) = u_{2R}, \qquad v(-R) = v_{2R} \tag{2.21}$$

on $[-R, \xi_0], R = 1, 2, \cdots$, where

$$\underline{u}_1(-R) \leq u_{2R} \leq \overline{u}_1(-R), \quad \underline{v}_1(-R) \leq v_{2R} \leq \overline{v}_1(-R). \tag{2.22}$$

Since the nonlinear terms in (2.3) are

$$f = u(a_1 - b_1 u - c_1 v), \qquad g = v(a_2 - b_2 u - c_2 v)$$

which satisfy $f_v \leq 0, g_u \leq 0$ in $[0, \max \overline{u}_1(\xi)] \times [0, \max \overline{v}_1(\xi)]$. It follows from (2.18),(2.19) and (2.21) that $\underline{u}_1, \overline{u}_1, \underline{v}_1$ and $\overline{v}_1$ are the upper and lower solutions of the problem (2.3),(2.17) and (2.21). According to the comparison existence theorem ([5],[6]) of the boundary value problem for a quasimonotone semilinear elliptic system on a bounded domain, we get the existence of solution of the problem (2.3),(2.17) and (2.21). Denote this solution by (u_R, v_R),then $0 \leq \underline{u}_1(\xi) \leq u_R(\xi) \leq \overline{u}_1(\xi)$, $0 \leq \underline{v}_1(\xi) \leq v_R(\xi) \leq \overline{v}_1(\xi)$.

Next we choose a convergent subsequence on $[-R, \xi_0]$ for every $R = 1, 2, \cdots$.

Consider the sequence $\{(u_R, v_R), R = 1, 2, \cdots\}$ on $[-1, \xi_0]$. Clearly they are uniformly bounded, i.e.

$$\|u_R\|_{L^\infty[-1,\xi_0]}, \quad \|v_R\|_{L^\infty(-1,\xi_0)} \leq M, \quad R = 1, 2, \cdots, \tag{2.23}$$

so we have

$$\|f(u_R, v_R)\|_{L^\infty(-1,\xi_0]}, \quad \|g(u_R, v_R)\|_{L^\infty[-1,\xi_0]} \leq M_1, \quad R = 1, 2, \cdots.$$

By means of the L^p estimate for a elliptic partial differential equation we have

$$\|u_R\|_{H^2[-1,\xi_0]}, \quad \|v_R\|_{H^2[-1,\xi_0]} \le C, \quad R = 1, 2, \cdots. \tag{2.24}$$

By means of the embedding theorem it follows from above formula that

$$\|u_R\|_{C^1[-1,\xi_0]}, \quad \|v_R\|_{C^1[-1,\xi_0]} \le C, \quad R = 1, 2, \cdots. \tag{2.25}$$

It follows that

$$\|f(u_R, v_R)\|_{C^1[-1,\xi_0]}, \quad \|g(u_R, v_R)\|_{C^1[-1,\xi_0]} \le C, \quad R = 1, 2, \cdots.$$

By using the Schauder estimate for the elliptic partial differential equations for any $0 < \alpha < 1$ we have

$$\|u_R\|_{C^{2,\alpha}[-1,\xi_0]}, \quad \|v_R\|_{C^{2,\alpha}[-1,\xi_0]} \le C, \quad R = 1, 2, \cdots. \tag{2.26}$$

It show that the sequence $\{(u_R, v_R)\}$ is compact in $C^2[-1, \xi_0] \times C^2[-1, \xi_0]$ so we can choose a subsequence $\{(u_{R_{1j}}, v_{R_{1j}})\}$ which has a limit in $C^2[-1, \xi_0] \times C^2[-1, \xi_0]$, i.e.

$$\lim_{j \to \infty} (u_{R_{1j}}, v_{R_{1j}}) = (u_{s1}, v_{s1})$$

and (u_{s1}, v_{s1}) satisfies (2.3) on $[-1, \xi_0]$ and the condition (2.17).

Without loss of generality we assume that $R_{1j} \ge 2$, and study $\{(u_{R_{1j}}, v_{R_{1j}})\}$ in $[-2, \xi_0]$. Repeating the same argument we get the convergent subsequence $\{(u_{R_{2j}}, v_{R_{2j}})\}$ in $C^2[-2, \xi_0] \times C^2[-2, \xi_0]$ and its limit is denoted by (u_{s2}, v_{s2}). Clearly (u_{s2}, v_{s2}) also satisafies (2.3) and (2.17). We recursively continue this process to obtain the function $(u_s(\xi), v_s(\xi))$ which is defined on $(-\infty, \xi_0)$ by the diagonal rule, and (u_s, v_s) satisfies (2.3) and (2.17) too. Since

$$\underline{u}_1(\xi) \le u_s(\xi) \le \overline{u}_1(\xi), \quad \underline{v}_1(\xi) \le v_s(\xi) \le \overline{v}_1(\xi), \tag{2.27}$$

it follows from (2.20) that $\lim_{\xi \to -\infty} u_s(\xi) = u^*$, $\lim_{\xi \to -\infty} v_s(\xi) = v^*$, and the Proposition is proved now.

LEMMA 1. *Let $c \le -c_0$, then there exists a solution of the problem (2.3),(2.4) and (2.17).*

Proof: By Proposition 1, the proof of existence for the problem (2.3),(2.4) and (2.17) attribute to construct the upper and lower solution pairs. Below we give $\underline{u}_1, \overline{u}_1, \underline{v}_1$ and $\overline{v}_1$ and check them by the conditions of Proposition 1. Let

$$\begin{aligned} \underline{u}_1(\xi) &= (u_1 - u^*)e^{k(\xi - \xi_0)} + u^*, \quad \overline{u}_1(\xi) = (u_1 - u^*)e^{k'(\xi - \xi_0)} + u^*, \\ \underline{v}_1(\xi) &= (v_1 - v^*)e^{k'(\xi - \xi_0)} + v^*, \quad \overline{v}_1(\xi) = (v_1 - v^*)e^{k(\xi - \xi_0)} + v^* \end{aligned} \tag{2.28}$$

where $k > -c$ and k' satisfies (2.16). It is obvious that

$$u_1 - u^* > 0, \quad v_1 - v^* < 0, \quad k > k' > 0,$$
$$\xi - \xi_0 \le 0, \quad e^{k(\xi-\xi_0)} \le e^{k'(\xi-\xi_0)}, \quad (\infty < \xi \le \xi_0)$$

Hence $\underline{u}_1(\xi) \le \overline{u}_1(\xi), \underline{v}_1(\xi) \le \overline{v}_1(\xi)$ and (2.19) and (2.20) hold. Now we check the differential inequalities. Note that $\underline{u}_1$ and $\overline{v}_1$ or $\overline{u}_1$ and $\underline{v}_1$ appear together in (2.18), so we have

$$\begin{aligned} &\underline{u}_1'' + c\underline{u}_1' + \underline{u}_1(a_1 - b_1\underline{u}_1 - c_1\overline{v}_1) \\ &\quad = (u_1 - u^*)e^{k(\xi-\xi_0)}(k^2 + ck) + \underline{u}_1(a_1 - b_1\underline{u}_1 - c_1\overline{v}_1) > 0, \end{aligned} \tag{2.29}$$

$$\begin{aligned} &\overline{v}'' + c\overline{v}_1' + \overline{v}_1(a_2 - b_2\underline{u}_1 - c_2\overline{v}_1) \\ &\quad < (v_1 - v^*)e^{k(\xi-\xi_0)}(k^2 + ck) < 0. \end{aligned} \tag{2.30}$$

Here we note that the points $(\underline{u}_1, \overline{v}_1)$ and $(\overline{u}_1, \underline{v}_1)$ on $(u, v)-$ plane are on the straightline which connects (u_1, v_1) and (u^*, v^*)(see Fig. 1).

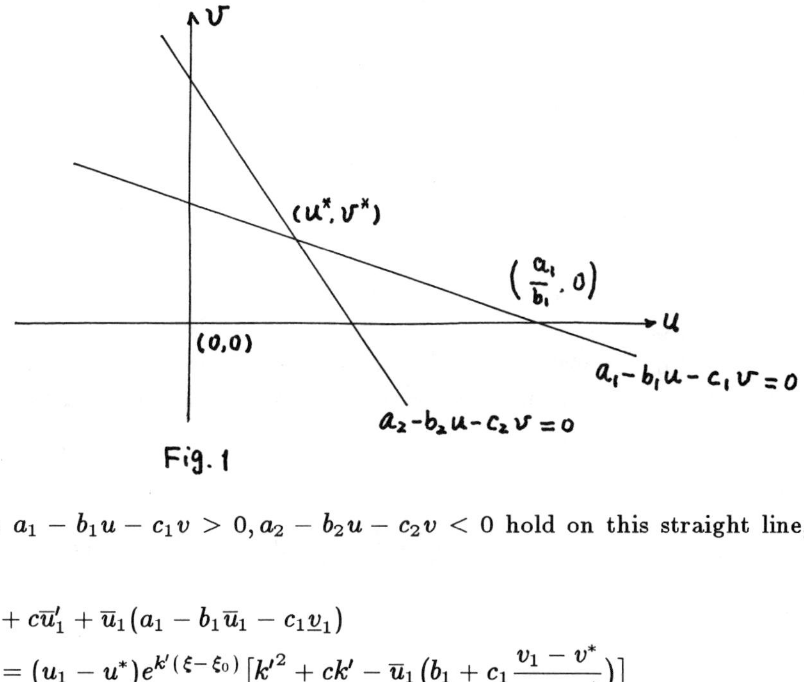

Fig. 1

Thus we have $a_1 - b_1u - c_1v > 0, a_2 - b_2u - c_2v < 0$ hold on this straight line, therefore

$$\begin{aligned} &\overline{u}_1'' + c\overline{u}_1' + \overline{u}_1(a_1 - b_1\overline{u}_1 - c_1\underline{v}_1) \\ &\quad = (u_1 - u^*)e^{k'(\xi-\xi_0)}\big[k'^2 + ck' - \overline{u}_1\big(b_1 + c_1\frac{v_1 - v^*}{u_1 - u^*}\big)\big] \\ &\quad \le (u_1 - u^*)e^{k'(\xi-\xi_0)}(k'^2 + ck' + c^*) < 0, \end{aligned} \tag{2.31}$$

$$\begin{aligned} &\underline{v}_1'' + c\underline{v}_1' + \underline{v}_1(a_2 - b_2\overline{u}_1 - c_2\underline{v}_1) \\ &\quad = (v_1 - v^*)e^{k'(\xi-\xi_0)}\big[k'^2 + ck' - \underline{v}_1\big(b_2\frac{u_1 - u^*}{v_1 - v^*} + c_2\big)\big] \\ &\quad \ge (v_1 - v^*)e^{k'(\xi-\xi_0)}(k'^2 + ck' + c^*) > 0. \end{aligned} \tag{2.32}$$

Here we have considered that k' satisfies (2.16). Hence we obtain the conclution by means of Proposition 1. We denote the solution of (2.3),(2.4) and (2.17) by $(u_s(\xi), v_s(\xi))$.

Remark 1 According to the expressions of the upper and lower solutions, we have

$$0 < k'(u_1 - u^*) \le u_s'(\xi) \le k(u_1 - u^*), \tag{2.33}$$

$$k(v_1 - v^*) \le v_s'(\xi_0) \le k'(v_1 - v^*) < 0. \tag{2.34}$$

To prove Theorem we need the following

Proposition 2. *Let $\bar{u}(\xi)$ and $\underline{v}(\xi)$ be in $C(-\infty,\infty) \cap C^2(-\infty,\xi_0) \cap C^2(\xi_0,\infty)$, and*

$$\bar{u}'(\xi_0 - 0) \ge \bar{u}'(\xi_0 + 0), \quad \underline{v}'(\xi_0 - 0) \le \underline{v}'(\xi_0 + 0). \tag{2.35}$$

Let $\underline{u}(\xi), \bar{v}(\xi) \in C^2(R)$ such that $\underline{u}(\xi) \le \overline{u}(\xi), \underline{v}(\xi) \le \overline{v}(\xi)$, which satisfy

$$\begin{aligned} \lim_{\xi\to-\infty} \underline{u}(\xi) \le u^* \le \lim_{\xi\to-\infty} \overline{u}(\xi), \\ \lim_{\xi\to-\infty} \underline{v}(\xi) \le v^* \le \lim_{\xi\to-\infty} \overline{v}(\xi), \end{aligned} \tag{2.36}$$

and

$$\begin{aligned} \lim_{\xi\to\infty} \underline{u}(\xi) \le \frac{a_1}{b_1} \le \lim_{\xi\to\infty} \overline{u}(\xi), \\ \lim_{\xi\to\infty} \underline{v}(\xi) = 0 \le \lim_{\xi\to\infty} \overline{v}(\xi). \end{aligned} \tag{2.37}$$

at $\xi = \pm\infty$. Moreover, $\underline{u}$ and $\overline{v}$, $\overline{u}$ and $\underline{v}$ satisfy the differential inequalities (2.18) on $(-\infty,\xi_0) \cup (\xi_0,\infty)$. Then there exists a solution of the problem (2.3),(2.4) and (2.5).

Proof: Similary, on $[\xi_0 - R, \xi_0 + R]$, $R = 1, 2, \cdots$, we consider system (2.3) with boundary conditions

$$\begin{aligned} u(\xi_0 - R) = u_{3R}, \quad u(\xi_0 + R) &= u_{4R}, \\ v(\xi_0 - R) = v_{3R}, \quad v(\xi_0 + R) &= v_{4R} \end{aligned} \tag{2.38}$$

here we require

$$\begin{aligned} \underline{u}(\xi_0 - R) \le u_{3R} \le \overline{u}(\xi_0 - R), \quad \underline{v}(\xi_0 - R) \le v_{3R} \le \overline{v}(\xi_0 - R), \\ \underline{u}(\xi_0 + R) \le u_{4R} \le \overline{u}(\xi_0 + R), \quad \underline{v}(\xi_0 + R) \le v_{4R} \le \overline{v}(\xi_0 + R) \end{aligned} \tag{2.39}$$

and

$$\begin{aligned} \lim_{R\to\infty} u_{3R} = u^*, \qquad \lim_{R\to\infty} u_{4R} = \frac{a_1}{b_1}, \\ \lim_{R\to\infty} v_{3R} = v^*, \qquad \lim_{R\to\infty} v_{4R} = 0. \end{aligned} \tag{2.40}$$

Since $\underline{u} \le \bar{u}, \underline{v} \le \overline{v}$ and (2.36),(2.37), u_{3R}, v_{3R}, u_{4R} and v_{4R} can be choosed, and by means of (2.35), $(\overline{u}(\xi), \underline{v}(\xi))$ and $(\underline{u}(\xi), \overline{v}(\xi))$ may be as generalized upper and lower solutions([7],[8]). We obtain the existence of solution for problem (2.3) and (2.38) by the comparison existence theorem on the bounded domain, repeat the same argument choosing subsequence in the proof of Proposition 1 again, then we get the functions $u(\xi)$ and $v(\xi)$ which are defined on whole $(-\infty,\infty)$. The functions $(u(\xi),v(\xi))$ satisfy (2.3) and we know from (2.40) that

$$\lim_{\xi\to\infty} u(\xi)=\frac{a_1}{b_1},\qquad \lim_{\xi\to-\infty} u(\xi)=u^*,$$

$$\lim_{\xi\to\infty} v(\xi)=0,\qquad \lim_{\xi\to-\infty} v(\xi)=v^*,$$

this complete the proof of the Proposition.

Now go back to the proof of the basic Theorem. We had obtained the local solution $u_L(\xi)$ and $v_L(\xi)$ on $[\xi_0,\infty)$ and the solution $(u_s(\xi),v_s(\xi))$ of the boundary value problem on semi-bounded interval $(-\infty,\xi_0)$. We define

$$\overline{u}(\xi)=\begin{cases} u_s(\xi), & (-\infty,\xi_0] \\ u_L(\xi), & [\xi_0,\infty), \end{cases} \tag{2.41}$$

$$\underline{v}(\xi)=\begin{cases} v_s(\xi), & (-\infty,\xi_0] \\ v_L(\xi), & [\xi_0,\infty). \end{cases} \tag{2.42}$$

It follows from (2.15),(2.33) and (2.34) that

$$\begin{aligned}\overline{u}'(\xi_0-0)&=u_s'(\xi_0)\\ &\geq k'(u_1-u^*)>\frac{1}{2}k'(\frac{a_1}{b_1}+u^*-2u^*)\\ &>u_L'(\xi_0)=\overline{u}'(\xi_0+0),\end{aligned} \tag{2.43}$$

$$\begin{aligned}\underline{v}'(\xi_0-0)=v_s'(\xi_0)&\leq k'(v_1-v^*)\\ &<-\frac{1}{2}k'v^*<v_L'(\xi_0)=\underline{v}'(\xi_0+0).\end{aligned} \tag{2.44}$$

Let $\underline{u}=u^*,\overline{v}=v^*$, then $(\overline{u}(\xi),\underline{v}(\xi))$ and $(\underline{u}(\xi),\overline{v}(\xi))$ satisfy the all conditions in Proposition 2, hence the existence of travelling wave solution is proved.

In order to prove the existence of monotone travelling wave solution we must reform the proof of above Propositions and Lemmas.

Proposition $1'$. *Assume that* $\underline{u}_1(\xi),\overline{u}_1(\xi),\underline{v}_1(\xi)$ *and* $\overline{v}_1(\xi)$ *satisfy the all conditions in Proposition 1 and* $\underline{u}_1(\xi)$ *and* $\overline{u}_1(\xi)$ *are monotone increasing functions, and* $\underline{v}_1(\xi)$ *and* $\overline{v}_1(\xi)$ *are decreasing functions. Moreover, we asuume that* $\underline{u}_1(\xi_0)=\overline{u}_1(\xi_0)=u_1$, $\overline{u}_1'(\xi_0)>0,\underline{u}_1'(\xi_0)>0$, $\underline{v}_1(\xi_0)=\overline{v}_1(\xi_0)=v_1$, $\underline{v}_1'<0$, $\overline{v}_1'(\xi_0)<0$. *Then the solutions* $u_s(\xi)$ *and* $v_s(\xi)$ *of the problem (2.3),(2.4) and (2.17) are non-decreasing and non-increasing functions respectively.*

Proof: Denote $m=\min_{(-\infty,\xi_0]}\underline{u}_1(\xi)$, $M=\max_{(-\infty,\xi_0]}\overline{u}_1(\xi)$, $n=\min_{(-\infty,\xi_0]}$ and $N=\max_{(-\infty,\xi_0]}\overline{v}_1(\xi)$. We can choose $M_1>0$ sufficiently large such that for $(u,v)\in$

$[m, M] \times [n, N]$, $F_1(u,v) = M_1 u + u(a_1 - b_1 u - c_1 v)$ becomes a increasing function for u and a decreasing function for v ; and $F_2(u,v) = M_1 v + v(a_2 - b_2 u - c_2 v)$ becomes a decreasing for u and a increasing function for v. Choose $u^{(0)} = \overline{u}_1(\xi)$, $v^{(0)} = \underline{v}_1(\xi)$ and recursively define a sequence by solving the following boundary value problem for $k \geq 1$:

$$-\frac{d^2 u^{(k)}}{d\xi^2} - c\frac{du^{(k)}}{d\xi} + M_1 u^{(k)} = F_1(u^{(k-1)}, v^{(k-1)}), \tag{2.45}$$

$$-\frac{d^2 v^{(k)}}{d\xi^2} - c\frac{dv^{(k)}}{d\xi} + M_1 v^{(k)} = F_2(u^{(k-1)}, v^{(k-1)}), \tag{2.46}$$

$$u^{(k)}(\xi_0) = u_1, \qquad v^{(k)}(\xi_0) = v_1, \tag{2.47}$$

$$u^{(k)}(-\infty) = u^*, \qquad v^{(k)}(-\infty) = v^*. \tag{2.48}$$

We know that $\underline{u}_1, \overline{u}_1, \underline{v}_1$ and $\overline{v}_1$ are the upper and lower solutions of the boundary value problem for the above linear system when $k = 1$, hence the solution $(u^{(1)}, v^{(1)})$ of the problem (2.45)-(2.48) is obtained by the method in Proposition 1, and by the generalized comparison principle we have

$$\underline{u}_1(\xi) \leq u^{(1)} \leq \overline{u}_1(\xi), \quad \underline{v}_1(\xi) \leq v^{(1)} \leq \overline{v}_1(\xi), \quad \xi \in (-\infty, \xi_0]. \tag{2.49}$$

Next we claim that $u^{(1)}$ and $v^{(1)}$ are monotone.

Since $\underline{u}_1(\xi_0) = \overline{u}_1(\xi_0), \underline{u}_1'(\xi_0) > 0$ and $\overline{u}'(\xi_0) > 0$, $\frac{du^{(1)}}{d\xi}(\xi_0) > 0$. Thus there is a $\delta_0 > 0$ such that $u^{(1)}(\xi_0 - \delta) \leq u^{(1)}(\xi_0)$ for $0 < \delta < \delta_0$. Denote $\tilde{u}^{(1)}(\xi) = u^{(1)}(\xi - \delta)$, $\tilde{u}^{(0)}(\xi) = \overline{u}_1(\xi - \delta)$ and $\tilde{v}^{(0)}(\xi) = \underline{v}_1(\xi - \delta)$, then we have

$$-\frac{d^2 \tilde{u}^{(1)}}{d\xi^2} - c\frac{d\tilde{u}^{(1)}}{d\xi} + M_1 \tilde{u}^{(1)} = F_1(\overline{u}_1(\xi - \delta), \underline{v}_1(\xi - \delta)), \tag{2.50}$$

$$\tilde{u}^{(1)}(\xi_0) \leq u_1, \tag{2.51}$$

$$\tilde{u}^{(1)}(-\infty) = u^*. \tag{2.52}$$

Since $\overline{u}_1(\xi)$ is increasing and $\underline{v}_1(\xi)$ is decreasing, we have

$$F_1(\overline{u}_1(\xi - \delta), \underline{v}_1(\xi - \delta)) \leq F_1(\overline{u}_1(\xi), \underline{v}_1(\xi)) = F_1(u^{(0)}, v^{(0)}). \tag{2.53}$$

Denote $w = u^{(1)} - \tilde{u}^{(1)}$ then (2.45), (2.47), (2.48), (2.50), (2.51), (2.52) and (2.53) imply

$$\begin{cases} -w'' - cw' + M_1 w \geq 0, \\ w(\xi_0) \geq 0, \qquad w(-\infty) = 0 \end{cases}$$

so $w \geq 0$ on $(-\infty, \xi_0]$ and it shows that $u^{(1)}(\xi)$ is a non- decreasing function.

Similarly we can prove $v^{(1)}$ is a non-increasing function. Moreover we can also check $\underline{u}_1, \overline{u}_1, \underline{v}_1$ and $\overline{v}_1$ are the upper and lower solutions of the problem (2.45)-(2.48) for $k = 2$. For any $k \geq 2$ we can recursively prove that $u^{(k)}$ is non- decreasing and $v^{(k)}$ is non-increasing. By using the methods of the constructing expansion intervals sequence and choosing diagonal subsequence we can obtain a convergent subsequence on $(-\infty, \xi_0]$ whose limit functions $u_s(\xi)$ is non-decreasing and $v_s(\xi)$ is non-increasing on $(-\infty, \xi_0]$ and (u_s, v_s) is a solution of the problem (2.3),(2.4) and (2.17), this complete the peoof of Proposition 1′.

Lemma 1′. *Let* $c \leq -c_0$. *Then there exists a monotone solution for the problem (2.3),(2.4) and (2.17).*

Proof: Since the functions $\underline{u}, \overline{u}, \underline{v}$ and $\overline{v}$ introduced in the proof of Lemma 1 satisfy the hypothesis of the Proposition 1, the conclution of the Lemma 1′.

Similar the reformation from Proposition 1 to Proposition 1′ we can reform Proposition 2 and establish Proposition 2′, we get that the solution of the problem (2.3),(2.4) and (2.5) is monotone if the upper and lower solutions are monotone. This complete the proof of the basic theorem.

REFERENCES

[1] Ya. I.Kanel, The existence of diffusion wave type solution for Belousov-Zhabotinskii system II, Siberie Math. J.,vol.32,No.3,47-59(1991).

[2] S.Ahmad and A.C.Lazer, An elementary approach to travelling front solution to a system of N competition-diffusion equations, Nonlinear Anal.TMA.vol.16,No.10,891-901(1991).

[3] R.A.Gardner, Existence and stability of travelling wave solution of competition models: A degree theoretic approach, J.D.E.,vol.44.No.3,343- 364(1982).

[4] Wei-Ming Ni, On the elliptic equation $\Delta u - K(x)u^{\frac{n+2}{n-2}} = 0$, its generalizations, and applications in Geometry, Indiana Univ.Math.J., vol.31, No.4,493-529(1982).

[5] L.Zhou and C.V.Pao, Asymptotic behavior of a competition-diffusion system in population dynamics, Nonlinear Anal.TMA,vol.6,No.11, 1163-1184(1982).

[6] Ye Qixiao and Li Zhengyuan, An introduction to reaction-diffusion equations (in Chinese), Beijing, Science Press, 1989.

[7] P.Fife, Mathematical Aspects of R-D equations, Lecture Notes in Biomath.,vol.28, Springer-Verlag, Berlin,1979.

[8] Zhou Li, The comparison method and its applications in Reaction- Diffusion equations, J.of H.U.S.T.,vol.13,N0.2,39-44(1985).

Department of Mathematics, Huazhong University of Science and Technology, Wuhan, China, 430074. Moscow Device Institute.

AMS/IP Studies in Advanced Mathematics
Volume 3, 1997

RESONANCES OF THE SCHWARZSCHILD METRIC

Agnès MOTET-BACHELOT

Abstract. This paper is devoted to the theoretical and computational investigations of the scattering frequencies of scalar, electromagnetic, gravitational waves around a spherical Black Hole. We develop a new procedure for the computation of resonances by spectral analysis of the transient scattered wave, based on Prony's algorithm.

This paper deals with the Scattering Frequencies of the *Regge–Wheeler* equation describing the perturbation of a massless field of spin s outside a Schwarzschild Black-Hole of radius 1 :

$$\partial_t^2 \Phi - \partial_x^2 \Phi + V(x)\,\Phi = 0, \quad t\in\mathbb{R}\,,\ x\in\mathbb{R}\quad , \tag{1}$$

$$V(x) = (1-\frac{1}{r})\,\{\,\frac{l(l+1)}{r^2} + \frac{1-s^2}{r^3}\,\},\ l\in\mathbb{N}\,,\ s\in\mathbb{R}\,, x = r + Log\,(\,r-1)\ ,\ 1<r<+\infty\ . \tag{2}$$

R. Phillips [8] has studied (1) when the potential V decays as $|x|^{-2-\varepsilon}$, $\varepsilon>0$. Since $V_{l,s}$ decays more slowly we cannot apply these results. *J. Dimock* [4] has investigated the scalar case ($s=0$). In [2] we have considered equation (1) as a perturbation of the free wave equation

$$\partial_t^2 \Phi_0 - \partial_x^2 \Phi_0 = 0, \quad t\in\mathbb{R}\,,\ x\in\mathbb{R}\quad ,$$

with more general assumption on the potential. We define the energy spaces

$$\mathscr{H}_o = BL^1(\mathbb{R})\times L^2(\mathbb{R})\,,\quad \mathscr{H} = H(\mathbb{R})\times L^2(\mathbb{R})\,,$$

which are respectively the closures of $C_o^\infty(\mathbb{R})\times C_o^\infty(\mathbb{R})$ with respect to the norms

$$E_o(f) = \int_{\mathbb{R}} |f_1'|^2 + |f_2|^2\,dx\,,\quad E(f) = \int_{\mathbb{R}} |f_1'|^2 + |f_2|^2 + V(x)\,|f_1|^2\,dx$$

We introduce the unitary groups $U_0(t)$ on $\mathscr{H}_0$ and $U(t)$ on $\mathscr{H}$ expressing the solutions of (1) at time t from the initial data, by putting

$$U(t)\,(\Phi(0), \partial_t \Phi(0)) = (\Phi(t), \partial_t \Phi(t)), \quad U_o(t)\,(\Phi_0(0), \partial_t \Phi_0(0)) = (\Phi_0(t), \partial_t \Phi_0(t))$$

We need some dense subspace $\mathscr{D}_0$ of $\mathscr{H}$ and $\mathscr{H}_0$, invariant by $U_0(t)$:

$$\mathscr{D}_o = \{ f = {}^t(f_1, f_2) \in C_o^\infty(\mathbb{R}) \times C_o^\infty(\mathbb{R}), \ \int f_2(x)\,dx = 0\}$$

We introduce the classical wave operators

$$W_\pm f = s - \lim_{t\to\pm\infty} U(-t)\,U_o(t)\,f \quad in \quad \mathscr{H}.$$

The Birman-Kato theorem and the invariance principle give the following

THEOREM 1. *For any f in $\mathscr{D}_o$, the limits $W_\pm f$ are well defined and satisfy*

$$E(W_\pm f) = E_o(f), \quad \overline{W_\pm(\mathscr{D}_0)} = \mathscr{H}.$$

W_+, W_- can be extended by continuity as isometries from $\mathscr{H}_o$ onto $\mathscr{H}$ and the Scattering Operator $S = W_+^{-1} W_-$ is an isometry from $\mathscr{H}_o$ onto $\mathscr{H}_o$.

The equivalence between the stationary scattering theory and the time dependent approach is expressed by the spectral representation that connects the Scattering Operator S and the Heisenberg matrix. We define the free spectral representation of $\mathscr{H}_0$ by putting for $f \in C_o^\infty(\mathbb{R}_x) \times C_o^\infty(\mathbb{R}_x)$:

$$\mathscr{R}_o f(\sigma, \omega) = E_o(f, \varphi_o(.\,, \sigma, \omega)),$$

$$\sigma \in \mathbb{R}^*, \ \omega \in \{-1, 1\}, \ \varphi_o(x, \sigma, \omega) = \frac{1}{\sigma\sqrt{2\pi}} \left(e^{-i\sigma x\omega}, \ i\sigma e^{-i\sigma x\omega}\right)$$

$\mathscr{R}_o$ can be extended as an isometry of $\mathscr{H}_o$ onto $L^2(\mathbb{R}_\sigma \times \{-1, 1\}_\omega)$. Then we prove that the Scattering Operator is unitarily equivalent to the Heisenberg matrix defined by the coefficients of transmission, T, and reflection, R_ω, given by

$$T(\sigma) = 2i\sigma\,[f_-(x, \sigma)\,\partial_x f_+(x, \sigma) - f_+(x, \sigma)\,\partial_x f_-(x, \sigma)]^{-1}, \quad \sigma \in \mathbb{R}^*$$

$$2i\sigma\,R_\omega(\sigma) = T(\sigma)\,[f_+(x, -\omega\sigma)\,\partial_x f_-(x, \omega\sigma) - f_-(x, \omega\sigma)\partial_x f_+(x, -\omega\sigma)].$$

Here, $f_\pm(x, \sigma)$ are the Jost functions, solutions of the integral equation:

$$\sigma \in \mathbb{R}^*, \ f_\pm(x, \sigma) = e^{\pm i\sigma x} + \int_{\pm\infty}^x \frac{\sin\sigma(x-y)}{\sigma}\,V(y)\,f_\pm(y, \sigma)\,dy$$

THEOREM 2. *For any $F \in L^2(\mathbb{R}_\sigma \times \{-1, 1\}_\omega)$ we have:*

$$[\mathscr{R}_o\,S\,\mathscr{R}_o^{-1}F]\,(\sigma, \omega) = T(\sigma)\,F(\sigma, \omega) + R_{-\omega}(\sigma)\,F(\sigma, -\omega), \quad \sigma \in \mathbb{R}^*, \ \omega \in \{-1, 1\}.$$

The meromorphic continuation of $T(\sigma)$ with respect to σ is connected with the analyticity of V with respect to x and we prove

THEOREM 3 . *We assume that assumptions* (5), (17) *are fulfilled. Then for any* $x \in \mathbb{R}$, $f_{\pm}(x, \sigma)$ *is analytic by respect to* σ *in* $\mathbb{C} \setminus i\mathbb{R}^{-}$.

The poles of $T(\sigma)$ are so called *resonances, scattering frequencies*, or *quasi–normal modes of the Black–Hole* [3]. The Lax-Phillips theory provides a time dependent characterisation of the resonances associated with a compactly supported perturbation of the wave equation in Minkowski space time: the solution has a spectral decomposition in terms of generalised eigenfunctions.

THEOREM 4 . *We assume V is a bounded compactly supported function; then* $T(\sigma)$ *is meromorphic on the whole complex plane; each horizontal strip contains a finite number of its poles; the poles with non–negative imaginary part are purely imaginary , of finite multiplicity, and their set is finite; the set of the poles with negative imaginary part is infinite.*

We arrange the frequencies σ_j *in decreasing order of their imaginary parts* :

$$Im\ \sigma_{j+1} \leq Im\ \sigma_j \ < 0 \leq Im\, \sigma_0 \leq Im\ \sigma_{-1} \leq \dots \leq Im\, \sigma_{-N} , \qquad 1 \leq j.$$

Then for each $n \geq -N$, $\varepsilon > 0$, $\rho > 0$, *there exists* $C(n, \varepsilon, \rho) > 0$, $C_j \in \mathbb{C}$ *such as*

$$|\Phi(t, x) - \sum_{j=-N}^{n} e^{-i\sigma_j t}\, C_j\, f_+(x, \sigma_j)| \leq C(n, \varepsilon, \rho).(1+|x|)^{1/2}\ |e^{(-i\sigma_{n+1}+\varepsilon)t}|$$

holds for any solution Φ *of* (1) *with compactly supported initial data f in* $\mathcal{H}$.

Now we prove that the resonances are the limits of the ones associated with cut-off and damped potentials: given $\gamma > 1$, $\varepsilon > 0$, $\rho > 0$, we put for V given by (2):

$$V^{\rho}_{\gamma,\varepsilon}(x) = \mathbf{1}_{[-\rho,\rho]}(x)\ .\ e^{-\varepsilon |x|^{\gamma}} .V(x)\ .$$

Let $\mathcal{R}(W, E)$ be the set of resonances in $E \subset \mathbb{C}$ for a potential W.

THEOREM 5 . *For any* $\eta > 0$, $\gamma > 1$, *and bounded open* Ω *satisfying*

$$\overline{\Omega} \subset \{\sigma \in \mathbb{C}^* \ , -\frac{\pi}{2\gamma} < Arg\ \sigma < \pi + \frac{\pi}{2\gamma}\ \} ,$$

there exists $\varepsilon_0 > 0$ *such as given* $\varepsilon \in]0, \varepsilon_0]$, *there exists* $R > 0$ *satisfying* :

$$\forall \rho > R,\ \ Card\, \mathcal{R}(V, \Omega) = Card\, \mathcal{R}(V^{\rho}_{\gamma,\varepsilon}\, , \Omega)$$

$$\forall \sigma \in \mathcal{R}\,(V,\ \Omega)\ \exists \sigma^{\rho}_{\gamma,\varepsilon} \in \mathcal{R}\,(V^{\rho}_{\gamma,\varepsilon}\, ,\ \Omega)\ ,\ |\sigma - \sigma^{\rho}_{\gamma,\varepsilon}| < \eta\ ;$$

If $\overline{\Omega} \subset \{\sigma \in C, Im\, \sigma > 0\ \}$ *we can take* $\varepsilon_0 = \varepsilon = 0$.

The numerical investigation of the resonances is very delicate because these poles are not characterised in a variational way and we know no error estimate. Hence the values obtained by computation can be some *artefact* ; therefore it is very important to compare the results given by different methods. All the methods used to find the Black Hole Resonances consist in solving the elliptic equation

$$- \partial_x^2 \Phi + V(x) \Phi = \sigma^2 \Phi$$

provided with the outgoing radiation condition [3], [7]. Instead, we solve the *time dependent* equation (1) and apply the *Prony procedure* already used by *G. Majda, W. Strauss, M. Wei*, for acoustic waves [6]; we refer to *M. Wei* [9] for a detailed investigation of this method. More precisely, we compute solution Φ of (1) by a finite difference scheme. According to Theorem 4 the asymptotic expansion

$$\Phi(t, x_0) \approx \sum_{j=1}^{N} C_j \, e^{-i\sigma_j t} ,$$

is valid for x_0 fixed and $t > t_0$ large enough. Then we choose a sample rate $\Delta T > 0$ and denote

$$f_k = \Phi(t_0 + k \, \Delta T, x_0) , \quad z_j = e^{-i\sigma_j \Delta T} .$$

Hence we have to solve the polynomial system

$$\sum_{j=1}^{N} C_j (z_j)^k = f_k , \quad k = 0, 1, \ldots$$

Following *Prony* this problem can be reduced to finding the zeros z_j of :

$$\sum_{k=0}^{N} \alpha_k z^k = 0 . \tag{3}$$

where the coefficients α_k are the solutions of the over determined linear system

$$\alpha_N = 1, \quad \sum_{j=0}^{N} \alpha_j f_{j+m} = 0 , \quad m = 0,\ldots,M-1, \; M > N. \tag{4}$$

This system is solved using the generalised inverse and the singular value decomposition. Since (4) is very ill-conditioned, this step is unstable with respect to the slight variations of f_k. Therefore the computation of Φ has to be very accurate. For instance we choose the spatial and temporal grid sizes of the finite difference scheme $\Delta t = \Delta x = 10^{-4}$ on the domain $[-40, 120]_x \times [0, 160]_t$. To solve (3) we use Muller's algorithm. The numerical experiment on super computer

CRAY2 shows the great accuracy of the Prony procedure for the computation of the first resonances and we constat an excellent agreement with the values obtained by stationary methods.

[1] A. BACHELOT, *Gravitational Scattering of electromagnetic field by Schwarzschild Black–Hole*, Ann.I.H.P. physique théorique. vol. 54, n°3, 1991, p. 261-320.

[2] A. BACHELOT, A. MOTET-BACHELOT, Les résonances d'un Trou Noir de Schwarzschild, *Ann.I.H.P. physique théorique*, vol.59, n° 1, 1993, p. 3-68.

[3] S. CHANDRASEKHAR, The mathematical theory of black-holes, *Oxford University Press, New–York*, 1983.

[4] J. DIMOCK, Scattering for the wave equation on the Schwarzschild metric, *Gen. Rel. Grav.* 17, 4, 1985, p.353-369.

[5] B. ENGQUIST, A. MAJDA, Radiation boundary conditions for acoustic and elastic wave calculations, *Comm. Pure Appl. Math.* 32, 1979, p.313-357.

[6] G. MAJDA, W. STRAUSS, M. WEI Numerical Computation of the Scattering Frequencies for Acoustic Wave Equations, *Comput. Phys.* 75, 2, 1988, p.345-358.

[7] H.P. NOLLERT, B.G. SCHMIDT Quasi Normal Modes of Schwarzschild Black-Holes-Defined and Calculated via Laplace Transformation, *to appear*.

[8] R. PHILLIPS, Scattering Theory for the Wave Equation with a Short Range Perturbation II, *Indiana Univ. Math. J.*, 33, 6, 1984, p.831-846.

[9] M. WEI, Numerical Computation of Scattering Frequencies, *Ph.D. thesis, Dept. of Applied Math., Brown Univ., Providence*, RI, 1986.

[10] M. ZWORSKI, Distribution of Poles for Scattering on the Real Line, *J. Funct. Anal.* 73, 1987, 277-296.

Département de Mathématiques Appliquées,
CeReMaB, unité associée au CNRS 226,
Université BORDEAUX 1,
F-33405 TALENCE cedex

AMS/IP Studies in Advanced Mathematics
Volume **3**, 1997

SCATTERING OPERATOR FOR THE KLEIN-GORDON EQUATION ON THE SCHWARZSCHILD METRIC

ALAIN BACHELOT

Abstract. We construct the wave operators describing the scattering of the Klein-Gordon field by a Schwarzschild black-hole. The scattering operator is implementable in Fock space.

The present paper is mainly devoted to the asymptotic completeness of the wave operators associated to the massive Klein-Gordon equation on the Schwarzschild metric. This work is the continuation of a program of rigorous mathematical studies on the fields theory around a black-hole. The previous investigations concerned the Maxwell system [1] [2], the black-hole resonance's [3], the non linear Klein-Gordon equation [4], the Dirac system [10].

The time dependent approach of the scattering of the scalar fields by a Schwarzschild black-hole was initiated by Dimock and Kay [6][7][8] who constructed the wave operators and, assuming the asymptotic completeness, investigated the quantum fields. This asymptotic completeness was proved by Dimock for the D'Alembertian [6]. In the case of the massive scalar fields, the situation is more complicated because the gravitational interaction is long range type and we must modify the defining of the wave operators. Finally our result allows to obtain the quantum Scattering Operator.

The Klein Gordon equation on the Schwarzschild background is given by

$$[\frac{\partial^2}{\partial t^2} - \frac{1}{r^2}\frac{\partial}{\partial r_*}r^2\frac{\partial}{\partial r_*} + (1-\frac{2M}{r})(-\frac{1}{r^2}\Delta_{S^2} + m^2)]\,\Psi = 0, \quad t\in\mathbb{R},\ r_*\in\mathbb{R},\ \omega\in S^2 \qquad (1)$$

where $m, M > 0$ are the mass of the field and the black-hole, Δ_{S^2} is the Laplacian on the Euclidean two sphere S^2 and $r > 2M$ is an implicit function of r_*:

$$r_* = r + 2M\,Log(\,r - 2M)\,,\ \ r > 2M\,.$$

We introduce the Hilbert space $\mathscr{H}$ of data's with finite energy, as the completion of $C_0^\infty(\mathbb{R}_{r_*}\times S^2_\omega)\times C_0^\infty(\mathbb{R}_{r_*}\times S^2_\omega)$ for the norm

$$\|{}^t(f,g)\|^2_{\mathscr{H}}=\int\{|\frac{\partial}{\partial r_*}f|^2+(1-\frac{2M}{r})[\frac{1}{r^2}|\nabla_{S^2}f|^2+m^2|f|^2]+|g|^2\}\,r^2dr_*d\omega .$$

The solution of (1) is given by a unitary group $U(t)$ on $\mathscr{H}$:

$${}^t(\Psi(t),\partial_t\Psi(t))=U(t)[\,{}^t(\Psi(0)\,,\partial_t\Psi(0))\,].$$

Near the *black–hole horizon* $\{r=2M\}\times S^2_\omega$ we compare the solutions of (1) with the plane waves solutions of

$$\partial_t^2\Psi_0-\partial_{r_*}^2\Psi_0=0, \tag{2}$$

by introducing the Hilbert spaces $\mathscr{H}_0^\pm$ as the completion of the sets of regular left/right going data's $\mathscr{D}_0^\pm$

$$\mathscr{D}_0^\pm=\{\,{}^t(f_1(r_*\,,\omega)\,,\ \pm\partial_{r_*}f_1(r_*\,,\omega))\,,\ f_1\in C_0^\infty(\mathbb{R}_{r_*}\times S^2_\omega)\,\}\,,$$

for the free energy norm

$$\|{}^t(f,g)\|^2_{\mathscr{H}_0^\pm}=4M^2\int\{|\frac{\partial}{\partial r_*}f|^2+|g|^2\}\,dr_*d\omega\,,$$

and the unitary group $U_0(t)$ on $\mathscr{H}_0^+\oplus\mathscr{H}_0^-$ defined by

$$F_0^\pm\in\mathscr{D}_0^\pm\ ,\ \ U_0(t)F_0^\pm(r_*\,,\omega)=F_0^\pm(\,\pm t+r_*\,,\omega)\,.$$

We construct an identification operator $\mathscr{I}_0$ between $\mathscr{D}_0^+\oplus\mathscr{D}_0^-$ and $\mathscr{H}$ by putting

$$F_0^\pm={}^t(f,g)\in\mathscr{D}_0^\pm\ ,\ \ (\mathscr{I}_0F_0^\pm)(r_*\,,\omega)={}^t(\chi_0(r_*)f(r_*\,,\omega)\,,\chi_0(r_*)g(r_*\,,\omega)),$$

where χ_0 is given such that

$$\chi_0\in C^\infty(\mathbb{R}_{r_*})\,,\ \ r_*<a\Rightarrow\chi_0(r_*)=1,\ \ r_*>b\Rightarrow\chi_0(r_*)=0,\ \ a<b.$$

We introduce the *horizon wave operators* $W_0^\pm$

$$F_0^\pm\in\mathscr{D}_0^\pm\ ,\ \ W_0^\pm F_0=s-\lim_{t\to\pm\infty}U(-t)\,\mathscr{I}_0\,U_0(t)F_0^\pm\ \ \textit{in}\ \mathscr{H}.$$

At the spatial infinity, the term $-2Mm^2r^{-1}$ in (1) is a long range type perturbation of the Klein-Gordon equation on the Minkowski space-time:

$$\frac{\partial^2}{\partial t^2}\Psi_\infty-\Delta_{\mathbb{R}^3_x}\Psi_\infty+m^2\Psi_\infty=0,\ \ \Delta_{\mathbb{R}^3_x}=\sum_{i=1}^3\frac{\partial^2}{\partial x_i^2}\ . \tag{3}$$

Following [8] we compare the solutions de (1) at the infinity with the fields evolving in the Minkowski space-time according to a Dollard-modified free dynamic. Let $\mathcal{H}_\infty$ be the Hilbert space

$$\mathcal{H}_\infty = \{ {}^t(f, g);\ f, \nabla_{\mathbb{R}^3_x} f, g \in L^2(\mathbb{R}^3_x, dx) \},$$

with the norm

$$\|{}^t(f, g)\|^2_{\mathcal{H}_\infty} = \int_{\mathbb{R}^3_x} \{ |\nabla_{\mathbb{R}^3_x} f(x)|^2 + m^2 |f(x)|^2 + |g(x)|^2 \} \, dx,$$

and the unitary propagator $U^D_\infty(t)$ given by

$$U^D_\infty(t) = \begin{bmatrix} cos(tB_\infty + D log\, t) & B^{-1}_\infty sin(tB_\infty + D log\, t) \\ -B_\infty sin(tB_\infty + D log\, t) & cos(tB_\infty + D log\, t) \end{bmatrix},$$

where

$$B_\infty = (-\Delta_{\mathbb{R}^3_x} + m^2)^{1/2}, \quad D = -Mm^2 \, |\nabla_{\mathbb{R}^3_x}|^{-1}, \quad log\, t \equiv sign(t)\, log|t| .$$

We also consider the non modified unitary group $U_\infty(t)$

$$U_\infty(t) = \begin{bmatrix} cos(tB_\infty) & B^{-1}_\infty sin(tB_\infty) \\ -B^{-1}_\infty sin(tB_\infty) & cos(tB_\infty) \end{bmatrix}.$$

We introduce a dense subspace $\mathcal{D}_\infty$ of $\mathcal{H}_\infty$

$$\mathcal{D}_\infty = \{ {}^t(f_1, f_2) \in \mathcal{H}_\infty ; \int_{\mathbb{R}^3_x} e^{-ix.\xi} f_j(x)\, dx \in C^\infty_0(\mathbb{R}^3_\xi \setminus \{0\}) \} .$$

Now we choose a cut-off function χ_∞ such that

$$\chi_\infty \in C^\infty(\mathbb{R}_{r_*}), \quad r_* < c \Rightarrow \chi_\infty(r_*) = 0, \quad r_* > d \Rightarrow \chi_\infty(r_*) = 1, \quad 0 < c < d,$$

and we define a (bounded) identification operator $\mathcal{J}_\infty$ between $\mathcal{H}_\infty$ and $\mathcal{H}$ by identifying the Euclidean norm of $x \in \mathbb{R}^3$ and $r_* \geq 0$ by putting

$$F_\infty \in \mathcal{H}_\infty, \quad 0 \leq r_* \Rightarrow (\mathcal{J}_\infty F_\infty)(r_*, \omega) = \chi_\infty(r_*) F_\infty(x = r_*\omega), \quad r_* \leq 0 \Rightarrow (\mathcal{J}_\infty F_\infty)(r_*, \omega) = 0.$$

We consider the time dependent modified wave operators $W^\pm_\infty$

$$F_\infty \in \mathcal{H}_\infty, \quad W^\pm_\infty F_\infty = s - \lim_{t \to \pm\infty} U(-t)\, \mathcal{J}_\infty U^D_\infty(t) F_\infty \quad in\ \mathcal{H}.$$

The main result is the following

THEOREM 1. *For all* $F_0^{\pm} \in \mathcal{D}_0^{\pm}$, $F_\infty \in \mathcal{D}_\infty$, *the limits* $W_0^{\pm} F_0^{\pm}$ *and* $W_\infty^{\pm} F_\infty$ *exist in* $\mathcal{H}$ *and are independent of* χ_0 *and* χ_∞ *and satisfy*:

$$\|W_0^{\pm} F_0^{\pm}\|_{\mathcal{H}} = \|F_0^{\pm}\|_{\mathcal{H}_0^{\pm}}, \quad \|W_\infty^{\pm} F_\infty\|_{\mathcal{H}} = \|F_\infty\|_{\mathcal{H}_\infty}.$$

Moreover, the strong asymptotic completeness holds: the range of $W_0^{\pm} \oplus W_\infty^{\pm}$ *is dense in* $\mathcal{H}$ *and the Scattering Operator* $S = (W_0^+ \oplus W_\infty^+)^{-1} . (W_0^- \oplus W_\infty^-)$ *is an isometry from* $\mathcal{H}_0^- \oplus \mathcal{H}_\infty$ *onto* $\mathcal{H}_0^+ \oplus \mathcal{H}_\infty$ *and satisfies the intertwining property*

$$\forall t \in \mathbb{R}, \ [U_0(t) \oplus U_\infty(t)] . S = S . [U_0(t) \oplus U_\infty(t)].$$

Finally we construct the quantum Scattering Operator:

THEOREM 2. *S is unitarily implementable in the free Fock–Cook representation of the quantized wave equation* (2) *and Klein–Gordon equation* (3).

We sketch the ideas of the proof. First we decouple the study near the horizon from the study at the infinity by comparing (1) with the Dirichlet problem for the same equation with homogeneous condition $\Psi = 0$ on the sphere $\{r_* = 0\}$; that is a short range perturbation of the initial problem and since $(1-2Mr^{-1})$ is exponentially decreasing as $r_* \to -\infty$, (2) is a short range perturbation of (1) on $]\infty, 0[_{r_*}$, the existence and asymptotic completeness of $W_0^{\pm}$ follow from the Kato-Birman theorem and the invariance principle. On the other hand the problem at the infinity is reduced to the study of the perturbation of Klein-Gordon equation (3) by the long range potential $-2Mm^2 |x|^{-1}$. We prove the existence and completeness of $W_\infty^{\pm}$ by using the results of Kitada [9] on the Schrödinger operators and the invariance principle for the long range potentials and we need the machinery of the two Hilbert spaces scattering theory to obtain :

$$(W_0^- \oplus W_\infty^-)(\mathcal{H}_0^- \oplus \mathcal{H}_\infty) = (W_0^+ \oplus W_\infty^+)(\mathcal{H}_0^+ \oplus \mathcal{H}_\infty) \equiv \mathcal{H}_{ac}.$$

To prove the strong asymptotic completeness $\mathcal{H}_{ac} = \mathcal{H}$, we decompose (1) on the spherical harmonics basis and we conclude by noting that by the analyticity of the gravitational interaction (see [3]) the spectrum of

$$-\partial_{r_*}^2 + (1 - 2Mr^{-1}) [\, l(l+1) r^{-2} + 2Mr^{-3} + m^2 \,]$$

is absolutely continuous on $L^2(R_{r_*})$. Finally, S is isometric thus orthogonal for the real fields; hence the implementability is a straightforward consequence of the Shale criterion.

[1] A. BACHELOT, Gravitational Scattering of electromagnetic field by Schwarzschild Black-Hole, *Ann.I.H.P. physique théorique*. vol. 54, n°3, 1991, p.261-320.

[2] A. BACHELOT, Scattering of electromagnetic field by De Sitter-Schwarzschild Black-Hole, in *"Non linear hyperbolic equations and field theory" Research Notes in Math.*253, 1992, Pitman.

[3] A. BACHELOT, A. MOTET-BACHELOT, Les résonances d'un trou noir de Schwarzschild, *Ann.I.H.P. physique théorique*, vol. 59, n°1, 1993, p. 3-68.

[4] A. BACHELOT, J.P. NICOLAS, Equation non linéaire de Klein-Gordon dans des métriques de type Schwarzschild , *C.R. Acad. Sci. Paris*, t.316, Série I, 1993, p. 1047-1050.

[5] P. DEIFT, B. SIMON, On the decoupling of finite singularities from the question of asymptotic completeness in two body quantum system, *J. Func. Anal.* 23, 1976, p.218-238.

[6] J. DIMOCK, Scattering for the wave equation on the Schwarzschild metric, *Gen. Rel. Grav.* 17, 4, 1985, p.353-369.

[7] J. DIMOCK, B.S. KAY, Scattering for massive scalar fields on coulomb potentials and Schwarzschild metrics, *Class. Quantum Grav.* 3, 1986, p.71-80.

[8] J. DIMOCK, B.S. KAY, Classical and Quantum scattering theory for linear scalar fields on the Schwarzschild metric I, *Ann. Phys.* 175, 1987, p.366-426.

[9] H. KITADA, Scattering theory for Schrödinger operators with long range potentials, *J. Math. Soc. Japan*, 30, 4, 1978, p.603-632.

[10] J.P. NICOLAS, Non linear Klein-Gordon equation and Dirac system on Schwarzschild-like metrics, *Ph.D. thesis*, *Bordeaux* 1 *University*, 1994.

Département de Mathématiques Appliquées,
CeReMaB, unité associée au CNRS 226,
Université BORDEAUX 1,
F-33405 TALENCE cedex

AMS/IP Studies in Advanced Mathematics
Volume **3**, 1997

PERIODIC BOUNDARY VALUE PROBLEM AND CAUCHY PROBLEM FOR GENERALIZED IMBq EQUATION WITH SEVERAL VARIABLES*

GUOWANG CHEN   ZHIJIAN YANG

Abstract

The existence and uniqueness of the classical local solutions of the periodic boundary value problem and Cauchy problem for the generalized IMBq equation with several variables are proved.

1. Introduction

We know that the equation (which we call Bq-equation)

$$\varphi_{tt} = \varphi_{xx} + (\varphi^2)_{xx} + \varphi_{xxxx} \tag{1}$$

was derived by J. Boussinesq in 1872 to describe shallow-water waves. The improved Bq-equation (which we call IBq) is

$$\varphi_{tt} - \varphi_{xx} - \varphi_{xxtt} = (\varphi^2)_{xx}. \tag{2}$$

A modification of the IBq equation analogous to the MKdV equation yields

$$\varphi_{tt} - \varphi_{xx} - \varphi_{xxtt} = (\varphi^3)_{xx}, \tag{3}$$

which we call the IMBq (see [1]). The IMBq equation with several variables is

$$\varphi_{tt} - \nabla^2\varphi_{tt} - \nabla^2\varphi = \nabla^2\varphi^3. \tag{4}$$

In the above equations, φ denotes the unknown function, subscripts x and t indicate partial derivatives and ∇ denotes the gradient operator.

In this paper, we study the following generalized IMBq equation with several variables

$$u_{tt} - A_0\nabla^2 u_{tt} - A_1\nabla^2 u - \nabla^2 g(u) - \overline{\nabla} f(\nabla u) - G(u) = 0, \tag{5}$$

where $u(x,t)$ denotes an unknown function of variables $x=(x_1,x_2,\cdots,x_n)\in R^n$ and $t \in R_+$, $\nabla=\left(\frac{\partial}{\partial x_1},\frac{\partial}{\partial x_2},\cdots,\frac{\partial}{\partial x_n}\right)$, $\overline{\nabla}=\sum_{i=1}^{n}\frac{\partial}{\partial x_i}$, A_0, $A_1>0$ are constants, g, f and G are the given nonlinear functions. The equation (5) is the nonlinear wave equation of higher order. The equation (5) contains not only the equations (2), (3) and (4) but the equation

$$u_{tt} - [a_0 + Na_1(u_x)^{N-1}]\, u_{xx} - a_2 u_{xxtt} = 0 \tag{6}$$

as well, where $a_0, a_2>0$ are constants, a_1 is an arbitrary real number, and N is an arbitrary natural number. In the study of the strain solitary waves in the nonlinear elastic rods there arises the longitudinal wave equation (6) (see [2]).

Let $\overline{\Omega}\subset R^n$ be an n-dimensional cube with width 2D(D>0) in each direction,

* This project is supported by the National Natural Science Foundation of China and partially by the National Natural Science Foundation of Henan Province.

that is $\overline{\Omega}=\{x=(x_1,x_2,\cdots,x_n)\,|\,|x_i|\leqslant D, i=1,2,\cdots,n\}$, $x+2De_i$ denotes $(x_1,\cdots,x_{i-1},x_i+2D,x_{i+1},\cdots,x_n)$ $(i=1,2,\cdots,n)$ and set $\overline{Q_T}=\{x\in\overline{\Omega},0\leqslant t\leqslant \overline{T}\}$.

For the equation (5), we discuss its periodic boundary value problem in (n+1)-dimensional cylindrical domain $Q_{\overline{T}}$,

$$\begin{aligned} u(x,t) &= u(x+2De_i,t), \qquad i=1,2,\cdots,n,\\ u(x,0) &= \varphi(x), \qquad u_t(x,0)=\psi(x), \end{aligned} \tag{7}$$

where $\varphi(x)$ and $\psi(x)$ are given functions of n-dimensional initial value, satisfying the periodic boundary conditions. For the equation (5), we also study Cauchy problem

$$u(x,0)=\varphi(x), \qquad u_t(x,0)=\psi(x), \tag{8}$$

where $\varphi(x)$ and $\psi(x)$ are given functions defined in R^n.

For simplicity, let $A_0=A_1=1$ in (5) and we here only prove the existence and uniqueness of the classical local solutions for the 2-dimensional periodic boundary value problem (5), (7) and Cauchy problem (5), (8), because we can prove the n-dimensional periodic boundary value problem (5), (7) and Cauchy problem (5), (8) by the method of the 2-dimensional case.

2. **Periodic boundary value problem** (5), (7)

Now, we establish an orthonormal base in $L_2(\Omega)$.

Let us consider the eigenvalue problem for the ordinary differential equation

$$\begin{cases} y''+\lambda y=0,\\ y(x_1+2D)=y(x_1). \end{cases}$$

We can obtain eigenvalue $\lambda_{1i}=\alpha_i^2=\left(\frac{i\pi}{D}\right)^2$ $(i=0,1\cdots)$ and the family of eigenfunctions $\{y_i(x_1)\}=\{\frac{1}{\sqrt{2D}},\frac{1}{\sqrt{D}}cos\alpha_ix_1,\frac{1}{\sqrt{D}}sin\alpha_ix_1, i=1,2,\cdots\}$, which composes an orthonormal base. Similarly, we can obtain eigenvalue $\lambda_{2j}=\beta_j^2=(\frac{j\pi}{D})^2$ $(j=0,1,\cdots)$ and the family of eigenfunctions $\{z_j(x_2)\}=\{\frac{1}{\sqrt{2D}},\frac{1}{\sqrt{D}}cos\beta_jx_2,\frac{1}{\sqrt{D}}sin\beta_jx_2, i=1,2,\cdots\}$ of the periodic boundary value problem $z''+\lambda z=0, z(x_2+2D)=z(x_2)$. $\{z_j(x_2)\}$ composes an orthonormal base. The family of functions $\{y_i(x_1)z_j(x_2), i,j=0,1,\cdots\}$ composes an orthonormal base in $L_2(\Omega)$ (See [3]).

Let $u(x,t)=\sum_{i,j=0}^{\infty}T_{ij}(t)y_i(x_1)z_j(x_2)$ be the solution of the periodic boundary value problem (5), (7). The initial value functions may be expressed $\varphi(x)=\sum_{i,j=0}^{\infty}\varphi_{ij}y_i(x_1)z_j(x_2)$ and $\psi(x)=\sum_{i,j=0}^{\infty}\psi_{ij}y_i(x_1)z_j(x_2)$. Substituting the expression of $u(x,t)$ into the equation (5), multiplying both sides by $y_i(x_1)z_j(x_2)$ and integrating on Ω, we get

$$\begin{aligned} &(1+\alpha_i^2+\beta_j^2)\ddot{T}_{ij}+(\alpha_i^2+\beta_j^2)T_{ij}-\int_\Omega\{\nabla^2 g(u)+\overline{\nabla}f(\nabla u)\\ &+G(u)\}y_i(x_1)z_j(x_2)dx=0, \quad i,j=0,1,\cdots. \end{aligned} \tag{9}$$

Substituting the expressions of $u(x,t)$, $\varphi(x)$ and $\psi(x)$ into the initial conditions (7), we obtain

$$T_{ij}(0) = \varphi_{ij},\ \dot{T}_{ij}(0) = \psi_{ij} \qquad (i,j = 0,1,\cdots), \tag{10}$$

where $\dot{T}_{ij}=\frac{d}{dt}T_{ij}(t)$ etc. Hence T_{ij} satisfy the infinite system of ordinary differential equations (9), (10). Let us now prove the existence of the solution for the initial value problem (9), (10). For this purpose, we shall first consider the existence of the solution for the following initial value problem of a finite system of ordinary differential equations

$$(1 + \alpha_i^2 + \beta_j^2)\ddot{T}_{Nij} + (\alpha_i^2 + \beta_j^2)T_{Nij} - \int_\Omega \{\nabla^2 g(u_N) + \overline{\nabla} f(\nabla u_N) + G(u_N)\} y_i(x_1) z_j(x_2) dx = 0, \tag{11}$$

$$T_{Nij}(0) = \varphi_{ij},\ \dot{T}_{Nij}(0) = \psi_{ij}, \quad i,j = 0,1,\cdots,N, \tag{12}$$

where $u_N(x,t)=\sum_{i,j=0}^{N} T_{Nij}(t) y_i(x_1) z_j(x_2)$. In order to use Leray-Schauder's fixed-point argument we are also going to consider the following inital value problem of the finite system of ordinary differential equations with θ

$$(1 + \alpha_i^2 + \beta_j^2)\ddot{T}_{Nij} + (\alpha_i^2 + \beta_j^2)T_{Nij} - \theta\int_\Omega \{\nabla^2 g(u_N) + \overline{\nabla} f(\nabla u_N) + G(u_N)\} y_i(x_1) z_j(x_2) dx = 0, \tag{13}$$

$$T_{Nij}(0) = \theta\varphi_{ij}, \quad \dot{T}_{Nij}(0) = \theta\psi_{ij}, \qquad i,j = 0,1,\cdots,N, 0\leqslant\theta\leqslant 1. \tag{14}$$

Lemma 1. *Suppose that the following conditions are satisfied.*

1. $g\in C^5$, $|g^{(i)}(s)|\leqslant K_i|s|^{m_1+(5-i)}$ $(i=1,2,3,4,5)$, (i) *denotes the order of derivatives.*

2. $f\in C^4$,

$$\begin{aligned} &|f(\nabla s)| = |f(s_{x_1},s_{x_2})| \\ &\leqslant K_6(|s_{x_1}|^{m_2+4} + |s_{x_1}|^{m_3+4}|s_{x_2}|^{m_4+4} + |s_{x_2}|^{m_5+4}), \\ &\left|\frac{\partial f}{\partial s_{x_1}^i}\right| \leqslant K_7(|s_{x_1}|^{m_2+(4-i)} + |s_{x_1}|^{m_3+(4-i)}|s_{x_2}|^{m_4+4}), \\ &\left|\frac{\partial f}{\partial s_{x_2}^i}\right| \leqslant K_8(|s_{x_1}|^{m_3+4}|s_{x_2}|^{m_4+(4-i)} + |s_{x_2}|^{m_5+(4-i)}), \quad (i = 1,2,3,4), \\ &\left|\frac{\partial^{i+j} f}{\partial s_{x_1}^i \partial s_{x_2}^j}\right| \leqslant K_9|s_{x_1}|^{m_3+(4-i)}|s_{x_2}|^{m_4+(4-j)} \quad (i,j = 1,2,3,\ i + j\leqslant 4). \end{aligned}$$

3. $G\in C^3$, $|G^{(i)}|\leqslant K_{10}|s|^{m_6+(3-i)}$ $(i=1,2,3)$.

4. $m_3+m_4>\max(m_1,m_2,m_5,m_6)$, *here* $m_i\geqslant 1$ $(i=1,2,\cdots,6)$, $K_i (i=0,1,\cdots,10)$ *are constants.*

Furthermore, suppose that $T(t)=(T_{Nij}(t),i,j=0,1,\cdots,N)$ *is the solution of the system* (13), *and let*

$$E_N(t) = \sum_{i,j=0}^{N}\Big(1 + \alpha_i^2 + \beta_j^2 + \sum_{\substack{l+k=8\\ l,k=0,2,4,6,8}} \alpha_i^l\beta_j^k + 2\sum_{\substack{l+k=10\\ l,k=2,4,6,8}} \alpha_i^l\beta_j^k + \alpha_i^{10} + \beta_j^{10}\Big)\dot{T}_{N_{ij}}^2 + \sum_{i,j=0}^{N}\Big(1 + \alpha_i^2 + \beta_j^2 + 2\sum_{\substack{l+k=10\\ l,k=2,4,6,8}} \alpha_i^l\beta_j^k + \alpha_i^{10} + \beta_j^{10}\Big)T_{Nij}^2 + 1,$$

then there is

$$\frac{dE_N(t)}{dt} \leqslant K_{11}(E_N(t))^{\frac{m_3+m_4+9}{2}}, \tag{15}$$

where $K_{11}>0$ *is a constant independent of* θ, D *and* N.

Lemma 2. *Under the conditions of Lemma 1, if* $\lim_{N\to\infty} E_N(0)=b=\sum_{i,j=0}^{\infty}\{1+\alpha_i^2+\beta_j^2 + \sum_{\substack{l+k=8\\ l,k=0,2,4,6,8}} \alpha_i^l\beta_j^k + 2\sum_{\substack{l+k=10\\ l,k=2,4,6,8}} \alpha_i^l\beta_j^k+\alpha_i^{10}+\beta_j^{10}\}\psi_{ij}^2 + \sum_{i,j=0}^{\infty}\{1+\alpha_i^2+\beta_j^2+2\sum_{\substack{l+k=10\\ l,k=2,4,6,8}}\alpha_i^l\beta_j^k+\alpha_i^{10}+\beta_j^{10}\}\varphi_{ij}^2+1<\infty$, *then* $E_N(t)$ *is uniformly bounded and independednt of* N *and* D *in any closed subinterval* $0\leqslant t\leqslant t_1<t_b$, *where* $t_b=2/[K_{11}(m_3+m_4+7)b^{(m_3+m_4+7)/2}]$.

Lemma 3. *Under the conditions of Lemma 2, there is a solution of the initial value problem* (11), (12) *of the finite system of ordinary differential equations in* $[0, t_1]$, *here* $0<t_1<t_b$.

Observe Lemma 2. We can prove Lemma 3 by Leray-Schauder's fixed-point theorem.

From Ascoli-Arzelà Theorem we have

Corollary. *Under the conditions of Lemma 3 and for the sequence* $\{T_{ij}\}_{i,j=0}^N$ $(N=1,2\cdots)$ *of the solutions for the initial value problem* (11), (12), *there is a convergent subsequence* $\{T_{N_kij}\}_{i,j=0}^{N_k}$. *As* $N_k\to\infty$, *then*

$$T_{N_kij}\to T_{ij},\quad \dot{T}_{N_kij}\to \dot{T}_{ij}\quad (i,j=0,1,\cdots)$$

uniformly in $[0,t_1]$.

Lemma 4. *Under the conditions of Lemma 2, the series* $\sum_{i,j=0}^{\infty}\dot{T}_{ij}^2$, $\sum_{i,j=0}^{\infty}\alpha_i^k\dot{T}_{ij}^2$ $(k=2, 8, 10)$, $\sum_{i,j=0}^{\infty}\beta_j^k\dot{T}_{ij}^2$ $(k=2,8,10)$, $\sum_{i,j=0}^{\infty}\alpha_i^k\beta_j^l\dot{T}_{ij}^2$ $(k,l=2,4,6,k+l=8)$, $\sum_{i,j=0}^{\infty}\alpha_i^k\beta_j^l\dot{T}_{ij}^2$ $(k,l=2, 4,6,8,k+l=10)$, $\sum_{i,j=0}^{\infty}T_{ij}^2$, $\sum_{i,j=0}^{\infty}\alpha_i^kT_{ij}^2$ $(k=2,10)$, $\sum_{i,j=0}^{\infty}\beta_j^kT_{ij}^2$ $(k=2,10)$ *and* $\sum_{i,j=0}^{\infty}\alpha_i^k\beta_j^lT_{ij}^2$ $(k,l =2,4,6,8,k+l=10)$ *are convergent and uniformly bounded in* $[0,t_1]$.

Corollary. *Under the conditions of Lemma 2, there exists a constant* $M_1>0$, *such that*

$$\| u \|_{H^5(\Omega)} + \| u_t \|_{H^5(\Omega)} \leqslant M_1,$$
$$\| u \|_{C^3(\Omega)} + \| u_t \|_{C^3(\Omega)} \leqslant M_1$$

in $[0,t_1]$, *here* $u(x,t)=\sum_{i,j=0}^{\infty}T_{ij}(t)y_i(x_1)z_j(x_2)$.

Lemma 5. *Under the conditions of Lemma 2,* $u_{N_k}\to u$, $u_{N_kx_i}\to u_{x_i}$ $(i=1,2)$ *and* $u_{N_kt}\to u_t$ $(N_k\to\infty)$ *uniformly hold in* Q_{t_1}, *where* $u_{N_k}(x,t)=\sum_{i,j=0}^{N_k}T_{N_kij}(t)y_i(x_1)z_j(x_2)$.

Theorem 1. *Under the conditions of Lemma 2, the functions* $T_{ij}(t)$ *are solutions of the initial value problem* (9), (10) *in* $0\leqslant t\leqslant t_1<t_b$, *where* $T_{ij}(t)=\lim_{N_k\to\infty}T_{N_kij}(t)$ *and* $T_{N_kij}(t)$ $(i=0,1,\cdots)$ *being solutions of the initial value problem* (11), (12).

Theorem 2. *Suppose that the conditions of Lemma 2 are satisfied and* $\varphi(x),\psi(x)\in H^6(\Omega)$, *then there exists a classical solution* $u(x,t)=\sum_{i,j=0}^{\infty}T_{ij}(t)y_i(x_1)z_j(x_2)$ *of the* *periodic boundary value problem* (5),(7) *in* Q_{t_1}, *where* $0<t_1<t_b=2/[K_{11}(m_3+m_4+7)b^{(m_3+m_4+7)/2}]$.

Theorem 3. *Suppose that* $g\in C^3$, $f\in C^2$, $G\in C^1$ *and* $\varphi(x),\tilde{\varphi}(x),\psi(x),\tilde{\psi}(x)\in H^6(\Omega)$. *Let* $u(x,t)$ *and* $v(x,t)$ *be two different classical solutions corresponding to* $\varphi(x),\psi(x)$ *and* $\tilde{\varphi}(x)$, $\tilde{\psi}(x)$ *respectively of the periodic boundary value problem* (5), (7) *in* Q_{t_1}. *Then for any* $\varepsilon>0$ *there exists a* $\delta>0$, *such that if* $\|\varphi-\tilde{\varphi}\|_{H^2(\Omega)}+\|\psi-\tilde{\psi}\|_{H^2(\Omega)}<\delta$, *we have*

$$\|u-v\|_{H^2(\Omega)}+\|u_t-v_t\|_{H^2(\Omega)}+\|u_{tt}-v_{tt}\|_{H^1(\Omega)}<\varepsilon,\quad 0\leqslant t\leqslant t_1.$$

Corollary. *Under the Conditions of Theorem* 3, *the solution of the periodic boundary value problem* (5), (7) *in* Q_{t_1} *is unique in generalized sense.*

3. **Cauchy problem** (5),(8)

Now, we are going to consider Cauchy problem (5), (8). The solution of Cauchy problem for the equation (5) with the initial condition (8) in the domain $[0,t_2]\times \mathbf{R}^2$ can be obtained by the limiting process of increasing the periodic of domain Ω to infinite.

Theorem 4. *Under the conditions of Lemma* 1, *if* $\varphi(x)$, $\psi(x)\in H^6(R^2)$, *then there exists a unique classical solution* $u(x,t)$ *of Cauchy problem* (5),(8) *in* $[0,t_2]\times R^2$, *where* $0<t_2<t_{\bar{b}}$,

$$\begin{aligned}\bar{b}=&\|\psi\|^2+\|\psi_{x_1}\|^2+\|\psi_{x_2}\|^2+\sum_{\substack{i+j=4\\ i,j=0,1,2,3,4}}\|\psi_{x_1^ix_2^j}\|^2+2\sum_{\substack{i+j=5\\ i,j=1,2,3,4}}\|\psi_{x_1^ix_2^j}\|^2\\ &+\|\psi_{x_1^5}\|^2+\|\psi_{x_2^5}\|^2+\|\varphi\|^2+\|\varphi_{x_1}\|^2+\|\varphi_{x_2}\|^2\\ &+2\sum_{\substack{i+j=5\\ i,j=1,2,3,4}}\|\varphi_{x_1^ix_2^j}\|^2+\|\varphi_{x_1^5}\|^2+\|\varphi_{x_2^5}\|^2+1,\end{aligned}$$

$$t_{\bar{b}}=2/[K_{11}(m_3+m_4+7)\bar{b}^{(m_3+m_4+7)/2}]$$

and $\|\cdot\|=\|\cdot\|_{L_2(R^2)}$.

References

[1] Makhankov, V. G. ,Dynamics of classical solitons (in non—integrable systems),Physics Reports ,A review section of Physics Letters (section C),**35**(1)(1978),1—128.

[2] Zhang Shanyuan,Zhuang Wei, Strain solitary waves in the nonlinear elastic rods, Acta Mechanica Sinica,**20**(1988),58—66(in Chinese).

[3] Michael Reed,Barry Simon, Method of Modern Mathematical Physics, Academic Press ,New York and London,1972.

Guowang Chen
Institute of Mathematics
Zhengzhou University
Zhengzhou,450052,P.R. China

Zhijian Yang
Department of Mathematics and Machanics
Zhengzhou Institute of Technology
Zhengzhou,450002,P.R. China

AMS/IP Studies in Advanced Mathematics
Volume **3**, 1997

A Degenerate Reaction-Diffusion Problem

ZUCHI CHEN AND XIANGDONG CHEN

ABSTRACT. We consider in this paper the existence, uniqueness and asymptotic behaviour of solutions of the mixed problem for a degenerate quasilinear parabolic equation which arises in population dynamics and reaction-diffusion problems .

1 INTRODUCTION

Let $D \subset R^n$ be a bounded domain with boundary $\partial D \in C^1$. We consider the non-negative solutions of the problem

$$\text{(P)} \qquad \begin{cases} w_t - L\Phi(w) = a(x)f(w), & (x,t) \in D \times (0,t) \\ w = \chi, & (x,t) \in \partial D \times [0,T) \\ w = w_0, & (x,t) \in \bar{D} \times \{0\} \end{cases}$$

where $L = \partial_i(a^{ij}\partial_j)$ with $a^{ij} = a^{ij}(x) = a^{ji}(x) \in C(\bar{D})$ is an uniformly elliptic operator, i.e.,there are constants Λ, λ such that $\Lambda \geq \lambda > 0$, and for any $\xi \in R^n$ it holds that $\Lambda|\xi|^2 \geq a^{ij}\xi_i\xi_j \geq \lambda|\xi|^2$. Here and throughout the notation ∂_i is used for $\frac{\partial}{\partial x_i}$, and the summation conviention over twice repeated indices is used. Suppose that

(H1) $\Phi \in C^1[0,\infty)$ is a monotonically increasing function satisfying that $\Phi(0) = \Phi'(0) = 0$, and Φ^{-1} is Hölder continuous;

(H2) $f \in C^1[0,\infty)$ is an increasing function with $f(0) = 0$;

(H3) $a \in C^0(\bar{D}), w_0 \in L^\infty(D), \chi \in C^1(\partial D \times (0,T))$ and $\chi = w_0$ on $\partial D \times \{0\}$.

Problems of this type arise in population dynamics and in reaction-diffusion processes. The investigation of this paper were motiveted by the special case of [1] and [2].

Definition 1.1 *We call $w \in L^\infty(D)$ an equilibrium solution of problem (P) if for any $\eta \in C^2(\bar{D})$ with zero value on ∂D, it holds that*

$$(1.1) \qquad -\int_D \Phi(w)L\eta dx + \int_{\partial D} \Phi(\chi)\frac{\partial \eta}{\partial \nu}ds = \int_D a(x)f(w)\eta(x)dx$$

where $\frac{\partial \eta}{\partial \nu} = a^{ij}\frac{\partial \eta}{\partial x_i}cos(n,x_j)$ is the oblique derivative of η. By an upper (lower) solution $\overline{w} \in L^\infty(D)(\underline{w} \in L^\infty(D))$ of (1.1) (i.e., by an equilibrium upper (lower) solution of (P)) we mean that $\overline{w}(\underline{w})$ satisfies (1.1) with the inequality sign $\geq$ ($\leq$), for any positive η as above.

If a^{ij} has Hölder continuous first derivatives and $\Phi \in C^2[0,\infty]$, we know that the solution of (1.1) is a classical one, and $u = \Phi(w) \in C(\bar{D})$ is a classical solution of the

Work supported by NSFC .

problem

$$(P_s) \qquad \begin{cases} Lu + a(x)g(u) = 0, & in\ D \\ u = \varphi, & on\ \partial D \end{cases}$$

where $g = f \circ \Phi^{-1}$, $\varphi = \Phi[\chi(\cdot)]$. We call such u a *stationary solution* of problem (P).u is called an *upper (lower) solution* of (P_s) if it satisfies

$$Lu + a(x)g(u) \leq (\geq) 0 \quad in\ D,\ u \geq (\leq)\varphi\ on\ \partial D.$$

Definition 1.2$w \in C([0,T]; L^1(D)) \cap L^\infty(Q_T)$ *is called a solution of (P) provided that for any $t \in [0,T]$ and any non-negative $\zeta \in C^2(\bar{Q}_t)$ with $\zeta = 0$ on $\partial D \times (0,T)$ the following holds*

$$\begin{aligned} \int_D w(x,t)\zeta(x,t)dx - \int_{Q_t} [w\zeta_t + \Phi(w)L\zeta]dxdt \\ + \int_0^t dt \int_{\partial D} \Phi(\chi)\frac{\partial \zeta}{\partial \nu}ds \\ = \int_D w_0\zeta(x,0)dx + \int_{Q_t} a(x)f(w)\zeta(x,t)dxdt \end{aligned} \tag{1.2}$$

where $Q_t = D \times (0,t)$ and $\frac{\partial}{\partial \nu}$ is as the one as in Definition 1.1. Instead of sign$'' ='' $ in (1.2) by $'' \geq (\leq)''$, then w is called an upper (lower) solution of (P) provided $\Phi \in C^2$ and $a^{ij} \in C^1$.

2 Existence and Uniqueness

Lemma 2.1 *Let $\underline{w}, \overline{w}$ be respectively the lower and upper solution of (P), then for any $t \in [0,T]$ the following holds*

$$\int_D [\underline{w}(x,t) - \overline{w}(x,t)]_+ dx \leq e^{kt} \int_D [\underline{w}(x,0) - \overline{w}(x,0)]_+ dx$$

where $r_+ = \max\{r, 0\}, k \geq 0$ is a constant.

By lemma 2.1 and conditions (H1) to (H3), using the upper-lower solution method, the following results can be obtained.

Theorem 2.1 *Suppose that (H1) to (H3) hold. Moreover,*

$$\lim_{s\to\infty} \frac{g(s)}{s} = 0, \quad \lim_{s\to 0+} \frac{g(s)}{s} = +\infty.$$

Then, for any $T \geq 0$ the problem (P) has an unique solution.

Theorem 2.2 *Let $\underline{w}_0$ and $\overline{w}_0$ be respectively the lower and upper solution of (1.1) satisfying $0 \leq \underline{w}_0 \leq \overline{w}_0$ a.e. in D, then,*

(a) If $\underline{w}_0 \leq w_0 \leq \overline{w}_0$ a.e. in D, then for any $t \geq 0$ we have $\underline{w}_0 \leq w(\cdot, t; w_0) \leq \overline{w}_0$ a.e. in D, where $w(\cdot, t; w_0)$ is a solution of (P) with the initial value of w_0;

(b) the mapping $t \to w(x,t;\underline{w}_0)$ is non-decreasing for a.e.$x \in D$; the mapping $t \to w(x,t;\overline{w}_0)$ is non-increasing for a.e.$x \in D$;

(c) $w(\cdot,t;\underline{w}_0)$ and $w(\cdot,t;\overline{w}_0)$ converge (in C^0 when $n = 1$, and in $L^p(D)$ with $p \geq 1$ when $n \geq 2$) respectively to the equilibrium solutions w and W at $t \to \infty$, where w and W are respectively the minimal and the maximal equilibrium solution of (P).

3 Stationary Case

In this section, we consider problem (P_s). Because of $g = f \circ \Phi^{-1}$, (H1) and (H2) the following holds:

(A1) $g \in C^{\alpha}([0,\infty)) \cap C^1((0,\infty)), \alpha \in (0,1)$, and $g(s) > 0$ when $s > 0, g(0) = 0$;

(A2) $g'(s) > 0$ when $s > 0$. Moreover, we suppose that

(A3) $h(s) = \int_0^s \frac{1}{g(\sigma)} d\sigma < +\infty, h(0) = 0$;

(A4) $g(s)$ is strictly concave in $(0,\infty)$. (A5) $\frac{g(s)}{s} \to 0$ *as* $s \to \infty$, $\frac{g(s)}{s} \to +\infty$ *as* $s \to 0^+$.

At first, we consider the unique solvability of problem (P_s). Define

$$D^+ = \{x \in D | a(x) > 0\},$$
$$D^- = \{x \in D | a(x) < 0\}, D^0 = int\{x \in D | a(x) = 0\}.$$

In view of the continuity of $a(x)$, D^+ consists at most of a countable number of connected components denoted by $D_k^+, k \in \mathbf{M} = \{1, 2, ..., r\}, r \leq \infty$. Then, we have

Lemma 3.1 *Let u be a non-negative solution of (P_s), then*

(a) Either $u \equiv 0$ or $u > 0$ in D_k^+;

(b) If $u > 0$ in D_k^+, then $u > 0$ in $\overline{D_k^+} \cap D$.

This lemma gives rise to the following classification of the positive solutions of (P_s).

Definition 3.1 *(a) For any subset $\mathbf{I} \subseteq \mathbf{M}$, let $\mathbf{S_I}$ be the class of solutions of (P_s) which are positive in $D_I^+ \equiv \cup_{k \in I} D_k^+$; (b) $\mathbf{N}_I$ denotes the set $\{u \in \mathbf{S}_I : u \equiv 0$ on $D^+ - D_I^+\}$.*

Then, we have an uniqueness result

Theorem 3.1. *Suppose (A1) to (A4) hold. Then, $\mathbf{N}_I$ has at most one element for any finite $I \subset \mathbf{M}$.*

The main result in this section is as follows

Theorem 3.2. *Assume (A1) and (A5) hold. Then for any $I \subset \mathbf{M}$ we have*

(a) $\mathbf{S}_I \neq \emptyset$;

(b) there exist a minimal solution u_I and a maximal solution U_I in $\mathbf{S}_I$ such that $u_I \leq u \leq U_I$ for the any other solution $u \in \mathbf{S}_I$;

(c) the maximal solution U_I coincides with the maximal solution in $\mathbf{S_M}$;

(d) if $u_I \neq U_I$, then $u_I \in \mathbf{N}_{I'}$, where $I' = \cap\{\tilde{I} | \tilde{I} \supseteq I, \mathbf{N}_{\tilde{I}} \neq \emptyset\}$.

The maximal solution $U \equiv U_{\mathbf{M}}$ can be constructed by means of a variational principle as follows. Consider the problem

$$\text{(3.1)} \qquad \begin{cases} Lv + a(x)g(v + h \quad) = 0 & in \quad D, \\ \qquad\qquad v = 0 & on \quad \partial D, \end{cases}$$

here $Lh = 0$ in D and $h = \varphi$ on ∂D. From §6 in [1] we know that (3.1) has an unique solution. Obviously, $u = v + h$ is a solution of (P_s). Let

$$J[w] = \int_D a^{ij} \frac{\partial w}{\partial x_i} \frac{\partial w}{\partial x_j} dx - 2 \int_D a(x) G(w + h) dx$$
$$G(s) = \int_0^s g(\tau) d\tau.$$

Based on the weak compactness of $H_0^1(D)$ and strong compactness of $L_2(D)$, using lower semi-continuity of $J[w]$, we can prove the following conclusion

Theorem 3.3 *Under the same assumption as for Theorem 2.2, we have*

(a) the problem

$$J[w] = \inf, \quad w \in H_0^1(D) \tag{3.2}$$

has a solution w_0;

(b) if in addition $r < +\infty$ and (A2) to (A4) hold, then (3.1) is uniquely solvable and $U \equiv w_0 + h$ is the maximal solution of (P_s).

4 Parabolic Case

In this section we consider the asymptotic behaviour of the solution of problem (P). The results are based on [3], and some methods in [3] and [4] are used.

Definition 4.1 *The interval $[w_1, w_2] = \{w \in L_+^\infty(D) | w_1 \le w \le w_2\}$ is called L^p – attractive if there exists a set $\chi \subseteq L_+^\infty(D)$ such that*

(a) $[w_1, w_2] \subseteq \chi$;

(b) for any $w_0 \in \chi$, there exists a solution $w(\cdot, t; w_0)$ of (P) such that

$$dist\{w(\cdot, t; w_0), [w_1, w_2]\} \to 0 \text{ in } L^p(D) \text{ as } t \to \infty.$$

Let us denote $w_I = \Phi^{-1}(u_I), W = \Phi^{-1}(U)$, where u_I and U are respectively the minimal and maximal solutions of (P_s) in $\mathbf{S}_I$. Because Φ is monotone , then w_I and W are respectively the minimal and maximal equilibrium solutions of (P) in D_I^+. From §2 we know that there exists uniquely a solution $w(x, t; w_0)$ of (P). Let us denote by $C_k (k = 1, 2, \ldots, l)$ the connected components of the set $\{x \in D | w(x, t; w_0) > 0\}$ which do not intersect Γ^+. We have

Theorem 4.1 *Suppose (A1) and (A5) hold, then*

(a) For all $w_0 \neq 0$ in $C_k \cap (\cup_{i \in I} D_i^+), k = 1, 2, \ldots l$, the interval $[w_I, W]$ attacts every solution $w(\cdot, t; w_0)$ of (P) in the L^p sense ($p \in [1, \infty)$ when $n \ge 2; p = \infty$ when $n = 1$);

(b) If $w_0 \le w_I$ and $w_0 \neq 0$ in $C_k \cap (\cup_{i \in I} D_i^+), k = 1, 2, \cdots, l$, then $w(\cdot, t; w_0) \to w_I$ in $L^p(D)$ when $t \to \infty$.

Theorem 4.2. *Assume $w(\cdot, t; w_0)$ is continuous in Q_∞. and (A1) to (A5) hold, $r < +\infty$. Then, for any initial condition w_0 the solution $w(\cdot, t; w_0)$ of (P) converges to an equilibrium solution as $t \to \infty$.*

References

1. P.De Mottoni, A.Schiaffino & A.Tesei,, *Attractivity properties of nonnegative solutions for a class of nonlinear degenerate parabolic problems,*, Ann. Mat. Pur.& Appl., **136**, (1984,), 35–48..
2. M.A.Pozio & A.Tesei,, *Support properties of solutions for a class of degenerate parabolic problems,*, Comm.in PDE., **12(1)** (1987,), 47–75..
3. J.Deuel & P.Hess,, *Nonlinear parabolic boundary value problems with upper and lower solutions,*, Isreal J.of Math., **29(1)** (1978,), 92-104..
4. T.Laetsch,, *Uniqeness for sublinear boundary value problems,*, J.Diff. Eqns, **13** (1973,), 13-23..

Department of Mathematics,University of Science

and Technology of China,Hefei,230026,China

AMS/IP Studies in Advanced Mathemati
Volume 3, 1997

CAUCHY PROBLEM FOR STRONGLY DAMPED NONLINEAR WAVE EQUATIONS*

ENGUI FAN[1] AND YUNBO WANG[2]

Abstract

In this paper, we consider the cauchy problem of the equations $u_{tt}-\Delta u-\upsilon\Delta u_t=F(\nabla u,\nabla_x\nabla u,\Delta u_t)$. Under the assumptions of power—type term $F(\hat{\lambda})=0(|\hat{\lambda}|^{1+\alpha})$, with integral number $\alpha>1$ and space dimensions $n>\frac{2}{\alpha-1}$, the global existence and uniqueness of the solutions to the equatoins with small initial data are obtained. The method is contracting mapping principle on the metric späce that is designed by energy estimates and decay estimate of the linear equations.

Key words and phrases Strongly damped wave equation, global classical solution, contraction mapping

Introduction.

In [1], by the countinuation extension principle, Ponce, G shows that if $F(\nabla u,\nabla_x\nabla u)$ is a smooth linear function with respect to the second derivatives in a neighborhood of the origin and the initial values is of certain smallness, then, the cauchy problem

$$\begin{cases} u_{tt}-\Delta u-\gamma\Delta u_t=F(\nabla u,\nabla_x\nabla u) & (1)\\ t=0:\ u=\varphi(x),\ u_t=\psi(x) & (2)\end{cases}$$

has a global classical solution.

In this paper, by the global iterative technique, we prove that the cauchy problem.

$$\begin{cases} u_{tt}-\Delta u-\gamma\Delta u_t=F(\nabla u,\ \nabla_x\nabla u,\ \Delta u_t), & (3)\\ t=0:\ u=\varphi(x),\ u_t=\psi(x) & (4)\end{cases}$$

has a global classical solution. Our method differs from that in [1], but we overcome the limitation that F is a smooth linear function with respect to the second derivatives in a neighborhood of origin and we add new terms Δu_t in F, improving the result in [1].

In problem (3)—(4), with the nonlinear term $F=F(\nabla u,\nabla_x\nabla u,\nabla u_t)$, we denote

$$\hat{\lambda}=((\lambda_i),i=0,1,\cdots,n;(\lambda_{ij}),i,j=0,1,\cdots,n,i+j\geqslant 1;(\lambda_{0ii},i=1,\cdots,n).$$

Let $F=F(\hat{\lambda})$ be sufficiently smooth in a neighborhood of $\hat{\lambda}=0$ and

* Supported by grants from the Inner Monglia Science Foundation

$$F(\hat{\lambda}) = 0(|\hat{\lambda}|^{1+\alpha}), \quad integer\ \alpha > 1, \tag{5}$$

$$:> \frac{2}{\alpha - 1} \tag{6}$$

Then we will prove the following main result.

Theorem 1 Assume that $F=F(\hat{\lambda})$ is sufficiently smooth and satisfies (5), the space dimensions satisfy (6). Then for any given integer $s \geqslant n+4$, there exists sufficiently small constants δ and E.

such that if

$$\varphi \in H^{s+3}(R^n) \cap W^{s,1}(R^n), \quad \psi \in H^{s+2}(R^n) \cap W^{s,1}(R^n)$$

and

$$\|\varphi\|_{s+3} + \|\varphi\|_{s,1} + \|\psi\|_{s+2} + \|\psi\|_{s,1} \leqslant \delta E$$

Cauchy problem (3)—(4) has a unique global classical solution $u \in X_{S,E}$, here

$$X_{S,E} = \{v = v(t,x) \mid D_s(v) \leqslant E, v(0,x) = \psi(x)\} \tag{7}$$

$$D_s(v) = \sup_{t \geqslant 0}(1+t)^{n/2} \|\nabla v\|_{s-n-2,\infty} + \sup_{t \geqslant 0} \|\nabla v\|_{s+2} + \left(\int_0^\infty \|\nabla_x \Delta u_t\|_s^2 dt\right)^{1/2}.$$

Our plan in this paper is as follows. In § 1 and § 2 we respectively construct the energy estimates and the decay estimate of the solution for the cauchy problem of the linear equation. In § 3 we give the proof of theorem 1.

Because of the limit of the length of article, we only give the results of lemmas and theorem without process of proofs.

§ 1 The energy estimates of the solution for the cauchy problem of the linear equations.

For the following typical cauchy problem.

$$\begin{cases} u_{tt} - \Delta u - \gamma \Delta u_t = 0, \\ t = 0: u = 0, \ u_t = \psi(x), \end{cases}$$

by Fourier transformation, we can obtain its solution

$$u(t,x) = F^{-1}\left(e^{-\gamma|\xi|^2 t}\left(\frac{e^{\Gamma t} - e^{-\Gamma t}}{2\Gamma}\right)\hat{\psi}(\xi)\right) \tag{1.1}$$

where $\Gamma = \sqrt{\gamma^2|\xi|^4 - 4|\xi|^2}$, we denote (1.1) in the form

$$u(t,x) = S(t)\psi$$

with

$$s(t): \psi \to u(t,)$$

Then by Duhamel principle, the solution of the cauchy problem

$$\begin{cases} u_{tt} - \Delta u - \gamma \Delta u_t = F(t, x) & (1.2) \\ t = 0: u = \varphi(x), \ u_t = \psi(x) & (1.3) \end{cases}$$

can be written in the form

$$u = s(t)\psi + \frac{\partial}{\partial t}(s(t)\varphi) + \int_0^t s(t-\tau)F(\tau, \cdot)d_\tau \tag{1.4}$$

By Galerkin method, we get

Lemma 1.1 For $\forall$ $T>0$, integer $s\geqslant 0$. Let

$$\varphi\in H^{s+1}(R^n),\psi\in H^s(R^n),F\in L^2(0,T;H^s(R^n))$$

then cauchy problem (1.2)—(1.3) has a unique solution $u=u(t,x)$ satisfying

$$u\in L^\infty(0,T,H^{s+1}(R^n)),$$
$$u_t\in L^\infty(0,T;H^s(R^n))\cap L^2(0,T;H^{s+1}(R^n)),$$
$$u_{tt}\in L^2(0,T;H^{s-1}(R^n)).$$

Lemma 1.2 For $\forall$ $T>0$, integer $s\geqslant 0$. Let

$$\varphi\in H^{s+3}(R^n),\psi\in H^{s+2}(R^n),F\in L^2(0,T;H^{s+1}(R^n)),$$

then the following energy estimates hold

$$\|\nabla u\|_s\leqslant\|\varphi\|_{s+1}+\|\psi\|_{s+1}+\int_0^t\|F\|_s dt,$$

$$\|\nabla_x\nabla u\|_s^2\leqslant c(\|\varphi\|_{s+2}^2+\|\psi\|_{s+1}^2+\int_0^t\|F\|_s^2 d\tau),$$

$$\|\nabla\Delta u\|_s^2+\gamma\int_0^t\|\nabla_x\Delta u_t\|_s^2 d\tau\leqslant c(\|\varphi\|_{s+3}^2+\|\psi\|_{s-2}^2+\int_0^t\|F\|_{s+2}^2 d\tau).$$

§2 Decay estimate of the solution for the cauchy problem of the linear equation

For the cauchy problem

$$\begin{cases}u_{tt}-\Delta u-\gamma\Delta u_t=0, & (2.1)\\ t=0:\ u=\varphi(x),\ u_t=\psi(x) & (2.2)\end{cases}$$

its solution can be written in the form

$$u=s(t)\psi+\frac{\partial}{\partial t}(s(t)\varphi), \tag{2.3}$$

then we have

Lemma 2.1 For the solution of (2.1)—(2.2), the following decay estimate holds

$$\|\nabla(s(t)\psi+\frac{\partial}{\partial t}(s(t)\varphi))\|_{N,\infty}\leqslant c(1+t)^{-\frac{n}{2}}(\|\varphi\|_{N+n+2,1}+\|\psi\|_{N+n+2,1}),\forall\, t\geqslant 0$$

where N is a positive integer, c is a positive constant,

§3 The global classical solution for the cauchy problem (3)—(4) of strongly damped nonlinear wave equations

Lemma 3.1[2] Assume that $F=F(w)$ is sufficiently smooth, with $w=(w_1,\cdots,w_N)$, and as $|w|<\gamma_0$, it holds

$$F(w)=0(|w|^{1+\alpha}),\ integer\ \alpha\geqslant 1 \tag{3.1}$$

If $w=w(x)$ satisfies

$$\|w\|_{L^\infty(R^n)}\leqslant\gamma_0 \tag{3.2}$$

then $\|F(w)\|_{s,r}\leqslant C\|w\|_{s,q}\|w\|_{0,p}\|w\|_{0,\infty}^{\alpha-1}$,

where

$$\frac{1}{r}=\frac{1}{p}+\frac{1}{q},\qquad 1\leqslant p,q,r\leqslant+\infty \tag{3.3}$$

Lemma 3.2[2] suppose that $F=F(w)$ is sufficienty smooth satisfying (3.1). $w=\overline{w}(x)$ and

$w=\overline{w}(x)$ satisfy (3. 2) respectively. Then we have

$$\| F(\overline{w}) - F(\overline{w} \|_{s,r} \leqslant c\{ \| w^* \|_{o,p}(\| \overline{w} \|_{s,q} + \| \overline{w}_{s,q} + \| w^* \|_{s,q}(\| \overline{w}_{o,p} + \| \overline{w} \|_{o,p})\}(\| \overline{w}_{0,\infty} + \| \overline{w} \|_{0,\infty})^{s-1},$$

where $w^* = \overline{w} - \overline{w}$, p, q, r satisfy (3. 3).

For any given $s \geqslant n+4$ and positive constant E, if we introduce the set of functions (7) and define a metric: $\forall\ \overline{v}, \overline{v} \in X_{S,E}$

$$\rho(\overline{v},\overline{v}) = D_s(\overline{v} - \overline{v}),$$

we get

Lemma 3. 3 Let $\| \varphi \|_{s+3} + \| \varphi \|_{s,1}$ be sufficiently small, then $X_{S,E}$ is a nonvoid complete metric space.

$\forall\ v \in X_{S,E}$, by solving the cauchy problem of following linear equation

$$\begin{cases} u_{tt} - \Delta u - \gamma \Delta u_t = F(\nabla v, \nabla_x \nabla v, \Delta v_t) \\ t = 0 : u = \varphi(x), u_t \doteq \psi(x) \end{cases}$$

we define a map

$$\hat{T} : v \rightarrow u = \hat{T} v$$

Then by lemma 1. 2, 2. 1, 3. 1—3. 2, we can prove

Lemma 3. 4 If δ and E are sufficently small, $\hat{T}$ maps $X_{S,E}$ in itself.

Lemma 3. 5 If δ and E are sufficiently small, $\hat{T}$ is contractive.

By the above two lemmas and Banach contractive mapping principle, we obtain the result of thorem 1 immediately.

Reference

1 Ponce, G. Nonlinear Analysis, Theory, Methods and Applications, 1985, 9(5): 399—418

2 TAsien Li and Yun mei Chen. Nonlinear evolution equations, Beijing: Science press, 1989, 3—48, 82—110

3 Nakao. M. Journal of Mathematical Analysis and Applications, 1990, 46: 217—240

4 Nakao. M. Mathematische zeitschrift, 1991, 206: 265—276

5 Andrews. G. Journal of Differential Equations, 1980, 35: 200—231

1. Department of Basic Subject
Zhang jiakou Agriculture college
Zhang jiakou 075131
and
2. Department of Mathe matics
Inner Mongolia university
Hohhot 010021

AMS/IP Studies in Advanced Mathematics
Volume **3**, 1997

SYSTEMS OF QUASILINEAR HYPERBOLIC EQUATIONS WITH QUADRATIC COUPLING

B. HANOUZET and R. NATALINI

Abstract. We consider weakly coupled systems of quasilinear hyperbolic equations. For some special coupling we are able, by using comparison methods, to show the global existence of entropy solutions.

This paper is concerned with the following class of systems of first order quasilinear hyperbolic equations:

$$\partial_t u_i + \partial_x f_i(u_i) = g_i(U) \ , \qquad i = 1, \ldots, N \ , \tag{1}$$

where $U = (u_1, \ldots, u_N)$ is the unknown function and f_i, $i = 1, \ldots, N$, $G = (g_1, \ldots, g_N)$ are some given locally Lipschitz continuous functions. Consider the Cauchy problem for the initial data

$$U(x, 0) = U^0(x) \ , \tag{2}$$

where U^0 is a given bounded measurable N–valued function. We are looking for the global existence of entropy solutions to (1)–(2). Observe that the L^∞–norms of the local solutions could blow up in finite time. The scalar quasilinear case was studied in [8] and [9] by using comparison methods. In the semilinear case with quadratic coupling, the problem was investigated in [3] and [1] by energy methods. Here, we extend comparison methods to quasilinear systems (1), which are weakly coupled. For more results and proofs see [4].

First we state a local existence result for the initial value problem (1)–(2). Let Σ_T denote the strip $\mathbb{R} \times (0, T)$ $(T > 0)$. We recall some standard definitions about the scalar problem

$$\partial_t u + \partial_x f(u) = g \ , \qquad \text{in} \ \ \Sigma_T \ , \tag{3}$$

$$u(x, 0) = u^0(x) \ , \tag{4}$$

where f is locally Lipschitz continuous and g is locally bounded in Σ_T. Denote by $[r]_\pm = \frac{1}{2}(|r| \pm r)$ and $H^\pm(r) = \frac{1}{2}(\operatorname{sgn} r \pm 1)$. Following [6] we say that a bounded function u is an *entropy subsolution* of (3)–(4) in Σ_T, if for any $\varphi \in C_0^\infty(\Sigma_T)$, $\varphi \geq 0$, and $k \in \mathbb{R}$, we have:

$$\iint \{[u-k]_+ \partial_t \varphi + H^+(u-k)(f(u)-f(k))\partial_x \varphi + H^+(u-k) g\varphi\}\, dx\, dt \geq 0 \; ; \tag{5}$$

and for any interval I

$$\operatorname*{ess\,lim}_{t\to 0+} \int_I [u(x,t) - u^0(x)]_+ \, dx = 0 \; . \tag{6}$$

An *entropy supersolution* of (3)–(4) is defined replacing in (5)–(6) $[\ \ ]_+$, H^+ by $[\ \ ]_-$, H^-. An *entropy solution* is defined as being both entropy subsolution and supersolution.

A bounded function U is an entropy solution of (1)–(2) if each component u_i of U is an entropy solution of $(1)_i - (2)_i$. Let us state our local existence result.

Theorem 1. *For each $U^0 \in L^\infty(\mathbb{R})^N$, there is $T > 0$ (depending only on $\|U^0\|_\infty$) such that there exists a unique (bounded) entropy solution U of (1)–(2) in Σ_T.*

We have also a continuation result.

Theorem 2. *Two possibilities occur: either $U \in L^\infty(\Sigma_T)^N$ for any $T > 0$ or there exists $T^* < +\infty$, such that for any $T < T^*$, U is defined on Σ_T and*

$$\lim_{T \to T^*_-} \|U\|_{L^\infty(\Sigma_T)^N} = +\infty \; .$$

In the first case the entropy solution U is said to be *global*.

To study the comparison properties of system (1) we have to deal with quasimonotone functions.

Definition 1. *Let Ω be an open, convex set in $\mathbb{R}^N$. A function $G : \Omega \to \mathbb{R}^N$ is said to be quasimonotone (nondecreasing) if each component g_i of G is nondecreasing in u_j for $j \neq i$.*

Concerning quasimonotone flows, see for ordinary differential equations and parabolic equations [10], [2] and references therein.

Consider the usual partial ordering generated by the positive cone of $\mathbb{R}^N$. Given $A \leq B$, we shall denote the corresponding interval by $[A, B]$.

Theorem 3. *Let $\underline{U}$ ($\overline{U}$) be an entropy subsolution (supersolution) of (1)–(2) in Σ_T with initial data $\underline{U}^0$ ($\overline{U}^0$). Assume there are $A, B \in \mathbb{R}^N$, $A \leq B$, such that:*

$$G \text{ is quasimonotone in } [A,B] \; ,$$
$$\underline{U}, \overline{U} \in [A,B] \text{ in } \Sigma_T.$$

If $\underline{U}^0 \leq \overline{U}^0$, *then* $\underline{U} \leq \overline{U}$ *in* Σ_T.

Let us notice that Theorem 3 is a consequence of the following local argument. Fix a real interval (a,b). Then, there exist two positive constants $L = L(A, B, f_i)$ and $M = M(A, B, G)$ such that, for each $t \in \left(0, \frac{b-a}{2L}\right)$, we have:

$$\sum_{i=1}^{N} \int_{a+tL}^{b-tL} [\underline{u}_i(x,t) - \overline{u}_i(x,t)]_+ \, dx \leq e^{tM} \sum_{i=1}^{N} \int_a^b [\underline{u}_i^0(x) - \overline{u}_i^0(x)]_+ \, dx \; .$$

This estimate shows that if $\underline{U}^0 \leq \overline{U}^0$ in some interval, then this inequality is established in a domain of determinacy. So Theorem 3 is really an hyperbolic result.

Example 1: Quasilinear Carleman's system.
Consider the 2×2 system (1) with the coupling term

$$g_1(U) = -g_2(U) = u_2^2 - u_1^2 \; .$$

In this case, for any nonnegative bounded initial data, there exists a unique global entropy solution (1)–(2). Moreover the following estimate holds true:

$$0 \leq \max(\|u_1\|_\infty, \|u_2\|_\infty) \leq \max(\|u_1^0\|_\infty, \|u_2^0\|_\infty) \; .$$

Notice that this result follows immediately by Theorem 3, because G is quasi-monotone. In the semilinear case this model was investigated in [5] and [11].

Similar arguments can work even if the coupling term is not quasimonotone. For example, consider two functions $\underline{G}, \overline{G}$, which are quasimonotone and Lipschitz continuous, such that

$$\underline{G}(U) \leq G(U) \leq \overline{G}(U) \; .$$

Let $\underline{U}$, U, $\overline{U}$ be a subsolution for $\underline{G}$, respectively a solution for G, a supersolution for $\overline{G}$. Assume that $\underline{U}^0(x) \leq U^0(x) \leq \overline{U}^0(x)$. Then

$$\underline{U}(x,t) \leq U(x,t) \leq \overline{U}(x,t) \quad \text{for} \quad t > 0 \; .$$

Example 2: Majda's model for combustion.
Consider the 2×2 system (1) with

$$g_1(U) = -g_2(U) = \varphi(u_1)u_2 \; .$$

Here φ is a Lipschitz continuous function, $0 \leq \varphi \leq 1$. For any bounded initial data, with $u_2^0 \geq 0$, there exists a unique global entropy solution to (1)–(2) and the following estimate holds true:

$$-\|u_1^0\|_\infty \leq u_1(x,t) \leq \|u_1^0\|_\infty + t\|u_2^0\|_\infty \; ;$$
$$0 \leq u_2(x,t) \leq \|u_2^0\|_\infty \; .$$

Observe that in the Majda's model $f_2 \equiv 0$ and the function φ is monotone, see [12], [7].

Example 3: Nonlinear optics.
Consider next the conservative coupling term [3]

$$g_1(U) = u_2^2 \; ; \qquad g_2(U) = -u_1 u_2 \; .$$

Now, for any bounded initial data, there exists a unique global entropy solution of (1)–(2) and we have, for any $t > 0$:

$$-\|u_1^0\|_\infty \leq u_1(x,t) \leq \|u_1^0\|_\infty + t\|u_2^0\|_\infty e^{2t\|u_1^0\|_\infty} \; ;$$
$$|u_2(x,t)| \leq \|u_2^0\|_\infty e^{t\|u_1^0\|_\infty} \; .$$

To obtain the above estimates, fix for any $T > 0$, the interval $[A, B]$, with

$$a_1 = -\|u_1^0\|_\infty, \; b_1 = \|u_1^0\|_\infty + T\|u_2^0\|_\infty e^{2T\|u_1^0\|_\infty}, \; b_2 = -a_2 = \|u_2^0\|_\infty e^{T\|u_1^0\|_\infty} .$$

We choose, as $\underline{G}$, $\overline{G}$, the lower, respectively upper, quasimonotone envelope of the function G in the interval $[A, B]$.

Observe that the bounds in the previous examples are independent on the form of the convection term. More generally, the one space dimension is not essential for our arguments. In fact we can easily extend Theorems 1, 2, 3 to several space variables, always assuming the weakly coupled form of the interactions.

The above results can also be extended to locally bounded entropy solutions as in [8] and [9].

References

[1] D. Aregba–Driollet and B. Hanouzet, Problème de Cauchy à données petites pour des systèmes hyperboliques semi–linéaires 1–D, *C.R. Acad. Sci. Paris*, t. **315**, Série I (1992), 1381–1383.

[2] J. Bebernes and D. Eberly, Mathematical problems from combustion theory, *Appl. Math. Sci.*, vol. **83**, Springer–Verlag, New York, 1989.

[3] D. Driollet and B. Hanouzet, Systèmes hyperboliques semi–linéaires conservatifs 1–D, *C.R. Acad. Sci. Paris*, t. **307**, Série I (1988), 231–234.

[4] B. Hanouzet and R. Natalini, Quaderno IAC n.28 (1993).

[5] I.I. Kolodner, On the Carleman's model for the Boltzman equation and its generalizations, *Ann. Mat. Pura Appl.* **63** (1963), 11–32.

[6] S.N. Kružkov, First order quasilinear equations in several independent variables, *Mat. Sb* **81** (1970), 285–255; *Math. USSR Sb.* **10** (1970), 217–243.

[7] A. Levy, On Majda's model for dynamic combustion, *Comm. P.D.E.* **17** (3 & 4) (1992), 657–698.

[8] R. NATALINI, Unbounded solutions for conservation laws with source, *Nonlin. Anal. T.M.A.* **21** (5) (1993), 349– 362.
[9] R. NATALINI and A. TESEI, Blow–up of solutions for a class of balance laws, *Quaderno IAC* n.1 (1993).
[10] H.L. SMITH, Systems of ordinary differential equations which generate an order preserving flow. A survey of results, *SIAM Review* **30** n. 1 (1988), 87–113.
[11] L. TARTAR, Some existence theorems for semilinear hyperbolic systems in one space variable, *Technical Summary Report 2164, University of Wisconsin–Madison, M.R.C.* (1981).
[12] Z. TENG and L. YING, Riemann problem for a reacting and convection hyperbolic system, *Approx. Theory Appl.* **1** (1984), 95–122.

Département de mathématiques appliquées,
CeReMaB, Unité associée au C.N.R.S. n° 226,
Université Bordeaux I, F–33405 Talence Cedex (France)

Istituto per le Applicazioni del Calcolo "M. Picone", C.N.R.,
Viale del Policlinico 137,
I–00161 Roma (Italy)

AMS/IP Studies in Advanced Mathematics
Volume 3, 1997

Large Time Step Glimm's and Godunov's Schemes and Their Entropy Consistency*

Jinyang Huang**

Abstract.

Large time step Glimm's and Godunov's schemes are introduced. Some results about the entropy consistency of the schemes, especially for isentropic equations of gas dynamics, are given.

1. Introduction

We are concerned with system of conservation laws of the form

$$u_t + f(u)_x = 0 \qquad -\infty < x < +\infty, \quad t > 0 \tag{1,1}$$

with initial condition

$$u(x,0) = u_0(x) \qquad -\infty < x < +\infty \tag{1,2}$$

where $u(x,t) \in R^m$ and $f(u)$ is a smooth mapping from a region Ω of R^m to R^m. We assume the system (1,1) is strictly hyperbolic, i.e. , the matrix df/du has real and distinct eigenvalues $\lambda_1(u), \cdots, \lambda_m(u)$.

Godunov's scheme and Glimm's scheme are semianalytical techniques for calculating discontinuous solutions to (1,1) and (1,2). To make the computation more efficient, it is reasonable to try a larger Courant number C, where $C = \max\limits_{i \le m}(|\lambda_i(u)|) \cdot h/\tau$ and h, τ are spatial and time step length respectively. But,if $C \ge 1/2$, neighboring waves may interact.It is difficult, even impossible to deal with the interactions exactly. Based on linearizing the interaction of nonlinear waves, R.LeVeque [5,6] (1984) introduced a large time step version of Godunov's scheme (L.T.S.Godunov's scheme for short), which is a conservative and consistent difference scheme for abitrarily large Courant number and convenient to compute,and illustrated that one could expect better numerical results with suitable Courant number (usually between 1 and 3). A parallel large time step version of Glimm's scheme and some similar results were given by R.LeVeque [6], Y.Brenier [2] and Jinghua Wang [9,10](1985-1989).In addition, Wang [10](1988) obtained consistency of L.T.S.Glimm's scheme with systems of conservation laws for any choice of time step and, furthermore, convergence of the scheme for the case that $C \le 1$, the total variation of $u_0(x)$ is sufficiently small and each eigenvalue of the system doesn't change its sign. Using Roe Riemann solver as building block, similar result was obtained by Brenier [2]. An error analysis of L.T.S. Godunov's scheme was performed by Lucier [7].

---

* Supported in part by the Natural Science Foundation of China and NSF Grant Number DMS-865731

Now we give descriptions of L.T.S. Glimm's scheme and Godunov's scheme. We discretize $R \times [0, +\infty)$ by choosing spatial mesh length h and time mesh length τ, and let $t_n = n\tau, n = 0, 1, 2, \cdots, x_k = kh, k = 0, \pm 1, \pm 2, \cdots$. Suppose that the approximate solution $u_h(x,t)$ to the initial value problem (1,1) and (1,2) has been constructed for $t < t_j$. In L.T.S. Godunov's scheme one sets

$$(1,3) \qquad u_h(x, t_j) = u_{ij} = \frac{1}{h} \int_{x_i}^{x_{i+1}} u_h(x, t_j - 0) dx \quad \text{for } x \in [x_i, x_{i+1})$$

In L.T.S. Glimm's scheme, one sets

$$(1,4) \qquad u_h(x, t_j) = u_{ij} = u_h(x_i + \theta_j h, t_j - 0) \quad \text{for } x \in [x_i, x_{i+1})$$

where $\{\theta_j | j = 1, 2, \cdots\}$ is an equidistributed sequence in (0,1]. Then for each $i \in Z$, we solve the Riemann problem

$$u_t + f(u)_x = 0$$

$$(1,5) \qquad u|_{t=t_j} = \begin{cases} u_{i-1,j} & x < x_i \\ u_{i,j} & x > x_i \end{cases}$$

and get the solution $u_{ij}(x,t)$ for $t > t_j$. Set

$$u_h(x,t) = u_h(x,t_j) + \sum_{i \in Z} (u_{ij}(x,t) - u_{ij}(x,t_j))$$

for $t_j \le t < t_{j+1}$ as the approximate solution to the problem (1,1) and (1,2) in the strip $S_j = \{(x,t) | -\infty < x < +\infty, t_j \le t < t_{j+1}\}$. To initiate the scheme at $j = 0$, we set

$$u_h(x, 0) = u_{i0} = \frac{1}{h} \int_{x_i}^{x_{i+1}} u_0(x) dx \quad \text{for } x \in [x_i, x_{i+1}), \quad i \in Z.$$

When $C \le 1/2$, L.T.S. Godunov's scheme and Glimm's scheme will reduce to Godunov's scheme and Glimm's scheme respectively. Therefore, they are generalizations of the corresponding schemes.

2 . Entropy Consistency

It is important to consider whether the L.T.S. Godunov's scheme and/or Glimm's scheme approximate an entropy solution. For scalar conservation laws with a convex flux function, Jinghua Wang and G. Warnecke[11] proved that both the schemes approximate the entropy solution if the Courant number ≤ 1. For the isentropic equations of gas dynamics with adiabatic exponent $\gamma = 3$, the same results were obtained by Guangshan Jiang and Jinghua Wang [13]. With adiabatic exponent $1 < \gamma < 3$, in the following we will give some similar results.

We are concerned with the initial value problem for isentropic equations of gas dynamics with adiabatic exponent $1 < \gamma < 3$ in Eulerian coordinates

$$(2,1) \qquad \begin{cases} \rho_t + (\rho u)_x = 0 \\ (\rho u)_t + (\rho u^2 + \rho^\gamma/\gamma)_x = 0 \end{cases}$$

$$(2,2) \qquad (\rho, u)(x,0) = (\rho_o(x), u_o(x))$$

We consider a weak solution $v = (\rho, u)$ to (2,1) and (2,2) to be admissible if it additionally satisfies an entropy condition of the form

$$(2,3) \qquad -\int\int_{R\times R+} (\eta(v)\varphi_t + q(v)\varphi_x)dxdt - \int_{-\infty}^{+\infty} \eta(v_0(x))\varphi(x,0)dx \leq 0$$

for all $\varphi \in C_o^\infty(R \times R+), \varphi \geq 0$. Here

$$\eta(v) = \rho u^2/2 + \rho^\gamma/\gamma(\gamma-1) \quad q(v) = \rho u^3/2 + \rho^\gamma u/(\gamma-1)$$

$\{\eta(v), q(v)\}$ is called mechanical energy entropy pair. The admissible solution is called entropy solution.

The following lemma shows that the entropy condition (2,3) rules out discontinuities which violate the entropy inequality,or equivalently Lax shock condition.

Lemma Let v be a piecewise smooth solution of (2,1) contains a discontinuity of left state v_l and right state v_r. Then the entropy inequality $u_r < u_l$ holds if and only if the entropy condition (2,3) holds.

The following theorem is our main result.

Theorem Suppose a bounded weak solution $v \in BV(R \times [0, +\infty))$ to the initial value problem (2,1) and (2,2) is approximated in L_1^{loc} sense by a subsequence of approximate solutions v_h of uniformly bounded variation. If v_h are constructed by L.T.S. Godunov's scheme with $C \leq 1$, then v satisfies the entropy conditon (2,3). If vh are constructed by L.T.S. Glimm's scheme using the average Riemann solvers (see [12] for their definition) for the rarefaction waves appearing in the Riemann problems (1,5) with $C \leq 1$ as $17/7 \leq \gamma \leq 3$ and $C \leq \min\{1, 6(3-\gamma)/(35-13\gamma)\}$ as $1 \leq \gamma \leq 17/7$, the same result holds.

Summary of the Proof:

It is sufficient to prove that

$$L_\varphi(v_h) \equiv -\int\int_{R\times R+}(\eta(v_h)\varphi_t + q(v_h)\varphi_x)dxdt - \int_{-\infty}^{+\infty} \eta(v_h(x,0))\varphi(x,0)dx \leq R(\tau)$$

for all $\varphi \in C_o^\infty(R \times R+), \varphi \geq 0$, where $R(\tau) \to 0$ as $\tau \to 0$.

Since v_h is piecewise continuous on $R \times [0, +\infty)$, we may rewrite $L_\varphi(v_h)$ as follows

$$L_\varphi(v_h) = \sum_{j=1}^{+\infty}\sum_{i\in Z} \int_{X_i}^{X_{i+1}} \varphi(x,t_j)[\eta(v_h(x,t_j+0)) - \eta((x,t_j-0))]dx + \sum_{j=1}^{+\infty}\sum_{i\in Z} \int_{t_j}^{t_{j+1}} \int_{x_i}^{x_{i+1}} [\eta(v_h)_t + q(v_h)_x]\varphi dxdt$$

where the integrals in the second term are in the sense of distribution. By making a Taylor expansion of φ to the second order at (x_i, t_j) in each $Y_{ij} = [x_i, x_{i+1})\times$
$[t_j, t_{j+1})$, then estimating carefully, we can get, for Godunov's scheme

$$L_\varphi(v_h) = \sum_{j=1}^{+\infty}\sum_{i\in Z}\varphi(x_i,t_j)\{h[\eta(\tfrac{1}{h}\int_{x_i}^{x_{i+1}} v_h(x,t_{j+1}-0)dx) - \eta(v_h(x_i,t_j))] + \int_{t_j}^{t_{j+1}}[q(v_h(x_{i+1},t)) - q(v_h(x_i,t))]dt\} + R_1(\tau)$$

where $R_1(\tau)\to 0$ as $\tau\to 0$, and, for L.T.S. Glimm's scheme

$$L_\varphi(v_h) = \sum_{j=1}^{+\infty}\sum_{i\in Z}\varphi(x_i,t_j)\{\int_{x_i}^{x_{i+1}}[\eta(v_h(x,t_{j+1}-0)) - \eta(v_h(x_i,t_j))]dx + \int_{t_j}^{t_{j+1}}[q(v_h(x_{i+1},t)) - q(v_h(x_i,t))]dt\} + R_2(\tau)$$

where $R_2(\tau)\to 0$ as $\tau\to 0$. By discussing each of four kinds of wave interaction and making complicated calculation and analysis, we can obtained with the Courant number $C\le 1$

$$h[\eta(\tfrac{1}{h}\int_{x_i}^{x_{i+1}} v_h(x,t_{j+1}-0)dx) - \eta(v_h(x_i,t_j))] + \int_{t_j}^{t_{j+1}}[q(v_h(x_{i+1},t)) - q(v_h(x_i,t))]dt \le 0$$

and, furthermore, if substitute the averge Riemann solvers for the rarefaction waves in the scheme, and take $C\le 1$ as $17/7\le\gamma\le 3$ and $C\le\min\{1, 6(3-\gamma)/(35-13\gamma)\}$ as $1\le\gamma\le 17/7$, we have

$$\int_{x_i}^{x_{i+1}}[\eta(v_h(x,t_{j+1}-0)) - \eta(v_h(x,t_j))]dx + \int_{t_j}^{t_{j+1}}[q(v_h(x_i,t)) - q(v_h(x_i,t))]dt = W_{ij}\le 0$$

Therefore, the theorem holds.

Remarks:

(1) For isentropic equations of gas dynamics in Lagrangian coordinates with the adiabatic exponent $\gamma\in[1,3]$, similar results hold.

(2) For the interaction between any pair of shock wave and shock wave or shock wave and rarefaction wave, under the condition $C\le 1$, we have $W_{ij}\le 0$. But for the interaction between any pair of rarefaction waves, $W_{ij} > 0$. It is the reason why we consider using average Riemann solvers instead of rarefaction waves.

(3) For Godunov's scheme with $C\le 1$, the interactions of waves are not difficult to be deal with (see [5]). But no same conclusion holds for Glimm's scheme.

REFERENCES

[1] Y. Brenier, Averaged multivalued solutions for scalar conservation laws. SIAM J. Numerical Anal. 21(1984), 1013-1037

[2] Y.Brenier, Stabilite du schema de Glimm Roe pour lois systems de conservation. Rapports de recherche 413. INIRA, Rocquencoourt(1985))

[3] J. Glimm, Solution in the large for nonlinear hyperbolic systems of equations. Comm. pure appl. math. 18(1965), 697-715

[4] S.K. Godunov, Bounds on discrepancy of approximate solutions constructed for the equations of gas dynamics, J. comput. math. and math. phys. 1(1961), 623-637

[5] R.J. LeVeque, Convergence of a large time step generalization of Godunov's method for conservation laws, Comm. pure appl. math. 37 (1984), 463-477

[6] R.J. LeVeque, A large time step generalization of Godunov's method for systems of conservation laws. SIAM J. numer. analy. 22 (1985) 1051-1073

[7] B.J. Lucier, Error bounds for the methods of Glimm, Godunov and LeVeque, SIAM J. numer. analy. 22(1985) 1047-1081

[8] J. Smoller, Shock waves and reaction- diffusion equations, Springer-Verlag, New York-Heidelberg-Berlin (1983)

[9] Jinghua Wang, Large time step generalizations of Glimm's scheme for systems of conservation laws. Chin. Ann. of Math. 913(1988),50-63

[10] Jinghua Wang, Large time step generalizations of random choice finite difference scheme for hyperbolic conservation laws, Acta Mathematia Scientia 9(1989), 33-43

[11] Jinghua Wang and G. Warnecke, On entropy consistency of L. T. S. Godunov's and Glimm's schemes, SIAM J. of numer. anal. 30(1993), 1229-1251

[12] Jinghua Wang and G. Warnecke, On the entropy consistency of large time step schemes using approximate Riemann solvers, SIAM J. Of numer. anal. 30(1993), 1251-1267

[13] Guangshan Jiang and Jinghua Wang, Entropy consistency of large time step schemes for isentropic equations of gas dynamics,(manuscript)

** Institute of Systems Science, Academia Sinica Beijing 100080, P.R. China

AMS/IP Studies in Advanced Mathematics
Volume **3**, 1997

NONLINEAR GALERKIN METHOD FOR NAVIER-STOKES EQUATIONS WITH STREAM FORM *

Li Kaitai, Ma Yichen, Feng Weibing

Abstract A finite element nonlinear Galerkin method is presented for the two dimensional incompressible Navier-stokes equations with stream form. The scheme is based on two finite element space X_H and X_h for the approximation of the stream function, defined respectively on one coarse grid with grid size H and one fine grid with grid size $h << H$. We prove that the difference between the solutions of new nonlinear Galerkin and the standard Galerkin methods is of the order of H^2 in stream function.

Key words: Navier-Stokes equation, stream function, nonlinear Galerkin method.

AMS(MOS) Subject Classification: 35A40, 65N60, 65M60.

1. Introduction

Let Ω be a bounded domain in R^2 assumed to have a Lipschitz continuous boundary Γ. We consider the time-dependent Navier-Stokes equations with stream form describing the flow of a viscous incompressible fluid confined in Ω

$$(1.1)\qquad \begin{cases} \frac{d(\Delta\psi)}{dt} - \nu\,\Delta^2\,\psi + \mathrm{rot}\psi\cdot\mathrm{grad}(\Delta\psi) = -\mathrm{rot}f & \text{in } \Omega,\ t>0 \\ \frac{d\psi}{dn} = 0 & \text{on } \Gamma,\ t>0 \\ \psi = 0 & \text{on } \Gamma,\ t>0 \\ \psi|_{t=0} = \psi_0 \end{cases}$$

We take $X = H_0^2(\Omega)$, $Y = H_0^1(\Omega)$, denote by $(\cdot,\cdot)$ the inner product on $L^2(\Omega)$ and $H_0^{-2}(\Omega)$ the dual space of $H_0^2(\Omega)$. We define the bilinear forms $a(\cdot,\cdot)$ and $r(\cdot,\cdot)$ on $X\times X$ and $Y\times Y$ respectively by

$$(1.2)\qquad \forall\psi,\ \phi\in Y \quad r(\psi,\phi) = (\nabla\psi,\nabla\phi)$$

$$(1.3)\qquad \forall\psi,\ \phi\in X \quad a(\psi,\phi) = \nu(\nabla\psi,\nabla\phi)$$

* **Partially supported by grants from National Science Foundation**

and trilinear form $a_1(\cdot,\cdot,\cdot)$ on $X \times X \times X$

$$\forall \psi, \phi, \eta \in X, \qquad a_1(\psi, \phi, \eta) = \left(\Delta\psi, \frac{d(\eta, \phi)}{d(x, y)}\right). \tag{1.4}$$

Also, we denote by $|\cdot|$ and $\|\cdot\|$

$$|\psi| = |(\nabla\psi, \nabla\psi)|^{\frac{1}{2}}, \qquad \|\psi\| = |(\Delta\psi, \Delta\psi)|^{\frac{1}{2}}$$

then we obtain the variational form of (1.1)

$$(Q) \qquad \begin{cases} \text{Find } \psi \in L^2(0,T,Y) \cap L^\infty(0,T,X) \text{ such that} \\ \frac{d}{dt}r(\psi,\phi) + a(\psi,\phi) + a_1(\psi,\psi,\phi) = (f, \text{rot}\phi) & \forall \phi \in X \\ r((\psi(0) - \psi_0), \phi) = 0 & \forall \phi \in Y \end{cases}$$

Let X_h and Y_h be two finite element subspaces of X and Y, then finite element approximative equation of (Q) is

$$(Q_h) \qquad \begin{cases} \text{Find } \psi_h \in L^\infty(X_h) \cap L^2(Y_h) \text{ such that} \\ \frac{d}{dt}r(\psi_h,\phi_h) + a(\psi_h,\phi_h) + a_1(\psi_h,\psi_h,\phi_h) = (f, \text{rot}\phi_h) & \forall \phi_h \in X_h \\ r((\psi_h(0) - \psi_0), \phi_h) = 0 & \forall \phi_h \in Y_h \end{cases}$$

For any fixed t, we define a projective operator $\pi_h : Y \to X_h$:

$$\begin{cases} \forall \psi \in Y, & \pi_h\psi \in X_h \text{ and} \\ \forall \phi_h \in X_h, & r(\psi - \pi_h\psi, \phi_h) = 0. \end{cases} \tag{1.5}$$

so, we have

$$\|\pi_h\psi\|_Y \le C\|\psi\|_Y \qquad \forall \psi \in Y \tag{1.6}$$

Assume that our finite element dissection is regular first degree element, according to classical interpolation theorem [2], the projective operator π_h defined by (1.5) satisfy following inequalities

$$\|\psi - \pi_h\psi\|_{1,\Omega} \le Ch\|\psi\|_{2,\Omega} \qquad \forall \psi \in H_0^2(\Omega) \tag{1.7}$$

$$\|\psi - \pi_h\psi\|_{0,\Omega} \le Ch\|\psi\|_{1,\Omega} \qquad \forall \psi \in H_0^1(\Omega) \tag{1.8}$$

and inverse inequality

$$\|\psi_h\|_{2,\Omega} \le Ch^{-1}\|\psi_h\|_{1,\Omega} \qquad \forall \psi_h \in X_h \tag{1.9}$$

Theorem 1.1 Let ψ be a solution of problem (Q), such that $\psi \in L^2(H_0^2(\Omega)) \cap C^0([0,T], H_0^1(\Omega))$, then for a sufficient small h, (1.6) holds, and (Q_h) possesses a unique solution, and we have error estimation

$$\forall (f, \psi_0) \in L^2(H_0^{-1}(\Omega)) \times H_0^{-2}(\Omega), \qquad \|\psi - \psi_h\|_{H^1(H_0^2(\Omega))} \le C(f, \psi_0)h. \tag{1.10}$$

The proof of this theorem is similar with [1,Thm 4.1], here, we omit this proof.

2. The nonlinear Galerkin method

Let h, H tend to zero, and $H > h > 0$. We consider two subspaces X_h and X_H, with $X_H \subset X_h$, and we write

$$X_h = X_H + X_h^H, \qquad \text{with } X_h^H = (I - \pi_H)X_h \tag{2.1}$$

where X_H and X_h^H are orthogonal with respect to the scaler product (grad$\cdot$, grad$\cdot$), X_H correspond to a space associated to a coarse grid. The following properties of X_h^H will be often used.

Lemma 2.1 Assume that (1.7)-(1.9) hold, then we have

$$\forall \chi \in X_h^H, \qquad |\chi| \le CH\|\chi\| \tag{2.2}$$

Next, we shall give the nonlinear Galerkin method of problem (Q_h). This method consists in looking for an approximate solution ψ^h of the form

$$\psi^h = \psi^H + \omega^h, \qquad \psi^h \in X_h, \;\; \psi^H \in X_H, \;\; \omega^h \in X_h^H$$

It is implemented starting from some given positive time t_0. On the interval of time $[0, t_0]$, we use a classical Galerkin method based on X_h, which gives the approximate solution ψ_h. For $t > t_0$, $\psi^h = \psi^H + \omega^h$ is given by the solution of

$$\frac{d}{dt}r(\psi^H, \phi) + a(\psi^H + \omega^h, \phi) + a_1(\psi^H, \psi^H, \phi) + a_1(\psi^H, \omega^h, \phi) \tag{2.3}$$

$$+ a_1(\omega^h, \psi^H, \phi) = (f, \text{rot}\phi) \qquad \forall \phi \in X_H$$

$$a(\psi^H + \omega^h, \chi) + a_1(\psi^H, \psi^H, \chi) = (f, \text{rot}\chi) \qquad \forall \chi \in X_h^H \tag{2.4}$$

$$\psi^H(t_0) = \pi_H \psi_0 \tag{2.5}$$

The existence and uniqueness of a solution ψ^h of (2.3)-(2.5) are given in the next theorem.

Lemma 2.2 Under assumptions (1.7)-(1.9). Let $f \in L^2(H_0^1(\Omega))$, $t_0 > 0$, $\psi_0 \in H_0^2(\Omega)$, then there exists a constant H_0 depending only of the data (Ω, ν, f), such that for h and H small enough, $h < H \le H_0$, we have $\pi_H X_h \ne \{0\}$.

Theorem 2.1 Under assumptions of lemma 2.2, problem (2.3)-(2.5) exists a unique solution ψ^h defined for $t > t_0$ with

$$\psi^h \in H^1(t_0, t_0 + T, X_h) \qquad \forall T > 0 \tag{2.6}$$

3. Error estimates

In this section, we aim to derive error estimates for the nonlinear Galerkin method presented by (2.3)-(2.5). We denote by $k(t)$ a constant depending on the data (Ω, ν, f), ψ_0, t_0 and t, which is continuous with respect to time on $[t_0, +\infty)$.

Theorem 3.1 Under assumptions (1.7)-(1.9), let $f \in L^2(H_0^{-1}(\Omega))$, $t_0 > 0$ and $\psi_0 \in H_0^2(\Omega)$, then, for h and H small enough, $h < H \leq H_0$, H_0 as in lemma 2.2, the solution ψ^h of problem (2.3)-(2.5) satisfies

$$\|\psi^h(t) - \psi_h(t)\| \leq k(t)H^2, \qquad t \geq t_0 \tag{3.1}$$

where $\psi_h(t)$ denotes the classical Galerkin solution of problem (Q_h)

In order to prove this theorem, we need do some preliminary work.

Similar to [4], the following lemma is easy to be proved.

Lemma 3.1 Let $\Omega \subset R^2$ is a C^2-classified bounded open set, then there exists a real constant C, such that

$$|a_1(\psi, \phi, \eta)| \leq C\|\psi\| \, \|\phi\| \, |\eta| \tag{3.2}$$

Lemma 3.2 Under the assumptions of theorem 3.1, the solution ψ_h of problem (Q_h) is such that $\omega_h = (I - \pi_H)\psi_h$ satisfies for $t \geq t_0$

$$|\omega_h^{(j)}| \leq CH\|\omega_h^{(j)}\| \leq k(t)H^2, \qquad j = 0, 1 \tag{3.3}$$

Now, we prove theorem 3.1

Proof: First, we set

$$\psi_h - \psi^h = e + E, \qquad e = \psi_H - \psi^H, \qquad E = \omega_h - \omega^h$$

where $\psi_H = \pi_H\psi_h$,, $\omega_h = (I - \pi_H)\psi_h$. By subtracting (2.7) and (2.8) from (Q_h), we obtain that

$$\frac{d}{dt}r(e, \phi) + a(e + E, \phi + \chi) + a_1(\psi_h, \psi_h, \phi) - a_1(\psi^h, \psi^h, \phi) + a_1(\omega^h, \omega^h, \phi) \tag{3.4}$$
$$+ a_1(\psi_h, \psi_h, \chi) - a_1(\psi^H, \psi^H, \chi) = -\frac{d}{dt}r(\omega_h, \chi) \qquad \forall \phi \in X_H, \ \chi \in X_h^H$$

by taking $\phi = e$, $\chi = E$, then

$$\frac{1}{2}\frac{d}{dt}|e|^2 + \nu\|e + E\|^2 \leq |r(\omega_{h,t}, E)| + |a_1(\psi_h, \psi_h, e) - a_1(\psi^h, \psi^h, e) \tag{3.5}$$
$$+ a_1(\omega^h, \omega^h, e) + a_1(\psi_h, \psi_h, E) - a_1(\psi^H, \psi^H, E)|$$

We aim to derive bounds for the terms in the right hand-side of (3.5). Using (3.3) and (2.2), we have

$$|r(\omega_{h,t}, E)| \leq |\omega_{h,t}| \, |E| \leq k(t)H^3\|E\| \tag{3.6}$$

Thanks to (3.2), (3.3) and (2.2), we get

$$|a_1(\psi_h,\psi_h,e) - a_1(\psi^h,\psi^h,e) + a_1(\omega^h,\omega^h,e) + a_1(\psi_h,\psi_h,E)$$
$$-a_1(\psi^H,\psi^H,E)| \leq k(t)|e|\,\|e+E\| + k(t)H^2\|e+E\|$$

So, we obtain

$$\frac{1}{2}\frac{d}{dt}|e|^2 + \nu\|e+E\|^2 \leq k(t)|e|\,\|e+E\| + k(t)H^2\|e+E\|$$
$$\leq k(t)|e|^2 + \frac{\nu}{4}\|e+E\|^2 + k(t)H^4 + \frac{\nu}{4}\|e+E\|^2$$

Hence

$$\frac{d}{dt}|e|^2 + \nu\|e+E\|^2 \leq k(t)|e|^2 + k(t)H^4 \tag{3.7}$$

Thanks to Gronwall inequality, we obtain $|e|^2 \leq k(t)H^4$. Hence, from (3.7), we have

$$\|e+E\|^2 \leq k(t)H^4$$

Thus, we complete the proof. □

References

[1] C. Bercovier and G. Raugel, A conforming finite element method for the time-dependent Navier-Stokes equations. **SIAM J. Numer. Anal. Vol.** 22 **No.** 2,June 1985, 455-473.

[2] P. G. Ciarlet, The finite element method for elliptic problem. **North-Holland, Amsterdam** 1978

[3] A. Ait Ou Ammi and M. Marion, Nonlinear Galerkin methods and mixed finite elements: two-grid algorithms for the Navier-Stokes equations. **Equiped' Analyse Numerique Lyon Saint-Etience Preprint** 1992

[4] P. Constantin and C. Foias, Navier-Stokes Equations. **University of Chicago Press, Chicago**, 1988

Research Centre for Applied Mathematics
Xi'an Jiaotong University
Xi'an, 710049, P. R. CHINA

AMS/IP Studies in Advanced Mathematics
Volume 3, 1997

A COMBINED SPECTRAL-FINITE ELEMENT METHOD FOR A TYPE OF NONLINEAR TWO-DIMENSIONAL SCHRÖDINGER EQUATIONS

XIAO-QING LIU SHENG-CHANG WU

Abstract

In this paper, a combined spectral-finite element method is proposed for a type of system of nonlinear, two-dimensional, semiperiodic Schrödinger equations. Some prior estimations are given and the convergence theorem is proved.

§1 Problem and Computation Algorithm

Let

$$I = \{x : 0 < x < 1\},$$
$$\tilde{I} = \{y : 0 < y < 2\pi\},$$
$$\Omega = I \times \tilde{I}.$$

We consider a type of system of nonlinear Schrödinger equations

$$\begin{aligned} &\frac{\partial U}{\partial t} = A(t)\Delta U + f(U), \quad in\ \Omega \times (0,T), \\ &U(0) = U_0, \quad in\ \Omega, \end{aligned} \tag{1.1}$$

where $U = U(x,y,t) = (U_1(x,y,t), U_2(x,y,t))^T$ has the period 2π for the variable y, and

$$U(0,y,t) = U(1,y,t) = 0, \quad \forall (y,t) \in \tilde{I} \times [0,T].$$

$f(U) = (f_1(U), f_2(U))^T$ is the nonlinear term. $A(t) = (a_{ij}(t))_{2\times 2}$ is a 2×2 real matrix satisfying the uniform positive definite condition, i.e. there exists a constant $k_1 > 0$, such that

$$\xi^T A(t)\xi \geq k_1 \|\xi\|^2, \quad \forall \xi \in \mathbf{R}^2.$$

Equations of the type above are found in theoretical physics and biological mathematics. Finite difference and finite element method are the classical numerical treatment for them. Since high order of precision can be reached in spectral approximation to smooth functions, we use spectral method in the y -direction if U is smooth enough with respect to y.

For convenience of expression, we introduce the following notations:

$$\begin{aligned}
\mathbf{C}_p^{\infty}(\Omega) &= \{\eta \in \mathbf{C}^{\infty}(\Omega) : \eta(x,y) = \eta(x, y+2\pi), \forall x \in I\},\\
\mathbf{C}_{0,p}^{\infty}(\Omega) &= \{\eta \in \mathbf{C}_p^{\infty}(\Omega) : \eta(0,y) = \eta(1,y) = 0, \forall y \in \tilde{I}\},\\
\mathbf{H}_p^{\mu}(\Omega) &= \overline{\mathbf{C}_p^{\infty}(\Omega)}^{\mathbf{H}^{\mu}(\Omega)},\\
\mathbf{H}_{0,p}^{\mu}(\Omega) &= \overline{\mathbf{C}_{0,p}^{\infty}(\Omega)}^{\mathbf{H}^{\mu}(\Omega)}.
\end{aligned}$$

$U \in [\mathbf{H}_{0,p}^1(\Omega)]^2$ is called the generalized solution of (1.1) if it satisfies

$$\begin{aligned}
&\frac{d}{dt}(U(t), v) + a(A(t)U(t), v) = (f(U(t)), v), \quad \forall v \in [\mathbf{H}_{0,p}^1(\Omega)]^2,\\
&U(0) = U_0,
\end{aligned} \tag{1.2}$$

where

$$a(u,v) = \iint_{\Omega} (\nabla u)(\nabla v) d\sigma.$$

Now we construct the computation algorithm. First, we divide I into N subintervals, with the nodes $0 = x_0 < x_1 < \cdots < x_N = 1$. Let $I_l = (x_{l-1}, x_l)$, $h = max\{x_l - x_{l-1} : 1 \le l \le N\}$, and $h' = min\{x_l - x_{l-1} : 1 \le l \le N\}$. We assume in addition that there exists a positive constant d independent of the divisions of I, such that $h/h' \le d$. For any integer $k \ge 0$, we denote by $\mathbf{P}_k$ the set of all the polynomials defined on $\mathbf{R}^1$ of degree $\le k$. We define the finite element subspace in the x-diretion as:

$$\begin{aligned}
\tilde{\mathbf{F}}_h^k &= \{\eta : \eta|_{I_l} \in \mathbf{P}_k, \quad 1 \le l \le N\}\\
\mathbf{F}_h^k &= \tilde{\mathbf{F}}_h^k \cap H_0^1(\Omega).
\end{aligned}$$

Let M be a positive integer, we define the subspace for Fourier spectral approximations as:

$$\mathbf{S}_M = Span\{e^{ijy} : |j| \le M\}.$$

and define

$$\mathbf{V}_\alpha = [\mathbf{F}_h^{k+1} \otimes \mathbf{S}_M]^2, \quad \alpha = (M, h, k).$$

Let $\mathbf{L}_M$ be the orthogonal projection from $\mathcal{L}^2(\tilde{I})$ onto $\mathbf{S}_M$. Π_h^k is the piecewise Lagrange interpolation of order k from $C(I)$ onto $\mathbf{F}_h^k$, i.e. for any $\eta \in \mathbf{C}(I)$, $\Pi_h^k \eta \in \mathbf{F}_h^k$ satisfies

$$\begin{aligned}
(\Pi_h^k \eta)(x_{l-1} + \frac{n}{k} h_l) =& \eta(x_{l-1} + \frac{n}{k} h_l),\\
& 0 \le n \le k, 1 \le l \le N.
\end{aligned}$$

It is obvious that $\mathbf{L}_M$ and Π_h^k are commutable, i.e.

$$\mathbf{L}_M \circ \Pi_h^k = \Pi_h^k \circ \mathbf{L}_M.$$

Let τ be the time step and $\mathbf{T}_\tau = \{t = l\tau : 0 \le l \le [\frac{T}{\tau}]\}$. Define

$$\eta_t(t) = \frac{1}{\tau}(\eta(t) - \eta(t-\tau)).$$

A fully discrete spectral-finite element scheme for (1.2) is to find $u(t) \in \mathbf{V}_\alpha$ for all $t \in \mathbf{T}_\tau$, such that

$$\begin{aligned} &(u_t(t), v) + a(A(t)u(t), v) = (\mathbf{L}_M \circ \Pi_h^k f(u(t)), v), \quad \forall v \in \mathbf{V}_\alpha, \\ &u(0) = \mathbf{L}_M \circ \Pi_h^{k+1} U_0. \end{aligned} \tag{1.3}$$

§2 A Prior Estimation

1.Estimation of $\|u(t)\|$

Suppose $\mathcal{B}$ is a Banach space, and $J \subset \mathbf{R}^1$is an interval. We define

$$\mathcal{L}^2(J;\mathcal{B}) = \{\eta : J \to \mathcal{B} \ : \ \|\eta\|_{\mathcal{L}^2(J;\mathcal{B})} = (\int_J \|\eta(t)\|_{\mathcal{B}}^2)^{1/2} < \infty\}.$$

Similarly, we define the spaces $\mathbf{C}(J;\mathcal{B})$ and $\mathbf{H}^\mu(J;\mathcal{B})$, etc.. For $r \ge 0$, and $s \ge 0$, we define

$$\mathbf{H}^{r,s}(\Omega) = \mathcal{L}^2(\tilde{I};\mathbf{H}^r(I)) \cap \mathbf{H}^s(\tilde{I};\mathcal{L}^2(I)),$$

with the norm

$$\|\eta\|_{\mathbf{H}^{r,s}(\Omega)} = (\|\eta\|^2_{\mathcal{L}^2(\tilde{I};\mathbf{H}^r(I))} + \|\eta\|^2_{\mathbf{H}^s(\tilde{I};\mathcal{L}^2(I))})^{1/2},$$

If $r \ge 1$, $s \ge 1$ then we define

$$\mathbf{M}^{r,s}(\Omega) = \mathbf{H}^{r,s}(\Omega) \cap \mathbf{H}^1(\tilde{I};\mathbf{H}^{r-1}(I)) \cap \mathbf{H}^{s-1}(\tilde{I};\mathbf{H}^1(I)),$$

with the norm

$$\|\eta\|_{\mathbf{M}^{r,s}(\Omega)} = (\|\eta\|^2_{\mathbf{H}^{r,s}(\Omega)} + \|\eta\|^2_{\mathbf{H}^1(\tilde{I};\mathbf{H}^{r-1}(I))} + \delta\|\eta\|^2_{\mathbf{H}^{s-1}(\tilde{I};\mathbf{H}^1(I))})^{1/2}.$$

Besides, we define

$$\begin{aligned} \mathbf{H}_p^{r,s}(\Omega) &= \overline{\mathbf{C}_p^\infty(\Omega)}^{\mathbf{H}^{r,s}(\Omega)}, \\ \mathbf{H}_{0,p}^{r,s}(\Omega) &= \overline{\mathbf{C}_{0,p}^\infty(\Omega)}^{\mathbf{H}^{r,s}(\Omega)}, \\ \mathbf{M}_{0,p}^{r,s}(\Omega) &= \overline{\mathbf{C}_{o,p}^\infty}^{\mathbf{M}^{r,s}(\Omega)}. \end{aligned}$$

Lemma 2.1 [1]. *If $r \ge \frac{1}{2}$, $s \ge 0$, $\bar{r} = min(r, k+1)$, then there exists a positive contant c independent of h and M, such that for all $\eta \in \mathbf{H}_p^{r,s}(\Omega)$*

$$\|\eta - \mathbf{L}_M \circ \Pi_p^k \eta\| \le c(h^{\bar{r}} + M^{-s})\|\eta\|_{\mathbf{H}^{r,s}(\Omega)}.$$

Lemma 2.2 [1]. *If $Mh \leq Const$, $r \geq 1$, $s \geq 1$, $\bar{r} = min(r, k+1)$, then there exists a positive constant c independent of h and M, such that for all $\eta \in \mathbf{M}_p^{r,s}(\Omega)$*

$$|\eta - \mathbf{L}_M \circ \Pi_h^k \eta|_1 \leq c(h^{\bar{r}-1} + M^{1-s})\|\eta\|_{\mathbf{M}^{r,s}(\Omega)}.$$

In particular, we have for all $\eta \in \mathbf{H}_p^1(\Omega)$ that

$$|\mathbf{L}_M \circ \Pi_h^k \eta|_1 \leq c\|\eta\|_1.$$

Lemma 2.3.

$$(u_t(t), u(t)) = \frac{1}{2}(\|u(t)\|^2)_t + \frac{\tau}{2}\|u_t(t)\|^2.$$

Theorem 1. *Suppose f is Lipschitz, i.e.*

$$\|f(u) - f(v)\|_1 \leq K\|u - v\|_1, \quad \forall u, v \in \mathbf{H}_1(\Omega).$$

We have the following estimation:

$$\|u(t)\| \leq c_1,$$

where c_1 is a constant independent of τ, h, M when τ, h, M^{-1} are sufficiently small.

2. Estimation of $|u(t)|_1$

By the inverse inequality in the finite element method [2], it is not difficult to prove that $|u(t)|_1 \leq (C_d h^{-2} + M^2)^{1/2}\|u(t)\|$, where C_d is a constant depending on the parameter d, but not on h and M. Here we are going to figure out a better estimation of $|u(t)|_1$.

Lemma 2.4. *$(A(t))^{-T}$ exists and satisfies the following:*
$\exists k_2 > 0$, such that

$$\xi^T (A(t))^{-T} \xi \geq k_2\|\xi\|^2, \quad \forall \xi \in \mathbf{R}^2.$$

Theorem 2. *Suppose f is Lipschitz. We have the following estimation:*

$$|u(t)|_1 \leq c_2,$$

where c_2 is a positive constant independent of τ, h, M.

§3 Convergence Theorem

Lemma 3.1(Gronwall's Inequality) [6]. *For functions $u(t)$ and $v(t)$, with $0 \le v(t) \le G$, if there exists a positive constant ρ, such that*

$$u(t) \le \rho + \tau \sum_{l=0}^{[\frac{t}{\tau}]} u(l\tau)v(l\tau), \quad 0 \le t \le T,$$

then

$$u(t) \le \rho e^{GT}, \quad 0 \le t \le T.$$

Theorem 3. *Suppose (1.2) has the solution $U \in \mathbf{C}(0,T;\mathbf{M}_{0,p}^{r,s}(\Omega)) \cap \mathbf{H}^1(0,T;\mathbf{H}^1(\Omega) \cap \mathbf{H}_p^{r-1,s-1}(\Omega)) \cap \mathbf{H}^2(0,T;\mathcal{L}^2(\Omega))$ with $r \ge 1$, $s \ge 1$. u is the solution of (1.3). τ, h, M^{-1} are sufficiently small with $hM \le$Const, $\bar{r} = min(r, k+2)$. If f is suitably smooth and Lipschitz, then there exists a positive constant C^*, such that*

$$\|u(t) - U(t)\|^2 \le C^*(\tau^2 + h^{2(\bar{r}-1)} + M^{2(1-s)}), \quad \forall t \in \mathbf{T}_\tau.$$

The convergence theorem above shows that the precision of the spatial discretization is mainly determined by $h^{2(\bar{r}-1)}$ when s is large, i.e. U is sufficiently smooth with respect to y. To reduce h in the method, we need refine the division only in one direction, the x-direction, while the classical finite element method requires refinement in two directions. Thus the computation time is saved.

References

[1] C.Camuto, Y.Maday, and A.Quarteroni, **Numer. Math., 39**(1982), 205-220.

[2] P.G.Ciarlet, The Finite Element Method for Elliptic Problems, North-Holland, Armsterdam, 1978.

[3] D.Gottlieb, Numerical Analysis of Spcetral Method, CBMS-NSF, Regional Conference Series in Applied Math. 26, 1977.

[4] B-Y.Guo and W-M Cao, **Journal of Computational Physics, 101**(1992), 375-385.

[5] S-C.Wu, **Science Exploration Vol.4 No.4**(1984), 39-48.

[6] Q-S.Chang, **Mathematica Numerica Sinica, 4**(1982), 373-384.

Institute of Applied Mathematics,
Academia Sinica,
Beijing 100080, China

AMS/IP Studies in Advanced Mathematics
Volume **3**, 1997

CAUCHY PROBLEM OF NONISENTROPIC GAS DYNAMICS EQUATIONS

Liu Fagui

Abstract In the paper, we obtain the existence and nonexistence theorems for C^1−solutionsof nonisentropic gas dynamics equations.

1 Introduction

For the following Cauchy problem :

$$u_t + A(u)_x = 0, \quad u(x,0) = u_0(x), \quad u \in R^n, \tag{1.1}$$

it is well−known that the C^1−solutions of the problem (1.1) exists only locally in time, even if $u_0(x)$ is smooth and small. Up to now, there are wonderful results as $n=2$ ([1],[2]), however, for $n \geqslant 3$, fewer results are obtained. Moreover, for Cauchy problem of the gas dynamics equations

$$u_t + p(v,s)_x = 0, \qquad v_t - u_x = 0, \qquad s_t = 0, \tag{H}$$

$$t = 0: \quad u = u_0(x), \quad v = v_0(x), \quad s = s_0(x), \tag{I}$$

the results of the existence or nonexistence are by no means satisfactory, where v, u, $p = k\exp(\frac{s}{c})v^{-\gamma}$, s denote respectively the specific volume, velocity, presure and entropy, $k > 0$, a constant, $c > 0$, the specific heat capacity, $\gamma > 1$, the adiabatic exponent. C. Z. Li & J. H. Wang have only considered the problem (H), (I) for a class of special initial value ([3]); Y. C. Zhao has given a sufficient condition existing global C^1 − solution ([4]), however, the restriction for $u_0(x)$ seems to be no ideal as $s_0(x) \not\equiv 0$.

In the paper, we consider Cauchy problem (H), (I), and obtain the global existence and nonexistence theorems of its C^1−solutions.

2 Boundedness

At first, the problem (H), (I) can be written as

$$u_t + p(v,s_0(x))_x = 0, \qquad v_t - u_x = 0, \tag{2.1}$$

$$t = 0; \quad u = u_0(x), \qquad v = v_0(x), \tag{2.2}$$

where $p(v,s_0(x))=K(x)v^{-\gamma}$, $K(x)=k\exp(\frac{s_0(x)}{c})>0$. The eigenvalues and Riemann invariants of (2.1) are respectively

$$\lambda_1=-\mu_1=-(K(x)\gamma)^{\frac{1}{2}}v^{-\frac{\gamma+1}{2}}<0, \tag{2.3}$$

$$z=u-B(v,x), \qquad w=u+B(v,x), \tag{2.4}$$

where $B(v,x)=\int_v^\infty\sqrt{-p_v(y,s_0(x))}dy$. Therefore, we get

$$\begin{cases} z_t+\lambda_1 z_x=Cs_{0'}(x)\lambda_1(w-z), \quad w_t+\mu_1 w_x=Cs_{0'}(x)\mu_1(w-z), \\ t=0: \quad z=z_0(x), \quad w=w_0(x), \end{cases} \tag{2.5}$$

where $C=\frac{1}{4c\gamma}$, $z_0(x)=u_0(x)-B(v_0(x),x)$, $-w_0(x)=u_0(x)+B(v_0(x),x)$.

Under the domain considered, we suppose that

(A) $w_0(x),z_0(x)\in C^1(R)$, $s_0(x)\in C^2$, $\|w_0(x)\|_{C^1}$、$\|z_0(x)\|_{C^1}$、$+\|s_0(x)\|_{C^2}$ is bounded, and $\delta=inf_x\, w_0(x)-sup_x\, z_0(x)>0$;

(B) $s_{0'}(x)\leqslant 0$, and there exists a constant $M_0>0$, such that

$$\sup_x w_0(x)-\inf_x z_0(x)\leqslant\frac{M_0}{4}, \quad C(\sup_x s_0(x)-\inf_x s_0(x))\leqslant\min(\frac{\delta}{M_0},\frac{1}{2}).$$

Theorem 2.1 *Under the hypothesis (A),(B), any C^1—solution of Cauchy problem (H), (I) satisfies*

$$0<\frac{\delta}{2}\leqslant(w-z)(x,t)\leqslant\frac{M_0}{2} \tag{2.6}$$

Proof Notice the monotonicity of characteristics the equations (2.5), we suppose that the first and the second characteristics are respectively

$$t=t_1(x,\beta),\ t_1(\beta,\beta)=0,\ t=t_2(x,\eta),\ t_2(\eta,\eta)=0. \tag{2.7}$$

Set $\frac{d}{d_1t}=\frac{\partial}{\partial x}+\frac{1}{\lambda_1}\frac{\partial}{\partial t}$, $\frac{d}{d_2x}=\frac{\partial}{\partial x}+\frac{1}{\mu_1}\frac{\partial}{\partial t}$, then by (2.5), we get

$$\begin{cases} \frac{dz}{d_1x}=z_x+\frac{1}{\lambda_1}z_t=-Cs_{0'}(x)(w-z), \\ \frac{dw}{d_2x}=w_x+\frac{1}{\mu_1}w_t=-Cs_{0'}(x)(w-z), \\ t=0: \quad z=z_0(\beta), \quad w=w_0(\eta) \end{cases} \tag{2.8}$$

Integrating (2.8) along (2.7), and noticing (A),(B), (2.6) is easily verified.

By the theorem 2.1, we can prove the following lemma.

Lemma 2.2 *Under the hypothesis (A) and (B), any C^1—solution of the problem (H),(I) satisfies*

$$0<v_1\leqslant v(x,t)\leqslant v^1<\infty, \qquad \|u(x,t)\|_{C^0}+\|s(x,t)\|_{C^0}\leqslant M,$$

where v_1、v^1、M are positive constants indepent of x, t.

3 The Main Results

Differentiating (2.5) with respect to t, and noticing (2.3), (2.4), we get

$$\begin{cases} (z_t)_t + a(x)\lambda(z_x)_x = \dfrac{\alpha}{w-z}(z_t - \dfrac{w-z}{\alpha}\mu A(x))(w_t - z_t), \\ (w_t)_t + a(x)\mu(w_x)_x = \dfrac{\alpha}{w-z}(-w_t - \dfrac{w-z}{\alpha}\mu A(x))(z_t - w_t), \\ w_t(x,0) = \mu_0(x)(A(x)(w_0(x) - z_0(x)) - a(x)w_{0'}(x)), \\ z_t(x,0) = \mu_0(x)(A(x)w_0(x) - z_0(x)) + a(x)z_{0'}(x)), \end{cases} \tag{3.1}$$

where $\mu_0(x)=(w_0(x)-z_0(x))^\alpha$, $\alpha=\dfrac{\gamma+1}{\gamma-1}$, $\lambda=-\mu=-(w-z)^\alpha>0$,

$$A(x) = \frac{K_{'}(x)}{\gamma-1}\left(\frac{\gamma-1}{4\sqrt{K(x)\gamma}}\right)^{\alpha+1}, \ a(x) = \left(\frac{\gamma-1}{4}\right)^\alpha (K(x)\gamma)^{\frac{1}{1-\gamma}} \tag{3.2}$$

By the lemma 2.2 and (B), there exist constants δ_1、δ_2—$(0<\delta_1<\delta_2)$, such that

$$\delta_1 s_{0'}(x) \leqslant -\frac{w-z}{\alpha}\mu A(x) \leqslant -\delta_2 s_{0'}(x) \tag{3.3}$$

Lemma 3.1 *Under the hypothesis (A) and (B), if*

$$0 \leqslant z_t(x,0)、w_t(x,0) \leqslant -\delta_1 s_{0'}(x), \tag{3.4}$$

then

$$0 \leqslant z_t(x,t)、w_t(x,t) \leqslant -\delta_1 s_{0'}(x). \tag{3.5}$$

Proof By (3.4) and the continuity of w_t and z_t, there exists constant $\delta>0$, such that as $t\in(0,\delta]$,

$$z_t - A(x)\mu\frac{w-z}{\alpha} \geqslant 0, \ -w_t - A(x)\mu\frac{w-z}{\alpha} \geqslant 0. \tag{3.6}$$

Therefore, by the proposition 3 ([4]), as $t\in(0,\delta]$, (3.5) is true.

The rest may be reduced by analogy. Thus, (3.5) is verified as $t\in[0,\infty)$.

By the theorem 2.1 and lemma 3.1, we can give the following main theorem.

Theorem 1 *Under the hypothesis (A) and (B), if $w_{0'}(x)$、$z_{0'}(x)$ satisfy*

$$0 \leqslant \mu_0(x)(A(x)(w_0(x) - z_0(x)) + a(x)z_{0'}(x)) \leqslant -\delta_1 s_{0'}(x),$$
$$0 \leqslant \mu_0(x)(A(x)(w_0(x) - z_0(x)) - a(x)w_{0'}(x)) \leqslant -\delta_1 s_{0'}(x),$$

then, Cauchy problem (H), (I) admits a unique global smooth solution on the upper—half plane $t\geqslant 0$.

Differentiating (2.5) with respect to x, and let $h = \frac{1}{2}\ln\mu$, then we get

Lemma 3.2 *Any smooth solution of the problem (H), (I) satisfies*

$$\begin{cases} \dfrac{dR}{d_1 t} = -A_1(x,t)R^2 - A_2(x,t)R + A_3(x,t), \\ \dfrac{dS}{d_2 t} = -A_1(x,t)S^2 + A_2(x,t)S - A_3(x,t), \end{cases} \tag{3.7}$$

where $\dfrac{d}{d_1 t}=\dfrac{\partial}{\partial t}+a(x)\lambda\dfrac{\partial}{\partial x}$, $\dfrac{d}{d_2 t}=\dfrac{\partial}{\partial t}+a(x)\mu\dfrac{\partial}{\partial x}$, $A_1(x,t)=\dfrac{\alpha a(x)}{w-z}\mu^{\frac{1}{2}}$,

$A_2(x,t)=\dfrac{(\alpha+1)(\alpha+6)A(x)\mu}{\alpha+2}$, $A_3(x,t)=(\dfrac{A_{'}(x)}{\alpha+2}+\dfrac{2(\alpha+1)^2A(x)}{(\alpha+2)^2})\mu^{\frac{1}{2}}(w-z)$,

$R(x,t)=\mu^{\frac{1}{2}}(z_x+\dfrac{(\alpha+1)A(x,t)(w-z)}{(\alpha+2)a(x)})$, $S(x,t)=\mu^{\frac{1}{2}}(w_x-\dfrac{(\alpha+1)A(x)(w-z)}{(\alpha+2)a(x)})$.

Under the hypothesis (A) and (B), by the theorem 2.1, there exist constants δ_0, β_1, β_0, such that

$$\beta_1 \leqslant \frac{A_1(x,t)}{2} \leqslant \delta_0, \quad \left|\frac{A_2(x,t)}{2A_1(x,t)} + A_3(x,t)\right| \leqslant \beta_1, \tag{3.8}$$

Therefore, by (3.7) and (3.8), we get

$$\frac{dR}{d_1 t} \leqslant -\beta_0 R^2 + \beta_1, \quad \frac{dS}{d_2 t} \leqslant -\beta_0 S^2 + \beta_1. \tag{3.9}$$

Lemma 3.3 *For the following initial value problem*

$$y,(t) = -\beta_0 \quad y^2 + \beta_1, \quad y(0) = R_0 = R_0(\beta). \tag{3.10}$$

If $R_0 < -\alpha_0 = -\sqrt{\frac{\beta_1}{\beta_0}}$, *then there exists a finite time* $T>0$, *such that*

$$y(t) \to \infty, \qquad \text{as} \quad t \to T^-. \tag{3.11}$$

Proof By (3.10), we get

$$y(t) = \frac{\alpha_0(1+E)\exp(-2\alpha_0\beta_0 t)}{1 - E\exp(-2\alpha_0\beta_0 t)}, \tag{3.12}$$

where $E = \frac{R_0 - \alpha_0}{R_0 + \alpha_0}$. Thus, if $R_0 < -\alpha_0$, then $E > 1$. Therefore, there exists finite time $T > 0$, such that

$$1 - E\exp(-2\alpha_0\beta_0 T) = 0. \tag{3.13}$$

Consequently, by (3.12), (3.11) is true.

Suppose that

$$R(x,0) = R_0(x) = \mu^{\frac{1}{2}}(v_0(x))(z_0,(x) + \frac{(\alpha+1)A(x)}{(\alpha+2)a(x)}(w_0(x) - z_0(x)),$$

$$S(x,0) = S_0(x) = \mu^{\frac{1}{2}}(v_0(x))(w_0,(x) - \frac{(\alpha+1)A(x)}{\alpha+2)a(x)}(w_0(x) - z_0(x)).$$

By the lemma 3.3, if there exists a point $0 \in R$, such that

$$R_0(\theta) < -\alpha_0 \quad \text{or} \quad S_0(\theta) < -\alpha_0, \tag{3.14}$$

then we can give the following nonexistence theorem.

Theorem 2 *Under the hypothesis (A) and (B), if the condition (3.14) is held, then Cauchy problem (H), (I) admits smooth solution in a finite time.*

REFERENCE

1 D. Q. Li, Proceeding of the Roy. Soc. of Edingburgh, 87A(1981), 225—261.

2 P. D. Lax, J. Math. Phys., 5(1964), 611—613.

3 C. Z. Li & J. H. Wang, Mathematica Acta Sci., 9(1989), 297—306.

4 Y. C. Zhao, Global smooth solutions of the gas dynamics, to appear.

* North China Inst. of Water Conservancy & Hydro. Power, Zhengzhou, P. R. China, 450045

AMS/IP Studies in Advanced Mathematics
Volume 3, 1997

Existence and Uniqueness of Global Solution for Certain Degenerate Reaction-Diffusion Systems

Lu Guo-fu

Abstract

In this paper we discuss the following nonlinear degenerate parabolic systems
$$u_{it} = \triangle a_i(u_i) + b_i(x,t,u_i)Du_i + f_i(x,t,u)$$
for $i = 1,2,\cdots,m$ and $u = (u_1,\cdots,u_m)$ is a vector function, with Dirichlet boundary condition. Under some structure conditions on a_i, b_i and f_i and initial data $u_i^0 \in L^{p_i}(\Omega)$ for some $p_i > p_i^0, i = 1,2,\cdots,m$, the result on existence and uniqueness of global solution is established.

1. Introduction

Consider the initial and boundary value problem
$$u_{it} = \triangle a_i(u_i) + b_i(x,t,u_i)Du_i + f_i(x,t,u), \qquad (x,t) \in Q_T \tag{1.1}$$
$$u_i(x,t) = 0, \qquad (x,t) \in \partial\Omega \times (0,T] \tag{1.2}$$
$$u_i(x,t) = u_i^0(x), \qquad (x,t) \in \Omega \times \{t = 0\} \tag{1.3}$$
where $i = 1,\cdots,m$ and any $T > 0$. $Du_i = \sum_{l=1}^{N} u_{ix_l}, u = (u_1,\cdots,u_m)$ and Ω is an open bounded domain in $\mathbf{R}^N$ with a smooth boundary $\partial\Omega$.

Our purpose is to prove the existence and uniqueness of locally bounded global solution in $\bar{\Omega} \times (0,T]$ for systems (1.1) — (1.3). The problem with $a_i(u_i) \equiv u_i^{m_i}$ and $b_i(x,t,u_i) \equiv 0$ has been studied by L. Maddalena [4]. As a result of unsuitable method being used by the author, the major estimate on the norm of L^∞ for solution is failure in [4]. Since considering the problem with the terms of first derivative and the data u_i^0 as some functions of $L^{p_i}(\Omega)$ for $p_i > p_i^0$ here, then we have some substantial difficulties for investigating the result on existence and unniqueness.

Actually, the existence program consists of the following: the initial and boundary data are approximated by smooth positive funcitions, and $f_i(x,t,u)$ is approximated by a sequence of linear maps for large $|u|$. Hence the problem turns to that the limit of the solutions for the approximating nondegenerate problem exists. Executing this procedure, one encounters two difficulties: one of them is to prove that the approximatting problems are nondegenerate and admit a global solution; the other is to prove uniform estimates for these solutions. Since $u_i \in L^{p_i}(\Omega)$, we get only interior estimate on L^∞ —norm of these solutions. Additional, with the terms of first order derivative, there must be some difficulty for proving the unique theorem. With accurate estimate on L^∞ —norm of solution u and boundary estimate on normal derivative for the solution of dual problem, we overcome these difficulties.

Acknowlehement The author is grateful to Professor Ye Qixiao and Professor Chen Yazhe for their guidance.

We assume that the coefficients of (1.1) are continuously differentiable with respect to their variables and satisfy the following structure conditions.

(**H.** 1) $a_i(0) = 0$ and there exist constants $m_i > 1$ and $\Lambda_l > 1, l = 0,1$ such that

$$\Lambda_0 u_i^{m_i - 1} \leqslant a'(u_i) \leqslant \Lambda_1 u_i^{m_i - 1}$$

for any $u_i \in \mathbf{R}^+$, $i = 1, \cdots, m$.

(**H.** 2) There exist constants $\Lambda_l > 0, l = 2,3,4$ such that

$$|b_i(x,t,u_i)| \leqslant \Lambda_2 + \Lambda_3 u_i^{\frac{m_i - 1}{2}}$$

and

$$|\sum_{l=1}^{N} \frac{\partial}{\partial x_l} b_i(x,t,u_i)| \leqslant \Lambda_4$$

for $(x,t) \in Q_T, u_i \in \mathbf{R}^+$ and $i = 1, \cdots, m$.

(**H.** 3) There exist constants $\alpha_{ij} > 0, \beta_i > 0$, functions $c_j(x,t), c_{ij}(x,t), \in L^\infty(Q_T)$ such that

$$f_i(x,t,u) \leqslant \sum_{j=1}^{m} c_{ij}(x,t)[a_j(u_j)]^{\alpha_{ij}} + c_i(x,t) \qquad \text{in } Q_T \times R_+^m$$

for $i = 1, \cdots, m$, with $\alpha_{ij} < 1$ whenever $mes\{c_{ij}(x,t) > 0\} > 0$.

(**H.** 4) That $f_i(x,t,u)$ are nonnegative, continuously differentiable with respect to u on $Q_T \times R_+^m$ and nondecreasing in u_j for all $j \neq i$ and $f_i(x,t,0) = 0$ for $i = 1, \cdots, m$.

(**H.** 4$'$) There exist constants α'_{ij} and nonnegative functions $c'_i(x,t), c'_{ij}(x,t) \in L^\infty(Q_T)$ such that

$$0 \leqslant \alpha'_{ij} < \alpha_{ij}, \sum_{j=1}^{m} |\frac{\partial}{\partial u_j} f_i(x,t,u)| \leqslant c'_i(x,t) + \sum_{j=1}^{m} c'_{ij}(x,t) u_j^{\alpha'_{ij} m_j} \quad \text{in } Q_T \times R_+^m$$

for $i = 1, \cdots, m$.

(**H.** 5) $u_i^0 > 0$ and $u_i^0 \in L^{p_i}(\Omega), i = 1, \cdots, m$, where $p_i \geqslant p_i^0 = \max_{1 \leqslant j \leqslant m} \{m_i + 1, m_j + \alpha_{ij} m_j, (N+2)\alpha_{ij} m_j - \frac{N}{2}(m_i - 1)\}$.

Dfeinition *A nonnegative vector function $u = (u_1, \cdots, u_m)$ (i. e. $u_i \geqslant 0, i = 1, \cdots, m$) is said to be a weak solution of problem* (1.1) — (1.3) *in $Q_T, T > 0$, if*

(*i*) *$\nabla u_i^{m_i}$ exists in the sense of distributions in Q_T and*

$$\iint_{Q_T} (u_i^{2m_i} + |\nabla u_i^{m_i}|^2) dx dt < +\infty$$

(*ii*) *u satisfies the identity*

$$\int_Q \varphi_i(x,T) u_i(x,T) dx = \iint_{Q_T} a_i(u_i) \Delta\varphi_i dx dt + \iint_{Q_T} B_i(x,t,u_i) D\varphi_i dx dt$$

$$+ \iint_{Q_T} [F_i(x,t,u)\varphi_i + \varphi_{it} u_i] dx dt + \iint_Q \varphi_i(x,0) u_i^0(x) dx$$

for every $\varphi_i \in C^{1,0}(\overline{Q}_T) \cap C^{2,1}(Q_T)$ such that $\varphi_i = 0$ on $\partial\Omega \times (0,T]$, where

$$B_i(x,t,u_i) = -\int_0^{u_i} b_i(x,t,s) ds$$

and

$$F_i(x,t,u) = f_i(x,t,u) - \sum_{l=1}^{N}\int_0^{u_i}\frac{\partial}{\partial x_l}b(x,t,s)ds$$

(iii) $u_i(x,t) = 0, (x,t) \in \partial\Omega \times (0,T)$ *in the sense of trace, for* $i = 1,\cdots,m$.

A vector function $u = (u_1,\cdots,u_m)$ *is a supersolution (subsolution) if we take the test function* $\varphi_i \geqslant 0$ *and* $\geqslant$ $(\leqslant)$ *instead of the equalities in (ii), (iii).*

The main results in the paper are the following theorems:

Theorem A *Assume that hypotheses (H. 1)—(H. 5) hold, then problem (1. 1)—(1. 3) at least possesses one weak solution* u *in* Q_T, *for any* $T > 0$.

Theorem B *Assume that hypothess (H. 1),, (H. 2) and (H. 4′) hold, problem (1. 1)—(1. 3) can't have two distinct weak solution in* Q_T, *for any* $T > 0$.

In the following sections, the outline of proofs on Theorem A and B is given.

2. Existence

To prove Theorem A we begin by regularizing problem (1. 1) — (1. 3) and get a priori estimate for the solution satisfying the regularized problem.

Let $\{R_i\}$ be a monotone increasing sequence of positive real numbers such that $\lim_{k\to+\infty} R_k = +\infty$ and f_{ik} be smooth function nondecresaing in $u_j, j \neq i$, which linearize for $|u| > R_k$ the function f_i and such that $0 \leqslant f_{ik} \leqslant f_i$ for $u_i \geqslant 0, i = 1,\cdots,m, k = 1,2,\cdots$.

To approximate the initial values u_i, pick $\{u_{ik}^0\}$ such that $u_{ik}^0 \in C_0^\infty(\Omega), u_{ik}^0 \geqslant 0$, $\| u_{ik}^0 \|_{L^{p_i}(\Omega)} \leqslant \| u_i^0 \|_{L^{p_i}(\Omega)}, u_{ik}^0 \to u_i^0$ *in* $L^{p_i}(\Omega)$.

Consider the following problem for every integer $k > 1$.

$$u_{ikt} = \Delta a_i(u_{ik}) + b_i(x,t,u_{ik})Du_{ik} + f_{ik}(x,t,u_k - \frac{1}{k}) \quad \text{in } \Omega \times (0,\infty) \tag{2. 1}$$

$$u_{ik}(x,t) = \frac{1}{k} \text{ on } \partial\Omega \times (0,\infty) \tag{2. 2}$$

$$u_{ik}(x,t) = u_{ik}^0 + \frac{1}{k} \text{ in } \Omega \times \{t = 0\} \tag{2. 3}$$

where $u_k - \frac{1}{k} = (u_{lk} - \frac{1}{k},\cdots,u_{mk} - \frac{1}{k})$.

Lemma 2. 1 *For every* $k \geqslant 1$, *problems (2. 1)—(2. 3) has a unique classical global solution* $u_k = (u_{lk},\cdots,u_{mk})$ *and* $u_{ik} \geqslant \frac{1}{k}$ *in* $\Omega \times (0,\infty)$.

Lemma 2. 2 *Under the assumptions of Theorem A, the follwing estimates hlod for the solutions* $u_k = (u_{lk},\cdots,u_{mk})$ *of (2. 1)—(2. 3) and any* $p \leqslant p_i$,

$$\| u_{ik} \|_{L^p(Q_T)} \leqslant C(p), \text{for } i = 1,\cdots,m.$$

where $C = C(p)$ *is a constant independent of* k.

Lemma 2. 3 *Under the assumptions of Theorm A, the following estimates hold for the solutions* $u_k = (u_{lk},\cdots,u_{mk})$ *of (2. 1)—(2. 3)*

$$\| u_{ik} \|_{L^\infty_{(Q_T^\tau)}} \leqslant C(\tau), \qquad \text{for } i = 1,\cdots,m.$$

where $C(\tau)$ *is a constant depending on* $\tau, N, \Omega, T, m_i, a_{ij}, C_i, C_{ij}, \| u_i^0 \|_{L^{p_i}}, m$ *and* Λ_l *for* $i,j = 1,\cdots,m, l = 0,1,\cdots,4, Q_T^\tau = \Omega \times [\tau,T]$.

Remark 1. By the proof of Lemma 2. 3, we can get the following conclusion:

Under the assumptions of Lemma 2. 3, there exist constants $0 \leqslant \gamma_i < \frac{1}{2}$ and $\sigma_i > 0$, such that

$$\sup_{\frac{1}{2}<s<t} \| u_{ik}(\cdot, s) \|_{L^\infty(\Omega)} \leqslant \frac{\sigma_i}{t^{\gamma_i}}$$

for any $0 < t < T$, where σ_i and λ_i depending on p_i and T, but independent of t and k. Furthermore if $p_i = \infty$, then we have $\gamma_i = 0$.

Lemma 2. 4 *Under the assumptions of Theorem A, we have*

$$\| a_i(u_{ik}) \|_{L^2(0,T,H'(\Omega))} \leqslant C_1$$

$$\sup_{[\tau,T]} \| \nabla a_i(u_{ik}) \|_{L^2(\Omega)} \leqslant C_2(\tau)$$

and

$$\| (a_i(u_{ik}))_t \|_{L^2(\tau,T,L^2(\Omega))} \leqslant C_3(\tau)$$

for every $0 < \tau < T$, and $C_1, C_2(\tau), C_3(\tau)$ are constants independent of k.

Due to Lemmas 2. 3 and 2. 4, we can complete the proof of Theorem A by weak convergence method.

Remark 2 As for the existence result, condition (H. 3) can't be weakened. If $\alpha_{ik} \geqslant 1$ for some i and j, there is no necessity of the existence for global solution of problem (1. 1) — (1, 3). Then Theorem A is false in this case. You can see readily in [3] about the case of $m = 1$.

3. Uniqueness

Let $u_k = (u_{1k}, \cdots, u_{mk})$ and $(\bar{u}_1, \cdots, \bar{u}_m)$ be two solutions respecively of problem (2. 1) — (2. 3) and of problem (1. 1) — (1. 3).

For simplicity, we suppose $T \leqslant 1$ first. It is sufficient to prove the result of uniqueness in this case. As for case $T > 1$, we can get the result in cylinder $\Omega \times (1, T]$ by repeating analogous discussion of this section.

We define for $i = 1, \cdots, m$ and $k = 1, 2, \cdots$

$$\psi_{ik}(x,t) = \begin{cases} \dfrac{a_i(u_{ik}) - a_i(\bar{u})_i}{u_{ik} - \bar{u}_i}, \text{if } u_{ik} \neq \bar{u}_i \\ a'(u_{ik}), \text{otherwise} \end{cases}$$

and

$$\tilde{\psi}_{ik}(x,t) = \begin{cases} \dfrac{B_i(x,t,u_{ik}) - B_i(x,t,\bar{u}_{ik})}{u_{ik} - \bar{u}_i}, \text{if } u_{ik} \neq \bar{u}_{ik} \\ -\, b_i(x,t,u_{ik}), \text{ otherwise} \end{cases}$$

where $B_i(s,t,u_{ik}) = \int_0^{u_{ik}} b_i(x,t,s)ds$. By the structure comditions (H. 1) and (H. 2), Lemma 2. 1 and Remark 1, we have

$$\frac{\Lambda_0}{m_i}\left(\frac{1}{2}\right)^{m_i-1} \leqslant \psi_{ik} \leqslant \Lambda_1\left(\frac{\sigma_i}{t^{\gamma_i}}\right)^{m_i-1} \tag{3. 1}$$

Let χ_{ik} be as characteristic function of $(u_{ik} - \bar{u}_i)^+(x,T)$. Now we can find three sequences of smooth functions $\psi_{ik}^{(n)}, \widetilde{\psi_{ik}^{(n)}}$ and $\chi_{ik}^{(n)} (n \geqslant 1)$ such that

$$\psi_{ik}^{(n)} \geqslant \psi_{ik}, (\widetilde{\psi_{ik}^{(n)}})^2 \leqslant \Lambda(1 + \psi_{ik}^{(n)}) \tag{3. 2}$$

and

$$\lim_{n\to\infty} \chi_{ik}^{(n)} = \chi_{ik}, \text{ uniformly in } \Omega \times \{t = T\}.$$

Consider the following adjoint nondegenerate boundary value problem in Q_T :

$$\varphi_{ikt}^{(n)} + \varphi_{ik}^{(n)}\Delta\varphi_{ik}^{(n)} + \tilde{\psi}_{ik}^{(n)}D\varphi_{ik}^{(n)} = \frac{\lambda}{t^{\gamma}}\varphi_{ik}^{(n)} \quad \text{in } Q_T \tag{3.4}$$

$$\varphi_{ik}^{(n)}(x,t) = 0 \quad \text{on } \partial\Omega \times (0,T] \tag{3.5}$$

$$\varphi_{ik}^{(n)}(x,t) = \chi_{ik}^{(n)} \quad \text{in } \Omega \times \{t = T\} \tag{3.6}$$

where positive constants λ and $\gamma \in [0,\frac{1}{2})$ are undetermined. It is well-known to us that problem (3. 4) - (3. 6) have a unique classical solution (Note $\gamma < \frac{1}{2}$, See [8]). Furthermore we have the following lemma.

Lemma 3. 1 *If $\varphi_{ik}^{(n)}$ is a solution of problem (3. 4)-(3. 6), then we have*

$$0 \leqslant \varphi_{ik}^{(n)} \leqslant e^{\lambda(t^{1-\gamma}-T^{1-\gamma})}$$

$$\iint_{Q_T} |\Delta\varphi_{ik}^{(n)}|^2 dxdt \leqslant C_6(k)$$

and

$$\iint_{Q_T} |\Delta\varphi_{ik}^{(n)}|^2 dxdt \leqslant C_7(k)$$

where $C_6(k)$ and $C_7(k)$ are constants independent of λ and n.

Lemma 3. 2 *For any $\varepsilon \in (0,1)$, we have*

$$0 \leqslant \frac{\partial\varphi_{ik}^{(n)}}{\partial\tilde{n}} \leqslant k^{m_i-1+\varepsilon}$$

for k large enough and $i = 1,\cdots,m$, where $\tilde{n}$ is the outer normal of $\partial\Omega$.

Lemma 3. 3 *Under the assumpions of (H. 1), (H. 2) and (H. 4), we can conclude that there exists constant $\tilde{\Lambda} > 0$ such that*

$$\sum_{i=1}^{m}\int_{\Omega}(\underline{u_i} - \overline{u_i})^{\pm}(x,T)dx \leqslant \tilde{\Lambda}\sum_{i=1}^{m}\int_{\Omega}(\underline{u_i^0} - \overline{u_i^0})^{\pm}dx$$

Where $\underline{u_i}$ and $\overline{u_i}$ are a subsolution and a supersloution of problem(1. 1)-(1. 3) respectively.

Consequently the conclusion of Lemma 3. 3 implies Theorem B.

References

[1] D. G. Aronson & L. A. Caffarelli, The initial trace of a solution of the porous medium equation, *Trans. Amer. Math. Soc.*, **280** (1983), 351-366.

[2] Chen Yazhe, Existence and uniquence of weak solutions for uniformly degenerate quasilinear parbolic equations, *Chin. Ann.* **6B** (1985), 131-146.

[3] Jiang Lishang, Existence of global for nonlinear parabolic equation $u_t = \Delta u^m + u^p$, *J. Partial Differential Equations*, Vol. 4, **3** (1991)

[4] L. Maddalena, Existence of global solution for reaction-diffusion systems with density dependent diffusion, *Nonlinear Analysis* **8** (1984), 1383-1394.

[5] A. Ladyzenskaja, V. A. Solonnikov & N. N. Uralceva, Linear and quasilinear equation of parabloic type, *Transl. of Math. Mongraphs*, Vol. 23, Amer. Math. Soc., Provisence, R. I., 1968.

[6] Q. X. Ye, Z. Y. Li, Introduction to reaction-diffusion equations (in Chinese), 1990, Science Press, Beijing.

Math. Dept. of Fuqing Branch, Fujian Normal Univ.,
Fujian, Fuqing, 350300, P. R. China

AMS/IP Studies in Advanced Mathematics
Volume **3**, 1997

ON THE SPHERICALLY SYMMETRIC MOTION OF SELF-GRAVITATING ISENTROPIC GAS SURROUNDING A SOLID BALL

Tetu Makino

1. Problem and Main result.

Let us consider the system of equations

$$\begin{aligned}\frac{\partial\rho}{\partial t}+u\frac{\partial\rho}{\partial r}+\rho\frac{\partial u}{\partial r}+\frac{2}{r}\rho u=0\ ,\\ \rho\left(\frac{\partial u}{\partial t}+u\frac{\partial u}{\partial r}\right)+\frac{\partial p}{\partial r}=-\frac{\rho}{r^2}\left(M+4\pi\int_1^r\rho(t,r)r^2dr\right)\ ,\end{aligned}\tag{1.1}$$

$$p=A\rho^\gamma\ ,\tag{1.2}$$

on $t\geq 0$ and $1\leq r<+\infty$. Here, M, A and γ are positive constants such that $1<\gamma\leq\frac{5}{3}$. This system governs the spherically symmetric isentropic motion of a self-gravitating atmosphere surrounding a solid star with radius 1 and mass M. The variable ρ means the density, u the outward velocity, and p the pressure. The problem is to find a solution of the system (1.1)(1.2) which satisfies the boundary condition

$$\rho u|_{r=1}=0\tag{1.3}$$

and the initial condition

$$\rho|_{t=0}=\rho^0(r)\ ,\quad u|_{t=0}=u^0(r)\ .\tag{1.4}$$

We are interested in the case where the support of $\rho^0(r)$ is compact. We have already solved the problem for the case without the self-gravitation in T.M. and S. Takeno, [8]. This note is an extension of that result. The underlying method is due to Ding Xiaxi, Chen Guiqiang and Luo Peizhu, [3]. Our main result is:

Theorem 1.1. *If ρ^0 and u^0 are bounded, $\rho^0\geq 0$ has a compact support, then there exists a weak solution (ρ,u) of (1.1)(1.2)(1.3)(1.4) on $0\leq t\leq T$ and $1\leq r<+\infty$ such that $0\leq\rho\leq C$ and $|u|\leq C$. Here T is a positive number depending upon the initial data and the parameters.*

2. Stationary solutions.

Our motivation is to establish an existence theorem for a class of initial data which contains stationary solutions of the form $\rho=\rho(r)$, $u=0$. These stationary solutions are given by the equation

$$\frac{dp}{dr}=-\frac{\rho}{r^2}\left(M+4\pi\int_1^r\rho(r)r^2dr\right)\ .\tag{2.1}$$

Given $\rho|_{r=1} > 0$, this equation has a solution $\rho = \rho(r)$.
At least if $M \leq (A\gamma/(r-1)^2)(\rho|_{r=1})^{\gamma-1}$ or if $\gamma \geq 4/3$, then there exists the first zero of the solution, that is, R such that $\rho(r) > 0$ for $0 \leq r < R$ and $\rho(R) = 0$. In fact, introducing the variable θ by $\rho = (A\gamma/4\pi(r-1))^{1/(2-\gamma)}\theta^{1/(\gamma-1)}$ and the parameter $\mu = (4\pi)^{(\gamma-1)/(2-\gamma)}((\gamma-1)/A\gamma)^{1/(2-\gamma)}M$, the equation (2.1) is reduced to the problem

$$\frac{d^2\theta}{dr^2} + \frac{2}{r}\frac{d\theta}{dr} + \theta^{\frac{1}{\gamma-1}} = 0\ , \quad \left.\frac{d\theta}{dr}\right|_{r=1} = -\mu\ . \tag{2.2}$$

Therefore if $\theta|_{r=1} = \alpha > 0$ is given, the first zero $R(\alpha,\mu)$ of the solution $\theta = \theta(r;\alpha,\mu)$ is finite if $\frac{d\theta}{dr}|_{r=1} \leq -\theta$ or of $\gamma \geq 4/3$. See [9] for the proof. However the quantity given beforehand in the physical context is not α but the total mass of the atmosphere $4\pi\int \rho(r)r^2dr$, or equivalently

$$\xi(\alpha,\mu) = \int_1^{R(\alpha,\mu)} \theta(r;\alpha,\mu)^{\frac{1}{\gamma-1}} r^2 dr\ . \tag{2.3}$$

Thus we have the problem: Given μ and β, how many α such that $\beta = \xi(\alpha,\mu)$ do there exist? As for this question, Song-Sun Lin and Wen-Chi Kuan have obtained a series of interesting results ([4]). They claim that in some cases there are plural stationary solutions with the same mass.

3. Approximate solutions

We write the equation (1.1) in the form

$$U_t + f(U)_r = H(U(t,r)) \tag{3.1}$$

where $U = {}^t(\rho, m)$ with $m = \rho u$, $f(U) = {}^t(m, m^2/\rho + p)$,
$H = {}^t(-2m/r, -2m^2/r\rho - (\rho/r^2)(M + \int_1^r \rho(t,r)4\pi r^2 dr))$. We construct approximate solutions by the Lax-Friedrichs scheme as in [8]. The Riemann invariants of the problem

$$U_t + f(U)_x = 0 \tag{3.1$_0$}$$

are $w_1 = u + \frac{2}{\gamma-1}c$, $w_2 = u - \frac{2}{\gamma-1}c$ with $c = \sqrt{dp/d\rho}$. We consider the invariant region $\Sigma(B) = \{(\rho,u)|\ 0 \leq \rho,\ w_1 \leq B,\ -B \leq w_2\}$. The mesh lengths $\Delta r = \Delta$ and Δt are supposed to satisfy $2\Lambda \leq \Delta r/\Delta t \leq 3\Lambda$,
where $\Lambda = \sup\{|u \pm c|\ |\ U \in \Sigma(B^{00})\}$, B^{00} being an arbitrary value greater than $B^0(\geq 1)$ such that $(\rho_0, u_0) \in \Sigma(B^0)$. We choose T such that

$$M\left(\frac{C(1+T)}{M}\right)^{\frac{(\gamma-1)}{2}} T + k\left(\left(\frac{C(1+T)}{M}\right)^{\frac{(\gamma-1)}{2}} B^0\right) < k\left(\left(\frac{C(1+T)}{M}\right)^{\frac{(\gamma-1)}{2}} B^{00}\right),$$

where C is a constant specified later and $k(x) = \int_0^x dx/(1 + x^{2/(\gamma-1)})$. Note that $K(+\infty)$ is finite.

Now the construction of approximate solutions $U^{\Delta}(r,t)$ is similar to that of [8]. Solving the Riemann problem for $(3.1)_0$ with the right and left initial values given by the average of $U^{\Delta}(r,(n-1)\Delta t-0)$, we get $U_0^{\Delta}(r,t)$ for $(n-1)\Delta t \le t < n\Delta t$. Then $U^{\Delta}(r,t)$ in this strip is given by

$$\begin{aligned}\rho^{\Delta} &= \rho_0^{\Delta} + \frac{2}{r}\rho_0^{\Delta}u_0^{\Delta}(t-n\Delta t)\\ u^{\Delta} &= u_0^{\Delta} + \frac{1}{r^2}\left(M + 4\pi\int_1^r \rho_0^{\Delta}(t,r)r^2dr\right)(t-n\Delta t)\,.\end{aligned}$$

Then $B_n = \sup\{w_1^{\Delta}, -w_2^{\Delta} \mid 1 \le r,\ (n-1)\Delta t \le t < n\Delta t\}$ is estimated by

$$B_{n-1} + \left(M + \sup\frac{4\pi}{r^2}\int_1^r \rho_0^{\Delta}(t,r)r^2dr + \frac{\gamma-1}{4}B_{n-1}^2\right)\Delta t\,.$$

On the other hand we see

$$\frac{1}{r^2}\int_1^r \rho_0^{\Delta}(t,r)r^2dr \le \sup\rho_0^{\Delta}\frac{1}{3}R[\rho_0^{\Delta}] \le \left(\frac{\gamma-1}{2\sqrt{A\gamma}}B_{n-1}\right)^{\frac{2}{\gamma-1}} \times \frac{1}{3}R[\rho_0^{\Delta}]\,,$$

where $R[\rho] = \sup\{r \mid \rho(r) \ne 0\}$. Since $\Delta r/\Delta t \le 3\Lambda$, we get $R[\rho_0^{\Delta}] \le R^0 + 3\Lambda T$. Thus we find a constant C such that

$$B_n \le B_{n-1} + \left(M + C(1+T)B_{n-1}^{\frac{2}{\gamma-1}}\right)\Delta t\,.$$

Note $2 < 2/(\gamma-1)$. Comparing B_n with the solution of

$$\frac{d\beta}{dt} = M + C(1+T)\beta^{\frac{2}{\gamma-1}}\,,\quad \beta(0) = B^0\,,$$

we get $B_n \le \beta(n\Delta t)$ as long as $n\Delta t \le T$. Since

$$\beta(t) = \left(\frac{M}{C(1+T)}\right)^{\frac{\gamma-1}{2}} k^{-1}\left(M\left(\frac{C(1+T)}{M}\right)^{\frac{\gamma-1}{2}} t + k\left(\left(\frac{C(1+T)}{M}\right)^{\frac{\gamma-1}{2}} B^0\right)\right)\,,$$

we have $\beta(T) \le B^{00}$.

4. Convergence of approximate solutions

By the compensated compactness method, we can show that there is a sequence $\Delta_j \to 0$ such that U^{Δ_j} tends to a limit U almost everywhere. (See [1] and [2].) In order to show that the limit U is a weak solution, we use Proposition 3 of [8]. The proof of these facts is the same as that of [8], so we omit it.

5. Concluding remarks.

Unfortunately the solution constructed here is temporally local. To establish the global existence remains an open problem. Although there are blowing up solutions for the case without solid ball at the center (see [5]), we conjecture that in this exterior problem we might construct global solutions. In fact when $\gamma = 1$, K. Mizohata et al [6], [7] constructed global weak solutions by using Glimm's method.

References

[1] Chen Guiqiang, *Convergence of the Lax-Friedrichs scheme for an isentropic gas dynamics (III),* Acta Math. Sci., 6(1986), 75-120.

[2] Ding Xiaxi, Chen Guiqiang and Luo Peizhu, *Convergence of the Lax-Friedrichs scheme for an isentropic gas dynamics (I), (II),* Acta Math. Sci., 5(1985), 415-432, 433-472.

[3] Ding Xiaxi, Chen Guiqiang and Luo Peizhu, *Convergence of the fractional step Lax-Friedrichs scheme and Godunov scheme for the isentropic system of gas dynamics,* Commun. Math. Phys., 121(1989), 63-84.

[4] Song-Sun Lin and Wen-Chi Kuan, *On the numbers of equilibria for self-gravitating isentropic gas surrounding a solid ball,* in preparation.

[5] T. Makino, *Blowing up solutions of the Euler-Poisson equation for the evolution of gaseous stars,* Transport Th. and Stat. Phys.,21(1992), 615-624.

[6] T. Makino, K. Mizohata and S. Ukai, *The Global Weak Solutions of the Compressible Euler Equation with Spherical Symmetry,* Japan J. Indust. Appl. Math., 9, (1992), 431-449.

[7] T. Makino, K. Mizohata and S. Ukai, *Global Weak Solutions of the Compressible Euler Equation with Spherical Symmetry (II),* preprint.

[8] T. Makino and S. Takeno, *Initial boundary value problem for the spherically symmetric motion of isentropic gas,* preprint.

[9] T. Makino, *On the existence of positive solutions at infinity for ordinary differential equations of Emden type,* Funkc. Ekv., 27(1984), 319-329.

Osaka Sangyo University
3-1-1, Nakagaito, Daito
Osaka 574, JAPAN

AMS/IP Studies in Advanced Mathematics
Volume **3**, 1997

Dynamics on the Attractor for Reaction-Diffusion Systems in Higher Space Dimensions

Yoshohisa Morita

Abstract

This report shows the existence and stability of spatially non-uniform periodic solution to reaction-diffusion systems in a domain with Neumann boundary condition. Moreover for a specific equation in a thin annulus domain it is given that its global attractor can be decomposed into a finite collection of compact invariant sets (which include periodic solutions) and connecting orbits between the invariant sets.

We are concerned with reaction-diffusion systems in a bounded domain $\Omega \subset R^n$ subject to Neumann boundary condition:

$$\begin{cases} \frac{\partial \mathbf{u}}{\partial t} = \nu D \Delta \mathbf{u} + f(\mathbf{u}) & \text{in} \quad \Omega \\ \frac{\partial \mathbf{u}}{\partial n} = 0 & \text{on} \quad \partial\Omega \end{cases} \tag{1}$$

where ν is positive constant,Δ denotes Laplacian,

$$\mathbf{u} = \begin{pmatrix} u_1 \\ \vdots \\ u_m \end{pmatrix}, \quad D = diag(d_1, \cdots, d_m), \quad d_i > 0,$$

and $\partial/\partial n$ denotes the outer normal derivative on the smooth boundary $\partial\Omega$. We assume that the function $f: R^m \longrightarrow R^m$ is so smooth that there is a unique solution with any initial data in an appropriate function space and that the semi-flow generated by the solutions admits a global attractor, the maximal compact invariant set attracting every solution. Here we consider Equation (1) under the situation that the ordinary differential equation

$$\frac{d\mathbf{u}}{dt} = f(\mathbf{u}) \tag{2}$$

has an asymptotically stable non-degenerate periodic solution,$\mathbf{u} = p(t)$, with period T,where *non-degnerate* means that zero Floquet exponent of the linearized equation around $\mathbf{u} = p(t)$ is simple. Then by virtue of the boundary condition Equation (1) also has the solution $\mathbf{u} = p(t)$ which is spatially uniform. Maginu [4] gave a sufficient condition for the occurrence of destabilization of the spatially uniform periodic solution.

Proposition1 (Maginu[4]) *Let μ_0 be so small that the equation*

$$(I+\mu D)\frac{d\mathbf{u}}{dt} = f(\mathbf{u}), \quad |\mu| < \mu_0 \tag{3}$$

has a periodic solution $\mathbf{p}^\star(t;\mu)$ with period $L(\mu)$ satisfying $\mathbf{p}^\star(t;0) = \mathbf{p}(t)$, $L(0) = T$. If $L'(0) < 0$, then the spatially uniform periodic solution $\mathbf{p}(t)$ to (1) is unstable for sufficiently small ν.

Equation (3) is a perturbed equation to (2), so we can evaluate

$$L'(0) = \frac{2\pi}{T}\int_0^T < D\dot{\mathbf{p}}(t), \mathbf{q}(t) > dt$$

where $< \cdot,\cdot >$ denotes the inner product in $\mathbf{R}^m$ and $\mathbf{q}(t)$ is th periodic solution to the adjoint system

$$\frac{d\mathbf{v}}{dt} = -\left(\frac{\partial f}{\partial \mathbf{u}}(\mathbf{p}(t))\right)^T \mathbf{v}$$

with the normalized condition

$$< \dot{\mathbf{p}}(t), \mathbf{q}(t) >= 1.$$

Maginu showed only the instablity of the spatially uniform periodic solution, while if ν is large enough, it is proved in [7] that there exists a stable spatially non-uniform periodic solution to Equation (1) in a dumbbell-shaped domain under the condition

$$\sigma := \int_0^T < D\dot{\mathbf{p}}(t), \mathbf{q}(t) > dt < 0 \tag{4}$$

Now we consider a specific system

$$\begin{cases} \frac{\partial \mathbf{u}}{\partial t} = \Delta \mathbf{u} + \begin{pmatrix} 1 & -a \\ a & 1 \end{pmatrix} \mathbf{u} - |\mathbf{u}|^2 \begin{pmatrix} 1 & -b \\ b & 1 \end{pmatrix} \mathbf{u} & in \Omega \\ \frac{\partial \mathbf{u}}{\partial n} = 0 & on \partial\Omega \end{cases} \tag{5}$$

where $\mathbf{u} = (u, v)^T, |\mathbf{u}|^2 = u^2 + v^2$ and a, b are real parameters. We easily see that Equation (5) has a spatially uniform periodic solution

$$\mathbf{u} = \begin{pmatrix} cos(a-b)t \\ sin(a-b)t \end{pmatrix} \tag{6}$$

for $a \neq b$. Moreover

$$\sigma = \int_0^T < \dot{\mathbf{p}}(t), \mathbf{q}(t) > dt = 1$$

and the linearized analysis tells that the solution (6) is always stable. Nevertheless Equation (5) possesses stable periodic solutions, which are spatially on-uniform, provided that one takes an appropriate domain.

Theorem 2 *Consider Equation (5) in an annulus $\Omega = \Omega(\varepsilon)$:*

$$\begin{aligned} \Omega(\epsilon) = & \quad \{x = (x_1, x_2) | R < \sqrt{x_1^2 + x_2^2} < R + \epsilon\} \\ = & \quad \{x = (r, \theta) | R < r < R + \epsilon, 0 \leq \theta < 2\pi\} \end{aligned} \tag{7}$$

If $R > (3m^2 - 1/2)^{1/2}$ for some positive integer m, then there are positive ϵ_1, b_1 such that for $0 < \epsilon < \epsilon_1, |b| < b_1$, there exist stable periodic solutions $\mathbf{u}_k^{\epsilon,b}(t, x), k = \pm 1, \cdots, \pm m$,satisfying

$$\mathbf{u}_k^{\epsilon,b} \longrightarrow (1 - k^2/R^2)^{1/2}(cos(at + k\theta), sin(at + k\theta))^T \quad as \quad \epsilon, b \longrightarrow 0.$$

We sketch key idea of the proof. First assume $b = 0$. Applying the transformation

$$\mathbf{v} = \begin{pmatrix} \cos at & \sin at \\ -\sin at & \cos at \end{pmatrix} \mathbf{u}$$

to Equation (5) yields

$$\mathbf{v}_t = \Delta \mathbf{v} + (1 - |\mathbf{v}|^2)\mathbf{v} \quad in\Omega(\epsilon), \quad \frac{\partial \mathbf{v}}{\partial n} = 0 \quad \partial\Omega(\epsilon) \tag{8}$$

The main result in [3] reveals that there are stable non-constant solutions to (8) written as

$$\mathbf{v} = a(r; \epsilon)(cos(at + k\theta), sink\theta)^T \tag{9}$$

where $a(\cdot; \epsilon) \longrightarrow (1 - R^2/k^2)^{1/2} \quad as \quad \epsilon \longrightarrow 0$. By this we obtain asymptotically stable periodic solutions

$$\mathbf{u} = a(r; \epsilon)(cos(at + k\theta), sin(at + k\theta))^T$$

to Equation (5) with $b = 0$. For $|b| << 1$ we can apply the standard perturbation theorem to the existence of the periodic solutions to Equation (5) (see [2]).

Next we discuss some dynamical structure, that is a Morse decomposition of the global attractor for Equation (5) in $\Omega(\epsilon)$. Recall the definition of the Morse decomposition which was given by Conely (see [5]).

Definition 3 *Let X be a compact metric space with a flow denoted $x \cdot t$ where $x \in X$ and $t \in (-\infty, \infty)$. A Morse decomposition of X is a finite ordered collection*

$$M(1) < M(2) < \cdots < M(K)$$

of disjoint compact invariant subsets of X such that if $x \in X$, then there exists $N \geq L$ such that

$$\alpha(x) \subset M(N) \quad \omega(x) \subset M(L)$$

and if $N = L$,then $x \cdot t \in M(N)$ for $t \in (-\infty, \infty)$, where $\alpha(x)$ and $\omega(x)$ denote the alpha- and omega-limit sets of the orbit through x. Combining the results given in [6] and [1], we obtain the following theorem:

Theorem 4 *Let the inner radius R of $\Omega(\epsilon)$ be in (m,m+1) and fix it. If ϵ and $|b|$ are sufficiently small, the global attractor for Equation (5) in $\Omega(\epsilon)$ has a Morse decomposition with Morse sets $\{M(j)\}_{j=0,1,\cdots,m+1}$ such that*

$$M(0) = \{\mathbf{u} = (cos(a-b)t, sin(a-b)t)^T\}, \quad M(m+1) = \{0\}$$

and $\mathbf{u}^{\epsilon,b}_{\pm j} \in M(j)(j = 1, \cdots, m)$.

References

[1] J.K.Hale and G.Raugel, *Reaction-Diffusion on Thin Domains, preprint*

[2] D.Henry, *Geometric Theory of semilinear Parabolic Equations, Springer-Verlag, New York 1981.*

[3] S.Jimbo and Y.Morita, *Stability of Non-constant steady state Solutions to Ginzburg-Landau Equation in Higher Space Dimensions, to appear in Nonlinear Analysis:T.M.A.*

[4] K.Maginu, *Stability of spatially Homogeneous Periodic Solutions of Reaction-Diffusion Equations,J.Differential Equations,* **31** *(1979) 130-138(1979).*

[5] *J.Mallet-Paret, Morse Decompositions for Delay-Differential Equations,Journal of Differential Equations,***72***(1988)270-315.*

[6] *K.Mischaikow and Y.Morita, Dynamical Strycture of Ginzburg-Landau Type system on* S^1*,in preparation.*

[7] *Y.Morita, Reaction-diffusion systems in nonconvex domains: Invariant manifold and reduced form, J.Dynamics and differential Equations* **2***(1990)69-115.*

Yoshihisa Morita

Rynkokn University

Dept. of Applied Mathematics and Informatics

Seta Ohtsu 520-21, Japan

AMS/IP Studies in Advanced Mathematics
Volume 3, 1997

FINITE DIMENSIONALITY OF ATTRACTORS OF MAGNETIC FLOW EQUATION

Qu Chaochun Chen Fengxin*

Abstract

In this paper, the Hausdorff dimension of attractor determined by magnetic flow equation is considered when n=2. We prove that the m-dimensional volume element in the phase space X decays exponentially as t approaches to infinity, and the Hausdorff dimension and the fractal dimension of the attractor are finite.

1. Introduction

The main concern in this paper is the Hausdorff dimension of the attractor associated with the following initial boundary value problem of magnetic flow equation (Abbr. M-F Eq.),

$$\begin{cases} \dfrac{\partial v}{\partial t} + (v \cdot \nabla)v = \nu\Delta v + \dfrac{1}{\rho}\nabla p - 2\Omega \times v + \dfrac{1}{\mu\rho}(\nabla \times b) \times b + f(x) \\ \dfrac{\partial b}{\partial t} = \lambda\Delta b + \nabla \times (v \times b) - \dfrac{1}{\mu}\nabla q + g(x) \\ div v = 0, \quad div b = 0, \\ v|_\Gamma = 0, b|_\Gamma = 0, v(0) = v_0, b(0) = b_0 \end{cases} \tag{1.1}$$

where $v = (v_1, v_2, v_3)$ is the velocity vector, $b = (b_1, b_2, b_3)$ is the magnetic field, p, q, are the pressure, ν, λ, μ, ρ are positive constants, $\Omega = (\Omega_1, \Omega_2, \Omega_3)$ is a constant vector, $f(x), g(x)$ are the volume forces, K is an open bounded domain in R^2 with boundary Γ of class C^2.

Let us denote $L^2(K)^3 = \{v = (v_1, v_2, v_3)|v_i \in L^2(K), i = 1,2,3\}$, with the usual norm, denote H_σ= the closure of $\{\phi = (\phi_1, \phi_2, \phi_3)| \in C_0^\infty(K)^3| \text{ div } \phi = 0. \text{ in } K\}$ in $L^2(K)^3$. H_π= the closure of $\{ \text{grad}\phi | \phi \in C^1(K)\}$ in $L^2(K)^3, V = \{u \in H_0^1(K)^3| \text{ div } v = 0.\}$, $X = H_\sigma \times H_\sigma, W = V \times V$. Let us denote

$$A = \begin{vmatrix} A_1 & 0 \\ 0 & A_2 \end{vmatrix} = \begin{vmatrix} -\nu P\Delta & 0 \\ 0 & -\lambda P\Delta \end{vmatrix}$$

$$M(u) = \begin{vmatrix} P[(\rho\mu)^{-1}(\nabla \times b) \times b - (v \cdot \Delta)v - 2\Omega \times v] \\ P[\nabla \times (v \times b)] \end{vmatrix}, F(x) = \begin{pmatrix} Pf(x) \\ Pg(x) \end{pmatrix}$$

where $u=(v,b)=(v_1,v_2,v_3,b_1,b_2,b_3)$, P is the prejection in $L^2(K)^3\times L^2(K)^3$ on X. Then (1.1) can be abbreviated as follows

$$\frac{\partial u}{\partial t}+Au=M(u)+F(x) \tag{1.2}$$

The existence of solution $u(x,t)\in L^\infty(0,T;X)\cap L^2(0,T;W)$ has already been proved in [3], and so has the existance of the attractor. In this paper, we prove that the dimension of the attractor determined by (1.1) is finite. The main theorem is

Theorem

(1) the m-dimansional volume element in the phase space X is exponentially decaying as $t\to\infty$;

(2) the Hausdorff dimension and the fractal dimension of the attractor are finite.

In the following, we equip X with the scalar product and norm

$$[u,v]=(u_1,v_1)+\frac{1}{\mu\rho}(u_2,v_2),|u|_0^2=|u|^2=(u_1,u_1)+\frac{1}{\mu\rho}(u_2,u_2)$$

and W with

$$\begin{aligned}((u,v)) &= [Au,v]=v(-\Delta u_1,v_1)+(\lambda/\rho\mu)(-\Delta u_2,v_2)\\ &= \nu\sum_{i,j}(\frac{\partial u_{1i}}{\partial x_j},\frac{\partial v_{1i}}{\partial x_j})+\frac{\lambda}{\rho\mu}\sum_{i,j}(\frac{\partial u_{2i}}{\partial x_j},\frac{\partial v_{2i}}{\partial x_j})\end{aligned}$$

$u=(u_1,u_2)\in D(A), v=(v_1,v_2)\in D(A)$. And

$$||u||^2=\nu\sum_{i,j}(\frac{\partial u_{1i}}{\partial x_j},\frac{\partial v_{1i}}{\partial x_j})+\frac{\lambda}{\rho\mu}\sum_{i,j}(\frac{\partial u_{2i}}{\partial x_j},\frac{\partial v_{2i}}{\partial x_j})$$

Clearly, $||u||$ is an equivalent norm of W.

2. Proof of the theorem

In this section, we follow the procedure introduced in [2] to estimate the dimension of the attractor for the dynamical systems determined by M-F equation.

First, we consider the existence of the solution of the first variation equation of (1.1). Suppose $u\in L^\infty(0,T;X)\cap L^2(0,T;W)$ $\forall, T>0$, is the solution of (0.1) [3]. Then the first variation equation of (1.1) is as follows

$$\begin{cases}\dfrac{\partial\phi}{\partial t}+(\phi\cdot\nabla)v+(v\cdot\nabla)\phi=\nu\Delta\phi-\dfrac{1}{\rho}\nabla p-2\Omega\times\phi\\ \qquad +\frac{1}{\mu\rho}(\nabla\times\psi)\times b+\frac{1}{\mu\rho}(\nabla\times b)\times\psi\\ \dfrac{\partial\psi}{\partial t}=\lambda\Delta\psi+\nabla\times(\psi\times b)+\nabla\times(v\times\psi)+\dfrac{1}{\mu}\nabla q\\ div\phi=0,\quad div\psi=0\end{cases} \tag{2.1}$$

We associate (2.1) with Dirichlet boundary condition $\phi|_\Gamma = 0, \psi|_\Gamma = 0$ and initial conditions $U(0) = (\phi(0), \psi(0)) = \xi, \xi \in X$ where we denote $U(x,t) = (\phi(x,t), \psi(x,t))$. Then we have following result.

Proposition 1.1. *If $u \in L^\infty(0,T;X) \cap L^2(0,T;W)$ is the solution of (1.1) $\forall T > 0$. Then the initial boundary value problem (2.1) possesses a unique solution*

$$U(x,t) = (\phi(x,t), \psi(x,t)) \in C(0,T;X) \cap L^2(0,T;W), \forall T > 0$$

and for every $t > 0$, the function $u(0) \to S(t)u(0)$ is Frechet differentiable in X at $u(0)$ with differential $L(t,u(0)) : \xi \in X \to U(t)$, where U is the solution of (2.1).

Let $U_1(t), \cdots, U_m(t)$ be m solutions of (2.1) corresponding to $\xi_1, \xi_2, \cdots, \xi_m$, $Q_m(\tau, u_0, \xi_1, \xi_2, \cdots, \xi_m)$ is the orthogonal prejector in X onto the space spanned by $U_1(t), \cdots, U_m(t)$. And let $\phi_1(t), \cdots, \phi_m(t)$ be an orthonormal basis of $Q_m(t)X$ at a given time, $\phi_j = (\phi_{1j}(t), \phi_{2j}(t))$. Since $U_1(t), \cdots, U_m(t) \in W, j = 1, \cdots, m$, and we have

$$\begin{aligned}
Tr(F'(S(\tau)u_0)Q_m(\tau)) &= \sum_{i=1}^m(-A\phi_i(\tau) + M(u(\tau)\phi_i, \phi_i) \\
= -\sum_{i=1}^m \|\phi_i\|^2 &+ \sum_{i=1}^m[(-(\phi_{1i}\cdot\nabla)v - (v\cdot\nabla)\phi_{1i} - 2\Omega\times\phi_{1i} + (\rho\mu)^{-1}(\nabla\times\phi_{2i})\times b \\
&+(\rho\mu)^{-1}(\nabla\times b)\times\phi_{2i}, \phi_{1i} + (\rho\mu)^{-1}(\nabla\times(\phi_{1i}\times b) + \nabla\times(v\times\phi_{2i}), \phi_{2i})] \\
= -\sum_{i=1}^m \|\phi_i\|^2 &+ \sum_{i=1}^m[(-(\phi_{1i}\cdot\nabla)v, \phi_{1i}) + (\rho\mu)^{-1}((\nabla\times b)\times\phi_{2i}, \phi_{1i}) \\
&+(\rho\mu)^{-1}(\nabla\times(v\times\phi_{2i}), \phi_{2i})]
\end{aligned}$$

since $(2\Omega\times\phi_{1i}, \phi_{1i}) = 0$. By the lemma in [3], we have

$$\sum_{i=1}^m((-(\phi_{1i}\cdot\nabla)v, \phi_{1i}) = \int_K \sum_{i=1}^m \sum_{j,k} \phi_{1i,k}\frac{\partial v_k}{\partial x_j}\phi_{1i,j} \tag{2.2}$$

and

$$\begin{aligned}
\sum_{i=1}^m((\nabla\times b)\times\phi_{2i}, \phi_{1i}) &= \int_K \sum_{i=1}^m\sum_{j,k}\phi_{2i,k}\frac{\partial b_j}{\partial x_k}\phi_{1i,j} - \int_K \sum_{i=1}^m\sum_{j,k}\phi_{2i,k}\frac{\partial b_j}{\partial x_k}\phi_{1i,j} \\
\sum_{i=1}^m(\nabla\times(v\times\phi_{2i}), \phi_{2i}) &= -\int_K \sum_{i=1}^m\sum_{j,k}\phi_{2i,k}\frac{\partial v_k}{\partial x_j}\phi_{2i,j}
\end{aligned} \tag{2.3}$$

So, we can estimate the terms above as follows

$$\begin{aligned}
|\sum_{i=1}^m((\phi_{1i}\cdot\nabla)v, \phi_{1i})| &\le (\int_K \sum_{j,k}(\frac{\partial v_k}{\partial x_j})^2 dx)^{\frac{1}{2}}(\int_K\sum_{j,k}(\sum_{i=1}^m \phi_{1i,k}, \phi_{1i,j})^2 dx)^{\frac{1}{2}} \\
&\le 3\|v\||\rho_1|_{L^2(K)}
\end{aligned} \tag{2.4}$$

where $\rho_1 = \sum_{i=1}^{m}\sum_{j=1}^{3} \phi_{1i,j}^2$. Similarly, we have

$$|\sum_{i=1}^{m}((\nabla \times b) \times \phi_{2i}, \phi_{1i})| \leq 3||b|||\rho_2|_{L^2(K)} \tag{2.5}$$

$$|\sum_{i=1}^{m}(\nabla \times (v \times \phi_{2i}), \phi_{2i})| \leq 3||v|||\rho_2|_{L^2(K)} \tag{2.6}$$

where $\rho_2 = \sum_{i=1}^{m}\sum_{j=1}^{3} \phi_{1i,j}^2 + \sum_{i=1}^{m}\sum_{k=1}^{3} \phi_{2i,k}^2$, $\rho_3 = \sum_{i=1}^{m}\sum_{k=1}^{3} \phi_{2i,k}^2$. From estimates above, we obtain that

$$Tr(F'(S(\tau)u_0)Q_m(\tau)) \leq -\sum_{i=1}^{m} ||\phi_i||^2 + C_0||u|||\rho|_{L^2(K)} \tag{2.7}$$

where $C_0 = 3 + 2/\mu\rho$. On the other hand, by the generalization of the Sobolev-Lieb-Thirring inequality, we have

$$|\rho_2|_{L^2(K)} \leq C_1(\sum_{i=1}^{m} |\phi_i|^2)^{\frac{1}{2}} \tag{2.8}$$

where C_1 depends only on n, K, and is independent of ψ_j and m. Hence , there exists a constant $C_2 = C_0C_1/2$, such that

$$\begin{aligned} Tr(F'(S(\tau)u_0)Q_m(\tau)) &\leq -\sum_{i=1}^{m} ||\phi_i||^2 + C_0||u||C_1(\sum_{i=1}^{m} ||\phi_i(\tau)||^2)^{\frac{1}{2}} \\ &\leq -\tfrac{1}{2}\sum_{i=1}^{m} ||\phi_i||^2 + C_2||u||^2 \end{aligned} \tag{2.9}$$

By the theorem obtained by Metivier [4] for the eigenvalues λ_j of Stokes operator $P\Delta$, as n=2, we have

$$\begin{aligned} \sum_{i=1}^{m} ||\phi_i||^2 &\geq min(\nu, \lambda)(\lambda_1 + \lambda_2 + \cdots + \lambda_m) \geq C\lambda_1[1 + 2 + \cdots + m] \\ &= C\lambda_1 \frac{m(m+1)}{2} \geq \frac{C\lambda_1}{2} m^2 \end{aligned} \tag{2.10}$$

Therefore

$$Tr(F'(S(\tau)u_0)Q_m(\tau)) \leq -\frac{C\lambda_1}{4} m^2 + C_2||u(\tau||^2 \tag{2.11}$$

Hence

$$\frac{1}{t}\int_0^t Tr(F'(S(\tau)u_0Q_m(\tau)) \leq -\frac{C\lambda_1}{4} m^2 + \frac{C_2}{t}\int_0^t ||u(\tau||^2 d\tau \tag{2.12}$$

Let E be the attractor for systems of M-F equation (0.1), which is existent by [3], and let

$$q_m = \lim_{t\to\infty} sup\, q_m(t) = \lim_{t\to\infty} \sup_{u_0\in E} \sup_{\substack{i \in H, |\xi_i| \leq 1 \\ i=1,\cdots,m}} \left(\frac{1}{t}\int_0^t Tr(F'(S(\tau)u_0)Q_m(\tau))d\tau\right) \tag{2.13}$$

Then by (2.12) we have

$$q_m \leq -K_1m^2 + K_2 \tag{2.14}$$

where $K_1 = C\lambda_1/4; K_2 = C_2 \sup_{u_0\in E} \limsup_{t\to\infty} \frac{1}{t}\int_0^t \|u(\tau)\|^2 d\tau$ which is bounded by [3]. So if we choose m such that

$$m = \sqrt{K_2/K_1} + 1 \tag{2.15}$$

then

$$\mu_1 + \mu_2 + \cdots + \mu_m \leq q_m \leq -k_1m^2 + k_2 \tag{2.16}$$

By application of proposition 5.2.1 and Theorem 5.3.3 in [2], we have proved the theorem. More precisely, we have

Theorem 2.1. *We consider the dynamical system associated with the MF equation (1.1). Let m be defined by (2.15). Then*

(1) the uniform Lyapunov exponents μ_j are majorized as in (2.16), $\forall m \in N$.

(2) the m-dimensional volume element in the phase space X is exponentially decaying as $t \to \infty$;

(3) the Hausdorff dimension of E is less than or equal to m, and its fractal dimension is less than or equal to $2m$.

References

[1] R. Hide " On planetary atomospheres and interiors", in mathematical problem in the geophysical science. I.W.H.Reid. ed. Amer. Math. Soc. Providence. R.I.(1971)

[2] R. Temam Infinite dimensional dynamical system in mechanics and physics. Springer-verlag, 1988

[3] Qu ChaoChun et. The solution of equation for the flow and the magnetic fields within the earth, (to appear in J. Mathe. Anal. Appl.)

[4] P.Constantin and C. Foias, Navier-Stokes Equations, The University of Chicago Press (1988).

* Department of Mathematics , Yunnan University, Kunming, 650091

* Yunnan Institute of Applied Mathematics, Kunming, 650091

** Supported By The Fund of The Yunnan Educational Committtee.

AMS/IP Studies in Advanced Mathematics
Volume **3**, 1997

ON A COMPLEX DYNAMICAL AND DISSIPATIVE STRUCTURE OF PARTIAL DIFFERENTIAL EQUATIONS

SZE KUI NG

1. Hamiltonian Systems

Let us first introduce a class of evolution equation which is considered as a unified mathematical model for mechanical systems with dissipation. For Hamiltonian mechanics let us introduce an evolution equation of the following form

$$-\frac{1}{ih+\beta}\frac{\partial\psi}{\partial t} = -\frac{1}{2}\frac{\partial^2\psi}{\partial x^2} + V(x)\psi \tag{1.1}$$

In this equation h and β are positive constants. We notice that when $\beta = 0$ this equation reduces to the Schrodinger equation and when $h = 0$ this equation reduces to a diffusion equation. We shall see that the constant β can describe the dissipation of energy. We emphasize that in this equation we require that both h and β must be positive.

Then, when $\beta << h$ we adopt this equation to describe the atomic world where quantum mechanics with Schrodinger equation is a suitable mathematical model for it. We here suppose that our model for atomic world with equation (1.1) with positive β is more realistic than quantum mechanics with Schrodinger equation. Then when $\beta << h$ and both β and h approaching 0 we adopt this evolution equation to describe the macro world where classical mechanics is a suitable mathematical model for it.

Then, when $h << \beta$ we adopt this evolution equation to extend the probability model based on the diffusion equation with $h = 0$. In this way we have physically observable wave amplitude to replace probability whose physical meaning is difficult to be defined and we then introduce complex dynamics into the probability model with the diffusion equation. The original probabilty model plays the role of background of this complex dynamical model and is important for this complex model to be well defined and to be a complete model including dissipation.

Then, when β and h are comparatively of the same magnitude we adopt this evolution equation to describe macro quantum system such as systems in chemical reaction where both quantum and thermal effects are important.

From the above cases we see that the evolution equation (1.1) is a unified mathematical model for systems described by quantum or classical mechanics.

2. The Second Law of Thermodynamics

Let us derive the second law of thermodynamics from the evolution equation (1.1). Let λ_n be eigenvalues of the Hamiltonian H of (1.1). Then we can write the solution of (1.1) in the form

$$\psi(t) = \sum_{n=1}^{\infty} c_n e^{(-ih-\beta)\lambda_n t}\phi_n \tag{2.1}$$

where for each n, ϕ_n is a normalized eigenfunction of H corresponding to λ_n. Then the total energy of the system at state $\psi(t)$ is given by

$$E_1 = \sum_{n=1}^{\infty} |c_n|^2 e^{-2\beta\lambda_n t}\lambda_n \tag{2.2}$$

Then we define

$$E_2 = \sum_{n=1}^{\infty} |c_n|^2 \lambda_n \tag{2.3}$$

Then we define the energy dissipated from the system at time t by

$$D = E_2 - E_1 \tag{2.4}$$

Then we define an entropy S_2 by

$$D = TS_2 \tag{2.5}$$

where T denotes the absolute temperature of the system and we assume for simplicity that T exists as a constant. Then on the other hand let us define the information entropy S_1 by

$$S_1 = -\sum_{n=1}^{\infty} p_n log p_n \tag{2.6}$$

where

$$p_n = \frac{|c_n|^2 e^{-2\beta\lambda_n t}}{\sum_{m=1}^{\infty} |c_m|^2 e^{-2\beta\lambda_n t}}$$

Then the total entropy of the system is defined by

$$S = S_1 + S_2 \tag{2.7}$$

Now we notice that $D > 0$ and we identify this fact as the second law of thermodynamics. The entropy S_2 is for this law and we notice that S_2 is an increasing function of t since T is a constant and D is an increasing function of t. On the other hand we also have the first law of thermodynamics which states that the free energy F of the system is given by

$$F = E_1 - TS_1 \tag{2.8}$$

We notice that this first law is in terms of the information entropy S_1 which is related to the evolution phenomena of the world. With the distinction of these two entropies S_1 and S_2 we then solve the paradox between the second law of thermodynamics and the evolution phenomena of the world.

3. Systems with Drift and Friction Terms

Let us extend the above evolution equation for Hamiltonian mechanics to include drift and friction. Let us consider an evolution equation of the form

$$-\frac{1}{ih+\beta}\frac{\partial\psi}{\partial t} = -\frac{1}{2}\frac{\partial^2\psi}{\partial x^2} + af\frac{\partial\psi}{\partial x} + bV\psi \tag{3.1}$$

where the term with first order $\frac{\partial}{\partial x}$ is for drift and friction and the constants a and b are in general of complex value and f and V are functions of x. This drift and friction term is analogous to the friction term which is proportional to velocity in the equation for harmonic oscillator in mechanics [1][2]. This evolution equation (3.1) is an extension of the Fokker-Planck equation. To illustrate damping and counter damping effect of the drift and friction term let us consider the following simple

Example.

$$-\frac{1}{ih+\beta}\frac{\partial\psi}{\partial t} = -\frac{1}{2}\frac{\partial^2\psi}{\partial x^2} + ax\frac{\partial\psi}{\partial x} \equiv L\psi \tag{3.2}$$

In this example eigenvalues λ_n and eigenfunctions ϕ_n of the operator L are given in the form

$$\phi_n = p_n(x)e^{-\frac{a}{2}x^2} \qquad \lambda_n = n + a \qquad n = 0, 1, 2, 3, \ldots \tag{3.3}$$

where p_n is some nth order polynomials in x for each n.

Then solution of equation (3.2) is of the form

$$\psi(t) = \sum_{n=0}^{\infty} c_n e^{(-ih-\beta)\lambda_n t}\phi_n \tag{3.4}$$

Then we consider

$$(ih+\beta)\lambda_n = \beta(n+\alpha) - hk + i(hn + h\alpha + k\beta) \tag{3.5}$$

where we write $a = \alpha + ik$ with real numbers α and k. Then when $\alpha > 0$ or $k < 0$ a damping term $\alpha\beta$ or $-hk$ appears in (3.4). Similarly when $\alpha < 0$ or $k > 0$ we have the counter damping term $\alpha\beta$ or $-hk$. We notice that since $\beta > 0$ the counter damping effect of $\alpha\beta$ and $-hk$ will be very small when n is very large because of the term βn. Also, since $h > 0$ we have that oscillation effect remains when n is large because of the term hn which will not be canceled by $h\alpha + k\beta$.

4. A Complex Dynamical and Dissipative Structure

Let us generalize the above examples of systems modeled by equations of mixed type of Schrodinger and diffusion as follows. We call a mathematical structure as a complex dynamical and dissipative structure if this mathematical structure is a complex structure which is obtained from a probability structure involving a time parameter by extending a real constant β of this probability structure to a complex constant $ih+\beta$ where by a probability structure we mean a probability space $(\Omega, \mathcal{F}, P)$ or an evollution equation whose states (or solutions) are probability density functions or probability distributions.

More generally we may suppose that such an evolution equation may be singular in the sense that it is the limit of nonsingular evolution equations which give probability density functions (For example the equation

$$\frac{dp}{ds} = -ap \qquad (a > 0)$$

is considered as limit of nonsingular equations

$$-\frac{dp}{ds} = -b\frac{d^2p}{dx^2} + ap \qquad (b > 0)$$

as $b \to 0$). Then this complex dynamical structure can include complex dynamical systems described by ordinary difference or differential equations.

5. The Burgers' Equation

As an example let us consider a complex dynamical and dissipative structure which is an extension of the Burgers' equation. Let Burgers' equation be given by

$$\frac{\partial p}{\partial t} = \frac{\sigma^2}{2}\frac{\partial^2 p}{\partial x^2} + p\frac{\partial p}{\partial x} \tag{5.1}$$

Then by transforming t to $(ih+\beta)t$ we have the following equation which is an extension of Burgers' equation

$$\frac{1}{ih+\beta}\frac{\partial \psi}{\partial t} = \frac{\sigma^2}{2}\frac{\partial^2 \psi}{\partial x^2} + \psi\frac{\partial \psi}{\partial x} \tag{5.2}$$

We regard (5.2) as a complex dynamical and dissipative structure where the dissipative structure comes from the Burgers' equation and the complex dynamical structure comes from the complex nature of this equation.

It is well known that by Hopf-Cole transformation solution of (5.1) is given by the form

$$p(x,t) = \sigma^2\frac{\partial ln\varphi(x,t)}{\partial x} \tag{5.3}$$

where $\varphi(x,t)$ is a solution of the heat equation

$$\frac{\partial \varphi}{\partial t} = \frac{\sigma^2}{2}\frac{\partial^2 \varphi}{\partial x^2} \tag{5.4}$$

From (5.3) we have that solution of (5.2) is given by the form

$$\psi(x,t) = \sigma^2\frac{\partial Log\psi_0(x,t)}{\partial x} \tag{5.5}$$

where ψ_0 is a solution of the following equation of mixed type of Schrodinger and diffusion:

$$\frac{1}{ih+\beta}\frac{\partial\psi^0}{\partial t} = \frac{\sigma^2}{2}\frac{\partial^2\psi_0}{\partial x^2} \tag{5.6}$$

and $Logz$ is a multivalued function of complex variable z with an analytic branch given by

$$Logz = iArcz + ln|z| \qquad 0 < Arcz < 2\pi \tag{5.7}$$

Thus ψ is of the form

$$\psi(x,t) = i\frac{\partial Arc\psi_0(x,t)}{\partial x} + \frac{\partial ln|\psi_0(x,t)|}{\partial x} \tag{5.8}$$

Let us investigate (5.8). The part

$$i\frac{\partial Arc\psi_0(x,t)}{\partial x} \tag{5.9}$$

of (5.8) comes from the complex nature (and the dissipative structure of Burgers' equation) of (5.2) and it is congruent periodic because of the property of $Arcz$. This gives a pseudo periodic space-time structure for ψ. From this we see that the graph of ψ which is as a function of space-time is rather complicate (It is more complicate than that of p of the Burgers' equation).

Thus while it is not easy to define chaos of partial differntial equations we propose that such a complex dynamical and dissipative phenomena for partial differential equations is analogous to the chaotic phenomena of ordinary differential equations. The space-time structure of (5.8) which is pseudo periodic reminds us to compare it with the space-time structure of the Belousov-Zhabotinskii reaction and to compare it with the dissipative structure of Prigogine [3].

References

1. L.D. Landau and E.M. Lifshitz, "Mechanics," Pergamon Press, New York, 1976.
2. P. Berge, Y. Pomeau and C. Videl, "Order within Chaos," Hermann, Paris, 1984.
3. I. Prigogine and I. Stengers, "Order Out of Chaos," Bantam Books, Inc., 1984.

Department of Mathematics, Hong Kong Baptist College, 224 Waterloo Road, Kowloon, Hong Kong

AMS/IP Studies in Advanced Mathematics
Volume 3, 1997

GLOBAL BEHAVIOR OF POSITIVE SOLUTIONS TO A QUASILINEAR PARABOLIC EQUATION WITH NONLINEAR BOUNDARY ABSORPTION*

Ming-Xin Wang

Abstract This note deals with the global behavior of positive solutions to a quasilinear parabolic equation with nonlinear boundary absorption. The necessary and sufficient conditions on the existence of global positive solutions are given.

1. Introduction

In this paper we deal with the global behavior of positive solutions to the following quasilinear parabolic equation

$$\begin{cases} (u^m)_t = \triangle u & \text{in } \Omega \times (0,T), \\ \dfrac{\partial u}{\partial n} = u^\alpha & \text{on } \partial\Omega \times (0,T), \\ u(x,0) = u_0(x) & \text{in } \Omega, \end{cases} \tag{1}$$

where $\Omega \subset I\!R^N$ is bounded domain with smooth boundary $\partial\Omega$, n is the unit outer normal vector. $m > 0$ and $\alpha \geq 0$. $u_0 \in L^\infty(\Omega)$ and $u_0(x) \geq \delta$ for some small $\delta > 0$. The nonlinear boundary condition can be physically interpreted as the nonlinear radiation law, which here is actually an absorption law. See [3].

Problem (1) has been discussed by many authors. If $m \leq 1$, we have proved in [5] that every solution exists globally if and only if $\alpha \leq m$. For the special case, $\Omega = (-1,1)$, $u_0(x)$ is radially symmetric $C^{2+\gamma}$ function and satisfies $u_{0x}(\pm 1) = \pm u_0^\alpha(\pm 1)$, Filo proved in [2] that for $m > 1$, if $2\alpha < m+1$ then every solution exists globally; if $2\alpha > m+1$ then every solution blows up in finite time. Wu complemented this problem in [7] and proved that every solution exists globally for $2\alpha = m+1$. Wolanski generalized these results to N-dimensional ball ($\Omega = B_R$) and general nonlinearities in [6]. However, we have to point out that the main theorem 2.2 of [6] fails, which concerns global existence of solutions, and the author did a misuse of maximum principle in his proof. Therefore, it is essential to generalize the results of [2] and [7] to general bounded domain $\Omega \subset I\!R^N$, i.e. to prove the following

* Project supported by the National Natural Science Foundation of China

Theorem Suppose $m > 1$, then every solution of (1) exists globally if and only if $2\alpha \le m+1$.

2. Local Solutions and Comparison Principle

Definition Let $T > 0$ and $Q_T = \Omega \times (0,T)$. A function $u(x,t)$, defined on $\overline{\Omega} \times [0,T)$, is called a upper (lower) solution of (1) on Q_T if the followings hold.

(i) for any $0 < \tau < T$, $u \in L^\infty(Q_\tau)$;

(ii) $u(x,0) \ge (\le) u_0(x)$ on $\overline{\Omega}$;

(iii) for every $\varphi \in C(\overline{\Omega} \times [0,T)) \cap C^{2,1}(\overline{\Omega} \times (0,T)) \cap V(Q_T)$ and $t \in (0,T)$,

$$\int_\Omega u^m(x,t)\varphi(x,t)dx \ge (\le) \int_\Omega u^m(x,0)\varphi(x,0)dx + \int_0^t \int_\Omega (u^m\varphi_t + u\Delta\varphi)dxds + \int_0^t \int_{\partial\Omega} (u^\alpha\varphi - u\frac{\partial\varphi}{\partial n})d\sigma ds$$

holds. Where $V(Q_T) = \{\varphi | \varphi \ge 0,\ \varphi_t, |\nabla\varphi|,\ \Delta\varphi \in L^2(Q_T)\}$.

A function $u(x,t)$, defined on $\overline{\Omega} \times [0,T)$, is called a solution of (1) on Q_T if it is both a upper solution and a lower solution.

By the theorems 2.1, 2.2 and 2.3 of [1] we have the following

Proposition 2.1 There exist $T > 0$ and a unique noncontinuable solution $u(x,t)$ of (1) on Q_T satisfying $u(x,t) \ge \delta$. Moreover, if $T < +\infty$ then

$$\lim_{t\to T^-} \sup \|u(\cdot,t)\|_\infty = +\infty,$$

i.e. u blows up in finite time, and T is called the blow up time of u.

Proposition 2.2 Assume that $\overline{u}(x,t)$ and $\underline{u}(x,t)$ is the upper and lower solution of (1) on Q_T respectively, and $\underline{u}(x,0) \ge \delta_1$ for some $\delta_1 > 0$. Then the solution $u(x,t)$ of (1) satisfies

$$\underline{u}(x,t) \le u(x,t) \le \overline{u}(x,t) \quad \text{on} \quad Q_T.$$

Especially, if $\overline{u}(x,t)(\underline{u}(x,t)) \in C(\overline{\Omega} \times [0,T)) \cap C^{1,0}(\overline{\Omega} \times (0,T)) \cap C^{2,1}(\Omega \times (0,T))$ and satisfies

$$\begin{aligned} &(\overline{u}^m)_t \ge \Delta\overline{u} \quad ((\underline{u}^m)_t \le \Delta\underline{u}) \quad \text{in} \quad Q_T, \\ &\frac{\partial\overline{u}}{\partial n} \ge \overline{u}^\alpha \quad (\frac{\partial\underline{u}}{\partial n} \le \underline{u}^\alpha) \quad \text{on} \quad \partial\Omega \times (0,T), \\ &\overline{u}(x,0) \ge u_0(x)(0 < \underline{u}(x,0) \le u_0(x)) \quad \text{in} \quad \Omega. \end{aligned}$$

Then we have $\overline{u} \ge u(\underline{u} \le u)$ on Q_T.

3. Proof of Theorem

We divide the proof into some lemmas. Remember $m > 1$ throughout this section.

Lemma 3.1 If $\alpha \le 1$, then every solution of (1) exists globally.

Proof. Let $M = \max_{\overline{\Omega}} u_0(x)$ and $\varphi(s)$ be the solution of

$$\varphi'(s) = \varphi^\alpha(s), \quad s > 0; \quad \varphi(0) = M.$$

Then $\varphi(s)$ is well defined on $[0, +\infty)$. Let $g(t)$ be the solution of

$$g'(t) = C\varphi^{\alpha-m}(g(t)), \quad t > 0; \quad g(0) = 0.$$

Since $\alpha \le 1 < m$, we have that $g(t)$ is well defined on $[0, +\infty)$. Take

$$\overline{u}(x,t) = \varphi(g(t) + h^*(x)), \tag{3.1}$$

where $h^*(x)$ is the positive solution of

$$\triangle h^* = k = |\partial\Omega|/|\Omega|, \quad x \in \Omega; \quad \frac{\partial h^*}{\partial n} = 1, \quad x \in \partial\Omega.$$

It is obvious that $\overline{u}(x,t)$ exists globally. Using $\alpha \le 1$ and $\varphi' > 0$, it can be proved that, for the suitable large constant C, $\overline{u}(x,t)$ defined by (3.1) is a upper solution of (1). Hence $u(x,t) \le \overline{u}(x,t)$, and in turn $u(x,t)$ exists globally. #

Lemma 3.2 Suppose $\alpha > 1$. If $2\alpha \le m+1$, then every solution of (1) exists globally.

Proof. First, by proposition 2.2 we know that, if $u_0(x) \ge 1$ then $u(x,t) \ge 1$. Therefore $u^\alpha \le u^{(m+1)/2}$. By proposition 2.2 again, we see that it is enough to prove lemma 3.2 for the special case $2\alpha = m+1$.

Choose $M \ge \|u_0(x)\|_\infty + 1$, $K > 0$ to be determined. Let

$$\varphi(s) = [M^{1-\alpha} - (\alpha-1)s]^{1/(1-\alpha)}, \quad g(t) = M^{1-\alpha}(1 - e^{(1-\alpha)Kt})/(\alpha - 1),$$

then we have, by $2\alpha = m+1$,

$$\varphi'(s) = \varphi^\alpha(s), \quad s > 0; \quad \varphi(0) = M, \tag{3.2}$$

$$g'(t) = K\varphi^{\alpha-m}(g(t)), \quad t > 0; \quad g(0) = 0. \tag{3.3}$$

Denote $h(x)$ the eigenfunction corresponding to the first eigenvalue λ_0 of $-\triangle$, i.e.

$$-\triangle h = \lambda_0 h \text{ in } \Omega, \quad h = 0 \text{ on } \partial\Omega.$$

Then $\lambda_0 > 0$, $h(x) > 0$ in Ω, and $\dfrac{\partial h}{\partial n} < 0$ on $\partial\Omega$. Without loss of generality we may consider $h(x) \le 1/2$. Let $C_1 = \max_{\overline{\Omega}} |\nabla h(x)|$, $C_2 = \min_{\partial\Omega}(-\dfrac{\partial h}{\partial n})$, $l = 1/C_2$, and

$$\overline{u}(x,t) = \varphi(g(t) + \epsilon(1-h(x))^{l/\epsilon}) = [M^{1-\alpha}e^{(1-\alpha)Kt} - (\alpha-1)\epsilon(1-h(x))^{l/\epsilon}]^{1/(1-\alpha)},$$

where $\epsilon = \theta M^{1-\alpha}$, $\theta > 0$ satisfies $(\alpha-1)\theta < 1$. Since

$$\begin{aligned} &M^{1-\alpha}e^{(1-\alpha)Kt} - (\alpha-1)\epsilon(1-h(x))^{l/\epsilon} \\ &\ge M^{1-\alpha}(e^{(1-\alpha)Kt} - (1-\alpha)\theta). \end{aligned}$$

We see that if

$$T_0 = -\ln(\theta(\alpha-1))/[2K(\alpha-1)], \tag{3.4}$$

then $\overline{u}(x,t)$ is well defined on $\overline{\Omega}\times[0,T_0]$. For $(x,t)\in\Omega\times(0,T_0]$, denote $\tau = g(t)+\epsilon(1-h(x))^{l/\epsilon}$, using (3.2) we have

$$\overline{u}_t = \varphi'(\tau)g'(t), \tag{3.5}$$

$$\begin{aligned}
\nabla\overline{u} &= -\varphi'(\tau)l(1-h(x))^{(l-\epsilon)/\epsilon}\nabla h,\\
\Delta\overline{u} &= \alpha l^2\varphi^{\alpha-1}(\tau)\varphi'(\tau)|\nabla h|^2(1-h)^{2(l-\epsilon)/\epsilon}\\
&\quad+\frac{1}{\epsilon}l(l-\epsilon)\varphi'(\tau)|\nabla h|^2(1-h)^{(l-2\epsilon)/\epsilon}+\lambda_0 l\varphi'(\tau)h(1-h)^{(l-\epsilon)/\epsilon}\\
&\le (\alpha l^2C_1^2+l^2C_1^2\varphi^{1-\alpha}(\tau)/\epsilon+\lambda_0 l\varphi^{1-\alpha}(\tau))\varphi'(\tau)\varphi^{\alpha-1}(\tau).
\end{aligned} \tag{3.6}$$

Since $\alpha>1$ and $\varphi'>0$, we see that

$$\varphi^{1-\alpha}(\tau)/\epsilon \le \varphi^{1-\alpha}(0)/\epsilon = M^{1-\alpha}/\epsilon = 1/\theta.$$

This inequality and (3.6) yield

$$\Delta\overline{u} \le ((\alpha+1/\theta)l^2C_1^2+\lambda_0 l)\varphi'(\tau)\varphi^{\alpha-1}(\tau). \tag{3.7}$$

Choose $K=[(\alpha+1/\theta)l^2C_1^2+\lambda_0 l]/m$. Because $\alpha-m=1-\alpha<0$, $\tau\ge g(t)$ and $\varphi'>0$, by (3.3), (3.5) and (3.7) we have

$$\begin{aligned}
(\overline{u}^m)_t &= m\overline{u}^{m-1}\overline{u}_t = m\varphi^{m-1}(\tau)\varphi'(\tau)g'(t)\\
&= mK\varphi^{\alpha-1}(\tau)\varphi'(\tau)\varphi^{m-\alpha}(\tau)\varphi^{\alpha-m}(g(t))\\
&\ge mK\varphi^{\alpha-1}(\tau)\varphi'(\tau)\varphi^{m-\alpha}(\tau)\varphi^{\alpha-m}(\tau)\\
&= \Delta\overline{u} \quad \text{in } \Omega\times(0,T_0],\\
\frac{\partial\overline{u}}{\partial n} &= \varphi^{\alpha}(\tau)l\left(-\frac{\partial h}{\partial n}\right) \ge \varphi^{\alpha}(\tau)=\overline{u}^{\alpha} \quad \text{on } \partial\Omega\times(0,T_0],\\
\overline{u}(x,0) &\ge \varphi(0)=M>u_0(x) \qquad \text{on } \overline{\Omega}.
\end{aligned}$$

By proposition 2.2 we have $u(x,t)\le\overline{u}(x,t)$ on $\overline{\Omega}\times[0,T_0]$, therefore $u(x,t)$ is well defined on $\overline{\Omega}\times[0,T_0]$. Since K is independent of M, from (3.4) we see that T_0 is independent of M. Using the *Continuation of Solutions Method* we get that $u(x,t)$ exists globally. Lemma 3.2 is proved. #

Lemma 3.3 Suppose $\alpha>1$. If $2\alpha>m+1$, then every solution of (1) blows up in finite time.

Proof Let $0<\epsilon\le\delta/2$ and $\varphi(s)=[\epsilon^{1-\alpha}-(\alpha-1)s]^{1/(1-\alpha)}$. There exists $s_1>0$ such that $\varphi(s)\le\delta$ for all $0\le s\le s_1$. Let $h_0(x)=x_1+\cdots+x_N$, $A>0$ such that $h_0(x)+A>0$ on $\overline{\Omega}$. Take $0<C_3\le 1/N$ such that $\max_{\overline{\Omega}}(h_0(x)+A)C_3\le s_1/2$. Let $g(t)$ be the solution of

$$g'(t)=\begin{cases} C_4\varphi^{\alpha-m}(g(t)), & \alpha\ge m,\\ C_4\varphi^{\alpha-m}(g(t)+s_1/2), & \alpha<m,\end{cases} \qquad g(0)=0,$$

where $C_4 > 0$ is small. Since $2\alpha > m+1$, we see that there exists $T_0 : 0 < T_0 < +\infty$, such that $g(T_0) = \epsilon^{1-\alpha}/(\alpha-1)$ if $\alpha \geq m$; $g(T_0) = \epsilon^{1-\alpha}/(\alpha-1) - s_1/2$ if $\alpha < m$. Let

$$\underline{u}(x,t) = \varphi(g(t) + \sigma + C_3(h_0(x) + A)),$$

where $\sigma \geq 0$ such that $\sigma + \max_{\overline{\Omega}}(h_0(x)+A)C_3 = s_1/2$. It is obvious that $\underline{u}(x,t)$ blows up in finite time, denote the blow up time by T. Thanks to $\varphi' > 0$ and $0 \leq \sigma + (h_0(x)+A)C_3 \leq s_1/2$, similar to the proof of lemma 3.2 we can verify that, for the suitable small constant $C_4 > 0$, $\underline{u}$ satisfies

$$\begin{cases} (\underline{u}^m)_t \leq \Delta \underline{u} & \text{in } \ \Omega \times (0,T), \\ \dfrac{\partial \underline{u}}{\partial n} \leq \underline{u}^\alpha & \text{on } \ \partial\Omega \times (0,T), \\ \underline{u}(x,0) \leq \varphi(s_1/2) \leq \delta \leq u_0(x) & \text{in } \ \Omega. \end{cases}$$

Therefore $u(x,t) \geq \underline{u}(x,t)$ on $\overline{\Omega} \times (0,T)$, and hence $u(x,t)$ blows up in finite time. the proof is completed. #

Now, our theorem is proved.

Remark 1 For the special case $\Omega = B_R$, the ball of $\mathbb{R}^N$, and $m \leq 1 < \alpha$. If $u_0(x) = u_0(r)$ is radially symmetric positive function, $u_0'(R) = u_0^\alpha(R)$, $u_0'(r) > 0$ for $0 < r \leq R$. Similar to the proof of [4] we can prove that the set of blow up points is $\partial\Omega$.

Remark 2. For the following problem

$$\begin{cases} (u \ln(1+u))_t = \Delta u & \text{in } \ \Omega \times (0,T), \\ \dfrac{\partial u}{\partial n} = u^p(\ln(1+u))^q & \text{on } \ \partial\Omega \times (0,T), \\ u(x,0) = u_0(x) & \text{in } \ \Omega, \end{cases}$$

where $\Omega \subset \mathbb{R}^N$ is bounded smooth domain, $u_0 \in L^\infty(\Omega)$ and $u_0(x) \geq \delta$ for some small $\delta > 0$, $p, q \geq 0$ are parameters. Similar to the proofs of lemmas 3.1 and 3.3 we have the following results.

(i) If $p > 1$, then every solution blows up in finite time;

(ii) If $p < 1$, then every solution exists globally;

(iii) If $p = 1$, then every solution exists globally if and only if $q \leq 1$.

Remark 3 For the general problem

$$\begin{cases} a(u)_t = \Delta u & \text{in } \ \Omega \times (0,T), \\ \dfrac{\partial u}{\partial n} = b(u) & \text{on } \ \partial\Omega \times (0,T), \\ u(x,0) = u_0(x) & \text{in } \ \Omega, \end{cases}$$

where $a(s), b(s) \in C^1(0,+\infty)$, $a(s), a'(s)$ and $b(s)$ are positive functions for $s > 0$. $u_0 \in L^\infty(\Omega)$ and $u_0(x) \geq \delta$ for some $\delta > 0, \Omega \subset \mathbb{R}^N$ is a bounded smooth domian. We can prove the following results.

(i) Assume that there exist positive function $f(s) \in C(0,+\infty)$ and a constant $M > 0$, such that $(b'(s)+1)/a'(s) \geq f(s)$ for $s > 0, f(s)$ is monotone (nonincreasing or nondecreasing) for $s \geq M$, and

$$\int^{+\infty} \frac{ds}{b(s)f(s)} < +\infty,$$

then every solution blows up in finite time.

(ii) Assume that $\int^{+\infty} ds/b(s) = +\infty$. If there exist positive function $f(s) \in C(0,+\infty)$ and a constant $M > 0$, such that $(b'(s)+1)/a'(s) \leq f(s)$ and $f(s)$ is monotone (nonincreasing or nondecreasing) for $s \geq M$, and

$$\int^{+\infty} \frac{ds}{b(s)f(s)} = +\infty,$$

then every solution exists globally.

References

[1] J.R. Anderson, Local existence and uniqueness of solutions of degenerate parabolic equations, Commu. in P. D. E., 16 (1) (1991), 105-143.

[2] J. Filo, Diffusivity versus absorption through the boundary, J. of Diff. Eqns., 99(1992), 281-305.

[3] H.A. Levine & L.E. Payne, Nonexistence theorems for the heat equation with nonlinear boundary conditions and for the porous medium equation backward in time, J. of Diff. Eqns., 16(1974), 319-334.

[4] J. López Gómez, V. Márquez & N. Wolanski, Blow up results and localization of blow up points for the heat equation with a nonlinear boundary condition, J. of Diff. Eqns., 92(1991), 384-401.

[5] M.X. Wang & Y.H. Wu, Global existence and blow-up problems for quasilinear parabolic equations with nonlinear boundary conditions, SIAM J. Math. Anal., 24(6) (1993), 1515-1521.

[6] N. Wolanski, Global behavior of positive solutions to nonlinear diffusion problems with nonlinear absorption through the boundary, SIAM J. Math. Anal., 24 (2) (1993), 317-326.

[7] Y.H. Wu, Remarks on the diffusivity versus absorption through the boundary, submitted to J. Systems Science and Math. Sciences.

Department of Mathematics
Henan University
Kaifeng 475001, P.R. China

AMS/IP Studies in Advanced Mathematics
Volume **3**, 1997

INTERACTION OF THREE PROGRESSING WAVES WITH DIFFERENT SINGULARITIES FOR SEMILINEAR WAVE EQUATIONS

Wang Weike*

Abstract. In this paper, we consider the interaction of triple of conormal waves with different singularities for semilinear wave equations. We will show that if three characteristic hyperplanes carrying different singularities intersect transversally at the origin, then the solution will be conormal with respect to the three hyperplanes, and a new singularity will be produced on the surface of the light cone at later times. We can also prove here that the strength of the new singularity will be dependent only on the weakest one and strongest one in the three hyperplanes.

For a solution to a nonlinear system, when more than a pair of characterristic hypersurface carrying conormal singularities intersect transversally in a lower dimensional manifold, a new singularity can form, even for a second order equation. An example showing that the new singularity can appear is due to J. Rauch and M. Reed (see [7]). The solution to a semilinear wave equation in two space dimensions is conormal in the past with respect to a triple of characteristic hyperplanes which intersect transversally at the origin, a new singularity is present at later times on the surface of the light cone over the origin. This result was established by J. M. Bony [2], M. Beals [1] and R. Melrose [6] individually. The general second order semilinear case was treated in J. Y. Chemin [4]. Recently, D. Y. Fang [5] considered the third order equations for a particular case. But in those paper mentioned above, authors only discuss the interaction of three waves with the same singularities.

In this paper, we shall study the interaction of three different singularities along three characteristic hyperplanes which intersect transversally at the origin. We will show that, in the future, the solution of semilinear wave equa-

* Supported by Foundation of the State Commission of Education and the National Natural Science Foundation of China .

tions has singularities not only on those hyperplanes, but also on the surface of the light cone over the origin at later times. And this singularity (on the light cone over the origin) satisfies the "sum law" for the weakest singularity and the strongest singularity. Our method mainly based on the second microlocalization techniques (see [2]) and the conormal space with respect to different singularities (see [8]).

Let u be a solution belonging to H^s , in an open neighborhood Ω of $\mathbf{R}^3$ containing origin, of

$$\Box u = f(t, x, y, u), \tag{1}$$

where $\Box = \frac{\partial^2}{\partial t^2} - \frac{\partial^2}{\partial x^2} - \frac{\partial^2}{\partial y^2}$, and f a real C^∞ function. Put $\Omega_\pm = \Omega \cap \{\pm t > 0\}$, let Σ_1, Σ_2 and Σ_3 be characteristic hyperplanes for $\Box$, intersecting transversally at 0, and let Γ be the full light cone, $\Gamma^\pm$ the two half light cone in $\Omega_\pm$.

Let S be either a smooth hypersurface or the union of hypersurfaces intersecting only two by two and transversally, the space of conormal distribution $H^{s,k}(S)$ is the set of those $u \in H^s_{loc}(\Omega)$ such that $X^I u \in H^s_{loc}(\Omega)$ for $|I| \leqslant k$, where X^I is the produce of $|I|$ smooth vector fields tangent to Σ. Our main result are

Theorem 1. *If u is a solution of (1) belonging to $H^s_{loc}(\Omega)$ $(s > \frac{3}{2} + 1)$, and assume that $u \in H^{s+l_j, m_j}(\Sigma_j)$ is near to Σ_j in Ω_- and $u \in H^{s+m}_{loc}(\Omega)$ away from $\Sigma_1 \cup \Sigma_2 \cup \Sigma_3$ in Ω_-, with $m = l_j + m_j (j = 1,2,3)$. Then for each $s' < s$, we have*

1) *$u \in H^{s'+m}_{loc}(\Omega)$ outside $\Sigma_1 \cup \Sigma_2 \cup \Sigma_3 \cup \Gamma^+$,*

2) *$u \in H^{s'+l_j, m_j}(\Sigma_j)$ near $\Sigma_j \backslash (\bigcup_{i \neq j} \Sigma_i \cup \Gamma^+)$,*

3) *$u \in H^{s+\tau, m-\tau}(\Gamma^+)$ near $\Gamma^+ \backslash (\Sigma_1 \cup \Sigma_2 \cup \Sigma_3)$, with $\tau = \min(m, [s + \bar{l} + \underline{l} - \frac{3}{2} + 1])$, $\bar{l} = \max(l_1, l_2, l_3)$, $\underline{l} = \min(l_1, l_2, l_3)$ ($[a]$ denote the integer part of a).*

Without loss of generality, we can assume that $l_1 \leqslant l_2 \leqslant l_3$. Let $\mathscr{M}_{\Sigma_1}$ and $\mathscr{M}_{\Sigma_1 \cup \Sigma_2}$ be the sets of smooth vector fields, tangent to Σ_1 and tangent to $\Sigma_1 \cup S_2$ respectively. As in J. M. Bony [2], let $\mathscr{M}$ be a Lie sub-algebra of Op $(\Sigma^{0,1})$ generated (over Op $(\Sigma^{0,0})$) by 1 and a finite number of singular vector fields $Z_1^g, \cdots, Z_m^g$, which are tangent, outside 0, to Σ_1, Σ_2, Σ_3 and Γ.

Definition 2. *We will say that u belongs to $N^{s,s'}_{k,p,q} = N^{s,s'}(\mathscr{M}, k; \mathscr{M}_{\Sigma_1}, p; \mathscr{M}_{\Sigma_1 \cup \Sigma_2}, q)$, if*

$$Z^{\alpha}M^{\beta}X^{\gamma}u \in N^{s,s'}, \quad \text{for } |\alpha| \leqslant k, |\beta| \leqslant p, |\gamma| \leqslant q, \tag{2}$$

where $Z^{\alpha} = Z_{j_1}\cdots Z_{j_{|\alpha|}}$, $M^{\beta} = M_{j_1}\cdots M_{j_{|\beta|}}$, $X^{\gamma} = X_{j_1}\cdots X_{j_{|\gamma|}}$, $Z_{j_i} \in \mathcal{M}$, $M_{j_i} \in \mathcal{M}_{\Sigma_1}$, $X_{j_i} \in \mathcal{M}_{\Sigma_1 \cup \Sigma_2}$, *and with* $N^{s,s'}$ *denoting 2-microlocal Sobolev space, namely*

$$N^{s,s'} = \{u;\ x^{\alpha}D^{\beta}u \in H^{s}_{loc}(\Omega),\ 0 \leqslant |\alpha| = |\beta| \leqslant s', s' > 0\}.$$

Taking a partition of unity $\{\chi_j\}$ homogeneous of degree zero and smooth away from the origin. With a sufficiently fine partition, it can be assumed that each of the functions χ_j has conic support intersecting at most two of the surface $\Sigma_1, \Sigma_2, \Sigma_3$ and Γ, except at the origin. It is easy to see

Lemma 3. *For any* $Z \in \mathcal{M}$, *we have*

$$Z = \sum_j \sum_l A_{j,l}\widetilde{Z}_{jl}), \tag{3}$$

where $\widetilde{Z}_{jl}$ *are the usual vector fields*, $A_{j,l} \in \mathrm{Op}(\Sigma^{0,0})$.

It is the simplest that there are the local coordinates (z_1, z_2, z_3), such that $\Box = \partial_1\partial_2 + \partial_2\partial_3 + \partial_3\partial_1 (\partial_j = \partial_{z_j})$ and $\Sigma_1 = \{z_1 = 0\}$, $\Sigma_2 = \{z_2 = 0\}$. In these coordinates, the natural time variable is $t = z_1 + z_2 + z_3$, and the surface of the light cone over the origin is defined by

$$\Gamma = \{\ 2(z_1z_2 + z_2z_3 + z_3z_1) - (z_1^2 + z_2^2 + z_3^2) = 0\ \}.$$

The generators of $\mathcal{M}_{\Sigma_1}$ and $\mathcal{M}_{\Sigma_1 \cup \Sigma_2}$ are $z_1\partial_1$, ∂_2, ∂_3 and $z_1\partial_1$, $z_2\partial_2$, ∂_3 respectively. As given in M. Beals [1], we denote

$$\left.\begin{aligned} W_0 &= z_1\partial_1 + z_2\partial_2 + z_3\partial_3, & W_1 &= (z_2 - z_3)\partial_1 + z_2\partial_2 - z_3\partial_3, \\ W_2 &= -z_1\partial_1 + (z_3 - z_1)\partial_2 + z_3\partial_3, & W_3 &= z_1\partial_1 - z_2\partial_2 + (z_1 - z_2)\partial_3, \end{aligned}\right\}$$

$$\left.\begin{aligned} V_1 &= (z_1 - (z_2 + z_3))\partial_1, & V_2 &= (z_2 - (z_1 + z_3))\partial_2, \\ V_3 &= (z_3 - (z_1 + z_2))\partial_3. & & \end{aligned}\right\}$$

From Proposition 2.4 and 2.9 of [1] and Lemma 3, we have

Lemma 4. 1) *Let K be an open conic neighborhood of* $\Sigma_1 \cap \Gamma^+$, *the collection* $\{W_0, W_2, W_3\}$ *generates (over* $\mathrm{Op}(\Sigma^{0,0})$) *all vector fields of* $\mathcal{M}$ *on K*.

2) *Let K be an open conic neighborhood of* $\Sigma_1 \cap \Sigma_2$, *the collection* $\{W_0, V_3, W_3\}$ *generates (over* $\mathrm{Op}(\Sigma^{0,0})$) *all vector fields of* $\mathcal{M}$ *on K*.

Analogous statements hold in an open conic neighborhood of $\Sigma_j \cap \Gamma^+$ $(j = 2, 3)$ *or* $\Sigma_i \cap \Sigma_3 (i = 1, 2)$.

Applying Lemma 4 we easily verify

Lemma 5. *If* $Z \in \mathcal{M}$, $M \in \mathcal{M}_{\Sigma_1}$, $X \in \mathcal{M}_{\Sigma_1 \cup \Sigma_2}$, *then*

$$[X, M] = \sum_j A_jM_j,\ [M, Z] = \sum_j B_jM_j,\ [X, Z] = \sum_j C_jX_j,$$

where $A_j \in C^\infty$, B_j, $C_j \in \mathrm{Op}(\Sigma^{0,0})$.

By Lemma 5, we know that the fields are commutable in (2), so Definition 2 is well definite. On the other hand, using Leibniz formula we can prove Lemma 6 below by induction on p and q.

Lemma 6. *Assume* $s > \frac{3}{2}$, $s + s' > \frac{3}{2}$, $k \geqslant 0$, $p \geqslant 0$, $q \geqslant 0$, *then* $N^{s;s'}_{k;p,q}$ *is an algebra. Moreover, if* $u_1, \cdots, u_m$ *belong to* $N^{s;s'}_{k;p,q}$ *and* F *is a* C^∞ *function of its arguments,* $F(z, u_1(z), \cdots, u_m(z))$ *belong to* $N^{s;s'}_{k;p,q}$

Below we shall work in either the original coordinates (t, x, y) or in the coordinates (z_1, z_2, z_3). By direct calculation, we have

Lemma 7. *Let* $M^g_j \in \{z_1\partial_1, \partial_2, \partial_3\}$ *and* $X^g_j \in \{z_1\partial_1, z_2\partial_2, \partial_3\}$, *and note* $M^I = M^g_{i_1} \circ \cdots \circ M^g_{i_p}$, $X^J = X^{g*}_{j_1} \circ \cdots \circ X^g_{j_q}$ *for* $I = (i_1, \cdots, i_p)$, $J = (j_1, \cdots, j_q)$. *Then there exist* $A_{I',J'} \in \mathrm{Op}(S^1_{1,0})$, $B_{I'',J''} \in C^\infty$ *(matrix value), such that*

$$[\Box, M^I X^J] = \sum_{|I'| \leqslant |I|} A_{I',J'} M^{I'} X^{J'} + \sum_{|I''| \leqslant |I|, |J''| \leqslant |J|} B_{I'',J''} M^{I''} X^{J''} \Box \tag{4}$$

Following (1) and (4), we have

$$\Box M^I X^J u = M^I X^J f(t, x, y, u) + \sum A_{I',J'} M^{I'} X^{J'} u + \sum B_{I'',J''} M^{I''} X^{J''} f(t, x, y, u).$$

Let U_{nl} denote the column vector, whose component are $M^I X^J u$ for $0 \leqslant |I| \leqslant n$, $0 \leqslant |J| \leqslant l$. For U_{nl} one has the system as follows

$$\Box U_{nl} + A_{nl} U_{nl} = F_{nl}(t, x, y, U_{nl}). \tag{5}$$

Here A_{nl} is a matric differential operator of order 1, F_{nl} is a column C^∞ function of its arguments. However, the next Theorem is proved by J. M. Bony for the equation (1) in [2]. As usual, the theorem of propagation singularities can be extended to the system which has diagonal principal part.

Theorem 8. *Assume that* U_{nl} *is a solution of* (5) *belonging to* $H^{3/2+\varepsilon}(\Omega)$ *and that* $U_{nl} \in N^{\sigma,-1/2}_k(\Omega, \mathscr{M})$ *in* Ω_-, *with* $\sigma > 2$ *and* $k \in \mathbf{N}$. *Then* $U_{nl} \in N^{\sigma',-1/2}_k(\Omega, \mathscr{M})$ *near* 0, *for each* $\sigma' < \sigma$.

Using Theorem 8 and by induction on p and q, we have

Proposition 9. *Assume that* u *is a solution of* (1) *belonging to* $H^{3/2+1+\varepsilon}(\Omega)$ *and that* $u \in N^{\sigma,-1/2}_{k,p,q}$ *in* Ω_- *with* $\sigma > 3$. *Then* $u \in N^{\sigma',-1/2}_{k,p,q}$ *near* 0, *for each* $\sigma' < \sigma$.

By Theorem 6.1 of [3], we easily prove

Proposition 10. *Let* $u \in N^{\rho,0}_{k,p,q}(\rho > \frac{3}{2})$ *is a solution of* (1). *If* $u \in H^{\rho+p+q+\tau',k-\tau'}(\Gamma^-)$ *near* $\Gamma^- \setminus (\Sigma_1 \cup \Sigma_2 \cup \Sigma_3)$, *then* $u \in H^{\rho+p+q+\tau',k-\tau'}(\Gamma^+)$ *near* $\Gamma^+ \setminus (\Sigma_1 \cup \Sigma_2 \cup \Sigma_3)$ *with* $\tau' = \min(k, [\rho - \frac{3}{2} - 1])$.

Finally, we come back to the proof of Theorem 1.

Proof of Theorem 1. Without loss of generality, we can assume that $\bar{l} = l_3$, $\underline{l} = l_1$. Let $\sigma = s + \frac{1}{2} + l_1$, $p = l_2 - l_1$, $q = l_3 - l_2$, $k = m_3$, it is clear that the assumptions of Theorem 1 implies that $u \in N^{s+1/2+l_1,-1/2}_{k,p,q}$ in Ω_-. By Proposition 9, we have $u \in N^{s'+1/2+l_1,-1/2}_{k,p,q}$ near 0, while a), b) of Theorem 1 are just simple consequences. But for c), we have $u \in H^{s'+l_3,m_3}(\Gamma^+)$ near $\Gamma^+ \setminus (\Sigma_1 \cup \Sigma_2 \cup \Sigma_3)$ only. Moreover, by Proposition 10, take $\rho = s + l_1$, $k = m_3$, then $\rho + p + q + \tau' = s + l_3 + \tau'$ and $\tau' = \tau - l_3$ (τ see Theorem 1), and we see that c) of Theorem 1 is valid. This completes the proof of Theorem 1.

REFERENCES

[1] M. Beals, Vector fields associated with the nonlinear interaction of progressing waves, Indiana Univ. Math., J. 37 (1988), 637—666.

[2] J. M. Bony, Second microlocalization and propagation of singularities for semilinear hyperbolic equations, Taniguchi Symp., Katata (1984), 11—49.

[3] J. M. Bony, Calcul symbolique et propagation des singularités pour les équations aux dérivées partielles non linéaires, Ann. Sci. Ec. Norm. Sup., (4)14, (1981), 209—246.

[4] J. Y. Chemin, Interaction de trois ondes les equations semilineaires strictement hyperboliques d'order 2, These Universite de Paris Centre D'Orsay, (1987).

[5] D. Y. Fang, Interaction of conormal waves for higher order semilinear hyperbolic equations, Doctoral Thesis of Fudan Univ., (1992).

[6] R. Melrose and N. Ritter, Interaction of nonlinear progressing waves, Annals of Math. (2), (1985), 187—213.

[7] J. Rauch and M. Reed, Singularities produced by the nonlinear interaction of three progressing waves; example, Comm. P. D. E., 7(1982), 1117—1133.

[8] W. K. Wang, Interaction of conormal waves with different singularities for semilinear equations, J. P. D. E., (4)4, (1991), 87—96.

Department of Mathematics
Wuhan University
Wuhan, 430072
P. R. China

AMS/IP Studies in Advanced Mathematics
Volume 3, 1997

THE BIDEMENSIONAL STEFAN PROBLEM WITH NONLINEAR CONVECTION

WANG XUEFENG*

1. INTRODUCTION

J.R.Cannon, E.DiBenedetto and G.H.Knightly discussed the bidementional Stefan problem with connections in paper[1],the connection is a linear Stocks equation there.In this paper we consider the Stefan problem with the nonlinear convection,i.e. the convection is a Navier-Stokes equation

$$\frac{\partial \vec{V}}{\partial t} - \nu\Delta\vec{V} + (\vec{V}\cdot\nabla)\vec{V} + \nabla_x p = \vec{f}(u) \quad \text{in} \quad \Omega_1 \tag{1.1}$$

here Ω_1 is a domain of liquid phase in $Q_T = \Omega\times(0,T], \Omega\subset R^2$ is a bounded domain. Let $\Omega_2 = Q_T\backslash\bar{\Omega}_1$ denote a domain of solid phase,$\vec{V} = \vec{V}(x,t)$ denote the velocity of liquid flew,$u = u(x,t)$ denote the quantity determined by the tempurature(see[1]), where $x = (x_1, x_2), t \in [0,T]$.Let$\Omega(t) = \Omega\times\{t\}, t \in [0,T]$.

As in [1], the problem reduce to (1.1)~(1.6)

$$\begin{cases} \frac{\partial u}{\partial t} - \Delta K(u) + \vec{V}\cdot\nabla_x u = 0 & (x,t)\in Q_T \\ -\nabla_x K(u)\cdot\vec{N} = g(x,t,u) & (x,t)\in S_T \\ u(x,0) = u_0(x) & x\in\Omega(0) \\ u(x,t) = 0 & (x,t)\in\Gamma \\ \{[\nabla_x K(u)]^+ - [\nabla_x K(u)]^-\}\cdot\nabla_x\Phi = L\Phi_t & (x,t)\in\Gamma \end{cases} \tag{1.2}$$

$\vec{V}$ satisfies conditions

$$\vec{V} = 0 \quad \text{on } S_1 \tag{1.3}$$

$$\vec{V}(x,0) = \vec{V}_0(x) \quad x\in\Omega_1(0), div\vec{V}_0(x) = 0 \tag{1.4}$$

$$div\vec{V} = 0 \tag{1.5}$$

$$\vec{V}(x,t) = 0 \quad a.e. \quad \text{on } \Omega_2 \tag{1.6}$$

To simplify,if there is no additional description,the notations are like as description in paper[1].

* Partially supported by grants from the National Science Foundation .

2. The Equation of the weak formulation

Let $\phi \in W_2^{1,1}(\Omega_T)$ such that $\phi(x,T)=0$.Then multiplying the first equation of (1.2) by ϕ and performing intergration by parts,we obtain (Note:$\nabla_x u\phi = (\nabla_x u)\phi$)

$$\begin{aligned}\iint_{\Omega_T} \{-\beta(u)\phi_t + \nabla_x K(u)\cdot\nabla_x\phi + \vec{V}\cdot\nabla_x u\phi\}dxdt \\ = -\int_{S_T} g(x,t,u)\phi ds + \int_{\Omega(0)} \beta(u_0)\phi(x,0)dx\end{aligned} \tag{2.1}$$

where $\beta(\cdot)$ is the maximal monotone graph

$$\beta(s) = \begin{cases} s, & \text{for } s>0 \\ [-L,0], & \text{for } s=0 \\ s-L, & \text{for } s<0 \end{cases} \tag{2.2}$$

Since the graph $\beta(\cdot)$ is multivalued,$\beta(u(x,t))$ has to be interpreted as a function $w(x,t) \subset \beta(u(x,t))$,the inclution being intended in the sense of the graphs.In order to simplify the symbolism we will keep the symbol $\beta(u(x,t))$, bearing in mind the way it has to be interpreted.Since $u_0 \neq 0$ except on $\Gamma, \beta(u_0(x))$ is unambigously a.e. defined in $\Omega(0)$.

To obtain a weak formulation of (1.1),(1.3)~(1.6),consider a smooth, divergence free vector value function $\vec{\psi}$ which is compactly supported in $\Omega_1(t) = \Omega_1 \cap \{t = t\}$,for all $t \in [0,T]$ and $\vec{\psi}(x,T) \equiv 0$.Take the "dot"product of (1.1) by $\vec{\psi}$ and integrate by parts in Ω_1. Routine calculations give

$$\begin{aligned}\iint_{\Omega_1} \{-\vec{V}\cdot\vec{\psi}_t + \nu\nabla_x\vec{V} : \nabla_x\vec{\psi} + (\vec{V}\cdot\nabla)\vec{V}\cdot\vec{\psi} - \vec{f}(u)\vec{\psi}\}dxdt \\ = \int_{\Omega_1(0)} \vec{V}_0(x)\vec{\psi}(x,0)dx\end{aligned} \tag{2.3}$$

where $\Omega_1 = \{(x,t) \in Q_T \mid u(x,t) > 0\}$.

Definition. *By a weak solution of (1.1)-(1.6),we mean a pair* $(u,\vec{V})$*such that*

(1) $\vec{V} \in V_2(\Omega_T), u \in V_2(\Omega_T) \bigcap C(\Omega_T)$

(2)$\vec{V} \in J_1^\infty(\Omega_T) = L^\infty(0,T;J(\Omega)) \bigcap L^2(0,t;J_1(\Omega))$

$\vec{V} = 0$ *a.e. on the set* $[u<0] \equiv \{(x,t) \in \Omega_T \mid u(x) < 0)\}$

(3) u and $\vec{V}$ *satisfy*

$$\begin{aligned}\iint_{\Omega_T} \{-\beta(u)\phi_t + \nabla_x K(u)\nabla_x\phi + \vec{V}\cdot\nabla_x u\phi\}dxdt \\ = -\int_{S_T} g(x,t,u)\phi d\sigma + \int_{\Omega(0)} \beta(u_0)\phi(x,0)dx\end{aligned} \tag{2.4}$$

for all $\phi \in W_2^{1,1}(\Omega_T)$ *such that* $\phi(x,T) \equiv 0$.*and*

$$\iint\limits_{[u>0]} \{-\vec{V}\cdot\vec{\psi}_t + \nu\nabla_x\vec{V} : \nabla_x\vec{\psi} + (\vec{V}\cdot\nabla)\vec{V}\cdot\vec{\psi} - \vec{f}(u)\vec{\psi}\}dxdt = \int\limits_{\Omega_1(0)} \vec{V}_0(x)\cdot\vec{\psi}(x,0)dx \tag{2.5}$$

for all $\vec{\psi} \in J_{1,1}(\Omega_T) = J_1(\Omega_T)\bigcap[W_2^{1,1}(\Omega_T)]^2$ *such that* $\nabla_x\vec{\psi} = 0, \vec{\psi}(x,T) \equiv 0$ *and* $supp\vec{\psi}(\cdot,t) \subset [u>0](t)$.

Remarks:
(1)Since we require u to be continuous,the set $[u > 0]$ is open in the relative topology of Q_T,and therefore the last integral identity is well defined.
(2)The integrals in the identity of the tempurature are well defined,under some basic assumptions listed below.

3.Assumptions and Statement of Results

With respect to the data $g(x,t,u), u_0(x), \vec{V}_0(x), \vec{f}(u)$ we assume the following

$[A_1]$ $u_0(x) \neq 0$ a.e. in $\Omega(0), u_0(x) > 0, x \in \Omega_1(0), u_0(x) < 0$ on $\Omega_2(0)$ .Moreover u_0 is essentially bounded and

$$\| u_0 \|_{\infty,\Omega(0)} < K_0.$$

where K_0 is a known positive constant.

$[A_2]$ The function $g(x,t,\xi)$ is continuous over $\bar{S}_T \times R$ and Satisfies the growth condition

$$| g(x,t,\xi) | < K_0 + K_1 | \xi |$$

where K_1 is a known positive constant .Moreover $\xi \rightarrow g(x,t,\xi)$ is monotone at the origin for all $(x,t) \in S_T$ i.e. $g(x,t,\xi)sign\xi \geq 0$.

$[A_3]$ With respect to $K(\cdot)$,we assume that $\xi \rightarrow K(\xi)$ is Lipschitz continuous in $R-\{0\}$ and is continuous at $\xi = 0, K(0) = 0$; moreover it is monotone increasing and satisfies

$$0 < \lambda_0 \leq K'(\xi) \leq \lambda_1 \quad a.e.\xi \in R - \{0\}.$$

where $\lambda_0 = \gamma_0\gamma_1^{-1}$ and $\lambda_1 = \gamma_1\gamma_0^{-1}$.

$[A_4]\vec{V}_0(x) \in J(\Omega)$ and $\vec{V}_0(0) = 0$ a.e. in $\Omega_2(0)$

$[A_5]\vec{f}$ is Lipschitz continuous in R and

$$| \vec{f}(s_1) - \vec{f}(s_2) | \leq K_2 | s_1 - s_2 |$$

for some constant K_2 and all $s_i \in R, i = 1,2$.We will also assume $\vec{f}(0) = 0$. This is no less of generality since $\vec{f}(0) = 0$ can always be realized by adition of a conservative force incorporated in the pressure term.

Theorem. *Under assumptions* $[A_1] \sim [A_5]$,*problem (1.1)* $\sim$ *(1.6) has a sulution Moreover*

(1) $U \in L_\infty(Q_T)\bigcap C(Q_T)$
(2) If $\xi \rightarrow K(\xi) \in C^\infty(R^-)$,*then* $u \in C^\infty([u<0])$

(3) If $\xi \to K(\xi) \in C^\infty(R^+), \xi \to \vec{f}(\xi) \in C^\infty(R)$ then $D_x^\alpha u \in L_{loc}^\infty(Q_T \bigcap [u > 0])$ for every multiindex α

$(a) D_x^\alpha u \in C_{loc}^{\sigma, \frac{\sigma}{2}}(Q_T \cap [u > 0]), \sigma = \sigma(\alpha) \in (0,1)$

$(b) x \to \vec{V}(x,t) \in C^\infty(\Omega \cap [u(\cdot,t) > 0]) a.e. t \in [0,T]$ $D_x^\alpha \vec{V} \in L_{loc}^\infty(Q_T \cap [u > 0])$

for every α

4. Proof of the theorem

The plan is to obtain the solution (u,$\vec{V}$) as a limit of solution sequence $(u_m, \vec{V}_m)$ of certain approximating problems solved in all Q_T. Since $\vec{V}$ must act ultimately only on set $[u > 0]$, we introduce in the approximating process a penalization acting, roughly apeaking,on $[u \leq 0]$.

For any natural namber m,find a pair $(u_m, \vec{V}_m), u_m \in V_2(Q_T), \vec{V}_m \in J_1^\infty(Q_T)$ satisfies

$$\iint_{\Omega_T} \{-\beta_m(u_m)\phi_t + \nabla_x K(u_m)\nabla_x\phi + \vec{V}_m \cdot \nabla_x u_m \phi\} dxdt = -\int_{S_T} g(x,t,u_m)\phi d\sigma + \int_\Omega \beta_m(u_0)\phi(x,0)dx \tag{4.1}$$

for all$\phi \in W_2^{1,1}(\Omega_T)$ such that $\phi(x,T) \equiv 0$ and

$$\iint_{\Omega_T} \{-\vec{V}_m \cdot \vec{\psi}_t + \nu\nabla_x\vec{V}_m : \nabla_x\vec{\psi} + (\vec{V}_m \cdot \nabla)\vec{V}_m \cdot \vec{\psi} + mH_m(u_m)\vec{V}_m \cdot \vec{\psi} - \vec{f}(u_m)\vec{\psi}\} dxdt = \int_\Omega \vec{V}_{0m}(x)\vec{\psi}(x,0)dx \tag{4.2}$$

for all $\vec{\psi} \in J_{1.1}(\Omega_T)$ such that $\vec{\psi}(x,T) \equiv 0$,where $mH_m(\cdot)$ is a penalty term defind as

$$H_m(s) = \begin{cases} 1 & -\infty < s \leq -\frac{2}{m} \\ -ms - 1 & -\frac{2}{m} < s \leq -\frac{1}{m} \\ 0 & -\frac{1}{m} < s < \infty \end{cases} \tag{4.3}$$

$\beta_m(\cdot)$ denotes a smooth sequence of funtions, such that $\beta_m(t) \to \beta(t)$,for all $t \in R\backslash\{0\}$ (it is uniformly convergence on compact subset) and satisfies

(1)β_m is monotone increasing

(2)$\beta'_m(s) \geq 1 \quad \forall s \in R$

(3)

$$\beta_m(s) = \begin{cases} s & \text{for } s \geq \frac{1}{m} \\ s - L & \text{for } s < -\frac{1}{m} \end{cases}$$

Suppose $\{\vec{V}_{0,m}\}$is a sequence in $J_1(\Omega)$ such that

$\vec{V}_{0,m}(x) \to \vec{V}_0(x)$ in $J(\Omega)$ $\vec{V}_0(x)$equals to zero in $\Omega_2(0)$

The crux of the matter lies in the proof of the following facts.

Proposition 4.1. *For all natural number m the system (4.1)-(4.2) admits a solution* $(u_m, \vec{V}_m)$. *Moreover there exists constants* C_0, C_1, C_2 *depending upon the data in assumptions* $[A_1] \sim [A_5]$ *but independent of m such that*

$$(a)\|u_m\|_{V_2(Q_T)}, \|\vec{V}_m\|_{J_1^\infty(Q_T)} \leq C_0$$

$$(b) m \iint\limits_{Q_T} H_m(u_m) \mid \vec{V}_m \mid^2 dx d\tau \leq C_1$$

$$(c)\|u_m\|_{\infty, Q_T} \leq C_2$$

Proposition 4.2. *The set* $\{u_m\}$ *of solutions is equicontinuous over compact subsets of* Q_T. *That is for every compact set* $K \subset Q_T$, *there exists a nondecreasing, continuous function* $\omega_k(\cdot) : R^+ \to R^+, \omega_k(0) = 0$, *dependent on* $dis(K, \partial Q_T)$ *but not upon m such that*

$$(4.4) \qquad \mid u_m(x_1, t_1) - u_m(x_2, t_2) \mid < \omega_k(\mid x_1 - x_2 \mid + \mid t_1 - t_2 \mid^{\frac{1}{2}})$$

for every pair $(x_i, t_i) \in K, i = 1, 2$.

Useing these propositions we can prove the theorem (omit).

Author would like to thank prof.Jiang Lishang for some information.

REFERENCES

1. J.R.Cannon,E.DiBenedetto & G.H.Knightly, *The bidimensional Stefan problem with time dependent case*, Comm.P.D.E. 8 no. 14 (1983), 1549-1604.
2. R.Temam, *Navier-Stokes equations theory and numerical analysis*, North-Holland- Amsterdam. New Yoek.Oxford,revised edition, 1984.
3. O.A.Ladyzhenskaya, *The mathematical problems of dynomics of vescous incompre- ssible flow (in Chinese)*, Shanghai Scientific and Technical Publish House,second edition, 1985.
4. J.R.Cannon & E.DiBenedetto, *On the existence of weak-solutions to an n-dimensional Stefan problem with nonlinear boundary conditions,*, SIAM J.Math.Anal. **11**, no. 4 (1985), 632-645.
5. O.A. Ladyzhenskaja,V.A.Solonniksv and N.N.Vral'ceva, *Linear and quasilinear equations of parabolic type,Amer.Math.Soc.Transl.Math.Mono. 23*, providence, RI, 1968.
6. J.R.Cannon,W.T.Ford and A.Lair, *Quasilinear parabolic systims,J.Differential Equation.*
7. J.R.Cannon and R.Ewing,, *Quasilinear parabolic systems with nonlinear boundary conditions, inverse and improperly posed problems in differential equations (Proc.Conf.Math.Numer.Meth.Halle, DDR,1979) ,35-43*, Math.kes.I. Akademic-Verlag,Bellin, 1979.
8. E.DiBenedetto, *Continuity of weak solutions to certain Singular parabolic equation*, Annali di Mat. pure et Applieate, **130** no. 4 (1982).
9. M.Struwe, *on the Holder Continuity of bounded weak solutions of quasi- linear parabolic systems*, Manuscripta Math. **35** (1981), 125-145.

Department of Mathematics
East China Normal University
Shanghai 200062 China

AMS/IP Studies in Advanced Mathematics
Volume 3, 1997

SOME PROBLEMS ON SEMICONDUCTOR EQUATIONS[①]

Wang Yuanming **Guan Ping**

MODELING EQUATIONS

The fundamental system of partial defferential equations that describes the physical process occuring in semiconductor devices is usually given in the following macroscopic form

$$\begin{aligned} & -\operatorname{div}(\varepsilon \nabla \varphi) = q(p - n + D) \\ & \operatorname{div}\mathbf{J}_n(\mathbf{B}) - q\partial n / \partial t = qR, \quad \text{in } Q_T = \Omega \times (0,T] \\ & \operatorname{div}\mathbf{J}_p(\mathbf{B}) + q\partial p / \partial t = -qR, \\ & \mathbf{J}_n(\mathbf{B}) = q(\mu_n n\mathbf{E} + D_n \nabla n) - \mu_n^{*} (\mathbf{J}_n(\mathbf{B}) \times \mathbf{B}), \\ & \mathbf{J}_p(\mathbf{B}) = q(\mu_p p\mathbf{E} - D_p \nabla p) + \mu_p^{*} (\mathbf{J}_p(\mathbf{B}) \times \mathbf{B}). \end{aligned} \tag{1}$$

where φ, the electrostatic potential; p, the concentration of positively charge holes; and n, the concentration of negatively charge conduction electrons, are the unknowns. $\mu_{n/p}$, the electron / hole mobilities, $D_{n/p}$, the electron / hole diffusion coefficients, and R, the carrier recombination rate are supposed to be known functions of $\nabla\varphi$ and n, p respectively. Most recombination generation phenomena can be explained by the Shockley–Read–Hall recombination model

$$R(n, p) = \frac{n_i^2 - np}{\tau_p(n + n_i) + \tau_n(p + p_i)}$$

here n_i denotes the intrinsic concentration; τ_n, τ_p are the carrier lifetimes as reciprocals of the corresponding capture rates per single carrier.

---

①Partially supported by National Natural Science Foundation

The initial boundary conditions can be written as

$$\begin{aligned}
&Q\varphi = \bar{\varphi} \qquad \text{on} \quad \Sigma = \Sigma_D \cup \Sigma_N \\
&n = \bar{n}, \quad p = \bar{p} \quad \text{on} \quad \Sigma_D \\
&\mathbf{J}_n(\mathbf{B}) \cdot \gamma = \mathbf{J}_p(\mathbf{B}) \cdot \gamma = 0 \quad \text{on} \quad \Sigma_N \qquad (2)\\
&n(0, x) = n_0(x), \quad p(0, x) = p_0(x), \quad x \in \Omega
\end{aligned}$$

where $Q\varphi = \begin{cases} \varphi & \text{on } \Sigma_D \\ \nabla\varphi \cdot \gamma + a\varphi & \text{on } \Sigma_N \end{cases}$, γ is outward unit normal at any point of $\partial\Omega$.

EXISTENCE

System (1) in two–dimentional version can be written in the following form when $\mathbf{B} = \mu\mathbf{K}$

$$\begin{aligned}
&-\operatorname{div}(\varepsilon\nabla\varphi) = q(p - n + D) \\
&-\operatorname{div}(A_1 e^{a\varphi}\nabla(e^{-a\varphi} n)) + \partial n / \partial t = -R(n, p) \qquad (3)\\
&-\operatorname{div}(A_2 e^{-a\varphi})) + \partial p / \partial t = -R(n, p)
\end{aligned}$$

1 STEADY STATE SOLUTIONS

We consider the steady state model of (3) subject to the following mixed boundary conditions

$$\begin{aligned}
&\varphi = \bar{\varphi} \quad n = \bar{n}, \quad p = \bar{p} \quad \text{on} \quad \partial\Omega_D \\
&\frac{\partial\varphi}{\partial\gamma} = \frac{\partial n}{\partial\gamma} = \frac{\partial p}{\partial\gamma} = 0 \quad \text{on} \quad \partial\Omega_N \qquad (4)
\end{aligned}$$

and assume that the functions $\varepsilon(x)$, $D(x)$, A_i etc satisfy hypothesis (H):

(H_1) $\varepsilon(x)$, $D(x)$ and $a_{ij}^{(k)}$ $(k = 1,2;\ i,j = 1,2,\cdots,n)$ are bounded measurable, $D(x) \in L^2\Omega)$, $a_{ij}^{(k)} \in H^1(\Omega)$ and

$$\varepsilon(x) \geqslant \varepsilon_0 > 0, \ \xi A_i \xi' \geqslant a_i |\xi|^2 \quad \forall x \in \Omega, \ \xi \in R^N, \ a_i > 0 \ (i = 1, 2)$$

where $A_k = \left(a_{ij}^{(k)}\right)$ $(k = 1, 2)$

(H_2) $\bar{\varphi}$, $\bar{n}$, $\bar{p}$, $\in C^2(\bar{\Omega})$, an $\bar{n} > 0$, $\bar{p} > 0$

Let $\alpha = \min(\inf \bar{n}e^{-a\bar{\varphi}}, \ \inf \bar{p}e^{a\bar{\varphi}})$ $\quad \beta = \max(\sup \bar{n}e^{-a\bar{\varphi}}, \ \sup \bar{p}e^{-a\bar{\varphi}})$

$\gamma = \min(\inf \bar{\varphi}, \ \gamma')$ $\quad \delta = \max(\sup \bar{\varphi}, \ \delta')$

here γ', δ' are constants satisfying

$$\beta e^{-a\gamma'} - \alpha e^{-a\gamma'} - \inf D \leqslant 0 \quad \text{and} \quad \alpha e^{-a\delta'} - \beta e^{-a\delta'} - \sup D \geqslant 0$$

THEOREM 1.1[2] Suppose that hypothesis (H) holds. Then there exists a weak solution (φ, n, p) in $H^1(\Omega) \cap L^\infty(\Omega)$ to equilibrium model correspending to problem (3), (4) and

$$\gamma \leqslant \varphi \leqslant \delta, \quad \alpha e^{a\gamma} \leqslant n \leqslant \beta e^{a\delta} \quad \text{and} \quad \alpha e^{-a\delta} \leqslant p \leqslant \beta e^{-a\gamma}$$

For the BVP with Dirichlet–oblique derivative boundary conditions or R with non–monotonicity in n, p, we have obtained the similar results.

2 TIME–DEPENDENT SOLUTIONS

To study the global existence and uniqueness of problem (2), (3), we impose the following hypotheses

(H_1) $A_i = A_i(t, x; \nabla\varphi)$ are bounded, nonsymmetric positive definite matrices and satisfy caratheodory condition. For any $\varphi \in L^\infty(0, T; H^2(\Omega))$ we have

(i) $A_i = \varepsilon_i I + B_i(t, x; |\nabla\varphi|) \qquad (i = 1, 2)$

where ε_i are posive constants, I is a unit matrix of order N and B_i satisfy

$$\left|B_i(t, x; |\eta|)\eta\right| \leqslant K,$$

$$\left|B_i(t, x; |\eta_1|)\eta_1 - B_i(t, x; |\eta_2|)\eta_2\right| \leqslant K\left|\eta_1 - \eta_2\right|$$

(H_2) (i) $D(x) \in L^\infty(\Omega)$

(ii) $0 \leqslant n_0(x), \ p_0(x) \leqslant K, \quad \forall x \in \Omega \ \ a.e.,$

$\bar{n}, \ \bar{p} \in H^1(\Omega) \cap L^\infty(\Omega), \ \bar{n}, \ \bar{p} \geqslant C_0 > 0$

THEOREM 1.2[3] Under the assumptions (H_1), (H_2), the weak solution (φ, n, p) to problem (2), (3) exists.

If we add some stronger conditions on A_i and elliptic BVP, we can obtain the uniqueness of the solution.

BEHAVIOR OF SOLUTIONS

We also proved the results of the asymptotic behavior and decay estimate of solutions to the problem (2), (3).

THEOREM 2.1[5] Suppose that (H_1), (H_2) in theorem 1.2 hold,

$$\bar{n}\ \bar{p} = n_i^2, \qquad \nabla(\ln\bar{n} - \bar{\varphi}) = \nabla(\ln\bar{p} + \bar{\varphi}) = 0$$

then

$$\lim_{t\to\infty}\left\{ \|n(t) - n^*\|_2 + \|p(t) - p^*\|_2 + \|\varphi(t) - \varphi^*\|_{H_1} \right\} = 0$$

THEOREM 2.2[6] In addition to the assumptions of theorem 2.2, we also assume that

$$n_0(x), \qquad p_0(x) \geqslant C_1 > 0$$

Then there exist positive constants C and β such that

$$\|n(t) - n^*\|_2 + \|p(t) - p^*\|_2 + \|\varphi(t) - \varphi^*\|_{H^1} \leqslant Ce^{-\beta t}, \qquad \forall t \in R^1$$

NUMERICAL METHODS

The classical model of semiconductor devices is a convection–diffusion model. When the convection term dominates the diffusive term wiggles will appear in the numerical solution if classical finite element, finite difference or finite box methods are used. The phenomenon is caused by the negative diffusion arising from the numerical discretisation method. In order to control wiggles and improve the accuracy of the numerical solution, several methods have been proposed and widely used, our method is based on variational principle. We consider the stationary model of semiconductor devices

$$-\triangle\varphi = \alpha(p - n + D) \tag{5}$$

$$\gamma_n \nabla \cdot [\mu_n(\nabla n - n\nabla\varphi)] = R(n,\ p) \tag{6}$$

$$\gamma_p \nabla \cdot [\mu_p(\nabla p + p\nabla\varphi)] = R(n,\ p) \tag{7}$$

$$\varphi|_{\Gamma_1} = n|_{\Gamma_1} = p|_{\Gamma_1} = 0$$

$$\left.\frac{\partial\varphi}{\partial y}\right|_{\Gamma_2} = \left.\frac{\partial n}{\partial y}\right|_{\Gamma_2} = \left.\frac{\partial p}{\partial y}\right|_{\Gamma_2} = 0$$

The electron current continuity equation (6) can be written as the variational form

$$(\gamma_n \nabla \cdot (\mu_n \nabla n),\ Q_n v) - (\gamma_n \mu_n \nabla\varphi \cdot \nabla n,\ Q_n v) - (\gamma_n \nabla \cdot (\mu_n \nabla\varphi) n,\ Q_n v)$$

$$= (R(n,\ p),\ Q_n v), \quad \forall v \in H_E^1(\Omega), \tag{8}$$

where $H_E^1(\Omega) = \left\{ v \in H^1(\Omega);\ v|_{\Gamma_1} = 0 \right\}$, Q_n is an introduced weight function to eliminate the numerical wiggles. To convert the convenction term to diffusive term, we let Q_n satisfies

$$\int_\Omega \gamma_n \mu_n [(\nabla\varphi \cdot \nabla n) Q_n + \nabla Q_n \cdot \nabla n] v dx = 0, \quad \forall v \in H_E^1(\Omega) \tag{9}$$

The equation (10) becomes

$$\gamma_n(\nabla \cdot (\mu_n \nabla n),\ Q_n v) + \gamma_n(\mu_n \nabla n \cdot \nabla Q_n,\ v) - \gamma_n(\nabla \cdot (\mu_n \nabla\varphi) n,\ Q_n v)$$

$$= (R(n,\ p),\ Q_n v), \ \forall v \in H_E^1(\Omega)$$

i.e. $\gamma_n(Q_n \mu_n \nabla n,\ \nabla v) + \gamma_n(\nabla \cdot (\mu_n \nabla\varphi) Q_n\ n,\ v)$

$$= -(RQ_n,\ v) \quad \forall v \in H_E^1(\Omega) \tag{10}$$

Similarly, the hole current continuity equation can be also expressed as

$$\gamma_p(Q_p \mu_p \nabla p,\ \nabla v) - \gamma_p(\nabla \cdot (\mu_p \nabla\varphi) Q_p\ p,\ v)$$

$$= -(RQ_p,\ v) \quad \forall v \in H_E^1(\Omega) \tag{11}$$

where Q_p is also the weight function which satisfies

$$\int_\Omega \gamma_p \mu_p [(\nabla\varphi \cdot \nabla p) Q_p - \nabla p \cdot \nabla Q_p] v dx = 0, \quad \forall v \in H_E^1(\Omega) \tag{12}$$

Functions Q_n and Q_p satisfied (9) and (12) respectively can be chosen easily, e.g.,

$$Q_n = e^{-\varphi}, \qquad Q_p = e^{\varphi}$$

System (6) (weak form), (10), (11) can be discretized and solved by the iteration method. The convergence has been proved[7].

We have recently given a novel method to control numerical solution ocsillation of the current continuity equations and a mixed finite element approximation to IVBP with magnetic influence (see [9] and [10]).

REFERENCES

[1] Y.M.Wang: Mathematical model and its analysis for the carrier transport in semiconductor devices, Appl. Math. J. of Chinese Univ., Vol.2, No.2, 228–240, 1987.

[2] Y.M.Wang & P.Guan: Periodic and stationary solutions to a semiconductor equation, Chinese Annals of Math. 14A: 2, 262–273, 1993

[3] P.Guan & Y.M.Wang: On existance of weak solution to a kind of semiconductor equation affected by a magnetic field, J. Southeast Univ., Vol.20, No.6, 79–87, 1990.

[4] Y.M.Wang & P.Guan: The existence and uniqueness of solutions to the mixed initial–boundary value problem of a parabolic–elliptic coupled system arising in semiconductor theory, Appl. Math. J. of Chinese Univ., Vol.4, No.3, 411–427, 1989.

[5] P.Guan: On asymptotic behavior of time–dependent solutions to a kind of semiconductor equations, J. Southeast Univ., Vol.19, No.1, 12–18, 1989.

[6] Y.M.Wang, Z.Y.Qiu & P.Guan: The decay estimate of the solutions to the magneto–semiconductor equations (In Chinese), J. of Southeast Univ. Vol.22, No.6, 30–38, 1992.

[7] H.Yie, G.X.Cao & Y.M.Wang: Novel method to control numerical solution oscillation of diffusion–drift equation, Electronics Letters, IEE, Vol.26, No.18, 1487–1489, 1990.

[8] Andor, L. et al: IEEE Trans, ED–32, 1224, 1985.

[9] Y.M.Wang & J.S.Fan: A novel technique to eliminate the drift term for semiconductor device modelling, to appear.

[10] Y.M.Wang, H.W.Wu & J. Zhu: A mixed finite element method in the simulation of semiconductor devices, to appear.

Southeast University, Nanjing, PRC

AMS/IP Studies in Advanced Mathematics
Volume 3, 1997

Existence and Nonexistence of Global Solution of Nonlinear Parabolic Equation with Nonlinear Boundary Condition

WU YONG-HUI WANG MING-XIN

Abstract

This paper deals with the existence and nonexistence of global positive solution of the following equation:

$$\begin{cases} u_t = \nabla(u^{q-1}\nabla u) - \alpha u^m, & x \in \Omega, \quad t > 0, \\ \dfrac{\partial u}{\partial n} = u^p, & x \in \partial\Omega, \quad t > 0, \\ u(x,0) = u_0(x) > 0, & x \in \bar{\Omega}, \end{cases}$$

where p, q, m, α are parameters with $q \geq 1$, $m, p > 0, \alpha \geq 0$. $\Omega \subset R^N$ is a bounded domain with $\partial\Omega$ smooth enough, $N \geq 1$. The sufficient and necessary conditions for the global existence of solution are obtained.

Key Words and Phases: Nonlinear parabolic equiation, nonlinear boundary condition, existence, blow-up, subsolution and supersolution.

§1. Introduction

This paper deals with the following problem:

$$\begin{cases} u_t = \nabla(u^{q-1}\nabla u) - \alpha u^m, & x \in \Omega \times (0,T), \quad (1.1) \\ \dfrac{\partial u}{\partial n} = u^p, & (x,t) \in S_T = \partial\Omega \times (0,T), \quad (1.2) \\ u(x,0) = u_0(x) > 0, & x \in \bar{\Omega}, \quad (1.3) \end{cases}$$

where p, q, m, α are parameters with $p, m > 0$, $q \geq 1$, $\alpha \in R$, n is the outward normal vector. (1.1) becomes the well-known porous medium equation for $\alpha = 0$, and it has a variety applications in physical, biological and engineering problems. Many mathematicians devote to the study of (1.1) with Dirichlet. Neumann or Rabin boundary conditions, and some beautiful results have been obtained, see [1] and the references there. However, just as professor H.A. Levine pointed out (in [2]) that few efforts are made when the nonlinearity occurs in the boundary, and the results are far more complete. To our knowledge, [3] first studies (1.1) $-$(1.3) with $\alpha = 0$, $q = 1$ and $\frac{\partial u}{\partial n} = b(u)$, and obtains that if $b(u) = |u|^{1+\epsilon} h(u)$ with $\epsilon > 0$, $h(u)$ is increasing, then (1.1)$-$(1.3) has no global solution. Later, [5] deals with the generalized nonlinear boundary condition and obtains that if $b(u)$ and $b'(u)$ are continuous, positive and increasing, then the solution of (1.1)$-$(1.3), with $\alpha = 0$, $q = 1$ and $\frac{\partial u}{\partial n} = b(u)$, blows up in finite time provided that $\int^{\infty} \frac{ds}{b(s)b'(s)} < \infty$, and it exists globally provided that $\int^{\infty} \frac{ds}{b(s)b'(s)} = \infty$. [8] considers the problem from another point of view, i.e. the nonlinear diffusion

$$\begin{cases} u_t = \nabla(a(u)\nabla u), & x \in \Omega, \quad t > 0, \\ \frac{\partial u}{\partial n} = 1, & x \in \partial\Omega, \ t > 0, \\ u(x,0) = u_0(x) > 0, & x \in \bar{\Omega}, \end{cases} \tag{1.4}$$

Let $a(u), a'(u)$ be positive, continuous and $\lim_{u \to \infty} \frac{a'(u)}{a(u)} \le M$. [8] proves that if $\int^{\infty} \frac{ds}{a(s)} < \infty$, then the solution of (1.4) blows up in finite time, if $\int^{\infty} \frac{ds}{a(s)} = \infty$, then the solution of (1.4) exists globally. Recently [7] deals the case with both nonlinear diffusion and nonlinear boundary condition,

$$\begin{cases} u_t = \nabla(a(u)\nabla u), & x \in \Omega, \quad t > 0, \\ \frac{\partial u}{\partial n} = b(u), & x \in \partial\Omega, \ t > 0, \\ u(x,0) = u_0(x) > 0, & x \in \bar{\Omega}, \end{cases} \tag{1.5}$$

Under the hypothesis that $a(s)$ and $b(s)$ are positive nondecreasing c' functions for $s > 0$, if $\int^{\infty} \frac{ds}{b(s)} < \infty$, then the solution of (1.5) blows up in finite time; if $\int^{\infty} \frac{ds}{b(s)} = \infty$, and $a(s) + a(s)b'(s) + a'(s)b(s)$ is nondecreasing, then there are two subcases: if $\int^{\infty} \frac{ds}{(a(s) + a(s)b'(s) + a'(s)b(s))b(s)} < \infty$, then the solution blows up in finite time; if $\int^{\infty} \frac{ds}{(a(s) + a(s)b'(s) + a'(s)b(s))b(s)} = \infty$, then the solution of (1.5) exists globally.

In this paper, we put three kinds of nonlinearity together, i.e., nonlinear diffusion, nonlinear reaction and nonlinear boundary condition. In this paper we assume that $\alpha > 0$ (For the case of $\alpha \le 0$, i.e. the reaction term is a source, we can deal with it in the same way as that of [7] and the results are trivial, we omit it). Here the reaction term is absorption. It is very important to realize the relations of the three nonlinear terms interact with each other.

In the case of $p+q \leq 2$ or $p+q > 2$ and $m \leq 1$, we can prove in the same way as that of [7] that the solution of (1.1)–(1.3) exists globally for $p+q \leq 2$, and it (with large initial data) blows up in finite time for $p+q > 2$ and $m \leq 1$. Here we focus our attention on the case of $p+q > 2$ and $m > 1$.

§2. Main Results and Preliminaries

A solution $u(x,t)$ of (1.1)–(1.3) means that

$$u(x,t) \in C^{2,1}(\Omega \times (0,T)) \cap C(\bar{\Omega} \times [0,T)),$$

and $u(x,t)$ satisfies (1.1)–(1.3), where $T \leq \infty$. If $T = \infty$, we say that $u(x,t)$ exists globally, if $T < \infty$ and there exists $t_n \nearrow T^-$, such that

$$\lim_{n\to\infty} \sup_{x\in\bar{\Omega}} |u(x,t_n)| = \infty,$$

we say that $u(x,t)$ blows up in finite time.

It is easy to prove that, if $T < +\infty$, then $u(x,t)$ blows up in finite time.

The main results of this paper are as follows.

Theorem 1. Suppose $p > 1$, $p+q > 2$ and $m > 1$, then the solution of (1.1) - (1.3) exists globally if $2p+q < m+2$, and the solutin blows up in finite time for large initial data if $2p+q \geq m+2$.

Theorem 2. Suppose $0 < p \leq 1$, $p+q > 2$, and $m > 1$. If $p+q < m+1$, then the solution of (1.1)–(1.3) exists globally; if $p+q > m+1$, then the solution of (1.1)–(1.3) blows up in finite time for large initial data. For the case of $p+q = m+1$, the situation is very subtle, the geometry of the domain Ω becomes very important. In this case, if $\frac{|\partial\Omega|}{|\Omega|} < 1$, then the solution of (1.1)–(1.3) exists globally; if $\frac{|\partial\Omega|}{|\Omega|} > 1$, then the solution of (1.1)–(1.3) blows up in finite time.

Our approach to the proof of the theorems is based upon the subsolution and supersolution method. $\bar{u}(x,t)$ is called a supersolution of (1.1)–(1.3) if $\bar{u}(x,t) \in C^{2,1}(\Omega \times (0,T)) \cap C(\bar{\Omega} \times [0,T))$ and satisfies

$$\begin{cases} \bar{u}_t \geq \nabla(\bar{u}^{q-1}\nabla\bar{u}) - \alpha\bar{u}^m, & x \in \Omega,\ t > 0, \quad (2.1) \\ \dfrac{\partial\bar{u}}{\partial n} \geq \bar{u}^p, & x \in \partial\Omega,\ t > 0, \quad (2.2) \\ \bar{u}(x,0) \geq u_0(x), & x \in \bar{\Omega}. \quad (2.3) \end{cases}$$

A subsolution $\underline{u}$ of (1.1)–(1.3) is defined in the same way but with each " $\geq$" replaced by " $\leq$".

For subsolution and supersolution, we have the following comparison theorem.

Lemma Let $\underline{u}$ and $\bar{u}$ be a subsolution and a supersolution of (1.1)–(1.3). Then $\underline{u}(x,t) \leq \bar{u}(x,t)$ for all $x \in \bar{\Omega}$ and $t \in (0, T_{\max})$, where $T_{\max}$ is the maximal time of existence of $\bar{u}$. Moreover, if one of the inequalities (2.1)–(2.3) is strict, then $\underline{u}(x,t) < \bar{u}(x,t)$ for all $x \in \Omega$ and $t \in (0, T_{\max})$.

Under the previous assumptions, problem (1.1)–(1.3) has a unique, local solution $u(x,t)$. Furthermore, if $u_0(x) > 0$ on $\bar{\Omega}$, then $u(x,t) > 0$ for $t > 0$ and $x \in \bar{\Omega}$. Here, we restrict our attention to the positive solution of (1.1)–(1.3). By the standard monotone method we know that if we can find a supersolution $\bar{u}$ with $u_0(x) \leq \bar{u}(x,0)$, and $\bar{u}$ exists globally, then problem (1.1)–(1.3) admits a unique global solution $u(x,t)$ satisfying

$$0 < u(x,t) \leq \bar{u}(x,t).$$

If there exists a subsolution $\underline{u}$ of (1.1)–(1.3) such that $\underline{u}(x,0) \leq u_0(x)$ and $\underline{u}(x,t)$ blows up in finite time, then $u(x,t)$ must blow up in finite time. Therefore, in the proofs of these theorems, our main works are looking for the appropriate super-solutions and sub-solutions. We shall give the detail proof in [9].

References

[1] Aronson, D.G., The porous medium equation, Lecture Notes in Math. 1224, Nonlinear Diffusion Problems, 1-47, Ed. A. Dold & B. Eckmann, Springer-Verlag.

[2] Levine, H.A., The role of critical exponents in blow-up theorems, SIAM REVIEW, Vol. 32, No.2, 262-288 (1990).

[3] Levine, H.A. & L.E. Payne, Nonexistence theorems for the heat equation with a nonlinear boundary conditions and for the porous medium equation backward in time. J.Diff. Eqs., 16, 319-334 (1974).

[4] Sack P., Qualitative behavior for a class of reaction-diffusion-Convection Equations, Nonlinear Diffusion Equations Their Equilibrium State II, Ed. W. M. Ni & L.A. Peletier, 245-252,1986, Springer-Verlag.

[5] Walter, W., On existence and nonexistence in the large of solutions of parabolic equations with a nonlinear boundary conditions, SIAM J. Math. Anal. 6(1), 85-90 (1975).

[6] Walter, W., Differential and Integral Inequalities, Ergebnisse der Mathematik Undiher Grenzgebiete Band 55, Springer-Verlag, 1970.

[7] Wang Ming-xin & Wu Yong-hui, Global solutions and blow-up problems for nonlinear parabolic equations with nonlinear boundary conditions, SIAM J. Math. Anal., 24(6), 1515-1521, 1993.

[8] Ying Hong-min, Blow-up versus global solvability for a class of nonlinear parabolic equations, Preprint.

[9] Wu yong-hui & wang Ming-xin, Existence and nonexistence of global solution of nonlinear parabolic equation with nonlinear boundary condition, preprint.

Institute of Systems Science, Academia Sinica,
BEIJING 100080, P.R. CHINA

Institute of Applied Mathematics, Academia Sinica,
BEIJING 100080, P.R. CHINA
DEPT. OF MATH., HENAN UNIV., KAIFENG 475001, P.R. CHINA

AMS/IP Studies in Advanced Mathematics
Volume **3**, 1997

Spirals for Riemann Problems in Multipieces for 2-D Euler Equations by Using MmB Schemes

Shuli Yang*

Abstract

In this paper, a class of Taylor-FVM MmB schemes, which have high resolution and nonoscillatory properties, are derived on regular triangular meshes for 2-D conservation laws, then spirals (or vortexes) are presented for some Riemann problems in multi-pieces for 2-D Euler equations by using the schemes.

Keywords: MmB schemes, 2-D conservation laws, spiral

1. Descriptions of MmB schemes for 2-D conservation laws

Firstly, we consider the initial value problem for 2-D scalar conservation law,

$$\begin{cases} \dfrac{\partial u}{\partial t} + \dfrac{\partial f}{\partial x} + \dfrac{\partial g}{\partial y} = 0 \\ u(x,y,t)|_{t=0} = u_0(x,y) \end{cases} \tag{1}$$

where $u_0(x,y)$ is a piecewise smooth function.

From [1], let w(p) be the neiborhood of point p in the plane $t = t_0$. By the characteristic relation of (1), consider the values of the solution u in w(p) depend on the values of the interval I(w(p)), then the following inequalities are satisfied

$$\begin{matrix} inf u(Q) & \leq & u(Q') & \leq & sup u(Q) \\ Q \in I(w(p)) & & Q' \in w(p') & & Q \in I(w(p)) \end{matrix} \tag{2}$$

where p' is a point in the plane $t = t_1$. (2) are basic properties of (1).

We locally present the regular triangular meshes as follows

* Partly supported by NSFC

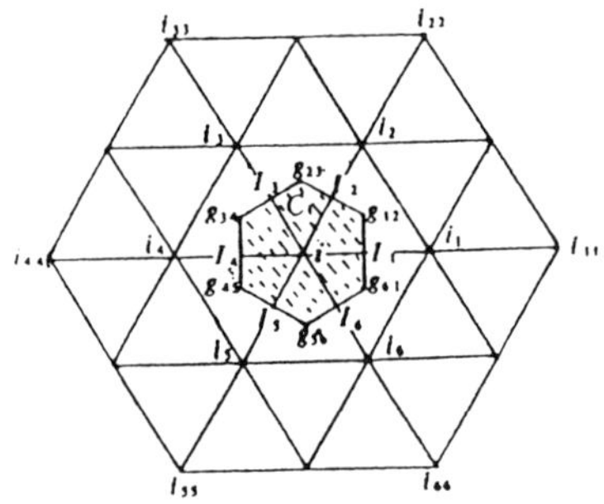

Fig. 1

Definition 1, A scheme

$$u_i^{n+1} = L_h u_i^n$$

is called a MmB scheme if

$$min(u_i, u_{i_1}, ..., u_{i_6}) \leq u_i^{n+1} \leq max(u_i, u_{i_1}, ..., u_{i_6})$$

The definition is equivalent to the form

$$u_i^{n+1} = C_i u_i^n + \sum_{j=1}^{6} C_{i_j} u_{i_j}, \quad C_i \geq 0, \quad C_{i_j} \geq 0$$

and

$$C_i + \sum_{j=1}^{6} C_{i_j} = 1$$

By Taylor expansion, we get the approximate equation which is second order accurate in time direction,

$$\begin{aligned} u^{n+1} &= u^n + \Delta t u_t^n + \tfrac{1}{2}\Delta t^2 u_{tt}^n \\ &= u^n - \Delta t(f_x + g_y) + \tfrac{1}{2}\Delta t^2[(f_u(f_x + g_y))_x + (g_u(f_x + g_y))_y] \end{aligned} \tag{3}$$

Integrate (3) on C_i, we have

$$\int\int_{C_i} u^{n+1} ds = \int\int_{C_i} u^n ds - \Delta t \int_l (f\nu^x + g\nu^y) dl + \frac{1}{2}\Delta t^2 \int_l (f_u\nu^x + g_u\nu^y)(f_x + g_y) dl \tag{4}$$

In (4), take u_i as the integral average u on C_i, that is

$$u_i = \frac{1}{Ar(c_i)} \int\int_{C_i} u ds$$

To discretize $\int_l (f_u\nu^x + g_u\nu^y)(f_x + g_y)dl$, now make transformation of the coordinates,

$$\begin{cases} x' = \nu^x x + \nu^y y \\ y' = \nu^x y - \nu^y x \end{cases}$$

then

$$\begin{cases} f_x = f_{x'}\nu^x - f_{y'}\nu^y \\ g_y = g_{x'}\nu^y + g_{y'}\nu^x \end{cases}$$

and

$$f_x + g_y = f'_{x'} + g'_{y'}$$

where

$$f' = f\nu^x + g\nu^y, \qquad g' = g\nu^x - f\nu^y$$

By constructing MmB schemes on rectangular meshes [2], modified schemes are derived in the following form,

$$\begin{aligned} u_i^{n+1} &= u_i^n - S_i \sum_{j=1}^{3} [a_{i_{j+3}}^+ (u_i - u_{i_{j+3}}) + a_{i_j}^- (u_{i_j} - u_i)] \\ &-\tfrac{1}{2} S_i \sum_{j=1}^{3} [a_{i_j}^+ (1 - a_{i_j}^+ \lambda) Q_{i_j}^+ (u_{i_j} - u_i) - a_{i_{j+3}}^+ (1 - a_{i_{j+3}}^+) Q_{i_{j+3}}^+ (u_i - u_{i_{j+3}}) \\ &-a_{i_j}^- (1 + a_{i_j}^- \lambda) Q_{i_j}^- (u_{i_j} - u_i) + a_{i_{j+3}}^- (1 + a_{i_{j+3}}^- \lambda) Q_{i_{j+3}}^- (u_i - u_{i_{j+3}})] \end{aligned} \tag{5}$$

where $\lambda = \frac{\Delta t}{h}$ and S_i, $Q_{i_j}^{\pm}$ and $Q_{i_{j+3}}^{\pm}$ are chosen as

$$Q_{i_j}^{\pm} = Q(r_{i_j}^{\pm}, \quad Q_{i_{j+3}}^{\pm} = Q(R_{i_{j+3}}^{\pm})$$

$$r_{i_j}^+ = \frac{a_{i_{j+3}}^+ (u_i - u_{i_{j+3}})}{a_{i_j}^+ (u_{i_j} - u_i)} \qquad r_{i_j}^- = \frac{a_{i_{j_j}}^- (u_{i_{j_j}} - u_i)}{a_{i_j}^- (u_{i_j} - u_i)}$$

$$r_{i_{j+3}}^+ = \frac{a_{i_{j+3_{j+3}}}^+ (u_{i_{j+3}} - u_{i_{j+3_{j+3}}})}{a_{i_{j+3}}^+ (u_i - u_{i_{j+3}})} \qquad r_{i_{j+3}}^- = \frac{a_{i_j}^- (u_{i_j} - u_i)}{a_{i_{j+3}}^- (u_i - u_{i_{j+3}})}$$

Theorem 3.1, Let $Q(r) \geq 0$, $Q(r) \equiv 0$ $(r \leq 0)$, if

$$Q_{i_j}^+ / r_{i_j}^+ \leq \frac{2}{1 - \lambda a_{i_j}^+} \qquad Q_{i_{j+3}}^- / r_{i_{j+3}}^- \leq \frac{2}{1 + \lambda a_{i_{j+3}}^-}$$

and

$$Q_{i_{j+3}}^+ \leq \frac{2}{1 - \lambda a_{i_{j+3}}^+} \qquad Q_{i_j}^- \leq \frac{2}{1 + \lambda a_{i_j}^-}$$

under the condition $\max\limits_{i,j} \lambda |a_{i_j}| \leq \frac{1}{12}$, scheme (7) is a MmB scheme.

In general way, we can generalize the MmB schemes to 2-D hyperbolic systems in conservation laws.

2. Numerical solutions of Riemann problems in multipieces for 2-D Euler equations

Consider the system of adiabatic flow

$$\begin{cases} \rho_t + (\rho u)_x + (\rho v)_y = 0 \\ (\rho u)_t + (\rho u^2 + p)_x + (\rho uv)_y = 0 \\ (\rho v)_t + (\rho uv)_x + (\rho v^2 + p)_y = 0 \\ (\rho(e + \frac{u^2 + v^2}{2}))_t + (\rho u(h + \frac{u^2 + v^2}{2}))_x + (\rho v(h + \frac{u^2 + v^2}{2}))_y = 0 \end{cases} \tag{6}$$

and $e = \frac{p}{(\gamma-1)\rho}$, $h = e + \frac{p}{\rho}$ where ρ, (u,v) and p denote density, velocity and pressure, respectively. In [3], Solutions of Riemann problems in four pieces have studied by using characteristic method. Here we give the numerical solutions of Riemann problems in multi-pieces, which only contain contact discontinuities, for (6).

From the following figures, we can find that the Taylor-FVM MmB schemes have high resolution nonoscillatory properties and the spirals are in accordance with the sources of the conjectures in paper [3].

a) Riemann problem in three pieces.

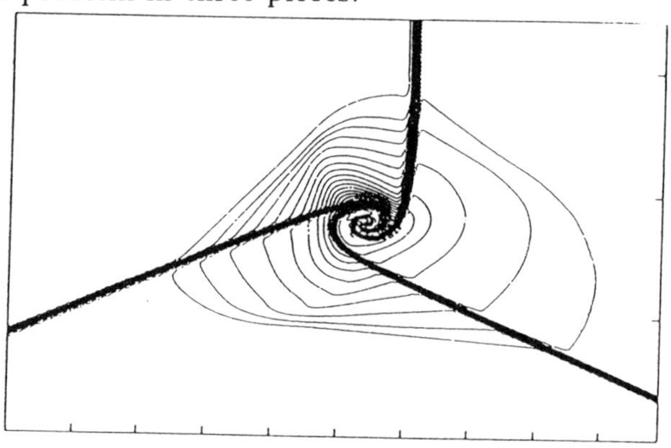

Fig. 2 Density Contour Lines

b) Riemann problem in five pieces.

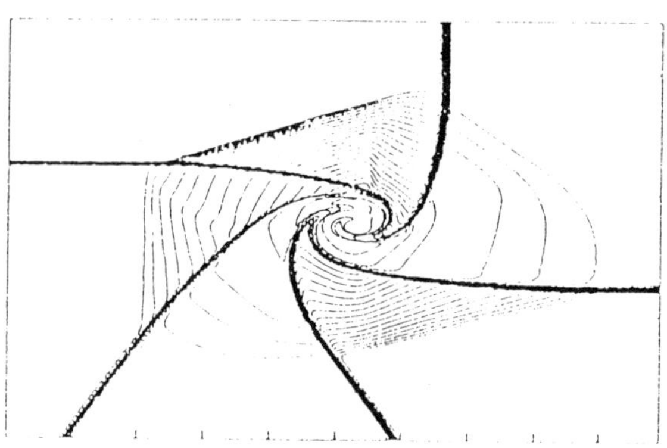

Fig. 3 Density Contour Lines

c) Riemann problem in six pieces

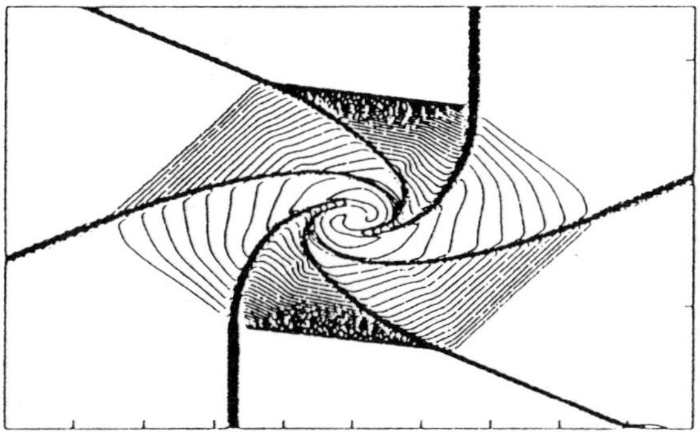

Fig. 4 Density Contour Lines

References

1. H-M. Wu and S-L. Yang, MmB-A new class of accurate high resolution schemes for conservation laws in two dimensions. IMPACT of Computing in Science and Engineering 1, 217-259(1989).
2. S-L. Yang, Second order accurate MmB finite difference and finite element schemes for initial value problems of 1-D and 2-D conservation laws. Ph.D. dissertation. Computing Center, Academia Sinica, Beijing, China, July, 1989.
3. T. Zhang and Y-X. Zheng, Conjecture on the structure of solutions of the gas dynamics systems. SIAM J. Math. Anal., 21, 593-630(1990).

Shuli Yang
Institute of Applied Mathematics
Academia Sinica
Beijing 100080, China

AMS/IP Studies in Advanced Mathematics
Volume **3**, 1997

Global Existence of Solutions for the Reaction-Diffusion Systems Which Don't Satisfy a Zero Sum Condition *

Yang Wanli

1.Introduction: It is well known that the reaction-diffusion systems have found a considerabale amount of interest in recent years. They arise naturally in a variety of models from thoretical physics,chemistry and biology [see, references].

The purpose of this work is to study the global existence for the following reaction-diffusion systems which don't satisfy a zero sum condition

$$\begin{cases} u_t = d_1\triangle u + divg_1(u) + f(u,v) & \\ v_t = d_2\triangle v + divg_2(v) + g(u,v) & t{\geq}0, x{\in}\Omega \\ \alpha u + (1-\alpha)\frac{\partial u}{\partial n} = 0 & \\ v = 0 & t{\geq}0, x{\in}\partial\Omega \\ u(0,x) = u_0(x), v(0,x) = v_0(x). & x{\in}\Omega \end{cases} \tag{1.1}$$

where Ω is a bounded domain in R^n with smooth boundary $\partial\Omega$.

This systems was studied (see [2]) in depth for the case where $\alpha = 1$ and $f = -g \leq 0$. In [3] S.Hollis, R.Martin & M.Pierre studied the case where $g_1{\equiv}g_2{\equiv}0$, and the boundary condition for v is $\alpha_2 v + (1-\alpha_2)\frac{\partial v}{\partial n} = \beta_2$. f and g satisfied the following bdsic hypotheses

$$f((\xi,\eta) + g(\xi,\eta){\leq}\mu_0(r), \eta{\geq}0, 0{\leq}\xi{\leq}r$$

In this present paper,we study the systems (1.1) in which $f(\xi,\eta) + g(\xi,\eta)$ don't be controlled by the function of ξ, using the Hö lder's inequality and duality technique. we obtain the Global existence of the solution for the systems (1.1), and extand the index control for $f(\xi,\eta) + g(\xi,\eta)$ to $\eta^{\frac{n+4}{n}}$.

Our paper is organized as follows: In the second section, we give the energy estimates of the solution;In the third section, using the Hölder's inequality and duality tecnique,we establish the$L_p$estimates;In the end , we obtain the Global

[0]* Supported in part by Doctor subject foundation of Chinese Education Council

existence of the solution, and give the following example for the chemical reaction

$$\begin{cases} u_t = d_1\triangle u + \varepsilon_1 div(u) - uv^2 + Bv & \\ v_t = d_2\triangle v + \varepsilon_2 div(v) + uv^2 + Cv & t\geq 0, x\in\Omega \\ u = v = 0, & t\geq 0, x\in\partial\Omega \\ u(0,x) = u_0(x), v(0,x) = v_0(x). & t = 0, x\in\Omega \end{cases} \tag{1.2}$$

We have the results of the global existence, for any $B\geq 0$.

2.The energy estimates

We consider the following Reaction-Diffusion systems

$$\begin{cases} u_t = d_1\triangle u + div g_1(u) + f(u,v) & \\ v_t = d_2\triangle v + div g_2(v) + g(u,v) & t\geq 0, x\in\Omega \\ \alpha u + (1-\alpha)\frac{\partial u}{\partial n} = 0 & \\ v = 0 & t\geq 0, x\in\partial\Omega \\ u(0,x) = u_0(x), v(0,x) = v_0(x). & t = 0, x\in\Omega \end{cases} \tag{2.1}$$

where, throughout this paper, the following basic hypotheses are assumed to hold:

($\mathbf{H_1}$) d_1, d_2, and α_1 are constants with $d_1, d_2\geq 0$; and $0\leq\alpha\leq 1$.

($\mathbf{H_2}$) $f(\xi,\eta)$ and $g(\xi,\eta)$ are locally Lipchitz functions from$[0,\infty)^2$to R with $f(0,\eta)\geq 0$and $g(\xi,0)\geq 0$ for all $\xi,\eta\geq 0$; also $g_1'(s)$ is a continuou function from R^+ to R, and $g_2'(s)$ is locally Lipchitz .

($\mathbf{H_3}$) Both u_0 and v_0 are measurable on Ω, and there is an $M_0\geq 0$ such that $0\leq u_0(x), v_0(x)\leq M_0$, for all $x\in\Omega$.

Under these assumptions, we have the following standard local existence result

Proposition 2.1 . Suppose that ($\mathbf{H_1}$)–($\mathbf{H_3}$) are satisfied. Then (2.1) has a unique noncontinuable (classical) solution (u,v) on $[0,T^*)\times\Omega$, and there are continuous increasing functions $N_1, N_2 : [0,T^*) \longrightarrow [0,\infty)$ such that

$$0\leq u(t,x)\leq N_1(t), 0\leq v(t,x)\leq N_2(t). \ for \ (t,x)\in[0,T^*)\times\Omega.$$

Moreover, if $T^*\leq\infty$ then

$$\lim_{t\longrightarrow\infty}\sup_{x\in\Omega}\{|u(t,x)| + |v(t,x)|\} = \infty.$$

This local existence result follows from the basic existence theory for abstract semilinear differential equations (see, e.g. [3] or [4]) . In general $T^*\neq\infty$, i.e. ($\mathbf{H_1}$)— ($\mathbf{H_3}$) are not sufficent to insure that solutions to (1,2) exist for all $t\geq 0$

In this paper , we consider the case which the u—component of the solution of (2.1) is "well- behaved" in the sense that it is bounded above by $\|u_0\|_\infty$ a priori (or, so does the v–component solution) . Hence , throughout this paper , we assume

($\mathbf{H_4}$) If $T^* \leq \infty$, then $N_1(t)$ is a bounded function

We introduc the control conditions as following

($\mathbf{H_5}$), $f(\xi,\eta) \leq C_1(r)\eta_2 + C_2(r)$, for all $\xi,\eta \geq 0$ and $0 \leq \xi \leq r$ where $C_1(r)$ and $C_2(r)$ are continuous increasing function

($\mathbf{H_6}$), There is a constand $R \geq \frac{1}{d_2}$ such that

$$R\xi^2 g_2'(\eta)^2 + (\eta+\xi)(g(\xi,\eta)+f(\xi,\eta)) \leq C_3(r)\eta^2 + C_4(r)$$

for all

$\xi,\eta \geq 0$ and $0 \leq \xi \leq r$ where $C_3(r), C_4(r) : R^+ \to R^+$ are continuous increasing functions

Remark 1: If there is a constand $R_1 \leq \frac{d_2}{2}$, such that

$$(\eta+\xi)[g(\xi,\eta)+f(\xi,\eta)] \leq -C_5(r)\eta^{m_1} + C_6(r)\eta^2 + C_7(r)$$

$$\xi^2 g_2'(\eta)^2 \leq R_1 C_5(r)\eta^{m_1} + C_8(r)\eta^2 + C_9(r)$$

for all

$$\xi,\eta \geq 1, and, 0 \leq \xi \leq r$$

Then (H_6) is satisfied .

For example 1: If $g_2(\eta) \equiv \eta$ then the inequality in (H_6) becomes as following

$$(g(\xi,\eta)+f(\xi,\eta)) \leq C_3(r)\eta + C_4(r)$$

hence if

$$g(\xi,\eta)+f(\xi,\eta) \leq C_{12}(r)\eta + C_{13}(r)$$

for all $\xi,\eta \geq 0$ and $0 \leq \xi \leq r$, Then the inequality in (H_6) is valid.

For functions g_2, g and f of the forms : $g_2(\eta) = \varepsilon\eta^m$, $f(\xi,\eta) = -C_0\xi^{\beta_1}\eta^\beta$, $g(\xi,\eta) = C_1\xi^{\beta_1}\eta^\beta$. where $C_o \geq C_1$, are constands,$\beta \geq 2m$. Then ,when ε is sufficiently small, (H_6) is valid.

Remark 2: For functions g_2, g, and f of the following forms

$$g_2(\eta) \equiv \eta, f(\xi,\eta) = -\xi^{\alpha_1}\eta^2 + B\xi^{\alpha_2}\eta,$$

$$g(\xi,\eta) = \xi^{\alpha_1}\eta^2 + C\xi^{\alpha_3}\eta,$$

Assumption ($\mathbf{H_6}$) is valid.In this case , the g_2, g and f don't satisfy the assumptions which is in [2].

Remark 3:For the functions g_2, g and f of the following forms :

$$g_2 = \eta^m where, m \leq \frac{n+2}{n}$$

$$f(\xi,\eta) = -\xi\eta^\beta + C\eta^{\alpha_2}$$

$$g(\xi,\eta) = \xi\eta^\beta + D\eta^{\alpha_1}$$

for all $\beta \geq 0$, $\alpha_i \leq \frac{n+6}{n}$.

Then the (H_6) holds.

We have the following theorem.

Theorem 2.2:Under the assumptions $(\mathbf{H}_1)$–$(\mathbf{H}_6)$. The solutions of (2.1) (u,v) possesses bounded energy in the following sense for $0 \leq T \leq T^*$

$$\max_{0 \leq t \leq T} \|u(t,x)\|^2_{2,\Omega} + \|u_x\|^2_{2,\Omega_T} \leq C(u_0, v_0, N_1(T))$$

and

$$u \in L_{\frac{2(n+2)}{n}}(\Omega_T)$$

$$\max_{0 \leq t \leq T} \|v(t,x)\|^2_{2,\Omega} + \|v_x\|^2_{2,\Omega_T} \leq C(u_0, v_0, N_1(T))$$

and

$$v \in L_{\frac{2(n+2)}{n}}(\Omega_T)$$

Where $C(u_0, v_0, r)$ is continuous increasing function from R^+ to R^+.

Throrem 2.3:Assume in addition to $(\mathbf{H_1}), (\mathbf{H_2}), (\mathbf{H_3}), (\mathbf{H_5}), (\mathbf{H_6})$ that hypothes $(\mathbf{H_4})$ is satisfied.then (u,v) possesces a priori estimates as following

$$\max_{0 \leq t \leq T^*} \int_\Omega v^2 + \int_{Q_T^*} |\bigtriangledown v|^2 \leq C(u_0, v_0, N_1(T^*))$$

and

$$\max_{0 \leq t \leq T^*} \int_\Omega u^2 + \int_{Q_T^*} |\bigtriangledown u|^2 \leq C(u_0, v_0, N_1(T^*))$$

Remark 4:For functions $g_2(\eta), f(\xi,\eta)$ and $g(\xi,\eta)$ of the forms:

$$g_2(\eta) = \varepsilon_2\eta^m, f = -v^p, g = u^p$$

If $p \geq 2m - 1$, with *varepsilon*$_2$ is sufficiently small . Then $(\mathbf{H_5}), (\mathbf{H_6})$ are satisfied.

3. The L_p -estimates

In order to obtain the L_p- estimates for v with $p \leq \infty$. we introduce the following control condition:

$(\mathbf{H_7})$ $g_2(\eta) \leq C\eta^{\frac{n+2+p_1}{n+2}}$; $f(\xi,\eta) + g(\xi,\eta) \leq C(r)\eta^{\frac{n+2+2p_2}{n}}$, for all $0 \leq \xi \leq r, \eta \geq 0$ where ,$2 \leq p_i \leq \frac{2(n+2)}{n}, i = 1, 2$.

Now we have the following theorem :

Theorem 3.1: Under assumption $(H_1) - - - (H_7), T^* \leq \infty$, and $p \geq 2$, Then we have

$$v \in L^p(Q_{T^*})$$

Our global existence results will require the following contrel growth conditions

$(\mathbf{H_8})$ $|g_2'(\eta)| \leq C\eta^{r_1}, r_1 \geq 1, for all \eta \geq 0;$

$|g(\xi,\eta)| \leq C(r) + C(r)\eta^{r_2}, r_2 \geq 1, for all 0 \leq \xi \leq 0, \eta \geq 0.$

$C(r)$ is a continuous function : $R^+ \longrightarrow R^+$

We have the following theorem

Theorem 3.2: If (H_1)–(H_8) hold ,then $T^* = \infty$

References

[**1**] S.L.Hollis,R.H.Martin and M.Pierre.

Global existence and boundedness in Reaction-Diffusion systems.

SIAM.J.Math. Anal.Vol.18.No.3.May 1987

[**2**] S.L.Hollis and J.Morgan.

Global existence and asymptotic decay for systems of convective Reaction-Diffusion equations

Nonlinear Analysis,T.M.A. Vol.17,No.8, 1991

[**3**] O.A.Ladyzenskaja,V.A.Solonnikov and N.N.Uralceva

Linear and Quasilinear Equations of Parabolic Type

Transl.math.Monogr.Am.math.Soc.23,Providence,RI 1968

[**4**] A.Pazy p(1 Semigroups of Linear Operators and Applications to Partial Differential Equations

Applied Math. Sciences 44. Springer,New York 1983

Department of mathematics
Lanzhou University
Gansu Province, 730000
P.R.of China

AMS/IP Studies in Advanced Mathematics
Volume 3, 1997

NONEXISTENCE OF GLOBAL SOLUTIONS TO A CLASS OF GENERALIZED IMPROVED BOUSSINESQ EQUATIONS

ZHIJIAN YANG* GUOWANG CHEN**

Abstract

In this paper, the existence and uniqueness of local solution and the nonexistence of global solution to the initial — boundary value problem of a class of generalized IBq (Improved Boussinesq) equations are studied and a few examples are given.

1. **Introduction.** In the present paper, making use of the contraction mapping principle and Fourier transform method, we study the existence of local solution and the nonexistence of global solution to a class of generalized IBq (Improved Boussinesq) equations

$$u_{tt} - u_{xx} - u_{xxtt} = a(u^p)_{xx} + bu^q \tag{1}$$

together with initial—boundary value conditions

$$\begin{aligned} &u_x(0,t) = 0,\ u_x(\pi,t) = 0, \\ &u(x,o) = \xi(x),\ u_t(x,o) = \eta(x), \end{aligned} \tag{2}$$

here a and b are arbitrary constants $(a \neq 0)$, $p(\geqslant 1)$ and $q(\geqslant 1)$ are integers. In particular, when $a=1$, $b=0$ and $p=2$, (1) becomes IBq equation

$$u_{tt} - u_{xx} - u_{xxtt} = (u^2)_{xx}. \tag{3}$$

IBq equation (3) was proposed in hydrodynamics to study shallow-water waves and ion-sound waves propogation [1], many authors have taken attention to its investigation and have derived some soliton solutions [1]; when $a=-1$, $p=1$, $b>0$ and $q \geqslant 2$, (1) becomes equation

$$u_{tt} - u_{xxtt} = bu^q. \tag{4}$$

According to the conclusion of paper [2], if initial value functions $\xi(x)$ and $\eta(x)$ satisfy

$$\int_0^\pi \xi^{q+1}(x)dx > [(q+1)/2b] \int_0^\pi [\eta^2(x) + (\eta'(x)^2)]dx, \tag{5}$$

then (4), (2) has no global solution, i. e., $\| u \|_{L_2[0,\pi]} + \| u_x \|_{L_2[0,\pi]} \to +\infty$ in finite time $\widetilde{T} < +\infty$. While we give the example of $\xi(x)$ and $\eta(x)$, which does not satisfy the condition (5), so the blowup theorems in paper [2] can not be used. But in this case according to our Th. 5, the solution of (4), (2) will blowup in finite time. For problem (3), (2), we also give the example of $\xi(x)$ and $\eta(x)$, for which the solution of (3), (2) will blowup in finite time.

2. Main Theorems and Conclusions.

Theorem 1. *Assume that $\xi(x)$, $\eta(x)\in D$. Then (1), (2) has a unique solution $u\in C^2([0,T_0);H^2[0,\pi])$, here $[0,T_0)$ is the maximal interval of existence of u, $D=\{v(x)\mid v\in H^2[0,\pi],\ v'(0)=v'(\pi)=0\}$.*

We define the Fourier transform $\hat{}\ : L_2[0,\pi]\to l^2$, $\forall\, f(x)\in L_2[0,\pi]$, $\hat{f}(k)=(2/\pi)\int_0^\pi f(x)\cos kx\,dx=f_k$, $k=0,1,\cdots$. Then $f(x)=f_0/2+\sum_{k=1}^{\infty}f_k\cos kx$ and $\|f\|^2=(\pi/2)[f_0^2/2+\sum_{k=1}^{\infty}f_k^2]$, here (and the following) $\|\cdot\|=\|\cdot\|_{L_2[0,\pi]}$, $\|\cdot\|_{H^k}=\|\cdot\|_{H^k[0,\pi]}$.

Theorem 2. *Assume that*

(i) *$a>0$, $2\leqslant p$ is an even; either $b\leqslant 0$, $0\leqslant q$ is an even or $b\geqslant 0$, $1\leqslant q$ is an odd.*

(ii) *$\xi(x)$, $\eta(x)\in D$, $\xi_0\leqslant -2(ap)^{-1/(p-1)}$, $\xi_k\leqslant 0$, $\eta_k\leqslant 0$, $k=0,1,\cdots$.*

(iii) *one of the following conditions* (H_1) *and* (H_2) *is valid,*

(H_1) *$b\neq 0$, $q\geqslant 2$, $\eta_0\leqslant -(|b|/2^{q-2}(q+1))^{1/2}(-\xi_0)^{(q+1)/2}<0$.*

(H_2) *$p\geqslant 4$ is an even, $\xi_p<0$, $\xi_1<-2(-p/a\xi_p)^{1/(p-2)}$, $\eta_1^2+\xi_1^2/2+a\xi_1^p\xi_p/2^{p-1}p\geqslant 0$.*

Then $\|u\|\to+\infty$ in finite time $\widetilde{T}<+\infty$, here u is the solution of (1), (2).

Especially for IBq equation (3), we obtain

Theorem 3. *Assume that $\xi(x)$, $\eta(x)\in D$, $\xi_0\leqslant -1$, $\xi_k\leqslant 0$, $\eta_k\leqslant 0$, $k=0,1,\cdots$ and $5\eta_2^2/8-\eta_1^2+\xi_2^2/2-(2-\xi_2)\xi_1^2/4\geqslant 0$, $\xi_2\leqslant 4\xi_1/\sqrt{10}<-3$, $\eta_1^2+\xi_1^2/2+(2/45)^{1/2}\xi_1^3\geqslant 0$. Then $\|u\|\to+\infty$ in finite time $\widetilde{T}<+\infty$. here u is the solution of (3), (2).*

Theorem 4. *Assume that*

(i) *$a<0$, $1\leqslant p$ is an odd and in particular $a\leqslant -1$ when $p=1$, either $b\leqslant 0$, $0\leqslant q$ is an even or $b\geqslant 0$, $1\leqslant q$ is an odd.*

(ii) *$\xi(x)$, $\eta(x)\in D$, $\xi_k\leqslant 0$, $\eta_k\leqslant 0$, $k=0,1,\cdots$ and in particular $\xi_0\leqslant -2(-ap)^{-1/(p-1)}$ when $p>1$.*

(iii) *one of the following conditions* (H_3) *and* (H_4) *is valid,*

(H_3) *the assumption* (H_1) *of Th. 2.*

(H_4) *$3\leqslant p$ is an odd, $\xi_1<-2(-(p+1)/2a)^{1/(p-1)}$ and $\eta_1^2+\xi_1^2/2+a\xi_1^{p+1}/$*

$2^{p-1}(p+1)\geqslant 0$.
Then $\| u \| \to +\infty$ *in finite time* $\widetilde{T}<+\infty$, *here* $u(x,t)$ *is the solution of* (1), (2).

Theorem 5. *Assume that*

(i) $a<0$, $b\geqslant 0$; $p\geqslant 1$, $q\geqslant 0$ *and in particular* $a\leqslant -1$ *when* $p=1$.

(ii) $\xi(x)$, $\eta(x)\in D$, $\xi_k\geqslant 0$, $\eta_k\geqslant 0$, $k=0,1,\cdots$ *and in particular* $\xi_0\geqslant 2(-ap)^{-1/(p-1)}$ *when* $p>1$.

(iii) *one of the following conditions* $(H_5)-(H_8)$ *is valid*,

(H_5) $b>0$, $q\geqslant 2$, $\eta_0\geqslant (b/2^{q-2}(q+1))^{1/2}\xi_0^{(q+1)/2}>0$.

(H_6) $p\geqslant 4$ *is an even*, $\xi_p>0$, $\xi_1>2(p/-a\xi_p)^{1/(p-2)}$, $\eta_1^2+\xi_1^2/2+a\xi_1^p\xi_p/2^{p-1}p\geqslant 0$.

(H_7) $p=2$, $\xi_2\geqslant 4\xi_1/\sqrt{10}>3/-a$, $\eta_1^2+\xi_1^2/2+(2/45)^{1/2}a\xi_1^3\geqslant 0, 5\eta_2^2/8-\eta_1^2+\xi_2^2/2-(2-a\xi_2)\xi_1^2/4\geqslant 0$.

(H_8) $p\geqslant 3$ *is an odd*, $\xi_1>2(-(p+1)/2a)^{1/(p-1)}$, $\eta_1^2+\xi_1^2/2+a\xi_1^{p+1}/2^{p-1}(p+1)\geqslant 0$.

Then $\| u \| \to +\infty$ *in finite time* $\widetilde{T}<+\infty$, *here* $u(x,t)$ *is the solution of* (1), (2).

3. The Proof of Main Theorems.

The Proof of Th. 1. Making use of the contraction mapping principle we can easily derive the conclusion of Th. 1.

The Proof of Th. 2. We deduce from $\xi(x)$, $\eta(x)\in D$ and Th. 1 that (1), (2) has a unique noncontinuable (on time t) solution $u(x,t)=\sum_{k=0}^{\infty}u_k(t)\cos kx$, and $\tilde{u}=(u_0(t),\cdots, u_k(t),\cdots)$ satisfies the following problem

$$\begin{gathered}(1+k^2)\ddot{u}_k+k^2u_k=f(t,u)_k,\\ u_k(0)=\xi_k,\ \dot{u}_k(0)=\eta_k,\ k=0,1,\cdots\end{gathered}\tag{7}$$

here $f(t,u)=a(u^p)_{xx}+bu^q$, $f(t,u)_k=\hat{f}(t,u)(k)=-2^{-(p-1)}ak^2\sum_{r_1+\cdots+r_p=k}u_{r_1}\cdots u_{r_p}+2^{-(q-1)}b\sum_{r_1+\cdots+r_q=k}u_{r_1}\cdots u_{r_q}$. Making use of Galerkin method we can prove that $u_k(t)\leqslant 0$, $\dot{u}_k(t)\leqslant 0$, $k=0,1,\cdots$, $t\in J^\circ$, here J° is the maximal inteval of existence of $\tilde{u}$.

1. If the condition (H_1) is valid, we deduce from (7) that $z(t)=-u_0(t)$ $(\geqslant 0)$ satisfies $\ddot{z}\geqslant (|b|/2^{q-1})z^q$. Multiplying the inequality by $\dot{z}(\geqslant 0)$, then integrating over $(0,t)$ we get easily

$$\dot{z} \geqslant (|b|/2^{q-2}(q+1))^{1/2} z,\ z(0) = -\xi_0 > 0. \tag{8}$$

We conclude from (8) that there exists $\tilde{T} < +\infty$ such that $z(t) = -u_0(t) \to +\infty$ as $t \to \tilde{T}^-$, and so $\| u \|^2 = (\pi/2)(u_0^2/2 + \sum_{k=1}^{\infty} u_k^2) \to +\infty$ as $t \to \tilde{T}^-$.

2. If the condition (H_2) is valid, we deduce from (7) that $y(t) = -u_1(t)$ ($\geqslant 0$) and $z(t) = -u_p(t)(\geqslant 0)$ satisfy $2\ddot{y} + y \geqslant (a/2^{p-1}) y^{p-1} z$. It follows from $z(0) = -\xi_p > 0$, $\dot{z}(t) \geqslant 0$ that $z(t) \geqslant -\xi_p$ for all $t \in J^\circ$. So from the above inequality we get easily

$$\dot{y} + y/\sqrt{2} \geqslant (-a\xi_p/2^{p-1}p)^{1/2} y^{p/2}, y(0) = -\xi_1 > 0. \tag{9}$$

From (9) we conclude that there exists $\tilde{T} < +\infty$ such that $y(t) \to +\infty$ as $t \to \tilde{T}^-$, and so $\| u \| \to +\infty$ as $t \to \tilde{T}^-$.

The Proof of Th. 3. Making use of the same method used in the proof of Th. 1, we get that $\tilde{u}(t) = (u_0(t), \cdots, u_k(t), \cdots)$ satisfies (7) (where $a=1, b=0$ and $p=2$) and $u_k(t) \leqslant 0$, $\dot{u}_k(t) \leqslant 0, k=0,1,\cdots, t \in J^0$. We deduce from (7) that $\tilde{y}(t) = -u_1(t)(\geqslant 0)$ and $\tilde{z}(t) = -u_2(t)(\geqslant 0)$ satisfy

$$2\ddot{\tilde{y}} + \tilde{y} \geqslant \tilde{y}\tilde{z}/2, \quad 5\ddot{\tilde{z}} + 4\tilde{z} \geqslant 2\tilde{y}^2. \tag{10}$$

Let $\bar{y} = \sqrt{10}\tilde{y}, \bar{z} = 4\tilde{z}$, we have

$$\ddot{\bar{y}} + \bar{y}/2 \geqslant \bar{y}\bar{z}, \ \ddot{\bar{z}} + 4\bar{z}/5 \geqslant \bar{y}^2. \tag{11}$$

First, let us think the initial value problem

$$\ddot{y} + y/2 = yz, \ \ddot{z} + 4z/5 = y^2, \tag{12}$$

$$y(0) = y_0, \dot{y}(0) = y_1, z(0) = z_0, \dot{z} = z_1, \tag{13}$$

here $y \geqslant 0, z \geqslant 0, \dot{y} \geqslant 0, \dot{z} \geqslant 0, y_0 = -\xi_1/\sqrt{10}, y_1 = -\eta_1/\sqrt{10}, z_0 = -\xi_2/4, z_1 = -\eta_2/4$. From the assumptions we know $z_0 \geqslant y_0 > 3/4$, and from which we can prove $z \geqslant y$ for all $t \in J^0$. So from (12) we conclude $\ddot{y} + y/2 \geqslant y^2$. Multiplying the inequality by $\dot{y}$, then integrating over $(0,t)$ we get easily

$$\dot{y} + y/\sqrt{2} \geqslant (2/3)^{1/2} y^{3/2}, \quad y(0) = y_0. \tag{14}$$

From (14) we know $z(t) \geqslant y(t) \to +\infty$ in finite time $\tilde{T} < +\infty$.

Making use of the comparison principle between ordinary differential equation and inequality we can obtain that $\bar{y}(t) \geqslant y(t), \bar{z}(t) \geqslant z(t)$ and so we get $\| u \| \to +\infty$ in finite time $\tilde{T} < +\infty$.

By virtue of the similar metheds used in the proof of Th. 2 and Th. 3, we can easily prove the Th. 4 and Th. 5.

4. Examples.

Example 1. For the problem (4), (2) (i. e., choosing $a=-1, p=1, b>0$ and $q\geqslant 2$ in equation (1)), if we choose $\xi(x)=1/2, \eta(x)=\sqrt{b}+\cos jx, j\geqslant 1$, then a simple computation shows that (5) is not valid and so the blowup theorems in paper [2] can not be used, but a simple varification shows that the assumptions (i), (ii) and (iii): (H_5) of Th. 5 hold. So $\|u\|\to+\infty$ in finite time $\widetilde{T}<+\infty$, here u is the solution of (4), (2).

Example 2. For initial boundary value problem (2) of IBq equation (3), if we choose $\xi(x)=-1-3\cos x-5\cos 2x$, $\eta(x)=-3\cos x-\sqrt{20}\cos 2x$, then a simple varification shows that the assumptions of Th. 3 hold. So $\|u\|\to+\infty$ in finite time $\widetilde{T}<+\infty$, here u is the solution of (3), (2).

References

[1] V. G. Makharav, *Dynamics of classical solition*, Physic Report A Review Section of Physics Letters (Section C), 35C: 1 (1978), pp. 1—128.

[2] H. A. Levien, *Instability and nonexistence of global solutions to nonlinear wave equations of the form $Pu_{tt}=-Au+F(u)$*, Trans. Amer. Math. Soc. 192(1974), pp. 1—21.

[3] BoB. Palais, *Blowup for nonlinear equations using a comparison principle in Fourier space*, Comm. Pure. Anal. Math. XLI (1988), pp. 165—196.

* Department of Mathematics and Mechanics,
Zhengzhou Institute of Technology,
Zhengzhou, 450002, P. R. China.

* * Institute of Mathematics,
Zhengzhou University,
Zhengzhou, 450052, P. R. China.

AMS/IP Studies in Advanced Mathematics
Volume 3, 1997

ZEROS OF SOLUTIONS OF HYPERBOLIC EQUATIONS WITH FUNCTIONAL ARGUMENTS

NORIO YOSHIDA

Oscillation theory for hyperbolic equations with functional arguments has been investigated by several authors. The existence of arbitrarily large zeros was studied by Mishev [2] and Mishev and Bainov [3, 4]. It is shown in Yoshida [6] that under some hypotheses there exists a bounded domain in which every solution to certain boundary value problems for the hyperbolic equation

$$u_{tt}(x,t) - (\Delta u(x,t) + \alpha \Delta u(x,t-\sigma)) + c(x,t,u(x,t),u(x,t-\tau)) = f(x,t)$$

has a zero. Kahane [1] has studied the oscillation of linear hyperbolic equations with variable coefficients. However, there appears to be no known results about zeros for nonlinear hyperbolic equations with variable coefficients and functional arguments.

We are concerned with the oscillatory behavior of solutions of the nonlinear hyperbolic equation with delays

$$(1)\quad u_{tt}(x,t) - (L[u(x,t)] + \alpha(t)L[u(x,t-\sigma)]) + c(x,t,u(x,t),u(x,t-\tau)) = f(x,t), \quad (x,t) \in \Omega \equiv G \times (0,\infty),$$

where G is a bounded domain of $\mathbf{R}^n$ with piecewise smooth boundary ∂G and L is the uniformly elliptic operator defined by

$$L[u] \equiv \sum_{i,j=1}^{n} \frac{\partial}{\partial x_i}\left(a_{ij}(x,t)\frac{\partial u}{\partial x_j}\right).$$

We assume throughout this paper that:

(H$_1$) σ, τ are nonnegative constants, $\alpha(t) \in C([0,\infty);[0,\infty))$ and $f(x,t) \in C(\overline{\Omega};\mathbf{R}^1)$;

(H$_2$) $c(x,t,\xi,\eta) \in C(\overline{\Omega} \times \mathbf{R}^1 \times \mathbf{R}^1;\mathbf{R}^1)$,
$$c(x,t,\xi,\eta) \geq K_0^2\xi \quad \text{for } (x,t) \in \Omega,\ \xi \geq 0,\ \eta \geq 0,$$
and
$$c(x,t,\xi,\eta) \leq K_0^2\xi \quad \text{for } (x,t) \in \Omega,\ \xi \leq 0,\ \eta \leq 0,$$
where K_0 is a nonnegative constant;

(H$_3$) $a_{ij}(x,t) \in C^1(\overline{G} \times (0,\infty);\mathbf{R}^1)$, $\frac{\partial}{\partial x_k}a_{ij}(x,t)$ $(i,j,k = 1,2,...,n)$ are uniformly bounded in $\overline{G} \times (0,\infty)$, and the matrix $(a_{ij}(x,t))$ is symmetric and uniformly positive definite in Ω.

We consider two kinds of boundary conditions:

$$(B_1)\qquad u = 0 \quad \text{on} \quad \partial G \times (0,\infty),$$

$$(\mathrm{B}_2) \qquad \frac{\partial u}{\partial \nu} + \mu u = \tilde{\psi} \quad \text{on} \quad \partial G \times (0, \infty),$$

where μ, $\tilde{\psi}$ are continuous functions on $\partial G \times [0, \infty)$, $\mu \geq 0$ on $\partial G \times [0, \infty)$ and $\frac{\partial}{\partial \nu}$ denotes the conormal derivative

$$\frac{\partial}{\partial \nu} = \sum_{i,j=1}^{n} a_{ij}(x,t)\gamma_j \frac{\partial}{\partial x_i},$$

γ_j being the j-th component of the unit exterior normal vector to ∂G.

Our objective is to obtain sufficient conditions guaranteeing that there are bounded domains in which every solution of the boundary value problem has a zero.

Definition 1. By a *solution* of the problem (1), (B_i) $(i = 1, 2)$, we mean a function $u(x,t) \in C^2(\overline{G} \times (-\sigma, \infty); \mathbf{R}^1) \cap C(\overline{G} \times (-\tau, \infty); \mathbf{R}^1)$ which satisfies (1), (B_i) $(i = 1, 2)$.

Definition 2. A solution u of the problem (1), (B_i) $(i = 1, 2)$ is said to be *oscillatory* in Ω if u has a zero in $G \times (t, \infty)$ for any $t > 0$.

In [1] Kahane showed that there are a positive constant M and a positive function $\Phi(x) \in \mathrm{C}^2(\overline{G})$ such that

$$(2) \qquad -L[\Phi(x)] \geq M^2 \Phi(x) \quad \text{in} \quad \Omega.$$

Theorem 1. *Assume that* (H_1)–(H_3) *hold. If there is a number* $s \geq T$ *such that*

$$(3) \qquad H(s) \equiv \int_s^{s+\pi/K} F(t) \sin K(t-s)\, dt = 0,$$

then every solution u of the problem (1), (B_1) *has a zero in* $G \times (s - T, s + \pi/K)$, *where* $K = (M^2 + K_0^2)^{1/2}$, $T = \max\{\sigma, \tau\}$ *and*

$$F(t) = \int_G f(x,t)\Phi(x)\, dx.$$

Proof. Suppose to the contrary that u is a solution of the problem (1), (B_1) which has no zero in $G \times (s - T, s + \pi/K)$. First we assume that $u > 0$ in $G \times (s - T, s + \pi/K)$. Multiplying (1) by $\Phi(x)$ and then integrating over G yields

$$(4) \qquad U''(t) - \int_G (L[u(x,t)] + \alpha(t)L[u(x,t-\sigma)])\, \Phi(x)\, dx + \int_G c(x,t,u(x,t),u(x,t-\tau))\Phi(x)\, dx = F(t),$$

where $U(t) = \int_G u(x,t)\Phi(x)\, dx$. Integrating by parts, we have

$$(5) \quad \int_G L[u(x,t)]\Phi(x)\, dx = \int_{\partial G} \left(\Phi(x)\frac{\partial u}{\partial \nu} - u\frac{\partial \Phi}{\partial \nu}(x) \right) dS + \int_G uL[\Phi(x)]\, dx.$$

Since $u > 0$ in $G \times (s-T, s+\pi/K)$ and $u = 0$ on $\partial G \times [0,\infty)$, we find that $\frac{\partial u}{\partial \nu} \le 0$ on $\partial G \times (s-T, s+\pi/K)$. Hence, the first term of the right hand side of (5) is nonpositive. Taking (2) into account, we see from (5) that

$$(6) \qquad \int_G L[u(x,t)]\Phi(x)\,dx \le -M^2 U(t), \quad t \in (s-T, s+\pi/K).$$

Analogously we obtain

$$(7) \qquad \int_G L[u(x,t-\sigma)]\Phi(x)\,dx \le -M^2 U(t-\sigma), \quad t \in (s-T, s+\pi/K).$$

It is easily seen that $u(x,t-\tau) \ge 0$ in $G \times [s, s+\pi/K)$. The hypothesis (H_2) implies that

$$(8) \qquad \int_G c(x,t,u(x,t),u(x,t-\tau))\Phi(x)\,dx \ge K_0^2 U(t), \quad t \in [s, s+\pi/K).$$

Combining (4), (6)–(8), we obtain

$$U''(t) + M^2 U(t) + \alpha(t) M^2 U(t-\sigma) + K_0^2 U(t) \le F(t), \quad t \in [s, s+\pi/K).$$

In view of the fact that $U(t-\sigma) \ge 0$ in $[s, s+\pi/K)$, we observe that $U(t)$ is a positive solution of

$$(9) \qquad U''(t) + \left(M^2 + K_0^2\right) U(t) \le F(t), \quad t \in [s, s+\pi/K).$$

However, the hypothesis implies that (9) has no positive solution in $[s, s+\pi/K)$ (see [5, Remark 2]). This is a contradiction. In the case where $u < 0$ in $G \times (s-T, s+\pi/K)$, we are also led to a contradiction. The proof is complete. □

Corollary 1. *Assume that* (H_1)–(H_3) *hold. If the function* $H(s)$ *given by* (3) *is oscillatory at* $s = \infty$, *then every solution* u *of the problem* (1), (B_1) *is oscillatory in* Ω.

Theorem 2. *Assume that* (H_1)–(H_3) *hold, and that* $K_0 > 0$. *If there is a number* $s \ge T$ *such that*

$$(10) \quad \widetilde{H}(s) \equiv \int_s^{s+\pi/K_0} \left(\widetilde{F}(t) + \widetilde{\Psi}(t) + \alpha(t)\widetilde{\Psi}(t-\sigma)\right) \sin K_0(t-s)\,dt = 0,$$

then every solution u *of the problem* (1), (B_2) *has a zero in* $G \times (s-T, s+\pi/K_0)$, *where*

$$\widetilde{F}(t) = \int_G f(x,t)\,dx,$$
$$\widetilde{\Psi}(t) = \int_{\partial G} \widetilde{\psi}(x,t)\,dS.$$

Proof. Suppose that u is a solution of the problem (1), (B_2) which has no zero in $G \times (s-T, s+\pi/K_0)$. We may assume that $u > 0$ in $G \times (s-T, s+\pi/K_0)$. Integrating of (1) over G and taking account of (H_2), (B_2), we find that $\widetilde{U}(t) = \int_G u(x,t)\,dx$ is a positive solution of

$$\widetilde{U}''(t) + K_0^2 \widetilde{U}(t) \le \widetilde{F}(t) + \widetilde{\Psi}(t) + \alpha(t)\widetilde{\Psi}(t-\sigma), \quad t \in [s, s+\pi/K_0).$$

Arguing as in the proof of Theorem 1, we are led to a contradiction. The proof is complete. □

Corollary 2. *Assume that* (H_1)–(H_3) *hold. If the function* $\widetilde{H}(s)$ *given by* (10) *is oscillatory at* $s = \infty$, *then every solution* u *of the problem* (1), (B_2) *is oscillatory in* Ω.

Remark 1. If $c(x,t,\xi,\eta) = K_0^2\xi + K_1(t)\eta^\gamma$, then $c(x,t,\xi,\eta)$ satisfies the hypothesis (H_2), where $K_1(t)$ is a nonnegative continuous function in $[0,\infty)$ and γ is a positive odd integer.

Remark 2. If $L = \Delta$ (Laplacian), we can choose λ_1 and $\Phi_1(x)$ instead of M and $\Phi(x)$, respectively, where $\lambda_1 > 0$ is the first eigenvalue of the eigenvalue problem

$$-\Delta w = \lambda w \quad \text{in} \quad G, \qquad w = 0 \quad \text{on} \quad \partial G,$$

and $\Phi_1(x)$ is the corresponding eigenfunction which may be chosen so that $\Phi_1(x) > 0$ in G.

Remark 3. In the case where $L = \Delta$ and $\alpha(t) = \alpha$ (positive constant), our results reduce to the results of [6].

Remark 4. We consider the boundary condition

$$(\mathrm{B}_3) \qquad u = \psi_1, \quad \frac{\partial u}{\partial \nu} = \psi_2 \quad \text{on} \quad \partial G \times (0,\infty),$$

where ψ_i $(i = 1,2)$ are continuous functions on $\partial G \times [0,\infty)$. Arguing as in the proof of Theorem 1, we conclude that Theorem 1 holds true if $F(t)$ and (B_1) are replaced by $F(t) + \Psi(t) + \alpha(t)\Psi(t-\sigma)$ and (B_3), respectively, where

$$\Psi(t) \equiv \int_{\partial G} \left(\Phi(x)\psi_2(x,t) - \psi_1(x,t)\frac{\partial \Phi}{\partial \nu}(x) \right) dS.$$

Remark 5. Our results can be generalized to the more general equation

$$u_{tt} - \left(L[u(x,t)] + \sum_{i=1}^{k} \alpha_i(t) L[u(x,t-\sigma_i)] \right)$$
$$+ c(x,t,u(x,t),u(x,t-\tau_1),\dots,u(x,t-\tau_m)) = f(x,t).$$

We conclude with two examples which illustrate Theorems 1 and 2.

Example 1. We consider the problem

$$(11) \qquad u_{tt}(x,t) - [u_{xx}(x,t) + u_{xx}(x,t-\pi)] + u(x,t-\pi) = -2\sin x \cdot \sin t,$$
$$(x,t) \in (0,\pi) \times (0,\infty),$$

$$(12) \qquad u(0,t) = u(\pi,t) = 0, \quad t > 0.$$

Here $n = 1$, $\alpha(t) = 1$, $\sigma = \tau = \pi$, $T = \pi$, $G = (0,\pi)$, $K_0 = 0$ and $f(x,t) = -2\sin x \cdot \sin t$. It is easy to see that $\lambda_1 = 1$, $K = 1$ and $\Phi(x) = \Phi_1(x) = \sin x$ and $F(t) = -\pi \sin t$. Since

$$H(s) = \int_s^{s+\pi} F(t)\sin(t-s)\,dt = -\frac{\pi^2}{2}\cos\, s,$$

we conclude that $H(s) = 0$ for $s = s_n = \frac{\pi}{2} + n\pi > \pi$ $(n = 1, 2, \ldots)$. Theorem 1 implies that every solution u of the problem (11), (12) has a zero in $(0,\pi)\times(s_n - \pi, s_n + \pi)$. One such solution is $u = \sin x \cdot \sin t$.

Example 2. We consider the problem

$$(13) \qquad u_{tt}(x,t) - [u_{xx}(x,t) + 3u_{xx}(x,t-\pi)] + u(x,t) = \sin x \cdot \cos 2t, \qquad (x,t) \in (0,\pi)\times(0,\infty),$$

$$(14) \qquad -u_x(0,t) + u(0,t) = -\cos 2t, \quad t > 0,$$

$$(15) \qquad u_x(\pi,t) + u(\pi,t) = -\cos 2t, \quad t > 0.$$

Here $n = 1$, $\alpha(t) = 3$, $\sigma = \pi$, $T = \pi$, $G = (0,\pi)$, $K_0 = 1$, $\mu = 1$ and $f(x,t) = \sin x \cdot \cos 2t$. It is easily verified that $\widetilde{\Psi}(t) = -2\cos 2t$, $\widetilde{\Psi}(t-\pi) = -2\cos 2t$ and $\widetilde{F}(t) = 2\cos 2t$, and therefore $\widetilde{F}(t) + \widetilde{\Psi}(t) + 3\widetilde{\Psi}(t-\pi) = -6\cos 2t$. Since

$$\widetilde{H}(s) = \int_s^{s+\pi} (-6\cos 2t)\cdot\sin(t-s)\,dt = 4\cos 2s,$$

we observe that $\widetilde{H}(s) = 0$ for $s = s_n = \frac{\pi}{4} + \frac{n\pi}{2} > \pi$ $(n = 2, 3, \ldots)$. It follows from Theorem 2 that every solution u of the problem (13)–(15) has a zero in $(0,\pi)\times(s_n - \pi, s_n + \pi)$. For example, $u = \sin x \cdot \cos 2t$ is such a solution.

References

1. C. Kahane, *Oscillation theorems for solutions of hyperbolic equations*, Proc. Amer. Math. Soc. **41** (1973), 183–188.
2. D. P. Mishev, *Oscillation of the solutions of hyperbolic differential equations of neutral type with "maxima"*, Godishnik VUZ Appl. Math. (to appear).
3. D. P. Mishev and D. D. Bainov, *Oscillation properties of the solutions of hyperbolic equations of neutral type*, Proceedings of the Colloquium on Qualitative Theory of Differential Equations, (1984), 771–780.
4. ———, *Oscillation properties of the solutions of a class of hyperbolic equations of neutral type*, Funkcial. Ekvac. **29** (1986), 213–218.
5. N. Yoshida, *On the zeros of solutions to nonlinear hyperbolic equations*, Proc. Roy. Soc. Edinburgh Sect. A **106** (1987), 121–129.
6. ———, *On the zeros of solutions of hyperbolic equations of neutral type*, Differential Integral Equations **3** (1990), 155–160.

Department of Mathematics, Faculty of Science, Toyama University, Toyama 930, Japan

AMS/IP Studies in Advanced Mathematics
Volume **3**, 1997

EXISTENCE AND CONVERGENCE OF SOLUTIONS TO HIGHER ORDER GENERALIZED KDV—BURGERS EQUATIONS

HUIJIANG ZHAO, CHANGJIANG ZHU AND ZHONG YU

Abstract

In this paper, by employing the energy method and theory of compensated compactness, we are concerned with the existence and convergence of solutions to the Cauchy problem of higher order generalized KdV—Burgers equations

(1) $u_t + Df(u) + \delta D^{2n+1}u = \varepsilon D^2 u$

(2) $u(0,x) = u_0(x)$

where $D = \frac{\partial}{\partial x}$.

1. Former Results

In the case of n=1, the Cauchy problem (1), (2) has been studied by many scholars([3—7]): For $\delta=0$, $\varepsilon>0$, the global existence results are well—known and Oleinik([6], 1957), Tartar([8], 1979) et al. proved the convergence of the solution to that of the invisid equation

(3) $u_t + Df(u) = 0$

with the same initial data (2); For $\delta>0$, $\varepsilon=0$, $f(u)=\frac{1}{2}u^2$, i. e. the KdV equation, global existence results had been obtained (Miura et al. [4], 1968—1969). For the convergence results, Lax P. D. and Levemore D. C. ([3], 1979) proved that the solution sequence $\{u^\delta(t,x)\}$, in general, does not converge to the solution of the invisid Burgers equation i. e. (3) with $f(u)=\frac{1}{2}u^2$.

Then the following question is natural: If one balances the size of δ and ε, how about the solutions of the Cauchy problem (1), (2)?

When n = 1, Schonbek ([7], 1981) gave the above problem a definite answer. Her results can be summarized in the following theorem:

Theorem. *Suppose that $n=1$, $f(u)\in C^2(R)$, $|f'(u)|\leqslant c(1+|u|)$. Let $u_\delta^\varepsilon(t,x):\Omega=[0,T]\times R\to R$ be a sequence of smooth solutions of (1), (2) which vanish at infinity and have initial data $u_\delta^\varepsilon(0,x)=u_0(x)$ which has compact support and satisfy $u_0(x)\in H^2(R)\cap L^4(R)$. If $\delta=0(\varepsilon^3)$, then there exists a subsequence (still labelled) $\{u_\delta^\varepsilon\}$ such that $u_\delta^\varepsilon\to u$, $f(u_\delta^\varepsilon)\to f(u_\delta^\varepsilon)$ in the sense of distributions and u is the solution of the Cauchy problem (1), (2). If in addition, $f''>0$, then $u_\delta^\varepsilon\to u$ strongly in $L^q(\Omega)$, $1<q<4$.*

In fact, Corollary 5. 1 in [7] allowed the flux function f to satisfy $|f'(u)|\leqslant c(1+|u|^2)$, however it's proofs seemed incomplete and the flux function f in Corollary 5. 2 ([7]) was too special.

In Schonbek's work, the following porblems remain unsolved:

(i). Under what conditions does the Cauchy problem (1), (2) admit a unique global smooth solution which satisfies certain regularity and asymptotic behaviors?

(ii). Can the growth condition of the flux function f be weakened?

(iii). Her proofs are too complicated, how can her proofs be simplified and generalized to the case of general $n(>1)$?

(iv). Is $f''>0$ necessary to guarantee the strong convergence of the solutions?

In this paper, we study the global existence and convergence of the solutions of the Cauchy problem (1), (2). When $n=1$, our results extend those of Schonbek's.

2. Main Results

Global existence results

Theorem 1. *Let $u_0(x)\in H^1(R)$, $f(u)\in C^2(R)$ and one of the following conditions holds:*

(i). $|f'(u)|\leqslant c(1+|u|^p)$, $0\leqslant p\leqslant 2$ *(When $p=2$, we ask $\|u_0\|_{L^2(R)}$ to be small in addition);*

(ii). $u_0(x)\in H^n(R)$, $(-1)^n f'(u)\geqslant 0$;

(iii). $u_0(x)\in H^n(R)$, $(-1)^n\int_0^u f(s)ds\geqslant 0$, $|f'(u)|\leqslant c(1+|u|^p)$, $0\leqslant p\leqslant 4n$ *(when $p=4n$ we ask $\|u_0\|_{L^2(R)}$ to be small in addition);*

(iv). $\|u_0\|_{L^2(R)}$ *is small.*

Then the Cauchy problem (1), (2) admits a unique global solution $u(t,x)\in C([0,\infty),H^1(R)\cap C^1([0,\infty),L^2(R)))$ and for each $k\in Z_+$, $u(t,x)\in C((0,\infty).H^k(R))\cap C^1((0,\infty),H^{k-1}(R))$.

Convergence results

For the convergence of solutions $\{u_\delta^\varepsilon\}$ as $\varepsilon\to 0_+$ and $\delta\to 0_+$, we have the following theorem:

Theorem 2. *Suppose that $u_0(x)\in H^{2n-1}(R)$, $f(u)\in C^{2n-1}(R)$, $|f'(u)|\leqslant c(1+|u|^p)$, $0\leqslant p<2$. Then, when δ,ε satisfy certain conditions, there exists a subsequence (still labelled) $\{u_\delta^\varepsilon\}$ satisfying $u_\delta^\varepsilon\to u$, $f(u_\delta^\varepsilon)\to f(u)$ in the sense of distributions and u is the generalized solution of (3), (2). Furthermore, if we assume further that there is no interval on which f is affine, then, $u_\delta^\varepsilon\to u$ strongly in $L^q(\Omega)$, $1\leqslant q\leqslant 2(p+1)$.*

For the convergence of solutions $\{u_\delta^\varepsilon\}$ as $\varepsilon\to 0_+$, we have the following results:

Theorem 3. *Suppose that $u_0(x)\in H^n(R)$, $(-1)^n f'(u)\geqslant 0$, then the solution sequence $\{u_\delta^\varepsilon\}$ of the Cauchy problem (1), (2) has a subsequence (still labelled) $\{u_\delta^\varepsilon\}$ which converges in the strong topology to a solution of the corresponding* *generalized higher order KdV equation*

(4) $u_t + Df(u) + \delta D^{2n+1}u = 0.$

3. Methods and Remarks

For the global existence results, we get our global existence results by the method of continuous extension of the local solution. But our methods to study the problem are different from that of the ordinary: First, for higher order equations, the normal tool to obtain the local existence result is the theory of analytic semigroups. But in our problem, the operator $Au = -\delta D^{2n+1}u + \varepsilon D^2$ is not an infinisimal generator of an analytic semigroup, hence the method cann't be used directly. In our paper, we use the method of Fourier analysis to get an integral representation of the solutions of the Cauchy problem (1), (2), then, by employing the contraction mapping principle, we get our local existence result; Next, we use two different methods to extend the local solution obtained above globally. The one reduces the global existence results to an algebraic inequality; The other is motivated by the work of Hoff and Smoller ([2]).

For the convergence results, since we can only obtain the weak convergence of the solutions, to obtain the weak continuity of the nonlinear functions, we need the theory of compensated compactness. The basic framework is well — known (see Schonbek [7] and the modified results given by Lu Yunguang [6], we adopt the latter in our analysis). The difficulty is to verify the $H^{-1}_{Loc}(\Omega)$ condition. Even in the case of $n=1$, Schonbek's proofs were too complicated. The techniques we used to prove the convergence of the solutions are different from those of Schonbek's in the following ways: (i). We find the proof of $\{\delta D^{2n} u^{\varepsilon}_{\delta}\} \in L^2_c(\Omega)$ is essential to the proofs of the others. The observation makes our analysis much more simple. (ii). The way to estimate $\| u^{\varepsilon}_{\delta}(t,x) \|_{L^{\infty}(R)}$ is different: In Schonbek's work, it's bounded by $\delta^{-1/3}$, but in our work, it's bounded by $\varepsilon^{-1/(2-p)}$. This makes it possible to generalize Schonbek's results ($n=1$) to general n.

Remarks. 1. It's easy to see that the same reults hold for the following equation

$$u_t + Df(u) - \delta D^{2n+1}u = \varepsilon D^2 u$$

2. The assumptions that $\| u_0 \|_{L^2(R)}$ is small in Theorem 1 and f satisfies certain growth and monotic conditions in Theorem 1, 2, 3 (especially $0 \leqslant p < 2$ in Theorem 2) are technical, how to remove or weaken them remains an open problem.

Acknowledgement We're grateful to Professor Ding Xiaxi for his encouragement and guidance.

References

1. Cassis B., Communi. in PDEs., 14(5), 619—634(1989).
2. Hoff D., Smoller J., J. Ann. IHP, Analyse non lineaire, 2(3), 213—235 (1985).
3. Lax P. D., Levemore C. D., Proc. Nat. Acad. Sci. USA, 76(8), 3602—3604(1979).
4. Miura R. M., J. Math. Phys., 9, 1202(1968).
5. Oleinik O. A., Amer. Math. Soc. Transl. Ser. 2, 26, 95—172(1957).
6. Lu Yunguang, Applicable Analysis, 31, 239—246(1989).
7. Schonbek M., Communi. in PDEs., 7(8), 959—1000(1981).
8. Tartar L., Nonlinear analysis and mechanics, Research Notes in Math., Heriot—watt Symposium, Vol. 4(Knop, R. J. Ed.), Pitman Press, New York(1979).

WUHAN INSTITUTE OF MATHEMATICAL SCIENCES
ACADEMIA SINICA
P. O. BOX 71007, WUHAN 430071
P. R. CHINA

AMS/IP Studies in Advanced Mathematics
Volume **3**, 1997

CONVERGENCE AND ERROR BOUND ANALYSIS OF A FINITE-DIFFERENCE SCHEME FOR THE ONE-DIMENSIONAL NAVIER-STOKES EQUATIONS

Jennifer Jing Zhao
Department of Mathematics & Statistics
University of Michigan-Dearborn, Dearborn MI 48128
David Hoff*
Department of Mathematics
Indiana University, Bloomington IN 47405

ABSTRACT. Convergence is proved and error bounds are derived for a finite difference approximation to discontinuous solutions of the Navier-Stokes equations for nonisentropic, compressible flow in one space dimension. The scheme is fully implicit and can be implemented under resonable mesh conditions. It is shown that the approximations converge at a rate of $O(\triangle x^r)(r < \frac{1}{12})$ when the initial velocity and energy are in L^2, and the initial density is piecewise H^1. The errors are measured in a norm which dominates the sup-norm of the error in the density, which in general is discontinuous.

Key words. difference scheme, Navier-Stokes equations, compressible flow

§ 1 INTRODUCTION.

We prove convergence and derive error bounds for finite difference approximations to discontinuous solution of the Navier-Stocks equations for nonisentropic, compressible flow in one space dimension:

$$
(1.1)\qquad \begin{cases} v_t - u_x = 0, \\ u_t + p(v,e)_x = (\frac{\varepsilon u_x}{v})_x, \\ e_t + p(v,e)u_x = \frac{\varepsilon u_x^2}{v} + (\frac{\lambda T(e)_x}{v})_x, \end{cases}
$$

*This research was supported in part by the NSF under grants DMS-9001606 and DMS-9201597

with initial data and boundary conditions:

$$(1.2) \qquad (v, u, e)(x, 0) = (v_0, u_0, e_0)(x), u(\pm 1, t) = e_x(\pm 1, t) = 0.$$

Here, v, u and e represent respectively the specific volume, velocity, and energy of the fluid, $p(v, e)$ and $T(e)$ are the pressure and temperature, and ε and λ are positive viscosity parameters. The equations (1.1)-(1.3) express conservation of mass and balance of momentum and energy. This system is formulated in Lagrangean coordinates, so that the lines $x = \mathit{constant}$ correspond to particle paths. (See Courant and Friedrichs [1].)

Of particular interest here is the case that the initial data and the corresponding solutions are not smooth. The solutions thus depend continuously on their initial data in a norm which dominates the L^2 norms of the perturbations in u and e, and the local variation of the perturbation in v. (See Hoff [2]-[4].)

Our goal in this paper is therefore to construct a scheme which can effectively approximate these solutions, and to understand the convergence properties of the scheme in a rigorous mathematical way. The scheme we study is fully implicit, so that a system of nonlinear equations must be solved at each time step. We show that this system can be solved by a simple iteration procedure, provided that appropriate mesh conditions are satisfied. We show that the error is bounded by $C\triangle x^r$ for any $r < 1/12$. These rather low rates of convergence are due to the fact that the error is measured in the unusually strong topology dictated by the continous-dependence theory; in particular, these error bounds dominate the sup-norm of the error for the discontinuous variable v.

We then define C_0 by

$$C_0 = \|u_0(x), \triangle e_0\|^2_{L^2(-1,1)} + \dagger\dagger \triangle v_0(x) \dagger\dagger^2_{H^1(-1,1)} + (TV(u_0, e_0))^2 + (\sum_{i=1} |[v_0(y_i)]|)^2$$

$$[v_0(y_i)] = v_0(y_i^+) - v_0(y_i^-), \dagger\dagger \cdot \dagger\dagger^2_{L^2(-1,1)} = \sum_{i=1}^{M-1} \| \cdot \|^2_{L^2(y_i, y_{i+1})}$$

and TV denotes the total variation of a function on (-1,1). C_0 is assumed to be small.

We let $\triangle x = K^{-1}$ be an increment in space, where K is a positive integer, and $\triangle t$ an increment in time. h will denote the pair $(\triangle x, \triangle t)$. We use k to denote an integer and j a half integer. We construct approximations $\{v_j^q, e_j^q, E_j^q = E(v_j^q) : j = -K + 1/2, \cdots, K - 1/2\}$ and $\{u_k^q : k = -K + 1, \cdots, K - 1\}$ to $\{v(x_j, t^q), e(x_j, t^q), E(x_j, t^q) : j = -K + \frac{1}{2}, \cdots, K - \frac{1}{2}\}$ and $\{u(x_k, t^q) : k = -K + 1, \cdots, K - 1\}$ for $q = 0, 1, 2, \cdots$, as follows:

$$(1.3) \qquad \frac{v_j^{q+1} - v_j^q}{\triangle t} = \delta u_j^{q+1}, j = -K + 1/2, \cdots, K - 1/2,$$

$$(1.4)\quad \begin{cases} \frac{u_k^{q+1}-u_k^q}{\Delta t} + \delta p_k^{q+1} = (\delta(\epsilon \rho^{q+1/2}\delta u^{q+1}))_k, \quad k=-K+1,\cdots,K-1 \\ u_{\pm K}^q = 0, \\ \rho_j^{q+1/2} = \begin{cases} \frac{E_j^{q+1}-E_j^q}{v_j^{q+1}-v_j^q}, v_j^{q+1} \neq v_j^q \\ E'(v_j^q), v_j^{q+1} = v_j^q, \end{cases} \\ p_j^{q+1} = p(v_j^{q+1}, e_j^{q+1}). \end{cases}$$

$$(1.5)\quad \begin{cases} \frac{e_j^{q+1}-e_j^q}{\Delta t} + p_j^{q+1}\delta u_j^{q+1} = \epsilon \rho_j^{q+1/2}\delta u_j^{q+1/2} + \lambda(\delta(\frac{\delta e^{q+1}}{v^{q+1}}))_j, \\ \qquad j=-K+1/2,\cdots,K-1/2, \\ \delta e_{\pm K}^q = 0.(\text{or} \quad e_{K-1/2}^q = e_{K+1/2}^q, \quad e_{-K+1/2}^q = e_{-K-1/2}^q) \end{cases}$$

Observe that the scheme (1.16)-(1.20) is nonlinearly implicit, so that a system of nonlinear algebraic equations must be solved at each time step. We define x_{k_i} to be the integer mesh point nearest to $y_i(x_{k_{i-1/2}} \leq y_i < x_{k_{i+1/2}})$, and we also assume that Δx is sufficiently small that there is at most one jump point y_i in each interval (x_j, x_{j+1}). We use G_0 to denote the discrete analog of C_0, where

$$G_0 \leq C(C_0^{1/2} + \Delta x),$$

C denotes a positive constant depending on $\underline{v}$,$\bar{v}$, and $\bar{e}$.

§ 2. A PRIORI ESTIMATES

In this section we derive various pointwise bounds and energy estimates for the solution to the difference scheme (1.3)-(1.5). These estimates will be used in section 3 to prove that the difference equations can be solved up to any finite time, and in section 4 in deriving estimates for the associated weak truncation errors.

We fix a positive constant $\theta \in (0, 1/4]$, and we assume that Δx and Δt satisfy the relation $\Delta t \leq \Delta x^2$; we also assume that there is an integer Q for which $1 \leq t^Q = Q\Delta t \leq 2$. We shall derive estimates for the following quantities:

$$A^n = \sup_{0\leq m\leq n} \{\sum_j (\Delta v_j^{m2} + \Delta e_j^{m2})\Delta x + \sum_k u_k^{m2}\Delta x + \sum_{k\neq k_j} \delta u_k^{m2} + (\sum_{i=1} |[v_{k_i}^m]|)^2\}$$
$$+ \sum_1^n (\sum_j \delta u_j^{m2} + \sum_{k\neq k_j} \delta v_k^{m2} + \sum_k \delta e_k^{m2})\Delta x \Delta t,$$

$$B^n = \sup_{1\leq m\leq n} \{(\tau^{m,Q})^{1/2}(\sum_j \delta u_j^{m2}\Delta x + \sum_k \delta e_k^{m2}\Delta x)\}$$
$$+ \sum_1^n (\tau^{m,Q})^{1/2+\theta} [\sum_k (\frac{u_k^m - u_k^{m-1}}{\Delta t})^2 \Delta x + \sum_{k\neq k_j} (\delta(\epsilon \rho^{m-1/2}\delta u^m)_k)^2 \Delta x$$
$$+ \sum_j (\frac{e_j^m - e_j^{m-1}}{\Delta t})^2 \Delta x + \sum_j (\delta(\frac{\delta e^m}{v^m})_j)^2 \Delta x + \sum_j \delta u_j^{m4} \Delta x]\Delta t,$$

$$F^n = \sup_{1\le m\le n} \{(\tau^{m,Q})^{3/2+\theta}[\sum_k (\frac{u_k^m - u_k^{m-1}}{\Delta t})^2 \Delta x + \sum_{k\neq k_j} (\delta(\epsilon\rho^{m-1/2}\delta u^m)_k)^2\Delta x$$
$$+ \sum_j (\frac{e_j^m - e_j^{m-1}}{\Delta t})^2\Delta x + \sum_j (\delta(\frac{\delta e^m}{v^m})_j)^2\Delta x]\}$$
$$+ \sum_{m=1}^n (\tau^{m,Q})^{3/2+\theta}\sum_j (\frac{\delta u_j^m - \delta u_j^{m-1}}{\Delta t})^2\Delta x\Delta t$$
$$+ \sum_{m=1}^n (\tau^{m,Q})^{\frac{3}{2}+\frac{5}{4}\theta}\sum_k (\frac{\delta e_k^m - \delta e_k^{m-1}}{\Delta t})^2\Delta x\Delta t,$$

where

$$\tau^{m,Q} = \begin{cases} t^m; & \text{if } \ t^m \le t^Q, \\ 1; & \text{if } \ t^m > t^Q. \end{cases}$$

The main results of this section are given in the following theorem.

Theorem 2.1 *Let $T > 0$ be given, and assume that the scheme (1.3)-(1.5) can be solved up to time t^n, where $0 < t^n \le T$, with $0 < \underline{v} \le v_j^m \le \bar{v} < \infty$, $0 < e_j^m \le \bar{e} < \infty$ for all $m \le n$, and furthermore that $e_j^m \ge \frac{1}{2}\tilde{e} > 0$, if $Q \le m \le n$. Then there exist positive constants $C(\underline{v}, \bar{v}, \bar{e}, \tilde{v}, \tilde{e}, \lambda, r, \varepsilon, \theta, \frac{1}{min_i|y_{i+1}-y_i|})$ and $h(G_0, C, T)$ such that, if $\Delta x \le h(G_0, C, T)$, then*

$$A^n + B^n + F^n \le C(G_0^{1/4} + \Delta x^{1/2}),$$
$$|u^m|_\infty \le |u^0|_\infty + C(G_0^{1/8} + \Delta x^{1/4}),$$
$$0 < e_j^m \le |e^0|_\infty + C(G_0^{1/8} + \Delta x^{1/4})$$

§ 3. SOLVABILITY OF THE SCHEME

In this section we show that the difference equations (1.3)-(1.5) can be solved forward in time by means of a simple iteration procedure, provided that certain conditions on the mesh are satisfied.

We let v^n, u^n, and e^n denote respectively the sequences $\{v_j^n\}_{j=-K+1/2,\cdots,K-1/2}$, $\{u_k^n\}_{k=-K,\cdots,K}$, and $\{e_j^n\}_{j=-K+1/2,\cdots,K-1/2}$. Then given$(v^n, u^n, e^n)$, we solve the difference equations (1.3)-(1.5) for $(v^{n+1}, u^{n+1}, e^{n+1})$ by applying the following fixed-point iteration: define

$$v_j^{(1)} = v_j^n, \quad u_k^{(1)} = u_k^n, \quad e_j^{(1)} = e_j^n, \tag{3.1}$$

and for $m = 1, 2, \cdots$, compute

$$v_j^{(m+1)} = v_j^n + \Delta t\delta u_j^{(m)}, \quad j = -K + 1/2, \cdots, K - 1/2, \tag{3.2}$$

$$\begin{cases} u_k^{(m+1)} = u_k^n - \Delta t\delta p(v_k^{(m+1)}, e_k^{(m)}) + \Delta t\delta(\epsilon\rho^{(m+1/2)}\delta u^{(m+1)})_k \\ \rho_j^{(m+1/2)} = \frac{\ln v_j^{(m+1)} - \ln v_j^{(m)}}{v_j^{(m+1)} - v_j^{(m)}} \\ u_{\pm K}^{(m)} = 0, \quad j = -K + 1/2, \cdots, K - 1/2 \end{cases} \tag{3.3}$$

$$(3.4)\quad \begin{cases} e_j^{(m+1)} = e_j^n + \Delta t\epsilon\rho_j^{(m+1/2)}\delta u_j^{(m+1)^2} - \Delta t p(v_j^{(m+1)}, e_j^{(m+1)})\delta u_j^{(m+1)} \\ \qquad +\Delta t\lambda\delta(\frac{\delta e^{(m+1)}}{v^{(m+1)}})_j, \\ \delta e_{\pm K}^{(m)} = 0, \qquad j = -K+1/2, \cdots, K-1/2. \end{cases}$$

In the following theorem we show that this iteration procedure converges, and that it can be applied to compute the solution of the scheme (1.3)-(1.5) up to any positive time T, provided that appropriate mesh conditions are satisfied. These mesh conditions depend only on the initial data and on T; it is therefore unnecessary to compute new mesh conditions at each time step.

THEOREM3.1 *Given positive numbers $\underline{v}, \bar{v}, \bar{u}, \bar{e}$, and given a points $(\tilde{v}, \tilde{e}) \in (\underline{v}, \bar{v}) \times (0, \bar{e})$, there are constants $C(\underline{v}, \bar{v}, \bar{e}, \tilde{v}, \tilde{e}, \lambda, \epsilon, r, \theta, \frac{1}{\min_i |y_{i+1}-y_i|})$, h_0 and $\overline{G_0}$, such that, if $T > 1$, if $0 < \underline{v} \le v_j^0 \le \bar{v} < \infty$, $0 < e_j^0 \le \bar{e} < \infty$, $|u_k^0|_\infty \le \bar{u}$, and $G_0 \le \overline{G_0}$, and if the mesh parameters satisfy some resonable conditions, then the scheme (1.3)-(1.5) can be solved for all $t^n \le T$ by means of the iteration procedure (3.1)-(3.4). The solution satisfies*

$$\begin{cases} 0 < \underline{v} \le v_j^n \le \bar{v} < \infty, \quad |u_k^n| \le \frac{3}{2}\bar{u}, \\ 0 < e_j^n \le \frac{3}{2}\bar{e}. \quad 0 < \frac{1}{2}\tilde{e} \le e_j^n, \ if\ t^n \ge t^Q. \end{cases}$$

§4. CONVERGENCE AND ERROR BOUNDS

The results of section 3 show that for any $T > 0$, the scheme can be solved up to T with $0 < \underline{v} \le v_j^n \le \bar{v} < \infty$,$|u^n|_\infty \le \frac{3}{2}\bar{u}$ and $0 < e_j^n \le \frac{3}{2}\bar{e} < \infty$, where $t^n \le T$. In this section, we use the solutions of our difference scheme to construct approximate solutions $(v^h(x,t), u^h(x,t), e^h(x,t))$ to the initial boundary value problem (1.1)-(1.2), where h denotes $(\Delta x, \Delta t)$. We then show that these approximate solutions converge in a appropriate norm.

We define v^h to be the piecewise bilinear interpolant of $\{v_j^m\}$, except at the boundaries and near the lines of discontinuity $x = y_i$, where v^h is defined to be constant in x. $p^h(x,t)$ and $E^h(x,t)$ are defined similarly. $u^h(x,t)$ and $e^h(x,t)$ are defined as the piecewise bilinear interpolatants of $\{u_k^m\}$ and $\{e_j^m\}$.

THEOREM 4.1*Let $T > 0$, $\bar{v} > \underline{v} > 0$, $\bar{e} > 0$, $\bar{u} > 0$, and $(\tilde{v}, \tilde{e}) \in (\underline{v}, \bar{v}) \times (0, \bar{e})$ be given, assume that all the hypotheses in Theorem 3.1 are in force. Then there exists a constant C depending on $\underline{v}, \bar{v}, \bar{e}, \tilde{v}, \tilde{e}, \lambda, r, \epsilon, \theta$ and $\frac{1}{\min_i |y_{i+1}-y_i|}$ and G_0, such that for any $\sigma \in (0,1)$, the following estimates hold:*

$$|\tau_{1\sigma}| \le CC_\sigma(\chi)(\Delta x + \Delta t^{-3/4-\theta/2}).$$
$$|\tau_{2\sigma}| \le CC_\sigma(\psi)\Delta x^{-1/4-\theta/2}$$
$$|\tau_{3\sigma}| \le CC_\sigma(\phi)(T-\sigma)^{1/2}\Delta x^{-\frac{3}{4}-\frac{5}{8}\theta},$$

where $\tau_{1\sigma}, \tau_{2\sigma}, \tau_{3\sigma}$ are the three weak truncation errors. For example:

$$\tau_{1\sigma}(\chi, h) = \int_\sigma^T \int_{-1}^1 (v_t^h - u_x^h)\chi dx dt.$$

$C_\sigma(\chi), C_\sigma(\psi), C_\sigma(\phi)$ are functionals of χ, ψ, ϕ. There exists a subsequence $\{h_l\}$ of $\{h\}$ and functions $v(x,t), u(x,t), e(x,t), \tilde{E}(x,t)$, and $\tilde{p}(x,t)$ such that, for all $\sigma \in (0,1), \{v^{h_l}, u^{h_l}, e^{h_l}, E^{h_l}, p^{h_l}$ converge to these functions in a reasonable sense.

Furthermore, for test functions $\phi \in C_0^\infty((-1,1) \times R), u, v, e$ satisfy

$$\int_{-1}^1 (v\phi)(x,t)dx - \int_{-1}^1 v_0(x)\phi(x,0)dx + \int_0^t \int_{-1}^1 (v_t^h - u_x^h)\chi dx dt = 0,$$
$$\int_{-1}^1 (u\phi)(x,t)dx - \int_{-1}^1 u_0(x)\phi(x,0)dx + \int_0^t \int_{-1}^1 (-u\phi_t - p\phi_x + \frac{\epsilon u_x}{v}\phi_x)dx dt = 0.$$

and if $\phi \in C_0^\infty(R \times R)$, then

$$\int_{-1}^1 (e\phi)(x,t)dx - \int_{-1}^1 e_0(x)\phi(x,0)dx$$
$$+ \int_0^t \int_{-1}^1 (-e\phi_t + pu_x\phi - \frac{\epsilon u_x^2}{v}\phi + \frac{\lambda e_x}{v}\phi_x)dx dt = 0, e_x(\pm 1, t) = 0, t \in (0,T];$$

where $R = (-\infty, +\infty)$.

The main result of this paper is given in the next theorem.

THEOREM 4.2

Let $T > 0, \underline{v}, \bar{v}, \bar{u}, \bar{e}, (\tilde{v}, \tilde{e}) \in (\underline{v}, \bar{v}) \times (0, \bar{e})$ be given for the equations (1.1)-(1.2) and assume that all the hypotheses in Theorem 3.1 are in force. Then the scheme (1.3)-(1.5) has solutions (v^h, u^h, e^h) which are constructed in the beginning of this section and defined up to T. There exists a constant C depending $\underline{v}, \bar{v}, \tilde{v}, \tilde{e}, r, \lambda, \epsilon, \theta, \frac{1}{\min_i |y_{i+1} - y_i|}$ and $\overline{G_0}$, such that: (v^h, u^h, e^h) converges to the unique solution (v, u, e) obtained in Theorem 4.1; and the error between (v^h, u^h, e^h) and (v, u, e) is bounded by

$$\xi(T) \leq C^T(\xi(0) + \Delta x^r), \quad r < \frac{1}{12}$$

where

$$\xi(t) = \sup_{s \in [0,t]} \{\|(u^h - u)(\cdot, s)\|_{L^2(-1,1)} + \|(e^h - e)(\cdot, s)\|_{L^2(-1,1)}$$
$$+ \dagger\dagger(v^h - v)(\cdot, s)\dagger\dagger_{H^1(-1,1)}\}.$$

REFERENCES

[1] R.Courant and K.O. Friedrichs, Supersonic Flow and Waves,Springer-Verlag, 1948.

[2] David Hoff, Discontinuous solutions of the Navier-Stokes Equations for compressible flow, *Archive Rational. Mech. Ana.*114(1991),15-46.

[3] David Hoff, Global well-posedness of the Cauchy problem for the Navier-Stokes equations of nonisentropic flow with discontinuous initial data, *Journal of Differential Equations* 95,No.1(1992),33-73.

[4] David hoff, Construction of solutions for compressible isentropic Navier-Stokes Equations in one space dimension with nonsmooth initial data, *Proc. Roy. Soc. Edinburgh* 03A(1986),301-315.

[5] Roger Zarnowski and David Hoff, A finite-diffence scheme for the Navier-Stokes equations of one-dimensional, isentropic,compressible flow, *SIAM J. Numer. Ana.*28(1991),78-112.

[6] J. Zhao, A convergent finite-difference scheme for the Navier-Stokes equations of nonisentropic, compressible flow, Ph.D dissertation, Indiana University, 1992.

AMS/IP Studies in Advanced Mathematics
Volume 3, 1997

VACUUM PROBLEM FOR THE DAMPED p—SYSTEM①

YONG—SHU ZHENG

Abstract

The author extends the results in [1,2]to the general p(v). For arbitrary large initial data, the global smooth solution of the initial value problem for the damped p—system is proved to be uniformly (i. e. ,independent of the time t)away from vacuum provided that the initial data are away from vacuum.

§ 1. Introduction

Consider the initial value problem for damped p—system

$$v_t - u_x = 0, \quad u_t + p(v)_x = -2\alpha u, \quad x \in R, \quad t > 0, \tag{E}$$

$$v(x,0) = v_0(x), \quad u(x,0) = u_0(x), \quad x \in R. \tag{I}$$

where α is a positive constant, $p(v) \in C^2(b,\infty)$, $p'(v)<0$, $p''(v)>0$, b is a finite number or $-\infty$. The system (E) has eigenvalue

$$\lambda = -\sqrt{-p'(v)}, \quad \mu = \sqrt{-p'(v)}.$$

Let r,s be the Riemann invariants, namely

$$r = u + \Phi(v), \quad s = u - \Phi(v) \tag{R}$$

where

$$\Phi(v) = \int_{\underline{v}}^{v} \mu(\tau)d\tau \quad (b < \underline{v} < \infty)$$

The Riemann invariants give a one to one smooth mapping from $\Omega=\{(u,v)\,|\,u\in R, b<v<\infty\}$onto $\Omega_1=\{(r,s)\,|\,2\Phi(b)<r-s<2\Phi(\infty)\}$;and transform (E),(I)into

$$r_t + \lambda r_x = -\alpha(r+s), \quad s_t + \mu s_x = -\alpha(r+s), \quad x \in R, \quad t > 0, \tag{E_1}$$

$$r(x,0) = r_0(x), \quad s(x,0) = s_0(x), \quad x \in R \tag{I_1}$$

where $r_0(x)=u_0(x)+\Phi(v_0(x))$, $s_0(x)=u_0(x)-\Phi(v_0(x))$.

---

① Projects supported by the Science Foundation of Fujian Province,PRC.

As $b<v<\infty$, then $\lambda<\mu$. Hence, the system (E) is strictly hyperbolic. But if $\lim_{v\to\infty} p'(v) = 0$, as $v=\infty$, then $\lambda=\mu=0$. Hence, it is not strictly hyperbolic; and is known as "vacuum state".

As it is well known, to solve the global C^1—solution for the initial value problem (E), (I) with large initial data will probably encounter the question that whether the vacuum state occurs in a finite time. For this reason, one supposes, in [3—5] that the initial data satisfy the following condition (A), in addition to the condition (C) below.

$$\{(r,s) \mid \inf_x r_0(x) \leqslant r \leqslant \sup_x r_0(x), \quad \inf_x s_0(x) \leqslant s \leqslant \sup_x s_0(x)\} \subset \Omega_1. \tag{A}$$

However, under the condition (A), the initial data are not really large. Consequently, the additional assumed condition (A) can be removed if the C^1-solution of the initial value problem (E), (I) in arbitrary domain of their existence can be proved to be away from vacuum under assumed condition (C) (namely, the initial data are away from vacuum). This is critical for studying the existence or nonexistence of global C^1-solution with really large initial data.

For certain special p-system, namely

$$p(v) = k^2 v^{-\gamma} \quad (v>0; \quad k>0, \quad 1\leqslant \gamma < 3)$$

the above conclusion was proved to be tenable in [1, 2]. In this paper, the author will extend the conclusion to the general p-system; and will confirm the conjecture in [3].

Suppose that $p(v)$ satisfies the following condition

$$\begin{cases} p(v)\in C^3(b,\infty), \quad p'(v)<0, \quad p''(v)>0, \quad 4p'(v)p'''(v) \geqslant 5[p''(v)]^2, \\ \qquad\qquad\qquad\qquad\qquad\qquad\qquad\qquad \forall\, v\in(b,\infty) \\ \lim_{v\to b_+} p(v)=\infty, \quad \lim_{v\to\infty} p'(v) = 0. \end{cases} \tag{P}$$

and the initial data satisfy the condition

$$\begin{cases} v_0(x), u_0(x) \in C^1(R), \quad b< v_{0*} \leqslant v_0(x) \leqslant v_0^*, \\ \| v_0(x) \|_{C^1} + \| u_0(x) \|_{C^1} \leqslant M \end{cases}. \tag{C}$$

where M can be an arbitrarily large positive constant.

Our main result is stated as follows:

Theorem Consider the p-system satisfying the condition (P). If the initial data satisfy the condition (C), then the initial value problem (E), (I) has the *a priori* estimate for C^1-solution in the domain where the C^1-solution exists,

$$v(x,t) \leqslant K$$

where K only depends on α, v_{0*}, v_0^*, M and $p(v)$, but independent of the time t, that is,

$$K = v_{0*} + 4(-p'_0)(p''_0)^{-1}\Big[-1+exp\{\frac{1}{4}p''_0(-p'_0)^{-\frac{5}{4}}M_1\}\Big], \tag{1}$$

where
$$p'_0 = p'(v_{0*}), p''_0 = p''(v_{0*}), \tag{2}$$

$$M_1 = max\{g(v_0^*), M_2/2\alpha\}, \tag{3}$$

$$M_2 = max\{0, \sup_x[\mu^{\frac{1}{2}}(v_0(x))\frac{dr_0(x)}{dx} + 2\alpha g(v_0(x))],$$
$$\sup_x[\mu^{\frac{1}{2}}(v_0(x))\frac{ds_0(x)}{dx} + 2\alpha g(v_0(x))]\}, \tag{4}$$

$$g(v) = \int_{v_{0*}}^{v} \mu^{\frac{1}{2}}(\tau)d\tau. \tag{5}$$

The theorem shows that, in the case $\alpha > 0$, the C^1-solution of the initial value problem (E), (I) is uniformly (i. e., independent of the time t) away from vacuum as long as there is no vacuum in the initial data. Obviously, the result is better than what is in the case $\alpha = 0$.

§ 2. Preliminary Results

Lemma 1. If $p(v)$ satisfies the condition (P), then b is a finite number.

As an extension of Lemma 1 in [3], this can be proved in the same way.

Consequently, there is no loss of generality in assuming $b = 0$.

Lemma 2. If $p(v)$ satisfies the condition (P), then

$$v \leqslant v_{0*} + 4(-p'_0)(p''_0)^{-1}[-1 + exp\{\frac{1}{4}p''_0(-p'_0)^{-\frac{5}{4}}g(v)\}], \quad \forall v \geqslant v_{0*} \tag{6}$$

where $p'_0, p''_0, g(v)$ are given respectively by expression (2), (5).

Proof. From expression (P), we have

$$\frac{p'''(v)}{p''(v)} \leqslant \frac{5}{4}\frac{p''(v)}{p'(v)}. \tag{7}$$

For given $v_{0*} \in (0, \infty)$, intergrating both sides of (7) from v_{0*} to v with $v \geqslant v_{0*}$, we get

$$[-p'(v)]^{-\frac{5}{4}}p''(v) \leqslant (-p'_0)^{-\frac{5}{4}}p''_0. \tag{8}$$

Intergrating both sides of (8) again from v_{0*} to v, we have

$$[-p'(v)]^{-\frac{1}{4}} \leqslant (-p'_0)^{-\frac{1}{4}} + \frac{1}{4}p''_0(-p'_0)^{-\frac{5}{4}}(v - v_{0*}). \tag{9}$$

From (9), we get

$$\frac{4(-p'_0)^{\frac{5}{4}}}{p''_0(v - v_{0*}) + 4(-p'_0)} \leqslant [-p'(v)]^{\frac{1}{4}} = \mu^{\frac{1}{2}}(v). \tag{10}$$

Finally, by intergrating both sides of (10) from v_{0*} to v, we make expression (6) tenable.

Lemma 3. Under the condition (P), (C), any C^1-solution $(r(x,t), s(x,t))$ of the initial value problem (E_1), (I_1) have the *a priori* estimate in the domain where the C^1-solution exists,

$$\mu^{\frac{1}{2}}(v(x,t))\frac{\partial r(x,t)}{\partial x} + 2\alpha g(v(x,t)) \leqslant M_2, \tag{11}$$

$$\mu^{\frac{1}{2}}(v(x,t))\frac{\partial s(x,t)}{\partial x}+2ag(v(x,t))\leqslant M_2, \tag{12}$$

where M_2 and $g(v)$ are given respectively by expression (4), (5).

Proof To differentiate (E_1) with respect to x, thus

$$(r_x)'=-\mu_s r_x^2+\mu_s r_x s_x-ar_x-as_x, \tag{13}$$

$$(s_x)`=-\mu_s s_x^2+\mu_s r_x s_x-ar_x-as_x, \tag{14}$$

Where $r_x=\frac{\partial r(x,t)}{\partial x}$, $s_x=\frac{\partial s(x,t)}{\partial x}$, $\mu_s=\frac{\partial\mu(r,s)}{\partial s}$, $'=\frac{\partial}{\partial t}+\lambda\frac{\partial}{\partial x}$, $`=\frac{\partial}{\partial t}+\mu\frac{\partial}{\partial x}$

Now from (E_1), (R), we get

$$s_x=v', \quad r_x=v`. \tag{15}$$

Next, from (13), (14) and (15), we obtain

$$(\mu^{\frac{1}{2}}r_x)'=-\mu_s\mu^{\frac{1}{2}}r_x^2-a\mu^{\frac{1}{2}}r_x-a\mu^{\frac{1}{2}}s_x, \tag{16}$$

$$(\mu^{\frac{1}{2}}s_x)`=-\mu_s\mu^{\frac{1}{2}}s_x^2-a\mu^{\frac{1}{2}}r_x-a\mu^{\frac{1}{2}}s_x. \tag{17}$$

and since

$$\mu_s=\frac{\partial\mu}{\partial v}\frac{\partial v}{\partial s}=-\frac{\mu_v}{2\mu}=\frac{p''(v)}{4[-p'(v)]}>0,$$

so that

$$(\mu^{\frac{1}{2}}r_x)'\leqslant-a\mu^{\frac{1}{2}}r_x-a\mu^{\frac{1}{2}}s_x, \tag{18}$$

$$(\mu^{\frac{1}{2}}s_x)`\leqslant-a\mu^{\frac{1}{2}}r_x-a\mu^{\frac{1}{2}}s_x, \tag{19}$$

Let

$$W=\mu^{\frac{1}{2}}r_x+2ag(v), \qquad Z=\mu^{\frac{1}{2}}s_x+2ag(v). \tag{20}$$

Due to (18), (19), (5) and (15), it reads that

$$W'\leqslant a(Z-W), \tag{21}$$

$$Z`\leqslant a(W-Z). \tag{22}$$

Based on (20), (21), (22) and citing Lin[6], Li[7], the *a priori* estimate (11), (12) can thus be completed.

§ 3. The Proof of Theorem

The proof of theorem. At first, from (11), (12) and (15), we have

$$\mu^{\frac{1}{2}}(v)v`+2ag(v)\leqslant M_2, \quad \mu^{\frac{1}{2}}(v)v'+2ag(v)\leqslant M_2. \tag{23}$$

It follows that by (5), (23),

$$[g(v)]'+2ag(v)\leqslant M_2, \quad [g(v)]`+2ag(v)\leqslant M_2. \tag{24}$$

Thus

$$[g(v)e^{2at}]'\leqslant M_2e^{2at}, \quad [g(v)e^{2at}]`\leqslant M_2e^{2at}. \tag{25}$$

Let (x,t) be any point in the domain where the C^1-solution exists, and $X=x_\lambda(\tau;x,t)$ denotes λ-characteristic curve passing through the point (x,t) and joining x-axis at $(0,\eta)$. Intergrating the first equation of (25) along the λ—characteristic curve from 0 to t, we get

$$g(v(x,t)) \leqslant g(v_0(\eta))e^{-2\alpha t} + \frac{M_2}{2\alpha}(1-e^{-2\alpha t})$$

$$\leqslant max\{g(v_0^*), M_2/2\alpha\} = M_1 \tag{26}$$

Combining (26) and Lemma 2, thus as $v \geqslant v_{0*}$, we can obtain

$$v(x,t) \leqslant v_{0*} + 4(-p'_0)(p''_0)^{-1}[-1+exp\{\frac{1}{4}p''_0(-p'_0)^{-\frac{5}{4}}M_1\}].$$

This yields the theorem.

References

[1] Y. S. Zheng, Vacuum problem of quasilinear system with dissipation, J. of Huaqiao Univ. (Natural Science) 13 : 4(1992), 442—447.

[2] Y. S. Zheng, Global smooth solutions for systems of gas dynamics with dissipation, Acta Math. Sci., 7 : 4(1987), 383—390.

[3] L. W. Lin and T. Yang, Existence and nonexistence of global smooth solutions for damped p—system with "really large" initial data, chinese J. Prat. Diff. Equa., 4 : 2 (1991), 45—51.

[4] L. W. Lin and Y. S. Zheng, Existence and nonexistence of global smooth solutions for quasilinear hyperbolic systems, Chin. Ann. of Math., 9B : 3(1988), 372—377.

[5] J. H. Wang and C. Z. Li, Global regularity and formation of singularities of solution for quasilinear hyperbolic systems with dissipation, Chin. Ann, of Math., 9A : 5 (1988), 509—523.

[6] L. W. Lin, On the occurrence of "vacuum states" for 2×2 quasilinear hyperbolic conservation laws, (to appear)

[7] C. Z. Li and Y. P. Zhang, A result on the existence of globally smooth solution to the thermally insulated gas dynamics system of equations, J. Chengdu Univ. of Sci. and Tech., (2)(1992), 67—73.

Address Department of Applied Mathematics, Huaqiao University,
Quanzhou, 362011, Fujian, P. R. China